Quantum Mechanics

Offering a detailed account of the key concepts and mathematical apparatus of quantum mechanics, this textbook is an ideal companion to both undergraduate and graduate courses. The formal and practical aspects of the subject are explained clearly alongside examples of modern applications, providing students with the tools required to thoroughly understand the theory and apply it. The authors provide an intuitive conceptual framework that is grounded in a coherent physical explanation of quantum phenomena, established over decades of teaching and research in quantum mechanics and its foundations. The book's educational value is enhanced by the inclusion of examples and exercises, with solutions available online, and an extensive bibliography is provided. Notes throughout the text provide fascinating context on the tumultuous history of quantum mechanics, the people that developed it, and the questions that still remain at its center. This title is also available as Open Access on Cambridge Core.

Ana María Cetto is a full professor at the Institute of Physics, National Autonomous University of Mexico. Specializing in the foundations of quantum physics, she has authored or coauthored 150 research papers and 25 books. Her awards and recognitions include Mexico's Woman of the Year in 2003, the UNESCO Kalinga Prize in 2023, the Oganesson Prize in 2024, and the AIP Tate Medal in 2025. Cetto was a corecipient of the Pugwash Nobel Peace Prize in 1995 and the IAEA Nobel Peace Prize in 2005.

Luis de la Peña holds the title of Professor Emeritus at the Institute of Physics, National Autonomous University of Mexico. Specializing in the foundations of quantum theory and stochastic processes, he has authored or coauthored more than 160 research papers and 10 books. His books include *Introducción a la Mecánica Cuántica* (1979), *The Quantum Dice* (1996), and *The Emerging Quantum* (2016). He was awarded the National Science Prize (Mexico) in 2002.

Quantum Mechanics
A Physical Approach

ANA MARÍA CETTO
National Autonomous University of Mexico

LUIS DE LA PEÑA
National Autonomous University of Mexico

Shaftesbury Road, Cambridge CB2 8EA, United Kingdom

One Liberty Plaza, 20th Floor, New York, NY 10006, USA

477 Williamstown Road, Port Melbourne, VIC 3207, Australia

314–321, 3rd Floor, Plot 3, Splendor Forum, Jasola District Centre,
New Delhi – 110025, India

103 Penang Road, #05–06/07, Visioncrest Commercial, Singapore 238467

Cambridge University Press is part of Cambridge University Press & Assessment,
a department of the University of Cambridge.

We share the University's mission to contribute to society through the pursuit of
education, learning and research at the highest international levels of excellence.

www.cambridge.org
Information on this title: www.cambridge.org/9781009679626

DOI: 10.1017/9781009679633

© Ana María Cetto and Luis de la Peña 2026

This publication is in copyright. Subject to statutory exception and to the provisions of relevant collective licensing agreements, with the exception of the Creative Commons version the link for which is provided below, no reproduction of any part may take place without the written permission of Cambridge University Press & Assessment.

An online version of this work is published at doi.org/10.1017/9781009679633 under a Creative Commons Open Access license CC-BY-NC 4.0 which permits re-use, distribution and reproduction in any medium for noncommercial purposes providing appropriate credit to the original work is given and any changes made are indicated. To view a copy of this license visit https://creativecommons.org/licenses/by-nc/4.0

When citing this work, please include a reference to the DOI 10.1017/9781009679633

First published 2026

A catalogue record for this publication is available from the British Library

A Cataloging-in-Publication data record for this book is available from the Library of Congress

ISBN 978-1-009-67962-6 Hardback

Additional resources for this publication at www.cambridge.org/cetto-delapena.

Cambridge University Press & Assessment has no responsibility for the persistence or accuracy of URLs for external or third-party internet websites referred to in this publication and does not guarantee that any content on such websites is, or will remain, accurate or appropriate.

For EU product safety concerns, contact us at Calle de José Abascal, 56, 1°, 28003 Madrid, Spain,
or email eugpsr@cambridge.org

Contents

Preface		*page* xi
1	**The Quantum World and Its Description**	1
	1.1 The Quantum World, Part of the Physical World	2
	1.2 Is Quantum Mechanics a Complete Theory?	5
	1.3 Main Issues in the Interpretation of Quantum Mechanics	8
	1.4 Completing the Quantum Ontology	12
	Bibliographical Notes	13
2	**Matter and Field**	14
	2.1 Time of Rupture	14
	2.2 Planck: The First Quantum Leap	15
	2.3 Einstein: Quantization as a Universal Phenomenon	22
	2.4 Early Attempts to Model the Atom	27
	2.5 Bohr's Quantum Atom Model	29
	2.6 The Early Quantization Rules	31
	2.7 De Broglie: Waves Associated with Corpuscular Motion	33
	2.8 Electron Beam Diffraction	35
	2.9 Back to Planck: Completing the Spectral Law with the Zero-Point Field	38
	2.10 Corollary: The Electrodynamic Essence of Quantum Mechanics	40
	Problems	41
	Bibliographical Notes	43
3	**Quantum Mechanics in the Heisenberg (Matrix) Picture**	44
	3.1 The Beginnings of Matrix Mechanics	44
	3.2 The Heisenberg Picture	45
	3.3 x- and p- Representations	49
	3.4 Laws of Motion in the Heisenberg Picture	53
	3.5 Matrix Description of the Harmonic Oscillator	54
	3.6 Generalization to Arbitrary Potentials	59
	3.7 State Vectors and the Hilbert Space. Pauli Matrices	59
	3.8 Why Operators Instead of Functions?	62
	Problems	63
	Bibliographical Notes	64

4	**Quantum Mechanics in the Schrödinger (Wave) Picture. Dirac's Notation**		65
	4.1	Wave Patterns of Quantum Particles	65
	4.2	The Continuity Equation	69
	4.3	The Stationary Schrödinger Equation	71
	4.4	Basic Operators and Their Expectation Values in Configuration Space	89
	4.5	General Properties of Quantum Operators and Eigenvalue Equations	91
	4.6	Dirac's Symbolic Notation	95
	Problems		104
	Bibliographical Notes		106
5	**One-Dimensional Potential Steps, Barriers, and Wells**		107
	5.1	General Approach	107
	5.2	The Step Potential	109
	5.3	The Square Well	113
	5.4	The Square Potential Barrier: Tunneling	120
	5.5	The Symmetric Double Well: Energy Splitting and Periodic Tunneling	126
	5.6	Corollary: Meaning of Particle-Wave Behavior	131
	Problems		131
	Bibliographical Notes		132
6	**The WKB Approximation. Electronic Properties of Solids**		133
	6.1	The Semiclassical Approach	133
	6.2	Transmission Across a Barrier. Alpha Decay	140
	6.3	Electronic Properties of Solids	147
	Problems		160
	Bibliographical Notes		161
7	**The Free Particle. The Time-Dependent Schrödinger Equation**		162
	7.1	The Free Particle	162
	7.2	The Time-Dependent Schrödinger Equation	177
	7.3	The Propagator in the General Case	184
	Problems		188
	Bibliographical Notes		190
8	**Dynamics of Quantum Systems**		191
	8.1	Dynamics in the Heisenberg Description	191
	8.2	The Schrödinger Description	192
	8.3	The Dynamics of Expectation Values. Ehrenfest's Theorem	193
	8.4	Elements for a Statistical Description	196
	8.5	The Heisenberg Inequalities and Their Meaning	199
	8.6	Quantum Canonical Transformations	204
	8.7	*Integrals of Motion and Symmetries: Noether's Theorem	215
	Problems		220
	Bibliographical Notes		221

Contents

Preface *page* xi

1 The Quantum World and Its Description 1
  1.1 The Quantum World, Part of the Physical World 2
  1.2 Is Quantum Mechanics a Complete Theory? 5
  1.3 Main Issues in the Interpretation of Quantum Mechanics 8
  1.4 Completing the Quantum Ontology 12
  Bibliographical Notes 13

2 Matter and Field 14
  2.1 Time of Rupture 14
  2.2 Planck: The First Quantum Leap 15
  2.3 Einstein: Quantization as a Universal Phenomenon 22
  2.4 Early Attempts to Model the Atom 27
  2.5 Bohr's Quantum Atom Model 29
  2.6 The Early Quantization Rules 31
  2.7 De Broglie: Waves Associated with Corpuscular Motion 33
  2.8 Electron Beam Diffraction 35
  2.9 Back to Planck: Completing the Spectral Law with the Zero-Point Field 38
  2.10 Corollary: The Electrodynamic Essence of Quantum Mechanics 40
  Problems 41
  Bibliographical Notes 43

3 Quantum Mechanics in the Heisenberg (Matrix) Picture 44
  3.1 The Beginnings of Matrix Mechanics 44
  3.2 The Heisenberg Picture 45
  3.3 x- and p- Representations 49
  3.4 Laws of Motion in the Heisenberg Picture 53
  3.5 Matrix Description of the Harmonic Oscillator 54
  3.6 Generalization to Arbitrary Potentials 59
  3.7 State Vectors and the Hilbert Space. Pauli Matrices 59
  3.8 Why Operators Instead of Functions? 62
  Problems 63
  Bibliographical Notes 64

4	**Quantum Mechanics in the Schrödinger (Wave) Picture. Dirac's Notation**	65
	4.1 Wave Patterns of Quantum Particles	65
	4.2 The Continuity Equation	69
	4.3 The Stationary Schrödinger Equation	71
	4.4 Basic Operators and Their Expectation Values in Configuration Space	89
	4.5 General Properties of Quantum Operators and Eigenvalue Equations	91
	4.6 Dirac's Symbolic Notation	95
	Problems	104
	Bibliographical Notes	106
5	**One-Dimensional Potential Steps, Barriers, and Wells**	107
	5.1 General Approach	107
	5.2 The Step Potential	109
	5.3 The Square Well	113
	5.4 The Square Potential Barrier: Tunneling	120
	5.5 The Symmetric Double Well: Energy Splitting and Periodic Tunneling	126
	5.6 Corollary: Meaning of Particle-Wave Behavior	131
	Problems	131
	Bibliographical Notes	132
6	**The WKB Approximation. Electronic Properties of Solids**	133
	6.1 The Semiclassical Approach	133
	6.2 Transmission Across a Barrier. Alpha Decay	140
	6.3 Electronic Properties of Solids	147
	Problems	160
	Bibliographical Notes	161
7	**The Free Particle. The Time-Dependent Schrödinger Equation**	162
	7.1 The Free Particle	162
	7.2 The Time-Dependent Schrödinger Equation	177
	7.3 The Propagator in the General Case	184
	Problems	188
	Bibliographical Notes	190
8	**Dynamics of Quantum Systems**	191
	8.1 Dynamics in the Heisenberg Description	191
	8.2 The Schrödinger Description	192
	8.3 The Dynamics of Expectation Values. Ehrenfest's Theorem	193
	8.4 Elements for a Statistical Description	196
	8.5 The Heisenberg Inequalities and Their Meaning	199
	8.6 Quantum Canonical Transformations	204
	8.7 *Integrals of Motion and Symmetries: Noether's Theorem	215
	Problems	220
	Bibliographical Notes	221

9	**The One-Dimensional Harmonic Oscillator**		222
	9.1 A Maximally Coherent Oscillating Packet with Minimal Dispersion		222
	9.2 Eigenfunctions and Eigenvalues of the Hamiltonian		226
	9.3 Transitions Between States: Selection Rules for the Harmonic Oscillator		231
	9.4 Raising and Lowering Operators		232
	9.5 Two Coupled Harmonic Oscillators		239
	Problems		242
	Bibliographical Notes		243
10	**The Physics Underlying Quantum Phenomena**		244
	10.1 Quantum Mechanics as a Stochastic Theory. The Nelson Theory		245
	10.2 Stochastic Electrodynamics. Basic Properties of the ZPF		251
	10.3 Statistical Approach: The Fokker–Planck Equation		259
	10.4 Transition to the Quantum Description in Terms of Operators		261
	10.5 Hydrodynamic Quantum Analogs		267
	Problems		268
	Bibliographical Notes		269
11	**Angular Momentum Theory**		270
	11.1 Orbital Angular Momentum. Spherical Harmonics		270
	11.2 Radial Hamiltonian for Central Forces		280
	11.3 Matrix Representation of the Angular Momentum. The Pauli Matrices		283
	11.4 Addition of Two Angular Momenta		291
	Problems		298
	Bibliographical Notes		299
12	**Central Potentials. The Hydrogen Atom**		300
	12.1 Reduction of the Two-Body Problem		300
	12.2 Angular Momentum Selection Rules for Dipole Transitions		305
	12.3 The 3D Rigid Rotor		306
	12.4 The Free Particle in 3D		308
	12.5 The Hydrogen-Like Atom		310
	12.6 The Hydrogen Spectrum		315
	12.7 The Atom in an Electromagnetic Field. The Zeeman Effect		320
	Problems		328
	Bibliographical Notes		330
13	**Stationary Perturbation Theory and the Variational Method**		331
	13.1 Introduction to Time-Independent Perturbation Theory		331
	13.2 Perturbation Theory for Nondegenerate States		334
	13.3 Perturbation Theory for Degenerate States		339
	13.4 The Stark Effect		346

13.5	The Dielectric Constant	348
13.6	*The Canonical Transformation Method	351
13.7	*The Feynman and Hellman Method	352
13.8	The Variational Method: Minimum Energy of the Quantum States	353
	Problems	359
	Bibliographical Notes	362

14 The Electron Spin. Entangled States — 363

14.1	Discovery of the Electron Spin	363
14.2	The Quantum Theory of Spin	366
14.3	Effects of Spin on Atomic Spectra	370
14.4	Modeling the Electron Spin	378
14.5	Quantum Correlations and Apparent Nonlocality	380
	Problems	389
	Bibliographical Notes	391

15 Identical-Particle Systems: Quantum Statistics. The Density Matrix — 392

15.1	Multipartite Systems	392
15.2	Bosons and Fermions. The Pauli Principle	398
15.3	Effects of Statistics on the Energy Spectrum	402
15.4	Collective Phenomena. Phonons and Bose-Einstein Condensates	408
15.5	The Density Matrix. Pure States and Mixtures	414
	Problems	434
	Bibliographical Notes	436

16 Atoms and Molecules — 437

16.1	Multielectron Atoms. The Hartree–Fock Method	437
16.2	The Periodic Table of the Elements	442
16.3	The Helium Atom	445
16.4	Molecular Structure	454
16.5	Effects of Nuclear Motion on Diatomic Molecules	465
16.6	Intermolecular Forces. Casimir and van der Waals Forces, Measurable Effects of the ZPF	471
16.7	Nuclear Structure. Extending the Periodic Table of the Elements	477
	Problems	480
	Bibliographical Notes	482

17 Time-Dependent Perturbations. Field Quantization and Second Quantization — 483

17.1	Time-Dependent Perturbation Theory	483
17.2	Radiative Transitions	493
17.3	Radiative Corrections: The Full Picture	506
17.4	The Radiative Corrections According to SED. Atomic Lifetimes and Lamb Shift	512

	17.5	Canonical Field Quantization Clarified	519
	17.6	*Second Quantization. An Introduction	522
	Problems		529
	Bibliographical Notes		530
18	**Elastic Scattering Theory**		**531**
	18.1	Collision Phenomena	531
	18.2	The Born Approximation	537
	18.3	Form Factors	544
	18.4	Partial-Wave Expansion. Introducing the S Matrix	547
	18.5	*Resonant Scattering	554
	Problems		558
	Bibliographical Notes		560
19	**Relativistic Equations. An Introduction**		**561**
	19.1	The Klein–Gordon Equation	561
	19.2	The Dirac Equation	566
	19.3	Properties of the Dirac Equation. The *Zitterbewegung*	572
	19.4	The Free Particle	576
	19.5	*The Dirac Particle in an External Electromagnetic Field	580
	19.6	Solution of Dirac's Equation for the Hydrogen-Like Atom	583
	19.7	Approximate Versions of Dirac's Equation	584
	Problems		586
	Bibliographical Notes		587
Appendix A	**Mathematical Tools**		**588**
	A.1	Solving Second-Order Linear Homogeneous Differential Equations: Special Functions	588
	A.2	Mathematical Identities	596
	A.3	Curvilinear Coordinates	596
	A.4	Dirac's Delta	599
	A.5	The Gamma Function	600
	A.6	Solving Eigenvalue Problems by Means of Algebraic Consistency Conditions	601
	Bibliographical Notes		604
	Physical Constants		605

References   606
Index   620

	17.5	Canonical Field Quantization Clarified	519
	17.6	*Second Quantization. An Introduction	522
	Problems		529
	Bibliographical Notes		530

18 Elastic Scattering Theory — 531
- 18.1 Collision Phenomena — 531
- 18.2 The Born Approximation — 537
- 18.3 Form Factors — 544
- 18.4 Partial-Wave Expansion. Introducing the S Matrix — 547
- 18.5 *Resonant Scattering — 554
- Problems — 558
- Bibliographical Notes — 560

19 Relativistic Equations. An Introduction — 561
- 19.1 The Klein–Gordon Equation — 561
- 19.2 The Dirac Equation — 566
- 19.3 Properties of the Dirac Equation. The *Zitterbewegung* — 572
- 19.4 The Free Particle — 576
- 19.5 *The Dirac Particle in an External Electromagnetic Field — 580
- 19.6 Solution of Dirac's Equation for the Hydrogen-Like Atom — 583
- 19.7 Approximate Versions of Dirac's Equation — 584
- Problems — 586
- Bibliographical Notes — 587

Appendix A Mathematical Tools — 588
- A.1 Solving Second-Order Linear Homogeneous Differential Equations: Special Functions — 588
- A.2 Mathematical Identities — 596
- A.3 Curvilinear Coordinates — 596
- A.4 Dirac's Delta — 599
- A.5 The Gamma Function — 600
- A.6 Solving Eigenvalue Problems by Means of Algebraic Consistency Conditions — 601
- Bibliographical Notes — 604
- Physical Constants — 605

References — 606

Index — 620

Preface

Textbooks are usually designed to introduce students to a new subject, to show them the ways in which it can be learned, and to give them the tools to use and apply it. The purpose of this book is also to provide the student with an adequate physical understanding of the subject – which, in the case of quantum mechanics, is particularly non-trivial.

Quantum physics has come a long way since its inception a little over a century ago. Its applications are expanding every day, many of them with important technological implications. For the new generations, life is unimaginable without the products of quantum physics, which have become part of everyday reality. We are often unaware that many devices that have revolutionized society through optical communications, high-speed internet, high-performance computing, materials by design, and imaging techniques, to name but a few, are based on quantum mechanical effects.

However, these successes come with a caveat: the difficulty – even for the expert – of obtaining a satisfactory physical understanding of quantum phenomena. The impossibility of direct access to the inside of the atom and other quantum constituents of matter has resulted in a theory that still relies on some abstract or mathematical postulates introduced by its founders, in lack of a causal explanation on physical grounds. Instead, a wide range of interpretations can be found, some of which may even contradict fundamental physical principles. This state of affairs is reflected in research papers, conference debates, online courses, … and textbooks. It even goes beyond the boundaries of science and contributes to the widespread view that "the quantum world is weird."

Our aim is to provide the reader with a picture of quantum phenomena that is at once as transparent and internally coherent as possible and compatible with the rest of physics, and that serves as a sound conceptual basis for more advanced courses in quantum theory or its many applications. It is not another interpretation, but a more complete picture that we, together with other colleagues in the field, have arrived at through decades of fruitful research into the physical foundations of the theory. Research that, by considering all the relevant physical ingredients that make up quantum systems, enables us to find a causal explanation for the theory's postulates.

This book is intended primarily for undergraduate and graduate courses in physics, chemistry, engineering, and materials science. Students in quantum information and communication, nanotechnology, quantum chemistry, and other physics-related fields, particularly philosophy of physics, may also find it useful as a reference or supplementary textbook. For this reason, the book covers the basic material of the university curriculum and includes material for more advanced courses or specific needs in sections marked with an asterisk in the Contents list. It emphasizes the theoretical and conceptual aspects and illustrates them with applications of current relevance. Where appropriate, it provides explanations or derivations not found in the conventional literature in order to give a more complete account of the underlying physics.

Some relevant historical, biographical, and bibliographical information is given throughout to allow the reader to appreciate the impressive development of quantum theory since its inception and the contributions of its main pioneers. Where necessary, background elements of classical physics and mathematics are also included. The text includes illustrative exercises throughout, as well as a selection of problems at the end of each chapter. Its educational value is enhanced by an accompanying online book with detailed solutions, which is available to students and instructors at https://www.cambridge.org/cetto-delapena. Due to space limitations, the number of solved exercises and problems is limited; the student is encouraged to take advantage of the wide variety of exercises and applications of quantum mechanics available on the internet.

The book may be of particular help to those interested in quantum theory who have been dissatisfied or confused by the conventional literature on the subject. We hope that our perspective on quantum mechanics can appeal to the curious mind, young or otherwise, to delve with confidence into the normally complex and mysterious depths of quantum nature, and to see the quantum world as part of the world that can be understood and explained.

We invite you to join us on this exciting journey of understanding.

This book is the product of many years of research and teaching on the subject. We are grateful to Fondo de Cultura Económica for allowing us to use material from an earlier textbook, *Introduccción a la Mecánica Cuántica*. We would also like to thank everyone who supported us in preparing this volume. Special thanks go to Alejandro Aguilar Sierra, who generously shared his Metagráfica program to draw the figures, and to Carlos López Nataren, who provided consistent, timely technical assistance. We are grateful to Cristo Sánchez Nicolás and Eduardo Varela Ortuño, our students, for their work on the figures and tables, extended figure descriptions, and thoughtful review of the manuscript. We would also like to thank our Ph.D. student and young colleague, José Francisco Pérez Barragán, for carefully reading the final proofs. Our colleague and friend John Bush from MIT provided helpful comments and suggestions on the topic of hydrodynamic quantum analogs. We are likewise grateful to the Instituto de Física for providing working conditions that enabled us to devote ample time to writing and preparing this volume. Lastly, we thank the staff at Cambridge University Press for their encouragement and valuable support throughout the entire editorial process.

1 The Quantum World and Its Description

Quantum mechanics (QM) is just over a century old. This is a short time for a fundamental theory of physics when you consider that it is expected to describe an important part of the world for centuries to come. Geometry, for example, is over two thousand years old and still going strong. Physical theories are much more recent, even Newtonian mechanics – the oldest branch of modern physics – is only three centuries old. Yet it is a well-established theory, so well established that it has successfully survived, one after the other, the unexpected birth of two mechanics, relativity and quantum mechanics, two whole new chapters of physics, both of which emerged in the twentieth century. The effect of the new theories on classical mechanics has essentially been to define its scope: we now know that it applies in general to systems that are not too small – that is, not atomic in size or less – and to speeds that are not too close to the speed of light. Newtonian mechanics is still used to describe the behavior of all sorts of objects and situations around us, as well as to design the most gigantic airplanes and ships, even spaceships, huge bridges, and all sorts of machines.

Quantum mechanics, for its part, has grown impressively in the first hundred years of its life, reaping success after success. It works flawlessly well in thousands of laboratories and millions of personal computers around the world, although its predictions are statistical in nature. However, like any scientific theory, QM is certainly not infallible; we just haven't reached the limits of its applicability yet.

This should not be a cause for concern: recognizing that there are limits to the application of a physical theory is neither unexpected nor extraordinary. On the contrary, identifying such limits is an important part of the daily task of science. When a new physical theory is born (and this applies to virtually all scientific endeavors), its limits of applicability – its *range*, in the language of T. A. Brody (1993)[1] – are unknown. In practice, much of the day-to-day work of research is, often unconsciously, to determine the range of the theory. Will it still work in my new application, at such high speeds, at low or high temperatures, over short or long distances? The scope of QM is not yet clearly defined, but its expansion into new domains and applications to more extreme scales will ultimately determine it.

Quantum mechanics is currently the most valuable tool we have for the study and understanding of the atomic and subatomic world. Quantum mechanics and the theories based on it have enabled us to explore not only the atomic nucleus, but also the structure of its constituent parts and their elementary components on the one hand,

---

[1] Thomas Alexander Brody (1922–1988) was a remarkable German–Mexican physicist, polymath, and polyglot. He was a central figure in the development of Mexican activity in physics, especially nuclear and statistical physics, as well as in computation and philosophy of science, always with a pronounced critical attitude.

and, seemingly paradoxically, the structure of matter from the nanoscale to the scale of the universe. In fact, today's cosmology recognizes that physics is ubiquitous and determines the large-scale dynamics of the cosmos, which emerges bottom-up from local dynamics at the small scale; there is no separate large-scale physics of the cosmos. Hence the importance of the continuing search for a quantum theory of gravity.

1.1 The Quantum World, Part of the Physical World

The three current branches of mechanics: classical mechanics, quantum mechanics, and the theory of relativity, each have to some extent their own domain of knowledge and application. The domain of QM is essentially (but not exclusively) the submicroscopic world, that is, the world of atoms, molecules, and their constituent elements, known globally as quantum particles, where the term "particle" must be used with due (quantum) care. The most important quantum particles are those that make up ordinary matter: electrons, protons, and neutrons, to which we should add – among others – the photon, the "particle" that represents the smallest exchangeable quantity of electromagnetic radiation of a given frequency. The first known submicroscopic particles, the electron and the proton, have gradually been joined by many others. Today, according to the so-called Standard Model, the list of fundamental quantum particles consists of the quarks and the gluons, the photon, the Higgs particle, and the leptons, as well as their antiparticles in all cases. If we leave the details to the specialists, we can consider about 30 particles – some elementary, the rest composite – and several hundred resonances. Quantum mechanics, in its diversity, applies to all of these particles and their possible combinations, although it is usually restricted to atoms, molecules, and systems composed of not too many particles, without excluding, however, some larger systems.

In fact, in our everyday world, QM extends very far beyond submicroscopic systems; there are even huge physical or technological systems whose basic functioning is quantum in nature. Well-known examples are the laser, a large crystal, a volume of superfluid, or even a superconducting magnet that can weigh tons. Quantum mechanics is a central theoretical corpus for a wide range of contemporary subjects and applications, starting at the level of today's most fundamental theory, quantum chromodynamics (QCD), with its "colored" quarks and gluons – the constituents of protons and neutrons, along with the rest of baryons and mesons – which in turn are the constituents of atomic nuclei, which along with electrons are the constituents of atoms, which in turn combine to form molecules. Quantum phenomena at the nanoscale are the focus of increasing attention and may even become a key to understanding phenomena as complex as the functioning of the human brain. Quantum mechanics is also having an astonishing impact on current developments in astrophysics, where it plays an important role in the study of the evolution of the universe.

In recent decades, QM has opened its doors to *information theory*, which as a field of knowledge dedicated to the study of the quantification, storage, and communication of digital information, is at the crossroads of probability theory, statistics, statistical mechanics, computer science, information engineering, and even electrical and electronic engineering. With the support of quantum theory, quantum information has found very important and extensive applications, of which quantum cryptography is the best known.

A word of caution, however: it is true that we are made up of protons, neutrons, and electrons, which are quantum particles, but that does not make us quantum persons. When systems made up of millions of particles reach the macroscopic scale (our scale), various factors intervene that cause their behavior to be governed by the laws of classical physics. A prime example of the need to make this distinction is the popular Schrödinger cat, discussed in Section 14.5.3.

So although QM has a relatively well-defined domain of knowledge, its applicability depends on the nature of the system, its size, its temperature, and so on, and even on the purpose of the study. Take a small piece of matter: in general, the energies involved in quantum processes are very small, so that as the temperature rises, thermal fluctuations may be sufficient to disrupt proper quantum behavior. For a system of macroscopic dimensions, on the other hand, classical physics is usually the natural choice; however, only a quantum treatment gives the correct optical and electronic properties of a crystal. The very existence of nanophysics warns us against the idea of a dichotomy with the quantum world in opposition to and separate from the classical world: there is a continuous spectrum in the physical world where different theories coexist, each with its own laws, overlapping here and there and contrasting in their demands.

1.1.1 The quantum toolkit

As with other branches of physics, dealing with the quantum world has required the development of a range of theoretical, experimental, and computational tools and techniques. Much of the success of QM lies precisely in these tools, which have reached a high degree of sophistication.

Immediately after the publication of the seminal papers by Heisenberg and Schrödinger, which laid the foundations for modern QM, a number of scientists, mainly physicists and mathematicians – including the founders themselves – began to test the basic equations and apply them to the simplest problems at hand. The application of algebraic methods developed by mathematicians furthered the systematic development of matrix mechanics, and methods of solving partial differential equations applied to physics were developed to deal with the Schrödinger equation. Since only a handful of simple problems were found to be amenable to an exact solution, a variety of approximation methods were developed to deal with problems that did not admit of an analytical solution or whose solution was too complex for practical application, perturbation theory being the most extended. Methods for dealing with composite or many-body systems, or systems in a highly excited state, or problems involving a time-dependent interaction, or statistical methods for dealing with an ensemble of similar quantum systems ... are now widely used. Most of these tools and methods, which form the core of the present text, are presented at an appropriate level in many QM textbooks, which the student is advised to consult regularly.

At the same time, new experimental techniques were developed to confirm the predictions of the nascent theory and to discover additional features of quantum systems that were promptly taken up by theorists for further elaboration and interpretation.

The result was a quantum effervescence around 1900 to 1927 – to define a period – that had a major impact on physics at the time and has not abated over the years. New instruments based on quantum phenomena, such as microscopes using the wave properties of electrons, quantum tunneling or atomic force microscopes, lasers based on the stimulated emission of radiation, optical tweezers based on nonlinear quantum optics, diagnostic tools based on NMR (nuclear magnetic resonance), the GPS (global positioning system) based on the astonishing precision of atomic clocks, related paraphernalia based on solid-state physics, and so on, have become the daily bread of modern laboratories and common technological and medical applications. In various branches of quantum physics, data-intensive calculations and computer simulations have become an important complementary tool.

1.1.2 Scope and purpose of this book

In principle, each of the preceding topics is a subject of study for QM. In practice, however, an undergraduate textbook is largely confined to the development of the basic theory, the exposition of quantum formalism, and its applications to the study of the lightest atoms and molecules, electrons bound or scattered by simple potentials, and elements of condensed matter, including crystals and metals. Depending on the authors' preferences or areas of research, some texts (the majority) focus on the formal aspects of the theory, others on the more fundamental and conceptual aspects, or on topics of current practical interest. Given the wide range of physics that QM touches on, it would be illusory to attempt to produce a truly comprehensive textbook, or to pretend that this one is. The present book is conventional in the sense that it introduces the student to the main formal and practical aspects of QM, but at the same time it is unconventional in the sense that it systematically strives to provide physical explanations that are as clear as possible and, where feasible, novel elements aimed at enabling the student to acquire a coherent picture of the physics underlying quantum formalism. In short, this book discusses standard QM, but from a nonstandard constructive perspective.

The student may have encountered the favored slogan "shut up and calculate"[2] or been told that their questions are "illegitimate questions." In science there are no illegitimate questions, there are only illegitimate answers. The student soon discovers that *understanding* QM is much more complex than *learning* QM; that he or she has learned to calculate without a true understanding of the underlying physics. Indeed, modern QM is in some ways cryptic and enigmatic, in stark contrast to the clarity and elegance of its formal apparatus. In preparing this textbook, the authors have been mindful of these difficulties and, rather than avoiding them, have tried to pave the way, as far as possible, to an understanding of the physics behind the mathematics. This aim is reflected both in the content and in the realistic and objective argumentation throughout the book.

---

[2] Concerned about attitudes toward QM and feeling uncomfortable with the Copenhagen interpretation (see Section 1.3.1), David Mermin (1989) wrote: "If I were forced to sum up in one sentence what the Copenhagen interpretation says to me, it would be 'Shut up and calculate!' " This meme has become part of modern quantum folklore because of its telling power.

1.2 Is Quantum Mechanics a Complete Theory?

As some physicists and philosophers of science have pointed out, despite its successes, current QM lacks adequate explanations for important observed atomic properties. Among those that have attracted the most attention are quantum fluctuations, the stability of atoms, atomic transitions, and even quantization itself. Let us briefly review these topics.

In the years 1924–1926, the present-day QM was formulated independently by L. de Broglie,[3] W. Heisenberg,[4] and E. Schrödinger,[5] who were separated by several hundred kilometers and by opposing points of view. At the time, Heisenberg, who had just given birth to his highly successful theory that would soon become known as *matrix mechanics*, was still tinkering with his strange (at least to physicists of the time) but successful algebra. Heisenberg's observations were highly significant: he realized that, in contrast to the classical case, his quantum description in terms of matrices – a form of operators, as we will see in Chapter 3 – for the dynamical variables of atomic electrons implied inevitable fluctuations. In the absence of an alternative explanation for such behavior, he decided to accept the fluctuations as inherent to quantum particles in general, since they were all described by the same kind of matrices. Thus, a few months after his careful investigation, he added the notion of *quantum fluctuations* to his theory as a *postulate*, namely that the quantum particles possess an *intrinsic* (*irreducible*) and *permanent* fluctuating motion. He then developed the subject further and arrived at a universal relationship between the position and momentum fluctuations of a quantum particle. This relationship became known as the *uncertainty principle* or *indeterminacy principle* (depending on personal preferences)[6] and was expressed as the "uncertainty relation" or "indeterminacy relation" – both popular names for the Heisenberg inequality. The immediate and widespread success of this work made Heisenberg an eminent public figure, and his uncertainties or indeterminacies have been immoderately extrapolated to other areas of knowledge, acquiring various scientific or philosophical, even mystical, connotations.

The general acceptance of quantum fluctuations as a basic *postulate* of QM has several major implications. At the time of its introduction, there was no evidence for a mechanism that could explain random motions in quantum systems (and this remained so for decades). Fluctuations of the quantum variables were observed, they were a fact; but the postulate that they were intrinsic – or innate, with no known explanation – turned QM into a noncausal theory, a theory that denied the existence of a possible cause for the fluctuations. Moreover, the theory under construction was

---

[3] Louis Victor Pierre Ramon de Broglie (1892–1987) was an eminent French theoretical physicist who was awarded the Nobel Prize in 1929 for his discovery of the wave properties of matter.

[4] Werner Karl Heisenberg (1901–1976) published his groundbreaking paper on quantum theory in 1925 (at the age of 24), for which he won the Nobel Prize in 1932. His scientific career was exceptional, although it was marred by his work for the German regime during the Second World War (1939–1945).

[5] Erwin Rudolf Josef Alexander Schrödinger (1887–1961) was an eminent Austrian theoretical physicist and philosopher, one of the creators of QM, for which he shared the 1933 Nobel Prize with P. A. M. Dirac.

[6] It should be noted that the terms "uncertainty principle" and "indeterminacy principle" have different connotations: the former refers to *our* impressions or to *our* knowledge and the latter implies an actual lack of determinism in the theory.

indeterministic because of the postulate of irreducible, uncertain, and unpredictable random motions. In short, Heisenberg's matrix mechanics led to a noncausal and indeterministic theory. However, this radical departure from central tenets of classical science came about through a postulate, not as the inevitable result of direct experimentation.

Let us look at the question from the atomic perspective. For simplicity, we consider the hydrogen atom. Its quantum mechanical description can be carried out in detail and with exact analytical procedures using the Schrödinger equation, as will be seen in Chapter 12. This equation contains the Coulomb force between the nucleus (a single proton) and the electron that make up the atom; no other force appears explicitly. Although this force alone cannot be held responsible for the quantum fluctuations, the fact remains that QM *correctly* predicts the energy spectrum of the hydrogen atom and several other of its important atomic properties. So the cause, the reason, the explanation of quantum fluctuations is hidden in the quantum formalism.

Next we consider the issue of atomic stability. Atomic electrons, being charged particles undergoing complicated motions, are expected to radiate as any charged particle does when accelerated. This means that the atom should lose energy and collapse after a short time. But atoms are known to be stable. Returning to the Schrödinger equation, we find that the Coulomb force, the only force that appears in the description of the atom, is also unable to explain the source of the extra energy that is constantly and continuously required to stabilize the system. Again, something is missing – or hidden in the formalism.

Related to the preceding is the question of atomic quantization. Formally, there is no problem: the quantum-mechanical formalism leads to the existence of quantized states. But *physically* it is a puzzle: what is the mechanism that allows some energy states but forbids others? There must be some (hidden) balancing process behind the well-defined orbits corresponding to the stationary quantized solutions so accurately predicted by Schrödinger's theory. Formal explanations along the lines of "each quantum state corresponds to a local minimum of the energy of the distribution of particles," may be interesting and correct, but they do not solve the puzzle. You will see that textbooks on QM usually consider the matter to be solved by mathematics, and in fact the mathematical explanation is the best one you can get – indeed the only one – as long as you give up trying to understand the phenomenon.

To this we can add the question of "spontaneous" transitions between atomic states, which QM does not account for; one has to resort to quantum electrodynamics (QED) to find an explanation. The fact is that these transitions take on a spontaneous appearance when an approximation is made to obtain QM from QED, which means that something important (the cause) contained in QED is lost in the process.

We conclude that although QM is correct in terms of its formal apparatus – and we may add its extended predictive capacity – it is a theory that has never failed – it is an *incomplete* theory in that it lacks one (or more?) crucial element(s) for its full understanding. In other words, its *ontology* – the collection of entities that make up the quantum world and their relationships – has yet to be fully deciphered.

1.2.1 On the seat of quantum phenomena

An important aspect that makes QM difficult to understand is that it expresses all its results as if they occurred in an abstract Hilbert space,[7] rather than in the spacetime of our everyday experience. Matrix mechanics was born as a particularly obscure theory, so much so that it was immediately qualified as not visualizable; H. A. Lorentz[8] spoke out against this obscurity at the Fifth Solvay Conference in 1927 – where QM as we know it today took root – and demanded a clear and visualizable description in our spacetime: "I should like to preserve this ideal of the past, to describe everything that happens in the world with distinct images. I am ready to accept other theories, on condition that one is able to re-express them in terms of clear and distinct images."[9] It is instructive to compare this demand with the following comment by W. Pauli[10] concerning the electron spin: "The lack of a concrete picture is the most satisfactory state of affairs."[11] These contrasting remarks by two of the most eminent theoretical Nobel-Prize-winning physicists illustrate well the two poles of theoretical thinking in Western philosophy under the influence of a realistic versus an idealistic outlook.

Somehow, a search along Lorentz's direction was proposed by H. Groenewold[12] and simultaneously and independently by J. E. Moyal,[13] each building on earlier (1934) ideas by E. Wigner.[14] At present, however, there is no satisfactory statistical phase-space description of QM, as there is, for example, of classical statistical mechanics; no quantum theory describes in detail in real spacetime what is happening, for example, inside an atom, what an electron is doing right now. This makes it difficult to think of QM today as a well-developed, finished statistical theory. Chapter 3 introduces the reader to Heisenberg's matrix mechanics, and Chapter 10 contains the explanation provided by stochastic electrodynamics (SED) for the emergence and physical meaning of the quantum operators underlying the Hilbert-space formalism. This, in turn, will explain the reason for the inability of current QM to provide a (visualizable) phase-space description.

---

[7] David Hilbert (1862–1943) was a highly influential German mathematician.
[8] Hendrik Antoon Lorentz (1853–1928) was an influential Dutch theoretical physicist, who shared the 1902 Nobel Prize with Pieter Zeeman for the Zeeman effect.
[9] Quotation taken from Bacciagaluppi and Valentini (2009).
[10] Wolfgang Ernst Pauli (1900–1958) was an eminent Austrian theoretical physicist and mathematician, a member of the group of young physicists – the *wunderkinder* – who developed quantum theory in its Copenhagen version, including quantum field theory and QED. He was awarded the Nobel Prize in 1945 for his contributions, in particular the *exclusion principle*.
[11] Quoted by B. L. van der Waerden (1960), p. 202.
[12] Hilbrand Johannes Groenewold (1910–1996) was a Dutch physicist, who in his PhD thesis in 1946 sought a phase-space description of QM.
[13] José Enrique Moyal (1910–1998) was an Australian mathematician and physicist.
[14] Eugene Paul Wigner (1902–1995) was an important Hungarian–US theoretical physicist, who was awarded the Nobel Prize in 1963 for his contributions to the development of quantum theory.

1.3 Main Issues in the Interpretation of Quantum Mechanics

The current QM is not only incomplete, but an unfinished theory. Any physical theory consists of at least two parts, one being its formal (logical) apparatus and the other being the set of specific meanings attributed to its nonlogical elements (the semantics). This is the process that gives the formalism a physical meaning; in short, that links mathematics and nature, identifying, for example, its symbols $m, f, a, E \ldots$ with mass, force, acceleration, electric field, and so on. The formal apparatus of a theory is well defined, but such identifications (correspondences) are freely assigned, which opens the door to more than one specific reading and allows it to change according to personal preferences or the trends of the time.

In a mature theory all identifications are well defined and consolidated by practice, so there is a single and universal interpretation, a fixed semantics whose meanings cannot be changed without changing the theory itself. Nobody pretends to change the identifications of Newtonian mechanics or Maxwellian electromagnetism. But this is not the case with QM, which is plagued by around two dozen interpretations.

Quantum phenomena normally take place on a scale of length and time that does not allow them to be observed directly: we cannot follow the trajectory of an atomic electron, for example, or its transition from one state to another, or the paths of individual particles in a double-slit array. The quantum formalism has been built on indirect information, extracted, for example, from atomic spectra, or the interference pattern registered on the detection screen, or the various physical properties of bulk matter. The lack of direct contact with the individual quantum objects thus leaves considerable room for interpretation.

And indeed, since its inception, a number of different interpretations of QM have emerged; some are more popular than others, but none is universally accepted. Philosophical precepts have an important influence on the overall interpretation of the quantum formalism and on the meaning assigned to the elements of the description. This makes the subject extremely rich and complex, so much so that a serious analysis of it would require at least one specialized volume. Nevertheless, considering the importance of the subject, we shall briefly discuss some of the salient points on which the various interpretations differ. The interested reader is invited to consult the relevant bibliography at the end of the chapter; in particular, Wikipedia (2025) contains a detailed discussion of some of the interpretations.

1.3.1 Realism and idealism

At the risk of unwarranted simplification – a risk that we are forced to take for the sake of brevity and simplicity – we can say that usually either a realistic or an idealistic (even mystical) position is at stake. Take, for example, the wave function ψ, which is used in QM to describe the state of a system. According to the realistic interpretation adopted throughout this book, ψ refers to a physical system – be it a particle, an atom or any other object – that has an existence of its own. By contrast, the view expressed by Pauli (1960) that "the concept of a material object, of a constitution and nature independent of the observer, is foreign to modern physics, which, forced by the facts, has had to renounce this abstraction," is idealistic, by implying that the object vanishes to be replaced by the world of ideas that appear in the mind of the researcher. Another example of the idealistic interpretation is the statement by Heisenberg (1958) that an elementary particle "is not a material formation in space and time, but in some way a symbol, adopted by which the laws of nature acquire a particularly simple form."

Furthermore, according to an objective view, ψ describes a system's physical state that follows its own laws, governed by an equation (the Schrödinger equation) without our intervention, that is, independent of us, our sensory experience, or our knowledge or ignorance of it. By contrast, according to the subjective view, ψ refers to our knowledge of the system. Thus it has been said that "quantum theory does not refer to nature, but to our knowledge of nature" (Heisenberg, 1951). Similarly, probabilistic statements in quantum theory are often interpreted as subjective judgments, as a measure of our ignorance of the system, rather than, for example, as statistical assessments of the fraction of favorable cases that will occur in the given situation, independent of any observer, who may or may not exist.

A scientist, especially a classical physicist, tends spontaneously to take a realistic and objective view when investigating nature. However, this view is neither the only possible nor the most common. In the mainstream empiricist interpretation, ψ is considered to refer only to properties of the observational data that relate to the system (which may or may not have its own independent existence, depending on the school of thought). In such a view, since the data can be influenced by our observations, it is the latter that define the state: without observations (without an observer), there is no definite state. The assumption that the observed system and the observer are logically inseparable at the quantum level is a central feature of the *orthodox interpretation* of QM, conventionally called the *Copenhagen interpretation*, with N. Bohr[15] (who lived and worked in Copenhagen) as one of its main founders, and Heisenberg, M. Born,[16] and P. A. M. Dirac[17] as its initial supporters. However, the term "orthodox" should not be taken to imply unanimity of opinion, but only the existence of certain features common to the various strands of this hegemonic school.

1.3.2 Individual vs statistical meaning of ψ

Another major issue is whether ψ refers to the state of a single system, or to that of a representative member of an ensemble of equally prepared systems. The latter is the

---

[15] Niels Henrik David Bohr (1885–1962) was a Danish physicist who made fundamental contributions to the understanding of quantum atomic structure and, more generally, to quantum theory; in particular, he proposed the first model of an atom with discrete energy levels, for which he was awarded the 1922 Nobel Prize in Physics. As a major promoter of scientific research, he succeeded in mainstreaming the Copenhagen version of QM.

[16] Max Born (1882–1970) was a German theoretical physicist and mathematician, a distinguished mentor, and key figure in the development of QM, for which he was awarded the Nobel Prize in 1954. During the early development of quantum theory, Born was a central figure in Göttingen, the *alma mater* of famous mathematicians and theoretical physicists.

[17] Paul Adrien Maurice Dirac FRS (1902–1984) was an English engineer, mathematician, and theoretical physicist, considered one of the most important scientists of the twentieth century. He made notable contributions to QM and quantum field theory, for which he shared the 1933 Nobel Prize with Schrödinger. It is commonly considered that his was the first thesis on quantum mechanics.

statistical or *ensemble interpretation* of QM, which was first proposed by J. C. Slater[18] in 1929 and has been supported over time by a range of physicists, including highly relevant figures such as A. Einstein,[19] Schrödinger, K. Popper,[20] D. I. Blokhintsev,[21] and others – but still enjoys limited popularity.

Much more widespread is the orthodox interpretation, according to which ψ describes the state of an *individual* system. This fundamental discrepancy gave rise to the famous polemic between Einstein, on the one hand, and Bohr and Born, on the other, which lasted for several decades, until Einstein's death in 1955, and is documented – albeit only partially – in several very instructive compilations (see the "Bibliographical Notes" section at the end of the chapter). For example, in his 1953 essay in honor of Born, Einstein showed that the classical limit of the wave function for particles bound to a potential well describes an ensemble of particles, not a single particle, and argued that since the character of the description cannot change by passing to the limit, one must conclude that in general ψ describes an ensemble. Similar reasoning has led to a number of results that are natural if one adopts the ensemble interpretation, but paradoxical if one adopts the orthodox one; we will encounter some of them throughout this book. The so-called *wave-function collapse* or *reduction* introduced by J. von Neumann,[22] which is discussed in Sections 14.5.3 and 15.5.7, is a notable example.

Within the ensemble interpretation of QM it is possible to overcome some obscure conceptual or philosophical points, and even others of a more straightforward physical nature. This is ultimately the reason for our choice. It is important to note, however, that this choice is not exempt from difficulties. An example that sums up the problem well is that although the quantum description is essentially statistical, it is generally incapable of providing a coherent statistical description in phase space, as was mentioned earlier.

Another question that has raised serious difficulties of interpretation relates to the nature of the quantum corpuscle: is it a wave, or a particle, or both in a complementary way, or a *sui generis* entity, neither wave nor particle, but possessing the properties of both? This topic will be discussed in detail in Section 2.8, where it will be argued that in the experiment of electron diffraction, the electrons *are and remain particles* all the time; they are simply distributed in space according to the structure of a diffraction pattern; in short the "wave" pattern is a multi-particle drawing.

The example just discussed shows that ψ does not fully describe the behavior of a single particle, that is, it can only give an incomplete description. To claim that the

---

[18] John Clarke Slater (1900–1976) was a US-born physicist, best known for developing the theory of the electronic structure of atoms, molecules and solids and for his contributions to microwave electronics.

[19] Albert Einstein (1879–1955). Undoubtedly the most famous scientist of the twentieth century, he was inter alia the creator of special and general relativity. Einstein was a major contributor to the foundation of quantum theory and a severe critic of the direction the theory took under the influence of the Copenhagen school.

[20] Karl Raimund Popper (1902–1994), born in Austria, was one of the most influential philosophers of science of the twentieth century, best known for his theory of empirical falsification.

[21] Dimitri Ivanovich Blokhintsev (1908–1979) was a prominent Soviet theoretical physicist, founder and first director of the Joint Institute for Nuclear Research in Dubna.

[22] John von Neumann (1903–1957) was a Hungarian–US mathematician, physicist, computer scientist, engineer and polymath. Among many other contributions, he pioneered the application of operator theory to quantum mechanics, which is contained in his influential work *Mathematical Foundations of Quantum Mechanics*, published in 1932.

description is complete would be tantamount to considering that nature possesses elements of unknowability that are beyond our investigation. This view, which is usually inspired or reinforced by philosophical considerations, contrasts with the view that follows from nonquantum physics and science in general, that the world is knowable in principle. This difference in perspective is of practical importance, since resolving the question in one direction or the other leads either to inquiring about the origin and nature of, for example, stochasticity and indeterminism, or to refraining from doing so because the question is considered meaningless. In turn, the conclusion that the description provided by ψ is incomplete raises the question: Can it be completed? Is it possible to refine the current quantum description and produce a deterministic theory that will allow us to fully predict the outcome of each and every experiment and get rid of the uncertainties? An informed discussion of this thorny question is left for Chapter 14, in the context of the Bell theorem,[23] as applied to the entangled spin state.

1.3.3 (A)causality and (non)locality

A related point of discrepancy between the various schools of thought is the issue of causality or acausality of quantum phenomena. A simple way of framing this dilemma is as follows. Once we agree that every electron behaves unpredictably at some point in time, it is natural to ask what is the cause of this random behavior. For the orthodox school, it is an irreducible phenomenon, the result of *intrinsic quantum fluctuations*; there is nothing more to explain. From a causal perspective, on the other hand, the phenomenon is in principle explainable, even if the description is indeterministic, at least for the time being – or simply for practical reasons. Although this point is of secondary importance for the exposition of QM in its present form, it is by no means so for future research. For if this behavior is thought to be caused by a physical agent – external or internal – such an agent requires careful physical investigation, which opens up a wide field of study.

The causal school of thought is often identified with the notion of hidden variables, according to which we should locate the cause of stochasticity in the existence of physical variables that are still unknown, that is, that remain hidden at the level of description of current QM. It has become commonplace to deny the possibility of their inclusion on the basis of several theoretical results, notably the Bell theorem and its variants, pre- or post-Bell. However, as will be discussed in Section 14.5, not only is the validity of these theorems controversial, but realistic hidden-variable theories that limit the kind of allowed hidden variables have been constructed, proving their possible existence, at least in principle.

Another issue in QM is that of nonlocality; since several parts of this book are concerned with the subject, we will limit ourselves here to a couple of observations.

The fundamental equation of QM, Schrödinger's equation, is a theoretical construct *sui generis*, as is explained in Section 4.4. The point is that this equation contains

---

[23] John Stuart Bell (1928–1990) was an influential Irish physicist who proposed the theorem on hidden variables that bears his name.

nonlocal elements that can affect any result derived from it. We can see this from its derivation in Section 4.4, where we find that the equation contains, though somehow hidden, the de Broglie relation $\lambda = h/p$, which is statistical and nonlocal in nature. Such properties can be transferred to any system it deals with; for example, by opening a new slit in a multislit diffraction experiment, the wave function $\psi(x)$ is modified everywhere, which means that the spatial distribution of particles $\rho(x)$, which is equal to $\psi^*\psi$, is also modified everywhere. Something similar happens with a vibrating string attached at both ends: by moving the position of one of the end points you change the shape of the whole string.

For the orthodox physicist, there is an additional, different kind of nonlocality associated with the measurement process, manifested in the collapse of the wave function. Such collapse involves the spontaneous reduction of an extended wave function to a simpler more localized one by an act of measurement or observation; this is an extreme example of a nonlocal quantum phenomenon. At present, some orthodox physicists are seriously trying to solve the collapse problem (or, more generally, the measurement problem) by introducing concepts such as spontaneous decoherence and the like (see Section 15.5.7).

Although the exposition in this book goes against the grain of the orthodox interpretation, it is useful for the reader to familiarize themself with it, as well as with other interpretations, on which there is an abundance of literature.[24]

1.4 Completing the Quantum Ontology

In Chapter 2 it will become clear that quantum matter is in constant contact with the electromagnetic radiation field. This field contains a random, nonthermal, space-filling component, the zero-point field (ZPF), discovered by M. Planck in 1911.[25]

The ZPF, as we know it today, is probably a product of the evolution of the universe. Considering that the background field is created and maintained by distant matter, it is natural to assume that it is of cosmological origin, and therefore that $\hbar$ is related to other universal constants. In a state of equilibrium, the charged particles radiate and contribute to the background field, while simultaneously absorbing energy from it. The cosmic microwave background (CMB) is the visible remaining part of the fossil radiation that was released shortly after the Big Bang and is absorbed by detectors; the ZPF is the *invisible* (nonphotonic) part of the background radiation. The field thus regenerated coincides on average with the stationary field at any frequency. This kind of

---

[24] As a rule, textbooks use the orthodox interpretation in one or other of its many variants. A notable exception is Sokolov, Loskutov, and Ternov's book *Quantum Mechanics* (1960), which uses the ensemble interpretation; unfortunately the arguments used throughout the original Russian text to justify their approach have been suppressed in the English translation. Other exceptions are the textbooks by de la Peña (1979), Bohm (1989), and Ballentine (1998).

[25] Max Karl Ernst Ludwig Planck (1858–1947) was a German theoretical physicist who made major contributions to theoretical physics. He is best known for his discovery of the quantization of energy, for which he was awarded the Nobel Prize in 1918. He introduced the term 'quantum'.

electromagnetic Mach principle[26] establishes a relationship between the cosmological and atomic constants and leads to (essentially) the correct value of the Planck constant.

As will be shown in Chapter 10 and elsewhere, the ZPF is key to understanding quantum fluctuations, wave-like behavior and apparent nonlocality, atomic stability, electron transitions between states, and quantization. In short, QM recovers the lost causality to become an intelligible causal stochastic theory. As the text progresses, we will have the opportunity to delve more deeply into such issues.

Bibliographical Notes

The following is a selection of texts on the foundations of QM, ranging from the elementary to the professional level.

Bacciagaluppi and Valentini (2009). An informative presentation of what happened during the pivotal Fifth Solvay Conference on quantum theory.
Ballentine (1970, 1972). A useful presentation of the arguments in favor of the ensemble interpretation of QM.
Born and Einstein (2005). A compilation of the Born–Einstein letters up to Einstein's death, with comments (and sometimes interpretations) by Born.
Brody (1993). A rich (high-level) discussion of some of the conceptual problems of QM, by the informed German-Mexican physicist-philosopher Thomas Brody.
Einstein (1948). Einstein's personal discussion of the EPR paper, Einstein, Podolsky, and Rosen (1935). Reproduced in Born and Einstein (2005).
Home (1997). A valuable discussion of the statistical interpretation.
Longair (2013). A textbook that provides a careful account of the conceptual development of QM during the first three decades of the twentieth century.
de la Peña and Cetto (1997). Estimate of Planck's constant based on cosmological parameters.
Ross–Bonney (1975). A discussion and proposal of the statistical interpretation of QM.
Selleri (1983). A realistic (Bohmian) discussion of the meaning and contents of QM. This book exists in several languages; the author himself was a polyglot.
Wick (1995). A highly recommended discussion of the supposed line separating quantum and classical worlds.
Wikipedia (2025): https://en.wikipedia.org/wiki/Interpretations_of_quantum#mechanics#Influential_interpretations. An informative discussion and comparison of some of the interpretations of QM.

---

[26] In a nutshell, for our purposes, Mach's principle can be taken to mean that *local physical laws are determined by the large-scale structure of the universe*. In the cosmological community, this is a relatively extended view.

2 Matter and Field

Quantum mechanics is significantly different from classical mechanics in a number of ways. To establish this theory, old ideas had to be discarded and well-established principles had to be modified or even abandoned. For example, it was first necessary to be convinced of the physical reality of the atomic structure of matter; then to show that Newtonian mechanics is not directly applicable to the study of the atom; later to show that Maxwell's[1] electrodynamics alone does not describe all the elementary processes of interaction between an atom and other quantum particles, and so on.

The moral is obvious, but takes time to grasp: in order to understand QM, we need to free our physical intuition from much of the thinking generated by everyday prequantum experience, and reread nature as it presents itself to us with new evidence and phenomena, in as receptive but critical and objective an attitude as possible to our claimed new knowledge. One of the first steps, then, is to determine *which* prequantum ideas needed to be adapted, adjusted, or replaced, and *in what sense* they needed to be modified or replaced. This is the purpose and essence of this chapter. The following chapters open the door to the two pillars on which the quantum edifice as we know it today has been built, namely Heisenberg's rather abstract formulation and the more accessible Schrödinger theory. Once the student is familiar with the basic conceptual content of QM, he or she can then naturally move on to the more formal treatment of the subject in the following chapters.

2.1 Time of Rupture

By the beginning of the twentieth century, a number of fundamental problems had accumulated for which physics had no answers. This was partly due to the fact that a large amount of spectroscopic evidence of atomic and molecular origin had been collected during the nineteenth century, which needed to be explained. In addition, by the end of the century several unexpected radiations had been discovered, including X-rays and some manifestations of radioactivity. Experimental physics (and serendipity) opened the doors to the future. We are not talking about more or less complex

---

[1] James Clerk Maxwell (1831–1879) was a Scottish theoretical physicist and mathematician, considered one of the greatest scientists of all time. He formulated the classical electromagnetic theory, which unified electricity, magnetism, and light as an electromagnetic phenomenon, and established fields as an essential part of nature, on a par with matter. He also contributed to statistical physics (the classical distribution is called the Maxwell–Boltzmann distribution).

problems that arose from the lack of a method, given the theory – as would be the case, for example, with the three-body gravitation problem, which is still alive and well – but about something more fundamental: problems that escaped the context of the physical theories that existed at the time, or for which at best (or at worst) incorrect solutions were obtained. In other words, it was understood that the physical laws known at the time were applicable to a certain (admittedly very broad) type of systems or phenomena, but that they left out much of physical reality, which became less and less explainable the more it was studied with the tools at hand.

This situation led to a questioning of the theories themselves. The intense crisis that ensued succeeded in shaking the conviction prevalent at the end of the century that physics was already a practically complete science, and forced recognition of the need for a fundamental change in part of the existing conception of the physical world.

The complexity of the problems encountered triggered a search that spanned several decades of intense and profoundly creative work, carried out by a large international (essentially central European) group of scientists – among them exceptionally brilliant figures – which led to the establishment of several new physical theories. For microsystems – atoms and molecules, and their structures as crystals, and so on – the new theory was quantum mechanics.

2.2 Planck: The First Quantum Leap

The first problem that revealed the need for a thorough revision of the classical theories on the interaction between matter and electromagnetic radiation was that of the blackbody. The blackbody is not a subject of great physical interest in itself, although it has many applications: blackbody-like materials are used for camouflage, as solar energy collectors or infrared thermal detectors, for example. Historically, however, the blackbody is of paramount importance because its solution revealed an emerging new situation in physics and paved the way for the formulation of the first quantum theory.

It is generally known that bodies absorb some of the incoming electromagnetic radiation; a perfect blackbody (a theoretical object) absorbs all of it. To make something that comes close to being a blackbody, it is enough to take a closed, hollow piece of metal at a fixed temperature T. If we open a very small hole in the cavity, practically all the radiation that enters through it will be trapped; this means that the cavity behaves (very approximately) like a perfect absorber, and the radiation that escapes through the hole can be regarded as blackbody radiation at that temperature (Fig. 2.1).

Several properties of blackbody radiation can be studied using classical thermodynamics. In particular, the energy density of the radiation follows the *Stefan–Boltzmann law*, derived by L. Stefan (1815–1863) and extended by L. Boltzmann,[2]

$$u(T) = aT^4, \qquad (2.1)$$

---

[2] Ludwig Eduard Boltzmann (1844–1906) was an eminent Austrian physicist and philosopher. His most notable achievement was the development of statistical mechanics, in particular the statistical explanation of the second law of thermodynamics, and the introduction of the concept of entropy.

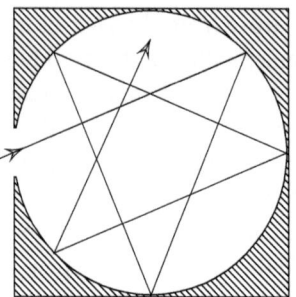

Figure 2.1 Diagram depicting a blackbody. Radiation entering through the small perforation is trapped inside the cavity after multiple reflections and has little chance of escaping. The cavity is therefore an almost perfect absorber.

where $a = 4\sigma/c$, and σ is a universal constant (independent of the material of the walls, their size or shape), called the Stefan-Boltzmann constant, with value $\sigma = 5.67 \times 10^{-5}$ erg cm^{-2} deg^{-4} sec^{-1}.

However, a serious problem arises when analyzing the spectral composition of u. The radiation field has components of all frequencies; if we filter only the frequencies between ω and $\omega + d\omega$, we obtain the *spectral energy density* $du = \rho(\omega, T)d\omega$, which when integrated over all frequencies gives the total energy density

$$u = \int_0^\infty \rho(\omega, T)d\omega. \tag{2.2}$$

Taking advantage of the fact that, according to Kirchhoff's law,[3] the spectral density of radiation from a blackbody is independent of the nature of the walls, Planck proposed to model the walls of the cavity as a collection of oscillators. This greatly simplifies the calculations, as will be shown in what follows.

Since each mechanical oscillator radiates when it vibrates, it is subject to a radiation force (called the *radiation reaction*); taking this force into account, the equation of motion of a charged oscillator with frequency ω is

$$\ddot{x} + \omega^2 x - \tau \dddot{x} = \frac{e}{m}\mathcal{E}_x, \tag{2.3}$$

where $\tau = 2e^2/3mc^3$ and $\mathcal{E}_x$ is the x-component of the radiation field in the cavity, which we take as a function of time only, independent of x. This so-called *long-wavelength approximation*[4] is justified by the fact that for the frequencies of interest (e.g., those of visible light), the wavelength of the radiation is $10^2 - 10^3$ times greater than the atomic radius, so that within the volume occupied by the atom the field $\mathcal{E}_x$ can be considered practically independent of position. Taking the Fourier transform of Eq. (2.3), we get

$$\tilde{x}(\omega') = \frac{e}{m} \frac{\tilde{\mathcal{E}}_x(\omega')}{\omega^2 - \omega'^2 + i\tau\omega'^3}, \tag{2.4}$$

where $\tilde{x}(\omega')$ is the Fourier transform of $x(t)$,

---

[3] Gustav Robert Kirchhoff (1824–1887) was a German (Prussian) physicist who among other things, derived important laws relating to linear electric circuits and blackbody radiation

[4] The long-wavelength approximation is equivalent to the electric dipole approximation, because in both cases the space-dependent coefficient of the electric field component e^{ikx} is replaced by 1.

$$\tilde{x}(\omega) = \frac{1}{2\pi} \int_{-\infty}^{\infty} x(t) e^{-i\omega t} dt, \qquad (2.5)$$

and similarly for $\tilde{\mathcal{E}}_x$. The energy of the oscillator averaged over a period is therefore

$$\overline{E(\omega,T)}^t = \overline{T+V}^t = \overline{2T}^t = m\overline{\dot{x}^2(t)}^t = \frac{e^2}{m} \int_{-\infty}^{+\infty} \frac{\omega'^2 |\tilde{\mathcal{E}}_x(\omega')|^2}{(\omega^2 - \omega'^2)^2 + (\tau \omega'^3)^2} d\omega'. \qquad (2.6)$$

It is easy to calculate an approximate value for this expression, noting that the integrand has a sharp maximum for frequencies close to ω because $\tau\omega \ll 1$; this allows one to replace ω' in the integral by ω, except, of course, where the difference $\xi = \omega - \omega'$ appears, and then extend the integration over ξ to the whole interval $(-\infty, \infty)$. This gives for the average energy of the oscillator in thermal equilibrium with the field in the cavity

$$\overline{E(\omega,T)}^t \simeq \frac{e^2}{4m} |\tilde{\mathcal{E}}_x(\omega)|^2 \int_{-\infty}^{\infty} \frac{d\xi}{\xi^2 + (e^2\omega^2/3mc^3)^2} = \frac{3\pi c^3}{4\omega^2} |\tilde{\mathcal{E}}_x(\omega)|^2. \qquad (2.7)$$

On the other hand, the average energy density of the field is given by

$$u = \frac{1}{4\pi} \overline{\mathcal{E}^2}^t = \frac{1}{4\pi} \overline{(\mathcal{E}_x^2 + \mathcal{E}_y^2 + \mathcal{E}_z^2)}^t = \frac{3}{4\pi} \overline{\mathcal{E}_x^2}^t$$

$$= \frac{3}{4\pi} \int_0^{\infty} |\tilde{\mathcal{E}}_x(\omega)|^2 d\omega, \qquad (2.8)$$

whence its spectral energy density is given according to Eq. (2.2) by

$$\rho(\omega) = \frac{3}{4\pi} |\tilde{\mathcal{E}}_x(\omega)|^2. \qquad (2.9)$$

Combining this expression with Eq. (2.7), we obtain Planck's formula (1899),

$$\rho(\omega, T) = \frac{\omega^2}{\pi^2 c^3} \overline{E(\omega,T)}^t, \qquad (2.10)$$

a remarkable result that we will use repeatedly. It tells us that the spectral energy density of blackbody radiation of frequency ω is proportional to the average energy of the mechanical oscillators with which it is in thermal equilibrium. The factor $\omega^2/\pi^2 c^3$ represents the *density of modes* of the field (or *density of states*, in its quantum version). Despite its simplicity, this factor – which had been calculated for an elastic medium around 1878 by Lord Rayleigh[5] – played an important role in the development of quantum theory, as it took some time to learn how to determine it for quantum systems.[6]

We now move on to calculate the average energy $\overline{E}$ of the oscillators using classical statistical mechanics. According to this theory, the average number of microsystems with energy E of a macrosystem in thermodynamic equilibrium at temperature T is given by the Boltzmann distribution

$$N(E) = N_0 e^{-\beta E}, \qquad (2.11)$$

---

[5] John William Strutt (1842–1919), known as Lord Rayleigh, was a British mathematician and physicist who made extensive contributions to science, particularly optics and fluid dynamics. for which he was awarded the Nobel Prize in Physics in 1904.
[6] Lord Rayleigh's calculation assumed a continuous exchange of energy, as does this text, which casts some doubt on its validity in the quantum case. Hence Einstein's concern to find an alternative, in principle unproblematic route to Planck's distribution; see Section 17.2.4.

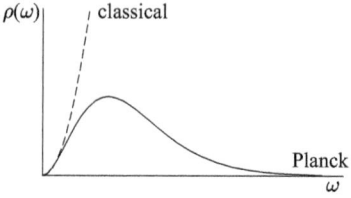

Figure 2.2 Spectral density of blackbody radiation as a function of frequency, at fixed temperature. The solid curve represents Planck's formula, which agrees with the experimental data, and the dashed curve represents the classical Rayleigh–Jeans formula.

where N_0 does not depend on energy nor temperature; β is the inverse temperature, $\beta = 1/kT$, with k the Boltzmann constant. The average energy obtained from this distribution is

$$\overline{E} = \frac{\int_0^\infty E e^{-\beta E} dE}{\int_0^\infty e^{-\beta E} dE} = -\frac{\partial}{\partial \beta} \ln \int_0^\infty e^{-\beta E} dE = \frac{\partial}{\partial \beta} \ln \beta = \frac{1}{\beta}, \qquad (2.12)$$

which leads to equipartition between the oscillators,

$$\overline{E} = kT, \qquad (2.13)$$

regardless of their nature and their frequency. In a system composed of a large number of independent similar oscillators (an ensemble), the result obtained by averaging over the ensemble of oscillators (as earlier) is expected to be equal to the time average of any one of them; this is the intuitive form of the *ergodic principle*. In substituting this expression in Eq. (2.10), we obtain the classical prediction

$$\rho(\omega, T) = \frac{\omega^2}{\pi^2 c^3} kT. \qquad (2.14)$$

But this result, which is the so-called *Rayleigh–Jeans distribution*,[7] agrees with the experimental data only for small frequencies, as is shown in Fig. 2.2. Furthermore, according to this formula, $u = \int_0^\infty \rho d\omega$ is infinite at any temperature $T > 0$, an absurd result that contradicts the Stefan–Boltzmann law, Eq. (2.1). This is the *ultraviolet catastrophe*, as P. Ehrenfest[8] called it.

2.2.1 Energy quantization

The difficulty with the ultraviolet catastrophe survived a few years and found a solution only when Planck proposed in 1900 a new way of looking at the problem, which implied a profound change in the prevailing physical conceptions at the time. Briefly stated, we can summarize Planck's theory as follows. Equation (2.13) for the average energy was derived based on the assumption that each oscillator can have an arbitrary energy, within a continuum. Planck found it necessary to introduce a hypothesis in the form

---

[7] James Hopwood Jeans (1877–1946) was an English physicist, astronomer, and mathematician, who made important contributions in several areas of physics and astronomy.
[8] Paul Ehrenfest (1880–1933) was a renowned Austrian-Dutch theoretical physicist who made important contributions to statistical physics and its relation to QM. In Chapter 8 the student will learn about Ehrenfest's theorem.

2.2 Planck: The First Quantum Leap

of a constraint, namely, that the material oscillators can only absorb multiples of a minimal amount of energy E_1 from the field,

$$E = nE_1, \quad n = 0, 1, 2, \ldots \tag{2.15}$$

where E_1, which Planck called *energy quantum*, depends on the frequency and perhaps other characteristics of the oscillator, but not on its temperature. This hypothesis violates a general principle of classical physical thought, that of energy as a continuum.

We now repeat the preceding calculation, but introduce Planck's postulate as a constraint, in the sense that the oscillators of a given frequency ω have an energy $E = nE_1$, with n an integer. Strictly speaking, according to Planck's argument, Eq. (2.16) refers to the (quantum) energy available for exchange with the field; the oscillators may have an additional (fixed) energy E_0 that does not participate in this exchange (see Section 2.6). The average energy is then obtained by introducing this constraint in Eq. (2.12),

$$\overline{E(\omega, \beta)} = \frac{\sum_{n=0}^{\infty} n E_1 e^{-\beta E_1 n}}{\sum_{n=0}^{\infty} e^{-\beta E_1 n}}. \tag{2.16}$$

To calculate the value of this expression, we introduce the auxiliary variable $x = e^{-\beta E_1}$. Deriving the formula $\sum_{n=0}^{\infty} x^n = (1-x)^{-1}$, we obtain

$$\frac{d}{dx} \sum_{n=0}^{\infty} x^n = \frac{1}{x} \sum_{n=0}^{\infty} n x^n = (1-x)^{-2}, \tag{2.17}$$

which allows us to write

$$\overline{E} = E_1 \frac{\sum_{n=0}^{\infty} n x^n}{\sum_{n=0}^{\infty} x^n} = E_1 \frac{x}{1-x} = \frac{E_1}{\frac{1}{x} - 1} = \frac{E_1}{e^{\beta E_1} - 1}. \tag{2.18}$$

whence from (2.6),

$$\rho(\omega, T) = \frac{\omega^2}{\pi^2 c^3} \frac{E_1}{e^{E_1/kT} - 1}. \tag{2.19}$$

The function $E_1(\omega)$ can be determined using Wien's[9] displacement law of thermodynamics as follows. By studying the adiabatic contraction of a perfectly reflecting body, Wien had shown that the spectral energy density must have the general form

$$\rho(\omega, T) = \omega^3 f(\omega/T), \tag{2.20}$$

where $f(\omega/T)$ is a universal function that could not be determined by classical theory. Comparing, we see that for Eq. (2.19) to obey this law, E_1 must be proportional to ω. Planck wrote this result in the form (we use modern notation)

$$E_1 = \hbar\omega, \tag{2.21}$$

where $\hbar$ is a constant to be determined with dimensions of action; it is the Planck constant divided by 2π. Equations (2.19) and (2.21) allow us to find the unknown function f; by combining them we get the famous *Planck distribution law*,

$$\rho(\omega, T) = \frac{\hbar \omega^3}{\pi^2 c^3} \frac{1}{e^{\hbar\omega/kT} - 1}. \tag{2.22}$$

---

[9] Wilhelm Wien (1864–1928) was a German theoretical physicist, who was awarded the Nobel Prize in 1911 for his discoveries concerning the laws of thermal radiation.

This result solves the problem,[10] because, on the one hand, it is free from the ultraviolet catastrophe and, on the other hand, it leads to the Stefan–Boltzmann law, Eq. (2.1),

$$u(T) = \int_0^\infty \rho(\omega, T) d\omega = \frac{\pi^2 k^4}{15 c^3 \hbar^3} T^4, \qquad (2.23)$$

thus providing a theoretical expression for a in terms of fundamental constants, since it follows from a comparison with Eq. (2.1) that

$$a = \frac{\pi^2 k^4}{15 c^3 \hbar^3}. \qquad (2.24)$$

Planck used this result, in reverse, to derive from it the numerical value of $\hbar$ given the empirical value of a, and obtained $\hbar = 1.05 \times 10^{-27}$ erg s. With this value of $\hbar$, Eq. (2.22) gives the solid curve shown in Fig. 2.2, which agrees with the experimental data. This theory also allowed Planck to obtain a number of collateral results, including a determination of the numerical value of the electric charge of the electron – by then a recent discovery by J. J. Thomson[11] – a result that convinced Rutherford[12] (and many other sceptics) of the validity of the atomic theory.

Suppose for a moment that in Eq. (2.22) we let the parameter $\hbar$ "tend to zero," keeping T and ω fixed (this is a purely formal procedure, since $\hbar$ is a constant of nature, not a parameter). As the exponent tends to zero, we can expand the exponential into a Taylor series, keeping only the first two terms of the expansion, to obtain

$$\rho(\omega, T) = \frac{\hbar \omega^3}{\pi^2 c^3} \frac{1}{\left(1 + \frac{\hbar \omega}{kT} + \ldots\right) - 1} \simeq \frac{\hbar \omega^3}{\pi^2 c^3} \frac{kT}{\hbar \omega} = \frac{\omega^2}{\pi^2 c^3} kT, \qquad (2.25)$$

which coincides with the classical Rayleigh–Jeans spectrum, Eq. (2.14). So we see that Planck's theory correctly reproduces the experimental results with $\hbar \neq 0$, and leads to the (incorrect or approximate) classical results with $\hbar = 0$. We conclude that the classical theory must be replaced by a new theory with $\hbar \neq 0$ (more precisely, with the numerical value given earlier).

Planck's constant $\hbar$ permeates all of modern physics, and its value has been determined by many different methods, so that it is now known with a high degree of accuracy. Since 2019, because of its central importance in QM, Planck's constant is taken to be exact by definition, along with other universal constants such as the elementary charge e, the speed of light in vacuum c, and the Boltzmann constant k.

---

[10] Based on plausibility arguments, Wien had published in 1896 the formula $\rho_W(\omega, T) = (\hbar \omega^3 / \pi^2 c^3) e^{-\hbar \omega / kT}$ for the spectral composition of the radiation, which, although approximate, gives excellent results for low temperatures (or high frequencies). The existence of this approximate formula was of great help to the later work of Planck and Einstein.

[11] Sir Joseph John Thomson FRS (1856–1940) was a distinguished British engineer and physicist, the discoverer of the electron, for which he won the Nobel Prize in 1906. He proposed the plum pudding model of the atom. He was an exceptional mentor to future scientists, half a dozen of whom went on to win the Nobel Prize in Physics or Chemistry.

The initials J. J. are important, firstly because there were several important quantum Thomsons, and secondly because he was popularly known as JJ.

[12] Ernest Rutherford (1871–1937) was an exceptional experimental New Zealand-born British physicist and chemist who discovered the atomic nucleus, for which he was awarded the Nobel Prize in Chemistry in 1908; he also identified and named the nuclear radiations α, β, γ, among many other contributions.

We have introduced the symbol $\hbar$ (h-bar) to denote the quantity $h/2\pi$, where $h = 6.6249 \times 10^{-27}$ erg s is the constant originally introduced by Planck; the frequent appearance of the reduced Planck constant $\hbar$ in this text (and everywhere) is associated with the common use of the (angular) frequency ω instead of $\nu = \omega/2\pi$. We will refer interchangeably to both h and $\hbar$ as Planck's constant where there is no danger of confusion.

Exercise E2.1. On the scale of $\hbar$

Estimate the number of energy quanta $E_1 = \hbar\omega$ contained in a macroscopic but small oscillator with mass $m = 1$ g oscillating with an amplitude a of 1 mm and frequency $\nu = 100$ Hz. Compare with the result obtained for an electron with mass $m \sim 10^{-27}$ g oscillating with the same amplitude and frequency,

Solution. The energy of the macroscopic oscillator is

$$E = \tfrac{1}{2} m\omega^2 a^2 = \tfrac{1}{2} \times 1.0 \times (2\pi \times 10^2)^2 \times 10^{-2} \text{ erg} \simeq 2000 \text{ erg}. \tag{2.26}$$

In terms of Planck's energy unit $E_1 = \hbar\omega$, this corresponds to

$$n = \frac{E}{\hbar\omega} = \frac{2 \times 10^3}{1.05 \times 10^{-27} \times 2\pi \times 10^2} \simeq 3 \times 10^{27}. \tag{2.27}$$

Assuming that we can measure changes in the energy of one part in a million, we would not be able to record changes in n smaller than

$$|\Delta n| = 3 \times 10^{21}. \tag{2.28}$$

It is clear that under these conditions it is impossible to observe variations in the system when n changes by a few units; any such change will appear to us to be continuous.

For the electron the energy is 2×10^{-25} erg, hence $n \simeq 3$, and minimal (discrete) changes in the energy will substantially modify its behavior. It is precisely the smallness of $\hbar$ that hides the quantum phenomenon outside the atomic scale. This explains why it was precisely when experimental physics reached the atomic level that the quantum phenomena became observable and contradictions with classical notions arose.

Planck's formula for the spectral energy density, Eq. (2.22), implies that the mean thermal energy of an oscillator is given by

$$\overline{E} = \frac{\hbar\omega}{e^{\hbar\omega/kT} - 1}, \tag{2.29}$$

as can be verified by combining Eqs. (2.18) and (2.21). An important consequence is that oscillators of the same frequency all share the same spectrum, regardless of their nature; this is in particular the case with the material oscillators and those of the blackbody radiation field. However, the explicit statement of this conclusion still required some years of maturation, as we will see later.

2.3 Einstein: Quantization as a Universal Phenomenon

Planck arrived at his blackbody formula after renouncing any attempt to explain the observed results in terms of a continuous spectrum, insisting that it should be applied only to the absorption of radiation by the walls of the thermal cavity. For his part, around 1905 Einstein began a novel analysis of Planck's results and concluded that the electromagnetic field itself had a discrete, corpuscular character. This was a drastic revision of the dominant wave (continuous) conception of light since the work of Young[13] and Fresnel[14] in the early nineteenth century. Much later, in 1926, the US physicist-chemist G. N. Lewis (1875–1946) proposed to call the quantum of the radiation field a *photon*, a name that is common today and that we will use from now on when appropriate (Einstein never used it).

Einstein reached his conclusion by studying the statistical behavior of a very faint monochromatic radiation field, showing that under these conditions things happen as if it were something similar to a dilute gas in which individual molecules, which are independent of each other, are responsible for the fluctuations. Convinced that $\hbar\omega$ represents the energy of each photon, he reinterpreted Planck's results by saying that the cause of the discrete behavior of the energy exchanged by Planck's oscillators is to be sought not in the oscillators themselves, but in the radiation field, thus making Planck's result the natural consequence of the discrete properties of this field, that is, a manifestation of its organization into photons.

As his work progressed, Einstein was able to show that the Planck distribution is consistent only with the hypothesis that when an atom emits energy $\hbar\omega$, a momentum of value $\hbar\omega/c$ (a pencil of radiation) is transferred to it in the opposite direction to that of propagation. This means that the emitted photon has a well-defined direction; on the contrary, if the emitted radiation had the structure of a spherical wave, as was usually assumed, the recoil momentum would be zero. To quote Einstein:

> Outgoing radiation in the form of spherical waves does not exist. During the elementary process of radiative loss, the molecule undergoes a regression of magnitude $h\nu/c$ in a direction that is determined by "chance", according to the current state of the theory.... These properties of elementary processes... make the formulation of a quantum theory of radiation inevitable.

Later it was shown that these quanta of the radiation field also have a defined angular momentum (polarization state), of magnitude $\hbar$ and always parallel or antiparallel to the direction of propagation (this corresponds to the spin of the photon).

It is easy to be convinced that a photon with energy $h\nu$ has a momentum of value $h\nu/c$, if we apply the well-known relativistic formula relating momentum and energy of a particle, $E = c\sqrt{m^2c^2 + p^2}$, to the case of the photon whose mass is zero, so

$$E = h\nu = cp. \tag{2.30}$$

---

[13] Thomas Young (1773–1829) was a British polymath who made notable contributions to human vision, mechanics of solids, energy, physiology, and even Egyptology. He established the wave theory of light and was described as "the last man who knows everything."

[14] Augustin-Jean Fresnel (1788–1827) was a French civil engineer and physicist whose research in optics led to the generalized acceptance of the wave theory of light.

Since the momentum is concentrated in the direction of propagation, we can rewrite this result in vector form,

$$\boldsymbol{p} = \frac{h\nu}{c}\hat{\mathbf{k}} = \frac{h}{\lambda}\hat{\mathbf{k}} = \hbar\boldsymbol{k}, \qquad (2.31)$$

where $\lambda = c/\nu$ is the length of the electromagnetic wave, which propagates in the $\hat{\mathbf{k}}$ direction; $\boldsymbol{k} = (2\pi/\lambda)\hat{\mathbf{k}}$ is the wave vector and its magnitude is the *wave number*. Formula (2.31) agrees with electromagnetic theory, which states that the momentum carried by a plane wave of energy E is $\boldsymbol{p} = \hat{\mathbf{k}}E/c$, but it was derived starting from the notion of photon, not from that of wave, which is an extended object, not a point entity.

2.3.1 The photoelectric effect

Einstein immediately applied his photonic conception of light to the study of various problems. Firstly, in his early work of 1905 he revised the problem of the *photoelectric effect* and showed that the new theory allowed him to address serious unresolved theoretical difficulties. Let us briefly review the situation.

In 1887, during the laboratory work that led him to demonstrate the existence of electromagnetic waves, H. Hertz[15] observed a new phenomenon that today we can describe as the emission of electrons when light falls on an alkali metal. The experimental study of this phenomenon soon led to results that were paradoxical to the classical theories. For example, the maximum energy of the electrons emitted does not depend on the intensity of the incident light, but on its frequency (its color); on the other hand, the number of photoelectrons emitted depends on the intensity of the light beam. It was also found that for each material there is a characteristic frequency of the incident radiation, below which there is no photoelectric effect.

All these peculiarities, inexplicable in classical physics, become natural in the photon theory, according to which a given atom will either absorb a photon of energy $\hbar\omega$ or nothing at all. In the first case the atomic electron receiving this energy will escape, overcoming the attraction of the material, which becomes charged (ionized). During its escape the electron will do some work W against the material it leaves (W is the so-called *work function* of the metal), so it will be emitted with a maximum energy given by

$$\tfrac{1}{2}mv^2 = \hbar\omega - W, \qquad (2.32)$$

which is indeed independent of the intensity of the light, but dependent on its frequency. If $\omega < W/\hbar$, no electron will be emitted. Furthermore, the number of emitted electrons can be expected to be proportional to the number of available photons, i.e., to grow with the intensity of the incident light.

Subsequent very fine experiments by Millikan[16] in 1914–1916 confirmed the validity of the theory and showed that the constant used in these formulae was exactly Planck's constant.

---

[15] Heinrich Rudolf Hertz (1857–1894) was an outstanding German experimental physicist who validated Maxwell's theory by building the first radio communication system in his laboratory (and unwittingly became the first radio amateur in history, although without a partner). The international unit of frequency, a lunar crater and an asteroid are named after him.

[16] Robert Andrews Millikan (1868–1953) was a US experimental physicist who was awarded the Nobel Prize in Physics in 1923 for his measurement of the electric charge of the electron and his verification of the quantum theory of the photoelectric effect.

2.3.2 The specific heat of solids

Inspired by Planck's work, Einstein also tackled the problem of the heat capacity of solids, again taking a radical stance. In 1906 he published another of his famous papers, in which he argued that we should expect other oscillators, such as those used in the theory of heat, to behave in a similar way to the blackbody. According to the classical law of energy equipartition, the specific heat (at constant volume) of a solid is a constant (this is the so-called *Dulong–Petit law*),

$$c_V = \left(\partial \overline{E}/\partial T\right)_V \sim \partial \left(kT\right)/\partial T = \text{const.} \tag{2.33}$$

However, already in 1900 there were indications – thanks to the work of W. Nernst[17] and others – that the specific heat of solids actually tends to zero as the absolute temperature approaches zero. The contradiction is resolved by adopting the viewpoint proposed by Einstein, that is, by treating each atom of the solid as an oscillator with a common frequency $\nu = \omega/2\pi$. The N atoms composing a mole of the solid contain the average energy

$$\overline{E} = \frac{3Nh\nu}{e^{h\nu/kT} - 1}, \tag{2.34}$$

as follows from Eq. (2.29) applied to the $3N$ degrees of freedom of the solid. Therefore, the specific heat at constant volume is

$$c_V = \left(\frac{\partial \overline{E}}{\partial T}\right)_V = 3R\left(\frac{h\nu}{kT}\right)^2 \frac{e^{h\nu/kT}}{(e^{h\nu/kT} - 1)^2}, \tag{2.35}$$

where $R = kN$ is the gas constant. This formula reproduces the classical result for large temperatures (with respect to the so-called *Debye temperature*, which is characteristic of the material and is of the order of 200 K for most solids), and shows that c_V does indeed go to zero with T,

$$c_V = 3R\left(\frac{h\nu}{kT}\right)^2 e^{-h\nu/kT} \quad (T \to 0). \tag{2.36}$$

Einstein's comparison of the experimental data for the heat capacity of diamond with his formula (2.35), shown in Fig. 2.3, confirms the decrease in the heat capacity at low temperatures. A more detailed description was later developed by P. Debye,[18] Born, T. von Kármán (1881–1963), and others, using elaborations of the Einstein model just discussed – which is now regarded as the beginning of solid-state quantum theory.

2.3.3 The Compton effect

In 1921–1923, photon theory received a new boost with the Compton effect, a phenomenon that A. H. Compton[19] discovered experimentally with the newly invented X-ray spectrometer and also explained theoretically.

---

[17] Walther Hermann Nernst Görbitz (1864–1941) was a German physicist and chemist who was awarded the Nobel Prize in Chemistry in 1920 for his studies of chemical affinity in accordance with the third law of thermodynamics. As early as 1916, he proposed that the mechanism for compensating for the energy lost by atomic electrons could be found in the ZPF (see Section 2.9).

[18] Peter Debye (1884–1966) was a distinguished Dutch physico-chemist who was awarded the Nobel Prize in Chemistry in 1936.

[19] Arthur Holly Compton 1892–1962 was a prominent US experimentalist who received the Nobel Prize in 1927 for his discovery, verification, and explanation of the Compton effect.

2.3 Einstein: Quantization as a Universal Phenomenon

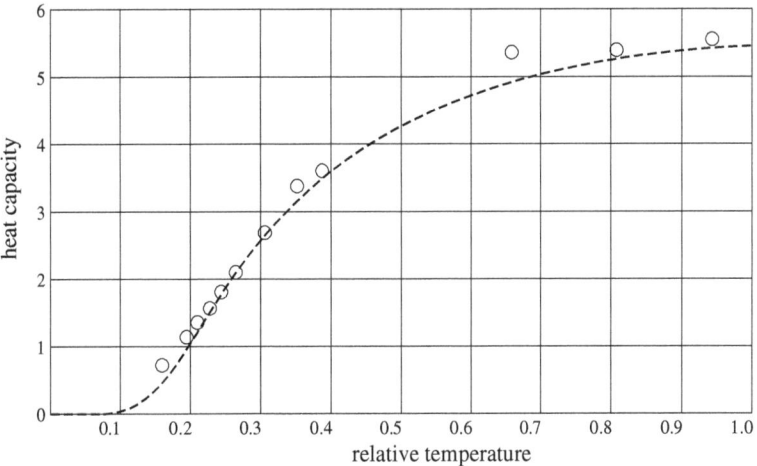

Figure 2.3 Variation of the heat capacity of diamond with temperature. Comparison of the predictions of Einstein's theory with experimental data (small circles) of the time. Figure adapted from Einstein (1907).

According to classical theory, the frequency of an electromagnetic beam does not change when it is scattered by an electron. However, using high-energy X-rays emitted by molybdenum and incident on graphite, Compton observed that part of the scattered beam had a frequency ω lower than the frequency ω_0 of the incident beam and that the difference $\omega_0 - \omega$ varied as a function of the scattering angle θ. The photon theory offers the following explanation.

Each scattering event is considered as an elastic collision between an electron at rest and a corpuscle (the photon) with speed c and initial energy $\hbar\omega_0$. Let us assume that the electron is free or loosely bound to an atom. Under these conditions, the conservation equations are, according to Eqs. (2.30) and (2.31),

$$\hbar\omega_0 + E_0 = \hbar\omega + E, \tag{2.37}$$

$$\hbar\boldsymbol{k}_0 + \boldsymbol{p}_0 = \hbar\boldsymbol{k} + \boldsymbol{p}. \tag{2.38}$$

For an electron initially at rest, $E_0 = m_0 c^2$ and $\boldsymbol{p}_0 = 0$; the moving electron has momentum $\boldsymbol{p} = m\boldsymbol{v}$ and energy $E = mc^2$, with $m = m_0/\sqrt{1 - v^2/c^2} \equiv m_0\gamma$. With these substitutions, the preceding system is rewritten as

$$\hbar\omega_0 - \hbar\omega = mc^2 - m_0 c^2, \tag{2.39}$$

$$\hbar\boldsymbol{k}_0 - \hbar\boldsymbol{k} = m\boldsymbol{v}. \tag{2.40}$$

It is left as an exercise to show that the electron at rest cannot absorb all the energy of the incident photon, so in all cases only part of it is absorbed. We now square the system of equations:

$$\omega_0^2 + \omega^2 - 2\omega_0\omega = (m - m_0)^2 c^4/\hbar^2, \tag{2.41}$$

$$k_0^2 + k^2 - 2k_0 k \cos\theta = m^2 v^2/\hbar^2, \tag{2.42}$$

eliminate k from the second, using the formula $\omega = kc$, and subtract the result from the first, which, after some simplifications, gives

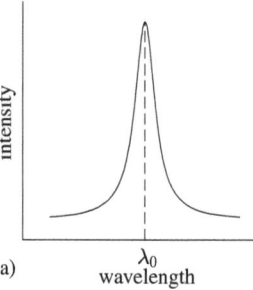 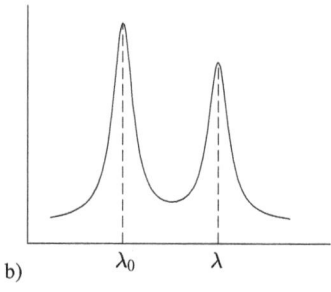

Figure 2.4 The Compton effect. Spectral distribution (a) of the incident X-rays and (b) of the scattered X-rays, with the first maximum at the original wavelength corresponding to Thomson scattering by tightly bound electrons and the second at a larger wavelength corresponding to Compton scattering by weakly bound electrons.

$$\omega_0\omega(1-\cos\theta) = \frac{c^4}{2\hbar^2}\left[\frac{m^2 v^2}{c^2} - (m-m_0)^2\right] = \frac{m_0 c^2}{\hbar}(\omega_0 - \omega). \quad (2.43)$$

Substituting $\omega = 2\pi c/\lambda$ and dividing the result by $\omega_0\omega$, we obtain the following expression for the change in wavelength of the scattered photon as a function of the scattering angle,

$$\Delta\lambda = \lambda - \lambda_0 = \frac{2\pi\hbar}{m_0 c}(1-\cos\theta) = 2\lambda_C \sin^2\frac{\theta}{2}. \quad (2.44)$$

The quantity

$$\lambda_C = \frac{2\pi\hbar}{m_0 c} = \frac{h}{m_0 c} \quad (2.45)$$

is an important parameter characteristic of the particle, known as the *Compton wavelength*. For the electron, its value is 2.4×10^{-10} cm.

From Eq. (2.44) we see that the difference between the initial and final wavelengths increases with the scattering angle. Now suppose that the incident radiation has a spectral distribution as shown in Fig. 2.4a. Since the scattering material has a large number of tightly bound atomic electrons oscillating about their equilibrium position, there will be a scattered component that does not change its wavelength; this primary radiation is correctly described by classical physics (it is the *Thomson scattering*) and corresponds to the first maximum in Fig. 2.4b. However, weakly bound (almost free) electrons will produce *Compton scattering* and cause the appearance of a secondary maximum centered at $\lambda > \lambda_0$, as shown in the same figure.

As the Compton wavelength of elementary particles is relatively small, Compton scattering is observed only at relatively small wavelengths, that is, with X- and γ-rays. For example, for visible light with $\lambda \sim 10^{-5}$ cm

$$\frac{\Delta\lambda}{\lambda} \sim \frac{\lambda_C}{\lambda} \sim 10^{-5}, \quad (2.46)$$

while for X-rays with $\lambda \sim 10^{-9}$ cm,

$$\frac{\Delta\lambda}{\lambda} \sim \frac{\lambda_C}{\lambda} \sim 10^{-1}. \quad (2.47)$$

Compton's work of 1926 is now regarded as the definitive confirmation of the photon theory of light. Of particular importance was the fact that it confirmed the validity of

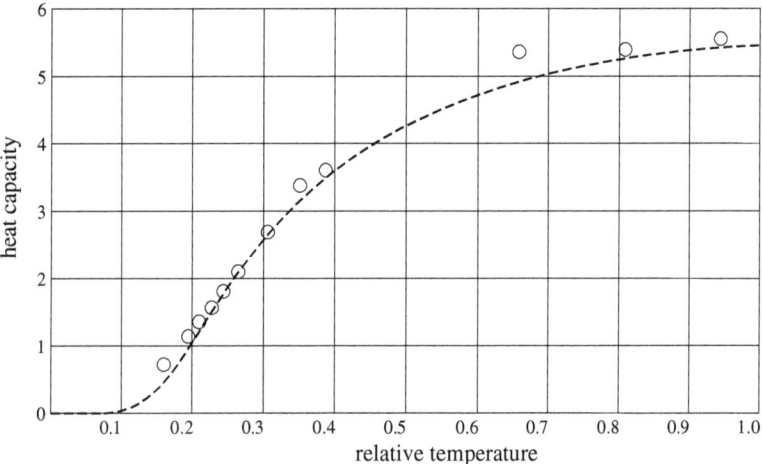

Figure 2.3 Variation of the heat capacity of diamond with temperature. Comparison of the predictions of Einstein's theory with experimental data (small circles) of the time. Figure adapted from Einstein (1907).

According to classical theory, the frequency of an electromagnetic beam does not change when it is scattered by an electron. However, using high-energy X-rays emitted by molybdenum and incident on graphite, Compton observed that part of the scattered beam had a frequency ω lower than the frequency ω_0 of the incident beam and that the difference $\omega_0 - \omega$ varied as a function of the scattering angle θ. The photon theory offers the following explanation.

Each scattering event is considered as an elastic collision between an electron at rest and a corpuscle (the photon) with speed c and initial energy $\hbar\omega_0$. Let us assume that the electron is free or loosely bound to an atom. Under these conditions, the conservation equations are, according to Eqs. (2.30) and (2.31),

$$\hbar\omega_0 + E_0 = \hbar\omega + E, \tag{2.37}$$

$$\hbar\mathbf{k}_0 + \mathbf{p}_0 = \hbar\mathbf{k} + \mathbf{p}. \tag{2.38}$$

For an electron initially at rest, $E_0 = m_0 c^2$ and $\mathbf{p}_0 = 0$; the moving electron has momentum $\mathbf{p} = m\mathbf{v}$ and energy $E = mc^2$, with $m = m_0/\sqrt{1 - v^2/c^2} \equiv m_0 \gamma$. With these substitutions, the preceding system is rewritten as

$$\hbar\omega_0 - \hbar\omega = mc^2 - m_0 c^2, \tag{2.39}$$

$$\hbar\mathbf{k}_0 - \hbar\mathbf{k} = m\mathbf{v}. \tag{2.40}$$

It is left as an exercise to show that the electron at rest cannot absorb all the energy of the incident photon, so in all cases only part of it is absorbed. We now square the system of equations:

$$\omega_0^2 + \omega^2 - 2\omega_0\omega = (m - m_0)^2 c^4/\hbar^2, \tag{2.41}$$

$$k_0^2 + k^2 - 2k_0 k \cos\theta = m^2 v^2/\hbar^2, \tag{2.42}$$

eliminate k from the second, using the formula $\omega = kc$, and subtract the result from the first, which, after some simplifications, gives

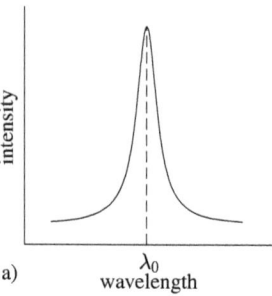
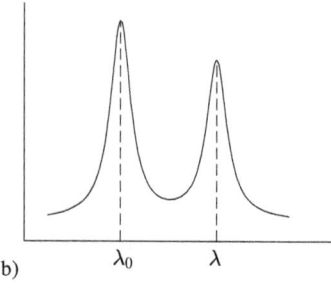

Figure 2.4 The Compton effect. Spectral distribution (a) of the incident X-rays and (b) of the scattered X-rays, with the first maximum at the original wavelength corresponding to Thomson scattering by tightly bound electrons and the second at a larger wavelength corresponding to Compton scattering by weakly bound electrons.

$$\omega_0\omega(1-\cos\theta) = \frac{c^4}{2\hbar^2}\left[\frac{m^2v^2}{c^2} - (m-m_0)^2\right] = \frac{m_0c^2}{\hbar}(\omega_0-\omega). \quad (2.43)$$

Substituting $\omega = 2\pi c/\lambda$ and dividing the result by $\omega_0\omega$, we obtain the following expression for the change in wavelength of the scattered photon as a function of the scattering angle,

$$\Delta\lambda = \lambda - \lambda_0 = \frac{2\pi\hbar}{m_0c}(1-\cos\theta) = 2\lambda_C\sin^2\frac{\theta}{2}. \quad (2.44)$$

The quantity

$$\lambda_C = \frac{2\pi\hbar}{m_0c} = \frac{h}{m_0c} \quad (2.45)$$

is an important parameter characteristic of the particle, known as the *Compton wavelength*. For the electron, its value is 2.4×10^{-10} cm.

From Eq. (2.44) we see that the difference between the initial and final wavelengths increases with the scattering angle. Now suppose that the incident radiation has a spectral distribution as shown in Fig. 2.4a. Since the scattering material has a large number of tightly bound atomic electrons oscillating about their equilibrium position, there will be a scattered component that does not change its wavelength; this primary radiation is correctly described by classical physics (it is the *Thomson scattering*) and corresponds to the first maximum in Fig. 2.4b. However, weakly bound (almost free) electrons will produce *Compton scattering* and cause the appearance of a secondary maximum centered at $\lambda > \lambda_0$, as shown in the same figure.

As the Compton wavelength of elementary particles is relatively small, Compton scattering is observed only at relatively small wavelengths, that is, with X- and γ-rays. For example, for visible light with $\lambda \sim 10^{-5}$ cm

$$\frac{\Delta\lambda}{\lambda} \sim \frac{\lambda_C}{\lambda} \sim 10^{-5}, \quad (2.46)$$

while for X-rays with $\lambda \sim 10^{-9}$ cm,

$$\frac{\Delta\lambda}{\lambda} \sim \frac{\lambda_C}{\lambda} \sim 10^{-1}. \quad (2.47)$$

Compton's work of 1926 is now regarded as the definitive confirmation of the photon theory of light. Of particular importance was the fact that it confirmed the validity of

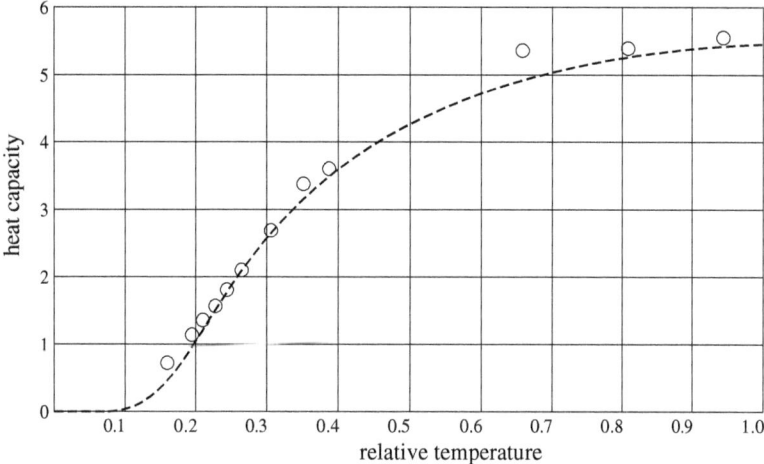

Figure 2.3 Variation of the heat capacity of diamond with temperature. Comparison of the predictions of Einstein's theory with experimental data (small circles) of the time. Figure adapted from Einstein (1907).

According to classical theory, the frequency of an electromagnetic beam does not change when it is scattered by an electron. However, using high-energy X-rays emitted by molybdenum and incident on graphite, Compton observed that part of the scattered beam had a frequency ω lower than the frequency ω_0 of the incident beam and that the difference $\omega_0 - \omega$ varied as a function of the scattering angle θ. The photon theory offers the following explanation.

Each scattering event is considered as an elastic collision between an electron at rest and a corpuscle (the photon) with speed c and initial energy $\hbar\omega_0$. Let us assume that the electron is free or loosely bound to an atom. Under these conditions, the conservation equations are, according to Eqs. (2.30) and (2.31),

$$\hbar\omega_0 + E_0 = \hbar\omega + E, \tag{2.37}$$

$$\hbar\boldsymbol{k}_0 + \boldsymbol{p}_0 = \hbar\boldsymbol{k} + \boldsymbol{p}. \tag{2.38}$$

For an electron initially at rest, $E_0 = m_0 c^2$ and $\boldsymbol{p}_0 = 0$; the moving electron has momentum $\boldsymbol{p} = m\boldsymbol{v}$ and energy $E = mc^2$, with $m = m_0/\sqrt{1 - v^2/c^2} \equiv m_0\gamma$. With these substitutions, the preceding system is rewritten as

$$\hbar\omega_0 - \hbar\omega = mc^2 - m_0 c^2, \tag{2.39}$$

$$\hbar\boldsymbol{k}_0 - \hbar\boldsymbol{k} = m\boldsymbol{v}. \tag{2.40}$$

It is left as an exercise to show that the electron at rest cannot absorb all the energy of the incident photon, so in all cases only part of it is absorbed. We now square the system of equations:

$$\omega_0^2 + \omega^2 - 2\omega_0\omega = (m - m_0)^2 c^4/\hbar^2, \tag{2.41}$$

$$k_0^2 + k^2 - 2k_0 k \cos\theta = m^2 v^2/\hbar^2, \tag{2.42}$$

eliminate k from the second, using the formula $\omega = kc$, and subtract the result from the first, which, after some simplifications, gives

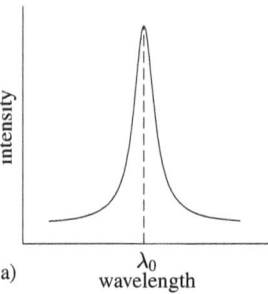 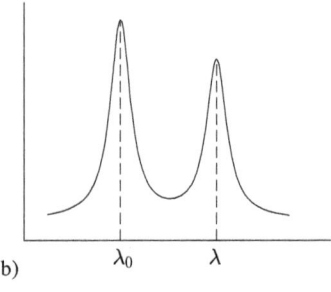

Figure 2.4 The Compton effect. Spectral distribution (a) of the incident X-rays and (b) of the scattered X-rays, with the first maximum at the original wavelength corresponding to Thomson scattering by tightly bound electrons and the second at a larger wavelength corresponding to Compton scattering by weakly bound electrons.

$$\omega_0\omega(1-\cos\theta) = \frac{c^4}{2\hbar^2}\left[\frac{m^2 v^2}{c^2} - (m-m_0)^2\right] = \frac{m_0 c^2}{\hbar}(\omega_0 - \omega). \tag{2.43}$$

Substituting $\omega = 2\pi c/\lambda$ and dividing the result by $\omega_0\omega$, we obtain the following expression for the change in wavelength of the scattered photon as a function of the scattering angle,

$$\Delta\lambda = \lambda - \lambda_0 = \frac{2\pi\hbar}{m_0 c}(1-\cos\theta) = 2\lambda_C \sin^2\frac{\theta}{2}. \tag{2.44}$$

The quantity

$$\lambda_C = \frac{2\pi\hbar}{m_0 c} = \frac{h}{m_0 c} \tag{2.45}$$

is an important parameter characteristic of the particle, known as the *Compton wavelength*. For the electron, its value is 2.4×10^{-10} cm.

From Eq. (2.44) we see that the difference between the initial and final wavelengths increases with the scattering angle. Now suppose that the incident radiation has a spectral distribution as shown in Fig. 2.4a. Since the scattering material has a large number of tightly bound atomic electrons oscillating about their equilibrium position, there will be a scattered component that does not change its wavelength; this primary radiation is correctly described by classical physics (it is the *Thomson scattering*) and corresponds to the first maximum in Fig. 2.4b. However, weakly bound (almost free) electrons will produce *Compton scattering* and cause the appearance of a secondary maximum centered at $\lambda > \lambda_0$, as shown in the same figure.

As the Compton wavelength of elementary particles is relatively small, Compton scattering is observed only at relatively small wavelengths, that is, with X- and γ-rays. For example, for visible light with $\lambda \sim 10^{-5}$ cm

$$\frac{\Delta\lambda}{\lambda} \sim \frac{\lambda_C}{\lambda} \sim 10^{-5}, \tag{2.46}$$

while for X-rays with $\lambda \sim 10^{-9}$ cm,

$$\frac{\Delta\lambda}{\lambda} \sim \frac{\lambda_C}{\lambda} \sim 10^{-1}. \tag{2.47}$$

Compton's work of 1926 is now regarded as the definitive confirmation of the photon theory of light. Of particular importance was the fact that it confirmed the validity of

the laws of conservation of momentum and energy for individual events of radiation–matter interaction (there had been doubts about their general validity in microphysics). It also confirmed that the emitted photons travel in a well-defined direction, like a radiation pencil, as established by Einstein, contrary to the previous popular belief that the emission has spherical symmetry. The reader should note the significance of these statements: the radiation pencil refers to a specific, isolated, one-particle event, whereas spherical symmetry results from a large number of similar events randomly distributed in all directions.

2.4 Early Attempts to Model the Atom

Already by 1912, many physicists were taking the quantum hypothesis seriously. The work of atomic spectroscopists and that of J. J. Thomson, Rutherford, and others on the electron and the atomic structure had provided sufficient experimental evidence to complement the field quantization results of Planck and Einstein. The conditions were ripe for the emergence of the first quantum theory of the structure of matter, namely Bohr's model of the hydrogen atom (1913), which gave rise to what is known as *primitive quantum mechanics*, which already considered quantized (stationary) atomic orbits.

The problem that initially concerned Bohr was merely that of constructing a model of the atom; with the development of his own work, Bohr turned his attention to a second problem closely related to the first: the structure of atomic absorption and emission spectra. Let us briefly review the situation of these two problems around 1912.

The first idea that came to mind is that the atom must be a static system, since according to electrodynamics an accelerated electric charge e radiates, emitting an average energy W_L per second given by Larmor's formula,

$$W_L = \frac{2e^2}{3c^3}\overline{\ddot{r}^2} \qquad (2.48)$$

(the bar indicates orbital (time) average). On this basis, J. J. Thomson had proposed in 1903 to visualize the atom as a uniform distribution of positive charge filling the entire atomic volume (with a radius of the order of 10^{-8} cm), with the (point) electrons in equilibrium inside, the so-called *plum pudding model*. However, later experiments on the scattering of α-particles (i.e., helium nuclei with two protons and two neutrons) by thin metal foils, conducted in Thomson's laboratory in Cambridge and by Rutherford and his colleagues H. Geiger (1862–1945) and E. Marsden (1889–1970) in Manchester, showed a relatively large number of particles scattered at large angles ($\sim 150°$).

These observations led Rutherford to propose an alternative model of the atom, with the positive charge of the protons concentrated in a very small central region (the nucleus), which carry virtually all the atomic mass, and sufficient orbital electrons to ensure the electrical neutrality of the atom. Positive particles that come very close to the nucleus are violently repelled and deflected by the strong central and quasi-point-like field (in fact a head-on collision would result in total reflection of the α-particle), which explains the frequent occurrence of large scattering angles. Using this model, Rutherford was able to correctly predict the fraction of scattered particles within the solid angle $d\Omega$, that is, to calculate the angular distribution of the scattered particles.[20] These results meant the end of Thomson's model and of other earlier alternatives, but introduced a new difficulty: the problem of the stability of the atom, since the orbiting electrons of a planetary model should radiate until they collapse on the nucleus.

Bohr – who had joined Rutherford's group in 1911 for a few months after a longer stay with Thomson – felt that the way out of this difficulty should be sought in Planck and Einstein's quantum theory, which, he hoped, might well be compatible with the idea of a nonradiating accelerated electron. A second difficulty with Rutherford's model, the fact that it did not assign a size to the atom, suggested to Bohr the idea that there should be an additional principle in the form of a *constraint* that would fix the atomic scale. A dimensional analysis shows that $\hbar$, e, and m can be used to construct an elementary length of appropriate value for the atom ($\hbar^2/me^2 \approx 10^{-9}$ cm), which is not possible with e and m alone, a fact that confirmed Bohr's suspicion that $\hbar$ plays a central role.

The concrete way of incorporating quantum theory into Rutherford's planetary model was suggested to Bohr by means of an analysis of the characteristics of atomic spectra. Atomic spectroscopy had reached a high level of development and the visible absorption and emission spectra (spectra consisting of series of characteristic lines) of many elements were known in sufficient detail. In 1885 J. Balmer (1825–1898) – a Swiss mathematician, numerologist, and elementary-school teacher – showed that the four visible lines of the emission spectrum of hydrogen have a wavelength λ given by the expression

$$\lambda = a\frac{n^2}{n^2 - 4}, \qquad (2.49)$$

where $a = 3645.6 \times 10^{-7}$ cm and $n = 3, 4, 5, 6$. Balmer later proved that the same formula correctly gave the other eight lines of the series, now called Balmer's series. He was also able to accurately predict – on the basis of purely numerological studies – the characteristic properties of the various series that could be obtained by replacing the number 4 by the square of a higher integer.

A few years later J. Rydberg (1854–1919) proved that all known series of the hydrogen spectrum, including Balmer's, can be obtained from a simple formula if the wave number $k = 2\pi/\lambda = \omega/c$ is used instead of the wavelength $\lambda = c/\nu$.[21] The formula he proposed, taking into account minor subsequent modifications, is

$$k = 2\pi R\left(\frac{1}{n_1^2} - \frac{1}{n_2^2}\right), \qquad (2.50)$$

where R is a universal, empirical constant, called the *Rydberg constant*, of value 109,737.3 cm^{-1}. The numbers n_1 and n_2 are positive integers, of which n_1 defines the series, while $n_2 > n_1$ defines the line of the series; the Balmer series corresponds to

---

[20] In Section 18.2 we will have the opportunity to study this problem with quantum methods. As we shall see, the formula derived by Rutherford using classical methods coincides exactly with the corresponding quantum version, a fortuitous fact that certainly facilitated the development of quantum theory.

[21] The Rydberg relation $k = 2\pi/\lambda$ can be rewritten in the form $\lambda = h/p$, with $p = \hbar k$; this is just the de Broglie relation in Equation (2.73). It will reappear in the next chapter associated with the Schödinger equation.

$n_1 = 2$. Equation (2.50) incorporates the *combination principle* proposed in 1908 by W. Ritz (1878–1909), which states that the frequency of each spectral line ($\omega = kc$) of any element can be expressed as the difference of two spectral terms, each containing an integer. The empirical law (2.50) finds no theoretical justification in any classical model, according to which the emitted frequency should have a continuous spectrum instead of the observed discrete one.

2.5 Bohr's Quantum Atom Model

It was precisely Eq. (2.50) that inspired Bohr's creativity by interpreting it as an expression of energy conservation. Bohr assumed that the electrons move in stationary orbits but can occasionally "jump" from one orbit n to another n', emitting a photon. Considering that a spectral line is produced by photons of (angular) frequency $\omega = ck$, and using Einstein's quantum relation $E = \hbar\omega$, he obtained

$$\hbar\omega = 2\pi\hbar cR \left(\frac{1}{n_1^2} - \frac{1}{n_2^2} \right) = E_1 - E_2, \tag{2.51}$$

whence

$$E_n = \frac{2\pi\hbar cR}{n^2}. \tag{2.52}$$

On the other hand, using the quantum rule for angular momentum L, obtained in 1912 on the basis of spectroscopic studies by J. W. Nicholson (1881–1955), and assuming circular orbits with $L_n = n\hbar$, Bohr obtained

$$r_n = \frac{n^2\hbar^2}{e^2 m}, \quad E_n = -\frac{e^2}{2r_n} = -\frac{e^4 m}{2\hbar^2 n^2}, \tag{2.53}$$

which led him to the formula for the Rydberg constant R (with $e' = e_0 = -e$),

$$R = \frac{me^4}{2c\hbar^3}. \tag{2.54}$$

This result is in excellent agreement with the empirical value.

The quantum model of the atom proposed by Bohr in 1913 can thus be summarized in two postulates:

I. The atom has stationary orbits – that is, the electrons orbiting the nucleus do not radiate – corresponding to discrete values of energy $E_n = E_1/n^2$. Changes in the energy of the system can occur only at transitions between orbits.

II. The radiation absorbed or emitted by the atom during a transition between two stationary states of energy E' and E'' is monochromatic and its frequency ω is given by the formula (for $E' > E''$)

$$\hbar\omega = E' - E''. \tag{2.55}$$

Bohr chose a form of Postulate I that was consistent with Rydberg's formula for the H atom, and the result was surprising: the ionization energy, the radius of the atom, the frequencies of the emission lines, etc., were immediately correct.

Armed with his theory, Bohr was able to draw a number of additional conclusions. For example, he deduced the existence of a series of lines in the infrared, corresponding to $n_1 = 3\, (n_2 = 4, 5, 6, \ldots)$, predicted by Ritz and observed in 1908 by F. Paschen (1865–1947), and another one in the ultraviolet, corresponding to $n_1 = 1\, (n_2 = 2, 3, 4, \ldots)$, observed the following year by T. Lyman (1874–1954). The series for $n_1 = 4$ and 5 were observed in 1922 and 1924, respectively. Furthermore, Bohr's theory found convincing experimental support with his own interpretation of a highly irregular series, discovered by E. Pickering (1846–1919) in the light of a star in 1896, as being due to He (see Section 12.6), and not to H, as was verified in the same year (1913).

Direct experimental confirmation of the existence of stationary and discrete atomic levels, as postulated by Bohr's theory, was provided by the famous work of J. Franck[22] and G. L. Hertz (1887–1975). By bombarding atoms (e.g., mercury vapor) with electrons, they found that as soon as the collision energy exceeded certain well-defined values, new emission lines appeared. The quantum interpretation of this phenomenon is straightforward: if the collisions between the electrons and the atoms in the vapor are energetic enough, they can induce transitions between levels, exciting the atoms; when an atom is de-excited, it emits the excess energy in the form of light. Using this method, it became possible to directly determine the atomic levels of many elements and verify the validity of Bohr's hypothesis.

Bohr's model enjoyed other great successes. One of the most important was Sommerfeld's[23] text *Atomic Structure and Spectral Lines* (1919), which was based on Bohr's theory and became the bible in its field until the advent of contemporary quantum theory. However, it also had major limitations, since it was based on a semiclassical theory developed specifically for the H atom, which does not allow for extension even to He. It also did not allow the intensity of the spectral lines to be determined. In short, it was clear that a new mechanics of atomic systems still had to be constructed, which would, of course, have to incorporate essentially the quantum aspect, while at the same time generalizing and explaining Bohr's postulates.

Another important shortcoming of Bohr's model is the following. The assumption of a stationary electron orbiting the nucleus contradicts electrodynamics, which shows that accelerated electric charges radiate and thus lose energy. Bohr *solved* the problem in a convenient way: he simply *postulated* that atomic electrons moving in closed orbits do not radiate, relying on the other apparent violations of electromagnetism known at the time. This is a historical problem that has haunted us for decades, the solution to which we will have the opportunity to discuss in Chapter 10 and elsewhere in this book.

---

[22] James Franck (1882–1964) was a German physicist, who shared the 1925 Nobel Prize in Physics with Gustav Hertz for the first electrical measurement that clearly showed the quantum nature of atoms.

[23] Arnold Johannes Wilhelm Sommerfeld (1868–1951) was a German physicist known for his pioneering work in quantum and atomic physics. He also served as a mentor to several scientists who excelled in their respective fields, including Pauli, Debye, Heisenberg, and Bethe. By his much sought-after high-quality physics textbooks, he alone could be considered a forerunner of Landau–Lifschitz.

2.6 The Early Quantization Rules

The Bohr model contains only a single quantum number n, which labels the energies of the stationary states. It soon was realized that more quantum numbers needed to be introduced to give a more complete picture of atomic structure and spectra. The establishment of general quantization rules therefore became a central problem. Over time several rules were proposed, generally approximate but very useful, one of which will be briefly discussed here.

Consider a system of N particles whose dynamical behavior is described in classical mechanics by the $3N$ generalized coordinates $q_1, q_2, \ldots, q_{3N}$ and their $3N$ conjugate canonical momenta $p_1, p_2, \ldots, p_{3N}$. The laws of motion are Hamilton's equations,

$$\dot{q}_i = \frac{\partial H}{\partial p_i}, \quad \dot{p}_i = -\frac{\partial H}{\partial q_i}, \tag{2.56}$$

where the Hamiltonian H of the system is the total energy expressed as a function of the variables q_i and p_i and time, when appropriate. For separable (i.e., integrable) systems with periodic motions, it is convenient to introduce the action variables J_i defined as

$$J_i = \oint p_i dq_i, \tag{2.57}$$

where the integration is carried out over a complete period (of oscillation or rotation, as the case may be); the generalized coordinate conjugate to J_i is the angle variable θ_i that satisfies the canonical equation

$$\nu_i \equiv \dot{\theta}_i = \frac{\partial H}{\partial J_i}, \tag{2.58}$$

where ν_i is the frequency associated with the periodicity of q_i. The variables J_i are also known as adiabatic invariants, because they are invariant to very slow (adiabatic) changes in the system parameters (lengths, electric charges, etc.).

The Bohr quantization postulate, in the form given to it by W. Wilson (1887–1948) in 1915 and Sommerfeld in 1916, known as the *Wilson–Sommerfeld rule*, consists in proposing that the numerical value of each of the adiabatic invariants is an integer multiple of Planck's constant:

$$J_i = hn_i = 2\pi\hbar n_i, \tag{2.59}$$

where the quantum number n_i can take any of the values $n_i = 1, 2, 3, \ldots$ The quantization postulate proposed by Bohr corresponds to the special case of the angular momentum of the orbital electron.

The Wilson–Sommerfeld quantization rule reproduces Bohr's formula (2.51), as shown in the following Exercise E2.2 for the case of circular orbits. Sommerfeld himself (1916) studied the more complicated case of elliptical orbits and, to his amazement, arrived at the same formula, showing that the energy depends only on the orbital quantum number n. This is a well-known example of *degeneracy of states*, the reason for which will be explained in detail in Section 12.5.

Exercise E2.2. Applying the Wilson–Sommerfeld quantization rule
Use the Wilson–Sommerfeld quantization rule to obtain the emission spectrum of the hydrogen-like atom.

Solution. We consider a hydrogen-like atom with Z protons in its nucleus and a single orbital electron. We denote by m the reduced mass of the system and by e_0 the charge of the proton; in the polar coordinate system in the plane of the orbit, the Hamiltonian reads

$$H = \frac{p_r^2}{2m} + \frac{p_\phi^2}{2mr^2} - \frac{Ze_0^2}{r}. \tag{2.60}$$

Hamilton's equations can be written as two equations of motion for the radial and angular momenta, respectively,

$$\dot{p}_r = -\frac{\partial H}{\partial r} = \frac{p_\phi^2}{mr^3} - \frac{Ze_0^2}{r^2}, \quad \dot{p}_\phi = -\frac{\partial H}{\partial \phi} = 0, \tag{2.61}$$

and two relations between velocities and generalized momenta,

$$p_r = m\dot{r} = m\frac{\partial H}{\partial p_r}, \quad p_\phi = mr^2\dot{\phi} = mr^2\frac{\partial H}{\partial p_\phi}. \tag{2.62}$$

Integrating the second Eq. (2.61) gives the law of conservation of angular momentum, $p_\phi = mr^2\dot{\phi} = \text{const}$. We will limit ourselves to the case of circular orbits for the sake of simplicity. In this case $\dot{r} = 0$, so $p_r = 0$ and from the first Eq. (2.61) it follows that

$$p_\phi^2 = Ze_0^2 mr. \tag{2.63}$$

The Hamiltonian is thus reduced to

$$H = -\frac{Ze_0^2}{2r}. \tag{2.64}$$

All the preceding results are classical; to make contact with the quantization methods, we restate them in terms of the adiabatic invariants. Since there is only one periodic degree of freedom in the present problem, there is only one adiabatic invariant, which is

$$J_\phi = \oint p_\phi d\phi = 2\pi p_\phi. \tag{2.65}$$

We now express the radius of the orbit, the angular frequency and the energy (the numerical value of the Hamiltonian) in terms of J_ϕ,

$$r = J_\phi^2 / 4\pi^2 Zme_0^2; \tag{2.66}$$

$$\dot{\phi} = 8\pi^3 Z^2 me_0^4 / J_\phi^3; \tag{2.67}$$

$$H = -2\pi^2 Z^2 me_0^4 / J_\phi^2 \tag{2.68}$$

To quantize these classical results, we apply the Wilson–Sommerfeld rule,

$$J_\phi = 2\pi \hbar n, \quad n = 1, 2, 3, \ldots \tag{2.69}$$

With this we obtain for the energy levels of the atom

$$E_n = -\frac{Z^2 e_0^4 m}{2\hbar^2} \frac{1}{n^2} \tag{2.70}$$

and for the corresponding atomic radius

$$r_n = \frac{\hbar^2}{Z e_0^2 m} n^2. \tag{2.71}$$

Finally, the emission spectrum of the atom is obtained by applying Bohr's Postulate II,

$$\omega_{nn'} = \frac{E_n - E_{n'}}{\hbar} = \frac{Z^2 e_0^4 m}{2\hbar^3}\left(\frac{1}{n'^2} - \frac{1}{n^2}\right). \tag{2.72}$$

The model correctly predicts the quantization of the hydrogen-like atom and the value of the Rydberg constant, justifying the quantization procedure. This was the first successful quantum model of an atom, and its success helped to establish the notion of both the atomic disposition of matter and its quantum nature.

2.7 De Broglie: Waves Associated with Corpuscular Motion

In 1923–1924, de Broglie published a series of articles related to his doctoral work, in which he proposed to give serious consideration to a new quantum feature he had discovered, namely a wave phenomenon accompanying the motion of quantum particles. Reconciling the wave and particle descriptions of the behavior of light had been his aim from the beginning of his studies, but he went much further by introducing the concept of matter waves.

Leaving aside the details, de Broglie's idea was as follows. As explained in Section 2.3, the combination of the relativistic formula relating the energy of a photon to its momentum, $E = cp$, with Einstein's formula, $E = \hbar\omega$, results in Eq. (2.30), $\hbar\omega = cp$. Furthermore, for a wave traveling at the speed c the equation $c = \lambda\omega/2\pi$ holds, so that λ and p are related by $\lambda = 2\pi c/\omega = h/p$ (see Eq. (2.31)). De Broglie skillfully combined arguments drawn from optics and special relativity to identify the particle's velocity v with the group velocity of a modulated wave packet that supposedly accompanies the particle. He concluded that the wavelength of this phase wave is in general given by $\lambda = h/p$, that is, that this relationship holds for *all* quantum corpuscles, whatever their mass,

$$\lambda = \frac{h}{p}. \tag{2.73}$$

This is the essence of de Broglie's major and original contribution to our knowledge of the quantum world. For de Broglie, all quantum particles behave in a similar way to the photon, regardless of whether they have zero mass or not, which means that there is always a *wave* phenomenon of wavelength λ associated with the particle's linear momentum p. The physical nature of this phase wave was not specified by de Broglie and has been the subject of much controversy ever since, embodied in the seemingly paradoxical wave–particle dual behavior of quantum objects. We will have several opportunities to return to this important issue, and in particular in Chapter 10 we will discuss a way out of the controversy.

Einstein reached a similar conclusion to de Broglie almost simultaneously, although he did so by a different route. This was based on an idea proposed to him by N. Bose.[24] Einstein began in 1925 to construct quantum statistics,[25] which he applied to the case of ideal gases, using his old method of energy fluctuations. Applied for simplicity to a gas of photons (considered as noninteracting molecules with zero chemical potential, of which there can be any number in any one state), the argument is basically as follows. The average energy of the gas at temperature $T = 1/k\beta$ is given by Eqs. (2.16) and (2.18),

$$\overline{E} = \frac{\sum_{n=0}^{\infty} n E_1 e^{-\beta E_1 n}}{\sum_{n=0}^{\infty} e^{-\beta E_1 n}} = \frac{E_1}{e^{\beta E_1} - 1}. \tag{2.74}$$

Further, it is a simple exercise to show that the mean square energy is given by (see also Exercise 15.44)

$$\overline{E^2} = \frac{\sum_{n=0}^{\infty} n^2 E_1^2 e^{-\beta E_1 n}}{\sum_{n=0}^{\infty} e^{-\beta E_1 n}} = \overline{E}(2\overline{E} + E_1). \tag{2.75}$$

Therefore, the dispersion (variance or mean square deviation) of the energy is given by

$$(\Delta E)^2 = \overline{E^2} - \overline{E}^2 = \overline{E}^2 + E_1 \overline{E}. \tag{2.76}$$

The first term on the right is dominant at high temperatures, and can therefore be interpreted as classical or wave-like; the second term is proportional to the average number of energy packets $\bar{n} E_1$ and can therefore be considered as corpuscular in nature. The presence of the corpuscular term is natural because we are dealing with energy packets, but the wave term suggested to Einstein that the molecules of a quantum gas also manifest wave properties and that it should therefore be possible to observe, for example, interference phenomena in molecular beams. As predicted by Einstein, experiments conducted at the end of the twentieth century using *Bose–Einstein condensates* (bosons cooled to near absolute zero, which tend to concentrate in the ground state, giving rise to a highly correlated and coherent state) revealed the predicted interference between two small but macroscopic volumes of atoms, suitably prepared. Einstein further noted that he was aware of similar results, which linked a wave phenomenon to the motion of quantum corpuscles, that had been proposed "by a young French physicist."[26]

Armed with his ideas, de Broglie reinterpreted the Wilson–Sommerfeld quantization rule $J = \oint p \, dq = nh$ as the condition for the electronic orbit to fit precisely an integer number of wavelengths. He explained that this was the only stable solution by arguing that any other orbit that might initially exist would tend to disappear due to the destructive interferences produced by the superposition of waves with different phases. In other words, de Broglie identified the quantization postulate with the condition of stability of the orbit, thereby adding a new content to Bohr's ideas.

---

[24] Satyendra Nath Bose (1894–1974) was an Indian polymath, best known for his work in the early 1920s in developing the foundations of Bose–Einstein statistics for bosons and the theory of the Bose–Einstein condensate. Considering the frequent reference to bosons in the normal quantum literature, Bose's name is probably the most directly or indirectly cited in the physics literature, and although he was often proposed for the Nobel Prize, he did not receive it.

[25] The quantum statistics describing the behavior of fermions is called Fermi–Dirac statistics. Both quantum statistics have the same limiting classical statistics at high temperatures (see Chapter 15).

[26] Paul Langevin, who was to chair de Broglie's examination committee, had sent Einstein a copy of de Broglie's doctoral thesis, requesting his opinion.

2.8 Electron Beam Diffraction

For de Broglie himself, the waves associated with corpuscular motion had a physical character; thus, in reply to J. Perrin[27] during his doctoral examination he stated that it should be possible to demonstrate the existence of such physical waves by an electron-diffraction experiment. Shortly afterwards, a series of electron diffraction experiments showed that there are indeed situations in which electrons behave in an analogous way to X-rays, with precisely the wavelength h/mv predicted by de Broglie's theory.

In 1927 C. J. Davisson,[28] together with L. Germer (1896–1971), observed – thanks to a lucky event correctly interpreted by Born's group – that a piece of nickel crystal produced electron scattering similar to that which would be obtained with X-rays. To reveal the wave properties of the electrons, their de Broglie wavelength must be of the order of the lattice constant of the crystal, i.e., $\lambda \sim 10^{-8}$cm $= 1$ Å, which is of the order of the wavelength of soft X-rays. Since, according to Eq. (2.77), the de Broglie wavelength is given by

$$\lambda = \frac{h}{mv} = \frac{h}{\sqrt{2meV}}, \qquad (2.77)$$

where V is the accelerating voltage, this must be of the order of a few tens of volts, a sufficiently high value to guarantee an essentially monoenergetic beam, but low enough for the electrons to be scattered only by the surface layer of the crystal.[29]

An electron-diffraction device is shown schematically in Fig. 2.5. Assume that the electron beam is incident perpendicular to the crystal surface, so that it can be represented by a plane wave. The atoms on the crystal surface will scatter the electrons, and

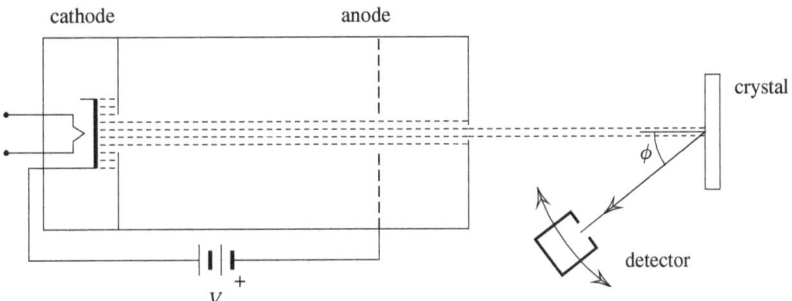

Figure 2.5 Diffraction of an electron beam by the surface of a crystal. A monoenergetic beam is emitted from the electron gun; the detector is placed at different angles ϕ to receive the scattered electrons.

---

[27] Jean Baptiste Perrin (1870–1942) was a French physicist who was awarded the Nobel Prize in Physics in 1926 for his experimental work on Brownian motion, which served to verify Einstein's theory of the phenomenon, confirming the atomic nature of matter (see Chapter 10).

[28] Clinton Joseph Davisson (1881–1958) was a US-born physicist who shared the 1937 Nobel Prize in Physics with G. P. Thomson for their discovery of electron diffraction.

[29] The Davisson-Germer experiment led to the analytical technique called low-energy electron diffraction, which is used to study crystal surfaces and the processes that take place on them. The electrons used have energies between 10 eV and 200 eV, corresponding to wavelengths of about 1 to 4 Å.

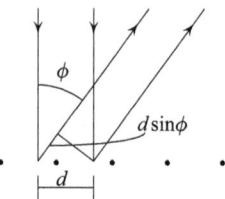

Figure 2.6 A collimated monoenergetic electron beam incident perpendicular to the crystal surface is scattered at different angles ϕ given by Eq. (2.78) with $n = 0, 1, 2, \ldots$ The most pronounced diffraction maximum is obtained for $n = 1$.

the accompanying wave at any point will be nonzero only in the directions in which there is constructive interference between the various components scattered to that point. As illustrated in Fig. 2.6, the condition for maximum constructive interference is that the optical path difference must be an integer multiple of the wavelength (von Laue, 1912),

$$d \sin \phi = n\lambda, \quad n = 0, 1, 2, \ldots, \tag{2.78}$$

where ϕ is the scattering angle. The wavelength obtained from the results of Davisson and Germer turned out to be in perfect agreement with that predicted by de Broglie's formula.

It is known that X-rays can be diffracted not only by single crystals, but also by polycrystals. G. P. Thomson[30] showed in 1927 that the same is true for electrons, by passing an electron beam with an energy of a few thousand eV through a very thin polycrystalline film whose crystals are randomly oriented. The analysis of the three-dimensional problem is quite complicated. However, it can be simplified by using the method proposed by W. L. Bragg,[31] which consists of considering the reflection of the beam from successive planes of atoms at equal distances d from each other. Maximum constructive interference occurs when the difference in optical paths between adjacent beams is an integer multiple of the wavelength, as shown in Fig. 2.7a. Therefore,

$$2d \sin \theta = n\lambda, \quad n = 0, 1, \ldots \tag{2.79}$$

where θ is the angle measured from the plane, determined by the orientation of the crystal, which varies randomly.

It is clear that among the multitude of crystals that make up the sample, there will be some oriented precisely in directions that satisfy Bragg's Law, producing luminous points on the screen (such as P in Fig. 2.7b). Since the orientation of the crystals is random, there will also be some whose orientation differs from that of the previous ones only by a rotation around the beam direction (i.e., the device has axial symmetry). These crystals form a diffraction ring, whose radius R can be calculated from the formula

$$\frac{R}{L} = \tan 2\theta, \tag{2.80}$$

---

[30] George Paget Thomson (1892–1975) was an English physicist, the son of J. J. Thomson. He shared the 1937 Nobel Prize in Physics with Davisson for his experiments on electron diffraction.
[31] William Lawrence Bragg (1890–1971) was a British physicist, who shared the 1915 Nobel Prize in Physics with his father William Henry Bragg for their analysis of crystal structure by means of X-rays.

2.8 Electron Beam Diffraction

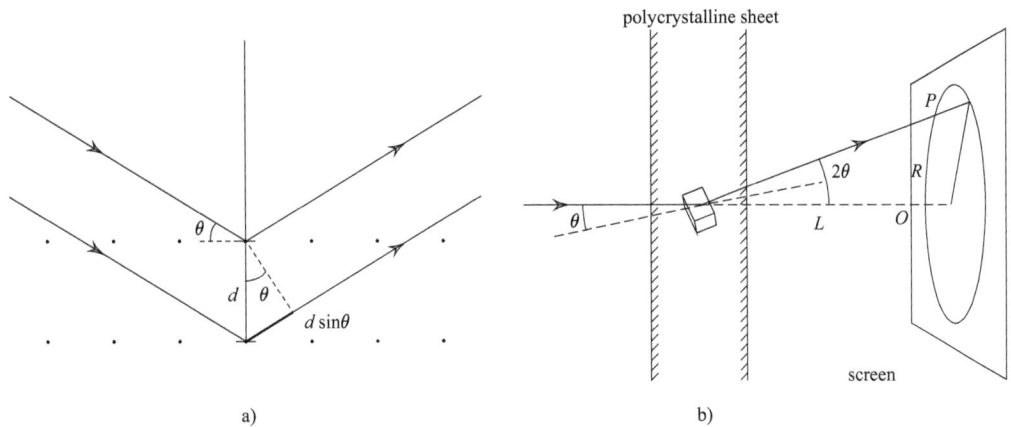

Figure 2.7 (a) Diagram of the reflection of a collimated monoenergetic electron beam by a three-dimensional lattice, leading to the Bragg equation (2.79). (b) Schematic of Thomson's electron diffraction experiment using a thin polycrystalline sheet. The orientation of the small crystals varies randomly, resulting in a circular diffraction pattern consisting of a series of concentric rings.

where L is the distance between the sample and the screen, as in Figure 2.7b. This equation was confirmed experimentally by Thomson using de Broglie's formula for the wavelength λ.

Davisson himself stated at a conference held in 1928:

> In the course of the last few years we have concluded that there are circumstances under which it is convenient, if not necessary, to regard electrons as waves rather than as particles, and we are making more and more use of terms such as diffraction, reflection, refraction and dispersion to describe their behavior.

Subsequently, it was shown experimentally that wave behavior is indeed shared by molecular beams of H or He (Estermann, Frisch and Stern, 1930–31), neutrons (Fermi and Marshall, 1947), and so on. Today it is accepted as a property associated with the motion of all corpuscles, as proposed by de Broglie, and is even used in the manufacture of various technological devices. For example, in electron or proton microscopes using electromagnetic lenses, the corpuscles are treated as a wave beam with a de Broglie wavelength much smaller than that of visible light and a resolution of the order of 10^3 times higher than that of conventional optical microscopes. An analogous principle is used in the construction of several other devices in modern electron optics.

The de Broglie wave is only important for particles on the atomic or subatomic scale, as its wavelength becomes negligible for macroscopic bodies. To get an idea of the scale involved, let us consider the case of the electron, which has a mass of about 9.1×10^{-28} g. If the electron moves at a speed of about one hundredth of the speed of light, the de Broglie wavelength is of the order of 10^{-8} cm, which is about the first Bohr orbit radius of H. In fact, we should expect that the wave properties of the corpuscles will manifest themselves only when the de Broglie wavelengths are comparable to the characteristic dimensions of the system. This helps us to understand why quantum phenomena remained hidden as long as physics was not directly concerned with the study of matter at the atomic or molecular scale.

In the usual interpretation of QM, the corpuscular and wave properties of quantum objects have often been regarded as mutually exclusive. It is now difficult to maintain this position, since experiments (with photons) have been carried out which show that, by small changes in the experimental setup, both aspects of their behavior can be manifested simultaneously. The first of these experiments was carried out by Grangier, Roger, and Aspect (1985); a more recent one is described by Mizobuchi and Ohtaké (1992). Strictly speaking, in diffraction experiments one ultimately observes both the (statistical) interference patterns and the corpuscular structure of the individual spots.

2.9 Back to Planck: Completing the Spectral Law with the Zero-Point Field

In 1911, while searching for a well-founded quantum theory for the blackbody problem,[32] Planck discovered the need to add an extra contribution to his spectral energy distribution law, in the form of a zero-point term carrier of an energy

$$E_0 = \frac{1}{2}\hbar\omega \qquad (2.81)$$

per mode of frequency ω. He found that the thermal field inside the blackbody was always accompanied by this nonthermal (i.e., *zero-point*) field, which meant that the reference level of the field energy had to be shifted by $\hbar\omega/2$. Therefore, the average energy per normal mode given by Eq. (2.29) should be completed with Eq. (2.81),

$$\overline{E} = \frac{1}{2}\hbar\omega + \frac{\hbar\omega}{e^{\hbar\omega/kT} - 1}, \qquad (2.82)$$

Planck's complete blackbody radiation formula reads

$$\rho(\omega, T) = \frac{\hbar\omega^3}{\pi^2 c^3}\left(\frac{1}{2} + \frac{1}{e^{\hbar\omega/kT} - 1}\right) \qquad (2.83)$$

instead of (2.22). The first term,

$$\rho_0(\omega) = \frac{\hbar\omega^3}{2\pi^2 c^3}, \qquad (2.84)$$

is the spectral energy density of the added ZPF, which is the product of the energy of the mode and the density of modes (see Section 10.2),

$$\rho_0(\omega, T) = \frac{\omega^2}{\pi^2 c^3} \cdot \frac{1}{2}\hbar\omega. \qquad (2.85)$$

Although Planck's theory was not to become generally accepted, the associated hypothesis of zero-point energy remained alive. In particular, it was vigorously defended by Nernst (1916), who argued that if radiation and matter are in statistical equilibrium with each other, then their respective mean energies must be described

---

[32] The most significant challenge for Planck arose from his prolonged rejection of the necessity for field quantization. It was Einstein who first proposed a correct and convincing derivation of the Planck distribution.

by the same law (i.e., a *single* law), namely by Planck's complete radiation law, which includes the zero-point term. Nernst regarded the ZPF as a fluctuating electromagnetic field of unknown origin, present everywhere and at all times. He postulated that interaction with this field was the mechanism responsible for atomic stability and the source of the quantum behavior of matter.

Nernst's proposal seems to have been well known in Germany, yet it did not enter the mainstream of physics at the time. With the advent of quantum formalism in 1925 the concept of ZPE – albeit of the mechanical oscillators – became necessary, as it is identified with their ground-state energy. It is commonly seen, both in textbooks and research papers, as a direct consequence of Heisenberg's uncertainty principle. But we can look at it the other way round: the Heisenberg uncertainty principle as a *consequence* of the ZPE, not as a principle in itself (see what follows).

The ZPE is not exclusive to the blackbody radiation; as we will see later (see in particular Chapters 9 and 17), QM has taught us that the minimum energy (the *ground-state energy*) of *any* harmonic oscillator of frequency ω is indeed $\hbar\omega/2$. More generally, current quantum field theory endows space with several "virtual" zero-point fields, including the electromagnetic (quantum) vacuum fluctuations. However, for Planck – and for Nernst – the ZPF was associated with the ground state of a *real* Maxwellian electromagnetic radiation field. Planck's (original) blackbody formula contains the thermal part of the radiation, that which is exchanged in the form of well-defined energy packets (photons) during atomic and molecular transitions. The nonthermal part does not participate in the (photonic) energy exchange, it constitutes the invisible (or dark) part of the radiation field. But its presence certainly has physical consequences, as is to be expected and as is shown throughout this text.

Exercise E2.3. Energy density of the zero-point field

Calculate the energy density of the ZPF contained in a large cavity surrounded by conducting walls, assuming a cutoff frequency ω_C.

Solution. Consider a rectangular cavity with dimensions $L_1 \times L_2 \times L_3 = V$. For the tangential component of the electric field on the conducting surfaces to vanish, the field modes contained in the cavity must be

$$u_x = A_1 \cos k_1 x \sin k_2 y \sin k_3 z,$$
$$u_y = A_2 \sin k_1 x \cos k_2 y \sin k_3 z,$$

and

$$u_z = A_3 \sin k_1 x \sin k_2 y \cos k_3 z, \qquad (2.86)$$

where $k_i = n_i \pi / L_i$ and n_i ($i = 1, 2, 3$) are nonnegative integers with at least two of them nonzero. The frequency of the mode is $\omega_k = c(k_1^2 + k_2^2 + k_3^2)^{1/2}$. The condition $\nabla \cdot \boldsymbol{E} = 0$ implies $\sum A_i k_i = 0$, which confirms that there are two independent polarizations for each wavenumber $\boldsymbol{k}$, except when one of the integers n_i is zero.

When the cavity is large, the allowed values of $\boldsymbol{k}$ can be considered to form a continuum and the density of modes in the positive $\boldsymbol{k}$-space is $2V/\pi^3$. The total energy density of the ZPF modes of frequency $\omega < \omega_C$ is therefore given by the integral over the full $\boldsymbol{k}$-sphere divided by 8,

$$\frac{2}{V}\sum_k \frac{1}{2}\hbar\omega_k \to \frac{2}{8\pi^3 c^3}\int_0^{\omega_C}\frac{1}{2}\hbar\omega\, 4\pi\omega^2 d\omega = \frac{\hbar\omega_C^4}{8\pi^2 c^3}. \tag{2.87}$$

It is clear that the energy density is dominated by the high-frequency modes. Taking a minimum wavelength of $\lambda_C = 0.4 \times 10^{-6}$ m, so as to include the visible light, we get an energy density of 23 J/m^3, which is 10^9 times that of a 100 W light bulb at a distance of one meter (Ballentine 2010, ch. 19). In Chapter 10 we return to the issue of the (in principle) infinite energy content of the ZPF, and discuss the Casimir effect as a result of the change in energy content due to the presence of material bodies.

2.9.1 Stochastic electrodynamics

Stochastic electrodynamics is the theory developed by taking the ZPF at face value, as a real fluctuating Maxwellian field that pervades the whole space the whole time. Some back-of-the-envelope calculations show that the properties of the stochastic motions impressed on the electrons by the ZPF are fully consistent with quantum fluctuations. In particular, Heisenberg's inequality is readily obtained by taking into account that the energy of an oscillator is limited from below by the ZPE, and by using the virial theorem to calculate the dispersions of p and x (for details see Chapter 8), to get

$$\frac{\langle(\Delta p)^2\rangle}{2m} = \frac{m\omega^2\langle(\Delta x)^2\rangle}{2} \geq \frac{\hbar\omega}{4}, \tag{2.88}$$

whence $\Delta p \Delta x \geq \hbar/2$, with $\Delta x = \langle(\Delta x)^2\rangle^{1/2}$ and similarly for p. This is a simple version of Heisenberg's inequality, albeit without any mystery. In Chapter 10 and elsewhere in the book, it will be shown that the constant interaction of the particle with the ZPF has indeed far-reaching consequences, in that it provides a physical basis for quantum formalism, both in the Schrödinger and in the Heisenberg descriptions. Moreover, as will be shown in Chapter 17, inclusion of the ZPF *ab initio* – as an irremovable integral component of the matter-field system – allows us to go beyond QM and make contact with the more complete theory of (nonrelativistic) *quantum electrodynamics* (QED), without the need to introduce additional elements or ad hoc assumptions.

2.10 Corollary: The Electrodynamic Essence of Quantum Mechanics

It should be clear by now that atomic matter does not live in isolation, but is in constant contact with electromagnetic radiation, among other possible vacuum fields – although the net effects of those other fields appear to be negligible compared to the electromagnetic one for the purposes of usual QM. As we have seen throughout this chapter, the development of QM has been marked from the beginning by numerous demonstrations of this contact and its decisive influence on both matter and field. In fact, what is generally called "quantum mechanics" is not just a mechanical theory; it is the *quantum theory of matter in its interaction with the radiation field*, even if the latter is not usually made explicit. Although this book focuses primarily on the quantum behavior of *matter*, in due course we will approach the study of the matter-field system from an integrated perspective in order to provide a more comprehensive understanding of quantum phenomena.

Problems

P2.1 Obtain the limit expressions of the Planck distribution for small and large frequencies, at a fixed temperature. What is the form of the function $f(\omega/T)$ that appears in Wien's law (2.20) for high frequencies, and why is it not possible to determine it classically? Discuss your results.

P2.2 Show that Planck's law predicts that the spectral density of blackbody radiation has a maximum for each temperature, which occurs at the wavelength

$$\lambda_n = \frac{2\pi c \hbar}{4.965} \frac{1}{k_B T}. \qquad (2.89)$$

Calculate ν_m and explain why $\nu_n \neq c/\lambda_n$. This formula – known as *Wien's displacement law* – shows that as the temperature of the blackbody rises, the maximum intensity of the radiation shifts toward shorter wavelengths.

P2.3 Construct a graph of the average energy of the Planck oscillators versus frequency and use it to show that Planck's postulate, $E_n = n\hbar\omega$, introduces a cutoff in frequency space. Find this cutoff frequency. This result shows that the postulate prevents arbitrarily high frequency modes from being excited at a given temperature, in agreement with experimental observations.

P2.4 Calculate the energy of a quantum of visible light of wavelength 6000Å. Calculate the number of quanta of this wavelength that are emitted per second by a 10-watt light source.

P2.5 Taking the Sun as a blackbody at 5700 K, find the fraction of its mass that it emits as electromagnetic radiation per year.

P2.6 Ultraviolet light of wavelength 3500Å is incident on a potassium surface. The maximum energy of the emitted photoelectrons is observed to be 1.6 eV. Calculate the work function of potassium, neglecting thermal corrections.

P2.7 A 100 MeV photon collides with an electron at rest and is scattered at 45° from the direction of incidence. Calculate the energy of each particle after the collision and determine the final direction of the electron.

P2.8 Assuming that the relevant classical laws apply, calculate the power emitted by an electron moving in a circular Bohr orbit characterized by the quantum number n.

P2.9 Deuterium is an isotope of hydrogen whose nucleus consists of a tightly bound proton and neutron, with atomic weight 2 (the masses of the proton and neutron are nearly equal). Determine: (a) the Rydberg constant for deuterium in terms of that for hydrogen; (b) the ratio of the wavelengths of the deuterium spectrum to those of the corresponding hydrogen transitions.

P2.10 Use the Wilson–Sommerfeld quantization rule to determine the emission spectrum of a particle moving in the potential

$$V(r) = V_0 \left(\frac{r}{a}\right)^k, \qquad (2.90)$$

with $k \gg 1$, assuming that it is possible to restrict oneself to the study of circular orbits.[33] Draw a representative graph of this potential, and compare your results with those of 2.22.

P2.11 Calculate the de Broglie wavelength of a point particle moving at speed $c/100$, in the following cases:

(a) an electron ($m_e = 9.1 \times 10^{-28}$ g),
(b) a proton ($m_p = 1\,836.1\, m_e$),
(c) a pebble ($m = 10$ g),
(d) the Earth ($M_T = 6 \times 10^{27}$ g).

Compare your results with the wavelength of yellow light and with the radius of the first Bohr orbit ($a = 0.529 \times 10^{-8}$ cm).

P2.12 Find the expression for the orbital speed of the electron of a hydrogen atom that is in its lowest energy state using the Bohr model. Since this state is stationary, it can be described by an ensemble of H atoms in their ground state. Use this observation to determine the wavelength associated with the corresponding orbital velocity, and compare it with the circumference of the orbit. Discuss the result.

P2.13 In traditional optics, an optical instrument cannot resolve details of an object smaller than the wavelength used for observation. For example, a virus 200 Å in diameter cannot be studied with an instrument that uses visible light in the region of thousands of angstroms. An electron microscope, however, makes it possible. Determine the acceleration potential required to obtain electrons with a de Broglie wavelength $10^2 - 10^3$ times smaller than the dimensions of the virus.

P2.14 Determine the wavelength of a beam of neutrons which undergoes first-order diffraction by a crystal. The beam is incident at an angle of 40^0 with respect to the set of planes of the crystal lattice, whose spacing is 2.85 Å. What is the kinetic energy of the incident neutrons?

P2.15 A technique for monochromatizing[34] a beam of slow neutrons consists of directing the entire beam toward a crystal of known structure and positioning the collector to receive the monochromatic beam of diffracted neutrons. Suppose that a crystal is used whose distance between successive layers is 1.2 Å. Considering only Bragg diffraction of order 1, at what angle with respect to

---

[33] Bertrand's theorem (1873) of classical mechanics establishes that the only central forces that can give rise to bounded orbits, *all* of them closed for classical problems, are those proportional to r (Hooke's law) or to r^{-2} (law of gravitation). Therefore, for arbitrary k in the corresponding classical problem, the orbits are generically open and it is artificial to apply the quantization rules. However, all of these central problems do have *circular* orbits; see José and Saletan (1998).

[34] By analogy with optics, the term monochromatic is used to refer to a beam whose particles all have the same de Broglie wavelength (if all the particles have the same mass, the beam is both monoenergetic and monochromatic).

the initial direction of the beam should the collector be oriented to select the neutrons with $\lambda = 0.8$ Å?

P2.16 Compare the zero-point energy of a radiation field mode of wavelength 6000Å, with the average thermal energy of the same field mode at room temperature.

P2.17 An electron in an antenna is subject to oscillations at a frequency $\nu = 10^{15}$ Hz. Calculate the mean amplitude Δx of the oscillations at zero temperature and at room temperature.

Bibliographical Notes

Planck's seminal work on blackbody radiation: Planck (1901).

On Hamiltonian mechanics, see, e.g., Landau and Lifshitz (1976) section 49, José and Saletan (1998) section 6.4, or Goldstein et al. (2011).

Weinberger (2006) revisits de Broglie's crucial contribution to the birth of QM.

Stone (2015) reveals the significance of Einstein's major contributions to quantum theory.

Planck introduced the ZPF in his second radiation theory by means of the concept of zero-point energy (Planck 1911, 1912, 1913). Most of Planck's articles on radiation and quantum theory are conveniently collected in Planck (1958).

3 Quantum Mechanics in the Heisenberg (Matrix) Picture

Our journey into QM begins, as has historically been the case, with the matrix formulation of the theory pioneered by Werner Heisenberg. This has the advantage of bringing us close to the tools and language used in contemporary developments and applications of QM, supported by the necessary conceptual background that facilitates a physical understanding of the subject.

After a brief introduction, we present the basic elements of the Heisenberg picture in terms of matrices (operators) representing the dynamical variables of the quantum system and derive their laws of motion. The harmonic oscillator, a central and particularly simple problem in QM, is studied to familiarize the student with the new physics and the operator language that describes it. To complete the theory, the state vectors on which the operators act are introduced. The chapter concludes with a hint at the origin of the puzzling substitution of operators for the continuous dynamical variables of classical mechanics. A first-principles derivation of the quantum operator formalism, which reveals its deep physical content, is left to Chapter 10.

3.1 The Beginnings of Matrix Mechanics

In 1925, when the old quantum theory and related efforts seemed unable to account for the experimental data, Heisenberg decided to take a radical approach. Enough evidence had accumulated to accept the quantization of dynamical quantities such as energy and angular momentum, but the (classical) *kinematic* content of the old quantum theory simply did not fit the scheme. So, with great insight, he developed a description in which the dynamical variables satisfied a rather strange algebra, but the structure of the dynamical laws remained that of the classical laws.

The basic elements of the (matrix) theory proposed by Heisenberg were substantially elaborated by him together with Bohr, Born, Jordan[1] and the ever-present help of Pauli. This led to the matrix formulation of QM, which immediately gained the support of a substantial part of the physics community, even philosophers of science, despite its strange peculiarities – indeed, it was somehow stimulated by them – and its mathematical difficulty, since matrices were then largely unknown to physicists, including Heisenberg himself. In 1927, Heisenberg added the *principle of indeterminism*, which

---

[1] Pascual Jordan (1902–1980) was a brilliant German mathematician who studied under Max Born in Göttingen; his collaboration with Born and Heisenberg over several years was of great importance for the development of QM, in particular matrix mechanics, and quantum field theory. The introduction of anticommutators in QM is due to him.

attracted much attention and became the subject of a variety of interpretations. The subsequent formalization of the matrix formulation allowed it to be extended to other important areas of physics, in particular quantum field theory.

For all its reworking, matrix mechanics remained what it was, an abstract theory, impenetrable and mathematically complex, at odds with known theoretical physics. Things changed radically in 1926 with the proposal by Schrödinger of an alternative, less obscure theory (discussed in the next chapter), which was later independently shown by Schrödinger himself and by Born and others to be formally equivalent to matrix mechanics. The two seemingly unrelated theories turned out to be two mathematical descriptions of the same quantum phenomena. Physicists then began to turn to the new version because it was expressed in more familiar mathematical terms. What was surprising, however, was that both versions of the theory, so different in nature, required the quantum variables to be described in terms of *operators*.

Heisenberg realized that his formalism could only be applied to very simple cases, such as the harmonic oscillator, the rotator, and the oscillator perturbed by a nonlinear term. With great mathematical skill, Pauli managed to solve the H-atom problem within matrix mechanics, and shortly afterwards Lucy Mensing[2] tackled the problem of diatomic molecules, with some success in quantizing angular momentum. Despite its practical limitations, Heisenberg's QM was (and is) of extraordinary value because, apart from opening the way for further research, it already contained the essence of today's QM.

3.2 The Heisenberg Picture

In Heisenberg's picture (*description*, *representation*) each dynamical variable is represented by a matrix, with as many columns as rows. This applies to variables that are quantized, that is, that have a denumerable set of eigenvalues (or proper values, see what follows). However, even continuous matrices are allowed as a matter of principle.

The matrix Q with elements Q_{kl} should be *Hermitian* to correspond to a real variable – or *observable* in Heisenberg's now conventional language[3] – which means that it must be equal to its adjoint,

$$Q^\dagger = Q^{T*} = Q, \tag{3.1}$$

or, in terms of the matrix elements, $Q_{nl} = Q^*_{ln}$, where * indicates complex conjugation and T transposition (i.e., exchange of the indices n and l, or, equivalently, transformation of rows into columns and vice versa).

In particular, based on the rich evidence of atomic spectra, Heisenberg introduced "position" x_{nl} and (linear) "momentum" p_{nl} as parameters related to the atomic tran-

---

[2] Lucy Mensing (1901–1995) was a German physicist who pioneered quantum mechanics. She was the first to validate the allowable discrete values for quantum angular momentum.

[3] Influenced by positivism, Heisenberg argued that what we perceive is all that exists, and that what does not exist should not be part of a scientific theory. In particular, since atomic orbits are not observed, they do not exist and cannot be part of the theory; only observables, such as the frequencies and intensities of spectral lines, are acceptable. This now largely abandoned positivist philosophy of science is still often found in the QM literature, including textbooks.

sition between states n and l (or rather to the spectral line of light emitted or absorbed during the transition $n \leftrightarrow l$). This was a very insightful step which has implications for the whole of contemporary quantum theory. In fact, the introduction of the parameter sets $\{x_{ln}\}$ and $\{p_{ln}\}$ was unknowingly equivalent to transforming the (canonical) phase-space variables of classical mechanics into operators (matrices).

The position and momentum variables turned out to satisfy a universal relation, not limited to any particular system or state of the system, and so Heisenberg postulated it as a general *quantization rule*,

$$\sum_l (x_{ml} p_{ln} - p_{ml} x_{ln}) = i\hbar \delta_{nm}. \tag{3.2}$$

When Born saw this principle, he identified it as the matrix expression of a commutation relation between *matrices* $\hat{x}$ and $\hat{p}$ (a caret ˆ is added to distinguish operators – q-variables in Dirac's language – from numbers or c-variables), namely,

$$\hat{x}\hat{p} - \hat{p}\hat{x} \equiv [\hat{x}, \hat{p}] = i\hbar \hat{I}. \tag{3.3}$$

It is common to omit the identity matrix $\hat{I}$ and write simply

$$[\hat{x}, \hat{p}] = i\hbar. \tag{3.4}$$

This is the basic (canonical) quantum commutator between the operators associated with the variables x, p, in 1D notation. The mathematical convention is that an operator acts on whatever is to its right. Therefore, Eq. (3.3) tells us that the result of operating first with $\hat{p}$ and then with $\hat{x}$ is different from the result of operating first with $\hat{x}$ and then with $\hat{p}$; the operators *do not commute*.

The extension of Eq. (3.4) to three dimensions is

$$[\hat{x}_i, \hat{p}_j] = i\hbar \delta_{ij}, \tag{3.5}$$

where $i, j = 1, 2, 3$. This equation – actually a postulate! – constitutes the building block of matrix mechanics. Both $\hat{x}_j$ and $\hat{p}_j$ are square matrices; more specifically, Hermitian operators, as corresponds to real quantities. The size of the matrices – the number of rows and columns – depends in each case on the number of possible values of the index n or l, that is, on the number of possible states of the system.

3.2.1 General quantum commutators

Heisenberg's matrices are a particular representation of operators. In the following we introduce some important concepts that will help the reader to get acquainted with the quantum operator formalism. In Chapter 4 we will derive general relations between operators and matrices using the Schrödinger formalism; but at this stage we do not need to distinguish between an operator and its matrix representation.

Firstly, as indicated in Eq. (3.1), an operator $\hat{Q}$ representing a *real* (not complex) dynamical variable $\hat{Q}(\hat{x}, \hat{p})$ (a space coordinate, a momentum, an energy, etc.) is Hermitian. But if $\hat{F}$ and $\hat{G}$ are two Hermitian operators, that is, $\hat{F} = \hat{F}^\dagger$ and $\hat{G} = \hat{G}^\dagger$, then in general their product $\hat{A} \equiv \hat{F}\hat{G}$ is not Hermitian, since

$$\hat{A}^\dagger = (\hat{F}\hat{G})^\dagger = \hat{G}^\dagger \hat{F}^\dagger = \hat{G}\hat{F} \neq \hat{A}. \tag{3.6}$$

The condition that the product be Hermitian, $\hat{A}^\dagger = \hat{A}$, implies $\widehat{FG} = \widehat{GF}$. So *the product of two Hermitian operators is Hermitian if and only if they commute* (this is an important result).

In general, two operators taken freely do not commute – two arbitrary square matrices of the same dimension generally do not commute. A convenient way to handle this property when referring to $\hat{F}, \hat{G}$ is to introduce their *commutator*,

$$[\hat{F}, \hat{G}] \equiv \widehat{FG} - \widehat{GF}. \tag{3.7}$$

Writing

$$[\hat{F}, \hat{G}] = i\hat{C}, \tag{3.8}$$

it is easy to prove that due to the i factor, $\hat{C}$ is Hermitian if $\hat{F}$ and $\hat{G}$ are Hermitian. The fact that operators in general do not commute gives the commutator a particularly high value as a tool for the study of quantum theory, in particular quantum dynamics.

Alongside the transformation of the dynamical variables into operators, the classical Poisson brackets are transformed into quantum commutators, according to the *correspondence rule* established by Dirac,

$$\sum_i \left[\frac{\partial F}{\partial q_i} \frac{\partial G}{\partial p_i} - \frac{\partial F}{\partial p_i} \frac{\partial G}{\partial q_i} \right] \equiv \{F, G\}_{qp} \mapsto \frac{1}{i\hbar} \left[\hat{F}, \hat{G} \right], \tag{3.9}$$

where the sum in the Poisson bracket extends over all degrees of freedom, that is over all the pairs of (classical) canonical variables (q_i, p_i).[4] The factor $\hbar$ is required for dimensional reasons and the i preserves hermiticity of the result, as mentioned before.

Let us use Dirac's correspondence rule to calculate some simple but important commutators. First we take any two Cartesian coordinates x, y (of the same or of different particles); their Poisson bracket is $\{x, y\} = 0$, from which we trivially obtain $[\hat{x}, \hat{y}] = 0$. More generally, by similar arguments, we can write

$$[\hat{x}_i, \hat{x}_j] = 0, \tag{3.10}$$

$$[\hat{p}_i, \hat{p}_j] = 0, \tag{3.11}$$

in addition to (3.5),

$$[\hat{x}_i, \hat{p}_j] = i\hbar \delta_{ij}. \tag{3.12}$$

These are general rules; they apply to the variables x_i and p_j of any quantum system, regardless of the type of system, its state, or its dimensions. They form the set of canonical commutators of QM.[5]

---

[4] *Canonical* means standard or fundamental, although its original meaning was that of being prescribed by a high father. In classical Hamiltonian physics, canonical transformations $(x, p) \to (X, P)$ are those that keep the Poisson brackets invariant, i.e., $\{F, G\}_{xp} = \{F, G\}_{XP}$. The Poisson brackets were introduced by S.D. Poisson (1781–1840, the man behind the Poisson distribution) as general structures that help express the laws of Hamiltonian dynamics.

[5] We note that the phase-space symplectic structure of classical mechanics, defined by the basic Poisson bracket $\{x_i, p_j\}_{x,p} = \delta_{ij}$, is inherited by QM. This is a specific example of the Dirac correspondence rule.

3.2.2 The Heisenberg inequality

In 1927 Heisenberg published a mathematical proof of his now famous inequality, based on a particular reading of the basic commutator (3.4): he interpreted his result as a limitation on the degree to which x and p can be determined simultaneously. For this reason, it is often referred to as the "uncertainty relation" or even elevated to the "uncertainty principle," meaning that we cannot define the state of a system with absolute precision, much less follow its detailed evolution. Heisenberg had his reasons: he had discovered, as discussed in Chapter 2, that quantum variables exhibit unpredictable behavior for which there is no known cause. A more extreme (but common) reading of the inequality sees uncertainty as intrinsic – a fact of nature, rather than an expression of our limited knowledge – and speaks of a "principle of indeterminism," implying that determinism and causality become meaningless at the microscopic level, and that quantum particles possess something like self-action (or even free will).

Here we derive the Heisenberg inequality by means of a purely algebraic procedure, leaving for later the important discussion of its physical meaning. For this purpose we construct two vectors u_n, v_n

$$u_n = x_{nl}\epsilon_l, \quad v_n = p_{nl}\epsilon_l, \tag{3.13}$$

with ϵ_l unit orthogonal vectors and $l \neq n$, and apply the Cauchy–Schwarz inequality to the scalar products,

$$\left|u_n \cdot u_n^*\right| \left|v_n \cdot v_n^*\right| \geq \left|u_n \cdot v_n^*\right|^2. \tag{3.14}$$

In terms of the matrix elements of x and p this inequality reads

$$\sum_{l \neq n} |x_{nl}|^2 \sum_{l' \neq n} |p_{nl'}|^2 \geq \left|\sum_{l \neq n} x_{nl} p_{nl}^*\right|^2. \tag{3.15}$$

Because $\hat{x}$ and $\hat{p}$ are Hermitian and must satisfy Eq. (3.4), we have

$$\left|\sum_{l \neq n} x_{nl} p_{nl}^*\right|^2 = \left(\frac{\hbar}{2}\right)^2, \tag{3.16}$$

whence (3.15) gives the anticipated result,

$$\sum_{l \neq n} |x_{nl}|^2 \sum_{l' \neq n} |p_{nl'}|^2 \geq \left(\frac{\hbar}{2}\right)^2. \tag{3.17}$$

3.2.3 Dispersion of dynamical variables

Quantum mechanics takes the definition given by statistical theory for the dispersion of a variable F, which is

$$\Delta F = \left[\overline{(F - \bar{F})^2}\right]^{1/2}, \tag{3.18}$$

where ¯ means average value, and translates it freely into the operator language. Thus, for a system in a state n, the standard quantum formula for the mean-square dispersion or *variance* of $\hat{F}$ is

$$\left(\Delta \hat{F}\right)_n^2 = \left[\left(\hat{F} - F_{nn}\right)^2\right]_{nn} = \left(\hat{F}^2\right)_{nn} - F_{nn}^2 = \sum_{l \neq n} |F_{nl}|^2, \qquad (3.19)$$

where the diagonal matrix elements $\left(\hat{F}^2\right)_{nn}$, F_{nn} are the average values of $\hat{F}^2$, $\hat{F}$, respectively, in that state. Applied to $\hat{x}$ and $\hat{p}$, this gives for their mean square dispersion in state n,

$$\left(\Delta \hat{x}\right)_n^2 = \sum_{l \neq n} |x_{nl}|^2, \quad \left(\Delta \hat{p}\right)_n^2 = \sum_{l \neq n} |p_{nl}|^2 \qquad (3.20)$$

and according to (3.17) their product gives $\left(\Delta \hat{x}\right)_n^2 \left(\Delta \hat{p}\right)_n^2 \geq (\hbar/2)^2$. By noting that this inequality holds for any state n, we can write it in the general form

$$\left(\Delta \hat{x}\right)\left(\Delta \hat{p}\right) \geq \frac{\hbar}{2}. \qquad (3.21)$$

From a purely algebraic perspective, the Heisenberg inequality represents thus a restriction on the minimum value of the product of the dispersions of $\hat{x}$ and $\hat{p}$ for any given state n, resulting from the noncommutativity of the respective operators.

It is a useful exercise to show that a similar procedure leads in the more general case of two noncommuting operators $\left[\hat{F}, \hat{G}\right] = i\hat{C}$, to the inequality

$$\left(\Delta \hat{F}\right)\left(\Delta \hat{G}\right) \geq \frac{1}{2}\left|\left[\hat{F}, \hat{G}\right]\right| = \frac{1}{2}\left|\langle \hat{C}\rangle\right|, \qquad (3.22)$$

where $\langle \hat{C}\rangle = |C|_n$ when Eq. (3.22) is calculated in the state n. We will return to these results and derive a more general inequality in Chapter 8; for the time being, we stress the fact that (3.22) is obtained without introducing any particular interpretation: it is merely an algebraic result for any pair of noncommuting variables.

3.3 x- and p- Representations

The fact that all variables in x-space commute with each other, as indicated in (3.10), allows us to use a specific *representation* in which x is a c-number and all dynamical quantities are either functions of x or operate on functions of x. To find the expression for the momentum operator $\hat{p}$ in the x-representation, consider an arbitrary (differentiable) function $\psi(x)$ and apply to it Eq. (3.5),

$$[x, \hat{p}_x]\psi = (x\hat{p}_x - \hat{p}_x x)\psi = i\hbar\psi. \qquad (3.23)$$

This expression is satisfied by taking $\hat{p}_x = -i\hbar(\partial/\partial x)$, as is easily verified, since

$$\hat{p}_x \hat{x} \psi = -i\hbar \frac{\partial}{\partial x}(x\psi) = -i\hbar(1 + x\frac{\partial}{\partial x})\psi = (-i\hbar + x\hat{p}_x)\psi,$$

and (3.12) follows. Therefore, $\hat{x}$ and $\hat{p}$ are represented in x-space – which is the usual representation in Schrödinger's formulation of QM, as we will see in Chapter 4 – by

$$\hat{x} = x, \quad \hat{p} = -i\hbar \frac{\partial}{\partial x}. \tag{3.24}$$

Alternatively, in the p-representation the roles of $\hat{x}$ and $\hat{p}$ are reversed; Eq. (3.4) still holds, now with

$$\hat{x} = i\hbar \frac{\partial}{\partial p}, \quad \hat{p} = p, \tag{3.25}$$

as the student can easily prove. This is an example of the fact that *the quantum laws are independent of the representation*. Notice, however, that there does *not* exist a joint xp-representation: the description can be made either in x-space or in p-space (or any other space resulting from a canonical transformation) but not in the joint phase space. This has important consequences, as we will have occasion to see later.

Equation (3.4) can be used to derive further interesting relations, as follows. Consider the commutator

$$\left[\hat{x}, \hat{p}^2\right] = \hat{x}\hat{p}^2 - \hat{p}^2\hat{x};$$

by adding and subtracting the quantity $\hat{p}\hat{q}\hat{p}$ and rearranging, one gets

$$\left[\hat{x}, \hat{p}^2\right] = \hat{x}\hat{p}^2 - \hat{p}^2\hat{x} = \hat{x}\hat{p}^2 - \hat{p}\hat{x}\hat{p} + \hat{p}\hat{x}\hat{p} - \hat{p}^2\hat{x}$$
$$= \left[\hat{x}\hat{p} - \hat{p}\hat{x}\right]\hat{p} + \hat{p}\left[\hat{x}\hat{p} - \hat{p}\hat{x}\right] = 2i\hbar\hat{p}.$$

Proceeding by iteration, one gets

$$\left[\hat{x}, \hat{p}^n\right] = i\hbar n \hat{p}^{n-1} = i\hbar \frac{d\hat{p}^n}{d\hat{p}}, \tag{3.26}$$

and

$$\left[\hat{x}^n, \hat{p}\right] = i\hbar n \hat{x}^{n-1} = i\hbar \frac{d\hat{x}^n}{d\hat{x}}. \tag{3.27}$$

It is an interesting exercise to show that if $\hat{F}(\hat{p})$ depends *only* on $\hat{p}$, and $\hat{G}(\hat{x})$ depends *only* on $\hat{x}$, and both functions are polynomials, analytic functions or series, then (using a Taylor series expansion),

$$\left[\hat{x}, \hat{F}(\hat{p})\right] = i\hbar \frac{d\hat{F}}{d\hat{p}}, \tag{3.28}$$

and

$$\left[\hat{p}, \hat{G}(\hat{x})\right] = -i\hbar \frac{d\hat{G}}{d\hat{x}}. \tag{3.29}$$

These results show that the commutator of $\hat{x}$ or $\hat{p}$ with a function of $\hat{p}$ or $\hat{x}$, respectively, acts in general as a derivative of the function.

It is important to keep in mind that due to the noncommutativity of $\hat{x}$ and $\hat{p}$, in the preceding equations $\hat{F}(\hat{p})$ cannot depend on $\hat{x}$, and similarly $\hat{G}(\hat{x})$ cannot depend on $\hat{p}$; the exceptions are *adding* to $\hat{F}(\hat{p})$ a function of $\hat{x}$ (because it commutes with $\hat{x}$), and similarly *adding* a function of $\hat{p}$ to $\hat{G}(\hat{x})$. An important example is the Hamiltonian operator $\hat{H}$

$$\hat{H} = \hat{T}(\hat{p}) + \hat{V}(\hat{x}), \tag{3.30}$$

where $\hat{T}$ stands for the kinetic energy and $\hat{V}$ for the potential energy (see the next section). In the x-representation, $\widehat{V}(\hat{x}) = V(x)$, and in the p-representation, $\widehat{T}(\hat{p}) = T(p)$. Throughout the text we will use Eqs. (3.5) and (3.10)–(3.12) to derive a number of other interesting algebraic properties of operators.

Exercise E3.1. Differential property of commutators

Show that the commutators possess a property similar to the derivation property of a product.

Solution. Adding and subtracting the product $\widehat{abc}$, we find that

$$[\hat{c}, \widehat{ab}] = [\hat{c}, \hat{a}]\hat{b} + \hat{a}[\hat{c}, \hat{b}] \tag{3.31}$$

for any trio of operators $\hat{a}, \hat{b}, \hat{c}$ (this is a result worth bearing in mind). This expression is formally equivalent to the derivation formula for a product ab of functions ab,

$$\hat{D}ab = (\hat{D}a)b + a(\hat{D}b). \tag{3.32}$$

We thus see that $[\hat{c}, \cdot]$ has algebraic properties analogous to those of the derivative operator $\hat{D}$, to which we must add $[\hat{c}, \hat{a} + \hat{b}] = [\hat{c}, \hat{a}] + [\hat{c}, \hat{b}]$, and $[\hat{c}, \text{const}] = 0$.

3.3.1 Quantum correspondence rules

As mentioned earlier, because of the noncommutativity of $\hat{x}$ and $\hat{p}$ the quantum formalism does not allow for a joint phase-space description, so to apply the formalism to functions containing products of $\hat{x}$ and $\hat{p}$, one must specify the order of the operators. For example, to the classical dynamical variable $A = x^2 p$ may correspond the quantum variables $\hat{x}^2\hat{p}, \hat{x}\hat{p}\hat{x}, \hat{p}\hat{x}^2$ or any linear combination of them, which are obviously not equivalent. To determine which operator $\widehat{A}(\hat{x}, \hat{p})$ corresponds to the classical variable $A(x, p)$, we need a set of rules, called the *correspondence rules* of QM. The physical origin of these rules and their physical interpretation are not always clear, although the discussions in Chapters 10 and 17 will allow us to gain some insight into this issue.

A correspondence rule is a recipe for establishing one-to-one correspondence between classical and quantum dynamical variables. For example, Weyl's correspondence rule – one of the best known – can be stated as follows. Let $\tilde{A}(\beta, \gamma)$ be the (double) Fourier transform of the classical dynamical variable $A(x, p)$,

$$A(x, p) = \frac{1}{2\pi} \int \tilde{A}(\beta, \gamma) e^{i\beta x + i\gamma p} d\beta d\gamma; \tag{3.33}$$

then the corresponding quantum variable is

$$\widehat{A}(\hat{x}, \hat{p}) = \frac{1}{2\pi} \int \tilde{A}(\beta, \gamma) e^{i\beta \hat{x} + i\gamma \hat{p}} d\beta d\gamma. \tag{3.34}$$

Note that both expressions contain the same nucleus (kernel) $\tilde{A}(\beta, \gamma)$; note also that the condition of Hermiticity of $\widehat{A}(\hat{x}, \hat{p})$ requires that $\tilde{A}(-\beta, -\gamma) = \tilde{A}^*(\beta, \gamma)$.

Exercise E3.2. Weyl's rule of correspondence

Use the Weyl rule to construct the operators corresponding to the classical expressions qp and qp^2, where q, p are two canonical variables satisfying $\{q, p\} = 1$.

Solution. Take first $A(q, p) = qp$ (classical variables); the integral formula for the Dirac delta function allows us to write

$$\begin{aligned} qp &= \int q' p' \delta(q - q') \delta(p - p') dq' dp' \\ &= -\frac{1}{4\pi^2} \int (-iq')(-ip') e^{i\beta(q-q')} e^{i\gamma(p-p')} dq' dp' d\beta d\gamma \\ &= -\frac{1}{4\pi^2} \int \left[e^{i\beta q} \frac{\partial}{\partial \beta} e^{-i\beta q'} \right] \left[e^{i\gamma p} \frac{\partial}{\partial \gamma} e^{-i\gamma p'} \right] dq' dp' d\beta d\gamma \\ &= -\int e^{i\beta q} \frac{\partial}{\partial \beta} \delta(\beta) e^{i\gamma p} \frac{\partial}{\partial \gamma} \delta(\gamma) d\beta d\gamma, \end{aligned} \quad (3.35)$$

that is,

$$A = qp = \int [i\delta'(\beta)][i\delta'(\gamma)] e^{i\beta q + i\gamma p} d\beta d\gamma, \quad (3.36)$$

where the prime indicates derivation with respect to the argument. Comparing with expression (3.33) we see that $\tilde{A}(\beta, \gamma) = 2\pi i \delta'(\beta) i \delta'(\gamma)$, so substituting this result into the quantum expression (3.34) and integrating by parts over β and γ, we get

$$\widehat{A}(\widehat{q}, \widehat{p}) = \int i\delta'(\beta) i\delta'(\gamma) e^{i\beta \widehat{q} + i\gamma \widehat{p}} d\beta d\gamma = -\left[\frac{\partial}{\partial \beta} \frac{\partial}{\partial \gamma} e^{i\beta \widehat{q} + i\gamma \widehat{p}} \right]_{\beta=\gamma=0}. \quad (3.37)$$

To evaluate this expression we expand the exponential in a Taylor series,

$$-\left\{ \frac{\partial}{\partial \beta} \frac{\partial}{\partial \gamma} \left[1 + i(\beta \widehat{q} + \gamma \widehat{p}) + \frac{i^2}{2!}(\beta \widehat{q} + \gamma \widehat{p})^2 + \frac{i^3}{3!}(\beta \widehat{q} + \gamma \widehat{p})^3 + \ldots \right] \right\}_{\beta=\gamma=0}. \quad (3.38)$$

When deriving, only the second-degree term contributes, leading to

$$\begin{aligned} \widehat{A}(\widehat{q}, \widehat{p}) &= \tfrac{1}{2} \left\{ \frac{\partial}{\partial \beta} \frac{\partial}{\partial \gamma} \left[\beta^2 \widehat{q}^2 + \beta \gamma \widehat{q} \widehat{p} + \beta \gamma \widehat{p} \widehat{q} + \gamma^2 \widehat{p}^2 \right] \right\}_{\beta=\gamma=0} \\ &= \tfrac{1}{2} [\widehat{q}\widehat{p} + \widehat{p}\widehat{q}]. \end{aligned} \quad (3.39)$$

The correspondence sought is thus

$$qp \to \tfrac{1}{2}(\widehat{q}\widehat{p} + \widehat{p}\widehat{q}). \quad (3.40)$$

Similarly,

$$qp^2 \to \tfrac{1}{3}(\widehat{q}\widehat{p}^2 + \widehat{p}\widehat{q}\widehat{p} + \widehat{p}^2\widehat{q}). \quad (3.41)$$

A more detailed analysis shows that Weyl's rule makes the product $q^n p^m$ correspond to a fully symmetric linear combination of terms, each with n factors $\widehat{q}$ and m factors $\widehat{p}$.

3.4 Laws of Motion in the Heisenberg Picture

Heisenberg's assumption that the structure of the quantum laws should correspond to that of classical physics proved to be highly profitable. In particular, for a classical particle subject to a conservative force, the time evolution of a variable $F(x, p)$ that does not depend explicitly on time is governed by the Hamiltonian $H = (p^2/2m) + V(x)$; applying the chain rule of derivation one gets, with $p_i = m\dot{x}_i$ and $f_i = -\partial_i V$ (summation over repeated indices is understood),

$$\frac{dF}{dt} = \frac{\partial F}{\partial x_i}\frac{p_i}{m} + \frac{\partial F}{\partial p_i}f_i(x) = \frac{\partial F}{\partial x_i}\frac{\partial H}{\partial p_i} - \frac{\partial F}{\partial p_i}\frac{\partial H}{\partial x_i} = \{F, H\}_{xp}. \tag{3.42}$$

Note that the time derivative of F becomes determined by the Poisson bracket of F and the Hamiltonian H. As in classical physics, also in quantum physics the Hamiltonian of the system determines the time evolution of any dynamical variable. Applying Dirac's rule of correspondence, Eq. (3.9), to (3.42), we get (note that time t is not an operator)

$$i\hbar\frac{d\hat{F}}{dt} = \sum_i \left[\frac{\partial \hat{F}}{\partial \hat{q}_i}\frac{\partial \hat{H}}{\partial \hat{p}_i} - \frac{\partial \hat{F}}{\partial \hat{p}_i}\frac{\partial \tilde{H}}{\partial \hat{q}_i}\right] = \left[\hat{F}, \hat{H}\right]. \tag{3.43}$$

This is Heisenberg's important general equation of evolution for a quantum variable that does not explicitly depend on time. The immediate generalization to the time-dependent case $F(x, p, t)$ is

$$i\hbar\frac{d\hat{F}}{dt} = i\hbar\frac{\partial \hat{F}}{\partial t} + \left[\hat{F}, \hat{H}\right]. \tag{3.44}$$

Note that by their mathematical structure the classical and quantum laws have exactly the same form, but what they say and what follows from them are quite different. In the classical case F and H are *functions* in the phase space spanned by q_i and p_i, whereas in the quantum case they are *operators* acting on an abstract space (see Section 3.7).

In Chapter 8 and elsewhere in the book we will have occasion to discuss Eq. (3.44) and to apply it to specific problems. At this point, by way of illustration, let us consider a 1D Hamiltonian operator of the form

$$\hat{H} = \hat{p}^2/2m + \hat{V}(\hat{x}). \tag{3.45}$$

By replacing $\widehat{F}$ in Eq. (3.28) and $\widehat{G}$ in Eq. (3.29) by $\widehat{H}$, we get

$$\frac{1}{i\hbar}\left[\hat{x}, \hat{H}\right] = \frac{1}{i\hbar}\left[\hat{x}, \frac{\hat{p}^2}{2m}\right] = \frac{1}{m}\hat{p}, \tag{3.46}$$

and

$$\frac{1}{i\hbar}\left[\hat{p}, \hat{H}\right] = \frac{1}{i\hbar}\left[\hat{p}, \hat{V}(\hat{x})\right] = \hat{f}(\hat{x}). \tag{3.47}$$

Note that for $\hat{F} = \hat{H}$ given by (3.45), Eq. (3.43) gives zero; thus this form of the equation of evolution corresponds to the case in which the Hamiltonian is a constant (independent of time). Applying now Eq. (3.43) to $\widehat{x}$ and $\widehat{p}$ successively gives the quantum version of Newton's equations of motion,

$$i\hbar\dot{\widehat{x}} = \frac{i\hbar}{m}\widehat{p} \quad \text{or} \quad \widehat{p} = m\dot{\widehat{x}}, \tag{3.48}$$

and

$$i\hbar\dot{\widehat{p}} = i\hbar f(\widehat{x}), \quad \text{or} \quad \dot{\widehat{p}} = f(\widehat{x}). \tag{3.49}$$

Combining these with Eqs. (3.46) and (3.47) we get the *Heisenberg equations of motion* for $\widehat{x}$ and $\widehat{p}$,

$$\dot{\widehat{p}} = -\frac{\partial\widehat{H}}{\partial\widehat{x}}, \quad \dot{\widehat{x}} = \frac{\partial\widehat{H}}{\partial\widehat{p}}, \tag{3.50}$$

which, by their structure, coincide formally with the corresponding Hamilton equations of motion of classical mechanics.

3.5 Matrix Description of the Harmonic Oscillator

The quantum harmonic oscillator is studied in detail in Chapter 9 by applying Schrödinger's equation. Here we treat the problem in terms of operators and use a purely algebraic procedure to construct the corresponding matrices, which allow us to find the energy eigenvalues, that is, the energy values associated with the stationary states of the oscillator.

3.5.1 HO energy eigenvalues and eigenstates

The Hamiltonian operator for the 1D harmonic oscillator with a natural oscillating frequency ω is

$$\widehat{H} = \frac{\widehat{p}^2}{2m} + \frac{1}{2}m\omega^2\widehat{x}^2, \tag{3.51}$$

so that from the dynamical equations (3.50), we get $\dot{\widehat{p}} = -m\omega^2\widehat{x}$, $\dot{\widehat{x}} = \widehat{p}/m$, or in terms of matrix elements, $\dot{p}_{ln} = -m\omega^2 x_{ln}$, $m\dot{x}_{ln} = p_{ln}$. The solution of these coupled equations is

$$x_{ln}(t) = x_{ln}e^{-i\omega_{nl}t}, \quad p_{ln}(t) = -im\omega_{nl}x_{ln}e^{-i\omega_{nl}t}, \tag{3.52}$$

with $\omega_{nl} = \pm\omega$. As mentioned in Section 3.2, these matrix elements connect the stationary states n and l, which means that the oscillator can make transitions from state n to state l, by either absorbing or emitting radiation of frequency ω and exchanging a corresponding amount of energy

$$\mathcal{E}_{nl} = \mathcal{E}_n - \mathcal{E}_l = \hbar\omega_{nl}, \tag{3.53}$$

according to Bohr's rule. The state of minimum energy corresponds to $n = 0$, this is the ground state of the oscillator; all states with $n > 0$ are excited states and there are no states with $n < 0$. When the transition is accompanied by an energy emission, the final state of the oscillator will be characterized by a lower quantum number, $l = n - 1$,

3.5 Matrix Description of the Harmonic Oscillator

and $\omega_{nl} = \omega$; when it is accompanied by an energy absorption, the final state will be $l = n + 1$ and $\omega_{nl} = -\omega$ (there are no intermediate states). From (3.52) we therefore get for the nonzero matrix elements for emissions (with $n > 0$)

$$x_{n-1,n}(t) = x_{n-1,n} e^{-i\omega t}, \quad p_{n-1,n}(t) = -im\omega x_{n-1,n} e^{-i\omega t}, \quad (3.54)$$

and for absorptions (with $n \geq 0$)

$$x_{n+1,n}(t) = x_{n+1,n} e^{i\omega t}, \quad p_{n+1,n}(t) = im\omega x_{n+1,n} e^{i\omega t}. \quad (3.55)$$

Note that because of the Hermiticity of $\hat{x}$ and $\hat{p}$,

$$x_{n-1,n} = x^*_{n,n-1}, \quad x_{n+1,n} = x^*_{n,n+1}, \quad (3.56)$$

and similarly for the matrix elements of $\hat{p}$. According to Eq. (3.4), for the fundamental commutator $[\hat{x}(t), \hat{p}(t)] = i\hbar$ to be satisfied at all times (including t = 0), $\hat{x} = \hat{x}(0)$ and $\hat{p} = \hat{p}(0)$ must be such that $[\hat{x}, \hat{p}] = i\hbar$, which leads to the sum rule (check it!):

$$\sum_{l=n\pm 1} \omega_{ln} |x_{nl}|^2 = \frac{\hbar}{2m}, \quad (3.57)$$

for $n = 0, 1, 2, \ldots$. This result defines the scale of the matrix elements (the amplitude of the oscillations).

Using Eqs. (3.51)–(3.57), we obtain for the diagonal elements of $\hat{H}$ (summation over the indices $l = n \pm 1$ is understood)

$$H_{nn} = \frac{p_{nl} p_{ln}}{2m} + \frac{1}{2} m\omega^2 x_{nl} x_{ln} = \frac{1}{2}\hbar\omega + 2m\omega^2 |x_{n-1,n}|^2. \quad (3.58)$$

It is left for the student to prove that the nondiagonal elements of $\hat{H}$ are all equal to zero. In other words, the Hamiltonian is represented by a diagonal matrix, which means that it does not connect different states n, l. The matrix element H_{nn} is a property of the state n; from (3.51) it is clear that this property is associated with the energy of the state. Its value, $H_{nn} \equiv \mathcal{E}_n$, is called an energy *eigenvalue*, and the corresponding (stationary) state is called an energy *eigenstate*, the names deriving from the German word *eigen* which means *own* or *proper*. Note that with $n = 0$ Eq. (3.58) gives the ground-state energy, $\mathcal{E}_0 = \hbar\omega/2$.

Using the sum rule (3.57) we can obtain one by one all the matrix elements of $\hat{x}$ and $\hat{p}$; this will allow us to calculate also the energy eigenvalues. Starting with $n = 0$, we get $\omega_{10} |x_{10}|^2 = \hbar/2m$, since there can only be upward transitions to $n = 1$, with $\omega_{10} = \omega$. Therefore,

$$|x_{10}|^2 = \frac{\hbar}{2m\omega}. \quad (3.59)$$

For $n = 1$, Eq. (3.57) combined with (3.59) gives $|x_{21}|^2 - |x_{01}|^2 = |x_{10}|^2$ or, taking into account that $|x_{01}|^2 = |x_{10}|^2$, $|x_{21}|^2 = 2|x_{10}|^2$. The same procedure gives after n iterations $|x_{n\,n-1}|^2 = n |x_{10}|^2$ or, since we are free to take x_{10} real,

$$x_{n\,n-1} = \sqrt{n} x_{10}, \quad x_{n\,n+1} = \sqrt{n+1} x_{10}, \quad (3.60)$$

and from Eqs. (3.55),

$$p_{n\,n-1} = im\omega\sqrt{n} x_{10}, \quad p_{n\,n+1} = -im\omega\sqrt{n+1} x_{10}. \quad (3.61)$$

Note that, since there is no upper limit to n, the matrices $\hat{x}$ and $\hat{p}$ are of infinite size. If we tag the first row and the first column with the lowest index, $n = 0$, we get

$$\hat{x} = \sqrt{\frac{\hbar}{2m\omega}} \begin{pmatrix} 0 & 1 & 0 & \cdots & 0 & \cdots \\ 1 & 0 & \sqrt{2} & \cdots & 0 & \cdots \\ 0 & \sqrt{2} & 0 & \cdots & 0 & \cdots \\ \cdots & \cdots & \cdots & 0 & \sqrt{n} & \cdots \\ 0 & 0 & 0 & \sqrt{n} & 0 & \cdots \\ \cdots & \cdots & \cdots & \cdots & \cdots & \cdots \end{pmatrix}, \qquad (3.62)$$

$$\hat{p} = -i\sqrt{\frac{\hbar m\omega}{2}} \begin{pmatrix} 0 & 1 & 0 & \cdots & 0 & \cdots \\ -1 & 0 & \sqrt{2} & \cdots & 0 & \cdots \\ 0 & -\sqrt{2} & 0 & \cdots & 0 & \cdots \\ \cdots & \cdots & \cdots & 0 & \sqrt{n} & \cdots \\ 0 & 0 & 0 & -\sqrt{n} & 0 & \cdots \\ \cdots & \cdots & \cdots & \cdots & \cdots & \cdots \end{pmatrix}. \qquad (3.63)$$

For the diagonal matrix elements of $\hat{H}$, that is, the successive energy eigenvalues, we use Eqs. (3.58), (3.59), and (3.60),

$$H_{nn} \equiv \mathcal{E}_n = \frac{\hbar\omega}{2}(1 + 2n) = \mathcal{E}_0 + n\hbar\omega, \qquad (3.64)$$

where

$$\mathcal{E}_0 = \frac{1}{2}\hbar\omega \qquad (3.65)$$

is the energy of the ground state. Notice that the kinetic and potential energy terms contribute an equal amount to the value of $\mathcal{E}_n$; this is the quantum version of the virial theorem applied to the HO.

Exercise E3.3. Heisenberg inequality for the harmonic oscillator

Use the results just obtained to derive the Heisenberg inequality for the case of the HO.

Solution. Using Eqs. (3.20) and (3.60), we get for the mean square dispersion of $\hat{x}$,

$$(\Delta\hat{x})_n^2 = \sum_{l \neq n} |x_{nl}|^2 = \frac{\hbar}{2m\omega}(1 + 2n), \qquad (3.66)$$

and similarly for the mean-square dispersion of $\hat{p}$,

$$(\Delta\hat{p})_n^2 = \sum_{l \neq n} |p_{nl}|^2 = \frac{\hbar m\omega}{2}(1 + 2n). \qquad (3.67)$$

The product of the root-mean-square dispersions is therefore

$$(\Delta\hat{x})_n (\Delta\hat{p})_n = \frac{\hbar}{2}(1 + 2n). \qquad (3.68)$$

This result shows that the Heisenberg inequality

$$(\Delta\hat{x})_n (\Delta\hat{p})_n \geq \frac{\hbar}{2} \qquad (3.69)$$

is satisfied for all n and takes its minimum value for the ground state, $n = 0$, for which the equal sign applies.

Note that although the total energy has a fixed value, both position and momentum are distributed (i.e., dispersive) variables. Throughout the text we will meet other examples of such unavoidable fluctuations. This result explains how it was possible for Heisenberg, who did not carry out any experiments, to speak of quantum fluctuations and evaluate some of their effects, soon after 1925.

3.5.2 Raising and lowering operators for the harmonic oscillator

The structure of the matrices $\hat{x}$ and $\hat{p}$ for the HO strongly suggests introducing linear combinations of them, such that either the upper off-diagonal or the lower off-diagonal lines vanish. This is easily achieved by observing that according to Eqs. (3.60) and (3.61),

$$p_{n\,n-1} = m\omega x_{n\,n-1} + i p_{n\,n-1} = 0, \qquad (3.70)$$

and

$$p_{n\,n+1} = m\omega x_{n\,n+1} - i p_{n\,n+1} = 0. \qquad (3.71)$$

We therefore define a new couple of operators $\hat{a}$ and $\hat{a}^\dagger$ as

$$\hat{a} = \frac{1}{\sqrt{2m\hbar\omega}} \left(m\omega\hat{x} + i\hat{p}\right), \qquad (3.72)$$

and

$$\hat{a}^\dagger = \frac{1}{\sqrt{2m\hbar\omega}} \left(m\omega\hat{x} - i\hat{p}\right). \qquad (3.73)$$

The corresponding matrices take the desired form: from Eqs. (3.62) and (3.63) we see that matrix $\hat{a}$ has off-diagonal elements only above the central diagonal, and its adjoint matrix $\hat{a}^\dagger$ has off-diagonal elements only below the said diagonal. The role of these operators is therefore to lower the state from n to $n - 1$ and to raise it to $n + 1$, respectively. For this reason they are called *lowering* and *raising* operators, respectively.[6]

The operators $\hat{a}, \hat{a}^\dagger$ are a most practical tool for describing the harmonic oscillator. In terms of them, the Hamiltonian and the canonical commutator become

$$\hat{H} = \tfrac{1}{2}\hbar\omega \left(\hat{a}^\dagger \hat{a} + \hat{a}\hat{a}^\dagger\right) = \hbar\omega \left(\hat{a}^\dagger \hat{a} + \tfrac{1}{2}\right), \qquad (3.74)$$

and

$$[\hat{a}, \hat{a}^\dagger] = 1, \qquad (3.75)$$

as the student may verify. It is important to note that the Hamiltonian (3.74) carries with it the term $\tfrac{1}{2}\hbar\omega$, which is the ground-state energy of the HO, (3.65). Note also that $\hat{H} - \tfrac{1}{2}\hbar\omega = \hbar\omega \hat{a}^\dagger \hat{a}$ becomes factorizable in this representation, which is a very useful property. The expression of $\hat{H}$ in terms of products $\hat{a}^\dagger \hat{a}$ and $\hat{a}\hat{a}^\dagger$ confirms that

---

[6] Their effect is similar (though not equivalent) to that of the lowering and raising angular-momentum operators $\hat{J}_\pm$, which are studied in Chapter 11.

the lowering effect of $\hat{a}$ is compensated by the raising effect of $\hat{a}^\dagger$ and vice versa, with the end result being that the operator $\hat{H}$ leaves the state of the oscillator unchanged. For this reason, $\hat{H}$ appears as a diagonal matrix, with elements given by (3.64).

The physics behind this algebraic exercise is very significant. As noted earlier, the operators $\hat{a}, \hat{a}^\dagger$ have the effect of lowering or raising the state of the oscillator; when this happens, the oscillator emits or absorbs radiation with the resulting energy loss or gain $\Delta E = \hbar \omega$. As will be seen in later chapters, it is precisely the coupling to the radiation field that induces these transitions. As Heisenberg brilliantly recognized, the matrix elements are the spectral evidence of such transitions.

3.5.3 Heisenberg's elementary oscillators

As we have seen in Section 3.3, a characteristic property of the Heisenberg picture (*representation* or *description*) is that the operators are generally time-dependent functions. This contrasts with the Schrödinger representation, which we will study in Chapter 4, where the wave function – the solution of a time-dependent differential equation – is the natural carrier of time dependence and the operators are generally time-independent functions.

As an example, we obtained the matrix elements (3.55), which are explicit expressions for the harmonic-oscillator operators as functions of time. It is important to note that each of these matrix elements is in itself an oscillator of frequency $\pm \omega$, depending on the sign of the transition. This feature is even more apparent in the expressions for the lowering and raising operators, which according to Eqs. (3.55), (3.72), and (3.73) are respectively given by

$$\hat{a}(t) = \hat{a}_0 e^{-i\omega t}, \quad \hat{a}_0 = \hat{a}(0), \tag{3.76}$$

and

$$\hat{a}^\dagger(t) = \hat{a}_0^\dagger e^{i\omega t}, \quad \hat{a}_0^\dagger = \hat{a}^\dagger(0). \tag{3.77}$$

In fact, by applying Heisenberg's equation of motion (3.43) to $\hat{a}, \hat{a}^\dagger$ with $\hat{H}$ given by (3.74):

$$[\hat{H}, \hat{a}] = -\hbar\omega\hat{a}, \quad [\hat{H}, \hat{a}^\dagger] = \hbar\omega\hat{a}^\dagger, \tag{3.78}$$

we get $\dot{\hat{a}} = -i\omega\hat{a}$, $\dot{\hat{a}}^\dagger = i\omega\hat{a}^\dagger$, with solutions given by (3.76), (3.77). Using Eqs. (3.72, 3.73) and expressing the initial conditions in terms of $\hat{x}_0$ and $\hat{p}_0$ instead of $\hat{a}_0$ and $\hat{a}_0^\dagger$ gives

$$\hat{x} = \hat{x}_0 \cos \omega t + \frac{\hat{p}_0}{m\omega} \sin \omega t, \tag{3.79}$$

and

$$\hat{p} = \hat{p}_0 \cos \omega t - m\omega \hat{x}_0 \sin \omega t. \tag{3.80}$$

These expressions are identical to the corresponding classical ones, although in this case they refer to operators that must always satisfy the commutation rule $[\hat{x}, \hat{p}] = i\hbar$; note therefore that the integration constants are noncommuting operators.

It is important to note that the results obtained do not depend on the *nature* of the oscillator, and are therefore applicable to any quantum oscillator, thus extending the theory beyond the purpose for which it was originally proposed. In particular, the present theory can be used to describe the oscillators of the quantum radiation field, as is done in Chapter 17 and in QED, and more generally, in any quantum field theory.

3.6 Generalization to Arbitrary Potentials

Let us now consider a particle bound by an arbitrary potential $V(x)$. A typical case would be an atomic electron bound by the Coulomb potential due to the nucleus, but for purposes of simplicity we continue to restrict the discussion to 1D.

In the general case of a nonlinear binding force, the solution of the Heisenberg equations of motion (3.50) can in principle also be expressed in terms of elementary oscillators of the type (3.52),

$$x_{nl}(t) = x_{nl} e^{i\omega_{nl} t}, \quad p_{nl}(t) = i m \omega_{nl} x_{nl} e^{i\omega_{nl} t}. \tag{3.81}$$

In practice, however, given an arbitrary potential, the algebraic problem can be quite complicated, or even exactly intractable. When it can be solved, the resulting matrix $\hat{x}(t)$ contains off-diagonal elements not only above and below the principal diagonal, as in the case of the harmonic oscillator studied previously, but everywhere. This expresses the fact that when in state n, the system can undergo transitions to different upper or lower levels l (not just $l = n \pm 1$), with an energy exchange given in each case by $\mathcal{E}_{nl} = \hbar \omega_{nl}$. In other words, the solution $\hat{x}(t)$ represents a collection of oscillators of *different* transition frequencies ω_{nl}. Only in the case of the linear oscillator are the frequencies restricted to $\pm \omega$, the natural frequency of the oscillator.

For the basic commutator $[\hat{x}(t), \hat{p}(t)] = i\hbar$ to be satisfied at all times, $\hat{x}$ and $\hat{p}$ must be such that $[\hat{x}, \hat{p}] = i\hbar$, which leads to the following generalization of Eq. (3.57), as the student can verify,

$$\sum_{l \neq n} \omega_{ln} |x_{ln}|^2 = \frac{\hbar}{2m}, \tag{3.82}$$

known as the *Thomas–Reiche–Kuhn sum rule*. The sum extends over all values of l representing the states that can be reached by a transition from state n; the term $l = n$ is obviously excluded from the sum, since $\omega_{nn} = 0$. Equation (3.82) will prove to be very useful in future applications.

As will be seen in Chapter 4 and elsewhere in the text, it is precisely the solution of the Heisenberg equations, contained in the full set of values for x_{ln} and the corresponding frequencies ω_{ln}, that allows one to establish contact with the spectroscopic data, that is, with the intensities and frequencies of the spectral lines. This is one of the great advantages of Heisenberg's formalism.

3.7 State Vectors and the Hilbert Space. Pauli Matrices

To complete the description provided by matrix mechanics, several important additional elements are needed. Throughout the text we will have the opportunity to become familiar with the formalism. At this stage we shall limit ourselves to introducing the states of the system on which the operators act.

In the language of matrices, states are represented by column vectors, i.e., one-column matrices with as many components as there are rows (and columns) in the matrices representing the operators. The (complex and normed) linear vector space in which the operators act is known as a *Hilbert space*.[7] The dimension of this vector space is open to the requirements, and can range from two to an infinite number. In the case of the HO, for example, the vectors are infinite in size, as shown in Section 3.5, which makes their explicit writing at least cumbersome and space (and time) consuming.

When state vectors represent energy eigenstates, as discussed earlier, they are called *energy eigenvectors*. More generally, an eigenvector represents a state with a well-defined value of the corresponding dynamical variable.

We will return to this important topic in later chapters; as a useful exercise, we consider here the simplest case of a two-dimensional vector space. This space can accommodate the description of many physical realizations of two-state systems, including the most important cases of qubits and the electron spin, which will be studied in detail in Chapter 14.

Exercise E3.4. Matrices and vectors in two-dimensional space

Find appropriate expressions for the vectors and matrices in two-dimensional Hilbert space.

Solution. A two-dimensional vector space can accommodate at most two independent vectors, which conveniently are taken to be orthonormal; we denote them with φ_+ and φ_- and write them as

$$\varphi_+ = \begin{pmatrix} 1 \\ 0 \end{pmatrix}, \quad \varphi_- = \begin{pmatrix} 0 \\ 1 \end{pmatrix}. \tag{3.83}$$

The basis (3.83) is orthonormal by construction (orthogonal and normalized to unity), since

$$|\varphi_+|^2 = \begin{pmatrix} 1 & 0 \end{pmatrix} \begin{pmatrix} 1 \\ 0 \end{pmatrix} = 1; \quad |\varphi_-|^2 = \begin{pmatrix} 0 & 1 \end{pmatrix} \begin{pmatrix} 0 \\ 1 \end{pmatrix} = 1, \tag{3.84}$$

and

$$\varphi_+^\dagger \varphi_- = \begin{pmatrix} 1 & 0 \end{pmatrix} \begin{pmatrix} 0 \\ 1 \end{pmatrix} = 0; \quad \varphi_-^\dagger \varphi_+ = \begin{pmatrix} 0 & 1 \end{pmatrix} \begin{pmatrix} 1 \\ 0 \end{pmatrix} = 0, \tag{3.85}$$

where the row matrices are the adjoint of the corresponding column vectors, In addition, the basis is complete, since any ψ state in this (two-dimensional) space can be written as a linear combination of the form

$$\psi = a\varphi_+ + b\varphi_- = \begin{pmatrix} a \\ b \end{pmatrix}. \tag{3.86}$$

---

[7] The Hilbert-space formalism was introduced by John von Neumann (1903–1957) in 1927–29 as a mathematical framework for QM, and proved to be very useful for its subsequent extension into quantum field theory. Here we have informally introduced Hilbert-space concepts, without pretending to be overly rigorous.

The coefficients a and b can be functions themselves, but in any case, for the state (3.86) to be normalized to 1, they must be such that

$$\begin{pmatrix} a^* & b^* \end{pmatrix} \begin{pmatrix} a \\ b \end{pmatrix} = a^*a + b^*b = 1. \tag{3.87}$$

With the preceding basis vectors and their adjoint, one can construct at most four (2×2) elementary operators $\hat{\sigma}_{ij}$ acting on this space,

$$\begin{pmatrix} 1 \\ 0 \end{pmatrix}\begin{pmatrix} 1 & 0 \end{pmatrix} = \begin{pmatrix} 1 & 0 \\ 0 & 0 \end{pmatrix}; \quad \begin{pmatrix} 0 \\ 1 \end{pmatrix}\begin{pmatrix} 0 & 1 \end{pmatrix} = \begin{pmatrix} 0 & 0 \\ 0 & 1 \end{pmatrix}; \tag{3.88}$$

$$\begin{pmatrix} 1 \\ 0 \end{pmatrix}\begin{pmatrix} 0 & 1 \end{pmatrix} = \begin{pmatrix} 0 & 1 \\ 0 & 0 \end{pmatrix}; \quad \begin{pmatrix} 0 \\ 1 \end{pmatrix}\begin{pmatrix} 1 & 0 \end{pmatrix} = \begin{pmatrix} 0 & 0 \\ 1 & 0 \end{pmatrix}. \tag{3.89}$$

The four operators just obtained, (3.88) and (3.89), form a complete basis in the two-dimensional Hilbert space. The latter have the effect of exchanging the states φ_+ and φ_-. The former, on the other hand, reduce a state vector to just one of its components,

$$\begin{pmatrix} 1 & 0 \\ 0 & 0 \end{pmatrix}\begin{pmatrix} a \\ b \end{pmatrix} = \begin{pmatrix} a \\ 0 \end{pmatrix}, \quad \begin{pmatrix} 0 & 0 \\ 0 & 1 \end{pmatrix}\begin{pmatrix} a \\ b \end{pmatrix} = \begin{pmatrix} 0 \\ b \end{pmatrix}; \tag{3.90}$$

for this reason they are called projection operators or *projectors*. They satisfy a *completeness relation*, that is, their sum is equal to 1:

$$\begin{pmatrix} 1 & 0 \\ 0 & 0 \end{pmatrix} + \begin{pmatrix} 0 & 0 \\ 0 & 1 \end{pmatrix} = \begin{pmatrix} 1 & 0 \\ 0 & 1 \end{pmatrix} = \mathbb{I}. \tag{3.91}$$

The Hermitian operators obtained by combining Eqs. (3.88)

$$\begin{pmatrix} 1 & 0 \\ 0 & 1 \end{pmatrix} \equiv \hat{\sigma}_0 = \mathbb{I}, \quad \begin{pmatrix} 1 & 0 \\ 0 & -1 \end{pmatrix} \equiv \hat{\sigma}_3, \tag{3.92}$$

and (3.89)

$$\begin{pmatrix} 0 & 1 \\ 1 & 0 \end{pmatrix} \equiv \hat{\sigma}_1, \quad \begin{pmatrix} 0 & -i \\ i & 0 \end{pmatrix} \equiv \hat{\sigma}_2, \tag{3.93}$$

are the unit operator $\mathbb{I}$ and the three *Pauli matrices* $\hat{\sigma}_k$ ($k = 1, 2, 3$). Note that the vectors of Eq. (3.83) are eigenvectors of the matrices $\hat{\sigma}_0$ and $\hat{\sigma}_3$, with eigenvalues given by $\sigma_0^\pm = 1; \sigma_3^\pm = \pm 1$.

The Pauli matrices are of importance in a number of applications; see in particular Chapters 11 and 14 for their application to the spin angular momentum. Any 2×2 matrix (any operator on a 2D space) can be written as a linear combination of the four matrices σ_s with appropriate (usually complex) coefficients. The general form of such a matrix $\hat{F}$ is thus

$$\hat{F} = a_0 \mathbb{I} + \sum a_k \hat{\sigma}_k = a_0 \mathbb{I} + \boldsymbol{a} \cdot \hat{\boldsymbol{\sigma}}$$

$$= \begin{pmatrix} a_0 + a_3 & a_1 - ia_2 \\ a_1 + ia_2 & a_0 - a_3 \end{pmatrix}. \tag{3.94}$$

Note that this 2×2 matrix contains all the information about a three-dimensional vector $\boldsymbol{a}$; this is called the Bloch representation of $\boldsymbol{a}$, and $\boldsymbol{a}$ is a Bloch vector.

3.8 Why Operators Instead of Functions?

The success of matrix mechanics comes with an undesirable shortcoming: the lack of a physical picture. Whereas in classical mechanics the dynamical variables describe trajectories in space-time where particles move and "live," in QM the corresponding operators "live" and act on vectors in an abstract Hilbert space. This is one of the most paradoxical results of theoretical physics: to start constructing a theory by introducing the notion of observables, and to end up with a theory characterized by its nonvisualizability.

The demand for visualization, voiced in the early stages of the theory by H. A. Lorentz and others, is as relevant today as it was then, and leads to a most fundamental question: what is behind the replacement of continuous phase-space functions by operators?

In looking for an answer to this question, it is appropriate to recall Heisenberg's observation on the failure of the (classical) *kinematic content* of the old quantum theory mentioned in Section 3.1. Heisenberg sought a radical change, not in the dynamics – which still fitted into the classical Hamiltonian scheme – but in the *kinematics*, that is, the objects used to describe the motions and the relations between them. And indeed, in his new mechanics the formal properties of the basic descriptive elements x, p changed radically. But behind this change in the mathematics lies a no less radical change in the physical meaning, which remains largely unaddressed: Heisenberg's postulates were and are taken for granted, because they work – because they "save" the phenomena.[8]

The answer, then, should be found by looking for a physical element that modifies the kinematics of the particle but keeping to the Hamiltonian approach. In Chapter 2 we saw that every typical quantum system consists of a matter part that is coupled to the radiation field, including the ZPF by default. In Chapter 10 we will show that this coupling radically changes the nature of the dynamical variables used for the description, leading to the replacement of the classical canonical phase space variables x, p by the corresponding operators $\hat{x}$, $\hat{p}$. Thus, although there is a formal one-to-one correspondence between variables and operators, they differ substantially in their physical meaning.

At this point we can already envisage why Heisenberg's mechanics leads to matrices instead of the classical kinematic variables. We have learned in Section 3.2 that x_{ln} and p_{ln} refer to the radiative transitions that give rise to the spectral lines, so that the postulate introduced by Heisenberg, without his being aware of it, is equivalent to considering the atom *in response to its coupling to the radiation field*; the atom coexists with an (undisclosed) electromagnetic field that induces the (observed) transitions. A first-principles derivation of the quantum operator formalism, revealing its deep physical content, is left to Chapter 10, where it will become clear that the new kinematic variables that constitute the core of Heisenberg's matrix mechanics express the *response* of the atom to the field modes that induce these transitions.

---

[8] Such was the argument for accepting the Ptolemaic description of the solar system: it worked, "it saved the phenomena." From the time of Plato's demand, this was the dominant astronomical attitude toward "explanation," in contrast to the cosmological position of the physicists of the time, who demanded an *understanding* of the phenomena.

Problems

P3.1 Demonstrate directly the following properties of the commutator:

$$[\hat{u},\hat{v}] = -[\hat{v},\hat{u}], \quad [\hat{u}+\hat{v},\hat{w}] = [\hat{u},\hat{w}] + [\hat{v},\hat{w}], \quad (3.95)$$

$$[\hat{u}\hat{w},\hat{v}] = \hat{u}[\hat{w},\hat{v}] + [\hat{u},\hat{v}]\hat{w}, \quad [\hat{u},[\hat{v},\hat{w}]] + [\hat{v},[\hat{w},\hat{u}]] + [\hat{w},[\hat{u},\hat{v}]] = 0. \quad (3.96)$$

The latter is the so-called *Jacobi identity*. Rewrite the last two results using the differential property of the commutator discussed in Exercise 3.11.

P3.2 Show that if $[\hat{F},\hat{G}] = 0$, then

$$[\hat{F}^n, \hat{G}^m] = 0, \quad [f(\hat{F},\hat{G}), g(\hat{F},\hat{G})] = 0 \quad (3.97)$$

for functions f, g that can be expanded in power series.

P3.3 Show that for two commuting operators $\hat{F}, \hat{G}$, $[\hat{F}^{-1}, \hat{G}] = [\hat{F}, \hat{G}^{-1}] = [\hat{F}^{-1}, \hat{G}^{-1}] = 0$.

P3.4 Show that for two arbitrary operators $\hat{F}, \hat{G}$, $e^{\hat{F}\hat{G}}\hat{F} = \hat{F}e^{\hat{G}\hat{F}}$.

P3.5 Show that

$$[\hat{p}, \sin \lambda x] = -i\lambda \cos \lambda x,$$
$$[\hat{p}, \cos \lambda x] = i\lambda \sin \lambda x, \quad (3.98)$$

and therefore,

$$\Delta \hat{p} \Delta \sin \lambda x \geq \frac{\lambda \hbar}{2} |\langle \cos \lambda x \rangle|,$$

and

$$\Delta \hat{p} \Delta \cos \lambda x \geq \frac{\lambda \hbar}{2} |\langle \sin \lambda x \rangle|. \quad (3.99)$$

Examine the limit $\lambda \to 0$ of this last equation.

P3.6 Let $\hat{A}$ be an operator satisfying the equation $\hat{A}^2 + 2a\hat{A} + 1 = 0$, with a a real constant. For what values of a is the operator $\hat{A}$ Hermitian?

P3.7 Let $\hat{B}$ be a Hermitian operator satisfying the equation $\hat{B}^3 = 4\hat{B}$. What are its eigenvalues?

P3.8 Show that

$$\left[\hat{A}, \frac{1}{\hat{B}}\right] = -\frac{1}{\hat{B}}\left[\hat{A}, B\right]\frac{1}{\hat{B}}. \quad (3.100)$$

P3.9 Consider a classical dynamical system with initial conditions or other relevant parameters distributed in such a way that it is convenient to make a statistical description of it. Defining averages as usual, show that for two arbitrary real dynamical variables, the inequality

$$\overline{(\Delta A)^2}\,\overline{(\Delta B)^2} \geq \left(\overline{AB} - \overline{A}\,\overline{B}\right)^2 \quad (3.101)$$

holds.

P3.10 Using Eqs. (3.60) and (3.61) for the harmonic oscillator in an eigenstate n, calculate the square of $\hat{H}$ and show that the dispersion of $\hat{H}$ is zero. What is the physical meaning of this result?

P3.11 Calculate the commutator of two Pauli matrices, $[\hat{\sigma}_i, \hat{\sigma}_j]$, with $i \neq j = 1, 2, 3$, and express the result in terms of Pauli matrices.

P3.12 Use the Pauli matrices to calculate $(\boldsymbol{a} \cdot \hat{\boldsymbol{\sigma}})^2 \begin{pmatrix} b \\ c \end{pmatrix}$, where $\boldsymbol{a}$ is a unit vector with Cartesian components a_1, a_2, a_3 and $\begin{pmatrix} b \\ c \end{pmatrix}$ is an arbitrary state vector. Comment on the outcome.

Bibliographical Notes

For the student interested in the origins of QM, a valuable survey is van der Waerden (1967).[9] See also the textbook by Longair (2013).

The seminal paper by Heisenberg (1925) is commented and explained by Aitchison, MacManus, and Snyder (2004).

The classic book by Dirac (2023) is not an introductory text, but a masterpiece for later. There are several revised editions of this now standard reading.

---

[9] Bartel Leendert van der Waerden (1903–1996) was a well-known Dutch mathematician and historian of mathematics.

4 Quantum Mechanics in the Schrödinger (Wave) Picture. Dirac's Notation

Just as Heisenberg's matrix mechanics represented a break with classical physics and the beginning of a new physics, Schrödinger's wave mechanics represented another, no less radical, break with classical physics and at the same time opened up another path to the quantum world. Together, these two complementary formulations paved the way for today's QM. Schrödinger's theory is all the more interesting by the fact that it makes substantial use of de Broglie's waves and related phenomena. The Schrödinger equation, being wave-like, constitutes mathematically an eigenvalue problem for bound systems and thus naturally leads to quantization, as will be illustrated here with a couple of simple examples.

The chapter concludes with Dirac's abstract formulation, which has the great advantage of being suitable for any description, be it Heisenberg's, Schrödinger's, or any other.

4.1 Wave Patterns of Quantum Particles

In 1926, prompted by the success of de Broglie's theory of waves associated with moving particles and Einstein's proposal of the wave properties of a beam of ideal gas molecules, Schrödinger took a momentous step: he proposed to start from a wave equation to adequately describe the behavior of electrons.[1] Before turning to the main aspects of this step, it is important to clarify as much as possible the wave element associated with quantum particles, since, as mentioned in Section 1.3, there is a good deal of confusion on this central point. To simplify the discussion, we will refer to electrons, although the conclusions are applicable to any quantum particle and context for which the de Broglie relation (2.73) holds.

Experience tells us that the electron can manifest wave properties (e.g., in diffraction experiments, as explained in Section 2.8), but it can also present itself as a simple corpuscle (e.g., in the cathode-ray experiments by which J. J. Thomson discovered the electron, or in Compton scattering, see Section 2.3.3). The crucial question is then: when and how do the electron's wave properties manifest themselves, and when do its corpuscular properties manifest themselves? In answer to this question, let us recall some simple but important experiments.

---

[1] In early December 1925, at the end of a colloquium given by Schrödinger on de Broglie's theory, Debye incidentally remarked that to deal properly with waves "one must have a wave equation." At another colloquium a few weeks later, Schrödinger happily announced that he "had one."

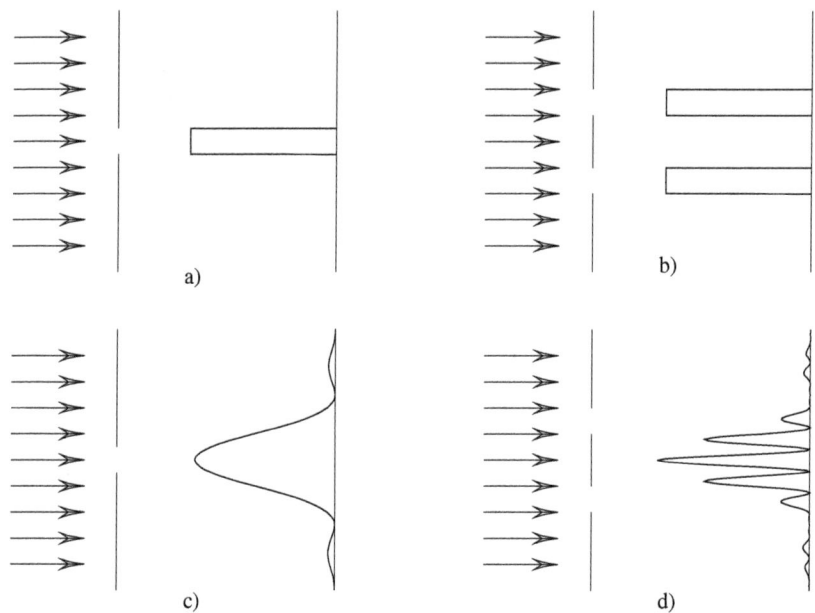

Figure 4.1 Distribution of a beam of particles on the screen after passing through one or two slits. (a) and (b) refer to macroscopic particles, which show corpuscular behavior; (c) and (d) refer to electrons, which show wavelike behavior.

In a first diffraction-like experiment, small classical particles (e.g., grains of fine sand) are sent toward a screen, all in the same direction and with the same momentum, forming a collimated, low-density uniform beam. A plate with a narrow slit is placed between the particle gun and the detecting screen, as shown in Fig. 4.1a. The nearly rectangular distribution recorded on the screen is due to the fact that the interposed plate does not affect the particles passing through the slit, except for the few that touch the edges. If the plate has several adjacent slits instead of a single one, the screen records an almost exact projection of each one, as shown in Fig. 4.1b. If any of the slits is covered, the corresponding particle distribution on the detecting screen simply ceases to appear on the screen, without affecting the rest. This is an example of classical corpuscular behavior described by Newtonian mechanics.

A second experiment consists of repeating the previous one, but this time, using electrons and covering the detection screen with a material that registers them (e.g., phosphor or photographic film). The result is surprising; for example, if we use a single sufficiently narrow slit, the electron distribution pattern on the screen is much smoother than in the previous case, with a large central maximum and a series of secondary maxima, similar to that shown in Fig. 4.1c. If we open more of the same slits sufficiently close to one another in the intermediate screen, we do not get a sequence of distributions similar to the previous one, offset from each other, but a more complicated and interesting distribution, with an alternating series of striking maxima and minima, the number of which is much greater than the number of slits, as shown in Fig. 4.1d.

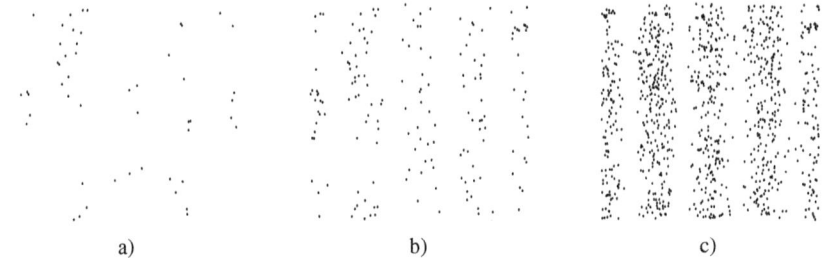

Figure 4.2 As the total number of electrons hitting the screen increases from (a) to (c), the diffraction pattern becomes more discernible and the granular structure becomes more masked.

The distribution of electrons on the screen is analogous to the one which we would obtain by performing a third experiment, this time with a monochromatic (i.e., with a well-defined wavelength) beam of light. When the light passes through a single slit, diffraction at the edges causes the beam to be bent, producing a Fraunhofer diffraction pattern on the far screen, as shown in Fig. 4.1c. However, when several slits are opened, the waves from the slits overlap on the detecting screen, creating a complex interference pattern – analogous to that of Fig. 4.1d – as each wave arrives with its own phase, resulting in alternating areas of constructive and destructive interference. In contrast, corpuscles in the classical mechanical sense cannot interfere destructively (it is impossible to reduce the number of corpuscles by adding more). For this reason, experiments of this type are considered to provide convincing evidence of the wave properties of electrons. Recall that it was precisely a similar (two-slit) experiment with light that led Young to discover (to observe) the wave nature of light in 1801.

Let us now repeat the experiment with electrons, this time reducing the intensity of the beam until no more than one electron flies toward the screen at a time. It turns out that, at sufficiently low intensities, the pattern gets a granular structure, associated with the fact that each incident electron produces a single bright spot on the screen. The electrons fall erratically on the screen, although they tend to fall more frequently in certain areas. We observe that in this case the diffraction pattern is formed by the superposition of independent spots, one for each electron, densely concentrated in the bright regions and few or none of them in the regions of low or no intensity, as shown in Fig. 4.2.

The experiment just discussed remained a *Gedankenexperiment* (thought experiment)[2] for several decades but has finally been carried out in several laboratories.[3] Note that these experiments were conducted long *after* the founding of QM; therefore, its founders had no way of knowing (or recognizing) the statistical origin of the wave patterns traced by the quantum corpuscles.

---

[2] *Gedankenexperiment* is the German term introduced by Einstein in creating his theory of relativity to refer to the unique approach of using imaginary or conceptual rather than actual experiments – an approach that Einstein himself repeatedly used with great success; see also Section 14.5.4.

[3] The first real experiment of this kind was performed by the German physicist Claus Jönsson (1930–2024) during 1961. See the bibliography at the end of the chapter.

4.1.1 Statistical meaning of the interference patterns

As we have just seen, although all the electrons are sent out under conditions that are as uniform and controlled as possible, their landing points vary randomly from one case to the next; there is therefore a random element in the behavior of the electrons (corresponding to the *quantum fluctuations* observed by Heisenberg), which prevents us from predicting their individual trajectories. However, each time the experiment is repeated under identical conditions and with a sufficiently large number of electrons – enough to hide the corpuscular structure of the pattern – the *same* interference pattern is reproduced. It does not matter whether the electrons are sent one by one over a long period of time or in a single bunch as long as the beam density is low enough to ensure that there is no interaction between electrons.

Therefore, the interference pattern is evidence for the existence of a statistical regularity. Although each electron has an unpredictable behavior, not everything is chaotic because the ensemble (the set of all electrons involved in the experiment) behaves in a reproducible way that can be predicted (with the help of QM). In short: *each particle gives rise to a random spot; it is the set of particles (the ensemble) that traces the "diffraction" pattern*. The statistics show the wave properties of the ensemble, characterized by the de Broglie wavelength. This is the essence of the wave properties of quantum particles.

This allows us to appreciate that Schrödinger's wave equation, which will be presented in Section 4.3, *describes the statistical behavior of the electrons* by referring precisely to the wave properties of the system. Unlike the classical dynamical equations, which apply to individual particles, the Schrödinger equation – the fundamental equation of QM – describes the behavior of an ensemble of particles, rather than that of any one particle.

There is an important consideration to this general statement. It was mentioned in Chapter 2, in the context of the thermal radiation distribution law, that the result obtained by averaging over the ensemble of oscillators is expected to be equal to the time average of any one of them. This *ergodic principle* can be considered to apply more generally to (quantum) systems undergoing periodic motion under conditions of stationarity, in which case averaging the individual motion over a very large number of periods is equivalent to averaging over a very large number of realizations. This applies in particular to stationary quantum states describing periodic motion.

By *ensemble* here we mean the same as in statistical mechanics: an infinitely large hypothetical collection of replicas of the physical system under consideration, each of which is in one of the possible dynamical states of the system consistent with the conditions of the problem. The study of the ensemble allows us to make statistical inferences about the system and to draw statistical conclusions – but only *statistical* ones – about the demeanor of any one member of the ensemble. An ensemble can only be approximated physically by a large number of similar independent experiments; if this number is large enough, we can expect the experimental statistical results to be close to the predictions made using the theoretical ensemble.

The reader should be aware that the ensemble perspective, although well established among many QM practitioners, is not the dominant one. More often than not, it is assumed in textbooks that the Schrödinger theory applies to individual particles (see the discussion in Section 1.3).

4.2 The Continuity Equation

To describe the results of electron experiments such as those discussed in the previous section, we need statistical methods. We start by determining the average number of electrons that fall into each small region of the screen. If da is a representative area element of the screen, then the average (or mean) number of electrons that can be expected to be in it, denoted by d^2n, can be written as

$$d^2n = \rho(\boldsymbol{x})da, \tag{4.1}$$

where $\rho(\boldsymbol{x})$ is the surface mean density of electrons at point $\boldsymbol{x}$ contained within the area da. Equation (4.1) does not say that each time we perform the experiment we will find exactly d^2n electrons in da, since the observed number will fluctuate around its mean value; what it says is that this *mean* value is exactly $\rho(\boldsymbol{x})da$. In a three-dimensional problem, (4.1) would be written

$$d^3n = \rho(\boldsymbol{x})d^3x, \tag{4.2}$$

where now $\rho(\boldsymbol{x})$ is the volume density of electrons. Since the number of required dimensions (i.e., appropriate to the specific problem) may vary from case to case, we will generally write in simplified form $dn = \rho(x)dx$, where dx represents the volume element in the corresponding space (one, two, three, ... $3N$ dimensions), and $\rho(x)$ is the density of particles in such space.

More generally, if the system consists of N particles, the number of dimensions of the configuration space is $3N$. For example, a lithium atom, with its three orbital electrons, is described by a wave function in 9-dimensional space if the nucleus is considered fixed, but 12-dimensional if the nuclear motion is taken into account. The total number of electrons involved in the experiment is then

$$N = \int \rho(x)dx, \tag{4.3}$$

where the integral extends over all the space allowed to the particles.

Now suppose that the electrons are moving with the local mean velocity $\boldsymbol{v}(\boldsymbol{x})$ – this is the *flux* or *flow* velocity – in a neighborhood of the point $\boldsymbol{x}$ at time t. The number of elementary charges passing through the unit area perpendicular to the motion per second is by definition the density of the electronic current. Since during a time dt all the dn electrons at a distance not greater than $d\boldsymbol{x} = \boldsymbol{v}dt$ pass through the area $d\boldsymbol{a}$, the average charge crossing the area during this time is $e\,dn = e\,\rho\,d\boldsymbol{a}\cdot d\boldsymbol{x} = e\,\rho\,v\,da\,dt$, so the electric current density is

$$j_e \equiv e\frac{d}{da}\frac{dn}{dt} = e\rho v, \tag{4.4}$$

or better

$$\boldsymbol{j}_e = e\rho(\boldsymbol{x})\boldsymbol{v}(\boldsymbol{x}) \equiv e\boldsymbol{j}(\boldsymbol{x},t)$$

where

$$\boldsymbol{j}(\boldsymbol{x},t) = \rho\boldsymbol{v}(\boldsymbol{x},t) \tag{4.5}$$

is the (particle) *current* or *flow density*.

Next, consider a fixed small macroscopic element of volume ΔV. Since electrons are conserved, the number of them $\Delta n = \int_{\Delta V} \rho \, d^3 x$ contained on average in ΔV can only change if the *net* current from ΔV is different from zero. If $d\boldsymbol{a}$ represents a surface element of this volume facing outwards, then the net current leaving ΔV is $\oint_{\Delta A} \boldsymbol{j}_e \cdot d\boldsymbol{a} = \int_{\Delta V} \nabla \cdot \boldsymbol{j}_e d^3 x$ (by the divergence theorem), where ΔA is the surface area enclosing the volume ΔV. This net outward current reduces the electric charge in ΔV by the amount $-\int \frac{\partial}{\partial t}(e\rho) d^3 x$ per unit time. Equating the two expressions gives

$$-e \int_{\Delta V} \frac{\partial \rho}{\partial t} d^3 x = \int_{\Delta V} \nabla \cdot \boldsymbol{j}_e d^3 x. \tag{4.6}$$

Since ΔV is arbitrary – it can be reduced to arbitrarily small values in principle – the preceding result can be written in the form of a continuity equation ($\boldsymbol{j}_e = e\boldsymbol{j}$)

$$\frac{\partial \rho}{\partial t} + \nabla \cdot \boldsymbol{j} = 0, \tag{4.7}$$

or alternatively, using Eq. (4.5),

$$\frac{\partial \rho}{\partial t} + \nabla \cdot \boldsymbol{v}\rho = 0. \tag{4.8}$$

These expressions are the differential, that is, the *local* form of the law of conservation of the number of electrons. Integrating (4.7) over the whole space and assuming that the current at the surface is zero at infinity – as is the case for any real or bounded system – gives

$$\frac{dN}{dt} = \frac{d}{dt} \int_V \rho d^3 x = \int_V \frac{\partial \rho}{\partial t} d^3 x = -\int_V \nabla \cdot \boldsymbol{j} d^3 x = -\oint \boldsymbol{j} \cdot d\boldsymbol{a} = 0, \tag{4.9}$$

so the total number N of electrons in the system is constant. This is the *global* version of the conservation law. Equation (4.9) applies to any finite system in which there is no particle creation or annihilation. In this book we will limit ourselves to the study of situations in which such essentially relativistic processes can be excluded a priori, so that the continuity equation will be applied without restriction and the number N will represent a constant of the system.

If the integration volume is finite, the previous result is still valid as long as the electrons do not cross the surface. This means that the number of electrons contained in the volume only changes due to those that escape through the surface. Turning now to the limit of an arbitrarily small volume, we recover the meaning of Eq. (4.7) as a local conservation law, which implies a global conservation law (N = const.), as we have seen.

The continuity equation is common to all nonrelativistic statistical problems for conserved particles, classical or quantum. However, it is clearly not sufficient for studying the behavior of electrons, since it contains the unknown velocity function $\boldsymbol{v}$. To make up for this deficiency we need to introduce a *dynamical law*; this will be precisely the Schrödinger equation, which determines all the functions describing the statistical behavior of the system, such as $\rho, \boldsymbol{v}$, and so on.

4.3 The Stationary Schrödinger Equation

In order to construct his wave equation, Schrödinger took as a starting point the general wave equation, well known in all areas of physics dealing with waves. For a scalar wave $\psi(x,t)$ propagating in free space with velocity v, oscillating in time with a (circular) frequency $\omega = 2\pi\nu$ and in space with a wavelength λ, this equation is

$$\nabla^2\psi - \frac{1}{v^2}\frac{\partial^2\psi}{\partial^2 t} = 0, \tag{4.10}$$

where, as holds for any propagating wave,

$$v = \lambda\omega/2\pi = \lambda\nu. \tag{4.11}$$

In attempting to develop his theory along this line, Schrödinger encountered difficulties, so he decided to first consider the equation for *standing* waves. A standing or stationary wave has the general form

$$\psi(\mathbf{x},t) = e^{-i\omega t}\varphi(\mathbf{x}). \tag{4.12}$$

The time-independent spatial factor $\varphi(x)$ in turn satisfies a stationary wave equation that is derived by combining the preceding equations,

$$\nabla^2\varphi + \frac{\omega^2}{v^2}\varphi = 0. \tag{4.13}$$

Introducing Eq. (4.11) we get

$$\nabla^2\varphi + \frac{4\pi^2}{\lambda^2}\varphi = 0, \tag{4.14}$$

or in terms of the *wave number* $k = 2\pi/\lambda$,

$$\nabla^2\varphi + k^2\varphi = 0. \tag{4.15}$$

This is the standard form of the stationary wave equation, with the parameter k taken as a measure of the wavelength. We now introduce a major step – the one required to leave classical physics and enter the quantum domain – by making the wavelength λ coincide with the de Broglie wavelength; with $p = |\mathbf{p}|$, this means

$$\lambda = \frac{h}{p} = \frac{2\pi}{k}, \quad \text{or} \quad \hbar k = p, \tag{4.16}$$

which shows that k is also the momentum p (in units of $\hbar$). Introducing this relation into the wave equation (4.15) gives

$$\nabla^2\varphi + \frac{\mathbf{p}^2}{\hbar^2}\varphi = 0. \tag{4.17}$$

This equation, together with the interpretation of ψ as a probability amplitude (see the next section)

$$\rho = \psi^*\psi = \varphi^*\varphi, \tag{4.18}$$

describes stationary quantum problems, for which ρ is time-independent. The monochromaticity of the wave function for ω fixed implies, according to the relation $E = \hbar\omega$, that the corresponding ensemble is characterized by a single energy, that is, that all the electrons that pertain to the ensemble have the same energy E. This is obviously an idealization, insofar as it is not possible to ensure in practice that each electron has exactly the same energy; but if the energy distribution is narrow enough, it can be considered to be essentially monoenergetic. In such a case it is convenient to express (4.17) in terms of the energy by writing $E = \boldsymbol{p}^2/2m + V(\boldsymbol{x})$, which leads to the *stationary Schrödinger equation*:[4]

$$\nabla^2 \varphi + \frac{2m}{\hbar^2}(E - V)\varphi = 0, \qquad (4.19)$$

or in terms of ψ, to include the time factor indicated in Eq. (4.12),

$$\nabla^2 \psi + \frac{2m}{\hbar^2}(E - V)\psi = 0. \qquad (4.20)$$

To arrive at this equation we have used relations characterizing the motion of corpuscles (such as the formula $E = \boldsymbol{p}^2/2m + V$ for the energy) and relations characterizing the de Broglie wave; thus the Schrödinger equation naturally combines corpuscular and wave elements and is expected to express the wave behavior of an ensemble of quantum corpuscles. We have not introduced any electron-specific elements, so it can be applied to any (nonrelativistic, spinless) quantum corpuscle.

Interestingly, the relation (4.18) was initially unknown to Schrödinger, having been proposed months later by Born; hence it is known as *Born's rule*. It specifies the probabilistic content of the Schrödinger equation and fixes the meaning of the wave function ψ (and of course of φ) as an *amplitude of probability*.

Equation (4.20) was put forward and studied by Schrödinger in a famous series of articles published in 1926. The enormous field of application of this equation has shown that it correctly describes any stationary quantum problem characterized by a fixed energy E and a potential V – which can be a function of any set of dynamical variables and not necessarily only of the position – and when relativistic or spin effects are negligible. Later we will study how to overcome these limitations and, in particular, how to generalize the differential equation to the time-dependent case in order to study the time evolution of any given quantum system (see Chapter 7).

It should be emphasized that Schrödinger's postulate has a *phenomenological* character,[5] since it takes up certain observed features of microsystems, but does not explain them physically on the basis of deeper principles, for example, related to the origin of their wave behavior or the quantum fluctuations. As a result, the nature of the wave function remains undefined. In addition, the use of de Broglie's relation $\lambda = h/p$, which contains the element λ that is arbitrarily extended in space, introduces features of nonlocality into the theory.

Schrödinger's aim was to show that it is possible to transform the general problem of quantization of a system into an eigenvalue problem, a well-defined mathematical

---

[4] In previous chapters, the symbol $\mathcal{E}$ was used for energy to distinguish it from the electric field; from now on, we will use the letter E to refer to energy, following the usual convention.

[5] In the context of this book, "phenomenological" refers to a procedure that takes into account characteristic features in the description without making their cause explicit.

problem. The first paper in the series was formal in style, without any clear theoretical elaboration. Only in his next paper, already convinced of the fundamental character of his equation, did Schrödinger try to derive it from physical arguments; specifically, by analogy with geometric optics. This analogy acquired a certain prestige and was studied for some time, but it has now evaporated. As in the case of matrix mechanics, the wave formulation of QM emerged as a mathematical formalism, to which far too many interpretations were subsequently added.

4.3.1 The quantum probability amplitude

The Schrödinger equation (4.20) is linear and homogeneous in ψ (or φ), so if ψ_1 and ψ_2 are two solutions of the same equation, the function $a\psi_1 + b\psi_2$ with a and b arbitrary constants is also a solution. This property is precisely the basis for interference phenomena, as those described in Section 4.1. It is usual to refer to this fundamental consequence of the linearity of the Schrödinger equation by saying that the *probability amplitude* φ satisfies the *superposition principle*. This is perhaps the most important general property of the wave function; we will have occasion to discuss it at length and to use it repeatedly.

To understand how interference occurs, recall from optics that if two coherent light beams are incident on a plate, the resulting amplitude at each point on the plate is the sum of the (complex) amplitudes of the two beams, each of which is a solution of the wave equation of electrodynamics. Writing these amplitudes in the form $\phi_1 = A_1 \exp i\theta_1$, $\phi_2 = A_2 \exp i\theta_2$, the amplitude Φ of the resulting wave is given by the superposition

$$\Phi = \phi_1 + \phi_2 = A_1 e^{i\theta_1} + A_2 e^{i\theta_2}. \tag{4.21}$$

A light spot on the plate corresponding to this amplitude has an intensity I proportional to $|\Phi|^2$,

$$I \sim |\Phi|^2 = |A_1 e^{i\theta_1} + A_2 e^{i\theta_2}|^2 = A_1^2 + A_2^2 + 2A_1 A_2 \cos(\theta_1 - \theta_2). \tag{4.22}$$

Since the phases θ_1 and θ_2 vary from one point on the plate to another if the beams come from different directions, $\cos(\theta_1 - \theta_2)$ oscillates between $+1$ and -1 (see Fig. 4.3). Therefore, the resulting intensity I shows alternating zones of maxima (when the phases coincide) and minima (when the phases differ by π), even though the amplitudes A_1 and A_2 vary little between nearby points.

The aforementioned similarity of the observed (quantum mechanical and optical) interference patterns suggests that we study the wave-like behavior of the electron density by means of the complex amplitude ψ, which has the following two properties:

a) it satisfies a linear and homogeneous wave equation;
b) the square of its modulus is proportional to the density of electrons at each point, $\rho(x) \sim |\psi|^2 = \psi^*\psi = \varphi^*\varphi$.

In this way, we ensure that the particle density is a nonnegative function that can manifest interference phenomena through the superposition of its constituent amplitudes, and that the diffraction of monoenergetic electrons produces a pattern similar to that of monochromatic light.

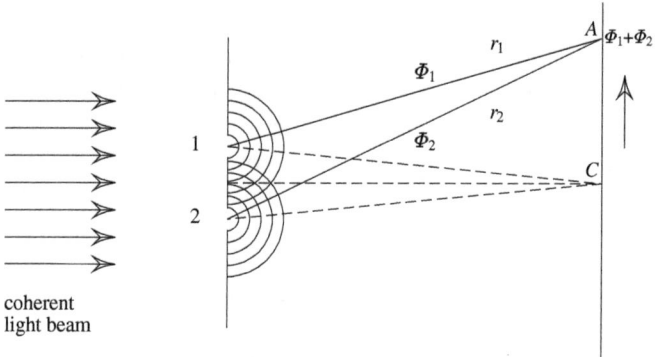

Figure 4.3 Interference of two coherent monochromatic light beams. The phase difference between Φ_1 and Φ_2 is due to the different distances traveled by the rays from slits 1 and 2 to reach point A on the screen. In C the two waves are in phase and their amplitudes add up constructively. As we move away from C, the phase difference $\theta_1 - \theta_2$ increases progressively and the interference term $2A_1 A_2 \cos(\theta_1 - \theta_2)$ changes sign, producing a pattern of alternating light and dark fringes.

4.3.1.1 Normalization of ψ

When fixing the proportionality factor in the expression $\rho(x) \sim |\psi|^2$, it is convenient that $\int |\psi|^2 dx$ integrated over the whole available space has a simple and, if possible, universal value. Therefore, taking advantage of the fact that the Schrödinger equation, being homogeneous, does not fix the amplitude of the solution, the function ψ is conveniently normalized to unity,

$$\int \psi^* \psi \, dx = 1. \tag{4.23}$$

If the system contains N particles, the total density ρ_N is divided by N.[6]

$$\rho \equiv \frac{\rho_N}{N} = \psi^* \psi, \tag{4.24}$$

and

$$\int \rho \, dx = 1, \quad \rho = \psi^* \psi. \tag{4.25}$$

We refer to ρ as the (normalized) *particle density* or *probability density*. Note, however, that these are two entirely different interpretations of this quantity: one as proportional to the local number of particles, the other as a probability. This distinction is not always made in the literature. In any case, Born's rule (4.25)[7] is

---

[6] This expression only makes sense if N is finite. There are theoretical descriptions that require N to be infinite, usually for unbounded problems; in such cases it is not possible to normalize ψ to unity. In Section 7.1 we will have the opportunity to see examples of this situation and study the methods for dealing with it.

[7] Many authors use the term "statistical interpretation" to refer to Born's interpretation of ψ as a probability amplitude. The reader should not be confused by this use of the term in QM. Born's interpretation of the amplitude is common to both the orthodox and the ensemble interpretations. Born used the term "statistical" within the orthodox theory because he was discussing an essentially statistical dispersion problem using the wave function; it was Pauli who suggested extending the "statistical" notion to all applications. In this text, probability will always be understood in an objective sense, i.e., a probability measures a property

universally accepted.[8] The ψ function, in turn, is variously called a wave function, a state function, a Schrödinger function, or a probability (density) amplitude, or more briefly, an amplitude.

It is appropriate to add a couple of comments here about the normalization of ψ. First, the normalization to unity – when feasible – is merely conventional. Second, the condition (4.23) determines ψ up to a constant phase. In fact, neither this condition nor the Schrödinger equation is altered by the change $\psi \to \psi' = \psi e^{i\alpha}$ with α a constant. This global phase α of the wave function is arbitrary and generally has no physical meaning, so it is customary to assign it the most convenient value in each case, which is usually zero.

The fact that the theory determines the wave function up to a numerical factor means that its absolute amplitude has no direct physical meaning. Practice strongly suggests that ψ is a mathematical entity, not a physical quantity. Compare this with the case of a real physical wave, such as a sound wave or an electromagnetic wave, whose magnitude is determined by the physics. In this text we will stick to the notion of ψ as a mathematical object – specifically an amplitude of probability – which does not correspond *directly* to a physical field, but expresses *real* (statistical) properties of the system described. It should be noted, however, that there are authors who assign the meaning of a physical field to the wave function; among them are Schrödinger (who directly identified ψ with the electron), de Broglie (who considered the ψ field as the element that guides the electron), and D. Bohm[9] (who postulated the ψ field as the source of the quantum phenomenon, in a context similar to that proposed by de Broglie, as a kind of guiding wave).

4.3.2 Quantization as an eigenvalue problem[10]

The procedure for solving the stationary Schrödinger equation differs from the usual procedure for solving a differential equation, where the general solution is usually sought first, and then particularized for the problem in question by selecting the specific values of the constants (or functions) of integration. In the quantum stationary case the process consists of first applying filters that select those solutions that satisfy a set of known conditions to be physically admissible. Among the mandatory conditions are (the prime denotes derivation):

(i) Both ψ and its spatial derivatives ψ' should be continuous functions;
(ii) ψ and ψ' must be single-valued at all points;
(iii) ψ and ψ' must be finite everywhere, and ψ square-integrable;
(iv) ψ and ψ' must satisfy the boundary conditions of the problem.

---

of the (state of the) system and *not*, for example, *our* degree of knowledge about it or *our* degree of confidence in *our* predictions.

[8] Often $\rho(x)$ is referred to as the "probability of finding the particle at position x." We prefer to avoid this language, because it involves the act of "finding," which is not part of the theoretical description.

[9] David Bohm (1917–1992) was an eminent US–Brazilian–British theoretical physicist. Among his many contributions to physics are his causal and deterministic interpretation of quantum mechanics and the Aharonov–Bohm effect discussed in Section 12.7.

[10] This is the title Schrödinger chose for his landmark series of four papers on wave mechanics of 1926. In the first of these he applied his equation to the H atom and showed that the quantum numbers emerge "in the same natural way as the integers specifying the number of nodes in a vibrating string."

All these conditions, which are typical of physically meaningful functions satisfying a wave equation, have their origin in the meaning of ψ and correspond to the requirement that the spatial density of particles be a well-behaved function. Furthermore, since the particle flow depends on both ψ and its spatial derivative (as we will see in Section 7.2.3), the continuity of the latter has been included in condition (i). These restrictions, in addition to possible problem-specific boundary and even initial conditions, allow us to select, among the – frequently infinite – number of solutions of a partial differential equation, the *single one* that satisfies *all* the conditions demanded by the physical situation.

Something similar occurs in other branches of physics in connection with differential equations, when the set of conditions that characterize the problem are sufficient to determine a unique solution, which is what is known in physics as *the solution* of the problem. However, in certain types of quantum problems – in particular, bound-particle problems – the conditions to be fulfilled are so strict that solutions exist only for certain values of the parameters appearing in the differential equation itself, which are generically called *eigenvalues* (or *proper*, or *characteristic values*). Typically, this is the case for a discrete (quantized) energy spectrum. This explains the title chosen by Schrödinger for his foundational papers and used for the title of this section. Consider the following elementary example of an analogous situation (albeit of a nonquantum nature), taken from classical mechanics.

Exercise E4.1. Stationary waves on a fixed string
Find the wavelength and frequency of the standing waves on a string of length L rigidly fixed at both ends, when the speed of propagation along the string is v.

Solution. The stationary oscillations of the string are governed by the differential equation

$$\frac{d^2 u}{dx^2} + k^2 u = 0, \tag{4.26}$$

where $u = u(x)$ represents the instantaneous local deviation of the string from rest, and $k^2 = \omega^2/v^2 = (2\pi/\lambda)^2$ is the wave number. The general solution is $u = A\cos kx + B\sin kx$. If one end of the string is fixed at $x = 0$ so $u(0) = 0$, the solution is reduced to $u(x) = B\sin kx$, where the value of B can be determined by the energy associated with the vibrations, or the maximum elongation of the string. If the other end of the string is fixed at $x = L$, the problem is overdetermined: in this case there are solutions only for the values of the parameter k given by

$$k = \frac{n\pi}{L}, \quad n = 1, 2, 3, \ldots \tag{4.27}$$

This condition is natural, since it means that stationarity is achieved only for vibrations that contain an integer number of half-wavelengths along the string. For any other wavelength, the destructive interference between the direct wave and the wave reflected at the fixed end cancels them out. The vibration frequencies of the string (its spectrum) are the harmonics given by the expression (4.27),

$$\nu_n = \omega_n/2\pi = v/\lambda_n = k_n v/2\pi = (v/2L)n. \tag{4.28}$$

The fundamental frequency v_1 is thus fixed by the length L of the string and the speed v, which depends on the material, its cross-section and, above all, its tension. This last property allows you to easily adjust the value of v_1 and its harmonics $v_n = nv_1$. Your guitar must be in tune!

Since the 1D stationary Schrödinger equation contains the energy E as the single free parameter, when the set of required conditions *overdetermines* the solution (when the number of conditions exceeds the number of integration constants at hand), it is the energy E itself that becomes conditioned to take certain discrete values E_n; these are the energy eigenvalues (proper values) for the problem.

There is, however, an important difference between the string and the quantum case. Whereas in the classical case for a given λ_n the frequency of vibration depends on v, and hence on the elastic properties of the string, in the quantum case the value of λ_n determines v (or the energy), and hence the state of motion of the system. The student should distinguish between the two cases: in acoustics nobody talks about quantization, only about harmonics; in quantum mechanics nobody talks of harmonics, but of quantum numbers.

Both the string problem and the bounded quantum problem give rise to an eigenvalue problem. In general, an eigenvalue equation can be written in the form

$$\hat{L}\psi_n = \lambda_n \psi_n, \tag{4.29}$$

where $\hat{L}$ is a suitable linear differential (or integral) operator and λ_n its eigenvalue corresponding to the solution ψ_n (not to be confused with the wavelength λ). For the string, in particular, $\hat{L} = -d^2/dx^2$, $\lambda_n = k_n^2$ and $k_n = n\pi/L$; whereas in the case of Eq. (4.19),

$$\left(-\frac{\hbar^2}{2m}\nabla^2 + V\right)\varphi_n = E_n\varphi_n, \tag{4.30}$$

we have $\hat{L} = (-\hbar^2/2m)\nabla^2 + V$, $\lambda_n = E_n$, and φ_n the eigenfunction. Solving the eigenvalue equation (4.29) means finding the eigenvalues λ_n (of the operator $\hat{L}$) and the corresponding eigenfunctions ψ_n (of such an operator) that are valid for the specific problem under the given conditions.[11] Schrödinger's aim was precisely to reduce quantization to an eigenvalue problem, a well-known mathematical technique.

4.3.2.1 Degeneracy of the solutions

When there is a one-to-one correspondence between the eigenvalues and the eigenfunctions, the solution is said to be *nondegenerate*. When, on the contrary, more than one eigenfunction corresponds to a given eigenvalue, the solution is *degenerate*. The order of degeneracy (usually denoted by the letter g) is given by the number of linearly independent eigenfunctions associated with the same eigenvalue.

---

[11] More precisely, when Neumann conditions ($\partial u/\partial n = 0$ at the boundary, where n represents the normal to the boundary surface) or Cauchy conditions (fixing u and $\partial u/\partial n$ on the boundary) are imposed on a *closed* surface, the solution becomes overdetermined. See the bibliography at the end of the chapter.

As we will see later, the absence of degeneracy is characteristic of 1D bounded states, while the stationary states of multidimensional problems are usually degenerate. We can understand this qualitatively by considering that, when there are multiple degrees of freedom, the energy is not sufficient to specify the state of motion, but additional constants are required, whose values must be fixed so as to correspond one to another. To each of the possible states belonging to the same energy corresponds a specific wave function and a set of quantum numbers that specify the value of the energy and the other dynamical variables.

The classical analogue may help to clarify this point: in 1D problems, energy is the only constant of motion; for problems with more degrees of freedom, additional integrals of motion appear, such as the angular momentum, some component of it, and so on. There can be different states of motion that correspond to the same energy but different values of angular momentum.

4.3.3 Continuous vs discrete spectrum

It is important to understand under which general conditions the energy spectrum is discrete and when it is continuous (nonquantized). The following general conclusions are drawn from studying the structure of the stationary Schrödinger equation.

When the potential in which the particles move is attractive and the system is bound (e.g., atomic electrons), the energy spectrum is discrete (see Fig. 4.4a). However, for a potential that does not constrain the motion, as in Fig. 4.4b, the differential equation has a solution for any admissible value of E (for any $E > 0$, in the case shown) and the energy spectrum is continuous; this is true even if the particles classically do not have access to a particular region of space, as long as the admissible space is unbounded. Thirdly, there are potentials of physical interest for which both cases can occur; an example of this situation is illustrated in Fig. 4.4c. In the classical case, for negative energies there are stationary bound states between the two turning points, and for positive energies the orbits are open. In the corresponding quantum case, the spectrum is discrete for negative energies and continuous for positive energies.

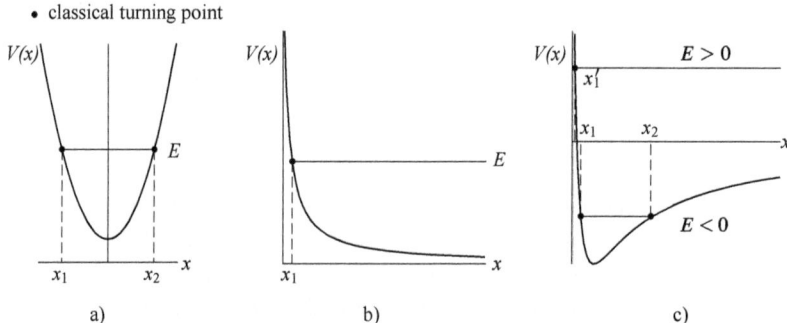

Figure 4.4 (a) An attractive potential (a potential well) forms a bounded system and gives rise to a discrete spectrum. (b) A repulsive potential produces open orbits and the spectrum is continuous. (c) There are potentials for which both situations can occur; in the example, for negative energies the spectrum is discrete, but for positive energies it is continuous. The notion of orbit used here is the classical one.

It follows that the index n (or the set of indices, when there are more degrees of freedom) numbering the solutions can be continuous, discrete, or partially continuous and partially discrete. Consider, for example, the case of an atomic nucleus and an electron attracted to it by a Coulomb potential. Under the right conditions, the system can form a hydrogen-like atom (single-electron bound states), which has a discrete energy spectrum (as predicted by Bohr's theory). However, if we throw the electron at the nucleus from the distance, the nucleus will deviate the electron without trapping it or altering its energy, which can obviously have any (positive) value.

Exercise E4.2. Asymptotic behavior of the wave function
Study the asymptotic behavior of ψ for a 1D problem under the hypothesis that $V(x)$ approaches a finite constant value as $|x| \to \infty$ (note that this condition may correspond to a real physical situation), and give a geometric interpretation of the result.

Solution. Let V_0 be the value of $V(x)$ at infinity. Given a value of E, there can be two cases, depending on the sign of $E - V_0$; we will study each one separately.

(a) $E - V_0 < 0$. In the asymptotic region we can approximate the Schrödinger equation by $\psi'' - q^2\psi = 0$ with q essentially constant and given by

$$q = \pm\frac{1}{\hbar}\sqrt{2m(V_0 - E)}. \tag{4.31}$$

The particular solutions of this equation are the exponentials $e^{\pm|q|x}$; therefore, the convergent asymptotic solution will be a decreasing exponential. Since ψ and ψ'' have the same sign, the curve is concave.

(b) $E - V_0 > 0$. Proceeding as in the previous case, we approximate the Schrödinger equation in the asymptotic region by means of the expression $\psi'' + k^2\psi = 0$ where k is taken as a constant given by

$$k = \pm\frac{1}{\hbar}\sqrt{2m(E-V_0)}. \tag{4.32}$$

The solutions of this equation are periodic functions of the general form $A\cos(kx + \theta)$. Since ψ and ψ'' have opposite signs, the curve is convex.

In a more realistic analysis, q and k are slowly varying functions rather than constant, which may change the details but not the qualitative behavior.

Exercise E4.3. Continuous and discrete spectra
Use the preceding results to establish the relationship between the nature of the energy spectrum (discrete or continuous) and the value of the energy. Also determine whether the system is in a bound state or not.

Solution. Assume that the potential is V_- at $x = -\infty$ and V_+ at $x = +\infty$, with $V_- > V_+$ (see Fig. 4.5). We will analyze three cases, depending on the value of the energy.

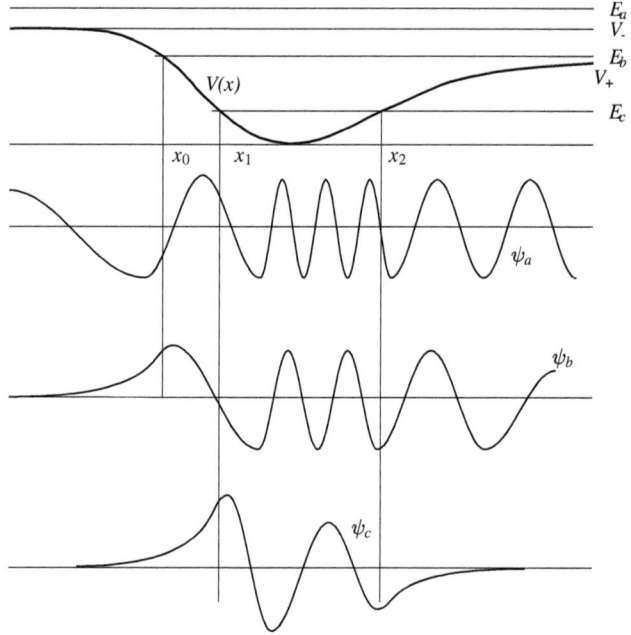

Figure 4.5 Different behavior of ψ depending on the energy of the system.

(a) $E = E_a > V_-$. In the Schrödinger equation $\psi'' + k^2(x)\psi = 0$, $k^2(x)$ is always positive, so ψ is oscillatory and finite for all x. Since this ψ is always acceptable, the spectrum is continuous. Furthermore, since ψ does not vanish at infinity, the system is not bounded (there is a significant probability that the particle is at any distance from the origin).

(b) $V_- > E = E_b > V_+$. Let x_0 be the point where $E_b = V(x)$; this is a classical turning point. To the right of x_0 the wave function is oscillatory (with variable wavelength, in general); to the left of x_0, $E - V$ is negative and the wave function is exponential; choosing the decreasing exponential as a particular solution, we obtain a wave function that is finite at all points. This wave function is acceptable and therefore corresponds to a continuous spectrum. As in the previous case, the system is not bounded.

(c) $E = E_c < V_+$. In this case there are two classical turning points, x_1 and x_2; between these two points ψ is oscillatory ($E > V$), but to the left of x_1 and to the right of x_2 it has an exponential-like behavior. If, under these conditions, it is possible to find continuous solutions, with continuous first derivative and which cancel at infinity, they will be acceptable and there will be stationary states which will be bounded (the number of particles at arbitrary distances is negligible).

4.3.4 Orthogonality and completeness of the eigenfunctions

In this section we will first show that the eigenfunctions of Eq. (4.19) are a set of mutually orthogonal functions, and clarify the meaning of this. For this purpose we write the equation for the proper *state n*, that is, for the eigenvalue E_n and the corresponding eigenfunction φ_n,

$$\nabla^2 \varphi_n + \frac{2m}{\hbar^2}(E_n - V)\varphi_n = 0. \tag{4.33}$$

The complex conjugate for another state m, given that the potential is a real function,[12] is

$$\nabla^2 \varphi_m^* + \frac{2m}{\hbar^2}(E_m - V)\varphi_m^* = 0. \tag{4.34}$$

Multiplying the first of these equations by φ_m^* and the second by φ_n and taking the difference gives

$$\varphi_m^* \nabla^2 \varphi_n - \varphi_n \nabla^2 \varphi_m^* + \frac{2m}{\hbar^2}(E_n - E_m)\varphi_m^* \varphi_n = 0. \tag{4.35}$$

We now define an auxiliary vector $\boldsymbol{B}$ as

$$\boldsymbol{B} = \varphi_m^* \nabla \varphi_n - \varphi_n \nabla \varphi_m^*. \tag{4.36}$$

Using the formulas $\nabla \cdot (f\boldsymbol{A}) = f \nabla \cdot \boldsymbol{A} + \boldsymbol{A} \cdot \nabla f$, $\nabla \cdot (\nabla f) = \nabla^2 f$, we see that

$$\nabla \cdot \boldsymbol{B} = \varphi_m^* \nabla^2 \varphi_n - \varphi_n \nabla^2 \varphi_m^*, \tag{4.37}$$

which, when introduced in Eq. (4.35), gives, after integrating over the whole space,

$$\int \left[\nabla \cdot \boldsymbol{B} + \frac{2m}{\hbar^2}(E_n - E_m)\varphi_m^* \varphi_n \right] d^3 x = 0. \tag{4.38}$$

For any bounded system, the first integral is zero, since by the divergence theorem it gives $\oint_A \boldsymbol{B} \cdot d\boldsymbol{a}$ and $\boldsymbol{B}$ is zero on the integration surface (there are no particles at infinity). Therefore,[13]

$$(E_n - E_m) \int \varphi_m^* \varphi_n dx = 0. \tag{4.39}$$

When the states are nondegenerate, that is, when for $n \neq m$ also $E_n \neq E_m$, we get

$$\int \varphi_m^* \varphi_n dx = 0, \quad n \neq m. \tag{4.40}$$

For $n = m$, the factor $E_m - E_n$ is zero and (4.39) is satisfied for any finite value of the integral; since φ_n is normalized to unity, we have

$$\int \varphi_n^* \varphi_n dx = 1. \tag{4.41}$$

With $\int \varphi_m^* \varphi_n dx$ defined as the dot (inner or scalar) product (φ_m, φ_n) in the space of square integrable functions (L_2), Eq. (4.40) tells us that the nondegenerate stationary

---

[12] This is a general rule; complex or imaginary potentials are in general part of a phenomenological model to simulate friction or particle loss.
[13] The integration volume will henceforth be denoted simply by dx instead of d^3x, as was done before. The advantage of this notation is not only its conciseness, but also that it does not specify the number of dimensions, which is irrelevant for our present purposes and may vary from problem to problem.

functions φ_n and φ_m^* are mutually orthogonal, while (4.41) says that they are normalized to unity.[14] Combining these two expressions into one with the help of the Kronecker function δ_{mn}, we write

$$\int \varphi_m^* \varphi_n dx = \delta_{mn}. \qquad (4.42)$$

It is usual to refer to (4.42) as the *property* (or *condition*) *of orthonormality*. In the case of degenerate states, using Schmidt's orthogonalization method, it can be shown that it is always possible to construct a system of mutually orthogonal degenerate functions by considering appropriate linear combinations of the degenerate solutions, so that (4.42) is general.[15]

It is also possible to show that the set of eigenfunctions is *complete*. In Section 7.1 it is shown that the condition of completeness can be expressed for a discrete set as

$$\delta(x, x') = \sum_n \varphi_n(x) \varphi_n^*(x'). \qquad (4.43)$$

This implies, in particular, that any function $f(x)$ that is sufficiently well behaved in the domain of the φ_n and satisfies the same boundary conditions can be written as the linear combination,

$$f(x) = \sum_n a_n \varphi_n(x), \qquad (4.44)$$

with constant coefficients a_n, which can be calculated by multiplying (4.44) by φ_k^* and integrating,

$$\int \varphi_k^*(x) f(x) dx = \sum_n a_n \int \varphi_k^* \varphi_n dx; \qquad (4.45)$$

using the orthonormality property (4.42), we get

$$a_k = \int \varphi_k^*(x) f(x) dx. \qquad (4.46)$$

The expansion (4.44) with the a_n given by (4.46) converges in the mean to $f(x)$ in the domain of definition; for continuous functions, the series and $f(x)$ are equal in that domain.

The property of completeness of the set of orthonormal functions, which is what allows us to do the expansion (4.44), is one of the most important mathematical properties of the Schrödinger eigenfunctions and will be fundamental for later developments. A well-known example of a series expansion in terms of a complete set of orthonormal functions is the Fourier expansion. The formulas for the Fourier expansion are just a particular case of (4.44) and (4.46), as the reader can easily check.

---

[14] As explained in more detail later, we consider the wave functions φ_n as the unit vectors of a linear vector space with the dot (the scalar) product as the norm. This is the specific Hilbert space of the problem at hand, constructed (defined) by the set $\{\varphi_n\}$. If φ_n is represented by a column vector, then to its adjoint corresponds a row vector. Thus, the vectors φ and $\varphi^\dagger$ belong to dual Hilbert spaces and are orthogonal.

[15] The student interested in these topics (including Schmidt's method of orthogonalization) may consult Margenau and Murphy (1956), section 8.2. This problem is studied in more detail in various texts on QM, such as Merzbacher (1998), section 8.4, or de la Peña and Villavicencio (2003).

4.3.4.1 The expansion theorem

The preceding results can be visualized by introducing a geometric interpretation, which generalizes the common description of vectors in 3D Euclidean space to n dimensions (with n possibly infinite). To do so, we have considered each φ_n as a unit vector (the norm = 1 follows from Eq. (4.41)) in the n-direction of an abstract linear complex normed vector space. Equation (4.44) then represents the expansion of the vector $f(x)$ in this space in terms of the orthogonal unit vectors φ_n. Since a_n now represents the component of f along $\boldsymbol{n}$, Eq. (4.46) can be interpreted as the *scalar product* of φ_n and f, defined as

$$(\varphi_n(x), f(x)) = \int \varphi_n^*(x) f(x) dx, \tag{4.47}$$

which is consistent with the definition given earlier of the inner product. This complex linear vector space, constructed on the basis of the eigenfunctions of the stationary Schrödinger equation, is an example of a Hilbert space as introduced in Section 3.4. Note that the set $\{\varphi_n\}$ spans its corresponding Hilbert space: for each given system of eigenfunctions there is a specific Hilbert space.

By way of illustration, we deduce an important theorem, called the *expansion theorem*. Let $f(x)$ and $g(x)$ be two arbitrary functions of x in $x_1 < x < x_2$, which we expand in terms of a complete set of orthonormal functions $\{\varphi_n\}$ in (x_1, x_2),

$$f(x) = \sum_n a_n \varphi_n(x), \qquad g(x) = \sum_n b_n \varphi_n(x). \tag{4.48}$$

According to (4.46),

$$a_n = \int_{x_1}^{x_2} \varphi_n^*(x) f(x) dx, \qquad b_n = \int_{x_1}^{x_2} \varphi_n^*(x) g(x) dx. \tag{4.49}$$

From these equations and using (4.42), we get

$$(f, g) = \int_{x_1}^{x_2} f^*(x) g(x) dx = \sum_{n,n'} a_n^* b_{n'} \int_{x_1}^{x_2} \varphi_n^*(x) \varphi_{n'}(x) dx = \sum_{n,n'} a_n^* b_{n'} \delta_{nn'}, \tag{4.50}$$

which leads to the expansion theorem,

$$(f, g) = \sum_n a_n^* b_n = (a_n, b_n) = \boldsymbol{a}^* \cdot \boldsymbol{b}. \tag{4.51}$$

In the particular case where the functions of (4.48) are Fourier expansions, this result is known as *Parseval's theorem*. This is an example of the invariance of the scalar product with respect to the representation, a topic that is discussed in detail in Chapter 8.

An alternative way of writing (4.51), frequently used in the literature, is obtained by rewriting Eqs. (4.49) in the form

$$b_n = (\varphi_n, g), \quad a_n^* = (\varphi_n, f)^* = (f, \varphi_n), \tag{4.52}$$

which, when introduced in (4.51), gives the elegant and transparent expression

$$(f, g) = \sum_n (f, \varphi_n)(\varphi_n, g). \tag{4.53}$$

An application of the theorem occurs when $f(x)$ represents a wave function $\psi(x)$,

$$\psi(x) = \sum_n a_n \varphi_n(x). \tag{4.54}$$

With $f = g = \psi$ in (4.51), we have

$$\int |\psi|^2 dx = \sum_n |a_n|^2, \tag{4.55}$$

and, using the normalization condition,

$$\sum_n |a_n|^2 = 1. \tag{4.56}$$

This result tells us that $|a_n|^2$ is the contribution from state φ_n to the density (or probability density) of particles in state ψ. In other words, a_n represents the probability amplitude of φ_n contained in state ψ. The consistency of this interpretation will be confirmed by subsequent developments.

Given the basis $\{\varphi_n(x)\}$ we can say that the set of coefficients a_n *represents* the function $\psi(x)$ in this basis, in the sense that knowing $\{a_n\}$ is equivalent to knowing $\psi(x)$ in the domain of definition. It is obvious that for a different basis the set of coefficients $\{a_n\}$ will be different, and will constitute a different *representation* of the same function. In particular, a rotation of the Hilbert space changes the vector basis to $\{\varphi'_n(x)\}$ without changing the function $\psi(x)$. The space does not change, it is only the description that changes.

4.3.5 Nondegeneracy of the one-dimensional bounded states

Let ψ_1 and ψ_2 be two stationary degenerate eigenfunctions corresponding to the same eigenvalue E of the energy; in the 1D case we have

$$E\psi_1 = \left(-\frac{\hbar^2}{2m}\frac{d^2}{dx^2} + V\right)\psi_1, \tag{4.57}$$

and

$$E\psi_2 = \left(-\frac{\hbar^2}{2m}\frac{d^2}{dx^2} + V\right)\psi_2. \tag{4.58}$$

Multiplying the first equation by ψ_2 and the second by ψ_1 and subtracting, we get

$$E\psi_2\psi_1 - E\psi_1\psi_2 = 0 = -\frac{\hbar^2}{2m}\left(\psi_2\frac{d^2\psi_1}{dx^2} - \psi_1\frac{d^2\psi_2}{dx^2}\right)$$

$$= -\frac{\hbar^2}{2m}\frac{d}{dx}\left(\psi_2\frac{d\psi_1}{dx} - \psi_1\frac{d\psi_2}{dx}\right). \tag{4.59}$$

From here it follows after an integration that

$$\psi_2\frac{d\psi_1}{dx} - \psi_1\frac{d\psi_2}{dx} = \text{const.} \tag{4.60}$$

If there exists a point at which ψ_1 and $d\psi_1/dx$ vanish simultaneously, the constant will be zero and we will have

$$\frac{1}{\psi_2}\frac{d\psi_2}{dx} = \frac{1}{\psi_1}\frac{d\psi_1}{dx} \tag{4.61}$$

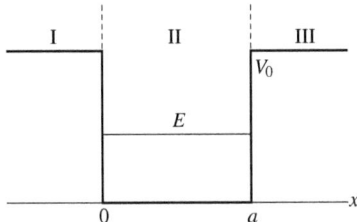

Figure 4.6 The finite square well. E is the energy of the trapped particle, which is less than the height of the well, V_0; a is the width of the well.

with solution

$$\psi_2 = a\psi_1, \tag{4.62}$$

where a is an arbitrary constant; then the two functions ψ_1 and ψ_2 are the same wave function after normalization. We conclude that in a 1D problem, if the wave function and its first spatial derivative vanish simultaneously at some point, the state is nondegenerate. In particular, for a bound system both ψ and $d\psi/dx$ vanish at infinity; thus *one-dimensional bound states* are nondegenerate.

In the case of more than one dimension, the states are distinguished by more than one index, one for each degree of freedom, and the energy is generally a function of these indices. For example, if E_n with $n = n_1 + n_2$, then the different possible combinations of n_1 and n_2 that give the same n correspond to the same energy, but to different wave functions: there is degeneracy.

In the following chapters we will have ample opportunity to use the Schrödinger equation to solve important quantum-mechanical problems. Here we apply it to a simple but physically appealing problem as a first illustration.

4.3.6 The square potential well

Imagine an electron trapped by a potential (usually called a *potential well*) as in Fig. 4.5. According to the preceding discussion, the stationary states will have a discrete energy spectrum. To determine the spectrum, we model the potential as a 1D square well, thereby simplifying the mathematical problem while retaining the essential physics (Fig. 4.6),

$$V(x) = \begin{cases} 0, & \text{if } x \in (0, a), \\ V_0, & \text{if } x \notin (0, a). \end{cases} \tag{4.63}$$

In order for the electrons to be trapped, we must take $E < V_0$.

The 1D stationary Schrödinger equation is

$$\varphi'' + \frac{2m}{\hbar^2}(E - V)\varphi = 0. \tag{4.64}$$

Due to the discontinuities of the derivative $V'(x)$ (the acting force), it is convenient to divide the x-space into three regions I, II, and III, as shown in Fig. 4.6. With the parameters k and q, both positive, defined as

$$k^2 = \frac{2m}{\hbar^2}E, \quad q^2 = \frac{2m}{\hbar^2}(V_0 - E), \quad 0 < E < V_0, \quad (4.65)$$

we have in region I ($x \leq 0$),

$$\varphi_I'' - q^2\varphi_I = 0; \quad \varphi_I = A_1 e^{-qx} + B_1 e^{qx},$$

in region II ($0 < x < a$),

$$\varphi_{II}'' + k^2\varphi_{II} = 0; \quad \varphi_{II} = A_2 \sin kx + B_2 \cos kx,$$

and in region III ($x \geq a$),

$$\varphi_{III}'' - q^2\varphi_{III} = 0; \quad \varphi_{III} = A_3 e^{q(x-a)} + B_3 e^{-q(x-a)}. \quad (4.66)$$

In each case we have written the general solution in a convenient form. Next we impose the appropriate conditions in each region. We must take $A_1 = 0$ to prevent φ_I from growing indefinitely; for a similar reason $A_3 = 0$. Therefore,

$$\begin{aligned}\varphi_I &= B_1 e^{qx}, \quad x \leq 0 \\ \varphi_{II} &= A_2 \sin kx + B_2 \cos kx, \quad 0 < x < a \\ \varphi_{III} &= B_3 e^{-q(x-a)}, \quad x \geq a.\end{aligned} \quad (4.67)$$

Now we need to *stitch*[16] the solutions together to ensure the continuity of φ and φ' at $x = 0$ and $x = a$, since the three pieces φ_K should be part of the same wave function. We therefore demand that

$$\varphi_I(0) = \varphi_{II}(0); \quad \varphi_{II}(a) = \varphi_{III}(a), \quad (4.68)$$

and

$$\varphi_I'(0) = \varphi_{II}'(0); \quad \varphi_{II}'(a) = \varphi_{III}'(a). \quad (4.69)$$

There are five conditions to be met, namely the four in Eqs. (4.68) and (4.69) and the normalization condition, but we have only four parameters at our disposal, B_1, A_2, B_2, B_3. The required fifth parameter follows from the conditioning of k, which leads to the quantization of the energy. The full solution to this and similar problems is treated in Chapter 5, so here we present the simplest possible case for illustration, which occurs when V_0 is infinite.

Exercise E4.4. The infinite square potential well
Determine the energy spectrum for a particle trapped in a rigid 1D potential well.

Solution. In the limit $V_0 \to \infty$, q becomes infinite (see Eq. (4.65)), so φ_I and φ_{III} are both zero. The wave function is then completely contained in the well (this is an exceptional situation),

$$\varphi = \begin{cases} 0, & x \leq 0 \\ \varphi_{II}, & 0 < x < a \\ 0, & x \geq a. \end{cases} \quad (4.70)$$

---

[16] Terms such as *sewing* and *splicing* are also used.

The continuity conditions for φ, Eqs. (4.68), reduce to

$$\varphi_{II}(0) = B_2 = 0; \quad \varphi_{II}(a) = A_2 \sin ka = 0, \quad \text{with } A_2 \neq 0 \quad (4.71)$$

whence k must take the values

$$k = \frac{n\pi}{a}, \quad n = 1, 2, 3, \ldots \quad (4.72)$$

The value $n = 0$ is excluded because the Schrödinger equation reduces for $k^2 = 0$ to $\varphi'' = 0$, whose solution $\varphi = Ax + B$ vanishes identically due to the boundary conditions. With (4.72) introduced into the first of Eqs. (4.65), we get for the energy eigenvalues

$$E_n = \frac{\pi^2 \hbar^2}{2ma^2} n^2. \quad (4.73)$$

The corresponding eigenfunctions are (after a change of name of the normalization constant)

$$\varphi_n = A_n \sin \frac{\pi n}{a} x, \quad (4.74)$$

with A_n fixed by the normalization condition $\int_{-\infty}^{\infty} |\varphi_n|^2 dx = 1$, which gives

$$|A_n|^2 \int_0^a \sin^2 \frac{n\pi}{a} x\, dx = \tfrac{1}{2} a |A_n|^2 = 1. \quad (4.75)$$

Taking A_n as real (i.e., setting to 0 the phase of the arbitrary factor $e^{i\alpha}$ in $A_n = |A_n| e^{i\alpha}$), we get

$$A_n = \sqrt{\frac{2}{a}}, \quad (4.76)$$

so the wave function of state n is

$$\varphi_n = \sqrt{\frac{2}{a}} \sin \frac{n\pi}{a} x \quad (4.77)$$

inside the well, and zero outside it. Due to the infinite discontinuity of $V'(x)$ at $x = 0$ and $x = a$, the continuity conditions on $\varphi'(x)$ at these points, Eqs. (4.69), do not apply. In Chapter 5 we will take a closer look at this situation and explain it physically.

The ground state of the system is by definition the state of minimum energy, namely for $n = 1$, with energy $E_1 = \pi^2 \hbar^2 / 2ma^2$. The corresponding electron density distribution,

$$\rho_1 = \varphi_1^2 = \frac{2}{a} \sin^2 \frac{\pi x}{a}, \quad (4.78)$$

is maximum at the center of the well.

An important lesson drawn from this exercise is that quantization results from the need to consistently meet the whole set of conditions that the wave function must satisfy. It is not just continuity and/or boundary conditions, as there may be other requirements. The discussion in Section A.1 of Appendix A may well be illustrative.

Note from Fig. 4.7 that the ground state has no nodes; this is a common property of the ground state, but not quite universal; see Section 15.3. Excited states φ_n for $n > 1$,

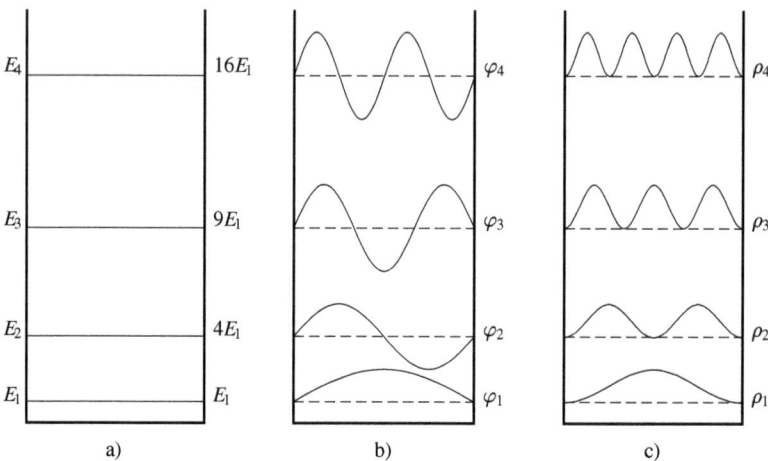

Figure 4.7 The infinite square well. (a) The first energy levels are shown in units of the ground-state energy; (b) shows the corresponding amplitudes; and (c) shows the density of electrons inside the well. Due to the repulsion from the perfectly reflecting walls, no electrons are found in their immediate vicinity.

have *nodes* (points where the electron density is zero), with the particles concentrating in the areas between them. That the number of nodes increases with the excitation of the state can be understood intuitively as follows. As a second derivative, the Laplacian term of the Schrödinger equation is proportional to the curvature of the wave function at each point, so the more nodes the wave function has, the greater the curvature. On the other hand, the Laplacian is proportional to the kinetic energy of the particle, given by $E - V$, so the greater the number of nodes, the greater the kinetic energy. Reversing the argument leads to the desired conclusion. As we move from one energy eigenstate to the next, the wave function acquires an additional node.

Exercise E4.5. On the "classical" limit $n \gg 1$

According to Bohr's *correspondence principle*,[17] the results of quantum theory agree with those of classical physics in the limit of very large quantum numbers. Show that when $n \to \infty$, the probability of the particle being in an infinite potential well at a point between x and $x + dx$ is independent of x, as is the case in classical physics.

Solution. For highly excited states ($n \gg 1$), ρ_n has a large number of nodes, and its envelope (a line segment) tends to merge with the continuum. However, the presence of nodes shows that the quantum distribution does *not* converge to the classical point-to-point uniform distribution. Nevertheless, by averaging the probability distribution over a small interval $\Delta x \ll a$ (i.e., making a coarse-grained description), one obtains

---

[17] In 1913, in an attempt to reproduce the old physics, Bohr postulated that QM reduces to classical mechanics for large quantum numbers n. The fact that Bohr's correspondence principle is generally inadequate to link classical and quantum theory has been widely discussed. However, it turns out that not necessarily the local densities, but the *locally averaged densities* of the eigenstates of the quantum system do resemble the corresponding classical density, as shown in the present exercise.

$$P(x_1 < x < x_1 + \Delta x) = \int_{x_1}^{x_1+\Delta x} |\psi(x)|^2 \, dx = \frac{2}{a} \lim_{n \to \infty} \int_{x_1}^{x_1+\Delta x} \sin^2 \frac{\pi n}{a} x \, dx$$

$$= \frac{2}{a} \int_{x_1}^{x_1+\Delta x} \frac{1}{2} dx = \frac{\Delta x}{a}, \tag{4.79}$$

which is the classical (uniform) result. From this we draw an important lesson: quantum results do not necessarily converge directly to classical ones in the limit of large quantum numbers, but a local coarse-grained description effectively corresponds to the classical result.

The solution (4.77) is an illustrative example of the wave description of the behavior of the ensemble of electrons: the two waves propagating and successively being reflected at the walls, e^{ikx} and e^{-ikx}, produce the standing wave $\sin kx$ where they superimpose and interfere with each other. To see this, rewrite Eq. (4.67) in the form $\varphi(x) = \frac{1}{2}(B_2 - iA_2)e^{ikx} + \frac{1}{2}(B_2 + iA_2)e^{-ikx} = Ae^{ikx} + A^*e^{-ikx}$. The wave function $\psi = e^{-i\omega t}\varphi(x)$ describes then the superposition of a wave $Ae^{-i(\omega t - kx)}$ propagating to the right with phase velocity ω/k and a wave $A^*e^{-i(\omega t + kx)}$ propagating to the left with the same phase velocity. As in the case of the vibrating string (formally analogous to the present one) the stationary states satisfy the condition $(\lambda/2)n = a$, that is, $k = p/\hbar = 2\pi/\lambda = \pi n/a$, which is Eq. (4.72).

The appearance of nodes with zero probability of particle presence is unexpected, at least from a classical point of view. From the (usual) orthodox perspective it points to the need to visualize the single electron as a (standing) wave distributed over the interior of the well. From an ensemble perspective, it can be understood as follows. The particles move with equal probability to the right or to the left, so that there is no effective flow at any point inside the well; the local flow velocity is zero, in line with Eq. (4.83). Since a particle cannot have both negative and positive momentum at the same time, the description must refer to different realizations of the ensemble of particles, or to an ensemble of different times in the motion of a single particle. Note that this is a particular application of the ergodic principle mentioned in Section 4.1.

4.4 Basic Operators and Their Expectation Values in Configuration Space

In the configuration-space description, the dynamical variable x is a c-number and the canonical momentum $\hat{p}$ must be replaced by the linear differential operator (see Section 3.3),

$$\hat{p} = -i\hbar \frac{\partial}{\partial x}. \tag{4.80}$$

Applied to the wave function ψ, this gives

$$\hat{p}(x)\psi = -i\hbar \frac{\partial \psi}{\partial x}, \tag{4.81}$$

which contains a real and an imaginary part,

$$\hat{p}\psi = m(v + iu)\psi, \tag{4.82}$$

where v is the flow velocity introduced in Eq. (4.5) in connection with the continuity equation,

$$v = \frac{i\hbar}{2m}\left(\frac{1}{\psi^*}\frac{\partial \psi^*}{\partial x} - \frac{1}{\psi}\frac{\partial \psi}{\partial x}\right), \tag{4.83}$$

and u is the diffusive velocity,

$$u = \frac{\hbar}{2m}\left(\frac{1}{\psi^*}\frac{\partial \psi^*}{\partial x} + \frac{1}{\psi}\frac{\partial \psi}{\partial x}\right) = \frac{\hbar}{2m}\frac{1}{\rho}\frac{\partial \rho}{\partial x}. \tag{4.84}$$

We leave the discussion of this component of the velocity to Chapter 10, except to note that it depends only on $\rho(x)$ and that it does not contribute to the mean (or expectation) value of the momentum,

$$\langle \hat{p} \rangle = -i\hbar \int \psi^* \frac{\partial \psi}{\partial x} dx = \int mv(x)\rho(x)\, dx = m\bar{v}. \tag{4.85}$$

More generally, for any dynamical variable $\hat{A}$ its expectation value is given by

$$\langle \hat{A} \rangle = \int \psi^* \hat{A} \psi\, dx. \tag{4.86}$$

This form coincides with the average value $\overline{A}$ when $\hat{A}$ depends only on position; however, as mentioned in Section 3.3.1, to calculate the correct expectation value of a dynamical variable that is a function of noncommuting factors, for example, $\hat{A}(x, \hat{p})$, one must first determine the order in which they appear.

4.4.1 Representation of some basic operators

We will use the notions just introduced to construct the expectation values of some fundamental dynamical variables in a consistent way. We start with the 3D Schrödinger equation (4.20), and note from Eq.(4.80) that the first term represents the *kinetic energy operator*,

$$\hat{T} = -\frac{\hbar^2}{2m}\nabla^2 = \frac{\hat{p}^2}{2m}, \tag{4.87}$$

and the second term represents the potential energy. Equation (4.20), multiplied from the left by ψ^* and integrated over the whole available space, yields

$$-\frac{\hbar^2}{2m}\int \psi^* \nabla^2 \psi\, dx + \int \psi^* V \psi\, dx = \int \psi^* E \psi\, dx. \tag{4.88}$$

The right-hand side is the expectation value of the energy,

$$\langle E \rangle = \int E\psi^* \psi\, dx = \int E\rho\, dx, \tag{4.89}$$

but it is also the energy eigenvalue E, since $\int \psi^* E\psi\, dx = E\int \rho\, dx = E$. This is consistent with the notion that in an energy eigenstate, each particle in the ensemble has the same energy E. As for the terms on the left-hand side of (4.88), they are the mean values of the kinetic and the potential energy, respectively, so that we can write

$$\langle \widehat{E} \rangle = \langle \widehat{T} \rangle + \langle \widehat{V} \rangle = \langle \widehat{T} + \widehat{V} \rangle. \tag{4.90}$$

The operator $\widehat{H}$ given by

$$\widehat{H} = \widehat{T} + \widehat{V} = -\frac{\hbar^2}{2m}\nabla^2 + \widehat{V} \tag{4.91}$$

is the Hamiltonian of the system, introduced in Section 3.4.

By writing the Hamiltonian operator as

$$\widehat{H} = \hat{p}^2/2m + V(\hat{x}), \tag{4.92}$$

the Schrödinger equation takes the form

$$\widehat{H}\psi = E\psi. \tag{4.93}$$

This is not just a compact way of writing the Schrödinger equation, but *a generalization of it* and *an expression of its meaning*, in that the Hamiltonian can take on a more general form, with the only restriction that it is independent of time.

4.5 General Properties of Quantum Operators and Eigenvalue Equations

We have been able to constructively deduce some important quantum operators, and we will encounter more as we go along. The ones we have already constructed ($\hat{x}, \hat{p}, \widehat{V}(x), \widehat{T}, \widehat{H}, \widehat{E}$) are all linear and Hermitian. Linear means that they satisfy the property

$$\widehat{F}(a\psi_1 + b\psi_2) = a\widehat{F}\psi_1 + b\widehat{F}\psi_2. \tag{4.94}$$

This guarantees the linearity of the theory, which in turn is necessary for the superposition of wave functions, as mentioned in Section 4.3.

Further, the operators can be multiplied (applied successively) and their multiplication is associative (as are the sums),

$$\widehat{F}\left(\widehat{G}\psi\right) = \left(\widehat{F}\widehat{G}\right)\psi = \widehat{FG}\psi. \tag{4.95}$$

As mentioned in Section 3.2, in this expression $\widehat{G}$ is applied first to ψ, followed by $\widehat{F}$. By inverting the order, instead of $\widehat{F}\widehat{G}$ one gets $\widehat{G}\widehat{F} = \widehat{F}\widehat{G} - i\hat{C}$, where $\hat{C} = -i\left[\widehat{F}, \widehat{G}\right]$.

We recall that Hermiticity means $\widehat{F}^\dagger = \widehat{F}$. The reason for this demand goes as follows. Let $\widehat{F}$ represent a variable with real average values; this means that

$$\langle \widehat{F} \rangle = \int \psi^* \widehat{F} \psi \, dx = \langle \widehat{F} \rangle^* = \int \psi \widehat{F}^* \psi^* \, dx. \tag{4.96}$$

A variable that satisfies this integral condition *for square-integrable functions* is Hermitian (or self-adjoint) by definition. The general definition of Hermicity is

$$\int \psi^* \widehat{F} \varphi \, dx = \int \varphi \widehat{F}^* \psi^* \, dx. \tag{4.97}$$

Quantum theory accepts also non-Hermitian operators; they refer to nonreal values of the variables. Non-Hermitian operators may be quite important; we will meet and use several of them. Some examples can be found in Chapters 9 and 11.

An important property of the eigenfunctions of a Hermitian operator corresponding to different eigenvalues is that they are orthogonal. In Section 4.3 we saw the example of the eigenfunctions of the Hamiltonian, which are the energy eigenvalues. More generally, we can consider two eigenstates m, n of a Hermitian operator $\widehat{F}$ and write the pair of (eigenvalue) equations

$$\widehat{F}\psi_n = f_n\psi_n, \quad \widehat{F}^*\psi_m^* = f_m\psi_m^*. \tag{4.98}$$

It follows that

$$\int \psi_m^* \widehat{F} \psi_n \, dx = f_n \int \psi_m^* \psi_n \, dx, \tag{4.99}$$

and that, being $\widehat{F}$ Hermitian,

$$\int \psi_m^* \widehat{F} \psi_n \, dx = \int \psi_n \widehat{F}^* \psi_m^* \, dx = f_m \int \psi_m^* \psi_n \, dx. \tag{4.100}$$

Taking the difference of the last two expressions gives $(f_n - f_m) \int \psi_m^* \psi_n \, dx = 0$. The solution of this equation is precisely the orthogonality condition,

$$\int \psi_m^* \psi_n \, dx = \alpha_n \delta_{mn}. \tag{4.101}$$

The constant α_n can be chosen to normalize the basis and is usually set to $\alpha_n = 1$.

Thus, we can use the set of eigenfunctions of any Hermitian operator $\widehat{F}$, which is complete and therefore constitutes a basis, to perform a "Fourier-type" expansion, as was done in Section 4.3 for the eigenfunctions of the Hamiltonian $\widehat{H}$. In particular, if $g(x)$ is a sufficiently continuous and square-integrable function and $\{\psi_n(x)\}$ is the set of eigenfunctions of $\widehat{F}$, we can write

$$g(x) = \sum_n c_n \psi_n(x), \quad \text{with} \quad c_n = \int \psi_n^*(x) g(x) \, dx. \tag{4.102}$$

The convergence properties of such expansions are similar to those of Fourier representations.

In principle, we can *prepare* a physical system to be in an eigenstate of the operator $\widehat{F}$. In QM, this preparation is interpreted as ensuring that each member of the ensemble has the same value as the dynamical variable F. Since we can in principle do this with any operator, there is an eigenvalue equation for every quantum dynamical variable.[18] Whether the system is in a given operator eigenstate or not is determined by the way it is set up and can be *partly* decided by the experimenter, as explained in Section 8.4. We will develop this topic further, but the previous results already show the central importance of eigenvalue equations in quantum theory.

---

[18] In modern QM, the term "observable" is formally used to denote any Hermitian operator that has a complete set of eigenfunctions. Therefore, physicists often speak of observables to refer to the operators that represent the dynamical variables. As explained in Chapter 3, this terminology contributes to the unwanted introduction of interpretive elements that are alien to the theory. To avoid possible confusion, we use this term in a very restricted way.

As a simple but important application of theorem (4.102), we expand the wave function ψ in terms of the orthonormal basis $\{\varphi_n\}$ in the following form (see Section 3.4)

$$\psi = \sum_n a_n \varphi_n, \qquad (4.103)$$

where the φ_n are the eigenfunctions of an operator $\widehat{F}$ (ψ is then said to be in the F-representation),

$$\widehat{F}\varphi_n = f_n \varphi_n. \qquad (4.104)$$

The coefficient $a_n = (\varphi_n, \psi)$ is the amplitude with which the state φ_n contributes to ψ. Substituting, we get

$$\langle \widehat{F} \rangle = \int \psi^* \widehat{F} \psi \, dx = \int dx \sum_n a_n^* \varphi_n^* \widehat{F} \sum_m a_m \varphi_m \qquad (4.105)$$

$$= \sum_{n,m} a_n^* a_m \int \varphi_n^* \widehat{F} \varphi_m \, dx = \sum_{n,m} a_n^* a_m f_m \int \varphi_n^* \varphi_m \, dx = \sum_{n,m} a_n^* a_m f_m \delta_{nm},$$

or finally,

$$\langle \widehat{F} \rangle = \sum_n |a_n|^2 f_n. \qquad (4.106)$$

This result says that the expectation value $\langle \widehat{F} \rangle$ is equal to the average of the eigenvalues of $\widehat{F}$, each taken with weight $|a_n|^2$. Remembering that $|a_n|^2$ is the probability that the state φ_n contributes to the state ψ, we see that (4.106) assigns to the eigenvalues f_n of the operator $\widehat{F}$ the meaning of the *contribution of the state n to the mean value* $\langle \widehat{F} \rangle$ with weight $|a_n|^2$, as suggested previously.

4.5.1 Matrix representation of operators and states

We will now apply Schrödinger's recipe for calculating a matrix element as follows.[19] Given an operator $\widehat{F}$, the matrix element F_{mn} in the vector basis $\{\varphi_n\}$ of some generic operator $\widehat{Q}$, that is, $\widehat{Q}\varphi_n = q_n \varphi_n$, is obtained by integrating the product $\varphi_m^* \widehat{F} \varphi_n$ over the domain,

$$F_{mn} \equiv \int \varphi_m^* \widehat{F} \varphi_n \, dx. \qquad (4.107)$$

One can thus construct a square matrix with elements F_{mn} representing the operator $\widehat{F}$ *in this* basis. We refer to this by saying that $\hat{F}$ is in the Q-representation. Note that each operator $\widehat{F}$ can be represented by a large number of different matrices, one for each Hermitian operator $\widehat{Q}$ that can be used to build the basis of the representation.

A particular but important result of formula (4.107) is obtained when $\widehat{F} = \widehat{Q}$, that is, when $\widehat{F}$ *is in its own representation*. Combining (4.107) with the eigenvalue equation for $\widehat{Q}$ and using $q_n = f_n$, it follows that

---

[19] It was Schrödinger who, in 1926, explained how to construct the Heisenberg matrices, thus bridging the gap between the two theories and demonstrating their mathematical equivalence. This was a crucial step, as physicists at the time were faced with two apparently very different and unrelated theories, and no elements to decide between them – personal inclinations notwithstanding.

$$F_{mn} \equiv \int \varphi_m^* \widehat{F} \varphi_n \, dx = f_n \int \varphi_m^* \varphi_n \, dx = f_n \delta_{mn}. \tag{4.108}$$

We see that $\widehat{F}$ is represented by a diagonal matrix, that is, *every operator is diagonal in its own representation*. This is a simple but important result. It means that finding the vector basis that diagonalizes the matrix $\widehat{F}$ is equivalent to solving the eigenvalue equation $\widehat{F}\varphi_n = f_n \varphi_n$.

It is easy to verify that the matrix representing a Hermitian operator is Hermitian. Indeed, combining the definition of the Hermitian operator, Eq. (4.96), and Eq. (4.107), which gives the matrix elements of the operator, it follows that

$$F_{mn} = \int \psi_m^* \widehat{F} \psi_n \, dx = \int \psi_n \widehat{F}^* \psi_m^* \, dx = \left(\int \psi_n^* \widehat{F} \psi_m \, dx \right)^* = F_{nm}^*, \tag{4.109}$$

or

$$F_{mn} = F_{nm}^*. \tag{4.110}$$

In matrix notation this means that

$$F = F^{T*} = F^{*T} = F^\dagger. \tag{4.111}$$

It is clear that with the given definitions, the algebraic relationships between operators are preserved when transcribed in terms of matrices. For example, the addition $\widehat{C} = \widehat{A} + \widehat{B}$ transforms into

$$C_{mn} \equiv \int \varphi_m^* \widehat{C} \varphi_n \, dx = \int \varphi_m^* \left(\widehat{A} + \widehat{B} \right) \varphi_n \, dx = A_{mn} + B_{mn}. \tag{4.112}$$

Since the relation is true for each matrix element, it is true for the matrix. The case of the product is an attractive exercise for the student.

Based on this equivalence, we will repeatedly manipulate relations between operators by following the rules of matrix algebra. For example, the adjoint of the operator $\widehat{A}\widehat{B}$ is $B^\dagger A^\dagger$, since the adjoint of a matrix product is the product of its adjoints in reverse order. Similarly, using the bases $\{\varphi_k\}$ and $\{a_k\}$, one gets

$$(\psi, \varphi) = \int \psi^* \varphi \, dx = \sum_{k,l} \int b_k^* \varphi_k^* a_l \varphi_l \, dx = \sum_{k,l} b_k^* a_l \delta_{kl} = \sum_k b_k^* a_k = \widehat{b}^\dagger \widehat{a}. \tag{4.113}$$

This is an example of application of the expansion theorem, see Section 4.3.4. In Chapter 3 we had an example of vectors and operators in a two-dimensional space; the same procedure applies to any number of dimensions.

To recapitulate, operators are represented by square matrices, and wave functions (vectors) are represented by column matrices, both in the same representation. The Hilbert space is the normed linear vector space spanned by a complete set of orthonormal complex vectors, with the inner product as defined in Section 4.3. In particular, $\{\varphi_k\}$ can be the set of eigenfunctions of some Hermitian operator of interest (such as the Hamiltonian), since they are then orthonormal. The Hilbert space constructed with the eigenstates of a quantum system is called the *state space of the system*. The operators act on this space, transforming one vector into another, that is, they represent a mapping of this space onto itself. It follows from the preceding that it is possible to construct different representations of this space by using different operators to define the basis.

4.6 Dirac's Symbolic Notation

We have discussed two different descriptions, one in purely algebraic terms, à la Heisenberg, the other in terms of wave functions in configuration space, à la Schrödinger. It is therefore natural to look for a description of the state vectors and their relations that does not rely on a particular representation. The result must contain exactly the desired information (that of the state); when necessary, we can resort to a specific representation to give concrete form to the expressions.

Such abstract formulation of QM, which is the idea behind the *theory of representations,* was initially developed by Dirac[20] and has achieved great importance (it is the natural choice in quantum field theory), because it represents a simpler and more transparent form of expression of the theory. Here we shall present it at a relatively informal introductory level; for a systematic and rigorous treatment, an advanced text on QM can be consulted.

4.6.1 The Dirac notation: States and operators

To construct the abstract description it is useful to refer to Heisenberg's representation in terms of matrices in a Hilbert space.[21] They will serve as a guide for the algebraic properties of the elements to be constructed. The resulting formalism must contain as basic elements those common to both Heisenberg's and Schrödinger's descriptions.

In the Dirac symbolic notation, the vector ψ in a Hilbert space $\mathcal{H}$ is denoted by the symbol $|\psi\rangle$ and is called a *ket*. The adjoint vector $\psi^\dagger$ is represented by the symbol $\langle\psi|$ and is called a *bra*; by definition $\langle\psi| = (|\psi\rangle)^\dagger = |\psi\rangle^\dagger$, with $\langle\psi|$ belonging to the dual space to $\mathcal{H}$. Then if ψ is represented by a column vector, say – as exemplified in Section 3.7 – its adjunct $\psi^\dagger$ becomes a row vector with conjugate elements; the product $\langle\varphi|\psi\rangle$ should result in a number (a scalar), i.e., it is a dot product. Translated into the new notation, the product of a bra by a ket (in *the order* bra-ket) is the *scalar product*,

$$(\varphi, \psi) \to \langle\varphi|\psi\rangle = \text{scalar}, \tag{4.114}$$

which, therefore, must satisfy the appropriate commutativity property. Since $|\psi\rangle^\dagger = \langle\psi|$, this implies that

$$\langle\varphi|\psi\rangle = (\langle\psi|\varphi\rangle)^* = \langle\psi|\varphi\rangle^*. \tag{4.115}$$

In particular, it follows that $\langle\varphi|\varphi\rangle$ is real and equal to the square of the norm of $|\varphi\rangle$. The appearance of the triangular bracket $\langle\cdot|\cdot\rangle$ is the origin of the terms "bra" and "ket," deliberately devoid of any unrelated meaning: a ket is a ket, and a bra is a bra.

---

[20] The first complete exposition of the quantum formalism was published by Dirac in 1930, in his classic book (Dirac, 1930). A couple of years earlier he had invented the compact and elegant *bra* and *ket* symbolic notation.

[21] In wave mechanics, the Hilbert space of interest is $L^2(\mathbf{R})$, constituted by the square-integrable complex functions in configuration space, and dot product $(f, g) = \int_\mathbf{R} dx f^*(x) g(x)$. In matrix mechanics, the Hilbert space is constituted by the space of infinite square-summable sequences, $l_2 = \left\{\vec{x} = (x_1, x_2, \ldots) \,|\, x_k \in \mathbf{C}, \sum_{k=1}^\infty |x_k|^2 < \infty\right\}$ with inner product $(\vec{x}, \vec{y}) = \sum_{k=1}^\infty x_k^* y_k$. In practice, the vectors used in QM are normally restricted to a subspace of the system's Hilbert space, but we won't be concerned about this.

Let us now consider the operators that act on the state vectors. An operator acting on a ket transforms it into another ket, both in the same Hilbert space; for instance,

$$|\psi\rangle = \widehat{F}|\varphi\rangle \equiv |\widehat{F}\varphi\rangle. \tag{4.116}$$

The corresponding expression for a bra is

$$\langle\psi| = \langle\varphi|\hat{F} = \langle\hat{F}^\dagger\varphi|. \tag{4.117}$$

Since $\widehat{F}|\varphi\rangle = |\widehat{F}\varphi\rangle$ is a ket, we can multiply it *scalarly* by another ket $|\chi\rangle$, in the form $\langle\chi|\psi\rangle = \langle\chi|F|\varphi\rangle$; in analogous form we can introduce other algebraic operations. When two bars meet, only one is written, $\langle\chi||\psi\rangle = \langle\chi|\psi\rangle$. In its turn, if necessary we can split the inner separator in two; for example, $\langle\varphi|\psi\rangle = \langle\varphi|1|\psi\rangle$, or $\langle\chi|\widehat{F}\varphi\rangle = \langle\chi|\widehat{F}|\varphi\rangle$.

A basis for the Hilbert space of interest consists of a complete set of kets, $\{|a_i\rangle\}$ which we will take as orthonormal. Therefore, for discrete indices we write

$$\langle a_i|a_j\rangle = \delta_{ij}, \tag{4.118}$$

while for continuous indices we must write

$$\langle q'|q''\rangle = \delta(q' - q''). \tag{4.119}$$

We are postulating that each state of the system is represented by a vector of the Hilbert space under consideration, but it is not necessary to postulate that each vector of this space corresponds to a physical state (although some authors do). Similarly, it is considered that every dynamical variable is represented by a self-adjoint operator, but not every Hermitian operator (i.e., a potential observable) necessarily represents a variable of physical interest.

A ket $|\psi\rangle$ can be expanded in terms of the basis vectors $\{|a_n\rangle\}$ in the usual way,

$$|\psi\rangle = \sum_k c_k |a_k\rangle. \tag{4.120}$$

To calculate the expansion coefficients, we multiply this expression by $\langle a_i|$ from the left and take advantage of the orthonormality of the basis to obtain

$$\langle a_i|\psi\rangle = \sum_k c_k \langle a_i|a_k\rangle = \sum_k c_k \delta_{ik} = c_i, \tag{4.121}$$

so in general, and as before,

$$c_k = \langle a_k|\psi\rangle. \tag{4.122}$$

We now insert (4.122) into (4.120) and rearrange to get

$$|\psi\rangle = \sum_k |a_k\rangle c_k = \sum_k |a_k\rangle\langle a_k|\psi\rangle. \tag{4.123}$$

Since this relation is valid for any ket $|\psi\rangle$, it means that

$$\sum_k |a_k\rangle\langle a_k| = \mathbb{I}. \tag{4.124}$$

($\mathbb{I}$ stands for the unit operator on the corresponding Hilbert space.) This is the *completeness* (or *closure*) relation (condition) for discrete variables in Dirac's symbolic notation; it is usually referred to as a *spectral decomposition of the unit*. In the case of a continuous variable q, the sum becomes an integral,

$$\int |q\rangle\langle q| dq = \mathbb{I}. \tag{4.125}$$

To convince ourselves of the validity of this expression, we multiply it from the right by $|q'\rangle$, to obtain

$$\int |q\rangle\langle q|q'\rangle dq = |q'\rangle; \tag{4.126}$$

since $\langle q|q'\rangle = \delta(q-q')$, this equation is identically satisfied.

Equation (4.124) is very helpful because it is possible to insert the quantity $\sum_i |\alpha_i\rangle\langle\alpha_i|$ into any expression without altering its value; this provides a practical means of transforming expressions from one basis to another. For example, to expand $|\psi\rangle$ in terms of the basis $|a\rangle$ we multiply it from the left by the unit written in the form $\sum_i |a_i\rangle\langle a_i|$, which gives, after rearranging,

$$|\psi\rangle = \left(\sum_i |a_i\rangle\langle a_i|\right)|\psi\rangle = \sum_i |a_i\rangle\langle a_i|\psi\rangle = \sum_i \langle a_i|\psi\rangle |a_i\rangle \equiv \sum_i c_i |a_i\rangle, \tag{4.127}$$

where the numbers $c_i = \langle a_i|\psi\rangle$ are the coefficients of the expansion; see Eq. (4.122). We can now consider the vector $|\psi\rangle$ to be represented by the (frequently infinite) set of numbers $\{c_0, c_1, c_2, \ldots\}$, which will play the role of coordinates in Hilbert space. This is the A-representation of $|\psi\rangle$, where $\hat{A}|a_i\rangle = \alpha_i |a_i\rangle$.

4.6.2 Change of representation: Transition to the Schrödinger description

To establish contact with Schrödinger's description, we compare the preceding results with their analogs written in terms of wave functions. In Dirac's version we have

$$|\psi\rangle = \sum_k c_k |a_k\rangle, \quad c_k = \langle a_k|\psi\rangle, \tag{4.128}$$

while in terms of wave functions in x space this is written

$$\psi(x) = \sum_k c_k \varphi_k(x), \quad c_k = \int \varphi_k^*(x)\psi(x)\,dx. \tag{4.129}$$

To relate the two expressions, in $c_k = \langle a_k|\psi\rangle$ we insert an expansion of the unit in terms of the basis $\{|x\rangle\}$, $\int dx |x\rangle\langle x| = \mathbb{I}$, which gives

$$c_k = \langle a_k|\psi\rangle = \int \langle a_k|x\rangle\langle x|\psi\rangle\,dx. \tag{4.130}$$

Equating this expression with the corresponding one in terms of the basis $\{\varphi_k(x)\}$ gives $c_k = \int \varphi_k^*(x)\psi(x)dx = \int \langle a_k|x\rangle\langle x|\psi\rangle\,dx$, which allows one to make the identifications (consider that $\langle x|\psi\rangle$ is a scalar, a function ψ at point x),

$$\psi(x) = \langle x|\psi\rangle, \tag{4.131}$$

$$\varphi_k^*(x) = \langle a_k|x\rangle. \tag{4.132}$$

Using Eq. (4.115), we see that $\langle a_k|x\rangle = \langle x|a_k\rangle^*$, so (4.132) can be written as a particular case of (4.131),

$$\varphi_k(x) = \langle x|a_k\rangle. \tag{4.133}$$

These important results establish the relationship between the ket $|\psi\rangle$ and the corresponding wave function $\psi(x)$ in x-space. From the abstract state vector $|\psi\rangle$ we have constructed the state wave function $\psi(x)$ in the x-representation as the scalar product $\langle x|\psi\rangle$; an abstract vector is thus transformed into a (scalar) function of x, which is precisely the wave function $\psi(x)$.

Equation (4.133) shows also how to construct the basis in the x-representation $\varphi_k(x)$ from the abstract basis $|a_k\rangle$. The set $\{\varphi_k(x)\}$ corresponds to $\{|a_k\rangle\}$ *for each value* of x. It is clear that if, instead of introducing the expansion $\mathbb{I} = \int dx|x\rangle\langle x|$ in the expression for c_k, we use the variable p, say, we obtain the wave function of the state ψ in the p-description (or representation),

$$\psi(p) = \langle p|\psi\rangle, \tag{4.134}$$

and, of course,

$$\varphi_k(p) = \langle p|a_k\rangle. \tag{4.135}$$

This shows how to obtain wave functions from state vectors in *any* representation, taking advantage of the fact that the latter do not imply a reference to a particular representation or basis. Wave functions, on the other hand, are specific representations of the state vectors, explicitly dependent on the variables and basis used for the description. By writing the Schrödinger equation from the beginning in the configuration space x, the x-description was selected a priori. This explains the observed asymmetry between the treatment given to x and other possible variables, such as the linear momentum p.

As a further example of calculus with the symbolic notation, we derive the important *expansion theorem* (see Section 4.3.4.1). Given two kets $|f\rangle$ and $|g\rangle$, we expand them in the same basis $|n\rangle$,

$$|f\rangle = \sum_n a_n|n\rangle; \quad |g\rangle = \sum_n b_n|n\rangle. \tag{4.136}$$

From here it follows that

$$\langle f|g\rangle = \sum_{n,m}\langle n|a_n^* b_m|m\rangle = \sum_{n,m} a_n^* b_m \langle n|m\rangle = \sum_n a_n^* b_n = (a,b), \tag{4.137}$$

which is precisely the expansion theorem

$$\langle f|g\rangle = (a,b). \tag{4.138}$$

In the preceding proof we used the fact that if $|q\rangle = a|n\rangle$, then $\langle q| = a^*\langle n|$, since $\langle q| = |q\rangle^\dagger$ The preceding result confirms that the value of the dot (scalar) product is independent of the representation.

Exercise E4.6. Two representations of the same state vector

As a complementary exercise to the previous one, find the equation relating two different representations of the same state vector $|f\rangle$.

Solution. Let the bases be $\{|a_l\rangle\}$ and $\{|b_k\rangle\}$, and write

$$|f\rangle = \sum_l \varphi_l |a_l\rangle = \sum_k \chi_k |b_k\rangle. \tag{4.139}$$

Each of the vectors $|b_k\rangle$ can in turn be expressed in the basis $\{|a_l\rangle\}$,

$$|b_k\rangle = \sum_l t_{lk} |a_l\rangle. \tag{4.140}$$

Combining, we get $|f\rangle = \sum_l \varphi_l |a_l\rangle = \sum_{k,l} \chi_k t_{lk} |a_l\rangle$; since this is valid for every $|a_l\rangle$, it follows that

$$\varphi_l = \sum_k t_{lk} \chi_k. \tag{4.141}$$

In terms of matrices, $\varphi = T\chi$. This expression, or, if you prefer, (4.141) gives the general relation between two representations of the same vector; it is the recipe for making a change of representation, with T being the operator that performs the transformation. Since according to Eq. (4.140), $b = Ta$, the matrix T transforms the basis a into b. An interesting and simple exercise left to the reader is to show that the matrix T is unitary, that is, that $T^\dagger = T^{-1}$. This important topic is discussed further in Section 8.6.

The student may mistakenly assume that because a ket can be re-expressed as a wave function in any space (x, or p, or whatever), the information contained in the ket is more than that contained in any of the corresponding wave functions. This is not the case: the Dirac symbolic notation is just that, a notation. It does not add extra physics to what is already contained in the wave functions, but it does make the algebra more transparent and easier to use. Likewise, the transition from one representation to another, as just explained, is similar to a change of coordinates: no *new* physical information is added.

4.6.3 Dirac's representation of operators

Consider an operator $\widehat{F}$ that transforms the ket $|a_j\rangle$, eigenvector of an operator $\widehat{A}$, into a new ket $|b_j\rangle$. Expressing the latter in terms of the basis a and denoting the coefficients with F_{kj} as before, we get $|b_j\rangle = \widehat{F}|a_j\rangle = \sum_k F_{kj} |a_k\rangle$. From this it follows that

$$\langle a_i | b_j \rangle = \langle a_i | \widehat{F} | a_j \rangle = \sum_k F_{kj} \langle a_i | a_k \rangle = F_{ij}, \tag{4.142}$$

whence the matrix elements of $\widehat{F}$ in the A-representation are given by

$$F_{ij} = \langle a_i | \widehat{F} | a_j \rangle. \tag{4.143}$$

This key result allows us to rewrite the $\widehat{F}$ operator itself in a form that is convenient for performing calculations and transformations. Introducing to the left and to the right of $\widehat{F}$ two spectral expansions of the unit in terms of the basis a, we obtain

$$\widehat{F} = \sum_{i,j} |a_i\rangle \langle a_i | \widehat{F} | a_j \rangle \langle a_j |, \tag{4.144}$$

and, using (4.143),

$$\widehat{F} = \sum_{i,j} F_{ij} |a_i\rangle \langle a_j|, \qquad (4.145)$$

which shows that the product of a ket by a bra *in this order*, $|\cdot\rangle\langle\cdot|$, *is an operator*, as we have already noted, and, moreover, that its elements

$$\widehat{O}_{ij} = |a_i\rangle\langle a_j|, \qquad (4.146)$$

can be used to express *any* operator acting on the *same* space in the form $\widehat{F} = \sum_{i,j} F_{ij} \widehat{O}_{ij}$; the set $\{\widehat{O}_{ij}\}$ constitutes thus a complete basis for the representation of the operators. The expansion coefficients F_{ij} are the matrix elements i,j of $\widehat{F}$ in the representation used, given by (4.143).

If the number of dimensions of the Hilbert space is finite, say N, there is only a finite number of linearly independent elementary operators $\widehat{O}_{ij}$, equal to N^2; thus in a Hilbert space of N dimensions, at most N^2 linearly independent dynamical variables can be introduced. Note that the powers and products of the elementary operators are expressed as linear combinations of themselves. For example,

$$\widehat{O}_{ij} \widehat{O}_{kl} = |a_i\rangle\langle a_j| a_k\rangle\langle a_l| = \delta_{jk} \widehat{O}_{il}, \quad \widehat{O}_{ij}^2 = \delta_{ij} \widehat{O}_{ij}. \qquad (4.147)$$

This shows, in particular, that for $i \neq j$, the operators $\widehat{O}_{ij}$ are square zero and have zero trace.

For $i = j$, the operators are idempotent, $\widehat{O}_{ii}^k = \widehat{O}_{ii}$ for any integer k; these are the N *projectors* or *projection operators*, which are conventionally denoted as

$$\widehat{P}_i \equiv \widehat{O}_{ii} = |a_i\rangle\langle a_i|, \qquad (4.148)$$

and are so called because acting on any vector $|\psi\rangle$, $\widehat{P}_i$ gives its component $\langle a_i|\psi\rangle$ in the direction $|a_i\rangle$,

$$\widehat{P}_i |\psi\rangle = |a_i\rangle\langle a_i|\psi\rangle = \langle a_i|\psi\rangle |a_i\rangle = c_i |a_i\rangle, \quad c_i = \langle a_i|\psi\rangle. \qquad (4.149)$$

Other features of projectors are discussed in Section 8.6.4.

Exercise E4.7. Dirac's notation in 2D Hilbert space

Express the 2D Hilbert space description introduced in Section 3.7 in terms of the Dirac notation.

Solution. The two basis vectors and their adjoint are denoted by

$$\begin{pmatrix} 1 \\ 0 \end{pmatrix} = |+\rangle, \quad \begin{pmatrix} 0 \\ 1 \end{pmatrix} = |-\rangle, \qquad (4.150)$$

$$(1\ \ 0) = \langle +|, \quad (0\ \ 1) = \langle -|. \qquad (4.151)$$

The orthonormality conditions take the form

$$\langle +|+\rangle = 1; \quad \langle -|-\rangle = 1, \qquad (4.152)$$

$$\langle +|-\rangle = 0; \quad \langle -|+\rangle = 0, \qquad (4.153)$$

and an arbitrary two-dimensional state vector is written as

$$|\psi\rangle = \begin{pmatrix} a \\ b \end{pmatrix} = a|+\rangle + b|-\rangle. \tag{4.154}$$

The projection operators (4.148) are given in the different notations by

$$\hat{P}_+ = |+\rangle\langle+| = \begin{pmatrix} 1 & 0 \\ 0 & 0 \end{pmatrix}; \quad \hat{P}_- = |-\rangle\langle-| = \begin{pmatrix} 0 & 0 \\ 0 & 1 \end{pmatrix}, \tag{4.155}$$

and their completeness relation is expressed as

$$|+\rangle\langle+| + |-\rangle\langle-| = \begin{pmatrix} 1 & 0 \\ 0 & 1 \end{pmatrix} = \mathbb{I}. \tag{4.156}$$

We now apply the Dirac notation to express the effect of the Pauli matrices $\hat{\sigma}_i$ ($i = 1, 2, 3$) introduced in Section 3.7,

$$\begin{pmatrix} 0 & 1 \\ 1 & 0 \end{pmatrix} \equiv \hat{\sigma}_1, \quad \begin{pmatrix} 0 & -i \\ i & 0 \end{pmatrix} \equiv \hat{\sigma}_2, \quad \begin{pmatrix} 1 & 0 \\ 0 & -1 \end{pmatrix} \equiv \hat{\sigma}_3. \tag{4.157}$$

Acting on the state vectors $|+\rangle$, $|-\rangle$, the $\hat{\sigma}_i$ produce the following results,

$$\hat{\sigma}_1|\pm\rangle = |\mp\rangle, \quad \hat{\sigma}_2|\pm\rangle = \mp i|\mp\rangle, \quad \hat{\sigma}_3|\pm\rangle = \pm|\pm\rangle. \tag{4.158}$$

While $\hat{\sigma}_1$ and $\hat{\sigma}_2$ change the state of the system from $|+\rangle$ to $|-\rangle$ and vice versa, $\hat{\sigma}_3$ gives us the two possible eigenvalues ($+1$ or -1); thus the vector basis $|+\rangle$, $|-\rangle$ is a good one for the operator $\hat{\sigma}_3$ (it is its proper basis).

4.6.4 Some basic theorems

When the operator $\widehat{F}$ is self-adjoint, $\widehat{F} = \widehat{F}^\dagger$, Eq. (4.143) gives

$$\widehat{F} = \sum_{i,j} \widehat{F}_{ij}|a_i\rangle\langle a_j| = \widehat{F}^\dagger = \sum_{i,j} F_{ij}^*|a_j\rangle\langle a_i| = \sum_{i,j} F_{ji}^*|a_i\rangle\langle a_j|, \tag{4.159}$$

where the last equality was obtained by exchanging the dummy indices i and j. Comparing coefficients we see that $F_{ij} = F_{ji}^*$, which shows that the matrix representing a Hermitian operator is Hermitian, as already noted in Chapter 3.

Let us assume we wish to solve the eigenvalue equation $\widehat{F}|\psi\rangle = \lambda|\psi\rangle$, with $|\psi\rangle = \sum c_k|a_k\rangle$ and $c_k = \langle a_k|\psi\rangle$. Using Eq. (4.145), we write

$$\sum_{i,j} F_{ij}|a_i\rangle\langle a_j|\psi\rangle = \sum_{i,j} F_{ij}c_j|a_i\rangle = \sum_i \lambda c_i|a_i\rangle, \tag{4.160}$$

which leads to the system of equations

$$\sum_j (F_{ij} - \lambda\delta_{ij})c_j = 0. \tag{4.161}$$

This shows that the eigenvalues λ are obtained by diagonalizing the matrix F, that is, changing from the basis $|a_k\rangle$ to a basis in which the matrix F is diagonal $F_{ij} = \lambda\delta_{ij}$; this is by definition the *proper representation* of the operator $\widehat{F}$. The diagonalization is done by solving the set of simultaneous equations (4.161) (one equation for each dimension). The eigenvalues λ are obtained from the condition that the determinant of the system (4.161) vanishes; this is the so-called *secular equation*. It is clear that finding an orthonormal vector basis that diagonalizes the matrix F can be an important task, because it gives us the eigenvalues of the operator and the corresponding eigenstates (eigenvectors). In Chapter 13 we will have the opportunity to use the diagonalization method to solve specific problems.

As just stated, an operator is in its own (proper) representation when its eigenvectors are used as the basis. Indeed, from the eigenvalue equation $\widehat{F}|n\rangle = f_n|n\rangle$ it follows that

$$F_{n'n} = \langle n'|\widehat{F}|n\rangle = \langle n'|f_n|n\rangle = f_n\langle n'|n\rangle = f_n\delta_{nn'}, \qquad (4.162)$$

and the matrix is diagonal, with the diagonal elements corresponding to the eigenvalues of the operator. This result can also be derived from Eq. (4.161), since with $|a_n\rangle = |n\rangle$ there is only one nonzero, for which (4.161) gives $F_{ij} = \lambda \delta_{ij} = f_i \delta_{ij}$.

The eigenvalues corresponding to different eigenstates of $\widehat{F}$ may be different or equal; in the latter case one speaks of *degeneracy of the state* (with respect to $\widehat{F}$). For a degeneracy of order g there are g different eigenfunctions of $\widehat{F}$ that correspond to the same eigenvalue. As will be seen in later chapters, in two- or three-dimensional problems and in systems with two or more equal particles it can happen that two or more different eigenstates of $\widehat{F}$ share the same eigenvalue; in 1D problems there is a one-to-one correspondence between the eigenvalues and the eigenstates of $\widehat{F}$.

More generally, for a given state $|\psi\rangle$ (not necessarily an eigenstate), the quantity

$$\langle \psi|\widehat{F}|\psi\rangle = \langle \widehat{F}\rangle \equiv \bar{F} \qquad (4.163)$$

is the *expectation value (expected value* or *mean value)* of $\widehat{F}$ in such a state. Therefore, the eigenvalues f_n coincide with the corresponding expectation values in the eigenvector (*proper*) basis. Note from (4.145) that any operator $\widehat{F}$ can be represented in terms of its eigenvectors and eigenvalues in the form

$$\widehat{F} = \sum_n f_n |n\rangle\langle n|, \qquad (4.164)$$

where $\widehat{P}_n \equiv \widehat{O}_{nn} \equiv |n\rangle\langle n|$ are the projectors defined in (4.148). This is the spectral decomposition of $\widehat{F}$. A particular case is obtained by taking $\widehat{F} = 1$, which gives the spectral decomposition of the unit, Eq. (4.124).

Now let $\widehat{F}$ be an operator acting on the variable x, which we indicate with an appropriate index, $\widehat{F}_x$. In the element $\langle x'|\widehat{F}_x|x\rangle$, x' behaves as a parameter, which allows us to write, using the orthonormality of the basis,

$$\langle x'|\widehat{F}_x|x\rangle = \widehat{F}_x \langle x'|x\rangle = \widehat{F}_x \delta(x' - x). \qquad (4.165)$$

The $\widehat{F}$ operator is in the x-representation; its matrix element between two states $\varphi_m(x')$ and $\varphi_n(x)$ is obtained by inserting the spectral expansions of the unit twice, $\int dx |x\rangle\langle x| = \mathbb{I}$, as follows,

$$\langle \varphi_m|\widehat{F}_x|\varphi_k\rangle = \int dx \int dx' \langle \varphi_m|x'\rangle\langle x'|\widehat{F}_x|x\rangle\langle x|\varphi_k\rangle \qquad (4.166)$$

$$= \int dx \int dx' \varphi_m^*(x')\widehat{F}_x \delta(x'-x)\varphi_k(x) = \int \varphi_m^*(x)\widehat{F}_x \varphi_k(x) dx,$$

in agreement with Eq. (4.143).

Finally, let us consider two operators $\widehat{A}$ and $\widehat{B}$ in the same Hilbert space; their product is obtained using (4.145),

$$\widehat{C} \equiv \widehat{A}\widehat{B} = \sum_{ij}\sum_{kl} A_{ij} B_{kl} |i\rangle\langle j|k\rangle\langle l| = \sum_{ij}\sum_{kl} A_{ij} B_{kl} \delta_{jk} |i\rangle\langle l|$$

$$= \sum_{il} \left(\sum_j A_{ij} B_{jl} \right) |i\rangle\langle l| \equiv \sum_{il} C_{il} |i\rangle\langle l|. \tag{4.167}$$

Therefore, the matrix elements of $\widehat{C}$ are calculated with the usual rule of matrix multiplication,

$$C_{ik} = \sum_j A_{ij} B_{jk}. \tag{4.168}$$

Exercise E4.8. Properties of commuting operators

Show that if two operators commute, the matrices representing them in any basis (in any representation) commute and have common eigenvectors.

Solution. Let $\widehat{A}$ and $\widehat{B}$ be the two operators. We denote the respective matrix elements in an arbitrary basis $\{|i\rangle\}$ by a_{ij}, b_{ij}. Applying Eq. (4.145), we obtain

$$[\widehat{A}, \widehat{B}] = \sum_{ij}\sum_{kl} a_{ij} b_{kl} \left(|i\rangle\langle j|k\rangle\langle l| - |k\rangle\langle l|i\rangle\langle j| \right) = \sum_{ijk} \left(a_{ik} b_{kj} - a_{kj} b_{ik} \right) |i\rangle\langle j|. \tag{4.169}$$

Therefore, the condition $[\widehat{A}, \widehat{B}] = 0$ leads to

$$\sum_k (a_{ik} b_{kj} - b_{ik} a_{kj}) = (ab)_{ij} - (ba)_{ij} = 0, \tag{4.170}$$

which shows that the matrices a, b commute. Now let $|n\rangle$ be the eigenvectors of $\widehat{A}$. In this basis we can write

$$\widehat{A} = \sum_n A_n |n\rangle\langle n|, \quad \widehat{B} = \sum_{n,m} B_{nm} |n\rangle\langle m|, \tag{4.171}$$

and the product $\widehat{A}\widehat{B}$ is therefore

$$\widehat{A}\widehat{B} = \sum_{l,n,m} A_n B_{lm} |n\rangle\langle n|l\rangle\langle m| = \sum_{n,m} A_n B_{nm} |n\rangle\langle m|, \tag{4.172}$$

while $\widehat{B}\widehat{A}$ is

$$\widehat{B}\widehat{A} = \sum_{l,n,m} B_{nm} A_l |n\rangle\langle m|l\rangle\langle l| = \sum_{mn} A_m B_{nm} |n\rangle\langle m|. \tag{4.173}$$

Taking the difference and setting the commutator to zero gives

$$[\widehat{A}, \widehat{B}] = 0 = \sum_{n,m} (A_n - A_m) B_{nm} |n\rangle\langle m|. \tag{4.174}$$

Multiplying on the left by $\langle k|$ and on the right by $|l\rangle$ and summing over k and l gives $(A_k - A_l) B_{kl} = 0$, the solution of which is $B_{kl} = B_k \delta_{kl}$, showing that $\widehat{B}$ is diagonal in the proper basis of $\widehat{A}$.

Problems

P4.1 Consider two functions ψ_1 and ψ_2, each corresponding to a Gaussian distribution of particles,

$$\psi_1 = A_1 e^{-(x-a_1)^2/4\sigma^2}, \quad \psi_2 = A_2 e^{-(x-a_2)^2/4\sigma^2}, \quad a_2 > a_1. \quad (4.175)$$

(a) Determine the normalization coefficients A_1 and A_2. Calculate $\bar{x}$ and $\overline{x^2}$ when the particle density is $\rho_1 = |\psi_1|^2$. What is the meaning of the parameters a and σ that appear in the normal distributions?

(b) Two new functions are constructed, ψ_+ and ψ_- defined as $\psi_\pm = A_\pm(\psi_1 \pm \psi_2)$. Determine the normalization constants A_+ and A_-. Construct and plot the particle densities $\rho_+ = |\psi_+|^2$, $\rho_- = |\psi_-|^2$. Discuss the two limiting cases $\sigma \to \infty, \sigma \to 0$.

P4.2 Consider a system with spherical symmetry, described by the probability amplitude $\psi(r) = N \exp(-r/a)$, where r is the radial coordinate, with $r = [0, \infty)$. Determine the value of the factor N so that ψ is normalized to unity and find the mean value of r and of r^2 for this distribution. Relate the results to the parameters of the distribution.

P4.3 A given function $\psi(x)$ is defined on the interval $(-\pi, \pi)$ in the form of a Fourier series

$$\psi(x) = \frac{A_0}{\sqrt{2\pi}} + \sum_{n=1}^{\infty} \left(\frac{A_n}{\sqrt{\pi}} \cos nx + \frac{B_n}{\sqrt{\pi}} \sin nx \right). \quad (4.176)$$

Using the orthogonality properties of the sine and cosine functions, obtain the coefficients A_0, A_n, and B_n for this $\psi(x)$.

P4.4 Determine the Fourier integral transform of the following functions:

(a) The square function $F(x) = \begin{cases} a, & |x| \leq d/2, \\ 0, & \text{otherwise.} \end{cases}$

(b) The wave packet $F(x) = \begin{cases} ae^{-iqx}, & |x| \leq d/2, \\ 0, & \text{otherwise.} \end{cases}$

(c) The Lorentzian distribution $F(x) = \frac{\delta}{\pi} \frac{1}{\delta^2 + x^2}$.

(d) The Gaussian distribution $F(x) = \frac{1}{\sqrt{2\pi\Delta^2}} e^{-x^2/2\Delta^2}$.

P4.5 Consider a function $f(x)$ that can be integrated twice, and $\tilde{f}(k)$ its Fourier transform. Express the Fourier transform of $df(x)/dx$ and of $xf(x)$ in terms of $\tilde{f}(k)$.

P4.6 Solving an eigenvalue equation $\hat{L}\psi = \lambda\psi$ means determining the eigenfunctions ψ that satisfy it, the corresponding eigenvalues λ, and fulfill certain requirements. Solve the following eigenvalue problems.

(a) $\hat{L} = i d/dx$, with the requirement that $\psi(x) = \psi(x + a)$, that is, that ψ is periodic with period a.

(b) $\hat{L} = d/dx$, under the constraint that ψ is finite. What happens if in addition $\psi(x) = \psi(x + s)$?

(c) $\hat{L}_c$ is such that $\hat{L}_c \psi(x) = \psi(-x)$.

(d) Examine the orthogonality of the eigenfunctions in the preceding three cases.

P4.7 An electron is enclosed in a 1D box with perfectly rigid walls, centered at the origin; the system is in a stationary state. Calculate the mean value of x and of x^2 as a function of energy.

P4.8 A 1D potential well of width a and infinite depth contains one electron; at a given moment, the electron probability density is triangular and symmetric.

(a) Determine the normalization constant and maximum particle density.

(b) Express the wave function in terms of the eigenfunctions obtained in Section 4.3.6 for the infinite well. Would you expect this state to be stationary? Why (not)?

P4.9 Consider an electron in a square potential well of infinite depth and width a.

(a) Calculate the expectation (average) values of $\hat{p}$ and $\hat{p}^2$ when the electrons are in an energy eigenstate n.

(b) Calculate the matrix elements p_{nm} and $(p^2)_{nm}$.

Discuss the meaning of your results.

P4.10 Three eigenfunctions ψ_1, ψ_2, and ψ_3 of some operator are linearly independent and degenerate, but not necessarily orthogonal. Construct three linear, orthogonal, and normalized combinations from them. Are the new functions eigenfunctions? Are they degenerate?

P4.11 An important problem in QM is that of a particle subjected to a linear force (e.g., the harmonic oscillator). The 1D stationary Schrödinger equation for this problem is

$$-\frac{\hbar^2}{2m}\varphi'' + \frac{1}{2}kx^2\varphi = E\varphi, \qquad (4.177)$$

where k is the oscillator constant. Solutions of the following types are proposed:

(a) $\varphi = A_1 \exp(ax^2) + A_2 \exp(-ax^2)$, with real and positive a,

(b) $\varphi = (B_1 + B_2 x) \exp(-bx^2)$, with real and positive b.

Find the values that the constants A_i, B_i, a, and b must have for these functions to be physically acceptable and normalized to unity. Find the value of E in each case, in terms of the parameters of the system.

P4.12 Using the Schrödinger equation, show that the greater the number of nodes in a wave function, the higher the energy and vice versa. by considering two different wave functions ψ_1, ψ_2 that intersect at a point P.

P4.13 Find an expression for the expectation value $\langle \hat{p}^2 \rangle$, in terms of the velocities v, u introduced in 4.4. Discuss your result.

P4.14 Given two operators $\hat{A}$, $\hat{B}$, express the matrix element C_{mn} of the product $\hat{C} = \hat{A}\hat{B}$ in terms of the matrix elements of $\hat{A}$ and $\hat{B}$. Now express the matrix element D_{mn} of the product $\hat{D} = \hat{B}\hat{A}$. How are the elements of matrices $\hat{C}$ and $\hat{D}$ related?

P4.15 Show that the expressions $\sum |n\rangle\langle n| = 1$, $\sum \varphi_n^*(x)\varphi_n(x') = \delta(x - x')$ are equivalent.

P4.16 (a) Calculate the average values of the Pauli matrices $\hat{\sigma}_i$ ($i = 1, 2, 3$) in the states $|+\rangle$ and $|-\rangle$.

(b) Calculate the average values of $\hat{\sigma}_i\hat{\sigma}_j$ ($i, j = 1, 2, 3$) in the states $|+\rangle$ and $|-\rangle$.

P4.17 Show that for any Hermitian operator $\widehat{A}$,

$$\langle n|(\Delta\widehat{A})^2|n\rangle = \sum_{n' \neq n} |\langle n|\widehat{A}|n'\rangle|^2, \tag{4.178}$$

where the deviation $\Delta\widehat{A}$ is defined as

$$\Delta\widehat{A} = \widehat{A} - \langle n|\widehat{A}|n\rangle. \tag{4.179}$$

Bibliographical Notes

It is recommended that the student read the pioneering work of Claus Jönsson on electron diffraction and study carefully the illustrations showing the experimental results using 1 to 5 slits [Jönsson (1961)]. A partial translation exists in English (1974) although the illustrations are poorer than the originals. Similar experiments were performed independently by Faget (1961).

These experiments could later be carried out more conveniently with an electron microscope; see Donati, Missiroli, and Pozzi (1973). Images of the experimental results from which Fig. 4.2 was constructed can be found in Merli, Missiroli, and Pozzi (1976, 1978).

For a very illustrative version of the experiment, see Tonomura et al.'s (1989) video, which can be found in Hitachi (23 October 2012). www.youtube.com/watch?v=mypzz99_MrM

An experiment on the simultaneous observation of particle and wave properties is that of Grangier et al. (1986). A more recent one is described in Mizobuchi and Ohtaké (1992).

The Schrödinger equation is a particular case of the Sturm–Liouville equation. This subject is discussed in some detail in various texts on QM, such as Merzbacher (1998) section 8.4. Accessible texts on differential equations dealing with the Sturm–Liouville equation are Butkov (1968) and Hassani (2013); Margenau and Murphy (1956), section 8.2 includes the Schmidt orthogonalization method.

5 One-Dimensional Potential Steps, Barriers, and Wells

To familiarize ourselves with some of the most basic features of quantum systems, in this chapter we will apply the Schrödinger equation to simple, one-dimensional problems. Fortunately, it is possible to obtain qualitatively correct conclusions about the behavior of quantum particles in the presence of more or less arbitrary potentials by replacing these with piecewise-constant potentials, which significantly simplifies the mathematics. For example, for many purposes an attractive potential can be represented by a square potential well, and a repulsive potential by a square barrier (Fig. 5.1). Although real devices do not have perfectly discontinuous interfaces, the micro- and nanostructures that are manufactured today are modeled, to a very good approximation, by square potentials. Thus, for example, multilayer structures combining metal, vacuum and semiconductor materials have nearly discontinuous interfaces and can therefore be represented by square potential barriers, wells and steps.

Many modern nanoelectronic devices are designed to prevent electrons from moving in certain directions. A *quantum dot* confines electrons to a small volume, so it is called a zero-dimensional system. *Quantum wires*, such as nanotubes, are considered one-dimensional because they confine electrons to move in their two opposite directions. A *quantum well* is considered two-dimensional because it confines electrons to a plane. All of these structures today can be made of carbon atoms. The confinement starts when the dimension in the direction of the constraint is comparable to the de Broglie wavelength of the carrier, which can be of the order of a nanometer (10^{-9} m). Also the strong nuclear binding potential, which has a range of a few femtometers (10^{-15} m), can be represented, to a first approximation, by a finite square-well potential. As already mentioned, this replacement of the real potential by a simpler one that retains its basic characteristics has the great advantage of allowing us to study the essence of the physical situation using simple mathematical methods, as the following examples will illustrate.

5.1 General Approach

To deal with simple potential problems in the easiest and most transparent way, we will use the time-independent 1D Schrödinger equation

$$-\frac{\hbar^2}{2m}\psi'' + V\psi = E\psi, \tag{5.1}$$

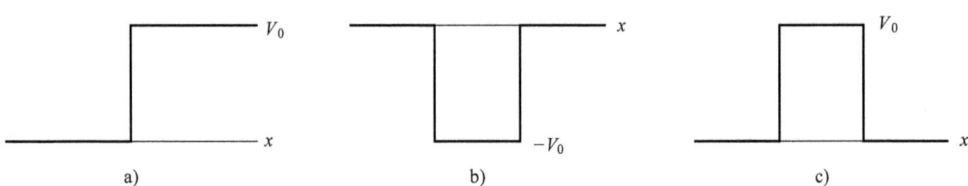

Figure 5.1 Piecewise constant potentials in one dimension. (a) The step potential, (b) The finite square well, (c) The square barrier.

which means that the solutions $\psi(x)$ will describe only stationary situations. This is an important restriction: it means that the system *has been given sufficient time to reach a stationary state*. With the introduction of the time-dependent Schrödinger equation in Chapter 7, a more complete and faithful description of quantum phenomena and their evolution in time will be possible.

Given that we will deal with potentials $V(x)$ that are piecewise constant, it is convenient to introduce two real parameters k and q, generally defined as

$$k^2 = \frac{2m}{\hbar^2}(E - V) \quad \text{when } E > V, \tag{5.2}$$

and

$$q^2 = \frac{2m}{\hbar^2}(V - E) \quad \text{when } E < V, \tag{5.3}$$

so that Eq. (5.1) is written as

$$\psi'' + k^2\psi = 0 \quad \text{when } E > V, \tag{5.4}$$

and

$$\psi'' - q^2\psi = 0 \quad \text{when } E < V. \tag{5.5}$$

Therefore, in those regions where $E > V$, the general solution will be of the form

$$\psi = Ae^{ikx} + Be^{-ikx}, \tag{5.6}$$

whereas in the regions where $E < V$, it will be of the form

$$\psi = Ce^{qx} + De^{-qx}, \tag{5.7}$$

with parameters A, B, C, D to be determined in each case by the conditions that must be satisfied for $\psi(x)$ and $\psi'(x)$ to be physical solutions. Specifically, we recall from Section 4.3 that ψ and ψ' must be finite, continuous, and single-valued functions at all points, in addition to satisfying the boundary conditions of the problem.

From Eqs. (5.6) and (5.7) it follows that the character of the solution depends drastically on whether the energy of the particles is greater or less than V: in the former case $\psi(x)$ is composed of oscillatory terms, with k^2 proportional to the (local mean) kinetic energy. From Eq. (4.83), we can identify the first term in Eq. (5.6) as representing the flow of particles to the right, with the velocity given by

$$v = \hbar k = \sqrt{2mE}, \tag{5.8}$$

and the second term as representing the flow of particles to the left with the opposite velocity. In contrast, when $E < V$, ψ is the sum of exponentially increasing and decreasing terms at a rate given by q, with q^2 proportional to the energy deficit.

5.2 The Step Potential

We begin by considering the scattering of an ensemble of free particles (electrons, say) by the step potential shown in Fig. 5.1 (a). For convenience we place the discontinuity of V at $x = 0$. Although it is possible to treat the problem more generally by assuming that the particles approach the discontinuity from both sides, to simplify the analysis, we will consider that they hit the step only from the left. Further, we assume a stationary situation, where the beam intensity is constant and all particles travel with the same velocity and direction. We can then speak of a *monoenergetic* or *monochromatic* beam, the latter term meaning that a single de Broglie wavelength h/mv is assigned to all the particles in the beam, regardless of whether they are sent one by one or in bulk. Although this is an idealized situation, it allows us to obtain some physically meaningful results, as will be seen in what follows.

5.2.1 $E < V_0$, total reflection

In the region $x < 0$ the particles move with kinetic energy E, while in the region $x > 0$ this energy is not sufficient to compensate for the potential, so the $E - V_0$ difference becomes negative. Classically, this region is inaccessible to particles. The quantum situation is somewhat more complex, since from Eqs. (5.6) and (5.7) we have

$$\psi = \begin{cases} Ae^{ikx} + Be^{-ikx}, & x \leq 0, \\ Ce^{-qx}, & x > 0. \end{cases} \quad (5.9)$$

with $k^2 = 2mE/\hbar^2$ and $q^2 = 2m(V_0 - E)/\hbar^2$, according to Eqs. (5.4) and (5.5). In the expression for $x > 0$ the term with positive exponential has been omitted, since it would make ψ increase indefinitely with x. For ψ and ψ' to be continuous at $x = 0$, it is required that $A + B = C$; $ik(A - B) = -qC$, whence

$$B = \frac{ik+q}{ik-q}A; \quad C = \frac{2ik}{ik-q}A, \quad (5.10)$$

which, when introduced in Eq. (5.9), gives

$$\psi = A\left[e^{ikx} + \frac{ik+q}{ik-q}e^{-ikx}\right], \quad x < 0, \quad (5.11)$$

$$\psi = \frac{2ik}{ik-q}Ae^{-qx}, \quad x > 0. \quad (5.12)$$

From the preceding expressions it follows that $|B| = |A|$. By writing $A = A_0 e^{-i\alpha/2}$, $B = A_0 e^{i\alpha/2}$ (which is a comfortable selection), with A_0 and α real, we obtain $C = A + B = 2A_0 \cos \alpha/2$. Therefore, from (5.9) we get

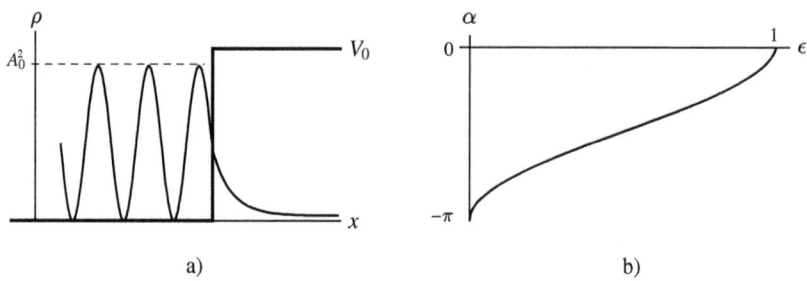

Figure 5.2 (a) Distribution of particles in the presence of a square potential step. (b) Phase shift α as a function of $\epsilon = E/V_0$.

$$\psi = 2A_0 \cos(kx - \alpha/2), \quad x < 0 \tag{5.13}$$

$$\psi = 2A_0 \cos(\alpha/2) e^{-qx}, \quad x > 0. \tag{5.14}$$

Since there is no net particle flow, we conclude that the number of particles reflected by the barrier is equal to the number of incident particles per unit time.

The particle density, given by the square of the expressions (5.13) and (5.14), is shown schematically in Fig. 5.2 (a). The effects of the interference of the concurring incident and reflected waves in the $x < 0$ region are clearly visible. The reflected wave is out of phase with the incident one by an amount

$$\alpha = \arccos \frac{k^2 - q^2}{k^2 + q^2} = \arccos(2\epsilon - 1) = -\arcsin 2\sqrt{\epsilon(1-\epsilon)}, \tag{5.15}$$

where $\epsilon = E/V_0$, $0 \leq \epsilon \leq 1$. Therefore, as $E \to 0$, $\alpha \to -\pi$, while as $E \to V_0$, $\alpha \to 0$; see Fig. 5.2 (b). It is interesting to compare these results with the optical case: we recall that when a light beam is reflected by a medium with lower refractive index there is no phase shift, while if the medium has a higher refractive index, the reflected wave is inverted with respect to the incident wave, that is, there is a phase shift of π. In the optical case, just as in the quantum case, at normal incidence (i.e., in 1D) the reflection from a mirror is total. In the $x > 0$ region, where there is no net particle transfer, one can speak of an *evanescent* solution, which decays exponentially with x according to Eq. (5.14).

This example serves to illustrate a remarkable phenomenon typical of QM. It is possible to change the phase shift α without changing the incident particle beam by altering the potential V_0 at $x = 0$; this modifies the distribution of particles in the entire region $x < 0$. From the point of view of the *description*, this is a nonlocal behavior, resulting from the interference of the two terms in the first of Eqs. (5.9). Other examples of such nonlocal behavior described by the wave function will be given in Section 4. The subject will be discussed in more detail in Sections 10.1.2 and 10.5.

5.2.2 $E > V_0$, partial transmission

In this case it is convenient to introduce the parameter k_1, defined as

$$k_1^2 = \frac{2m}{\hbar^2}(E - V_0), \tag{5.16}$$

so that the general solution of the Schrödinger equation can be written as

$$\psi = Ae^{ikx} + Be^{-ikx}, \quad x < 0 \tag{5.17}$$

$$\psi = Ce^{ik_1 x} + De^{-ik_1 x}, \quad x > 0, \tag{5.18}$$

with $k_1 < k$. In the particular case we are analyzing, with the particles incident from the left, it is clear that the flow in the region $x > 0$ is exclusively to the right, so we must take $D = 0$. From the continuity conditions at $x = 0$ we get

$$B = \frac{k - k_1}{k + k_1} A, \quad C = \frac{2k}{k + k_1} A, \tag{5.19}$$

whence

$$\psi = A \left[e^{ikx} + \frac{k - k_1}{k + k_1} e^{-ikx} \right], \quad x < 0; \tag{5.20}$$

$$\psi = A \frac{2k}{k + k_1} e^{ik_1 x}, \quad x > 0. \tag{5.21}$$

Using this wave function, we can calculate the net flux of particles both to the left and to the right of the barrier. From Eq. (4.83) we obtain the following expression for the current density $j = \rho v$, for $x < 0$,

$$j_< = \frac{i\hbar}{2m} \left(\psi \frac{d\psi^*}{dx} - \psi^* \frac{d\psi}{dx} \right) \bigg|_{x<0} = j_{\text{inc}} - j_{\text{refl}}, \tag{5.22}$$

where the incident and reflected current densities are given respectively by

$$j_{\text{inc}} = \frac{\hbar k}{m} |A|^2, \quad j_{\text{refl}} = \frac{\hbar k}{m} |B|^2. \tag{5.23}$$

Similarly, for $x > 0$ we obtain for the transmitted current density

$$j_> = j_{\text{trans}} = \frac{\hbar k_1}{m} |C|^2. \tag{5.24}$$

Since the boundary conditions guarantee that the flux density is continuous, both expressions (5.22) and (5.24) – whose value is constant – must be equal. It is not difficult to be convinced that the equality is in fact fulfilled by the values determined for B and C. Its physical meaning is clearly revealed by rewriting such an equality in the form

$$k|A|^2 = k|B|^2 + k_1 |C|^2, \tag{5.25}$$

which says that the particles impinging on the barrier are either reflected or transmitted, so in this case there is partial reflection. This suggests introducing the *reflection coefficient R* and the *transmission coefficient T*, defined as follows,

$$R \equiv \frac{j_{\text{refl}}}{j_{\text{inc}}} = \frac{k|B|^2}{k|A|^2} = \frac{|B|^2}{|A|^2} = \left(\frac{k - k_1}{k + k_1} \right)^2, \tag{5.26}$$

$$T \equiv \frac{j_{\text{trans}}}{j_{\text{inc}}} = \frac{k_1 |C|^2}{k|A|^2} = \frac{4kk_1}{(k + k_1)^2}. \tag{5.27}$$

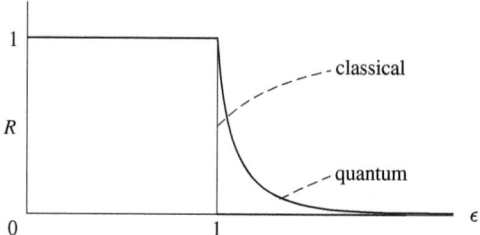

Figure 5.3 The reflection coefficient R as a function of $\epsilon = E/V_0$, for a beam of particles scattered by a square potential step.

It is clear that these coefficients satisfy the relation

$$R + T = 1, \tag{5.28}$$

which indicates that those particles that are not reflected are transmitted, that is, particles are neither lost nor created.

Figure 5.3 shows R as a function of $\epsilon = E/V_0$. As long as E is less than V_0 the reflection is total, $R = 1$, as we saw in the previous section. When $E > V_0$, particles can be transmitted according to Eq. (5.27), and they do so with increasing probability as E increases; when $E \gg V_0$, then $R \to 0$ and $T \to 1$, as expected. On the other hand, classical particles with $E > V_0$ would slow down, but all would be transmitted. This abrupt change of the reflection coefficient from 1 to 0 in the classical case is replaced by a continuous transition in the quantum case, as shown in Fig. 5.3.

It is easy to see that if we make the substitution $k \to -k_1$, $k_1 \to -k$, which corresponds to the case where the particle beam is sent from the right toward an attractive potential step, R and T are unchanged. Once again, the optical analogy is suggestive: according to a theorem of geometric optics, the transmission and reflection coefficients of a light beam passing through a medium with refractive index n_1 into a medium with index n_2 are equal to those obtained when the same beam passes from medium 2 into medium 1. Parameters like k, k_1, and so on, may therefore be considered as effective "refractive indices" – measured in suitable units – in the wave analog of QM.

Exercise E5.1. A two-dimensional barrier

Consider particles constrained to move in the xy-plane. The right half-plane $x > 0$ has the potential $V_0 > 0$, while the left half-plane $x < 0$ is held at zero potential. The particles can hit the boundary at $x = 0$ obliquely; the kinetic energy of the motion along the x-axis is $E_1 \leq V_0$, but that due to the motion along y is arbitrary. Determine the particle flux and analyze the results.

Solution. At the boundary there is total reflection in the x-direction, while the y-component travels freely. The wave function can be factorized in the form $\psi = \psi_1(x)\psi_2(y)$, where $\psi_2(y)$ is a plane wave with $E = E_2$ and $k = k_2$; $\psi_1(x)$ can be written as $\psi_1(x) = Ae^{ik_1x} + Be^{-ik_1x}$ for $x < 0$, and as $\psi_1(x) = Ce^{-qx}$ for $x > 0$, with $k_1^2 = 2mE_1/\hbar^2$ and $q = \sqrt{2m(V_0 - E_1)}/\hbar$. Therefore,

$$\psi_I(x, y) = (Ae^{ik_1x} + Be^{-ik_1x})e^{ik_2y}, \quad x < 0 \tag{5.29}$$

$$\psi_{II}(x,y) = Ce^{-qx}e^{ik_2 y} \quad x > 0. \tag{5.30}$$

By stitching the solutions at $x = 0$, we get

$$\frac{B}{A} = \frac{k_1 - iq}{k_1 + iq} = e^{-2i\alpha}, \quad \frac{C}{A} = \frac{2k_1}{k_1 + iq} = 2e^{-i\alpha}\cos\alpha, \tag{5.31}$$

with

$$\tan\alpha = \frac{q}{k_1} = \sqrt{\frac{V_0}{E_1} - 1}. \tag{5.32}$$

Note that $|B| = |A|$, as would be expected from total reflection; this makes the current density along the x-axis equal to zero. In the y-direction we get

$$j_I = 2\frac{\hbar k_2}{m}|A|^2(1 + \cos 2(k_1 x + \alpha)) = \frac{\hbar k_2}{m}\rho_I(x), \tag{5.33}$$

$$j_{II} = 4\frac{\hbar k_2}{m}|A|^2 e^{-2qx}\cos^2\alpha = \frac{\hbar k_2}{m}\rho_{II}(x). \tag{5.34}$$

To the left of the barrier the incident wave and the reflected wave interfere, as shown by the oscillatory term. The current density in region II decays exponentially along the x-direction; this current vanishes when $\cos\alpha \to 0$, that is, $(V_0/E_1) \to \infty$. In electrodynamics, this condition would be equivalent to a perfect conductor, and the finite penetration for finite V_0 corresponds to the skin effect; the solution is usually called an evanescent wave, *although it carries no energy*.

Note that changing V_0 with k_1 and k_2 fixed only changes the phase of the interference pattern by modifying the relative phase between the incident and reflected waves. This shows the importance (and observability) of the relative phases, even though the global phase remains arbitrary.

5.3 The Square Well

We consider a square potential well like the one shown in Fig. 5.4, of width a and depth V_0, and divide the x-space into three regions I, II, and III, as in Chapter 4 (compare with Figure 4.6). To follow established conventions, we take the potential outside the well to be zero, so that a bound particle has negative energy, while particles with positive energy are scattered by the potential, but do not remain bound to it.

5.3.1 $-V_0 < E < 0$, bounded states

For negative energies, the particles are bound by the potential. In terms of the parameters

$$q^2 = -\frac{2mE}{\hbar^2} = \frac{2m}{\hbar^2}|E|, \tag{5.35}$$

$$k^2 = \frac{2m}{\hbar^2}(V_0 - |E|) \equiv \frac{2m}{\hbar^2}E', \tag{5.36}$$

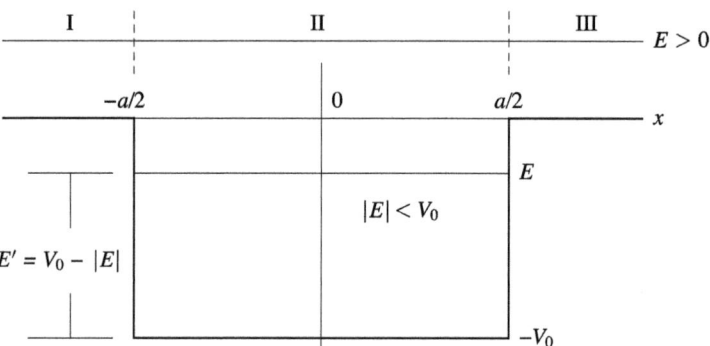

Figure 5.4 The finite-depth square well. For symmetry considerations, the origin of the x-coordinate is fixed at the center of the well.

the acceptable solutions of the Schrödinger equation are

$$\begin{aligned}\psi_I &= A_1 e^{qx}, & x &< -a/2, \\ \psi_{II} &= B\cos kx + C\sen kx, & -a/2 &< x < a/2, \\ \psi_{III} &= A_3 e^{-qx}, & x &> a/2.\end{aligned} \quad (5.37)$$

When writing Eqs. (5.37) we have taken into account the requirement that ψ be bounded at every point. The solution inside the well has been written in terms of trigonometric functions (instead of imaginary exponentials), which have definite (odd or even) parity. Given that $V(x)$ is an even function of x, changing the sign of x will either keep the Schrödinger equation the same or change its sign; therefore the parity of ψ must be either even or odd. This means that there is a family of even solutions and a family of odd solutions, which are given respectively by

$$\psi_I = Ae^{qx}, \quad \psi_{II} = B\cos kx, \quad \psi_{III} = Ae^{-qx}, \qquad \psi(x) = \psi(-x), \quad (5.38)$$

$$\psi_I = Ae^{qx}, \quad \psi_{II} = C\sin kx, \quad \psi_{III} = -Ae^{-qx}, \qquad \psi(x) = -\psi(-x). \quad (5.39)$$

The conditions of continuity of ψ and ψ' at the boundaries $x = a/2$ and $x = -a/2$ give respectively

$$Ae^{-\frac{a}{2}q} = B\cos \tfrac{1}{2}ak, \quad qAe^{-\frac{a}{2}q} = kB\sin \tfrac{1}{2}ak, \qquad \psi(x) = \psi(-x), \quad (5.40)$$

$$Ae^{-\frac{a}{2}q} = -C\sin \tfrac{1}{2}ak, \quad qAe^{-\frac{a}{2}q} = kC\cos \tfrac{1}{2}ak, \qquad \psi(x) = -\psi(-x). \quad (5.41)$$

To simplify writing, we introduce the dimensionless variable $y = \tfrac{1}{2}ak$ and the parameter, also dimensionless, $y_0 = \tfrac{1}{2}\sqrt{2ma^2 V_0/\hbar^2}$, so that

$$\tfrac{1}{2}aq = \sqrt{y_0^2 - y^2}. \quad (5.42)$$

5.3.1.1 Even and odd solutions

It is convenient to analyze each family separately. For even solutions, the first two equations (5.41) give

$$y\tan y = \sqrt{y_0^2 - y^2}. \quad (5.43)$$

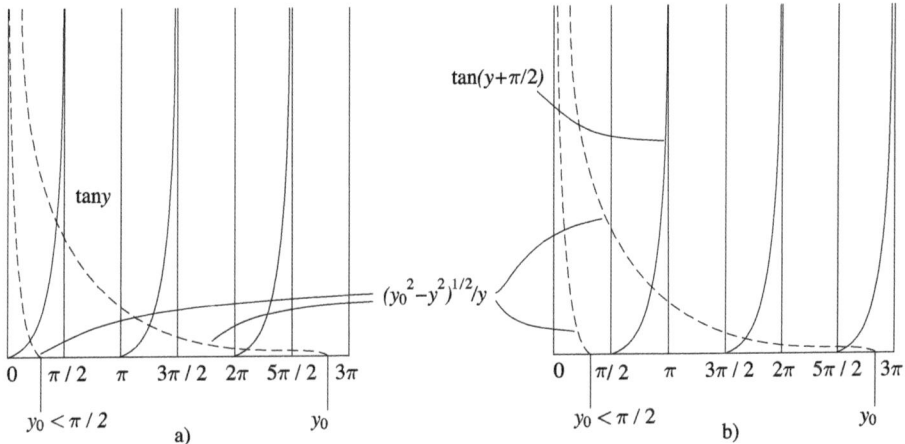

Figure 5.5 Solutions of the eigenvalue equation for the finite square well, for different values of the parameter y_0. (a) Even solutions, (b) odd solutions.

This is a transcendental equation, the solution of which can be obtained graphically from the intersections of the curves $\tan y$ and $\sqrt{y_0^2 - y^2}/y$, as shown in Fig. 5.5 (a). No matter how small the value of the parameter y_0, there exists at least one solution, meaning that a 1D well has at least one bounded state. In particular the ground-state solution is an even function. As y_0 grows, the number of intersections increases; that is, as the product $V_0 a^2$ increases, so does also the number of bounded states that the well can contain. In the limit where this product is infinite (the limit discussed in Section 4.3.6) the points of intersection become equidistant at the values $y = \pi/2, 3\pi/2, \ldots$ Therefore, for a potential well that is very deep or very wide or both, so that $y_0 \gg 1$, the energy eigenvalues are approximately given by the formula

$$y = \frac{a}{2\hbar}\sqrt{2mE'_n} = \pi\left(m + \tfrac{1}{2}\right), \tag{5.44}$$

where m is an integer and according to Eq. (5.36), E'_n is the energy measured from the bottom of the well. This expression, rewritten in the form

$$E'_n = \frac{\hbar^2 \pi^2}{2ma^2}(2m+1)^2, \tag{5.45}$$

corresponds to the solutions of the infinite well with odd n, given by

$$E'_n = \frac{\hbar^2 \pi^2}{2ma^2}n^2, \quad n \text{ odd}. \tag{5.46}$$

The odd solutions are obtained following an analogous procedure; from the second pair of equations (5.41) the eigenvalue equation becomes

$$-y \cot y = \sqrt{y_0^2 - y^2}, \tag{5.47}$$

and its graphical solutions are shown in Fig. 5.5(b). We see that for $y_0 \to \infty$, the solutions correspond to $y = \pi, 2\pi, 3\pi, \ldots$, the values given by

$$E'_n = \frac{\hbar^2 \pi^2}{2ma^2}n^2, \quad n \text{ even}. \tag{5.48}$$

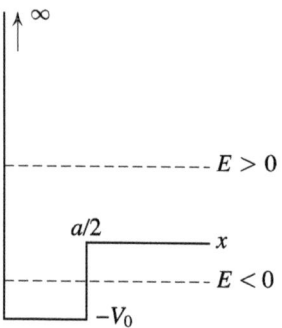

Figure 5.6 Effective square potential well of depth V_0 for central problems in two or three dimensions. The radius of the well is $a/2$.

If $y_0 < \pi/2$, there is no intersection, which means that for a 1D square well to have at least one odd bounded state, it is required that

$$a^2 V_0 > \frac{\pi^2 \hbar^2}{2m}. \tag{5.49}$$

Note that the odd solutions obtained here are also solutions of the potential shown in Fig. 5.6, since they cancel at $x = 0$ (precisely because they are odd), as required by the infinitely repulsive potential of this well at $x = 0$. As we shall see in Section 12.1, this is the type of effective potential for 2D and 3D central problems. From this we conclude that an attractive potential in two or three dimensions does not necessarily allow for bounded states; a condition of the type given by Eq. (5.49) must be satisfied for a first bounded state to occur.

In summary, we have shown that a 1D well always has bounded states that form a discrete spectrum; furthermore, the number of bounded states that fit in the well increases as a function of its effective width and depth.

Exercise E5.2. Quantum dots

A quantum dot can be thought of as electrons confined in a very small 3D box, a kind of artificial atom or zero-dimensional system. Find the eigenfunctions and eigenvalues for electrons found in an infinite 3D rectangular well and determine under what conditions the energy levels are degenerate.

Solution. Let a_1, a_2, a_3 be the x, y, z sides of the rectangular well. Solving Schrödinger's equation in a rectangular coordinate system, we obtain

$$\psi_{n_1 n_2 n_3} = \sqrt{\frac{8}{a_1 a_2 a_3}} \sin\left(\frac{\pi n_1}{a_1} x\right) \sin\left(\frac{\pi n_2}{a_2} y\right) \sin\left(\frac{\pi n_3}{a_3} z\right), \tag{5.50}$$

where the quantum numbers n_1, n_2, n_3 take any of the values $1, 2, 3, \ldots$ The total energy is given by

$$E_{n_1 n_2 n_3} = \frac{\pi^2 \hbar^2}{2m} \left(\frac{n_1^2}{a_1^2} + \frac{n_2^2}{a_2^2} + \frac{n_3^2}{a_3^2}\right). \tag{5.51}$$

It is not difficult to see that if the relation between two sides is an integer k, $a_2 = ka_1$, then there is a degeneracy for the pair of quantum numbers n_1, n_2 whenever n_2/k is an integer. Indeed, if n_1 is replaced with n_2/k and n_2 with $n_1 k$, the same energy is obtained as for n_1 and n_2. In particular, if two sides of the well are equal, there is a degeneracy of order $g = 2$ whenever $n_1 \neq n_2$. It remains for the student to find the energy degeneracy as function of the quantum numbers when the well is cubic.

As quantum dots have very few electrons, their relative simplicity has allowed them to be used successfully for detailed theoretical-experimental studies of the effects of electron–electron interactions, transport effects, and so on. Applications of low-dimensional systems in the fabrication of solid-state devices are already numerous (e.g., very low-noise amplifiers for satellite and communication antennas) and will certainly develop further.

5.3.2 $E > 0$, resonant transmission and resonant scattering

For positive energies, the solutions are oscillatory in the three regions,

$$\begin{aligned}
\psi_I &= A_1 e^{ikx} + B_1 e^{-ikx}, & x < -a/2 \\
\psi_{II} &= A_2 e^{ik_1 x} + B_2 e^{-ik_1 x}, & -a/2 < x < a/2 \\
\psi_{III} &= A_3 e^{ikx} + B_3 e^{-ikx}, & x > a/2,
\end{aligned} \quad (5.52)$$

with

$$k^2 = \frac{2mE}{\hbar^2}, \quad k_1^2 = \frac{2m}{\hbar^2}(E + V_0). \quad (5.53)$$

We write the continuity conditions of ψ and ψ', taking into account that when the particles approach the well from the left, in the region $x > a/2$ there are only transmitted particles flowing to the right, i.e., once again we should take $B_3 = 0$. Thus,

$$A_1 e^{-\frac{ia}{2}k} + B_1 e^{\frac{ia}{2}k} = A_2 e^{-\frac{ia}{2}k_1} + B_2 e^{\frac{ia}{2}k_1} \quad (5.54)$$

$$A_2 e^{\frac{ia}{2}k_1} + B_2 e^{-\frac{ia}{2}k_1} = A_3 e^{\frac{ia}{2}k} \quad (5.55)$$

$$k\left[A_1 e^{-\frac{ia}{2}k} - B_1 e^{\frac{ia}{2}k}\right] = k_1 \left[A_2 e^{-\frac{ia}{2}k_1} - B_2 e^{\frac{ia}{2}k_1}\right] \quad (5.56)$$

$$k_1 \left[A_2 e^{\frac{ia}{2}k_1} - B_2 e^{-\frac{ia}{2}k_1}\right] = k A_3 e^{\frac{ia}{2}k}. \quad (5.57)$$

We are left with five free parameters, so the four equations plus the normalization condition – which in this case is determined by the amplitude of the incoming beam A_1 – are satisfied without imposing any condition on the energy, and the spectrum is continuous, as it should be, since we are free to choose the energy of the incident beam.

To determine the reflection and transmission coefficients we must obtain the value of the coefficients B_1 and A_3. The calculation is somewhat laborious, so it is not included here; however, using the preceding equations it is easy to check the following results, which are plotted in Fig. 5.7,

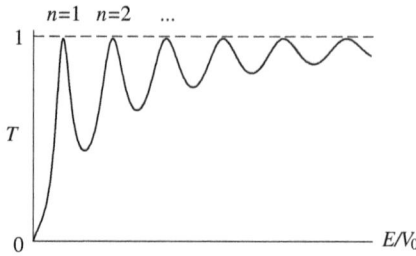

Figure 5.7 Transmission coefficient for free particles scattered by a square well, as a function of the ratio E/V_0.

$$R = \left|\frac{B_1}{A_1}\right|^2 = 1 - T; \tag{5.58}$$

$$T = \left|\frac{A_3}{A_1}\right|^2 = \left[1 + \frac{1}{4}\left(\frac{k_1}{k} - \frac{k}{k_1}\right)^2 \sin^2 k_1 a\right]^{-1}. \tag{5.59}$$

If $E \to \infty$, $k_1/k \to 1$, $(k_1/k) - (k/k_1) \to 0$ and $T \to 1$, as could be expected. However, also for $\sin k_1 a = 0$, i.e., for energy values such that

$$E + V_0 = \frac{\pi^2 \hbar^2}{2ma^2} n^2 \tag{5.60}$$

with integer n, the transmission coefficient takes the value $T = 1$. At exactly these energies all particles are transmitted (as would happen in the classical case for any positive energy). Here we have a 1D example of the phenomenon of *resonant transmission*, which will be discussed below and in more detail in Section 18.5.

The result just obtained can serve as a rudimentary model to explain phenomena such as the *Ramsauer–Townsend effect*.[1] This effect is observed in the scattering of low-energy electrons (< 1 eV) by noble gases, when at certain well-defined bombardment energies the electrons pass through the gas without being scattered. For example, argon is essentially transparent to electrons with energy 0.4 eV, but strongly scatters electrons with energy above or below this value. In wave language, we can explain such resonant transmission as being due to constructive interference between the left and right waves crossing the well, generated by multiple reflections at the discontinuities. Indeed, the maxima occur when the distance $2a$ travelled by the waves on their round trip is equal to an integer number of de Broglie wavelengths, as is clear from writing the expression $k_1 a = \pi n$ in the alternative form

$$\lambda = \frac{2\pi}{k_1} = \frac{2a}{n}. \tag{5.61}$$

Note that formula (5.60) agrees, except for the different convention on the origin of energy, with the expression (5.46) for the energy of the bounded states of the infinite well; for this reason, such levels are often called *virtual* energy levels.

Another interesting phenomenon exemplified by the previous results is that of resonances, or quasi-bounded states, which are so important, for example, in particle

---

[1] The effect is named after Carl Wilhelm Ramsauer (1879–1955) and Sir John Sealy Townsend, who independently studied the collisions between noble-gas atoms and low-energy electrons in 1921.

theory. This occurs when the reflection coefficient takes a maximum value, which, as follows from Eqs. (5.58) and (5.59), occurs approximately at energy values such that

$$k_1 a = \left(n + \tfrac{1}{2}\right)\pi, \qquad (5.62)$$

corresponding to the minima of the curve for T in Fig. 5.7. When this condition is met, we speak of a *resonance*. In this case the residence time turns out to be long and for this reason, such states are considered to be *quasi-bounded*.

5.3.3 Scattering matrix for one-dimensional problems

It is instructive, especially for future work, to consider the scattering problem from a more general perspective. For this purpose we introduce the important notion of *scattering matrix*, limited here to the 1D case. For an arbitrary potential vanishing at infinity on both sides, we can construct an incoming wave function in terms of the coefficients A_1 and B_3, as in Eqs. (5.52). It is convenient to represent the incident wave function as a column matrix,

$$\psi_{\text{inc}} = \begin{pmatrix} A_1 \\ B_3 \end{pmatrix}, \qquad (5.63)$$

and similarly the outgoing wave function,

$$\psi_{\text{out}} = \begin{pmatrix} A_3 \\ B_1 \end{pmatrix}. \qquad (5.64)$$

Note that at infinity, where $V(x) = 0$, the solutions of the Schrödinger equation are precisely the functions ψ_I and ψ_{III} in Eqs. (5.52), so knowledge of the incoming and outgoing matrices gives us the complete wave function in the asymptotic regions. Since the coefficients A_i and B_i are linearly related, see (5.54)–(5.57), we can establish a linear relationship between ψ_{inc} and ψ_{out} of the form

$$\psi_{\text{out}} = S\psi_{\text{inc}}, \qquad (5.65)$$

where S is a 2×2 matrix. Explicitly, we shall write

$$\begin{pmatrix} A_3 \\ B_1 \end{pmatrix} = \begin{pmatrix} S_{11} & S_{12} \\ S_{21} & S_{22} \end{pmatrix} \begin{pmatrix} A_1 \\ B_3 \end{pmatrix} = \begin{pmatrix} S_{11}A_1 + S_{12}B_3 \\ S_{21}A_1 + S_{22}B_3 \end{pmatrix}. \qquad (5.66)$$

The matrix S defined in this way, which transforms the incoming wave function into the outgoing one, is commonly called the *S-matrix* or *scattering matrix*. It can be shown that under the condition of conservation of flow, namely $T + R = 1$, the S-matrix is unitary, that is, it satisfies the condition $S^\dagger = S^{-1}$, or explicitly $S_{ij}^* = S_{ji}^{-1}$.

The S-matrix method provides a clear separation between the problem of determining the elements of S, which depend on the nature and dynamics of the system, and the initial conditions, which are arbitrary and depend on the specific problem. Thus, the study of the S matrix is equivalent to the general study of the system's dynamical behavior, independently of circumstantial or discretionary initial details. In the 1D case, the S-matrix allows us to determine the reflection and transmission coefficients, which contain the information about the properties of the scattering potential; the general 3D case is studied in Chapter 19.

5.4 The Square Potential Barrier: Tunneling

The potential to be studied is a square repulsive barrier of width a and height V_0. As in the classical case, the solution depends drastically on whether the energy of the incident particles is higher or lower than the barrier potential. In the quantum case, however, the behavior of the beam has peculiar properties, as will be shown later.

5.4.1 $E < V_0$, the tunneling effect

As before, we consider only the case where the particles hit the barrier from the left. The problem is similar to the one studied in connection with the potential step in Section 5.2; however, its detailed study is interesting because in this case a new phenomenon arises, with important applications, namely the tunneling (or tunnel) effect. We will show that a fraction of particles – which can be appreciable – manages to pass the barrier and move freely on the other side of it. In the classical case this possibility is totally forbidden by the conservation of energy.

From Eqs. (5.6) and (5.7), the general solution in each of the three regions (with the barrier at $(0, a)$), is

$$
\begin{aligned}
&I, x < 0: \quad \psi_I'' + k^2 \psi_I = 0, \quad &\psi_I = A_1 e^{ikx} + B_1 e^{-ikx}, \\
&II, 0 < x < a: \quad \psi_{II}'' - q^2 \psi_{II} = 0, \quad &\psi_{II} = A_2 e^{qx} + B_2 e^{-qx}, \\
&III, x > a: \quad \psi_{III}'' + k^2 \psi_{III} = 0, \quad &\psi_{III} = A_3 e^{ik(x-a)} + B_3 e^{-ik(x-a)}.
\end{aligned}
\tag{5.67}
$$

If particles are incident only from the left, $B_3 = 0$. Following the procedure used in the previous case, we obtain from the continuity conditions on ψ and ψ' that the coefficients can be written in the form

$$
\begin{aligned}
B_1 &= -i \frac{q^2 + k^2}{2kq} A_3 \sinh qa, \\
A_2 &= \frac{q + ik}{2q} A_3 e^{-qa}, \quad B_2 = \frac{q - ik}{2q} A_3 e^{qa}, \\
A_3 &= \frac{1}{\cosh qa + i \frac{1}{2}\left(\frac{q}{k} - \frac{k}{q}\right) \sinh qa} A_1,
\end{aligned}
\tag{5.68}
$$

with the usual definitions for k and q. The transmission coefficient is given by $|A_3/A_1|^2$,

$$
T = \left[1 + \tfrac{1}{4}\left(\frac{k}{q} + \frac{q}{k}\right)^2 \sinh^2 qa \right]^{-1}.
\tag{5.69}
$$

For $E = V_0$ we have $q = 0$, $\sinh qa = 0$, but $\lim_{q \to 0}\left(\frac{k}{q} + \frac{q}{k}\right)^2 \sinh^2 qa = k^2 a^2$, so that

$$
T = \frac{1}{1 + \tfrac{1}{4}k^2 a^2} = \frac{1}{1 + \frac{mV_0 a^2}{2\hbar^2}} \quad (E = V_0).
\tag{5.70}
$$

As E decreases, so does T, but it is still nonzero, meaning that there is partial transmission of the beam through the barrier: this is the tunneling effect. Classically we should expect total, not partial, reflection of the incident beam, so tunneling is considered to be a quantum mechanical feature with no classical counterpart.

The transmission coefficient acquires its minimum value when E approaches 0 (hence $k \to 0$); for such small energies, if $qa \gg 1$, we can approximate the sinh with an exponential,

$$\sinh qa = \tfrac{1}{2}(e^{qa} - e^{-qa}) \approx \tfrac{1}{2}e^{qa} \quad (qa \gg 1), \tag{5.71}$$

and T reduces to

$$T = \left(\frac{4kq}{k^2 + q^2}\right)^2 e^{-2qa}. \tag{5.72}$$

The behavior of this function is dominated by the exponential, so we can approximate the preceding expression by

$$T \approx e^{-2qa} = \exp\left[-\frac{2a}{\hbar}\sqrt{2m(V_0 - E)}\right]. \tag{5.73}$$

Due to the critical dependence of T on qa, it is necessary to develop calculational techniques that allow us to deal with more realistic potentials. In Chapter 6 we will present a method that is especially useful in these cases. Suffice it to say that the tunneling effect is a widespread phenomenon that gives rise to spontaneous decay and is used both in laboratory applications (such as the tunneling microscope) and in technology (chips, transistors, tunnel diodes, etc.).

5.4.1.1 An intriguing phenomenon

The tunneling effect allows us to understand many aspects of conduction in metals and semiconductors, as well as other phenomena such as nuclear α decay, which will be discussed in Chapter 6. But the tunneling phenomenon itself is not well understood. The solution of Schrödinger's equation is usually taken as *the* explanation, and the name associated with it seems to obviate the need for further investigation. A word of caution, however: in reality, the particles do not simply tunnel through the barrier. Inside the barrier, p is imaginary, corresponding to an evanescent solution with no particle flow. So how is it that particles are transmitted and appear on the other side of the barrier, moving at the original speed?

It is difficult to explain the physical origin of the tunnel effect within the usual QM scheme, where the incident electrons are described by a function of energy that is the same for all electrons; the description in terms of fixed energy states does not take fluctuations into account. A plausible, more physically oriented explanation is offered by stochastic theories (see Chapter 10), which see QM as providing an excellent statistical, but not exhaustive description; the *fine* behavior of the electrons is lost in the coarse averaging. The transmission coefficient is an excellent, albeit indirect, measure of the probability of individual electrons crossing the barrier. An illustrative and explanatory set of detailed predictions using Nelson's stochastic version of QM (see Section 10.1) can be found in McClendon and Rabitz (1988).

Another point worth noting is that the wave function ψ_I, which describes the particles moving in region I, i.e., in front of the barrier, carries information about the width a and the height V_0 of the barrier. This is a nonlocality due to the stationary, statistical

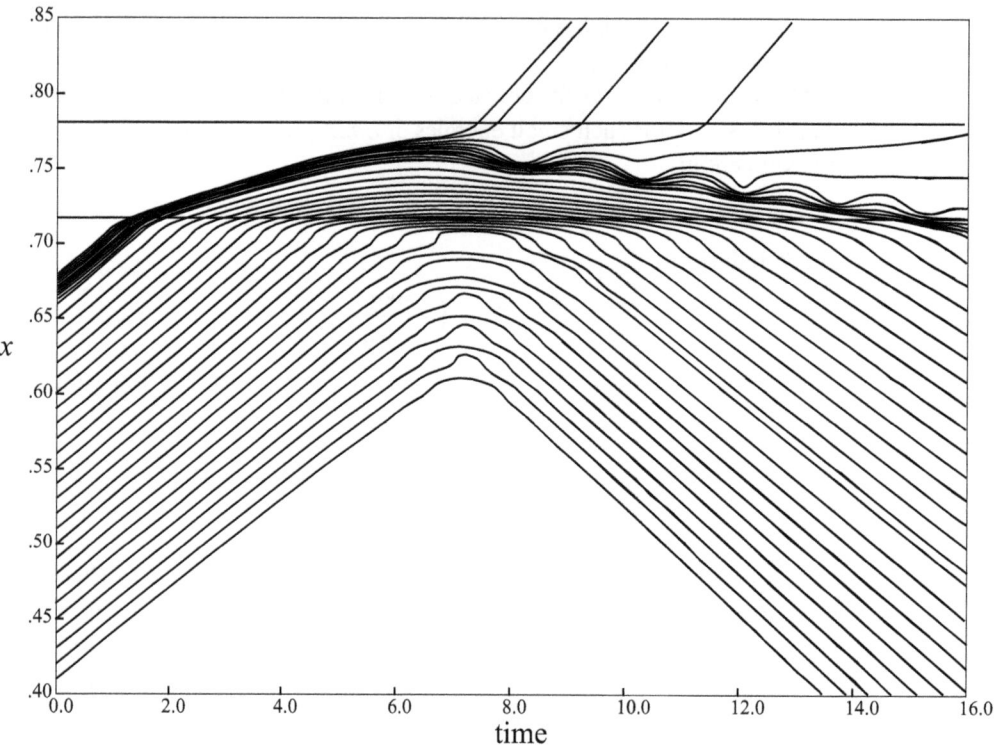

Figure 5.8 A collimated, monochromatic beam of particles scattered by a barrier, according to Bohm's theory (see Section 10.1). Note that many particles are scattered before reaching the barrier. Figure adapted from Dewdney and Hiley (1982).

description of the system. An impressive manifestation of this phenomenon is obtained by studying the trajectories of the particles (e.g., obtained from numerical simulations according to Bohm's theory, see Section 10.1.2), which clearly shows that a significant fraction of the incident particles are affected in their trajectory *before* they reach the barrier, as shown in Fig. 5.8.

5.4.1.2 Phase shift of the transmitted wave

It is instructive to analyze in more detail the transmitted wave $\psi_{\text{trans}} = \psi_{III}$. For this purpose we rewrite Eq. (5.68) as

$$\frac{A_3}{A_1} = \frac{\cosh qa - i\frac{1}{2}\left(\frac{q}{k} - \frac{k}{q}\right) \sinh qa}{\cosh^2 qa + \frac{1}{4}\left(\frac{q}{k} - \frac{k}{q}\right)^2 \sinh^2 qa}$$

$$= \frac{e^{i\delta}}{1 + \frac{1}{4}\left(\frac{q}{k} + \frac{k}{q}\right)^2 \sinh^2 qa} = \sqrt{T} e^{i\delta}, \qquad (5.74)$$

with the phase shift δ given by

$$\delta = \arctan \frac{1}{2}\left(\frac{q}{k} - \frac{k}{q}\right)\tanh qa. \tag{5.75}$$

Therefore, the transmitted wave is

$$\psi_{\text{trans}} = \sqrt{T}e^{ik(x-a)+i\delta}. \tag{5.76}$$

This result shows that the barrier produces two effects on the transmitted beam. The most evident one is the reduction of the wave amplitude, given by the factor $\sqrt{T}$; another effect is the phase factor $e^{i\delta}$ derived from the delay that the wave suffers when passing the barrier. In the limit where the barrier disappears we obtain $\delta = -\arctan ka \to 0$, which confirms that δ is effectively due to the presence of the barrier. (It is easier to go to this limit when $E > V_0$; then q becomes imaginary and $\tanh qa$ in Eq. (5.75) becomes $\tan qa$.) For this reason δ is called the *phase shift*. In Section 6.2.2 we return to this problem to determine the time delay produced by the barrier and in Section 18.4 it will be shown that particle scattering processes can be described in terms of these phase shifts.

5.4.1.3 Spontaneous decay

To gain some experience with the tunneling effect, we will use it as a basis for explaining the spontaneous decay of a system, for example, a radioactive nucleus. To do this, we modify the preceding potential by adding an infinitely rigid wall at $x = 0$, creating a well from which particles can only tunnel out to the right (Fig. 5.9). In this case the variable x represents the radial distance measured from the center of the well: the width of the well (i.e., its radius) is l and the width of the barrier is a. We write the solution of the Schrödinger equation as follows, taking A_1 real,

$$\begin{aligned}\psi_I &= A_1 \sin kx, \\ \psi_{II} &= A_2 e^{-q(x-l)} + B_2 e^{q(x-l)}, \\ \psi_{III} &= A_3 e^{ik(x-l-a)} + B_3 e^{-ik(x-l-a)}.\end{aligned} \tag{5.77}$$

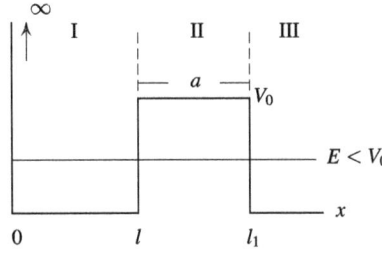

Figure 5.9 Schematic diagram of the potential well of width l, enclosed by a barrier of width a and height V_0, to show the phenomenon of spontaneous decay of a system as particles tunnel to the right. The number of particles inside the well decreases exponentially with time.

As before, $k^2 = 2mE/\hbar^2$, $q^2 = 2m(V_0 - E)/\hbar^2$. After stitching together the solutions at the boundaries, we get

$$A_2 = \tfrac{1}{2}A_1\left(\sin kl - \frac{k}{q}\cos kl\right),$$

$$B_2 = \tfrac{1}{2}A_1\left(\sin kl + \frac{k}{q}\cos kl\right),$$

$$A_3 = \tfrac{1}{2}e^{-qa}\left(1 + \frac{iq}{k}\right)A_2 + \tfrac{1}{2}e^{qa}\left(1 - \frac{iq}{k}\right)B_2,$$

$$B_3 = \tfrac{1}{2}e^{-qa}\left(1 - \frac{iq}{k}\right)A_2 + \tfrac{1}{2}e^{qa}\left(1 + \frac{iq}{k}\right)B_2. \tag{5.78}$$

These equations are satisfied for any energy $E < V_0$; therefore, the spectrum is continuous in principle. However, if we impose additional constraints, the spectrum can become discrete. In particular, this happens when the particles are initially confined in the well, so that in region III there are only particles escaping. In this case we must take $B_3 = 0$, from which it follows that

$$\left(\tan kl + \frac{k}{q}\right)\left[1 + \frac{k - iq}{k + iq}e^{-2qa}\right] = \frac{2k}{q}\frac{k - iq}{k + iq}e^{-2qa}. \tag{5.79}$$

This expression determines the discrete values for the energy of the quasi-confined particles. Comparing with Eq. (5.47), we see that the presence of a barrier of finite width modifies the energy levels of the particles inside the well; moreover, the solutions of Eq. (5.79) for k are complex, which implies a complex energy E. To see the meaning of this, we must refer to the complete expression for the wave function, including the exponential time factor, which is given by $\psi(x)e^{-iEt/\hbar}$. Therefore, the particle density inside the well is

$$\rho_I = |\psi_I e^{-\frac{iE}{\hbar}t}|^2 = |\psi_I|^2 e^{-\frac{i}{\hbar}(E - E^*)t} = |\psi_I|^2 e^{\frac{2\mathrm{Im}E}{\hbar}t}, \tag{5.80}$$

that is,

$$\rho_I = |\psi_I|^2 e^{-\lambda t}, \tag{5.81}$$

where the decay rate, λ, is given by

$$\lambda = -\frac{2}{\hbar}\mathrm{Im}E. \tag{5.82}$$

These equations show that the population inside the well decreases exponentially with time if the particles have complex energy ($\lambda \neq 0$), in other words, leakage through the barrier causes the population to decrease exponentially over time. The inverse of the parameter λ is the *mean* lifetime of the system, which means that over a period of time $t = 1/\lambda$ the population will decrease to the fraction $1/e \approx 35\%$ of its initial value. The reason this is called *spontaneous decay* is that there is no external factor causing the decay.

The conclusion is that the real part of the correction of k given by Eq. (5.79) with respect to the limit $a \to \infty$ (square well) represents a shift of the energy levels inside the well, while the imaginary part indicates that the states are no longer stationary and a decay takes place.[2] Note that while the population of region I decreases with time, that of region III must increase. In the present case the decay is due to the leakage of electrons trapped in the well; in the nuclear case, it will be the escape of α particles from a nucleus, and so on.

Exercise E5.3. Decay rate

Find an approximate formula to first order for the decay rate and its relationship to the transmission coefficient, assuming that the exact solution is close to the solution found for the infinitely wide barrier. Consider the case where the potential is large so that $ql \gg 1$, and the well is narrow so that $k_0 l \ll 1$.

Solution. To determine the decay (or escape) rate λ exactly, we would have to solve Eq. (5.79). We can solve it in an approximate way, considering that for an infinitely wide barrier ($a \to \infty$), the equation reduces to Eq. (5.47),

$$\tan kl + \frac{k}{q} = 0. \tag{5.83}$$

We write $E = E_0 + \delta E$, with E_0 the solution of this equation and δE the (complex) correction to the energy, and expand (5.79) in Taylor series around E_0. Since

$$k^2 = (k_0 + \delta k)^2 = \frac{2m}{\hbar^2} E = \frac{2m}{\hbar^2}(E_0 + \delta E), \tag{5.84}$$

the first-order terms give $2k_0 \delta k = (2m/\hbar^2)\delta E$, whence

$$k = k_0 + \delta k = k_0 + \frac{m}{\hbar^2 k_0}\delta E + \ldots, \tag{5.85}$$

Expanding (5.79) to first order in δk we obtain

$$\delta k(1 + ql \sec^2 k_0 l) = 2k_0 \frac{k_0 - iq}{k_0 + iq} e^{-2qa}. \tag{5.86}$$

If $k_0 l \ll 1$ (low energy, narrow well), we may take $\sec^2 k_0 l \simeq 1$. If, in addition, the potential is large so that for $ql \gg 1$, we can neglect unity against ql and write the above result in the form

$$\delta k = \frac{2k_0}{lq} \frac{k_0^2 - q^2 - 2ik_0 q}{k_0^2 + q^2} e^{-2qa}. \tag{5.87}$$

From this expression we obtain both the energy shift and the decay rate. In particular, using the relationship between the corrections to the wave number k and the energy E, we see that

$$\operatorname{Im} E \simeq \frac{\hbar^2 k_0}{m} \operatorname{Im} \delta k = -\frac{4\hbar^2 k_0}{ml} \frac{k_0^2}{k_0^2 + q^2} e^{-2qa}, \tag{5.88}$$

which when introduced in (5.82), gives

$$\lambda = \frac{8v}{l} \frac{k_0^2}{k_0^2 + q^2} e^{-2qa}, \tag{5.89}$$

---

[2] Note that by treating this problem as a stationary one, an inconsistency appears in the solution for ψ inside the well, due to the imaginary part of δk. To obtain the correct wave function one must solve the time-dependent Schrödinger equation, which is studied in Chapter 7.

where $v = p/m = \hbar k_0/m$ stands for the velocity of the particles inside the well. Noting that in the preceding approximation, $q^2/(k_0^2 + q^2) \approx 1$, and using Eq. (5.72) for the transmission coefficient through the barrier, we write the preceding result in the form

$$\lambda = \frac{v}{2l} T. \tag{5.90}$$

This is the desired relationship between the decay rate and the transmission coefficient. The meaning of this formula can be understood by considering that $v/2l$ is the average number of times the particle hits the wall per unit time; this value multiplied by the probability T that the particle escapes each time gives the escape rate per unit time.

5.4.2 $E > V_0$, resonant transmission

This problem is analogous to those studied earlier. Although its algebra is somewhat more complicated, it leads to similar results and, in particular, to resonant transmission at energies defined by the condition $\sin qa = 1$, that is, for $qa = n\pi$. Specifically, the transmission and reflection coefficients are given by

$$T = 1 - R = \frac{1}{1 + \left[\frac{1}{4}\left(\frac{q}{k} + \frac{k}{q}\right)^2 - 1\right]\sin^2 qa}. \tag{5.91}$$

If q and k are very different, the resonance peaks can be very sharp and noticeable.

5.5 The Symmetric Double Well: Energy Splitting and Periodic Tunneling

The tunneling effect studied earlier shows that the stationary energy levels of a well are altered by the presence of neighboring potentials; in other words, the behavior of particles in a given region of space depends not only on the local potential in that region, but also on the potential that exists at every point in the available space. This is an important observation. For example, if we place another potential well next to the first, separated from it by a finite barrier, the introduction of the second well will modify the levels of the first, and vice versa.

Even more important than the level shift is the change in the number of possible stationary (or quasi-stationary, in the sense of the previous section) levels contained in each well. For example, we will see that in the case studied in this section, each level of one of the wells is split by the presence of the other well, producing a pair of closely spaced levels. This result reflects a general regularity: if a number of similar wells are placed side by side, separated by finite barriers, each original level will decompose into as many neighboring levels. In the next chapter we will see how this phenomenon allows us to understand in a simple way a number of properties of periodic structures, such as those of a crystal.

5.5 The Symmetric Double Well: Energy Splitting and Periodic Tunneling

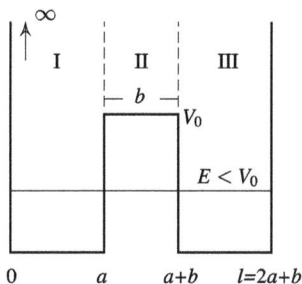

Figure 5.10 The symmetrical double well of total width $l = 2a + b$. Because the outer walls are impenetrable, there are stationary states inside.

Physically we can understand the phenomenon as follows. If we first imagine that each well is isolated by infinitely rigid walls and that the electrons are in a stationary state, then if we reduce the height of the barriers separating them, the boundary conditions of each well are modified, leading to an energy shift. Furthermore, the presence of an intermediate finite barrier allows particles to tunnel from one well to another, and the energy eigenvalues are consequently modified, as we saw in Section 4.

In the case of two wells separated by a finite barrier, as shown in Fig. 5.10, one well loses electrons by leakage and, on the other hand, it gains electrons that escaped from the second well. This double effect produces a level splitting. Clearly, the argument can be extended directly to the case of n similar wells. If the multi-well chain is bounded at both ends by infinitely repulsive potentials, as in Fig. 5.10, there are stationary states in the system, which occur when the flow of electrons escaping from one well is compensated by those entering from neighboring wells.

To find the energy levels of the double well we write the stationary solutions of the Schrödinger equation in the following form, limiting ourselves to the case $E < V_0$,

$$\begin{aligned} \psi_I &= A_1 \sin kx, \\ \psi_{II} &= A_2 e^{qx} + B_2 e^{-qx}, \\ \psi_{III} &= A_3 \sin k(l - x), \end{aligned} \quad (5.92)$$

with $l = 2a+b$. In writing Eqs. (5.92) the boundary conditions at $x = 0$ and $x = l$ have already been imposed. The remaining boundary conditions give the following relation between the constants,

$$\begin{aligned} A_1 \sin ka &= A_2 e^{qa} + B_2 e^{-qa}, \\ A_2 e^{q(a+b)} + B_2 e^{-q(a+b)} &= A_3 \sin ka, \\ k A_1 \cos ka &= q[A_2 e^{qa} - B_2 e^{-qa}], \\ q[A_2 e^{q(a+b)} - B_2 e^{-q(a+b)}] &= -k A_3 \cos ka. \end{aligned} \quad (5.93)$$

Since we have only three free constants to satisfy this system of four equations (one of them, A_1, for example, we fix by normalization), there is a condition between the parameters, which determines the allowed values for the energy E. To derive this quantization condition we first eliminate A_2 and B_2 from the previous system, thus obtaining the pair of equations

$$A_1\left(\tan ka + \frac{k}{q}\right) = A_3\left(\tan ka - \frac{k}{q}\right)e^{-qb}, \tag{5.94}$$

$$A_1\left(\tan ka - \frac{k}{q}\right) = A_3\left(\tan ka + \frac{k}{q}\right)e^{qb}. \tag{5.95}$$

Dividing the first equation by the second gives the quantization condition

$$\left(\tan ka + \frac{k}{q}\right)^2 = \left(\tan ka - \frac{k}{q}\right)^2 e^{-2qb}, \tag{5.96}$$

that is,

$$\tan ka + \frac{k}{q} = \pm\left(\tan ka - \frac{k}{q}\right)e^{-qb}. \tag{5.97}$$

Taking the limit $b \to \infty$ in this expression, we recover the quantization condition (5.47) for the potential well shown in Figure 5.6. Equation (5.97) confirms our expectations, since for finite b there is a correction with a double sign, that is, there are two possible values for ka. Therefore, if k_0 is a value of k corresponding to the well of Figure 5.6, the modified energies in the double well of Figure 5.10 are obtained by adding to k_0 the corrections calculated using (5.97). To estimate these first-order corrections, we take as a reference the value of k_0 given by (5.47),

$$\tan k_0 a + \frac{k_0}{q_0} = 0, \quad q_0 = \frac{2m}{\hbar^2}V_0 - k_0. \tag{5.98}$$

With $k = k_0 + \delta k$ and expanding $\tan ka$ in Taylor series around k_0, we obtain (for $k_0 a \ll 1$ and $q_0 a \gg 1$)

$$\tan ka + \frac{k}{q} \approx \tan k_0 a + a\delta k \sec^2 k_0 a + \frac{k_0}{q_0} + \frac{\delta k}{q_0} \simeq a\delta k\left(1 + \frac{1}{aq_0}\right) \simeq a\delta k \tag{5.99}$$

and

$$\tan ka - \frac{k}{q} \simeq \tan k_0 a + \frac{k_0}{q_0} - \frac{2k_0}{q_0} = -\frac{2k_0}{q_0}. \tag{5.100}$$

Introducing these values into (5.97), we obtain to first order

$$\delta k \approx \mp \frac{2k_0}{aq_0}e^{-q_0 b}. \tag{5.101}$$

It is left to the student to investigate the stationary wave functions for the lowest levels, both in the case of symmetric and antisymmetric solutions. In particular, it can be shown that the even wave function (with respect to the axis of symmetry of the potential) corresponds to the minimum energy of the system (with $\delta k < 0$), while a slightly higher energy corresponds to the odd function (with $\delta k > 0$). These two energies, denoted E_- and E_+, respectively, are given by the following expression, which follows directly from Eq. (5.101) and the definition of k, Eq. (5.6),

$$E_\pm = E_0 \pm \delta E = E_0 \pm \frac{4E_0}{aq_0}e^{-q_0 b}. \tag{5.102}$$

The respective wave functions $\psi_\pm$ are shown schematically in Figure 5.11 (a).

5.5 The Symmetric Double Well: Energy Splitting and Periodic Tunneling

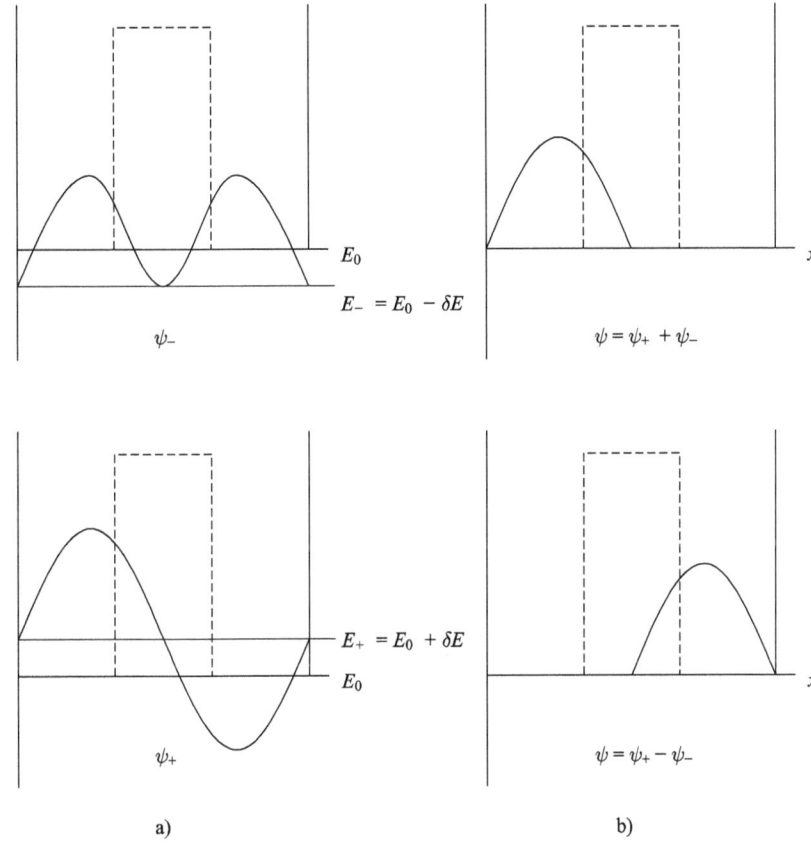

Figure 5.11 The symmetrical double well. (a) Symmetric and antisymmetric wave functions ψ_- and ψ_+ corresponding to the lowest energy levels, the first with a slightly lower energy than the second. (b) Distribution of particles in states $\psi_+ + \psi_-$ and $\psi_+ - \psi_-$.

5.5.1 Masers and molecular clocks

Suppose now that we use the stationary wave functions $\psi_\pm$ (which describe delocalized states, that is, with presence in both wells as shown in Figure 5.11a) to construct the following wave function (which is not an energy eigenfunction), by multiplying each (stationary) solution by its time factor,

$$\psi(x,t) = a\psi_+(x)e^{-\frac{iE_+}{\hbar}t} + b\psi_-(x)e^{-\frac{iE_-}{\hbar}t}. \tag{5.103}$$

We can select the coefficients a and b to satisfy any reasonable initial condition; in particular, for $t = 0$ we have

$$\psi(x, t = 0) = a\psi_+ + b\psi_-. \tag{5.104}$$

As can be seen from Figure 5.11 (b), if $a = b$, the resulting wave function has a large maximum in the left well and is practically zero in the right well, that is, the particles are initially (almost) all located in the left well. With this initial condition, the wave function (5.103) becomes

$$\psi(x,t) = a\left[\psi_+ e^{-\frac{i\delta E}{\hbar}t} + \psi_- e^{\frac{i\delta E}{\hbar}t}\right]e^{-\frac{iE_0}{\hbar}t} \qquad (5.105)$$

$$= ae^{-\frac{iE_0}{\hbar}t}\left[(\psi_+ + \psi_-)\cos\frac{\delta E}{\hbar}t - i(\psi_+ - \psi_-)\sin\frac{\delta E}{\hbar}t\right].$$

At $t = 0$ the particles are indeed in the left well; but a time T later, given by $(\delta E/\hbar)T = \pi/2$, they will have migrated to the right well. Therefore, the wave function (5.105) represents a packet that periodically tunnels from one well to the other, with the period given by

$$2T = \frac{\pi\hbar}{\delta E} = \frac{\pi\hbar}{4E_0}aq_0 e^{q_0 b}. \qquad (5.106)$$

It is possible not only to observe this effect, but to use it in the construction of technological devices. A typical example is provided by the ammonia molecule NH_3, which has a pyramidal structure with the three H atoms at the base and the N at the apex on either side of the plane defined by the three H atoms. Therefore, the intermolecular potential has a minimum at these two positions, that is, there are two wells separated by a barrier. For a qualitative analysis we can model this interatomic potential with rectangular wells. If we place ammonia in an electromagnetic field that oscillates with a period close to $2T$ (which corresponds to frequencies in the microwave band, with a wavelength close to 1.25 cm), the absorption of radiation by the molecule stimulates transitions between the two stationary states ψ_+, ψ_- separated by an energy $2\delta E$.

This phenomenon was exploited for the construction of the maser,[3] which essentially consists of a device containing NH_3 molecules initially in the ψ_+ state. When irradiated with the faint microwave field of frequency $\omega = 2\delta E/\hbar$ captured by a radio telescope, transitions to the ψ_- state are induced. The energy released is used to detect the presence of the radiation. The device functions as a microwave amplifier by stimulated emission of radiation, hence the name.

Clocks have also been built based on these transitions, to take advantage of the extraordinary stability of the operating frequency. Molecular clocks, which can be modelled as double potential wells, run with an accuracy of up to one second over several hundred years. This high precision made it possible to construct independent but synchronized clocks, which were used to verify some predictions of the theory of relativity to first order in v/c. In contrast, the usual interference experiments, which use a single clock and therefore require closed trajectories, only allow the observation of second-order effects, proportional to $(v/c)^2$.[4]

---

[3] Maser is an acronym for *microwave amplification by stimulated emission of radiation*. With further technological developments and the use of suitable transitions in other materials, light amplification by stimulated radiation was achieved, giving rise to the laser (see Section 17.2). These important devices act today both as amplifiers and generators. The initial work on masers was carried out independently in 1952–1954 by the Soviet physicists Nikolai Gennadievich Basov (1922–2001) and Alexandr Mijailovivh Prokhorov (1916–2002), and the American physicist Charles Hard Townes (1915–2015) and collaborators; all three were awarded the Nobel prize in 1964 for their contributions.

[4] In the 1990s, molecular clocks were replaced by atomic clocks, which can run to an accuracy of one second over 300 million years. This has led to the internationally agreed definition of the second based on the frequency of a particular transition of the caesium 133 atom, which is 9192631770 s^{-1}. Today, one of the most important applications of the amazing accuracy of atomic clocks is the GPS (Global Positioning System), used not only in telecommunications and transport, which was the original motivation, but also as a central element of mobile phones and other portable devices.

5.6 Corollary: Meaning of Particle-Wave Behavior

Although all the problems considered in this chapter have involved the motion of particles subject to simple potential forces, the solutions obtained are reminiscent of the behavior of waves, such as string waves, water surface waves, sound waves or light waves. We have been forced to use the language of waves to describe the statistical behavior of particles. We talk about standing waves, or waves that travel in one direction or another, and we associate them with a wave number and a wavelength. We observe interference between these waves. Phase shifts and resonances occur here and there. In the tunneling effect the wave "penetrates" the barrier, just as an evanescent "wave" does in the optical case. And yet the particles represented by these waves never cease to be particles.

It is important to remember that just as the particles always retain their individuality and smallness, the statistics of the ensemble also retain their distributed properties. We see corpuscular behavior when attention is focused on the particle, and when attention is focused on the ensemble – the pattern traced by the large collection of particles – we may recognize what we think of as a wave. And indeed, it is the wave-like distribution just predicted by Schrödinger's "wave" equation. We are then in the presence of a most subtle stratagem of nature – as if it were "determined" to confuse us. In Chapter 10, we will come back to this topic and discuss it in the context of stochastic electrodynamics.

Problems

P5.1 A particle moves in a symmetric potential $V(x) = V(-x)$, such that the spectrum is discrete for $E < 0$ and continuous for $E > 0$. Draw a graph showing the general form of this potential if it is known that

(a) there are an infinite number of bounded states,
(b) there are a small number of bounded states.

Explain your argument. What would happen if the potential were not symmetric?

P5.2 Show that the principle of invariance against time reversal implies that, for a given energy, the coefficients of transmission and reflection by a potential step are the same, whether the particles are incident from the right or from the left.

P5.3 Calculate the reflection coefficient for particles of mass m and energy $E < 0$, incident from the left on the complex potential step.

$$V(x) = \begin{cases} 0, & x \leq 0, \\ (1+ia)V_0, & x > 0, \end{cases} \quad (5.107)$$

where a is real and $V_0 > 0$.

P5.4 Show that a 1D square well has exactly $n+1$ even bounded states if

$$\frac{2\pi^2\hbar^2}{ma^2}n^2 \leq V_0 < \frac{2\pi^2\hbar^2}{ma^2}(n+1)^2 \quad (5.108)$$

and $n+1$ odd bound states if

$$\frac{2\pi^2\hbar^2}{ma^2}\left(n+\tfrac{1}{2}\right)^2 \le V_0 < \frac{2\pi^2\hbar^2}{ma^2}\left(n+\tfrac{3}{2}\right)^2. \qquad (5.109)$$

P5.5 Show that an extremely narrow and deep attractive potential that can be represented as a Dirac delta, $V(x) \simeq -\delta(x)/a$, contains a single bounded state, and find the corresponding energy eigenvalue. *Hint:* solve the Schrödinger equation for $x \ne 0$ and note that ψ' is not continuous at $x=0$.

P5.6 Extend the study of the finite rectangular well to the 3D case using Cartesian coordinates. Show that if the well is very deep, the results of Exercise 5.22 are recovered.

P5.7 Study the energy degeneracy as a function of the quantum numbers for a cubic potential well with impenetrable walls.

P5.8 Show that the S-matrix for particles through a 1D potential is unitary. Use the result to prove the following relations,

$$|S_{11}|^2 + |S_{12}|^2 = 1, \quad |S_{21}|^2 + |S_{22}|^2 = 1, \quad S_{11}S_{12}^* + S_{21}S_{22}^* = 0. \qquad (5.110)$$

P5.9 Determine the general form of the elements S_{ij} of the S-matrix describing the dispersion of particles through a 1D square well with $E > 0$.

P5.10 Determine the scattering matrix for the potential $V(x) = -a\delta(x-b)$.

P5.11 Derive in detail the formula (5.91) for the transmission coefficient of particles through a rectangular barrier when $E > V_0$.

Bibliographical Notes

A detailed and excellent textbook in two volumes is Cohen-Tannoudji et al. (1977).
Other excellent, advanced textbooks are Gasiorowicz (2003) and Davydov (2013).
For the evolution of a wave packet, see Goldberg et al. (1967).
The website https://commons.wikimedia.org/wiki/Category:Animations_of_quantum _tunneling provides a list of 12 links to different versions of animations of quantum tunneling. You might try making your own animations for the problems discussed in this chapter.

6 The WKB Approximation. Electronic Properties of Solids

In real situations, the potentials are usually more complicated than those studied in the previous chapter. However, exact solutions of the Schrödinger equation for such potentials do not generally exist, so it is necessary to develop specific solution methods. Depending on the nature of the problem and the purpose of the study, it is usually convenient to use an approximation method that provides a sufficiently simple and satisfactory solution. Several approximation methods exist, and in this book we will have the opportunity to draw on some of them.

In the first part of this chapter we present an approximation technique developed simultaneously and independently in 1926 by G. Wentzel (1896–1978), H. A. Kramers (1894–1952), and L. Brillouin (1889–1969), from whose surnames the acronym WKB is derived. The name WKB refers specifically to the method when applied to QM. More generally, it is a procedure for constructing approximate solutions to linear differential equations with spatially varying coefficients, known in mathematics as the Liouville–Green[1,2] method. In QM, the WKB method is extremely useful for 1D problems, although it is occasionally used for problems in more than one dimension. It is particularly suitable when the particle is in a sufficiently energetic state that its behavior can be considered to be close to classical almost everywhere, in the sense explained in what follows. For this reason, it is called a *semiclassical approximation* method. The meaning of the term semiclassical in this context should be clearly distinguished from its meaning in the context of *semiclassical theory*, where quantized matter is studied in interaction with classical (nonquantized) radiation.

The rest of the chapter is devoted to a discussion of the basic electronic properties of solids and some important applications of these properties, which are easily explained using the results obtained in the first part.

6.1 The Semiclassical Approach

To solve the stationary Schrödinger equation for the wave function ψ

$$-\frac{\hbar^2}{2m}\nabla^2\psi + V\psi = E\psi \tag{6.1}$$

---

[1] Joseph Liouville (1809–1882) was a notable French mathematician who made important contributions to several branches of mathematics.
[2] George Green (1793–1841) was an outstanding British mathematical physicist who, among several other contributions, introduced the concept of Green's functions and applied mathematics to the study of electricity and magnetism.

with the WKB method, we write ψ in the form[3]

$$\psi = e^{iS/\hbar}, \tag{6.2}$$

where $S(x)$ is a function of x with dimensions of action, usually complex. To determine S we substitute (6.2) into (6.1). This gives

$$\nabla^2 \psi = \nabla \cdot (\nabla \psi) = \nabla \cdot \left(\frac{i}{\hbar} \psi \nabla S\right) = \frac{i}{\hbar}(\nabla^2 S)\psi - \frac{(\nabla S)^2}{\hbar^2}\psi, \tag{6.3}$$

so Eq. (6.1) becomes

$$\frac{(\nabla S)^2}{2m} - \frac{i\hbar}{2m}\nabla^2 S + V = E. \tag{6.4}$$

Here we introduce the simplifying consideration that the system is in a state such that

$$\hbar|\nabla^2 S| \ll (\nabla S)^2, \tag{6.5}$$

in which case Eq. (6.4) takes the approximate form

$$\frac{(\nabla S)^2}{2m} + V \doteq E, \tag{6.6}$$

where the $\hbar$ has disappeared; this explains the name "semiclassical" for the procedure. Equation (6.6) reminds us of the Hamilton–Jacobi[4,5] equation of classical mechanics for the action $S(x)$, whose gradient determines the momentum of the particle

$$\boldsymbol{p} = \nabla S. \tag{6.7}$$

It is sufficient to substitute this last formula into the Hamilton–Jacobi equation to obtain the correct expression for the energy in a (classical) conservative problem. This means that if the condition (6.5) is satisfied, the function S is approximately given by Eq. (6.7).[6] The semiclassical approach starts from the assumption that under appropriate conditions (basically, that condition (6.5) is met) the quantum problem can be solved approximately by finding quantum corrections to the classical action and using this corrected action to determine the wave function.

Since the corrections we are looking for are caused by the term containing $\hbar\nabla^2 S$, which is proportional to $\hbar$, they in turn must depend on $\hbar$. Based on this, we propose that the solution of Eq. (6.4) can be expressed as an expansion of S in powers of $\hbar$,[7]

$$S = S_0 + i\hbar S_1 + (i\hbar)^2 S_2 + \ldots, \tag{6.8}$$

---

[3] The solution of the stationary wave equation is often written in optics in the form $\exp iS(x)$, where S is known as the *eikonal*, from the Greek word for icon (image). By extension, the WKB method is sometimes referred to as an expansion in terms of the eikonal, that is, a term corresponding to a wave, either stationary or propagating.

[4] Sir William Rowan Hamilton (1805–1865) was an eminent British mathematician, physicist, and astronomer, who made many contributions, including Hamiltonian mechanics.

[5] Carl Gustav Jacob Jacobi (1804–1851) was an eminent German mathematician; it would take a long list to recall all the achievements that bear his name, including a crater on the Moon.

[6] A word of caution, however: while the Hamilton–Jacobi equation of classical mechanics describes the deterministic motion of a single particle, Eq. (6.4) and its approximate version (6.6) refer to an ensemble of particles. The analogy is therefore strictly formal.

[7] This assumes that the solution can be expressed by an analytical function of Planck's constant, which is not always the case.

where none of the functions S_j, $j = 0, 1, 2, \ldots$ contains $\hbar$. Substitution of the gradient of this expression into (6.4) gives

$$\frac{1}{2m}[(\nabla S_0)^2 + 2i\hbar \nabla S_0 \cdot \nabla S_1 - 2\hbar^2 \nabla S_0 \cdot \nabla S_2 - \hbar^2(\nabla S_1)^2 + \ldots]$$
$$- \frac{i\hbar}{2m}[\nabla^2 S_0 + i\hbar \nabla^2 S_1 + \ldots] + V - E = 0. \qquad (6.9)$$

By canceling the coefficients of each power of $\hbar$ separately, we obtain a hierarchy of differential equations for the S_j, $j = 0, 1, 2, \ldots$,

$$\frac{(\nabla S_0)^2}{2m} + V = E, \qquad (6.10)$$

$$\nabla S_0 \cdot \nabla S_1 - \tfrac{1}{2}\nabla^2 S_0 = 0, \qquad (6.11)$$

$$2\nabla S_0 \cdot \nabla S_2 + (\nabla S_1)^2 - \nabla^2 S_1 = 0, \qquad (6.12)$$

and so on. The first of these equations coincides with the (classical) Hamilton–Jacobi equation, therefore we take for S_0 the classical action of the system, which is obtained from an integration of the classical momentum,

$$\nabla S_0 = \mathbf{p}. \qquad (6.13)$$

Once S_0 is determined, Eq. (6.11) should allow us to obtain S_1; substituting S_0 and S_1 in Eq. (6.12), we can determine S_2 in principle, and so on. Normally it is sufficient to know S to the first order in $\hbar$, which is achieved by solving only the first two equations of the hierarchy.

In two- or three-dimensional problems, additional considerations are required to solve Eqs. (6.10)–(6.12), hence the limited applicability of the method in such cases. In the following we restrict ourselves to the one-dimensional case, where Eqs. (6.10) and (6.11) take the simple form

$$(S_0')^2 = 2m(E - V) = p^2, \qquad (6.14)$$

$$S_1' = \frac{S_0''}{2S_0'} = \tfrac{1}{2}\frac{d}{dx}\ln S_0' = \frac{d}{dx}\ln \sqrt{p}. \qquad (6.15)$$

Integrating, we get

$$S_0 = \pm \int^x p\, dx; \quad S_1 = \ln \sqrt{p} = -\ln \frac{1}{\sqrt{p}}. \qquad (6.16)$$

Therefore, up to linear terms in $\hbar$ (classical solution plus first quantum correction) and to a normalization factor, the wave function has the general form

$$\psi = e^{(i/\hbar)(S_0 + i\hbar S_1)} = e^{iS_0/\hbar - S_1} = \exp\left[\pm(i/\hbar)\int^x p\, dx + \ln \frac{1}{\sqrt{p}}\right], \qquad (6.17)$$

that is,[8]

$$\psi = \frac{1}{\sqrt{|p|}} \exp\left[\pm i/\hbar \int^x p\, dx\right]. \qquad (6.18)$$

To establish the range of validity of this approximation, we note that Eqs. (6.10) and (6.13) allow us to write the magnitude of p in its classical form,

$$p = \sqrt{2m(E-V)}, \qquad (6.19)$$

and from (6.5) and (6.13) it follows that the semiclassical method is applicable only if

$$\hbar\left|\frac{dp}{dx}\right| \ll p^2. \qquad (6.20)$$

From Eq. (6.19) we get

$$\frac{dp}{dx} = -\sqrt{\frac{m}{2(E-V)}}\frac{dV}{dx} = \frac{mF}{p}, \qquad (6.21)$$

where $F(x)$ represents the external force acting on the particle; substituting in (6.20), we get

$$|\hbar m F/p^3| \ll 1. \qquad (6.22)$$

This condition is only fulfilled in regions of x where the force F is "small" and the classical momentum p is "large." From this we conclude that quantum effects are particularly noticeable: (a) where the potential V varies strongly; (b) in regions close to the classical turning points, for which $V = E$ and classical particles stop, since they cannot move to regions where $V > E$. In such regions, the semiclassical formulae are not valid.

Figure 6.1 shows a smoothly varying potential, to which we can apply the WKB method. The shaded zones around the turning points x_1 and x_2 are to be excluded according to the preceding discussion. To determine the width of the exclusion zones we proceed as follows. Near the turning point x_1 we can approximate the potential with a first-order Taylor expansion:

$$V(x) = V(x_1) + (x-x_1)V' = E - (x-x_1)F, \qquad (6.23)$$

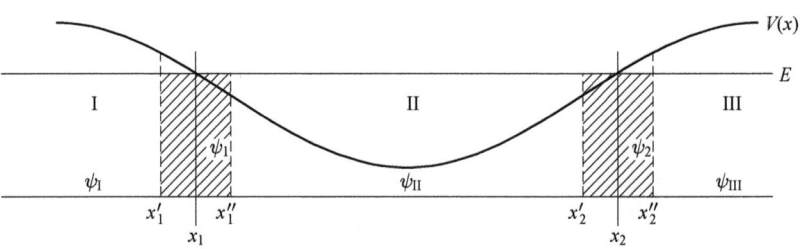

Figure 6.1 In regions I, II, and III the WKB functions are applicable; the shaded areas represent the exclusion zones.

---

[8] Note that in the present approximation the "classical action" $\int^x p\, dx$ is everywhere measured with respect to Planck's constant, a most fundamental reminder of the role of $\hbar$. Some authors refer to the semiclassical limit as the limit $\hbar \to 0$; however, $\hbar$ is a universal constant with a fixed value. Expressions that condition the validity of the approximation, such as (6.20) or (6.22), must be read as constraints on the dynamics, not on $\hbar$.

since at $x = x_1$, $V = E$. Substituting this expression in Eq. (6.19), we obtain for $x > x_1$ (to the right of the turning point),

$$p = \sqrt{2mF(x - x_1)}, \qquad (6.24)$$

and substituting this result in (6.22), we obtain

$$|x - x_1| \gg \tfrac{1}{2} \left| \frac{\hbar^2}{mF} \right|^{1/3}, \qquad (6.25)$$

which means that the semiclassical solution applies outside this zone, $x > x_1''$ in Fig. 6.1. Using this criterion we eliminate from our analysis the neighboring zones of x_1 and x_2, now the WKB approach can be safely applied in the regions marked I, II, and III.

The solution (6.18) applies to the case $E > V$, since p was taken to be real (see Eq. (6.19)); hence it corresponds to the wave function of region II, ψ_{II}. To construct this function, we note that the general solution is the sum of the two imaginary exponential solutions with arbitrary coefficients, which can be rewritten in terms of a trigonometric function using an amplitude and a phase as integration constants, with the origin for the latter at x_1,

$$\psi_{II}(x) = \frac{A_2}{\sqrt{p}} \sin \left[\frac{1}{\hbar} \int_{x_1}^{x} p(x') \, dx' + \alpha_2 \right], \quad x_1'' \le x \le x_2'. \qquad (6.26)$$

The factor $\sqrt{p}$ in the denominator of ψ_{II} means that the fraction of particles $\rho dx \sim \left(|A_2|^2 / p \right) dx$ that are on average in the interval dx becomes smaller as the flux velocity becomes larger. This is intuitively clear from the classical perspective, where the average time a particle spends in the interval $(x, x + dx)$ is inversely proportional to the average velocity at x.

In regions I and III, $V > E$ and p is imaginary, so the exponentials are real; therefore we must write for $x \ge x_2''$

$$\psi_{III} = \frac{1}{\sqrt{|p|}} \left[A_3 \exp\left(-\frac{1}{\hbar} \int_{x_2}^{x} |p(x')| \, dx' \right) + B_3 \exp\left(\frac{1}{\hbar} \int_{x_2}^{x} |p(x')| \, dx' \right) \right] \qquad (6.27)$$

and an analogous expression in region I for $x \le x_1'$. In these expressions $|p|$ is

$$|p| = \sqrt{2m(V - E)}. \qquad (6.28)$$

These equations show that in the classically allowed regions, the WKB wave function is oscillatory, while it tends to decay exponentially in the classically forbidden regions.

We now need to stitch the piecewise solutions together to ensure that they are part of the same wave function. The problem is not so simple, because the preceding solutions are not valid at the turning points. But just near the turning points the potential can be approximated with a first-order Taylor expansion, like (6.23), which, when introduced into the Schrödinger equation, gives a solution in terms of Bessel functions. Then, for example, the solution ψ_1 valid in the exclusion zone $x_1' < x < x_1''$, is tied to ψ_I at x_1' and to ψ_{II} at x_1'', which establishes a relation between the respective coefficients A_i, B_i. Similarly, the solution ψ_2 valid for $x_2' < x < x_2''$, is tied to ψ_{II} at x_2' and to ψ_{III} at x_2'' (see Fig. 6.1). Since this is a purely technical subject that can be consulted in the literature, we will omit it here and only present the final results. The pair of solutions tied to both sides of x_2, obtained under the hypothesis that in region III the solution is exponentially decreasing, is

$$\psi_{II}(x) = \frac{A}{\sqrt{p}} \sin\left[\frac{1}{\hbar}\int_x^{x_2} p(x')\,dx' + \frac{\pi}{4}\right], \quad x < x_2', \tag{6.29}$$

$$\psi_{III}(x) = \frac{A}{2\sqrt{|p|}} \exp\left[-\frac{1}{\hbar}\int_{x_2}^x |p(x')|\,dx'\right], \quad x > x_2''. \tag{6.30}$$

A similar result is obtained for the solutions around the turning point x_1, but with the integration limits reversed, so that the lower limit is always less than the upper one.

Exercise E6.1. Quantization in a potential well

Apply the WKB formulae to a particle bound by an arbitrary but smoothly varying potential well, similar to that shown in Fig. 6.1, to derive the action quantization condition and an (approximate) expression for the normalized wave function inside the well.

Solution. We can write the WKB wave function inside the well in two equivalent forms,

$$\psi = \frac{A'}{\sqrt{p}} \sin\left[\frac{1}{\hbar}\int_{x_1}^x p\,dx + \frac{\pi}{4}\right], \tag{6.31}$$

$$\psi = \frac{A''}{\sqrt{p}} \sin\left[\frac{1}{\hbar}\int_x^{x_2} p\,dx + \frac{\pi}{4}\right]. \tag{6.32}$$

From the second expression it follows that

$$\begin{aligned}\psi &= \frac{A''}{\sqrt{p}} \sin\left[\frac{1}{\hbar}\left(\int_{x_1}^{x_2} - \int_{x_1}^x\right) p\,dx + \frac{\pi}{2} - \frac{\pi}{4}\right] \\ &= -\frac{A''}{\sqrt{p}} \sin\left[\frac{1}{\hbar}\int_{x_1}^x p\,dx + \frac{\pi}{4} - \frac{1}{\hbar}\int_{x_1}^{x_2} p\,dx - \frac{\pi}{2}\right],\end{aligned} \tag{6.33}$$

and for this to coincide with Eq. (6.31) it must hold that

$$\frac{1}{\hbar}\int_{x_1}^{x_2} p\,dx + \frac{\pi}{2} = \pi(n+1), \quad n = 0, 1, 2, \ldots \tag{6.34}$$

$$A' = (-)^n A''. \tag{6.35}$$

In writing (6.34) we took into account that $\int_{x_1}^{x_2} p\,dx > 0$, so the right-hand side cannot be zero. The quantization condition (6.34) shows that bound particles have a discrete spectrum regardless of the shape of the well. This is a general rule, of which we have found examples in Chapter 5 and will find others later, an important one being that of the atom. Since p has a different sign on the forward and the return paths, we can write the quantization rule as a closed integral over a complete period,

$$\oint p\,dx = 2\pi\hbar\left(n + \tfrac{1}{2}\right). \tag{6.36}$$

This result (which is exact up to terms of order $\hbar$) agrees with the Wilson–Sommerfeld rule (see Section 2.6), except for the presence of the additional term $\frac{1}{2}h$, which is reminiscent of the zero-point energy.[9]

Equations (6.31) and (6.34) show that the phase of ψ changes from $\frac{1}{4}\pi$ at $x = x_1$ to $\pi(n + \frac{3}{4})$ at $x = x_2$, so the quantum number n determines the number of nodes of the wave function inside the well. Furthermore, by eliminating F from Eqs. (6.24) and (6.25), we obtain

$$|x - x_1| \gg \frac{\hbar}{2p}, \tag{6.37}$$

which confirms that the semiclassical approach applies to states characterized by a large quantum number n.

To determine the normalization constant we consider that since outside the well the wave function decays exponentially, we can neglect its contribution to the normalization integral. We then normalize the ψ given by (6.31) by approximating the square of the sine function, which is a rapidly oscillating function (remember that $n \gg 1$), by its mean value $\frac{1}{2}$, thus

$$|A'|^2 \int_{x_1}^{x_2} \sin^2\left(\frac{1}{\hbar}\int_{x_1}^{x} p(x')\,dx' + \frac{\pi}{4}\right)\frac{dx}{p} \simeq \frac{1}{2}|A'|^2 \int_{x_1}^{x_2} \frac{dx}{p}$$
$$= \frac{|A'|^2}{2m}\int_{x_1}^{x_2}\frac{dx}{v} = \frac{|A'|^2}{4m}\oint \frac{dx}{v} = \frac{|A'|^2}{4m}T = \frac{|A'|^2}{4m}\frac{2\pi}{\omega} = 1. \tag{6.38}$$

The period of oscillation $T = \oint dx/v$ has been re-expressed in terms of the frequency $\omega = 2\pi/T$. Taking the normalization constant to be real and positive, we get $A' = \sqrt{2m\omega/\pi}$. The normalized WKB wave function inside the well is thus

$$\psi = \sqrt{\frac{2\omega}{\pi v}} \sin\left(\frac{1}{\hbar}\int_{x_1}^{x} p\,dx + \frac{\pi}{4}\right), \tag{6.39}$$

subject to the quantization condition (6.36).

Exercise E6.2. Quantization in a multidimensional well

Obtain the quantization conditions for particles in a multidimensional potential well using the WKB method.[10]

Solution. We initially write the WKB solution in the form $\psi = Ae^{iS}$. For this ψ to be satisfactory, we must require that A be single-valued; the action S in turn may be multivalued, provided that the difference between two of its values satisfies the condition

$$\Delta S \equiv \oint_C \nabla S\, ds = 2\pi n, \quad n = 1, 2, 3, \ldots \tag{6.40}$$

---

[9] When applied to central potentials, the results of the WKB approximation are often improved by the replacement $l(l + 1) \to (l + \frac{1}{2})^2$, where l represents the angular momentum.

[10] The WKB problem in more than one dimension, if not separable, is considerably more complicated than the 1D case and involves the topological properties of phase space. Due to the complications of the full solution, this Exercise is given an informal treatment.

for any closed contour C of integration in x-space. However, it often happens that the function A is not single-valued on certain surfaces. In these cases we can proceed as follows. Let A and $A + \Delta A$ be two possible values of the function A at a given point. We rewrite the wave function in the form

$$\psi = (A + \Delta A)e^{iS} = Ae^{i(S - i\Delta \ln A)}, \qquad (6.41)$$

so that we have translated the change in A into a change in phase. Condition (6.40) is now rewritten in the form

$$\int_C \nabla S \, ds = 2\pi(n + \frac{i}{2\pi}\Delta \ln A). \qquad (6.42)$$

Considering the general results of the WKB method, which show that $\int_C \nabla S ds = (1/\hbar) \oint_C p_i dx_i$, we see that the quantization rules have the general form

$$\oint p_i \, dx_i = 2\pi \hbar(n + \frac{i}{2\pi}\Delta \ln A). \qquad (6.43)$$

There are as many quantization integrals as there are nonreducible contours in the problem (two contours are reducible if it is possible to continuously transform one into the other without crossing any singularity of the integrand). Contours that are reducible to a point can be eliminated because the integral over them is zero.

As an example of Eq. (6.43) we consider the case where A changes sign, as happens in the case of a perfectly reflecting wall. Since

$$\frac{i}{2\pi}\Delta \ln A = \frac{i}{2\pi}(\ln(-A) - \ln A) = \frac{i}{2\pi}\ln(-1) = \frac{i}{2\pi}\ln e^{-i\pi} = \tfrac{1}{2}, \qquad (6.44)$$

the quantization rule in this case contains a half-integer.

6.2 Transmission Across a Barrier. Alpha Decay

Consider a smoothly varying repulsive potential, as illustrated in Fig. 6.2. We shall use the WKB method to study the behavior of a beam of particles approaching the barrier from the left. Our first objective is to determine the reflection and transmission coefficients of particles with energy E less than $V_{\max}$. The following presentation is rather condensed; it is recommended that the student fill in the omitted steps.

To solve this problem, we need the connection formulae between the solutions in regions I and II and II and III, respectively. These formulae are obtained by following the same procedure used to establish the pair of equations (6.29)–(6.30). The result for regions I and II is

$$\psi_I = \frac{1}{\sqrt{p}}\left[A_1 e^{i\left(\frac{1}{\hbar}\int_x^{x_1} p\,dx + \frac{\pi}{4}\right)} + B_1 e^{-i\left(\frac{1}{\hbar}\int_x^{x_1} p\,dx + \frac{\pi}{4}\right)}\right], \quad x < x_1 \qquad (6.45)$$

$$\psi_{II} = \frac{1}{\sqrt{|p|}}\left[(A_1 + B_1)e^{\frac{1}{\hbar}\int_{x_1}^{x}|p|dx} + \frac{i}{2}(A_1 - B_1)e^{-\frac{1}{\hbar}\int_{x_1}^{x}|p|dx}\right], \quad x > x_1 \qquad (6.46)$$

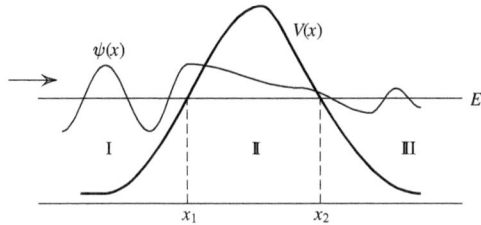

Figure 6.2 A smoothly varying potential barrier. The general form of the solution is shown for $E < V_{max}$.

Since in region III the particles travel only to the right, for the second pair of regions we get

$$\psi_{III} = \frac{1}{\sqrt{p}} A_3 e^{i\left(\frac{1}{\hbar}\int_{x_2}^{x} p\,dx + \frac{\pi}{4}\right)}, \quad x > x_2 \tag{6.47}$$

$$\psi_{II} = \frac{A_3}{\sqrt{|p|}}\left[e^{\frac{1}{\hbar}\int_x^{x_2}|p|dx} - \tfrac{1}{2}ie^{-\frac{1}{\hbar}\int_x^{x_2}|p|dx}\right], \quad x < x_2 \tag{6.48}$$

With the integral in the exponents written as

$$\int_x^{x_2}|p|dx = \int_{x_1}^{x_2}|p|dx - \int_{x_1}^{x}|p|dx \tag{6.49}$$

and the definition as

$$J \equiv \frac{1}{\hbar}\int_{x_1}^{x_2}|p|dx = \frac{\sqrt{2m}}{\hbar}\int_{x_1}^{x_2}\sqrt{V-E}\,dx, \tag{6.50}$$

ψ_{II} can be rewritten as

$$\psi_{II} = \frac{A_3}{\sqrt{|p|}}\left[e^{J-\frac{1}{\hbar}\int_{x_1}^{x}|p|dx} - \tfrac{1}{2}ie^{-J+\frac{1}{\hbar}\int_{x_1}^{x}|p|dx}\right]. \tag{6.51}$$

Comparing the coefficients in this expression with those in (6.46), we see that

$$A_1 + B_1 = -\tfrac{1}{2}iA_3 e^{-J}; \quad \tfrac{1}{2}i(A_1 - B_1) = A_3 e^{J}, \tag{6.52}$$

from which it follows that

$$A_1 = -iA_3(e^J + \tfrac{1}{4}e^{-J}), \quad B_1 = iA_3(e^J - \tfrac{1}{4}e^{-J}). \tag{6.53}$$

Using the definitions for the transmission and reflection coefficients introduced in Chapter 5, we obtain

$$T = \frac{|j_{\text{trans}}|}{|j_{\text{inc}}|} = \frac{|A_3|^2}{|A_1|^2} = (e^J + \tfrac{1}{4}e^{-J})^{-2}, \tag{6.54}$$

$$R = \frac{|j_{\text{ref}}|}{|j_{\text{inc}}|} = \frac{|B_1|^2}{|A_1|^2} = \left(\frac{e^J - \tfrac{1}{4}e^{-J}}{e^J + \tfrac{1}{4}e^{-J}}\right)^2. \tag{6.55}$$

Usually $J \gg 1$ since the WKB approximation is satisfactory only for wide barriers, so we can approximate the transmission coefficient as

$$T \approx e^{-2J} = e^{-\frac{2}{\hbar}\int_{x_1}^{x_2}\sqrt{2m(V-E)}\,dx}. \tag{6.56}$$

This is a very useful formula that generalizes the expression of T for arbitrary but smooth potentials. It shows that the transparency of a barrier to a particle depends very strongly on its energy and its mass; for example, if electrons are replaced by protons of the same energy, the transmission coefficient decreases by a factor of the order of $e^{\sqrt{1840}} \sim e^{42} \sim 10^{18}$. We will see some examples in what follows that illustrate the importance of this phenomenon.

Note that Fig. 6.2 again shows the existence of a nonlocal feature: the incident wave function is distorted to adapt to the wave function in region II. This adaptation involves a small zone *outside* the potential, as if to announce to the incoming wave that a change in the potential function is imminent. We must bear in mind that, as discussed in Section 5.4, the wave function provides a statistical, time-independent description, when the ensemble has had sufficient occasion to reach a stationary distribution.

6.2.1 Nuclear alpha decay

Some atomic nuclei decay spontaneously (without any external intervention) by emitting α particles, that is, ^{4}He nuclei, with two protons and two neutrons. The energy with which the α particles are emitted from the radioactive nucleus is characteristic of each isotope and is of the order of a few MeV, that is, much lower than the Coulomb repulsion energy between the α particle and the $Z - 2$ charged daughter nucleus (Z is the charge of the parent nucleus). This led E. Condon,[11] Gurney,[12] and, independently, Gamow,[13] in 1928 to understand α decay as a phenomenon of particle transmission through a potential barrier. The theory based on this hypothesis was not only the first physical application of quantum tunneling and one of the first verifications of Schrödinger's theory, but also showed that, with care, QM could be extended even to the scale of atomic nuclei.

To study nuclear alpha decay, we model the potential as shown in Fig. 6.3. R is the radius of the daughter nucleus (of charge $Z_2 = Z - 2$); for $r < R$ the particle is attracted by the nucleus with a potential assumed to be rectangular for simplicity, and for $r > R$ there is the Coulomb repulsion between the α particle (of charge $Z_1 = 2$) and the daughter nucleus. Assuming that the particle is in its ground state with energy E, the transmission coefficient is

$$T = \exp\left[-\frac{2\sqrt{2M}}{\hbar}\int_R^{R1}\sqrt{V-E}\,dr\right], \qquad (6.57)$$

[11] Edward Uhler Condon (1902–1974) was a US nuclear physicist and pioneer of QM. During the McCarthy era, when efforts were made to root out communist sympathizers in the United States, Condon was targeted by the House Un-American Activities Committee for being a "follower" of a "new revolutionary movement," namely QM. Condon defended himself with a notorious commitment to physics and science. A lunar crater is named after him.
[12] Ronald Wilfrid Gurney (1898–1953) was a British theoretical physicist who initiated the study of radioactive decay.
[13] Gueorgiy Antonovich Gamow (1904–1968) was a prominent Soviet–US polymath, theoretical physicist, and cosmologist, author of a large collection of popular science books, including *One Two Three ... Infinity* and the *Mr Tompkins* series, still in print after more than fifty years. Gamow made important scientific contributions, such as the liquid-drop model of the atomic nucleus.

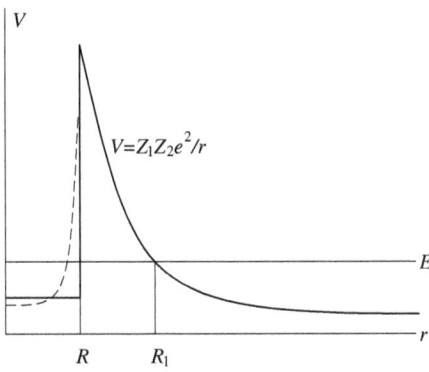

Figure 6.3 Idealized potential barrier for alpha decay. The dotted line inside the nucleus shows a more realistic representation of the barrier.

where M is the mass of the α particle and V is given by

$$V = \frac{Z_1 Z_2 e^2}{r}, \quad R < r < R_1. \tag{6.58}$$

The turning point at R_1 is given by the condition $V(R_1) = Z_1 Z_2 e^2 / R_1 = E$. The integral can be easily calculated with the change of variable $r/R_1 = \sin^2 \varphi$,

$$\int_R^{R_1} \sqrt{\frac{Z_1 Z_2 e^2}{r} - E}\, dr = \sqrt{E} \int_R^{R_1} \sqrt{\frac{R_1}{r} - 1}\, dr = \tfrac{1}{2} R_1 \sqrt{E}(\pi - 2\varphi_0 - \sin 2\varphi_0), \tag{6.59}$$

where $R/R_1 = \sin^2 \varphi_0$. By making the approximation $\sin \varphi_0 \approx \varphi_0$, which is valid for wide potential barriers, we obtain

$$\int_R^{R_1} \sqrt{V - E}\, dr = R_1 \sqrt{E}\left[\frac{\pi}{2} - 2\sqrt{\frac{R}{R_1}}\right]. \tag{6.60}$$

Substituting this result in Eq. (6.57), we get with $Z_1 = 2$, $\ln T = -2\pi e^2 Z_2 \sqrt{2M/\hbar^2 E} + 8e\sqrt{Z_2 MR/\hbar^2}$, which can be written as

$$\ln T = c - \frac{a}{\sqrt{E}}, \tag{6.61}$$

with the parameters a and c given by

$$a = \frac{2\pi e^2}{\hbar} \sqrt{2M} Z_2, \quad c = \frac{8e}{\hbar} \sqrt{Z_2 MR}. \tag{6.62}$$

An estimate of the decay rate λ is obtained by considering that, since the particle is confined in a "box" of width R and has the lowest possible energy, only half of a de Broglie wavelength is accommodated within the box, giving $R \sim h/p$ or, equivalently, $v \sim h/MR$, from which

$$\lambda \simeq \frac{\hbar}{MR^2} T. \tag{6.63}$$

Taking the logarithm and combining with the previous results, we get

$$\ln \frac{1}{\lambda} = -b + \frac{a}{\sqrt{E}}, \tag{6.64}$$

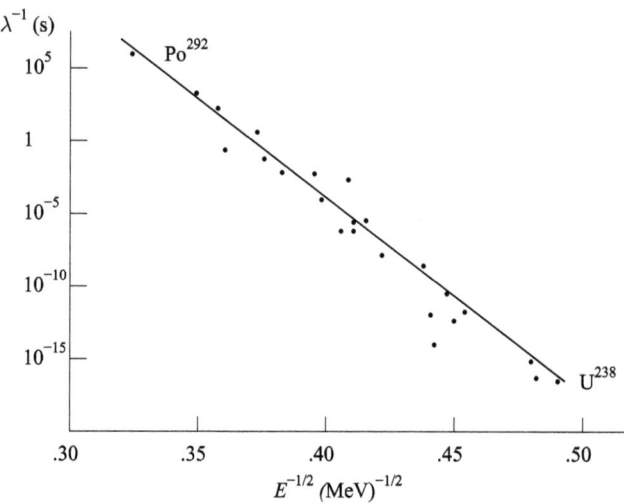

Figure 6.4 Decay constant for a wide range of radioactive isotopes as a function of alpha particle emission energy.

with b given by

$$b = \ln \frac{\hbar}{MR^2} + \frac{8e}{\hbar}\sqrt{Z_2 M R}. \tag{6.65}$$

The analysis of the parameters a and b calculated for different nuclei shows that b varies very slowly with the atomic number Z_2 (in a first approximation it can be taken as a constant), while a varies practically linearly with Z_2, as in Eq. (6.62). In this way, it is verified that, despite the many simplifications made, the theory is able to give a satisfactory qualitative and quantitative explanation of the phenomenon of nuclear α decay.

Note that according to Eq. (6.64), the mean life varies strongly with energy, due to the exponential dependence. For example, in Fig. 6.4 we see that as $1/\sqrt{E}$ goes from about 0.34 (E in MeV) to about 0.5 (less than twice), it changes by 23 *orders of magnitude*. This explains why some radioactive elements are much less stable than others, rendering them very useful for radiometric dating of a wide variety of materials of biological, geological or archaeological interest. For example, C^{14} has a mean life of roughly 8,000 years, while its other isotopes are stable; this makes it very suitable for dating materials (biological or mineral) with a high natural carbon content, since their C^{14}/C^{12} ratio decreases clearly with time.

6.2.2 Time delay by a potential barrier

In the first part of this section we studied the behavior of low-energy particles incident on a repulsive barrier for $E < V_{\max}$, and we saw that most of them are reflected by the potential, although some manage to pass through the barrier. If we increase the incident energy E, from the moment it is equal to some maximum potential, resonance phenomena similar to those studied in the previous chapter begin to appear. The main difference is that quantities such as $\sin qx$ are now replaced by $\sin \int^x q(x)dx$, and so on. This is easy to understand if we think of the potential $V(x)$

as a sequence of a large number of very narrow rectangular barriers placed side by side (see Problem P6.1).

If the incident energy is increased even more, so that $E \gg V_{\max}$, almost all the particles pass over the barrier. Since in this case the scattering is small, we can assume that a high-energy incident packet is transmitted almost without distortion. Under these conditions, the effect of the potential will essentially manifest itself in a delay of the packet. We will show that the delay predicted by quantum theory is given, to a first approximation, by the corresponding classical expression.

This example allows us to study a case of incident packet scattering that deviates from the oversimplified description of free particles described by plane waves. It is interesting to compare this description with that of the example in Section 5.4 in terms of the phase shift of the transmitted wave function, where it was the position of the packet in the asymptotic region that was shifted; here the focus is on the time delay of the arrival of the packet centroid.

Consider a potential that varies smoothly and differs from zero only in a region of width a; from the left and from a very large distance (which is eventually made infinite), particles are launched, forming a narrow packet with mean energy E_0 and initial amplitude $\psi(x,0)$. At a very large distance to the right of the potential (which is eventually considered infinite), there will be a packet of similar shape to the incident one, but with a certain phase that depends on the potential, shifted with respect to the phase it would have in the absence of the potential. To obtain the delay time Δt we first establish its relationship with the associated phase shift δ.

The incoming packet can be expressed in terms of free-particle solutions, which is equivalent to performing a Fourier analysis,

$$\psi(x,t) = \int A(\omega)e^{i(kx-\omega t)}\,dk, \tag{6.66}$$

where ω represents as usual the energy measured in units of $\hbar$,

$$\omega = \frac{E}{\hbar} = \frac{\hbar k^2}{2m}. \tag{6.67}$$

We will consider that the packet consists of particles that are launched from position x_0 with energies distributed around a sharp maximum centered at $E_0 = \hbar\omega_0 = \hbar k_0/2m$,

$$\psi(x,0) = 2\pi g(\omega - \omega_0)\delta(x - x_0), \tag{6.68}$$

where the function $g(z)$ is narrow and has a sharp maximum at $z = 0$; the numerical factor 2π is introduced for convenience. On the other hand, from (6.66) with $t = 0$ we obtain

$$\psi(x,0) = \int A(\omega)e^{ikx}\,dk. \tag{6.69}$$

Taking the inverse transform

$$A(\omega) = \frac{1}{2\pi}\int \psi(x,0)e^{-ikx}\,dx \tag{6.70}$$

and substituting the value of $\psi(x,0)$ given by (6.68), we get

$$A(\omega) = \int g(\omega-\omega_0)\delta(x-x_0)e^{-ikx}\,dx = g(\omega-\omega_0)e^{-ikx_0}. \tag{6.71}$$

We introduce this result in (6.66) to obtain

$$\psi_{\text{inc}}(x,t) = \int g(\omega - \omega_0) e^{ik(x-x_0) - i\omega t} dk. \tag{6.72}$$

This is the packet that hits the barrier.

To construct the transmitted packet we do as follows. We have seen in the previous sections and in Chapter 5, that when a plane wave is transmitted by the barrier, its amplitude is modified, but not its frequency. The amplitude of the transmitted wave packet can therefore be written from (6.72) by modifying the amplitude of each component by a factor $S(\omega)$, whose module obviously cannot exceed unity, that is, $|S(\omega)| \leq 1$,

$$\psi_{\text{trans}}(x,t) = \int g(\omega - \omega_0) S(\omega) e^{ik(x-x_0) - i\omega t} dk. \tag{6.73}$$

It is clear that $S(\omega)$ plays the role of a transmission amplitude of the frequency component ω, in the sense that

$$T(\omega) = |S(\omega)|^2. \tag{6.74}$$

Reversing this expression, we can write S in the form

$$S(\omega) = \sqrt{T(\omega)} e^{i\delta(\omega)}. \tag{6.75}$$

This formula says that $\sqrt{T}$ is the factor by which the amplitude of the component of frequency ω is reduced, while $\delta(\omega)$ is the phase shift produced by its delay (if there is no potential, $T = 1$ and $\delta = 0$). We will limit ourselves to the case where the magnitude T varies slowly with the energy, so that we can write approximately $T(\omega) \simeq T(\omega_0)$. We will also assume that the phase can be approximated up to linear terms in the variable ϵ,

$$\epsilon \equiv \omega - \omega_0. \tag{6.76}$$

With these considerations, we write

$$S(\omega) \simeq \sqrt{T(\omega_0)} e^{i\left(\delta(\omega_0) + \epsilon \frac{\partial \delta}{\partial \omega}\right)} = S(\omega_0) e^{i \frac{\partial \delta}{\partial \omega} \epsilon}. \tag{6.77}$$

In turn, as a function of the variable ϵ and up to linear terms, k is written as

$$k = \frac{\sqrt{2mE}}{\hbar} = \sqrt{\frac{2m\omega}{\hbar}} = \sqrt{\frac{2m\omega_0}{\hbar}} \left(1 + \frac{\epsilon}{2\omega_0}\right) = k_0 + \frac{m}{p_0} \epsilon. \tag{6.78}$$

We now introduce the approximation (6.77) into (6.73); using (6.76) and (6.78), we expand around k_0 and write

$$\psi_{\text{trans}} = S(\omega_0) \int e^{i \frac{\partial \delta}{\partial \epsilon} \epsilon} g(\epsilon) e^{i(k_0 + \frac{m}{p_0}\epsilon)(x-x_0) - it(\omega_0 + \epsilon)} \frac{m}{p_0} d\epsilon$$

$$= S(\omega_0) e^{ik_0(x-x_0) - i\omega_0 t} \int g(\epsilon) e^{\frac{im}{p_0}\left[x - x_0 - \frac{p_0}{m}\left(t - \frac{\partial \delta}{\partial \epsilon}\right)\right] \epsilon} \frac{m}{p_0} d\epsilon. \tag{6.79}$$

The factor preceding the integral is the central component of the energy packet; we see that its amplitude has indeed been reduced by the factor $S(\omega_0)$. Much more interesting, however, is the integral, since it describes the envelope of the incident packet, but with the time shifted by the quantity

$$\Delta t = \frac{\partial \delta}{\partial \epsilon} = \hbar \frac{\partial \delta}{\partial E}. \tag{6.80}$$

This gives the desired relationship between the phase shift δ and the delay time Δt.

To calculate δ we use the WKB solution for the transmitted wave function

$$\psi_{\text{trans}} = \frac{A}{\sqrt{p}} e^{\frac{i}{\hbar} \int_{x_0}^{x} p\,dx} \qquad (6.81)$$

and rewrite the integral of the exponent as

$$\int_{x_0}^{x} p\,dx = \int_{x_0}^{x} (p_0 + p - p_0)\,dx = p_0(x - x_0) + \int_{x_0}^{x} (p - p_0)\,dx, \qquad (6.82)$$

so that the transmitted packet takes the form

$$\psi_{\text{trans}} = B(p_0) e^{i \int_{x_0}^{x} (p - p_0)\,dx/\hbar}. \qquad (6.83)$$

In the absence of the scattering potential, we have $p = p_0$ at every point and the previous expression reduces to $\psi_{\text{trans}} = B(p_0)$; therefore, the factor of $B(p_0)$ represents the additional phase introduced by the potential along the path from x_0 to x. Since the potential is assumed to be localized, we can extend the limits of integration to infinity on both sides to obtain

$$\delta(E) = \int_{-\infty}^{\infty} \frac{p - p_0}{\hbar} dx = \frac{\sqrt{2m}}{\hbar} \int_{-\infty}^{\infty} (\sqrt{E - V} - \sqrt{E})\,dx. \qquad (6.84)$$

The delay time is obtained by plugging this result into formula (6.80), which gives

$$\Delta t = \sqrt{2m}\frac{\partial}{\partial E} \int_{-\infty}^{\infty} \left(\sqrt{E - V} - \sqrt{E}\right) dx = \int_{-\infty}^{\infty} \left(\sqrt{\frac{m}{2(E - V)}} - \sqrt{\frac{m}{2E}}\right) dx, \qquad (6.85)$$

i.e.

$$\Delta t = \int_{-\infty}^{\infty} \left(\frac{1}{v(x)} - \frac{1}{v_0}\right) dx, \qquad (6.86)$$

where $v(x)$ is the classical velocity at x. This is the same result that classical theory gives: the delay time is the total travel time $t_v = \int_{-\infty}^{\infty} dx/v(x)$ in the presence of the potential minus the travel time $t_0 = \int_{-\infty}^{\infty} dx/v_0$ that the free particle would need. In other words, in this approximation, the potential slows down the packet as a whole, as if it were a rigid structure. This is a direct consequence of the assumption that the real part of the reduction factor $S(\omega)$ is the same for the whole package.

Several different definitions for parameters such as delay time, transmission time, and so on are known in the specialized literature, the one used here to arrive at Equation (6.86) is one of them.

6.3 Electronic Properties of Solids

The mechanical, acoustic, thermal, or electrical properties of a body, be it a gas, a liquid or an ordered or disordered solid, are strongly dependent on the atomic or molecular content and structure. In particular, the electrical properties of solids are essentially determined by the dynamics of the electrons in their interior, leading to their classification as conductors, semiconductors, or insulators, depending on their ability to transfer charge and energy.

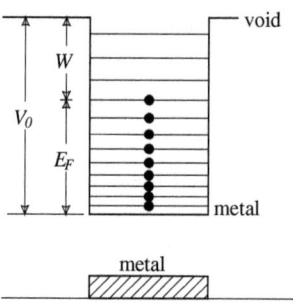

Figure 6.5 Elementary model of a metal. E_F is the Fermi energy, determined by the highest occupied level; W is the work function, and V_0 is the depth of the well. At $T = 0$ each of the lower levels is occupied by two electrons.

We begin the second part of this chapter by considering a simple model for conduction electrons in a metal, and continue with a discussion of the physics of electrons in a periodic structure, which allows us to explain important electronic properties of conductors and semiconductors and various combinations thereof, which form the basis of most of today's electronics.

6.3.1 The free-electron-gas model

About 80% of the 118 elements in the modern periodic table are metals.[14] The metals usually have 1, 2, or 3 electrons in their valence shell (i.e., the outer shell, see what follows), so they can donate electrons very easily. The high (electrical and thermal) conductivity of metals is due to the large number – of the order of 10^{22}/cm^3 – of electrons that can move (almost) freely in the material. This allows us to think of these conduction electrons as a "free-electron gas" inside the metal (a brief explanation of the physical origin of this phenomenon is given in Chapter 15). However, these electrons cannot escape from the material because work is required to extract them (the so-called work function W of the metal), which suggests that the metal should be represented in a first approximation as a finite potential well, which for simplicity will be assumed to be rectangular in the following.

Our first task is to estimate the depth of the well, for which we proceed as follows. Due to the large number of atoms in the metal, there is a very large number of stationary energy levels within the well – as will be shown below – so the electronic energy scheme of the metal is similar to that shown in Fig. 6.5. In the ground state (i.e., at absolute zero temperature), the electrons trapped in the well fill the lowest energy levels; each level can be occupied by a maximum of two electrons, due to the Pauli exclusion principle, as we will have occasion to study in Section 15.2. This means that if the metal contains N free electrons, at $T = 0$ they will occupy the $N/2$ lowest energy levels, producing a rectangular distribution. As the temperature increases, the higher levels begin to fill up; the distribution is no longer rectangular and becomes smoother, as shown in Fig. 6.6. This is the *Fermi–Dirac distribution*, which we will discuss later.

---

[14] This fraction is not a well-defined quantity, as the boundary between metals and nonmetals varies due to the lack of universally accepted definitions.

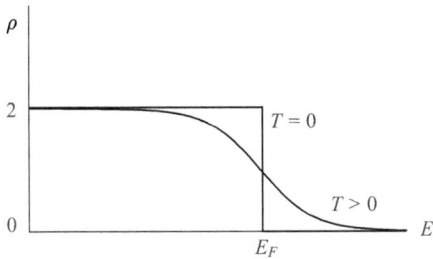

Figure 6.6 Electron level occupancy in a metal as a function of energy and temperature, according to the Fermi–Dirac distribution.

The highest energy level occupied determines the *Fermi energy* E_F,[15] and the surface in momentum space with $E = E_F$ is called the *Fermi surface*. To calculate the Fermi energy at zero temperature, we consider the metal as a cube with side L, so that the free electrons have momentum p_i ($i = x, y, z$) given by

$$p_i = \hbar k_i = \frac{2\pi \hbar}{L} n_i, \quad n_i = 0, 1, 2, \ldots n_{i\,\text{max}}. \quad (6.87)$$

Increasing n_i by one unit will give the smallest possible increase in momentum. Since, for a piece of metal, L is very large compared to the size of the atoms, this change in momentum is very small, so we can write $\Delta n_i \to L dp_i / 2\pi \hbar$. The number of states contained in the interval $(p, p + dp)$ is therefore

$$\Delta^3 n = \Delta n_x \Delta n_y \Delta n_z \to \frac{L^3}{(2\pi \hbar)^3} d^3 p. \quad (6.88)$$

Since $2\Delta^3 n$ electrons fit into each interval, if p_{max} is the momentum of the electrons on the Fermi surface, the total number N of free electrons becomes

$$N = 2 \sum \Delta n_x \Delta n_y \Delta n_z \to \frac{2L^3}{(2\pi \hbar)^3} \int_0^{p_{\text{max}}} d^3 p$$

$$= \frac{L^3}{4\pi^3 \hbar^3} 4\pi \int_0^{p_{\text{max}}} p^2 dp = \frac{L^3 p_{\text{max}}^3}{3\pi^2 \hbar^3}, \quad (6.89)$$

and for the density ρ_0 of free electrons in the metal we get $\rho_0 \equiv N/V = p_{\text{max}}^3 / 3\pi^2 \hbar^3$, hence

$$p_{\text{max}} = \hbar (3\pi^2 \rho_0)^{1/3}. \quad (6.90)$$

It follows that the Fermi energy is

$$E_F = \frac{p_{\text{max}}^2}{2m} = \frac{\hbar^2}{2m} (3\pi^2 \rho_0)^{2/3}. \quad (6.91)$$

---

[15] Enrico Fermi (1901–1954) was an exceptional Italian–US theoretical and experimental physicist, one of the great physicists of the twentieth century. He made a long list of contributions to quantum theory, nuclear physics, elementary particles, and statistical physics, and developed the first nuclear reactor (pile) in Chicago. He received the Nobel Prize in 1938 for his discovery of induced radioactivity. The (artificial) radioactive element $Z=100$ is named after him as *fermium*.

This allows us to estimate the Fermi energy for a representative metal. For example, assuming that there is one free electron per atom in silver (which is a realistic assumption), we can calculate the electron density as the number of silver atoms per unit volume. Since silver has a density of 10.5 g cm^{-3} and an atomic weight of 107.9 u (unified atomic mass units), with Avogadro's number (number of atoms in one gram-atom) equal to 6.022×10^{23}, we get

$$\rho_0 \approx \frac{10.5}{107.9} \times 6.022 \times 10^{23} \approx 6 \times 10^{22} \text{ cm}^{-3}. \tag{6.92}$$

Substituting this value into Eq. (6.91), we get $E_F \approx 8 \times 10^{12}$ erg ≈ 5.5 eV. Since the work function of silver is 4.7 eV, the depth of the potential well is 10.2 eV (see Fig. 6.5).

Equation (6.91) allows us to understand another property of metals that is not classically explained. When the electron-gas theory emerged as a model for metals, it was considered, based on classical arguments, that each electron contributes an amount $\frac{3}{2}k$ (k is Boltzmann's constant) to the specific heat, due to the equipartition theorem. However, experimentally it is observed that the free electrons do not contribute to the specific heat of the metal. This contradiction is easily resolved by the quantum theory just formulated. In fact, we can roughly calculate the average energy of the free electrons at low temperatures as

$$\overline{E}_{\text{elect}} = \frac{2}{N} \sum E(n) \Delta^3 n \rightarrow \frac{2}{\rho_0 L^3} \int \frac{p^2}{2m} \frac{d^3 p}{(2\pi\hbar)^3} L^3 = \tfrac{3}{5} E_F. \tag{6.93}$$

As the temperature rises, the electrons with an energy close to (but less than) E_F start to absorb energy and populate the upper levels close to E_F one by one. At low temperatures, the energy corrections due to these absorptions are negligible, which means that their contribution to the specific heat of the metal is effectively zero,

$$c_v^{\text{elect}} \equiv \left(\frac{\partial \overline{E}_{\text{elect}}}{\partial T} \right)_{T=0} = 0. \tag{6.94}$$

A calculation at higher temperatures allows us to verify that $c_v^{\text{elect}} \ll \tfrac{3}{2}k$, because the electron-level occupancy is determined by the Fermi–Dirac distribution, as is illustrated in Fig. 6.6 and will be discussed in Chapter 15.

6.3.1.1 Electron tunneling in metals

The electron-gas model can also explain various processes of electron emission from a metal. In general, it is possible to extract electrons by adding some extra energy to the metal. If this is done by increasing the temperature of the metal, thermionic emission occurs. This method is used to release electrons from an electrode (usually a filament) in various electronic devices, such as vacuum tubes, and also for power generation or cooling. For significant thermionic emission to occur, the metal must be heated to a temperature above 10^3 K to overcome its work function W.

It is also possible to extract electrons by applying a sufficiently strong uniform external electric field (*cold* or *field emission*). In this case, the potential energy diagram is similar to that shown in Fig. 6.7, which consists of the superposition of the metal's potential well of depth V_0 and the external potential $-e_0 \mathcal{E} x$, where $\mathcal{E}$ is the electric

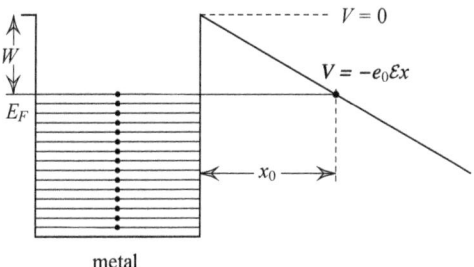

Figure 6.7 Field emission: the tunneling of electrons from a metal surface in the presence of an electric field. Tunneling occurs mainly with electrons that have an energy $E \simeq E_F$.

field strength. This creates a triangular potential barrier through which electrons can escape. The width of the barrier, x_0, is given by the intersection at $W = e_0 \mathcal{E} x_0$, because at normal temperatures the tunneling electrons have an energy close to the Fermi level. Equation (6.56) thus gives for the transmission coefficient

$$T = e^{-\frac{2}{\hbar}\sqrt{2m}\int_0^{x_0} \sqrt{W-e_0\mathcal{E}x}\,dx}. \tag{6.95}$$

that is,

$$T = e^{-\mathcal{E}_0/\mathcal{E}}, \tag{6.96}$$

where $\mathcal{E}_0$ is given by

$$\mathcal{E}_0 = \frac{4\sqrt{2m}\,W^{3/2}}{3\hbar e_0}. \tag{6.97}$$

An estimate of the value of this parameter under normal conditions is 10^6 V/cm. Equation (6.96) shows that fields of this intensity produce relatively strong emission currents.

Both thermionic and field emission had already been produced in the late nineteenth century, but could only be explained with the advent of QM. In fact, the quantum tunneling of electrons is considered one of the theory's early triumphs. Since then, electron tunneling has found a number of important applications. One example is the *scanning tunneling microscope* (STM), which is used to image surfaces at the atomic level. A very sharp conductive tip is brought close to the surface to be examined and a voltage is applied between the two, causing electrons to pass through the gap between them (see Fig. 6.8). The resulting tunnel current, which depends on the position of the tip and the applied voltage according to Eqs. (6.96) and (6.97), is monitored as the tip traverses the surface, and is usually displayed as an image.

6.3.1.2 The contact potential

Another example of the use of the simple metal model is the *contact potential*. This phenomenon, discovered by Volta,[16] occurs when the surfaces of two different metals

---

[16] Alessandro Volta (1745–1827) was an Italian physicist and chemist, inventor of the electrochemical battery (the common pile, 1799), and discoverer of methane gas. The unit of voltage of electric strength, the *volt*, is named after him.

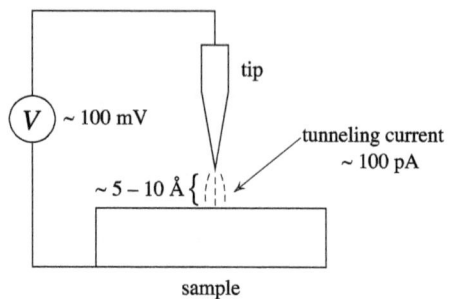

Figure 6.8 Schematic of a scanning tunneling microscope.

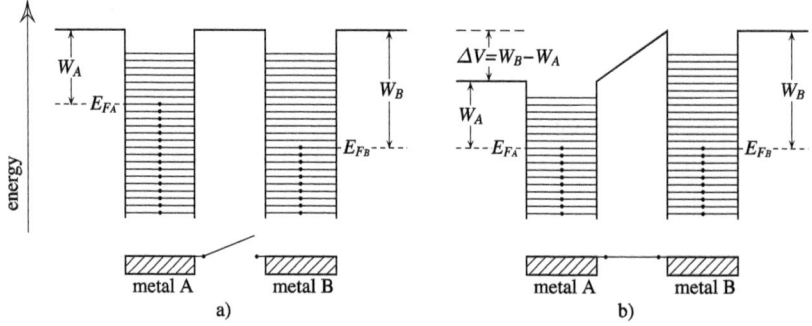

Figure 6.9 (a) Metals A and B having different work functions W_A, W_B. (b) The metals are brought into electrical contact Tunneling occurs from A to B until the Fermi levels are equalized.

come into contact, creating an electric potential difference between them. As long as the two metals are electrically isolated, the wells representing them will be at the same upper level, corresponding to the (common) vacuum potential, as shown in Fig. 6.9(a). When brought into electrical contact, the free electrons of each metal are separated from those of the other by a potential barrier. If the Fermi energy of metal A is higher than that of metal B, the more energetic electrons of metal A will tunnel into B, because there are holes in B for them to occupy. Since the reverse process does not occur, the net effect is a transfer of electrons from metal A to metal B, until the Fermi levels of the two metals are equalized, as shown in Fig. 6.9(b).

The number of electrons transferred is a very small fraction of the total number of free electrons, so that the Fermi energies are not significantly altered; therefore, the potential difference created between the metals is essentially given by the difference in work functions,

$$\Delta V = W_B - W_A. \tag{6.98}$$

Note that the explanation given is independent of the assumed shape of the well; in other words, the result (6.98) is essentially independent of the specific model.

In a chemical cell (an elementary battery) made of different metals, the potential difference between the electrodes is essentially equal to the corresponding contact potential given by (6.98), which also determines the direction of the current; the energy is extracted from the chemical processes taking place at the electrodes, involving the ions in the electrolyte.

The work function of a metal changes slightly with temperature, so the contact potential also changes with temperature. Thus, when two strips of different metals are joined at their ends to form a ring (a device that constitutes a *thermocouple*) and one of the junctions is heated, an electromotive force is generated, known as the thermoelectric potential; this phenomenon was discovered by J. Seebeck (1770–1831) in 1821. The inverse phenomenon discovered by J. Peltier (1785–1845) in 1834, which is also reversible, occurs when an electromotive force is applied to one of the junctions of the thermocouple: one of them cools down and the other heats up.

6.3.2 Conductors and semiconductors

Although the model of the metal as a simple potential well has allowed us to explain a number of important properties, it has the serious disadvantage that it takes no account of the internal structure of the material, which produces fundamental effects that cannot be understood without it. The physics of electrons in a solid is, strictly speaking, a many-body problem because the electrons interact with each other, yet most of the relevant features can be derived by assuming that they move independently in an effective potential. Therefore, to introduce us to the study of these effects, we will model the material as a periodic sequence of similar wells separated by barriers, with the center of each well corresponding to the position occupied by one of the positive ions in the lattice.

6.3.2.1 The Kronig–Penney model

To keep the study as simple as possible, we consider only the 1D case, with an infinite number of rectangular wells and barriers, as shown in Fig. 6.10. This periodic single-electron model, proposed and studied by R. Kronig (1904–1995) and W. G. Penney (1909–1991) in 1930, has an analytical solution, as we shall see, and therefore allows simple calculations.

The Kronig–Penney potential is defined by three parameters: the lattice periodicity a, the well's width b and its height V_0. The position of a point x in well n is given by the coordinate x_n measured from the left internal wall of this well,

$$x_n = x - na = x \bmod a, \tag{6.99}$$

and similarly for the neighboring wells. Note that point B has the coordinate $x_n = a$, which is equivalent to $x_{n+1} = 0$. The wave function inside the well n can be written as

$$\psi_n = A_n \sin kx_n + B_n \cos kx_n, \quad 0 \leq x_n \leq b, \tag{6.100}$$

while inside the barrier n it is

$$\psi_n = A'_n \sinh q(a - x_n) + B'_n \cosh q(a - x_n), \quad b \leq x_n < a, \tag{6.101}$$

with k and q given by Eqs. (5.2) and (5.3), respectively. Likewise, the wave function inside the well $n + 1$ is evidently

$$\psi_{n+1} = A_{n+1} \sin kx_{n+1} + B_{n+1} \cos kx_{n+1}, \quad 0 \leq x_{n+1} \leq b. \tag{6.102}$$

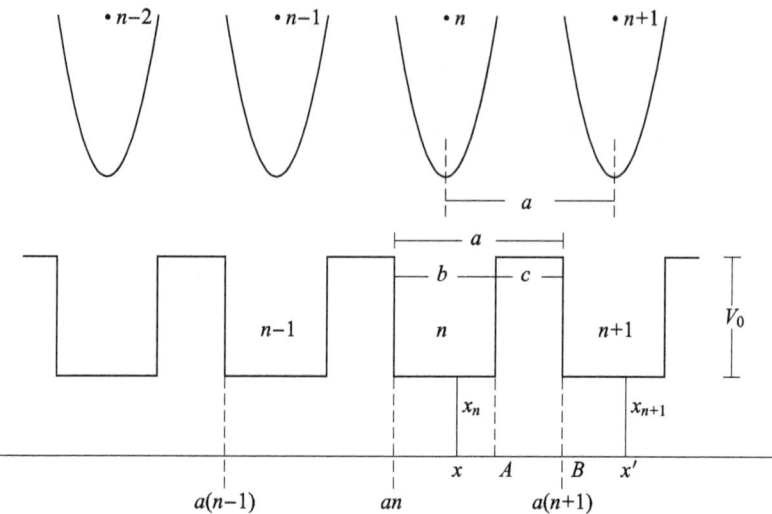

Figure 6.10 The Kronig–Penney model. Electrons move in a one-dimensional crystal under the influence of a periodic potential $V(x) = V(x + a)$, where a is the lattice parameter. In the lower diagram, b is the width of the well and $c = a - b$ is the width of the barrier of height V_0.

The continuity conditions at point A give ($x_n = b$; $a - b = c$)

$$A_n \sin kb + B_n \cos kb = A'_n \sinh qc + B'_n \cosh qc, \qquad (6.103)$$

$$k[A_n \cos kb - B_n \sin kb] = -q[A'_n \cosh qc + B'_n \sinh qc]. \qquad (6.104)$$

Analogously, the continuity conditions at point B give ($x_n = a$, $x_{n+1} = 0$)

$$B'_n = B_{n+1}, \quad -qA'_n = kA_{n+1}. \qquad (6.105)$$

We use these two expressions to eliminate A'_n and B'_n from Eqs. (6.103) and (6.104), thus obtaining

$$A_n \sin kb + B_n \cos kb = -(k/q)A_{n+1} \sinh qc + B_{n+1} \cosh qc, \qquad (6.106)$$

$$A_n \cos kb - B_n \sin kb = A_{n+1} \cosh qc - (q/k)B_{n+1} \sinh qc. \qquad (6.107)$$

To simplify the solution of this system of equations, we take the limit in which $V_0 \to \infty$ and at the same time the barriers become narrower and narrower, $c \to 0$, with the product cV_0 remaining constant (this limit is essentially that of a periodic succession of repulsive delta potentials). We write this constant in the form

$$cV_0 = \frac{\hbar^2 \lambda}{ma}, \qquad (6.108)$$

where λ is a (real and positive) dimensionless parameter. In this limit, $kb = ka$, $qc = \sqrt{2m\lambda c/\hbar^2} \to 0$ and $q^2 c = 2\lambda/a$. On the other hand, since qc is very small, we can write $\sinh qc \simeq qc$ and $\cosh qc \simeq 1$. With all these approximations, Eqs. (6.106) and (6.107) reduce to

$$A_n \sin ka + B_n \cos ka = B_{n+1}, \qquad (6.109)$$

$$A_n \cos ka - B_n \sin ka = A_{n+1} - \frac{2\lambda}{ka} B_{n+1}. \tag{6.110}$$

An inspection of this system of equations leads one to propose a general solution of the form

$$A_n = C_1 f^n, \quad B_n = C_2 f^n, \tag{6.111}$$

where f is a function to be determined and C_1 and C_2 are two constants, also to be determined. Substituting these expressions in the previous system, one obtains

$$C_1 \sin ka + C_2(\cos ka - f) = 0, \tag{6.112}$$

$$C_1(\cos ka - f) - C_2 \left(\sin ka - \frac{2\lambda}{ka} f \right) = 0. \tag{6.113}$$

For solutions with $C_1, C_2 \neq 0$ to exist, the determinant of the system must vanish. As a result, we get

$$f^2 - 2\left(\cos ka + \frac{\lambda}{ka} \sin ka \right) f + 1 = 0. \tag{6.114}$$

At this point it is convenient to define a new (real or complex) parameter $\varkappa$ such that

$$\cos \varkappa a \equiv \cos ka + \frac{\lambda}{ka} \sin ka, \tag{6.115}$$

in terms of which Eq. (6.114) takes the form

$$f^2 - 2f \cos \varkappa a + 1 = 0. \tag{6.116}$$

The solutions of this equation, $f = e^{\pm i \varkappa a}$, substituted into Eqs. (6.112) and (6.113), give for C_1 and C_2

$$C_1 = A \frac{f - \cos ka}{\sin ka}, \quad C_2 = A, \tag{6.117}$$

where A is the normalization constant. The double sign of the solutions indicates that the electrons can drift to the right and to the left. Taking (without loss of generality) only the positive sign, we determine A_n and B_n with the help of Eqs. (6.111); this gives

$$A_n = A \frac{f - \cos ka}{\sin ka} e^{i \varkappa na}, \quad B_n = A e^{i \varkappa na}. \tag{6.118}$$

Substituting these values into Eq. (6.100) for ψ and writing the product na in terms of the difference $x - x_n$, Eq. (6.120), we obtain, after some elementary trigonometric transformations,

$$\psi_n = A \frac{f - \cos ka}{\sin ka} e^{i \varkappa na} \sin kx_n + A e^{i \varkappa na} \cos kx_n$$

$$= A e^{i \varkappa x} \frac{e^{i \varkappa (a - x_n)} \sin kx_n + e^{-i \varkappa x_n} \sin k(a - x_n)}{\sin ka}, \tag{6.119}$$

that is,

$$\psi_n = A e^{i \varkappa x} u_n, \tag{6.120}$$

where u_n is given by

$$u_n = \frac{e^{i \varkappa (a - x_n)} \sin kx_n + e^{-i \varkappa x_n} \sin k(a - x_n)}{\sin ka}. \tag{6.121}$$

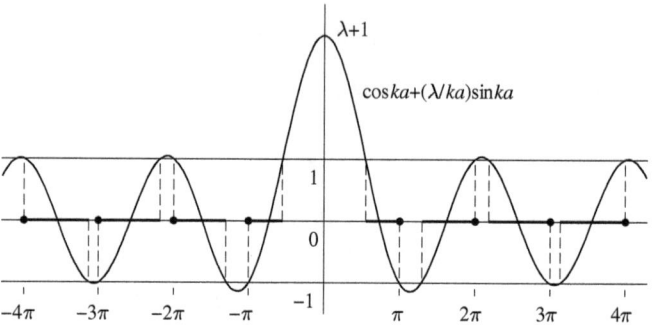

Figure 6.11 Graphical determination of the solutions of Eq. (7.92). Electrons with κ real, i.e., $|\cos\kappa a| \leq 1$, can move freely through the crystal; the regions corresponding to these allowed values of ka are marked with thick lines on the graph. As the barrier height increases, the allowed regions become narrower, until they disappear in the limit $\lambda \to \infty$ (thick dots on the graph, corresponding to $ka = n\pi$).

The fundamental equation (6.120), known as the *Bloch wave function*, was obtained here for the particular case of rectangular wells, but any periodic potential produces wave functions of this general form, a result that is known as *Floquet's theorem*. This theorem states that linear differential equations with periodic coefficients have solutions of the form $e^{\pm ikx} u(x)$, where $u(x)$ is a periodic function of x.[17] Bloch waves are the specific case applied to Schrödinger's equation, also when the periodic potential is not that of Kronig and Penney.

6.3.2.2 Electronic band structure

According to Eq. (6.120), when $\varkappa$ is real, the electrons move freely in the crystal. In this case the eigenfunction ψ_n given by Eq. (6.120) represents a plane wave times a periodic function, with the energy of the electrons determined by the value of $\varkappa$. To find the possible eigenvalues of the energy we have to study the solutions of Eq. (6.115) for real κ. This requires that ka is such that

$$-1 \leq \cos ka + \frac{\lambda}{ka} \sin ka \leq 1. \quad (6.122)$$

Figure 6.11 shows the function $\cos \varkappa a = \cos ka + (\lambda/ka) \sin ka$ for a given value of the parameter λ; the allowed values of ka (Brillouin zones) are marked with thick lines on the graph. We see that the solution defines a sequence of continuous bands, whose origin and position we can understand as follows. If for a moment we imagine the crystal as a set of wells isolated by impenetrable barriers (λ infinite), the solutions of the eigenvalue equation (6.115) would be precisely those corresponding to a single infinite well with a discrete energy spectrum, $ka = n\pi$, as can be seen in Fig. 6.12(a). By lowering the barriers or making them narrower (λ finite), the wells become interconnected and each energy level breaks down into N close levels, as we saw in Section 5.5. Since N is infinite in the model used, this decomposition produces an infinite number of infinitely close levels. This results in a series of bands, each with a continuum of

---

[17] A proof of Floquet's theorem can be found in Margenau and Murphy (2009), section 2.17.

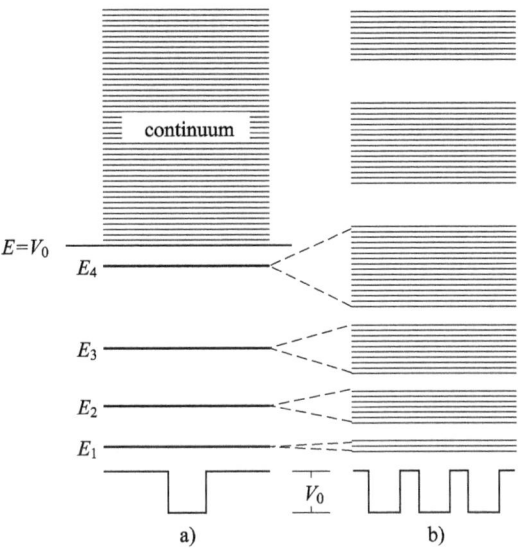

Figure 6.12 (a) Discrete energy levels for electrons in a single finite potential well for $E < V_0$ and continuous energy spectrum for $E > V_0$. (b) When the potential wells connect, continuous energy bands appear, separated by forbidden energy regions.

levels around the solution for an isolated well, as shown in Fig. 6.12(b). Note that the bands can also occur in regions that originally correspond to the continuum of one of the isolated wells.

In a real crystal, the bands are not strictly continuous, but are formed by a succession of discrete levels in close proximity, since the number of ions in the lattice is finite. The latter is the model we will use in the following to discuss qualitatively some important issues in conduction theory; more complete and quantitative analyses can be found in the literature on solid-state theory.

Consider a metal that contains one valence electron per atom (e.g., an alkali; the meaning of this term is discussed in Chapter 16). Since each electronic state can be occupied by up to two electrons, in the ground state the electrons occupy half the levels of the lower band (the band has N levels, where N is the number of atoms). Half of the electrons move in one direction and half in the opposite direction, so there is no net current. If we apply a voltage to the metal, more electrons will move in the opposite direction to the field; since the lower $N/2$ levels are already occupied, the electrons from the extra flow have to move to higher energy levels. Therefore, for electrical conduction to occur, the lower band must fulfil one of two conditions: a) have a sufficient number of vacancies (unoccupied levels), as in the alkali metal example, or b) be in contact with the second band. For this reason, the second band is called the *conduction band*, while the first is called the *valence band*.

In an insulator, on the other hand, all levels of the valence band are occupied, those of the conduction band are empty, and there is a separation of several eV between the two bands. This band gap is large enough to prevent electrons from passing into the conduction band when an external electric field is applied, so the material does not conduct.

Note the essential role played in the preceding argument by the fundamental fact that each electron level can be occupied by a maximum of two electrons (with different spin states); without this limitation, all crystals would be conductors. We can therefore say that the basic theory of conduction provides an excellent verification of this fundamental principle (or, more generally, of the fact that electrons satisfy the Fermi–Dirac statistics, as we shall see in Chapter 15).

In some materials it happens, however, that the energy gap (the zone between the valence and conduction bands) is comparatively very narrow. For example, for Si it is 1.14 eV and for Ge only 0.67 eV. These crystals remain insulators at $T = 0$, but their conductivity increases rapidly with temperature, as some thermally excited electrons pass into the conduction band. At room temperature ($kT \approx 10^{-2}$ eV) these materials reach conductivities intermediate between those of a dielectric and a conductor, and are therefore generically called *semiconductors*. Electrons that pass into the conduction band leave vacancies in the valence band, called *holes*.

The intrinsic conductivity of a semiconductor, that due to thermal excitation, has two contributions. On the one hand, electrons in the conduction band can produce a current in the usual way; on the other hand, when an electron in the valence band occupies one hole, but leaves another, a migration of vacancies occurs. It can be shown that this movement of electrons in a nearly full valence band (with few holes) can be consistently considered as the migratory movement – in the opposite direction – of the holes themselves. The current due to the movement of electrons is called *n-type*, while that due to the movement of vacancies is called *p-type*.

6.3.2.3 Extrinsic semiconductors

The conductivity of a semiconductor can be increased in various ways. For example, photons in the red or infrared range already have sufficient energy to transfer electrons into the conduction band; that is, some semiconductors are *photoconductors*. This property is widely used in the manufacture of photosensitive semiconductor devices.

A more important mechanism for increasing the conductivity of a semiconductor is to add impurities (or *dopands*) to the crystal lattice. For example, suppose that a small amount of arsenic is added during the preparation of a germanium crystal. Since As has five valence electrons compared to Ge's four, when an As atom takes the place of a Ge atom in the lattice, one valence electron is (almost) free and is easily transferred to the conduction band by thermal excitation. A significant effect on the conductivity of the material due to these substitutions is achieved at densities as low as one impurity per 10^6 atoms. Impurities that contribute electrons to the conduction band are called *donors* and the resulting semiconductor is *n*-type (with an excess of free electrons in the conduction band).

Impurities can also be introduced to create an electron deficit (presence of vacancies) in the valence band. These impurities are called *acceptors* and the resulting semiconductor is called *p*-type (with an excess of holes in the valence band). An example of this is when Ga is added to Ge; as Ga has only three valence electrons, it creates a hole in the band.

The energy levels produced by donors and acceptors are in the forbidden zone, very close to the conduction or valence band, respectively, as illustrated in Fig. 6.13.

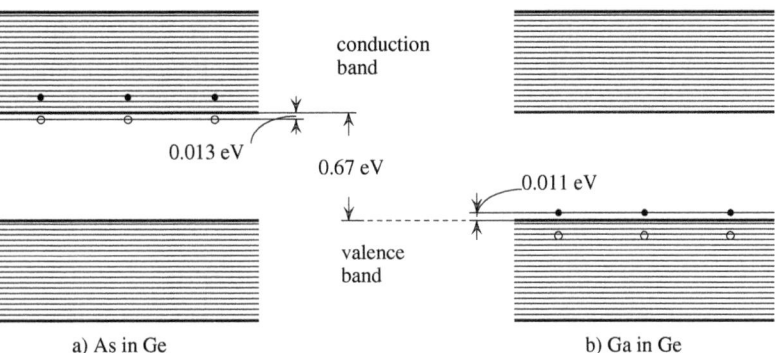

Figure 6.13 The valence and conduction bands of a doped Ge crystal. (a) Energy levels of a donor impurity (As) appear close to the conduction band (n-type doping); (b) energy levels of an acceptor impurity (Ga) appear close to the valence band (p-type doping).

The properties of semiconductors can vary over a wide range by controlling the nature and proportion of added impurities. It is these fundamental principles that are used in the construction of a variety of semiconductor electronic devices such as diodes, transistors, rectifiers, modulators, photocells, integrated circuit chips, in addition to LEDs and semiconductor lasers.

A semiconductor *diode* is formed by joining a *p*-type semiconductor with an *n*-type semiconductor; this is called a *p–n* junction. At the junction, electrons accumulate in the *p* region and holes accumulate in the *n* region, which is why it has some similarities to a charged capacitor. The device has resistances of very different orders of magnitude for currents flowing in one direction or the other, so it behaves like a rectifier (an element that allows electric current to flow in only one direction).

A *tunnel diode* is a *p–n* junction with a high concentration of impurities; in this case, the situation is such that applied voltages can modify and control the passage of electrons from one side of the junction to the other, so that for certain external voltages the net current decreases with the voltage, i.e., the device has negative resistance. This feature makes the tunnel diode extremely useful in the construction of many different types of electronic generators and switching circuits, mainly because of its very fast response. For example, it can be used to make generators that operate at frequencies above 10^{11} Hz.

The LED (acronym for *light-emitting diode*) is an electroluminescent heavily doped *p-n* junction diode. When current flows through the diode, electrons recombine with holes at the junction, releasing energy in the form of light, the color of which is determined by the energy band gap of the semiconductor used.

A *transistor* is obtained by juxtaposing two junctions, creating an *n–p–n* or *p–n–p* combination. These devices are used to control currents, so they can act as power, current or voltage amplifiers, as oscillators, and so on. Their great versatility, economy, smallness, durability, low working voltages, and so on, have made them indispensable in contemporary electronic technology, displacing thermionic tubes (bulbs) and allowing the construction of ultra-compact circuits (as, for example, in modern computers and cell phones).

Recent attention to low-dimensional systems that have electrons and holes confined to move in a plane, along a wire or within a quantum dot, forming structures with submicron resolution, has given rise to new physical concepts and applications, distinct from their three-dimensional analogs. Interested students are recommended to consult the specialized bibliography for more information on the subject.

Problems

P6.1 Derive Eq. (6.56) for the transmission coefficient T from the expression of the wave function for a rectangular barrier obtained in Chapter 5, considering the barrier of arbitrary profile as built up by a succession of rectangular barriers that become infinitely narrow in the limit.

P6.2 A particle moves in a conical potential that vanishes for $|x| > a$, has the value $-V_0$ at $x = 0$ and goes linearly from $-V_0$ to 0 in the intervals $|x| \leq a$. Study the bound states of the system using the WKB method.

P6.3 An electron with energy $E > V_0$ moves in the potential

$$V(x) = V_0 \cosh^{-2}(x/a), \quad V_0 > 0. \tag{6.123}$$

Determine the conditions under which the WKB method is applicable and obtain the transmission coefficient through the barrier.

P6.3 Use the WKB method to find the quantization conditions for particles moving in a well, one of the walls of which is impenetrable.

P6.4 Use the WKB method to find the energy of the stationary states of particles moving inside a box with impenetrable walls. Analyze the results and compare them with the exact ones.

P6.5 Determine the transmission coefficient for a monochromatic electron beam of energy 1000 eV and intensity 1 mA incident on a rectangular barrier of width 10 Å and height of 200 eV. The barrier height is now increased to 2000 eV; determine the new transmission coefficient and compare the results.

P6.6 Use the WKB method to determine the energy levels of a one-dimensional harmonic oscillator. Discuss your result.

P6.7 Use the WKB method to find the ratio between the width and depth of a one-dimensional square well that is required for no bound state to exist.

P6.8 Use the semiclassical approximation to determine the average value of the kinetic energy of a stationary state.

P6.9 Apply the reasoning used to arrive at Eq. (6.91), to obtain the formulae for the density of states as a function of the maximum energy level:

a) for a 2D layer,
b) for a 1D wire.

P6.10 Consider an infinite 1D periodic potential $V(x) = V(x+a)$. Starting from the Schrödinger equation for ψ_k as an eigenfunction corresponding to the energy E_k, obtain the Schrödinger equation for $u_k(x)$.

P6.11 Find the energy bands for the periodic potential

$$V(x) = g \sum_{n=-\infty}^{\infty} \delta(x - na), \quad g, a > 0. \tag{6.124}$$

P6.12 Estimate the minimum energy of the first allowed band for the previous problem, when $\lambda = 8$ and $g/a = 1$ eV.

Bibliographical Notes

For a detailed derivation of the WKB connection formulas, see Langer (1937), p. 669.
For a proof of Floquet's theorem, see Margenau and Murphy (2009), section 2.17.
Introductory studies on low-dimensional quantum systems can be found in Harrison (2000).
For a detailed treatment of semiconductors, see for example, Grundmann (2021).

7 The Free Particle. The Time-Dependent Schrödinger Equation

So far, we have been able to address a range of quantum problems using the time-independent Schrödinger equation, which has proved to be a powerful method for describing the behavior of a quantum system under stationary conditions. In the second part of this chapter we will extend the Schrödinger description to the time-dependent case, which must be applied whenever the state of the system evolves with time. In preparation for this, we will first consider the free particle, which might be thought to be the simplest problem of all; however, a detailed treatment serves to highlight some surprising features and to introduce powerful tools that are of wider application, notably the Green function and the Feynman propagator.

7.1 The Free Particle

A free quantum particle is characterized by a constant value of its energy, E, which is purely kinetic energy. To simplify the mathematics we will restrict ourselves to 1D (the generalization to 3D is immediate), in which case the Schrödinger equation is simply

$$\frac{d^2\varphi}{dx^2} + \frac{2m}{\hbar^2} E\varphi = 0. \tag{7.1}$$

In terms of the wave number

$$k = \sqrt{2mE}/\hbar = p/\hbar, \tag{7.2}$$

its general solution reads

$$\varphi = Ae^{ikx} + Be^{-ikx}, \tag{7.3}$$

where the constants A and B may depend on k. This is just the expression we used in Chapter 5 to study the dispersion of particles by simple potentials. We now complete the wave function by introducing the time factor (as was done in Section 5.4), $\psi(x,t) = e^{-i\omega t}\varphi(x)$, so that

$$\psi(x,t) = Ae^{-i(\omega t - kx)} + Be^{-i(\omega t + kx)}. \tag{7.4}$$

The first term of the right-hand side represents a plane wave propagating to the right with phase velocity $v = \omega/k$; the wave in the second term propagates to the left. It follows that, restricting ourselves to the case of particles moving to the right, we must take $B = 0$, so that the wave function becomes

$$\psi = Ae^{-i\omega t + ikx}. \tag{7.5}$$

With $e^{i\omega t}$ instead, the beam propagating to the right would correspond to $A = 0$, and instead of Eq. (7.5) we would have

$$\psi'(x,t) = A'e^{i\omega t - ikx}. \tag{7.6}$$

But ψ' is the complex conjugate of the ψ above (with $A'^* = A$), so the choice between the two alternatives is reduced to a simple change of name, whereby what we now call ψ becomes ψ^* and vice versa. Since this mere change of notation is physically insubstantial, we are free to adopt the time factor $e^{-i\omega t}$. Note that the wavelength associated to (7.5) is, according to Eq. (7.2),

$$\lambda = \frac{2\pi}{k} = \frac{h}{p}, \tag{7.7}$$

which is precisely de Broglie's wavelength.

The solution (7.5) represents a constant and uniform distribution of particles in time and space,

$$\rho = \psi^*\psi = |A|^2 = \text{const.} \tag{7.8}$$

It is clear that we have oversimplified the description of the physical situation: an infinite space cannot be uniformly filled with particles, since this would require the number of particles, the total momentum, the total energy, and so on, to be infinite – and forever. In the real world, the particle beam is finite in space and time. To confine the beam, some device (e.g., a shutter) is needed to allow or prevent the passage of particles, which necessarily affects at least the particles at both ends of the beam. This and other unavoidable effects – quantum or external – result in a dispersion of energy and momentum around their central values E and p, which becomes more pronounced the narrower or shorter the beam is.

The preceding observations indicate that the wave function (7.5) can reasonably describe the physical situation inside the real beam, but definitely not at the boundaries or outside it, where the correct wave function is essentially zero. A simple way to construct a more realistic function is to superpose components corresponding to all physically significant values of momentum and energy, taking each component with the appropriate amplitude and phase. Such a superposition must guarantee that the component waves interfere constructively within the volume occupied by the particles, but destructively outside it. But then the function ψ of interest is not an eigenfunction of the stationary Schrödinger equation, but a time-dependent superposition of solutions for different values of the parameter k (and corresponding values of the energy). The time factor is needed to specify the relative phase of each component of the solution.

Assume that the particle momentum is distributed in the interval (k_1, k_2) with amplitude $A(k)$; since for free particles $\omega = E/\hbar = \hbar k^2/2m$, the wave function is then

$$\psi(x,t) = \int_{k_1}^{k_2} A(k) e^{-\frac{i\hbar k^2}{2m}t + ikx} \, dk. \tag{7.9}$$

Such a structure is called a *wave packet*. Note that this packet evolves over time, so its shape changes as it spreads out, because it contains slower and faster than average particles. In short, the wave packet undergoes *dispersion*.[1]

---

[1] We recall that in his early work, Schrödinger proposed to interpret the wave function as a direct representation of the electron, which is distributed in each case in space as expressed by the wave function itself.

In Eq. (7.9), $A(k)$ is the amplitude with which the plane wave of momentum $\hbar k$ and energy $\hbar^2 k^2/2m$ contributes to the packet. The equation thus represents a spectral analysis of ψ in momentum space. If ψ were reduced to a monochromatic plane wave extending over the whole space, $A(k)$ would be nonzero only for the corresponding value of the variable k; however, for any finite packet, many components (in principle a continuum of them) are required to synthesize it by a superposition of plane waves. This means that $A(k)$ is a more or less wide function of k; the wider $A(k)$ is, the narrower $\psi(x)$ is, and vice versa. Since each Fourier component has well-defined momentum and energy, any finite packet necessarily has components with momentum and energy distributed with nonzero width. In general, $A(k)$ can be determined from the conditions of the problem.

Equation (7.5) is much simpler than its more realistic counterpart, (7.9), and is therefore preferred in applications whenever possible, as it greatly facilitates calculations. But its use introduces a mathematical difficulty, because given the uniformity of the distribution, the wave function (7.5) is not square integrable, that is, it is not normalizable,

$$\int |\psi|^2 dx = |A|^2 \int_{-\infty}^{\infty} dx = \infty. \tag{7.10}$$

If, on the other hand, $A(k)$ were reasonably selected, the number of particles associated with the packet would be finite and the corresponding wave function would be normalizable. We see that the problem arises from the extreme idealization that occurs when free particles are represented by plane waves extending through space (and time). To solve this problem caused by the approximations, we will take advantage of the fact that the normalization condition has a certain degree of arbitrariness. In fact, two different solutions to this problem are used, that is, two alternative ways of normalizing the plane wave, to which we will now turn our attention. The choice of one or the other method is usually dictated by reasons of convenience or custom.

7.1.1 The Born normalization

The basic idea of Born's method is to consider that the space in which the plane wave propagates has periodicity properties, that is, that it can be conceived as a set of equal boxes, side by side, inside each of which precisely what happens in the laboratory (represented by one of the boxes) is repeated. The boxes are usually taken to be cubic in shape, of edge L, and are called normalization cubes. After whatever calculations the problem requires, we let L tend to infinity. Since we are really only interested in what happens inside the normalization cube, imposing restrictions on what happens outside it has no physical consequences. Moreover, by taking the limit $L \to \infty$, the choice of boundary conditions becomes irrelevant, so that, in particular, they can be taken as periodic. Since the physical results are usually independent of the normalization volume if L is chosen large enough, the operation of taking the limit $L \to \infty$ is usually trivial. Note that the particle density $\rho = 1/L$ ($1/L^3 = 1/V$ in the three-dimensional

---

The fact that wave packets spread out continuously in space was the reason why this interpretation did not find support.

case, where V is the normalization volume) depends on L, but is normalized to unity in all cases, as required.

In the 1D case, the periodicity condition is written as

$$\varphi(x) = \varphi(x+L). \tag{7.11}$$

Applying this condition to the plane wave, Eq. (7.5), we get

$$Ae^{ikx} = Ae^{ik(x+L)}, \tag{7.12}$$

which means that $e^{ikL} = 1$. This condition is satisfied if k has one of the values

$$k = \frac{2\pi}{L}n, \quad n = 1, 2, 3, \ldots \tag{7.13}$$

Therefore $p = \hbar k$ and E are discrete, instead of forming a continuum. However, it is clear that in the limit $L \to \infty$ the discrete spectrum of these variables merges with the continuous one. For example, the increment of k between two successive values of n becomes $\delta k = \lim_{L \to \infty} (2\pi/L)$. It is important to note that this procedure eliminates possible effects due to the presence of boundaries in the vicinity. For example, if the free particles to be described are conduction electrons in a metal, the latter is modeled as having infinite dimensions.

The normalization condition is now imposed on the wave function inside the normalization cube (ψ is null outside),

$$\int_{-L/2}^{L/2} |\psi|^2 dx = |A|^2 \int_{-L/2}^{L/2} dx = |A|^2 L = 1. \tag{7.14}$$

For convenience, we take A real and positive, resulting in $A = 1/\sqrt{L}$. With this convention, the free-particle wave function traveling to the right is

$$\psi(x,t) = \frac{1}{\sqrt{L}} e^{-i\frac{E_n}{\hbar}t + i\frac{p_n}{\hbar}x}, \tag{7.15}$$

where

$$p_n = \hbar k_n = \frac{2\pi\hbar}{L}n, \quad E_n = \frac{\hbar^2 k_n^2}{2m} = \frac{2\pi^2\hbar^2}{mL^2}n^2 = E_1 n^2. \tag{7.16}$$

It is easy to verify that the wave functions $\varphi_n(x)$,

$$\varphi_n(x) = \frac{1}{\sqrt{L}} e^{i\frac{2\pi n}{L}x}, \tag{7.17}$$

are orthonormal in $(-L/2, L/2)$, since, by performing the integration, we get

$$\int_{-L/2}^{L/2} \varphi_{n'}^* \varphi_n \, dx = \frac{1}{L} \int_{-L/2}^{L/2} e^{2i\pi(n-n')x/L} dx$$
$$= \frac{\sin \pi(n-n')}{\pi(n-n')} = \begin{cases} 1, & n = n' \\ 0, & n \neq n'. \end{cases} \tag{7.18}$$

Identifying the final result with the Kronecker delta, we see that it is possible to represent the latter in the form

$$\delta_{nn'} = \frac{\sin \pi(n-n')}{\pi(n-n')}. \tag{7.19}$$

7.1.2 The Dirac normalization: Properties of the delta function

In preparation for the Dirac normalization method we will informally discuss some important properties of the Dirac delta "function," which can be seen as a generalization of the Kronecker delta to the case of continuous indices. Strictly speaking, the Dirac delta is not a function but a distribution, since it is not well behaved and only makes sense within an integral sign. The mathematical details can be found in Hassani (2013).

Let $\{\varphi_n\}$ be a complete set of orthonormal functions on the interval (a, b); we expand the arbitrary square-integrable function $f(x)$, $a < x < b$, in terms of the basis $\{\varphi_n\}$,

$$f(x) = \sum_n a_n \varphi_n(x) \tag{7.20}$$

$$a_n = \int_a^b f(x)\varphi_n^*(x)\,dx. \tag{7.21}$$

We now introduce the second expression into the first and interchange the order of the sum and the integral to get

$$f(x) = \sum_n \varphi_n(x) \int_a^b \varphi_n^*(x')f(x')\,dx' = \int_a^b dx'\, f(x') \sum_n \varphi_n(x)\varphi_n^*(x'). \tag{7.22}$$

Defining

$$\delta(x, x') = \sum_n \varphi_n(x)\varphi_n^*(x'), \tag{7.23}$$

we finally get

$$f(x) = \int_a^b \delta(x, x')f(x')\,dx'. \tag{7.24}$$

This result holds for any $f(x)$, which implies that the function δ defined in (7.23) must possess peculiar properties; we can understand its basic structure as follows. Let $\Delta(x - x')$ be a function with value $\Delta_0(x')$ if $|x - x'| < \epsilon$ and with value zero if $|x - x'| > \epsilon$, where ϵ is an arbitrarily small positive number; an example of such a function is illustrated in Fig. 7.1. Then for x, x' in (a, b),

$$\int_a^b \Delta(x - x')f(x')\,dx' = \int_{x-\epsilon}^{x+\epsilon} f(x')\Delta_0(x')\,dx' = f(x)\int_{x-\epsilon}^{x+\epsilon} \Delta_0(x')\,dx' + O(\epsilon); \tag{7.25}$$

to write the last equality, a Taylor expansion of $f(x')$ around x was done. Demanding that the area under the function Δ be unity,

$$\int \Delta_0(x')\,dx' = 1, \tag{7.26}$$

we obtain

$$\lim_{\epsilon \to 0} \int_a^b \Delta(x - x')f(x')\,dx' = f(x). \tag{7.27}$$

Comparing with (7.24) we see that we can make the identification

$$\delta(x, x') \equiv \delta(x - x') = \lim_{\epsilon \to 0} \Delta(x - x'). \tag{7.28}$$

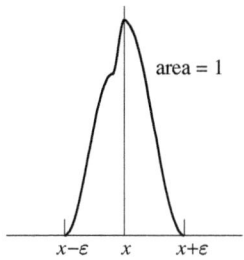

Figure 7.1 A possible function $\Delta(x - x')$. If the area under the curve is 1, in the limit $\sigma \to 0$ this function tends to Dirac's $\delta(x - x')$.

We have written δ here as a function of the argument $x - x'$, since it follows from the preceding reasoning that this function is nonzero only if its two arguments coincide. We can visualize this function as a pulse of zero width and infinite height, with area equal to unity; how to arrive at this limit is irrelevant, so the Dirac δ function can take many different explicit forms (we will have occasion to see some of them throughout the book). Using Eq. (7.28), we will henceforth write (7.24) in the form

$$f(x) = \int_a^b f(x')\delta(x - x')\,dx', \quad a < x < b, \tag{7.29}$$

which is the most important integral property of the δ function. Taking $f = 1$ we get the condition that the area under the curve is unity,

$$\int_a^b \delta(x - x')\,dx' = 1. \tag{7.30}$$

Note that when the variable x is dimensional, the $\delta(x)$ has the dimensions of x^{-1}.

The formula (7.23), which expresses the Dirac δ function in terms of a complete orthonormal set of functions, is the completeness relation. It means that there exists a unit vector for each direction in space, which in turn means that there exists a vector for each and every component of any function to be represented.

It can be shown that $\delta^{(n)}(x - a)$, the nth derivative of $\delta(x - a)$ with respect to x, has the property that

$$\int f(x)\delta^{(n)}(x - a)\,dx = (-1)^n f^{(n)}(a). \tag{7.31}$$

Moreover, if x_i represents a real root of the equation $y(x) = 0$ included in the integration interval, then

$$\int f(x)\delta[y(x)]\,dx = \sum_i \frac{f(x_i)}{\left|\left(\frac{dy}{dx}\right)_{x=x_i}\right|}, \tag{7.32}$$

as follows immediately from the fact that

$$\delta(y)\,dy = \delta(x)\,dx. \tag{7.33}$$

In a space of more than one dimension, the delta function is defined as the product of the $\delta(x_i)$. For example,

$$\delta(\mathbf{x} - \mathbf{a}) = \delta(x_1 - a_1)\delta(x_2 - a_2)\delta(x_3 - a_3). \tag{7.34}$$

If non-Cartesian coordinates ξ_i are used, it follows from a change of integration variables that

$$\delta(\mathbf{x} - \mathbf{x}') = \frac{1}{|J(x_i, \xi_i)|} \delta(\xi_1 - \xi_1')\delta(\xi_2 - \xi_2')\delta(\xi_3 - \xi_3'), \tag{7.35}$$

where $J(x_i, \xi_i)$ is the Jacobian of the coordinate transformation. The relation (7.33) is the one-dimensional case of this formula.

It is easy to be convinced that (7.29) corresponds to the generalization to the continuous case of Kronecker's δ. To see this, suppose that the variable x can only take n discrete values; (7.29) reduces in this case to

$$f(n) = \sum_{n'=n_a}^{n_b} f(n')\delta(n - n'), \tag{7.36}$$

which is precisely the defining property of $\delta_{nn'}$.

Exercise E7.1. Fourier representation of the delta function

Construct an explicit expression for Dirac's δ, using the free-particle solutions obtained in the previous section.

Solution. We use the complete set $\{\varphi_n\}$ of free-particle solutions,

$$\varphi_n(x) = \frac{1}{\sqrt{L}} e^{i\frac{2\pi n}{L}x}, \tag{7.37}$$

and from the completeness relation (7.23) applied to this set, we get

$$\delta(x - x') = \frac{1}{L} \sum_n e^{i\frac{2\pi n}{L}(x-x')}, \tag{7.38}$$

where the sum extends over all possible values of n, positive and negative.

Another important result is obtained by taking the limit of continuous n ($L \to \infty$), which we do as follows. Since $k_n = 2\pi n/L$, the minimum increment of k, corresponding to $\Delta n = 1$, is $\Delta k = 2\pi/L$; thus

$$\lim_{L \to \infty} \frac{1}{L} = \lim_{\Delta k \to 0} \frac{\Delta k}{2\pi} = \frac{1}{2\pi} dk. \tag{7.39}$$

Plugging this result into (7.38) and changing the sum to an integral gives

$$\delta(x - x') = \frac{1}{2\pi} \int_{-\infty}^{\infty} e^{ik(x-x')} dk. \tag{7.40}$$

This important formula clearly shows the structure of the δ function: for $x - x' \neq 0$ the integrand oscillates periodically and symmetrically, and the resulting net area is zero, but when $x - x' = 0$, the right-hand side reduces to $(2\pi)^{-1} \int_{-\infty}^{\infty} dk$, which is infinity. Using different sets of orthogonal functions it is possible to construct other integral representations of Dirac's δ.

As an example of the possibilities offered by this construction, we can use (7.40) to construct the Fourier integral transform $\tilde{f}(k)$ of a function $f(x)$. Indeed it is sufficient to write

$$f(x) = \frac{1}{\sqrt{2\pi}} \int_{-\infty}^{\infty} \tilde{f}(k) e^{ikx} dk, \tag{7.41}$$

to get

$$\tilde{f}(k) = \frac{1}{\sqrt{2\pi}} \int_{-\infty}^{\infty} f(x) e^{-ikx} \, dx. \qquad (7.42)$$

In particular, comparing (7.41) with (7.40) we see that the Dirac's δ is the Fourier transform of the unit. This direct procedure can be used to obtain any required integral transform. Observe that a Fourier series is the same, only for a discrete representation instead of the continuum.

7.1.2.1 The Dirac normalization

We take the nonnormalized free-particle wave function, Eq. (7.5), labeling it with the parameter p,

$$\varphi_p(x) = A e^{i\frac{p}{\hbar}x}. \qquad (7.43)$$

Using Eq. (7.40), we see that

$$\int_{-\infty}^{\infty} \varphi_{p'}^*(x) \varphi_p(x) \, dx = |A|^2 \hbar \int_{-\infty}^{\infty} e^{i\frac{x}{\hbar}(p-p')} d\frac{x}{\hbar} = 2\pi\hbar |A|^2 \delta(p-p'). \qquad (7.44)$$

For $p = p'$, the result is infinite; this is precisely the singularity obtained earlier when trying to normalize the wave function (7.43). But there is an essential difference between this result and the previous one, namely the divergence now has the form of a δ function, a quantity that we know how to handle with the tools developed previously. Therefore, we can now consider that the infinity implicit in (7.44) is acceptable. This result leads to the normalization method proposed by Dirac.

According to this method, (7.44) must be taken as the orthonormality condition of the free-particle functions, so all that remains is to give it a suitable form, which is achieved by assigning to the normalization constant A the value

$$A = \frac{1}{\sqrt{2\pi\hbar}}, \qquad (7.45)$$

to reduce the orthonormality condition to its simplest form,

$$\int_{-\infty}^{\infty} \varphi_{p'}^*(x) \varphi_p(x) \, dx = \delta(p-p'). \qquad (7.46)$$

This expression extends the orthonormality condition $\int_{-\infty}^{\infty} \varphi_{n'}^*(x) \varphi_n(x) \, dx = \delta_{n'n}$ to the continuous-spectrum case. Note that with this method, we are accepting as an integral part of the theory wave functions that in the strict sense are not square-integrable;[2] this will be the only exception allowed. On the other hand, by reinterpreting $\varphi_p(x)$ as a function of p with parameter x,

$$\phi_x(p) = \frac{1}{\sqrt{2\pi\hbar}} e^{i\frac{x}{\hbar}p}, \qquad (7.47)$$

---

[2] The inclusion of functions with continuous spectra extends the space considered; with them we leave the Hilbert space to enter a Banach space. However, it is usual (but not rigorous) to keep on calling it a Hilbert space, a practice we will follow.

Equation (7.46) takes the form

$$\int_{-\infty}^{\infty} \phi_x^*(p')\phi_x(p)\,dx = \delta(p-p'), \tag{7.48}$$

which is the extension to continuous variables of the completeness condition (7.23). Equations (7.46) and (7.48) provide a means to extend the methods used in the discrete-spectrum case to free-particle functions in the continuum.

In short, the wave function describing a free particle normalized à la Dirac is

$$\varphi_p(x) = \frac{1}{\sqrt{2\pi\hbar}} e^{i\frac{p}{\hbar}x}. \tag{7.49}$$

A very common form of the wave function for free particles is obtained by introducing the time dependence and moving to 3D,

$$\psi(\boldsymbol{x},t) = (2\pi\hbar)^{-3/2} e^{-i\frac{E}{\hbar}t + i\frac{\boldsymbol{p}}{\hbar}\cdot\boldsymbol{x}}. \tag{7.50}$$

7.1.3 The free-particle propagator

In Section 7.1 we saw that a general 1D free-particle packet has the form (see Eq. (7.9))

$$\psi(x,t) = \int A(k) e^{-i\hbar k^2 t/2m + ikx}\,dk, \tag{7.51}$$

with $k = p/\hbar$ and $\hbar k^2/2m = E/\hbar$. This expression allows us to study the time evolution of the packet with arbitrary initial conditions. Assuming that the particles are distributed at $t = 0$ with the amplitude

$$\psi(x, t=0) = \psi(x,0) = \psi_0(x) \tag{7.52}$$

and allowed to evolve freely, we can find $\psi(x,t)$ for all $t > 0$ as follows. Since the wave function is continuous, it must satisfy Eq. (7.51) at $t = 0$,

$$\psi_0(x) = \int A(k) e^{ikx}\,dk, \tag{7.53}$$

which shows that the amplitude $A(k)$ is the Fourier transform of the initial wave function,[3]

$$A(k) = \frac{1}{2\pi} \int \psi_0(x) e^{-ikx}\,dx. \tag{7.54}$$

Note that the distribution of the momentum does not change as the packet evolves. Plugging this expression into (7.51) we get

$$\psi(x,t) = \frac{1}{2\pi} \int_{-\infty}^{\infty} dk \int_{-\infty}^{\infty} dx'\,\psi_0(x') e^{-i\hbar k^2 t/2m - ik(x'-x)}, \tag{7.55}$$

which can be rewritten in the revealing form

$$\psi(x,t) = \int_{-\infty}^{\infty} K(x,t|x',0)\psi_0(x')\,dx', \tag{7.56}$$

---

[3] The numerical factor that multiplies the integrals in the Fourier transform is defined in different ways, depending on the author. Although we normally use the definitions (7.41) and (7.42), we will refer to the Fourier transform or its inverse with precision up to a numerical factor. Something analogous happens with the sign in the exponential.

7.1 The Free Particle

with

$$K(x,t|x',t') = \frac{1}{2\pi} \int_{-\infty}^{\infty} e^{ik(x-x') - i\hbar k^2 (t-t')/2m} dk. \tag{7.57}$$

Putting here $p = \hbar k$ and using (7.50), this expression can be rewritten in the following form, which turns out to be very important and representative of the general case,

$$K(x,t|x',t') = \int_{-\infty}^{\infty} \psi_p(x,t) \psi_p^*(x',t') dp. \tag{7.58}$$

According to Eq. (7.56), $\psi(x,t)$ is the sum of all the contributions from the initial distribution $\psi_0(x')$ that propagate from each point x' at time t' to the given x at time t. This means that the function $K(x,t|x',t')$ must be interpreted as the probability amplitude of the free particle moving from position x' at t' to position x at time t. This transition amplitude is known as the *Feynman*[4] *propagator* (for the free particle, in the present case; see Section 7.4 for the more general case). Mathematically, the propagator is a Green function, that is, the kernel of a linear integral transformation connecting the initial state to the final state, as follows from (7.56).

It is not difficult to realize that the quantum propagator is a particular wave function. Consider the case where the initial amplitude describes particles concentrated at x_0,

$$\psi_0(x') = \delta(x' - x_0); \tag{7.59}$$

introducing this condition in (7.56) gives

$$\psi(x,t) = \int K(x,t|x',0) \delta(x' - x_0) dx' = K(x,t|x_0,0), \tag{7.60}$$

which verifies that the propagator is a solution of the Schrödinger equation; in fact, being a Green function, it is a fundamental solution, which has the property of reducing to the Dirac's δ function for $\Delta t = 0$. This mathematical point, characteristic of Green functions, will be clarified in general in the next section. For the moment, we see from (7.40) and (7.57) that this is indeed the case for $t = t'$,

$$K(x,t|x',t) = \frac{1}{2\pi} \int_{-\infty}^{\infty} e^{ik(x-x')} dk = \delta(x - x'). \tag{7.61}$$

Finally, the explicit expression for the free-particle propagator is obtained by computing the integral indicated in (7.57), which gives

$$K(x,t|x_0,t_0) = \sqrt{\frac{m}{2\pi i \hbar (t-t_0)}} e^{\frac{im(x-x_0)^2}{2\hbar(t-t_0)}}. \tag{7.62}$$

Here we have set an arbitrary initial t_0, for generality. Taking the limit $t \to t_0$ in this expression gives another representation of Dirac's δ.

Because of its fundamental nature, the propagator contains all the information about the dynamics of the system under study; this has led to its extensive use in the study of

---

[4] Richard Phillips Feynman (1918–1988) was a famous US theoretical physicist, considered one of the most important physicists of all time. He made important contributions to quantum mechanics, quantum electrodynamics – for which he shared the 1965 Nobel Prize with J. Schwinger and S. Tomonaga – the physics of superfluids, and elementary particles. He introduced the important concepts of partons and nanotechnology, and carried out research into quantum computation.

quantum systems, including quantum field theory. It is therefore difficult to overstate the importance of the concept.

Exercise E7.2. Free-particle propagator in terms of the Hamiltonian

Show that it is possible to write the free-particle propagator in the form

$$K(x,t|x',t') = e^{-i\widehat{H}(t-t')/\hbar}\delta(x'-x), \tag{7.63}$$

where $\widehat{H}$ is in the x-representation.

Solution. Equation (7.57) can be written in the form $p = \hbar k$:

$$K(x,t|x',t') = \frac{1}{2\pi}\int_{-\infty}^{\infty} e^{-ikx'}e^{-i\frac{E_p}{\hbar}(t-t')}e^{ikx}dk \tag{7.64}$$

$$= \int_{-\infty}^{\infty}\varphi_p^*(x')e^{-\frac{iE_p}{\hbar}(t-t')}\varphi_p(x)dp = \int \varphi_p^*(x')e^{-\frac{i\widehat{H}}{\hbar}(t-t')}\varphi_p(x)dp,$$

where $\varphi_p(k)$ is given by (7.49); since $\widehat{H} = \widehat{H}_x$ acts only on the variable x, we can write

$$K(x,t|x',t') = e^{-i\frac{\widehat{H}}{\hbar}(t-t')}\int \varphi_p^*(x')\varphi_p(x)dp = e^{-i\frac{\widehat{H}}{\hbar}(t-t')}\delta(x'-x), \tag{7.65}$$

which is the desired result. For $t = t'$ we recover Eq. (7.61).

Introducing this result into (7.56) with $t = 0$ replaced by t', we get

$$\psi(x,t) = \int_{-\infty}^{\infty} e^{-i\frac{\widehat{H}}{\hbar}(t-t')}\delta(x'-x)\psi(x',t')dx' = e^{-i\frac{\widehat{H}}{\hbar}(t-t')}\psi(x,t'), \tag{7.66}$$

which shows that $S(t,t') \equiv e^{-i\frac{\widehat{H}}{\hbar}(t-t')}$ is the evolution operator. This result is generally valid when H does not explicitly depend on time. In Section 7.3 we will present a more general derivation of these results.

7.1.4 Green functions and the Dirac's delta function

Dirac's δ function can be used advantageously in the construction of the Green functions of linear differential equations. Although this is not the place to develop the subject fully, its extensive use, particularly in quantum theory, makes it convenient to review some of the techniques involved. To this end, we will consider some simple and interesting physical problems and discuss them in the course of their solution.

Suppose we wish to construct the solution of Poisson's equation, that is, to determine the potential Φ produced by a charge distribution ρ, subject to certain boundary conditions. The equation to be solved is

$$\nabla^2\Phi = -4\pi\rho(r). \tag{7.67}$$

Due to the linear nature of the problem, the potential generated by ρ (ρ is the *source* of the potential Φ) can be written as the sum of the potentials produced by each point charge; it is therefore convenient to study the latter case first.

A point charge e located at the origin generates the charge density

$$\rho(r) = e\delta(r). \tag{7.68}$$

This density is indeed zero at every point except the origin; furthermore, the total charge associated with it is

$$\int \rho(r) d^3r = e \int \delta(r) d^3r = e. \tag{7.69}$$

Therefore, the Poisson equation that determines the potential produced by a point charge of magnitude e placed at the origin is

$$\nabla^2 G(r) = -4\pi e \delta(r). \tag{7.70}$$

The potential G that solves this equation is, by definition, a Green function of Poisson's equation. More generally, if $\hat{L}_r$ is a linear differential operator, the Green function of $\hat{L}_r$ is the solution of the equation

$$\hat{L}_r G(r, r') = -4\pi \delta(r - r') \tag{7.71}$$

that satisfies the required boundary conditions.

The Green function allows us to construct immediately the solution of the inhomogeneous equation

$$\hat{L}_r \Phi(r) = -4\pi \rho(r) \tag{7.72}$$

for an arbitrary source ρ. Specifically, the solution is

$$\Phi(r) = \int G(r, r') \rho(r') d^3r', \tag{7.73}$$

where Φ and G satisfy the same boundary conditions. To show that the Φ given by this equation is a solution of (7.72), we apply to both sides of (7.73) the operator $\hat{L}_r$, obtaining

$$\hat{L}_r \Phi(r) = \int \rho(r') \hat{L}_r G(r, r') d^3r' = -4\pi \int \rho(r') \delta(r - r') d^3r' = -4\pi \rho(r), \tag{7.74}$$

which is precisely Eq. (7.72).

Returning to the problem of constructing the Green function for Poisson's equation, we propose to express it in terms of its Fourier transform,

$$G(r) = \frac{1}{(2\pi)^{3/2}} \int \tilde{G}(k) e^{i k \cdot r} d^3k. \tag{7.75}$$

Applying the operator ∇^2, we obtain

$$\nabla^2 G(r) = \frac{-1}{(2\pi)^{3/2}} \int k^2 \tilde{G}(k) e^{i k \cdot r} d^3k. \tag{7.76}$$

On the other hand, we can express the function δ in the form (see Eqs. (7.34) and (7.40))

$$\delta(r) = \frac{1}{8\pi^3} \int e^{i k \cdot r} d^3k. \tag{7.77}$$

Substituting these expressions in (7.70) and rearranging, we obtain

$$\int \left[\frac{k^2}{(2\pi)^{3/2}} \tilde{G} - \frac{4\pi e}{8\pi^3} \right] e^{i k \cdot r} d^3k = 0. \tag{7.78}$$

Since this condition must be satisfied for any r, the integrand must vanish, which demands that we take

$$\tilde{G}(k) = (2\pi)^{3/2} \frac{e}{2\pi^2 k^2}. \tag{7.79}$$

Plugging this result into Eq. (7.75) we get

$$G(r) = \frac{e}{2\pi^2} \int \frac{e^{i k \cdot r}}{k^2} d^3k. \tag{7.80}$$

The integral can easily be performed on a spherical coordinate system with the z-axis in the r direction; the result,

$$G(r) = \frac{e}{r}, \tag{7.81}$$

is the Coulomb potential of a point charge. Although this was predictable, the calculation illustrates well the ideas behind the procedure. Equation (7.81) gives the Green function that must be plugged into (7.73) to obtain the solutions of (7.67).

Exercise E7.3. Green functions for the Helmholtz equation
Generalize the preceding results to obtain the Green functions for the Helmholtz equation.

Solution. The Helmholtz equation is the time-independent part of the wave equation, frequently used in theoretical physics,

$$(\nabla_r^2 + k^2) G(r, r') = -4\pi \delta(r - r'). \tag{7.82}$$

Following the procedure used in the previous example, the Fourier transform of G can be easily shown to be

$$\tilde{G}(q) = \frac{(2\pi)^{3/2}}{2\pi^2} \frac{1}{q^2 - k^2}. \tag{7.83}$$

Taking the inverse Fourier transform, we get

$$G(r) = \frac{1}{2\pi^2} \int \frac{e^{i q \cdot r}}{q^2 - k^2} d^3q. \tag{7.84}$$

To carry out the integration, we express q in spherical coordinates, with the z-axis in the direction of r so that $q \cdot r = qr \cos\theta$. Integrating over the angular variables, we get

$$G(r) = \frac{1}{\pi} \int_0^\infty \frac{q}{ir} \frac{1}{q^2 - k^2} [e^{iqr} - e^{-iqr}] dq. \tag{7.85}$$

In the second integral we make the change of variable $q \to -q$ to obtain

$$G(r) = \frac{1}{i\pi r} \int_{-\infty}^\infty \frac{q}{q^2 - k^2} e^{iqr} dq. \tag{7.86}$$

We calculate this integral using the residue technique, moving to the complex q-plane. The integrand has simple poles at $q = k$ and $q = -k$. Only the integral $\int_{-\infty}^\infty$ contributes to the closed integral over the contour C^+ shown in Fig. 7.2a, since the integrand vanishes over the semicircle at infinity. Since this contour contains only the pole at $+k$, there is only one residue and we get

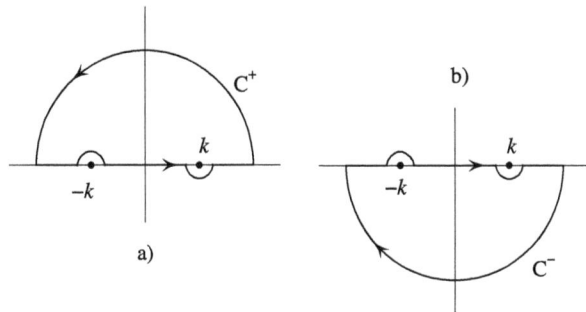

Figure 7.2 Possible integration contours for the calculation of the Green function for the stationary wave equation. The contour in (a) corresponds to an outgoing spherical wave centered at the origin, and the contour in (b) corresponds to an incoming wave.

$$G_+(r) = \frac{2\pi i}{i\pi r} \operatorname{Res}|_{q=k} \frac{q e^{iqr}}{(q-k)(q+k)} = \frac{e^{ikr}}{r}. \tag{7.87}$$

The amplitude of G_+ decreases with $1/r$, and the corresponding intensity decreases with $1/r^2$, corresponding to a spherical wave. Multiplying by the time factor $e^{-i\omega t}$, we see that the phase of this spherical wave is $kr - \omega t$, which is a wave leaving the origin. Therefore, Green's function (7.87) corresponds to the outgoing wave.

In a similar way we can construct the Green function for the incoming wave, integrating over the contour C^- (Fig. 7.2b), thus obtaining

$$G_-(r) = \frac{e^{-ikr}}{r}. \tag{7.88}$$

The general Green function of the problem is a linear combination of the two previous solutions. With $k = 0$ we recover the solution (7.81) of Poisson's equation.

To conclude, note that by applying the Laplace operator to Eq. (7.81) and combining with (7.70),

$$\nabla^2 \Phi(\mathbf{r}) = e \nabla^2 \frac{1}{r} = -4\pi e \delta(\mathbf{r}), \tag{7.89}$$

we get

$$\nabla^2 \frac{1}{r} = -4\pi \delta(\mathbf{r}), \tag{7.90}$$

which shows that the Laplacian of r^{-1} vanishes everywhere except at the origin, where it has a singularity.

7.1.5 Free-particle densities

The following densities are frequently used in applications; it is therefore useful to know them and have them at hand. Consider the 3D wave function for the free particle,

$$\psi = A^{3/2} e^{-\frac{iE}{\hbar}t + \frac{i\mathbf{p}}{\hbar}\cdot\mathbf{x}}, \tag{7.91}$$

where the constant A is L^{-1} for Born's normalization and $(2\pi\hbar)^{-1}$ for Dirac's normalization. The spatial density is constant, $\rho = A^3$, and with $\nabla\psi = (i/\hbar)\mathbf{p}\psi$ one gets for the current density

$$\mathbf{j} = \frac{i\hbar}{2m}A^3\left(-\frac{2i}{\hbar}\right)\mathbf{p} = \frac{\mathbf{p}}{m}\rho, \tag{7.92}$$

which is just the relation $\mathbf{j} = \rho\mathbf{v}$, with $\mathbf{v} = \mathbf{p}/m$.

N is the total number of particles in the ensemble, each one with charge e; the absolute density of particles is $\rho_{\text{abs}} \equiv N\rho$, where ρ, as usual, describes the relative particle density. Thus,

$$\int \rho_{\text{abs}} d^3x \equiv N \int \rho d^3x = N; \qquad \rho_{e,\text{abs}} = eN\rho = eNA^3. \tag{7.93}$$

With Born's normalization, this gives eN/L^3, so the total charge $Q = Ne$ is uniformly distributed in the normalization volume $V = L^3$. With Dirac's normalization, the quantity $(2\pi\hbar)^3$ plays the role of normalization volume *in phase space*. For a better understanding, we proceed as follows. The particle momentum is $p_n = (2\pi\hbar/L)n$, so the number of free-particle quantum states contained in the momentum interval $(p, p+dp)$ is

$$dn = \frac{L}{2\pi\hbar}dp. \tag{7.94}$$

In the 3D case this gives

$$d^3n = \left(\frac{L}{2\pi\hbar}\right)^3 d^3p = \frac{L^3 d^3p}{h^3}. \tag{7.95}$$

Therefore, since $L^3 d^3p = \int_V d^3x\, d^3p$ is the total phase-space volume, each phase-space volume equal to h^3 can contain only a single quantum free-particle state. This is usually expressed by saying that the quantum phase space is composed of cells of volume h^3, each containing at most one quantum state.[5] It follows that the number of cells (and of quantum states) in the interval d^3p is

$$d^3n = \frac{1}{h^3}\int_V d^3x\, d^3p = \frac{V}{(2\pi\hbar)^3}d^3p, \tag{7.96}$$

and the *space density of free-particle states* is then

$$\frac{d^3n}{V} = \frac{d^3p}{(2\pi\hbar)^3}. \tag{7.97}$$

This is an important result that we will have occasion to draw on.

Exercise E7.4. Spectral density of the radiation field

Determine the density of modes of the free radiation field that appears in Planck's spectral distribution law.

Solution. For an isotropic radiation field, the density of modes of frequency ω propagating in all directions is (recall that $p = E/c = \hbar\omega/c$)

---

[5] The interpretation of h^3 as an elementary volume of phase space first appeared in a paper by Planck in 1913, and was subsequently taken up by Bose.

$$\frac{d^3n}{V} = \frac{1}{(2\pi\hbar)^3}\int_{\Omega_p} d\Omega_p p^2\,dp = \frac{4\pi}{8\pi^3\hbar^3}\frac{\hbar^3}{c^3}\omega^2\,d\omega = \frac{\omega^2}{2\pi^2 c^3}\,d\omega. \qquad (7.98)$$

This gives for the spectral density of modes

$$\frac{1}{V}\frac{d^3n}{d\omega} = \frac{\omega^2}{2\pi^2 c^3}. \qquad (7.99)$$

Multiplying this mode density by the ground-state energy $\hbar\omega/2$ of a quantum oscillator of frequency ω and by 2 to account for both polarization states, gives the spectral energy density of the ZPF, (2.84),

$$\rho_0(\omega) = \hbar\omega^3/2\pi^2 c^3. \qquad (7.100)$$

7.2 The Time-Dependent Schrödinger Equation

So far we have been working with the stationary Schrödinger equation, except for the sections on the propagator. However, it is precisely these sections that give us a clue as to how to generalize the description to allow time to play its significant role. We should expect that in most situations – though not in most of the problems considered in this book – the system, left to itself, will evolve with time. How to introduce time into the stationary equation is far from obvious, but fortunately we already have at hand two conditions that the time-dependent Schrödinger equation must satisfy under all circumstances.

First, the generalized equation should in all cases reduce to the stationary one for time-independent problems. Second, and much more informative, is the general solution of the free-particle equation (7.51), which is already time-dependent.

$$\psi(x,t) = \int A(k)e^{-\frac{i\hbar k^2}{2m}t+ikx}\,dk, \qquad (7.101)$$

This should be a solution for $V=0$ of the generalized equation under construction. To proceed, we first take the second derivative of the preceding equation with respect to x, which gives

$$\frac{\partial^2 \psi(x,t)}{\partial x^2} = -\int k^2 A(k)e^{-\frac{i\hbar k^2}{2m}t+ikx}\,dk. \qquad (7.102)$$

In turn, deriving once the same expression now with respect to time, we get

$$\frac{\partial \psi(x,t)}{\partial t} = -\frac{i\hbar}{2m}\int k^2 A(k)e^{-\frac{i\hbar k^2}{2m}t+ikx}\,dk. \qquad (7.103)$$

Comparing results, we see that $\frac{\partial \psi}{\partial t} = \frac{i\hbar}{2m}\psi''$ or, by passing to 3D with the substitution $\psi'' \to \nabla^2\psi$,

$$i\hbar\frac{\partial \psi}{\partial t} = -\frac{\hbar^2}{2m}\nabla^2\psi. \qquad (7.104)$$

This is the time-dependent Schrödinger equation for a free particle. In the stationary situation the equation should be

$$E\psi = -\frac{\hbar^2}{2m}\nabla^2\psi, \tag{7.105}$$

which, by combining the last two results, means that for stationarity to be satisfied, it must be true that

$$E\psi = i\hbar\frac{\partial \psi}{\partial t}. \tag{7.106}$$

Applying this condition to the stationary Schrödinger equation for an arbitrary potential,

$$E\psi = -\frac{\hbar^2}{2m}\nabla^2\psi + V\psi, \tag{7.107}$$

it follows that, for consistency, the *time-dependent Schrödinger equation* should be

$$i\hbar\frac{\partial \psi}{\partial t} = -\frac{\hbar^2}{2m}\nabla^2\psi + V\psi. \tag{7.108}$$

This is the fundamental *postulate* of QM. The wave function is no longer restricted to stationary states and can evolve in time from arbitrary initial conditions. We will now use this postulate in order to draw some first conclusions that should strengthen or correct it accordingly.

The general solution of Eq. (7.108) can be expressed in terms of the eigenfunctions of the corresponding stationary equation for time-independent potentials. This is done by writing $\psi(x, t)$ as a coherent superposition of monochromatic (plane) waves, as follows

$$\psi(x, t) = \sum_n c_n e^{-\frac{iE_n}{\hbar}t}\varphi_n(x), \tag{7.109}$$

where the $\varphi_n(x)$ are the eigenfunctions of the corresponding stationary problem and the E_n their eigenvalues,

$$-\frac{\hbar^2}{2m}\nabla^2\varphi_n + V\varphi_n = E_n\varphi_n. \tag{7.110}$$

In fact, we can write under these conditions

$$i\hbar\frac{\partial \psi}{\partial t} = i\hbar\sum_n c_n e^{-\frac{iE_n}{\hbar}t}\frac{E_n}{i\hbar}\varphi_n = \sum_n c_n e^{-\frac{iE_n}{\hbar}t}\left(-\frac{\hbar^2}{2m}\nabla^2 + V\right)\varphi_n$$

$$= \left(-\frac{\hbar^2}{2m}\nabla^2 + V\right)\sum_n c_n e^{-\frac{iE_n}{\hbar}t}\varphi_n = \left(-\frac{\hbar^2}{2m}\nabla^2 + V\right)\psi. \tag{7.111}$$

To obtain the third equality we have considered that the term enclosed in parentheses does not depend on the index n, so it has been extracted from the sum. The last expression is just the time-dependent Schrödinger equation satisfied by $\psi(x, t)$, and thus Eq. (7.109) is its general solution.

An important fact is that the c_n in (7.109) are completely free, so they can be determined by the usual procedure allowing for any valid initial condition.

7.2.1 Coherent superposition of states

Since each of the constants c_n in (7.109) is multiplied by the corresponding oscillating factor $e^{-iE_n t/\hbar}$, the coefficients of φ_n can be taken as oscillators, so that $\psi(\mathbf{x}, t)$ acquires the form

$$\psi(\mathbf{x}, t) = \sum_n a_n(t)\varphi_n(\mathbf{x}), \quad \text{with} \quad a_n(t) = c_n e^{-\frac{iE_n}{\hbar}t}. \tag{7.112}$$

Therefore, if the system is not in an energy eigenstate and its wave function is given by a superposition with at least two nonzero coefficients c_n, the local particle density varies continuously with time. The general solution (7.109) of the Schrödinger equation represents a *coherent superposition of states* (i.e., they interfere with each other). The contribution of state φ_n to the relative probability density at a given point, however, remains constant, since $|a_n|^2 = |c_n|^2$.

As an illustration, consider the case of Eq. (7.112) in 1D with only two coefficients c_1 and c_2 different from zero, which we take to be real; for simplicity we also assume that φ_1 and φ_2 are real. The wave function can then be written in the form

$$\psi = e^{-i\omega_1 t}\left[c_1\varphi_1(x) + c_2 e^{-i(\omega_2-\omega_1)t}\varphi_2(x)\right], \tag{7.113}$$

and the particle density is

$$\rho = |\psi|^2 = c_1^2 \rho_1 + c_2^2 \rho_2 + 2c_1 c_2 \varphi_1 \varphi_2 \cos(\omega_2 - \omega_1)t, \tag{7.114}$$

with $\rho_1 = |\varphi_1|^2$, $\rho_2 = |\varphi_2|^2$. In this case the density at each point x oscillates periodically in time with frequency $\omega_{12} = |E_1 - E_2|/\hbar$ between the two extremes, $\rho = (c_1\varphi_1 + c_2\varphi_2)^2$ and $\rho = (c_1\varphi_1 - c_2\varphi_2)^2$, due to the interference term.

Note that the system cannot be said to be in state 1 or state 2 or in both states at the same time; it is in the quantum state characterized by the coherent (in time) superposition.

7.2.2 Specificity of the Schrödinger equation: Pure states and mixtures

The structure of the time-dependent Schrödinger equation (7.108) is *sui generis*, with no parallel in the rest of physics. It is the only fundamental differential equation in physics that has an imaginary coefficient; moreover, we are used to the wave equation being of second order in time, but this one is of first order in time. We can understand how the novel elements combine by noting that the cosine in Eq. (7.114) comes from the imaginary exponential, whose i is just the i in the $i\hbar(\partial\psi/\partial t)$ term of the Schrödinger equation. It is precisely the i that gives this equation a *wave* character (hyperbolic and reversible), despite being of first order in t. If the i were not present, the time factor in Eq. (7.111) would be of the form $e^{+\alpha t}$ or $e^{-\alpha t}$, where $\alpha > 0$ is a real constant. The demand that the density $\rho(x, t)$ remains finite would suffice to eliminate the first of these solutions, but we would then have $\rho(x, t) \sim e^{-2\alpha t} \to 0$ when $t \to \infty$, which corresponds to an *irreversible process*, similar to that described, for example, by the heat equation, or by a diffusion equation, which are also of first order in time. Herein lies the crucial role played by the i-factor: its presence *transforms* a parabolic *time-irreversible* equation into a hyperbolic *reversible wave* equation. Formally, we can say that the operator "$i\,\partial/\partial t$" plays the role of "$\partial^2/\partial t^2$". At the same time, this distinctive and unique combination indicates that the Schrödinger wave by its nature differs substantially from any other wave in physical theory – which explains the frequent difficulties in accommodating the "quantum wave" into a convincing customary classical conceptual scheme.

Note that interference can also occur at a fixed time, as can be seen by writing the wave function in the general form

$$\psi(x) = \sum_n a_n \varphi_n(x). \tag{7.115}$$

For constant a_n, it is the *spatial* coherence that determines the interference, since

$$\begin{aligned}\rho(x) &= \sum_{n,n'} a_n^* a_{n'} \varphi_n^*(x) \varphi_{n'}(x) \\ &= \sum_n |a_n|^2 |\varphi_n(x)|^2 + \sum_{n \neq n'} a_n^* a_{n'} \varphi_n^*(x) \varphi_{n'}(x),\end{aligned} \tag{7.116}$$

where the last sum contains the interference terms. For $a_n = c_n e^{-i\omega_n t}$ (general time-dependent case), both *spatial* and *temporal* coherence produce interference. We thus conclude that wave functions of the general type given by Eq. (7.115), where the coefficients may or may not depend on time, describe *coherent quantum states*.

There are, of course, situations of physical interest for which the superposition is *not coherent*; in such cases the particle density does not contain interference terms. For two incoherent states, for example, the density is given by

$$\rho = |a_1|^2 \rho_1 + |a_2|^2 \rho_2. \tag{7.117}$$

Obviously, in this case the state of the system cannot be described by a wave function such as (7.115); it is not factorizable and the description must be done by other means. The usual method in such a case is to introduce the so-called *density matrix* or *density operator*, a mathematical construction that allows one to integrate the possible incoherent coexistence of states and their quantum behavior in a single formalism. This important topic is dealt with in Chapter 15, so here we add only one remark on the terminology used to distinguish between the two cases just discussed. States of quantum systems described by wave functions of the type (7.109) or (7.115) (coherent superpositions) are known as *pure states*; states that require a density matrix (containing an *incoherent* combination of states and their statistical weights) for their description are called *mixtures*. Coherent states show interference; mixtures do not.

7.2.3 Flow velocity and current density

In Section 4.2 we saw that the particle density $\rho = \psi^*\psi$ satisfies the continuity equation

$$\frac{\partial \rho}{\partial t} + \nabla \cdot \boldsymbol{j} = 0, \tag{7.118}$$

where

$$\boldsymbol{j} = \rho \boldsymbol{v} \tag{7.119}$$

is the current or particle flow density and v is the corresponding flow (or flux) velocity. According to Eq. (4.83), the flow velocity is given by

$$v(x) = \frac{i\hbar}{2m}\left(\frac{\nabla\psi^*}{\psi^*} - \frac{\nabla\psi}{\psi}\right), \qquad (7.120)$$

whence

$$j(x) = \frac{i\hbar}{2m}(\psi\nabla\psi^* - \psi^*\nabla\psi). \qquad (7.121)$$

Now that we have the complete (time-dependent) Schrödinger equation, we can generalize these important expressions to include the time dependence. We start from $\rho = \psi^*\psi$, differentiate it with respect to time and introduce $\partial\psi/\partial t$ and $\partial\psi^*/\partial t$ obtained from the Schrödinger equation (7.108) and its complex conjugate,

$$\frac{\partial\rho}{\partial t} = \psi^*\frac{\partial\psi}{\partial t} + \psi\frac{\partial\psi^*}{\partial t} = \psi^*\left(\frac{i\hbar}{2m}\nabla^2\psi + \frac{V}{i\hbar}\psi\right) + \psi\left(\frac{-i\hbar}{2m}\nabla^2\psi^* - \frac{V}{i\hbar}\psi^*\right)$$

$$= \frac{i\hbar}{2m}(\psi^*\nabla^2\psi - \psi\nabla^2\psi^*) = -\frac{i\hbar}{2m}\nabla\cdot(\psi\nabla\psi^* - \psi^*\nabla\psi) = -\nabla\cdot j. \qquad (7.122)$$

Comparing with Eq. (7.118) we get (7.121). To this expression for j one can in principle add a vector of the form $\nabla \times A$ with arbitrary A, since $\nabla \cdot (\nabla \times A) \equiv 0$; this allows, in particular, the introduction of the potential A in the presence of an electromagnetic field (see in particular Chapters 12 and 17).

We recall that in Section 4.3, the continuity condition on ψ' was introduced to guarantee the continuity of the particle flow; from Eqs. (7.120) and (7.121) we see that if ψ and ψ' are indeed continuous, then v and j are also continuous, thus justifying our statement. In general, however, it is not necessary to impose this condition separately, since if the potential and ψ are both continuous, then also ψ' is continuous. This is easy to see as follows (in 1D, for simplicity). Let V and ψ be continuous on (a,b); integrating the Schrödinger equation over this interval, we get

$$E\int_a^b \psi\,dx - \int_a^b V\psi\,dx = -\frac{\hbar^2}{2m}\left[\frac{\partial\psi}{\partial x}\right]_a^b. \qquad (7.123)$$

For $b \to a$, the left-hand side tends to zero and ψ' is continuous, as stated. On the other hand, if V has an infinite discontinuity within the interval, the second integral on the left becomes indeterminate and ψ' is not necessarily continuous; in particular, it can be seen that if $V \to \infty$ when $b \to a$ such that $A = V(b-a)$ remains finite, then ψ' has a discontinuity given by $(2mA/\hbar^2)\psi$.

The previous results show that ψ', and therefore the velocity, can only have discontinuities at points where the potential is discontinuous, that is, where the force applied to the particle is infinite. Such behavior is analogous to the classical one. For example, a particle colliding perpendicularly with a rigid wall at velocity v is reflected at velocity $-v$, that is, there is a discontinuity in the velocity of magnitude $2v$. This explains the origin and physical significance of ψ'-discontinuities in problems such as the infinite well.

7.2.3.1 Examples of current density

It is instructive to calculate the current density in some simple cases.

(a) *Stationary case* in general. Inserting the general form of the stationary solution $\psi = e^{-i\omega t}\varphi(x)$ in the previous formulas, with $\varphi(x)$ a solution of the stationary Schrödinger equation, it follows that $\rho = \varphi^*(x)\varphi(x)$,

$$j = \frac{i\hbar}{2m}(\varphi\nabla\varphi^* - \varphi^*\nabla\varphi), \qquad (7.124)$$

$$v = \frac{i\hbar}{2m}\left(\frac{\nabla\varphi^*}{\varphi^*} - \frac{\nabla\varphi}{\varphi}\right). \qquad (7.125)$$

We see that in a stationary state there can be a flow, provided it is constant (as in the 1D potential problems studied in Chapter 5).

(b) *Stationary states with real $\varphi(x)$*. The φ_n corresponding to 1D bound states, such as the infinite potential well, are real; in such cases $v = 0$. This result does not imply that each particle is at rest; it simply means that the *net* flow velocity, which is a local mean net velocity of the particles of the ensemble, is zero. That one should expect the existence of particle motion in such cases can be seen by writing the wave function for the example of the square well in the form

$$\psi_n(x,t) = A_n e^{-iE_n t/\hbar}\sin k_n x = \frac{A_n}{2i}e^{-iE_n t/\hbar}\left(e^{ip_n x/\hbar} - e^{-ip_n x/\hbar}\right), \qquad p_n = \hbar k_n. \qquad (7.126)$$

Each term in this expression separately describes a beam of free particles, one traveling to the right with momentum p_n and the other to the left, with momentum $-p_n$; the superposition of these waves (defined only inside the well) must be interpreted as describing the coherent superposition of flows traveling in both directions, which confirms that the result $v = 0$ in such stationary cases does not mean there is no motion.

This observation can be very important for applications. Consider, for example, the ground state of the H atom, which is stationary. Half of the members of the ensemble are orbiting in one direction, and the other half in the other, so $\langle \hat{v} \rangle = 0$; that is, the electrons are in motion.

(c) *Free particle*. We write the 3D free-particle wave function as

$$\psi = A^{3/2}e^{-\frac{iE}{\hbar}t + \frac{i\mathbf{p}}{\hbar}\cdot\mathbf{x}}, \qquad (7.127)$$

where the constant A is L^{-3} if the Born normalization is adopted and $(2\pi\hbar)^{-3}$ according to the Dirac normalization. We have $\rho = A^3$, and $\nabla\psi = (i/\hbar)\mathbf{p}\psi$ gives for the current density

$$\mathbf{j} = \frac{i\hbar}{2m}A^3\left(-\frac{2i}{\hbar}\right)\mathbf{p} = \frac{\mathbf{p}}{m}\rho, \qquad (7.128)$$

so that the relation (7.119) holds regardless of normalization, with $\mathbf{v} = \mathbf{p}/m$.

Exercise E7.5. Schrödinger equation with particle absorption or creation

Show that a complex potential can be used to simulate (or model) the existence of particle sources or sinks. If the potential is of the form $V = V_r + iV_i$, show that $V_i < 0$

implies particle absorption. Further, show that for $V_i = -\hbar\lambda/2$, λ can be interpreted as the absorption coefficient.

Solution. The Schrödinger equation for a complex potential is

$$i\hbar\frac{\partial \psi}{\partial t} = -\frac{\hbar^2}{2m}\nabla^2\psi + (V_r + iV_i)\psi. \tag{7.129}$$

Multiplying this equation by ψ^* and its complex conjugate by ψ, and taking the difference of the two, we get

$$i\hbar\frac{\partial \psi^*\psi}{\partial t} = -\frac{\hbar^2}{2m}(\psi^*\nabla^2\psi - \psi\nabla^2\psi^*) + 2iV_i\psi^*\psi. \tag{7.130}$$

If we introduce $\rho = \psi^*\psi$, and $\boldsymbol{j}$ given by Eq. (7.121), we get on the left side the continuity equation and on the right side a source term that depends on the particle density,

$$\frac{\partial \rho}{\partial t} + \nabla \cdot \boldsymbol{j} = 2V_i\rho/\hbar. \tag{7.131}$$

This solves the first part of the problem. Integrating this equation over the whole space and identifying $\int \rho\, d^3x$ with the total number of particles N, we obtain

$$\frac{dN}{dt} = \frac{2}{\hbar}\int V_i\rho\, d^3x. \tag{7.132}$$

Particle absorption implies $dN/dt < 0$; for V_i constant, this means $V_i < 0$. In this case there exists a solution of the Schrödinger equation of the form

$$\psi = e^{-i\alpha t}\varphi(\boldsymbol{x}) \tag{7.133}$$

with α a complex constant, $\alpha = \omega + i\sigma$. Substituting, we obtain that the following must be fulfilled (we write $V_i = -\hbar\lambda/2$)

$$\hbar\alpha\psi = \hbar(\omega + i\sigma)\psi = -\frac{\hbar^2}{2m}\nabla^2\psi + V_r\psi - i\frac{\hbar}{2}\lambda\psi. \tag{7.134}$$

We choose the imaginary component σ so that the term containing it cancels the imaginary part of the potential $\sigma = -\lambda/2$. The resulting Schrödinger equation for ψ is the conventional one for the stationary case, but the wave function is

$$\psi = e^{-\frac{\lambda}{2}t - i\omega t}\varphi. \tag{7.135}$$

We see that the particle density now decreases exponentially with time,

$$\rho = e^{-\lambda t}|\varphi|^2, \tag{7.136}$$

where λ represents the absorption intensity and determines the decay time of the system (the time $\tau = 1/\lambda$ it takes for ρ to decay to $1/e$ its initial value). A problem analogous to the present one (but with real potentials) was studied in Section 6.2.

7.2.4 The Schrödinger equation in Dirac notation

To conclude this section, we write the Schrödinger equation in the Dirac notation introduced in Section 4.6. In preparation for it, we express the general state function $\psi(x,t)$ in terms of some suitable orthonormal basis $\{\varphi_k(x)\}$ in the form

$$\psi(x,t) = \sum_k a_k(t)\varphi_k(x). \tag{7.137}$$

Note that the time evolution is carried by the coefficients, since the basis is fixed. Substituting into the Schrödinger equation, we get

$$i\hbar\frac{\partial \psi}{\partial t} = i\hbar \sum_l \frac{da_l(t)}{dt}\varphi_l(x) = \widehat{H}\psi = \sum_l a_l(t)\widehat{H}\varphi_l(x).$$

Multiplying by $\varphi_k^*(x)$ from the left and integrating we get

$$i\hbar\sum_l \frac{da_l(t)}{dt}\int \varphi_k^*(x)\varphi_l(x)dx = i\hbar\frac{da_k(t)}{dt} = \sum_l a_l(t)\int \varphi_k^*(x)\widehat{H}\varphi_l(x)dx = \sum_l H_{kl}a_l(t). \tag{7.138}$$

The result obtained,

$$i\hbar\frac{da_k(t)}{dt} = \sum_l H_{kl}a_l(t), \tag{7.139}$$

is the Schrödinger equation in matrix form, with $\psi(x,t)$ represented by the column vector with elements $a_k(t)$, For the original vector $|\psi\rangle$, it reads

$$i\hbar\frac{d|\psi\rangle}{dt} = \widehat{H}|\psi\rangle. \tag{7.140}$$

7.3 The Propagator in the General Case

We will now derive an expression for the propagator of the wave function for a time-independent potential. As already explained, the propagator in QM is a Green function of the time-dependent Schrödinger equation; closed formulas such as (7.145) are obtained only in a few relatively simple cases, in particular for linear problems (the free particle, the uniform external field, the harmonic oscillator or their linear combinations).

As before, we start from the general expression for the wave function $\psi(x,t)$ in terms of stationary functions with time-dependent coefficients, Eq. (7.109),

$$\psi(x,t) = \sum_n c_n e^{-i\omega_n t}\varphi_n(x) = \sum_n a_n(t)\varphi_n(x), \tag{7.141}$$

where $\omega_n = E_n/\hbar$. To determine the coefficients c_n (which are independent of time) we consider the preceding expression at a fixed time t' and use the property of orthonormality of φ_n,

$$a_n(t') = \int \varphi_n^*(x')\psi(x',t')\,dx', \tag{7.142}$$

7.3 The Propagator in the General Case

thus obtaining

$$c_n = a_n(t')e^{i\omega_n t'} = \int \varphi_n^*(x')\psi(x',t')e^{i\omega_n t'}\,dx'. \tag{7.143}$$

Substituting this result in the initial equation (7.141), we get an integral expression for $\psi(x,t)$, which can be written in the form

$$\psi(x,t) = \int K(x,t|x',t')\psi(x',t')\,dx', \tag{7.144}$$

with the propagator K given by

$$K(x,t|x',t') = \sum_n e^{-i\omega_n(t-t')}\varphi_n^*(x')\varphi_n(x) = \sum_n \psi_n^*(x',t')\psi_n(x,t). \tag{7.145}$$

This is the general expression for the propagator. In particular, if here we let $t = t'$, the propagator reduces to $\delta(x - x')$, as follows from the completeness condition (7.23),

$$K(x,t|x',t) = \sum_n \varphi_n(x)\varphi_n^*(x') = \delta(x - x') \tag{7.146}$$

This is an expected result: introduced in Eq. (7.144) it means that for $t = t'$ the wave function remains the same. From the propagator K we can construct the causal Feynman propagator (or causal Green function) K_c, which is the propagator consistent with the condition that the effects propagate only into the future. To this end, we assume that at time $t' = t_0$ the system is in a certain state $\psi_0(x)$; for $t > t_0$, $\psi(x,t)$ is given by (7.144), where we must put $\psi(x',t') = \psi(x',t_0) = \psi_0(x')$; further, for $t < t_0$ we must have $\psi(x,t) = 0$ (there is no propagation of particles into the past, as demanded by causality). The Green function that satisfies these conditions is thus

$$K_c(x,t|x_0,t_0) = \begin{cases} K(x,t|x_0,t_0), & t > t_0 \\ 0, & t < t_0. \end{cases} \tag{7.147}$$

We can verify that the propagator coincides with the wave function that evolves from an initial delta distribution of particles. Indeed, with $\psi_0(x) = \delta(x - x_0)$, we get from Eq. (7.143), taking $t' = 0$,

$$c_k = \int \varphi_k^*(x)\delta(x - x_0)\,dx = \varphi_k^*(x_0), \tag{7.148}$$

which, when introduced in (7.141), gives

$$\psi(x,t) = \sum_n e^{-i\omega_n t}\varphi_n^*(x_0)\varphi_n(x) = K(x,t|x_0,0). \tag{7.149}$$

Let us analyze the meaning of the initial condition $\psi_0(x)$. To fully specify the function $\psi_0(x)$, we must determine both its magnitude and its phase; we can do this if we know two independent physical functions, such as the initial particle distribution $\rho_0(x)$ and the initial current $j_0(x)$ (or, if preferred, the initial local flow velocity, $v_0(x)$). Specifically, we write

$$\psi_0(x) = \sqrt{\rho_0}e^{iS_0}, \tag{7.150}$$

which gives

$$S_0(x) = \frac{m}{\hbar} \int_{x_0}^{x} v_0(x') dx'. \tag{7.151}$$

When we write the initial amplitude as a real function (e.g., a delta function), we are assuming that the initial local flow is zero. Such a situation can be achieved by assuming that the particles are at rest at $t = t_0$, but also, much more generally and correctly, by proposing an initial symmetric distribution of velocities, such that at time $t = t_0$, at a given point x, as many particles are moving to the right with an arbitrary speed v as are moving to the left with speed $-v$. It is clear that the appropriate choice must be made in each case, depending on the specific situation.

Exercise E7.6. Wave packet in a uniform field

Study the evolution of a packet of quantum particles moving in a uniform external field, such as the gravitational field near the Earth's surface ($V = mgy$) or the electric field between two flat electrodes ($V = eEx$).

Solution. To be less specific, we write the potential in the form $V = qx$. The Schrödinger equation to be solved is then

$$-\frac{\hbar^2}{2m}\psi'' + qx\psi = i\hbar\frac{\partial \psi}{\partial t}. \tag{7.152}$$

Partial differential equations of the type of Eq. (7.152) are encountered repeatedly in the course, so it is useful to study a method of solving them.

First, we do a Fourier transformation on the x-space to eliminate the double spatial derivative; specifically, we write

$$\psi(x,t) = \int \varphi(k,t)e^{ikx} dk, \quad \varphi(k,t) = \frac{1}{2\pi}\int \psi(x,t)e^{-ikx} dx. \tag{7.153}$$

The relation between the initial amplitudes is

$$\varphi_0 \equiv \varphi(k,0) = \frac{1}{2\pi}\int \psi_0(x)e^{-ikx} dx; \quad \psi_0(x) \equiv \psi(x,0). \tag{7.154}$$

It follows that

$$\frac{\partial \varphi}{\partial k} = -\frac{i}{2\pi}\int x\psi e^{-ikx} dx; \tag{7.155}$$

therefore, multiplying Eq. (7.152) by $(2\pi)^{-1}e^{-ikx} dx$, integrating over x and combining with the previous equations gives a first-order differential equation,

$$\frac{\partial \varphi}{\partial t} - \frac{q}{\hbar}\frac{\partial \varphi}{\partial k} + \frac{i\hbar}{2m}k^2\varphi = 0. \tag{7.156}$$

To solve it we use the method of characteristics (see the bibliography at the end of this chapter). We consider the system of subsidiary equations

$$\frac{dt}{1} = -\frac{\hbar dk}{q} = -\frac{2m}{i\hbar k^2}\frac{d\varphi}{\varphi} \tag{7.157}$$

with solutions

$$\varphi = \varphi_1 e^{i\hbar^2 k^3/6mq}, \quad k - k_0 = -\frac{q}{\hbar}t, \tag{7.158}$$

where φ_1 and k_0 are constants. Considering φ_1 as a function of k_0, it follows that

$$\varphi_1 \equiv \varphi_1(k_0) = \varphi_1\left(k + \frac{q}{\hbar}t\right) = \varphi e^{-i\hbar^2 k^3/6mq}. \tag{7.159}$$

In particular, for $t = 0$ we have that

$$\varphi_1(k) = \varphi_0 e^{-i\hbar^2 k^3/6mq}. \tag{7.160}$$

Eliminating φ_1 from these two equations gives

$$\varphi = \varphi_1\left(k + \frac{q}{\hbar}t\right)e^{i\hbar^2 k^3/6mq} = \varphi_0 \exp i(\hbar^2/6mq)\left(-3qtk^2/\hbar - 3q^2t^2k/\hbar^2 - q^3t^3/\hbar^3\right)$$

$$\equiv \varphi_0 K(k, t). \tag{7.161}$$

Plugging this expression into the Fourier transform of ψ and writing the coefficient $\varphi_0(k + \frac{q}{\hbar}t)$ in terms of ψ_0, we get

$$\psi = \int \varphi_0 K(k,t) e^{ikx}\, dk = \frac{1}{2\pi}\int\int \psi_0(x') e^{-ix'(k+qt/\hbar)} K(k,t) e^{ikx}\, dx'dk. \tag{7.162}$$

As

$$\int dk\, e^{i(x-x')k} K(k) = \int dk\, \exp\left[-(i\hbar t/2m)k^2 - (iqt^2/2m)k + i(x-x')k - iq^2t^3/6\hbar m\right]$$

$$= \frac{\text{const}}{\sqrt{t}} \exp\left[-iq^2t^3/6m\hbar + (im/2\hbar t)(x - x' - qt^2/2m)^2\right]$$

$$\tag{7.163}$$

(it is not necessary to specify the constant, since it will be absorbed in the normalization), we can write this in the form

$$\psi(x,t) = \frac{A}{\sqrt{t}} \int dx'\, \psi_0(x') e^{(im/2\hbar t)(x-x'-qt^2/2m)^2 - iq^2t^3/6m\hbar - iqx't/\hbar}, \tag{7.164}$$

which is the general solution of the problem.

Let us consider the particular case in which the particles are initially at a single point, $\psi_0(x') = \delta(x' - x_0)$; we obtain the propagator,

$$K(x,t|x_0,t_0)$$

$$= \sqrt{\frac{m}{2\pi i \hbar T}} \exp\left[(im/2\hbar T)(x - x_0 - qT^2 m)^2 - iqx_0 T/\hbar - iq^2 T^2/6m\hbar\right]$$

$$= \sqrt{\frac{m}{2\pi i \hbar T}} \exp\left[(im/2\hbar T)(x - x_0)^2 - iqT((x + x_0)/2\hbar - iq^2 T^3/24m\hbar\right], \tag{7.165}$$

where $T = t - t_0$ and A was determined with the normalization condition.

We use the preceding expression to calculate the current of particles,

$$j = \frac{i\hbar}{2m}(\psi\psi^{*\prime} - \psi^*\psi') = \rho\frac{x - x_0 - q(t-t_0)^2/2m}{t - t_0} = \rho v. \tag{7.166}$$

This result shows that the center of the packet moves according to classical laws, with an acceleration q/m, as dictated by the Ehrenfest theorem (see Section 8.3).

Problems

P7.1 Prove in detail the following properties of the Dirac's delta function:
a) $\delta(x) = \delta(-x)$,
b) $a\delta(ax) = \delta(x)$, $a > 0$,
c) $\delta(x^2 - a^2) = \frac{1}{2a}[\delta(x-a) + \delta(x+a)]$,
d) $\delta[f(x)] = \sum_i \frac{\delta(x-x_i)}{|f'(x_i)|}$, where x_i are the roots of $f(x_i) = 0$.

P7.2 Show that the normal distribution

$$\rho(x,\sigma) = \frac{1}{\sqrt{2\pi}\sigma} e^{-(x-a)^2/2\sigma^2} \tag{7.167}$$

has the property $\lim_{\sigma \to 0} \rho(x,\sigma) = \delta(x-a)$.

P7.3 When the Jacobian of a transformation from x_i to ξ_i vanishes, the transformation is no longer one-to-one. For example, at the origin of a polar system, $x = y = 0$ and $r = 0$, but θ has an arbitrary value. A coordinate that has no determinate value at a singular point of a transformation (where the Jacobian vanishes) is called a cyclic or ignorable variable. Show that if there are ignorable variables, the transformation equation should be changed to

$$\delta(x_1 - \alpha_1)\ldots\delta(x_n - \alpha_n) = \frac{\delta(\xi_1 - \beta_1)\ldots\delta(\xi_n - \beta_n)}{|J_k|}, \tag{7.168}$$

where $\xi_i, i = 1, 2, \ldots k$ are the nonignorable variables, and $J_k = \int \ldots \int J d\xi_{k+1} \ldots d\xi_n$, where J is the Jacobian of the transformation and the integral is performed over the ignorable variables.

P7.4 Show that in the plane we can write

$$\delta(x - x_0)\delta(y - y_0) = \frac{\delta(r - r_0)\delta(\theta - \theta_0)}{r} \quad (r_0 \neq 0), \tag{7.169}$$

where r and θ are the polar variables. At the origin, θ is ignorable. Show that at this point we must write $\delta(x)\delta(y) = \delta(r)/2\pi r$.

P7.5 Consider a free-particle wave function that at $t = 0$ has the form $\varphi(x)e^{ip_0x/\hbar}$, where $\varphi(x)$ is real and differs from 0 only for values of x in the interval $(-\delta, \delta)$. Find the interval of x in which the wave function is significantly different from zero at time t.

P7.6 Find the wave function at time t for free particles whose amplitude at $t = 0$ is

$$\psi(r,0) = \frac{1}{(\pi\sigma^2)^{3/4}} e^{-r^2/2\sigma^2 + i\mathbf{p}_0 \cdot \mathbf{r}/\hbar}. \tag{7.170}$$

P7.7 Using the general solution of the Schrödinger equation for a free particle

$$\psi(x,t) = Ae^{-iEt/\hbar + ipx/\hbar} + Be^{-iEt/\hbar - ipx/\hbar}, \tag{7.171}$$

show that the initial conditions can be chosen such that:

(a) $\psi(x,t)$ is an eigenfunction of the operator $\hat{p} = -i\hbar\partial/\partial x$,
(b) the probability density is independent of time, that is, it represents a standing wave.

P7.8 Solve the free-particle problem in cylindrical polar coordinates.

P7.9 Consider a packet of free particles with a Gaussian momentum distribution, $A(k) = e^{-(k-k_0)^2/q^2}$.

(a) Find $\psi(x,t)$ for this packet at time t;
(b) At what speed does the center of mass of the packet move?
(c) What is the rate at which the spatial dispersion of the particle increases? Can this speed be greater than the speed of light? Explain your answer.

P7.10 Show that if the coherent superposition (7.109) involves a finite number $N > 2$ of energy eigenstates n, m, the oscillations of ψ are not periodic unless the frequencies $\omega_{nm} = |E_n - E_m|/\hbar$ are all related by rational numbers.

P7.11 Show that if ψ_1 and ψ_2 are solutions of the time-dependent Schrödinger equation, the equation

$$\frac{\partial}{\partial t} \psi_1^* \psi_2 + \frac{i\hbar}{2m} \nabla \cdot (\psi_2 \nabla \psi_1^* - \psi_1^* \nabla \psi_2) = 0 \tag{7.172}$$

generalizes the continuity equation.

P7.12 A physical system is initially (at $t = 0$) in a state which is a coherent superposition of the eigenfunctions φ_1 and φ_2 of a Hamiltonian with eigenenergies E_1 and E_2, respectively. The φ_1 state is three times more probable than the φ_2 state. Write the most general possible initial wave function $\psi_0(x)$ consistent with the preceding data and determine $\psi(x,t)$ for all $t > 0$. Is the system in a stationary state? Does this state possess any time-invariant properties?

P7.13 Show that a nonstationary state cannot have a separable wave function of the form $\psi(x,t) = \chi(t)\varphi(x)$.

P7.14 Consider the relativistic Klein–Gordon equation for the free particle (Section 19.1):

$$\left(\nabla^2 - \frac{1}{c^2}\frac{\partial^2}{\partial t^2}\right)\psi(x,t) = \frac{m^2c^2}{\hbar^2}\psi(x,t). \tag{7.173}$$

Show that a conservation law similar to Eq. (7.118) is satisfied, with

$$\mathbf{j}(x,t) = \frac{i\hbar}{2m}\left(\psi\nabla\psi^* - \psi^*\nabla\psi\right). \tag{7.174}$$

What is now the value of $\rho(x,t)$? From this result discuss why the Klein–Gordon equation is not a good candidate to replace the Schrödinger equation for electrons in the relativistic case.

P7.15 A particle is in its ground state in a one-dimensional infinite square well of width L. Suddenly, at $t = 0$, the right wall of the well moves from $x = L$ to $x = 2L$. Is the particle still in a stationary state? Determine the probability of the new ground state.

P7.16 Show that the propagator given by Eq. (7.145) has the following integral property:

$$K(x_1, t_1 | x_2, t_2) = \int K(x_1, t_1 | x, t) K(x, t | x_2, t_2) \, dx, \quad t_1 < t < t_2. \tag{7.175}$$

Bibliographical Notes

A classic textbook on mathematical methods is Morse and Feshbach (1953).

On the theory of the relation between Green functions and the Dirac's delta function, see the excellent books by Friedman (1956) and Butkov (1968). A more recent alternative text is Hassani (2013).

For the method of characteristics to solve differential equations, see, for example, Webster (2016) or Ebert and Reissig (2018).

8 Dynamics of Quantum Systems

This chapter is devoted to the study of the dynamics of quantum systems, based on the formalism presented in Chapter 7 and previous chapters. Given the discussion in Section 4.6 we will use the operator notation and its matrix representation without distinction, moving freely from one to the other as convenient. This will allow us to study the system's evolution from different perspectives, depending on the description adopted. The chapter includes a discussion of the statistical information contained in the quantum formalism and a formal derivation of the Heisenberg inequalities, followed by an introduction to quantum canonical transformations. More advanced topics, which provide a valuable complement, include the relationship between the symmetry properties and the conserved dynamical variables of the system.

8.1 Dynamics in the Heisenberg Description

We will start from the Heisenberg equation of evolution for an operator $\hat{F}$ representing a generic dynamical variable F, introduced in Chapter 3, Eq. (3.43),

$$i\hbar \frac{d\hat{F}}{dt} = i\hbar \frac{\partial \hat{F}}{\partial t} + \left[\hat{F}, \hat{H}\right]. \tag{8.1}$$

The description in terms of this equation is called the *Heisenberg description*. Multiplying this equation from the left by the bra $\langle m|$ and from the right by the ket $|n\rangle$ we obtain

$$\frac{d}{dt}\langle m|\hat{F}|n\rangle = \langle m|\frac{\partial \hat{F}}{\partial t}|n\rangle + \frac{1}{i\hbar}\langle m|[\hat{F}, \hat{H}]|n\rangle. \tag{8.2}$$

Note that in this description, the matrix elements F_{mn} evolve because the *operator* $\hat{F}$ evolves in time according to Eq. (8.1), while the state vectors $|m\rangle$ and $|n\rangle$ do not change. Equation (8.2), on the other hand, contains the complete information about the evolution of $\hat{F}(x, p, t)$, encoded in its matrix elements in an *unspecified* description.

Exercise E8.1. Heisenberg's elementary oscillators

Using the basis of energy eigenstates, show that the matrix elements of any time-independent dynamical variable oscillate in time.

Solution. We apply Eq. (8.2) to a dynamical variable F that does not explicitly depend on time,

$$i\hbar\frac{d}{dt}\langle m|\widehat{F}|n\rangle = \langle m|[\widehat{F},\widehat{H}]|n\rangle, \tag{8.3}$$

and consider that the ket $|n\rangle$ represents an energy eigenstate, with eigenvalue E_n. Then

$$\langle m|[\widehat{F},\widehat{H}]|n\rangle = \langle m|\widehat{F}\widehat{H}|n\rangle - \langle m|\widehat{H}\widehat{F}|n\rangle$$
$$= E_n\langle m|\widehat{F}|n\rangle - E_m\langle m|\widehat{F}|n\rangle = -(E_m - E_n)\langle m|\widehat{F}|n\rangle, \tag{8.4}$$

which when substituted in (8.3) gives

$$\frac{d}{dt}\langle m|\widehat{F}|n\rangle = -\frac{1}{i\hbar}(E_m - E_n)\langle m|\widehat{F}|n\rangle. \tag{8.5}$$

In terms of the Bohr transition frequency

$$\omega_{mn} = \frac{E_m - E_n}{\hbar}, \tag{8.6}$$

Eq. (8.5) reads

$$\frac{d}{dt}\langle m|\widehat{F}|n\rangle = i\omega_{mn}\langle m|\widehat{F}|n\rangle. \tag{8.7}$$

with solution

$$F_{mn}(t) = \langle m|\widehat{F}|n\rangle = F_{mn}e^{i\omega_{mn}t}, \tag{8.8}$$

We see that the matrix $\widehat{F}$ in the energy representation has constant elements on the main diagonal and elements oscillating at a Bohr transition frequency outside it. This complements the results obtained for $x_{mn}(t)$ and $p_{mn}(t)$ in Section 3.5.3.

8.2 The Schrödinger Description

To study the quantum dynamics in the Schrödinger description we take into account that in this description the evolution of the (time-independent) variable F is expressed in the state vectors; in other words, we propose that (8.2) holds because the state vectors $|n\rangle$ and $|m\rangle$ evolve in Hilbert space. In this case, for an operator $\widehat{F}$ that does not depend explicitly on time, we can write from (8.2)

$$i\hbar\langle m|\widehat{F}|\frac{d|n\rangle}{dt} + i\hbar\frac{d\langle m|}{dt}\widehat{F}|n\rangle = \langle m|\widehat{F}\widehat{H}|n\rangle - \langle m|\widehat{H}\widehat{F}|n\rangle. \tag{8.9}$$

We observe that this equation is identically satisfied if

$$i\hbar\langle m|\widehat{F}\frac{d|n\rangle}{dt} = \langle m|\widehat{F}\widehat{H}|n\rangle. \tag{8.10}$$

The rest of Eq. (8.9) gives then

$$-i\hbar\frac{d\langle m|}{dt}\widehat{F}|n\rangle = \langle m|\widehat{H}\widehat{F}|n\rangle, \tag{8.11}$$

which is merely the adjoint of the previous one, with the exchange of the indices m and n; so the equation is indeed satisfied. Writing the solution in the form

$$\langle m|\widehat{F}\left[i\hbar\frac{d|n\rangle}{dt} - \widehat{H}|n\rangle\right] = 0, \tag{8.12}$$

it follows that for any $\langle m|$, the equation

$$i\hbar\frac{d|n\rangle}{dt} = \widehat{H}|n\rangle \tag{8.13}$$

must hold. This is precisely the time-dependent Schrödinger equation for a state $|n\rangle$, written in Dirac notation.[1] It corresponds to the description of the dynamics studied in Chapter 7, in which the time evolution of the system is expressed in the state vectors, the operators associated to the dynamical variables remaining fixed. For obvious reasons, this form of the theory is called the *Schrödinger representation* or *Schrödinger description*.

The above results confirm that Eq. (8.2), which implicitly contains any description, can be taken as the general evolution equation, abstracting from any reference to one or the other representation. In practice, however, a specific description is chosen as convenient, which may even be different from the two previous ones. An important example is the so-called *interaction description* studied in Section 8.6.6, which is a hybridization of the previous two, very useful for the study of complicated problems of systems in interaction.

8.3 The Dynamics of Expectation Values. Ehrenfest's Theorem

The evolution of the expectation value of a time-independent dynamical variable is described by (8.3) with $m = n$,

$$i\hbar\frac{d\langle\widehat{F}\rangle}{dt} = \langle[\widehat{F},\widehat{H}]\rangle, \tag{8.14}$$

with independence of the description used. In particular, for any $\widehat{F}$ commuting with the Hamiltonian, $d\langle\widehat{F}\rangle/dt = 0$ and the quantity $\langle\widehat{F}\rangle$ is conserved. This is of course the case for the Hamiltonian itself, so if the Hamiltonian does not explicitly depend on time, the energy $E = \langle\widehat{H}\rangle$ is an integral of motion of the system. This is the quantum version of the classical law of conservation of energy.

Equation (8.14) applied to different dynamical (time-independent) variables F can be used to obtain a series of particular results, each of which appears in the literature under the name of Ehrenfest's equation; for this reason it is appropriate to call Eq. (8.14) itself *Ehrenfest's theorem*. Applying it to the position operator (in 1D, for simplicity) $\frac{d\langle\widehat{x}\rangle}{dt} = -\frac{i}{\hbar}\langle[\widehat{x},\widehat{H}]\rangle$, with $H = (p^2/2m) + V$, we get

$$\langle\widehat{p}\rangle = m\frac{d\langle\widehat{x}\rangle}{dt}. \tag{8.15}$$

If we *define* $\langle\widehat{p}\rangle \equiv m\langle\widehat{v}\rangle = m\langle\dot{\widehat{x}}\rangle$, it follows that

$$\frac{d\langle\widehat{x}\rangle}{dt} = \left\langle\frac{d\widehat{x}}{dt}\right\rangle = \langle\dot{\widehat{x}}\rangle, \tag{8.16}$$

---

[1] When going from symbolic states $|n\rangle$ to functions $\langle x|n\rangle = \varphi_n(x)$, it is necessary to make the substitution $d/dt \to \partial/\partial t$.

and the operations of differentiating with respect to time and taking the expectation value are interchangeable. This is general, as can be seen by combining the expectation value of (8.1) (for time-independent F) with (8.14),

$$\left\langle \frac{d\widehat{F}}{dt} \right\rangle = \frac{1}{i\hbar} \langle [\widehat{F}, \widehat{H}] \rangle = \frac{d\langle \widehat{F} \rangle}{dt}. \qquad (8.17)$$

We now take $\widehat{F} = \widehat{p}$ in (8.14) to obtain $\frac{d\langle \widehat{p} \rangle}{dt} = -\frac{i}{\hbar}\langle [\widehat{p}, \widehat{H}] \rangle$, whence

$$\frac{d\langle \widehat{p} \rangle}{dt} = \langle -\left(\frac{\partial \widehat{V}}{\partial x}\right) \rangle = \langle f(\widehat{x}) \rangle. \qquad (8.18)$$

Combining Eqs. (8.15) and (8.18), we get

$$m\frac{d^2 \langle \widehat{x} \rangle}{dt^2} = \langle f(\widehat{x}) \rangle. \qquad (8.19)$$

The preceding results show that $\langle \widehat{x} \rangle$ and $\langle \widehat{p} \rangle$ satisfy *classical-like* relations in accordance with Ehrenfest's theorem. However, Eq. (8.19), which can be seen as the quantum version of Newton's second law, differs from it in an important way. Indeed, for the coordinate $\langle \widehat{x} \rangle$ of a classical particle we must write

$$m\frac{d^2 \langle x \rangle}{dt^2} = f(\langle x \rangle) \qquad (8.20)$$

instead of (8.19). The force $f(\langle x \rangle)$ and the *mean* force $\langle f(x) \rangle$ are different quantities, since the former is a local function at $\langle x \rangle$, whereas $\langle f(x) \rangle$ depends on the value of f in the whole accessible space. The difference can be made explicit by writing x in terms of its mean value and its deviation Δx,

$$x = \bar{x} + \Delta x, \quad \bar{x} \equiv \langle x \rangle, \qquad (8.21)$$

and expanding $f(x)$ around $\bar{x}$,

$$f(x) = f(\bar{x} + \Delta x) = f(\bar{x}) + \Delta x f'(\bar{x}) + \tfrac{1}{2}(\Delta x)^2 f''(\bar{x}) + \ldots \qquad (8.22)$$

Taking the mean value of this expression gives, with $\langle \Delta x \rangle = 0$,

$$\langle f(x) \rangle = f(\bar{x}) + \tfrac{1}{2}\langle (\Delta x)^2 \rangle f''(\bar{x}) + \ldots \qquad (8.23)$$

and substituting in (8.19) we get

$$m\frac{d^2 \bar{x}}{dt^2} = f(\bar{x}) + \tfrac{1}{2}\langle (\Delta x)^2 \rangle f''(\bar{x}) + \ldots, \qquad (8.24)$$

which differs from Eq. (8.20) by the presence of terms that depend on higher-order moments $\langle (\Delta x)^n \rangle$. In other words, stochasticity makes the average quantum trajectory of the particles differ in general from the corresponding classical one. For the case of *linear* forces $f = a + bx$ (free particle, uniform electric field, harmonic oscillator and their linear combinations) the second and higher derivatives of the external force are all zero and the quantum expression (8.24) reduces to the classical one (8.20); however, the analogy is only formal, since in the latter case the absence of fluctuations is central.

The quantum Eq. (8.19) describes the acceleration of the average value $\langle x \rangle$ as determined by the value of the force $f(x)$ averaged over the entire accessible space. When referred to a single particle (located in the mean at $\langle x \rangle$), Eq. (8.20) introduces a nonlocal element in the description. In Chapter 5 we found examples of this kind of nonlocalities when studying the dispersion of particles by a potential barrier. Such (apparent) nonlocalities are the result of reading (interpreting) the statistical description given by Eq. (8.19) in nonstatistical terms, as if it referred to a single particle, as is suggested by the Copenhagen point of view. Their appearance in the quantum description is one of the several arguments in favor of the ensemble interpretation of QM, as discussed in Section 10.1.2 in connection with the (nonlocal) quantum potential.

Exercise E8.2. Quantum virial theorem

Derive the quantum version of the virial theorem of classical mechanics for a particle subject to a potential V.

Solution. The virial theorem of classical mechanics is usually derived by analyzing the behavior of the time average of the quantity $\sum_i \mathbf{r}_i \cdot \mathbf{p}_i$, where the sum extends over all particles in the system.[2] For the derivation of a quantum version of this theorem, the same idea can be used, but applied to the average over the ensemble in a stationary state.

For a system in a stationary state, whose motion is restricted to a finite region of space, it must be true that

$$\frac{d}{dt}\langle \mathbf{r} \cdot \hat{\mathbf{p}} \rangle = 0. \tag{8.25}$$

On the other hand, from Eq. (8.14) it follows that

$$i\hbar \frac{d}{dt}\langle \mathbf{r} \cdot \hat{\mathbf{p}} \rangle = \langle [\mathbf{r} \cdot \hat{\mathbf{p}}, \hat{H}] \rangle. \tag{8.26}$$

From

$$[x\hat{p}_x, \hat{H}] = [x\hat{p}_x, \frac{\hat{p}_x^2}{2m}] + [x\hat{p}_x, V] = [x, \hat{p}_x^2]\frac{\hat{p}_x}{2m} + x[\hat{p}_x, V] = i\hbar \frac{\hat{p}_x^2}{m} - i\hbar x \frac{\partial V}{\partial x}, \tag{8.27}$$

we get

$$[\mathbf{r} \cdot \hat{\mathbf{p}}, \hat{H}] = i\hbar \left(\frac{\hat{\mathbf{p}}^2}{m} - \mathbf{r} \cdot \nabla V \right), \tag{8.28}$$

which when substituted in the preceding equation of motion gives

$$\frac{d}{dt}\langle \mathbf{r} \cdot \hat{\mathbf{p}} \rangle = 2\langle \hat{T} \rangle - \langle \mathbf{r} \cdot \nabla V \rangle. \tag{8.29}$$

For the bounded stationary case, condition (8.25) gives the virial theorem[3]

$$2\langle \hat{T} \rangle = \langle \mathbf{r} \cdot \nabla V \rangle. \tag{8.30}$$

Several particular cases are of interest. For example, for a central potential

$$2\langle \hat{T} \rangle = \langle r \frac{\partial V}{\partial r} \rangle, \tag{8.31}$$

---

[2] See, e.g., Goldstein et al. (2011).
[3] The name reflects the fact that the quantity $\mathbf{r} \cdot \nabla V$ is called virial (from Latin *vis*, force).

and if in addition the potential is of the form $V = ar^n$, then

$$\langle \widehat{T} \rangle = \frac{n}{2} \langle V \rangle = \frac{n}{n+2} E. \tag{8.32}$$

For a harmonic oscillator ($n = 2$) we have $\langle \widehat{T} \rangle = \langle V \rangle = \frac{1}{2} E$, while for a Coulomb potential ($n = -1$), we have $\langle T \rangle = -\frac{1}{2} \langle V \rangle = -E$. If the quantum average is interpreted as a time average – which is equivalent to assuming that each member of the ensemble reproduces the ensemble over time, that is, that the system has ergodic properties – all the above results are in agreement with the corresponding classical ones.

8.4 Elements for a Statistical Description

In the following, we will discuss the most important statistical properties of a quantum system.

8.4.1 The meaning of eigenvalues

We have seen in a number of examples throughout the book that when a physical system is in an eigenstate of an operator, the expectation value of that operator coincides with its eigenvalue, that is,

$$\widehat{F}|n\rangle = F_n|n\rangle \quad \Rightarrow \quad F_{nn} = \langle n|\widehat{F}|n\rangle = F_n. \tag{8.33}$$

Expressed in the Schrödinger representation, if

$$\widehat{F}\psi_n = F_n \psi_n, \tag{8.34}$$

the expectation value of $\widehat{F}$ in the state ψ_n is

$$\langle \widehat{F} \rangle_n = \int \psi_n^* \widehat{F} \psi_n dx = F_n. \tag{8.35}$$

The statistical explanation for this is that the ensemble described by ψ_n has been constructed in such a way that each and every one of its members has exactly the value F_n of the variable F, so that by taking the average we get F_n as a result. For example, the solution of the stationary Schrödinger equation $\widehat{H}\psi_n = E\psi_n$ (which is an eigenvalue equation), is a state function ψ_n corresponding to particles that all have the same energy E_n, as we saw in Chapter 4. A second example is given by the momentum equation $\hat{p}\psi = -i\hbar(\partial\psi/\partial x) = p\psi$; the solution of this differential equation, $Ae^{ipx/\hbar}$, describes a stationary plane wave with well-defined momentum p.

It is only possible to assign a single, well-defined value to a dynamical variable F if the system is in an eigenstate of the corresponding operator $\hat{F}$. In the general case where ψ is not an eigenstate of $\hat{F}$, the value of the corresponding variable is distributed among the elements of the ensemble.

The preparation process to obtain a desired eigenstate is, of course, a matter of specific manipulation (preparation) of the ensemble, a very important subject but beyond our present scope; see, however, the comment in Section 8.4.4.

8.4.2 Basic statistical elements

In Section 3.2.3 we introduced the variance or mean-square dispersion of $\widehat{F}$, defined as

$$\langle(\Delta\widehat{F})^2\rangle = \langle\widehat{F^2}\rangle - \langle\widehat{F}\rangle^2. \tag{8.36}$$

It is usual to also define the *standard deviation* (root mean square or *rms*), which is the square root of the variance and is sometimes written as ΔF or $\Delta_\psi F$ to make the dependence on the state explicit,

$$\Delta F \equiv (\Delta\widehat{F})_{\text{rms}} \equiv \sqrt{\langle(\widehat{F} - \langle\widehat{F}\rangle)^2\rangle} = \sqrt{\langle(\Delta\widehat{F})^2\rangle}. \tag{8.37}$$

When there is no risk of confusion, it is usual to write the average value of $\widehat{F}$ as $\langle\widehat{F}\rangle = \bar{F}$. Since, when applied to functions, the variance of F ($= \int (F - \bar{F})^2 \rho(x)dx$) is a sum of squares of deviations, any deviation (positive or negative) of the variable F from its mean value $\bar{F}$ adds to the result; this means that the variance of a variable is zero if and only if the variable has *exactly* the same value for each member of the ensemble. Even though in the quantum case we are dealing with operators (and $(\Delta\widehat{F})^2$ is not necessarily a positive quantity), it is common to consider (8.36) as the definition of variance. Note, however, that the definition of the powers of an operator in the form

$$\widehat{F^2} = \hat{F}^2, \tag{8.38}$$

which is an integral part of the quantum formalism, is not the only one possible in principle, and the following conclusions depend crucially on it.[4]

8.4.3 Simultaneous well-defined variables

Suppose we set up a system in an eigenstate of $\hat{H}$, so that the energy has a well-defined value. Under these conditions, what other variables could be well defined? A well-known example of a wave function that is a simultaneous eigenfunction of two operators ($\hat{H}$ and $\hat{p}$ in this case) is the free particle described by a plane wave. We then have $(\hat{p}^2/2m)\varphi_p(x) = E\varphi_p(x)$, and $\hat{p}\varphi_p(x) = p\varphi_p(x)$. These eigenvalue equations have the common solution $\varphi_p(x) = Ae^{ipx/\hbar}$; the particles described by this solution have well-defined energy E and momentum p.

The general answer is that it is possible to set up the system so that two (or more) of its dynamical variables F, G have well-defined values *if and only if their respective operators commute*.

---

[4] Note that (8.38) expresses a relation between operators, not between the values assigned to them. Indeed, if one wanted to apply a similar relation between the values V that a dynamical variable $\hat{A}$ can take, it should be possible to write

$$V_\psi(F(\hat{A})) = F(V_\psi(\hat{A})),$$

which (hypothetically) states that in the given state ψ, the value of a function of A is equal to the function evaluated at the value of the variable over that state. However, the *Kocher and Specker theorem* (1967) states that such a function V_ψ *does not exist* if the corresponding Hilbert space has more than two dimensions. An immediate implication of this result is that in a Hilbert space with more than two dimensions (which means that there can be more than four linearly independent dynamical variables) there is no state for which it is possible to simultaneously assign definite values to each of the dynamical variables of the system.

Here we show that the condition is necessary. By hypothesis, both $\widehat{F}$ and $\widehat{G}$ have well-defined values,

$$\widehat{F}\psi_n = F_n\psi_n, \quad \widehat{G}\psi_n = G_n\psi_n, \tag{8.39}$$

so that by applying $\widehat{G}$ from the left to the first equation

$$\widehat{G}\widehat{F}\psi_n = \widehat{G}F_n\psi_n = F_n\widehat{G}\psi_n = F_nG_n\psi_n, \tag{8.40}$$

and applying $\widehat{F}$ to the second equation

$$\widehat{F}\widehat{G}\psi_n = G_n\widehat{F}\psi_n = G_nF_n\psi_n, \tag{8.41}$$

we get $\widehat{F}\widehat{G}\psi_n = \widehat{G}\widehat{F}\psi_n$. Since this must hold for every ψ_n, we conclude that $\widehat{F}\widehat{G} = \widehat{G}\widehat{F}$. It is left to the student to prove that the condition is sufficient.

8.4.4 Well-defined and distributed variables

The preceding results suggest that once a starting operator has been selected, the remaining operators can be divided into two groups: those that commute with the selected operator and among themselves, and the rest of the operators. The first group is the *maximal set of commuting operators* (MSCO). Each operator can belong to one or the other group, this strongly depends on which was the initial selection; however, we know from the quantum canonical commutator $[x, p] = i\hbar$ that not all operators can belong to the MSCO, that is, the noncommuting group is never empty. In fact, the number of variables in an MSCO is generally very small: 3, 4..., not much more.

The classification is not unambiguous, since it is based on an initial arbitrary choice of the reference operator. However, in each particular case we can add appropriate criteria to define the two groups uniquely; we will assume that this has been done. Under these conditions, it is possible to prepare the system – at least in principle; the "preparation" of the state is usually a practical task – in a common eigenstate of all the MSCO operators, with the result that all these variables are dispersion-free. This means that the remaining dynamical variables (the non-MSCO variables) will be distributed without exception. In other words, *every quantum state is necessarily dispersive*.

Each dynamical variable belonging to the MSCO group gives rise to a *quantum number*; the corresponding operator satisfies an eigenvalue equation and the corresponding eigenvalue (or a number derived from it) becomes a quantum number of the state. Therefore, we can expect a physical system to have a small amount of quantum numbers in general.

As an example, suppose that the chosen initial variable is the Hamiltonian for a central-force problem, with eigenvalue n; the angular momentum L is then added, with eigenvalue l, and the projection l_z with eigenvalue m completes the MSCO: the system has been prepared in state $|nlm\rangle$, which precises the basis for the description. If, given such a common eigenstate, we were to make a series of measurements to determine the numerical value of any variable of the MSCO group, we would obtain the same value each time (apart from the inevitable experimental errors).

On the other hand, a series of measurements on a given variable of the non-MSCO group, would give different results each time, unpredictable from case to case. If the series of measurements is large enough, a law of distribution for that variable will eventually emerge. We see here one of the inevitable effects of quantum randomness: it is impossible to prepare a quantum system so that each of its dynamical variables has a well-defined value. This is a key feature that distinguishes quantum systems from classical ones and confirms the inevitability of the statistical approach.

8.4.4.1 Examples of well-defined and distributed variables

a) Consider first the case of particles moving in a potential $V(x)$. Then the momentum and the Hamiltonian operators do not commute, $[\hat{p}, \hat{H}] \neq 0$. So if we give all the particles the same energy, their momentum will be distributed; they will form a wave packet in momentum space.

b) A second example of great interest is that of an arbitrary pair of canonical variables. The fact that, for example, $[x, p] = i\hbar$ means that a quantum system cannot be prepared so that all the members of the ensemble have exactly the same momentum and are simultaneously at the same point, that is, the product of the dispersions of x and p_x cannot vanish. In our idealization of the description of the free particle with a plane wave of well-defined momentum, the particles were uniformly distributed in space; only by distributing the momentum within a packet of plane waves can the system acquire finite dimensions in x-space.

c) It could be argued that even if we accept that most of the variables are distributed (those that do not belong to the MSCO), this may not be important in practice if the system is prepared so that the dispersions are made small enough as to be considered negligible. In the next section we will show that this is not possible, since the dispersions are usually significant when compared to the characteristic values of the corresponding variables.

8.5 The Heisenberg Inequalities and Their Meaning

To give quantitative form to the preceding discussion, we begin with two simple auxiliary results of general use. First, we show that $\langle \hat{A}^\dagger \hat{A} \rangle_\psi$ is nonnegative for any operator (not necessarily Hermitian) and any state $|\psi\rangle$. We use the representation $\hat{A} = \sum_k a_k |k\rangle\langle k|$ (the coefficients a_k can be complex for $\hat{A}$ not Hermitian) and by inserting a pair of unit expansions we get

$$\langle \hat{A}^\dagger \hat{A} \rangle_\psi = \langle \psi | \hat{A}^\dagger \hat{A} | \psi \rangle = \sum_{k,l} \langle \psi | a_k^* | k \rangle \langle k | a_l | l \rangle \langle l | \psi \rangle = \sum_{k,l} \langle \psi | a_k^* a_l | k \rangle \delta_{kl} \langle l | \psi \rangle$$

$$= \sum_k |a_k|^2 \langle \psi | k \rangle \langle k | \psi \rangle = \sum_k |a_k|^2 |\langle k | \psi \rangle|^2 \geq 0, \quad (8.42)$$

hence (recall that $\hat{A}^\dagger$ contains the operation $\hat{A}^*$)

$$\langle \hat{A}^\dagger \hat{A} \rangle \geq 0. \quad (8.43)$$

Now let $\hat{A}$ and $\hat{B}$ be two operators with commutator $i\hat{C}$,

$$[\hat{A}, \hat{B}] = i\hat{C}. \quad (8.44)$$

The commutator of the deviations $\Delta\widehat{A}$ and $\Delta\widehat{B}$ is $[\Delta\widehat{A}, \Delta\widehat{B}] = [\widehat{A}-\overline{A}, \widehat{B}-\overline{B}] = [\widehat{A}, \widehat{B}]$; therefore,

$$[\Delta\widehat{A}, \Delta\widehat{B}] = i\widehat{C}. \tag{8.45}$$

This is our second observation; we now turn to the proof of Heisenberg's inequality, by considering the function $J(\alpha)$,

$$J(\alpha) = \langle (\alpha\widehat{A} + i\widehat{B})^\dagger (\alpha\widehat{A} + i\widehat{B}) \rangle. \tag{8.46}$$

where α is a real parameter. From (8.43) applied to $\alpha\widehat{A} + i\widehat{B}$ it follows that

$$J(\alpha) \geq 0. \tag{8.47}$$

We will take $\widehat{A}$ and $\widehat{B}$ to be Hermitian, so $(\alpha\widehat{A} + i\widehat{B})^\dagger = \alpha\widehat{A} - i\widehat{B}$; substituting and expanding, we get

$$J(\alpha) = \alpha^2 \langle \widehat{A}^2 \rangle + \langle \widehat{B}^2 \rangle + i\alpha \langle \widehat{A}\widehat{B} - \widehat{B}\widehat{A} \rangle \geq 0. \tag{8.48}$$

As a function of α, J describes a parabola in the upper half-plane, which takes its minimum value for $\alpha = \alpha_m$ and $J'(\alpha_m) = 0$ with

$$\alpha_m = -\frac{i\langle \widehat{A}\widehat{B} - \widehat{B}\widehat{A}\rangle}{2\langle \widehat{A}^2\rangle} = \frac{\langle \widehat{C}\rangle}{2\langle \widehat{A}^2\rangle}. \tag{8.49}$$

This minimum is

$$J_{\min} = J(\alpha_m) = \frac{\langle \widehat{C}\rangle^2}{4\langle \widehat{A}^2\rangle} + \langle \widehat{B}^2\rangle - \frac{\langle \widehat{C}\rangle^2}{2\langle \widehat{A}^2\rangle} = \langle \widehat{B}^2\rangle - \frac{\langle \widehat{C}\rangle^2}{4\langle \widehat{A}^2\rangle} \geq 0. \tag{8.50}$$

Multiplying the result by $\langle \widehat{A}^2 \rangle$ and reorganizing gives

$$\langle \widehat{A}^2 \rangle \langle \widehat{B}^2 \rangle \geq \tfrac{1}{4} \langle \widehat{C} \rangle^2, \tag{8.51}$$

or using (8.45),

$$\langle (\Delta\widehat{A})^2 \rangle \langle (\Delta\widehat{B})^2 \rangle \geq \tfrac{1}{4} \langle \widehat{C} \rangle^2. \tag{8.52}$$

This is a general version of the Heisenberg inequality, a particular version of which was derived in Section 3.6 for the variances of $\widehat{x}$ and $\widehat{p}$,

$$\langle (\Delta\widehat{x})^2 \rangle \langle (\Delta\widehat{p})^2 \rangle \geq \tfrac{1}{4} \hbar^2. \tag{8.53}$$

In the general case of conjugate canonical variables A, B we have $|\langle \widehat{C} \rangle| = \hbar$ and the Heisenberg inequalities take the form

$$\langle (\Delta\widehat{A})^2 \rangle \langle (\Delta\widehat{B})^2 \rangle \geq \tfrac{1}{4} \hbar^2. \tag{8.54}$$

Equation (8.52) makes clear the meaning and content of Heisenberg's inequality: the product of the dispersions is bounded from below by the mean value of the commutator. This is an inviolable theorem of QM, relating to objective statistical properties of the system.

Exercise E8.3. A stronger Heisenberg inequality
Derive an inequality that is stronger than (8.52) by expressing the product $\widehat{A}\widehat{B}$ in terms of their commutator and anticommutator. *Hint*: use the Schwarz inequality.

Solution. Proceeding as before, we write for $\overline{A} = \overline{B} = 0$, which is always possible; then

$$\langle(\Delta\widehat{A})^2\rangle\langle(\Delta\widehat{B})^2\rangle = \left[\int \left(\psi^*\hat{A}\right)\left(\hat{A}\psi\right)dx\right]\left[\int \left(\psi^*\hat{B}\right)\left(\hat{B}\psi\right)dx\right]$$

$$= \left[\int \left(\hat{A}\psi\right)^\dagger \left(\hat{A}\psi\right)dx\right]\left[\int \left(\hat{B}\psi\right)^\dagger \left(\hat{B}\psi\right)dx\right]$$

$$= \left[\int \left|\hat{A}\psi\right|^2 dx\right]\left[\int \left|\hat{B}\psi\right|^2 dx\right]$$

$$\geq \left|\int \left(\hat{A}\psi\right)^\dagger \left(\hat{B}\psi\right)dx\right|^2 = \left|\int \psi^*\hat{A}\hat{B}\psi\, dx\right|^2, \quad (8.55)$$

where we have used the Schwarz inequality. Introducing now the identity

$$\hat{A}\hat{B} = \tfrac{1}{2}\left[\hat{A},\hat{B}\right] + \tfrac{1}{2}\left\{\hat{A},\hat{B}\right\}, \quad (8.56)$$

where $\left\{\hat{A},\hat{B}\right\} \equiv \hat{A}\hat{B} + \hat{B}\hat{A}$ is the anticommutator, we get

$$\langle(\Delta\widehat{A})^2\rangle\langle(\Delta\widehat{B})^2\rangle \geq \left|\int \psi^*\hat{A}\hat{B}\psi\, dx\right|^2 = \left|\langle\hat{A}\hat{B}\rangle\right|^2 = \tfrac{1}{4}\left|\langle[\hat{A},\hat{B}]\rangle + \langle\{\hat{A},\hat{B}\}\rangle\right|^2. \quad (8.57)$$

For Hermitian operators $\langle[\hat{A},\hat{B}]\rangle^\dagger = -\langle[\hat{A},\hat{B}]\rangle$; this gives, after some simplification, the stronger relation

$$\langle(\Delta\widehat{A})^2\rangle\langle(\Delta\widehat{B})^2\rangle \geq \tfrac{1}{4}\left|\langle[\hat{A},\hat{B}]\rangle\right|^2 + \tfrac{1}{4}\left|\langle\{\hat{A},\hat{B}\}\rangle\right|^2. \quad (8.58)$$

It is easy to show that the equal sign holds only if $\hat{A}$ and $\hat{B}$ are related by a constant of proportionality, that is, if they refer to the same state. The preceding result can also be obtained using the previous derivation, taking α as a complex parameter.

8.5.1 What do the Heisenberg inequalities tell us?

As already mentioned, the Heisenberg inequalities are commonly referred to in the literature as *uncertainty* or *indeterminacy relations*. To avoid unnecessary and misleading allusions, we prefer not to add such qualifiers, which imply specific (and different) interpretations, the former often subjective and the latter indeterministic.

Sometimes we read in texts on QM that Eq. (8.53) tells us that particle trajectories do not exist – as Heisenberg himself concluded. It is true that the quantum description does not contain the notion of trajectories and does not describe them; but it does allow us to determine some of their properties, such as the variances and other statistical moments of the position and momentum variables. According to Heisenberg's inequality, paths of equally prepared electrons differ from electron to electron in such a way that the product of the dispersions of x and p always satisfies (8.53), and similarly for any conjugate pair $\widehat{A}, \widehat{B}$; since (8.52) has a statistical meaning, it does not apply to a single case, but to an ensemble of realizations. This is true whether we observe them or not, it is a corollary of the formalism which does not include the act of observation.

Note that the Heisenberg inequality does not imply that as the standard deviation of a variable decreases, the standard deviation of the corresponding conjugate canonical variable necessarily increases; in fact, the opposite may be true. To see this, consider the case of conjugate canonical variables,

$$\Delta_\psi q \, \Delta_\psi p \geq \tfrac{1}{2}\hbar, \tag{8.59}$$

for a particle subject to a binding potential. The more excited the bound state ψ is, the higher are the values of $\Delta_\psi q$ and $\Delta_\psi p$. When the excitation is reduced, it usually happens that both standard deviations decrease simultaneously, but their product still satisfies Heisenberg's inequality. For example, for the 1D harmonic oscillator studied in Chapter 9,

$$\langle(\Delta\widehat{x})^2\rangle = (n+\tfrac{1}{2})\frac{\hbar}{m\omega}; \quad \langle(\Delta\widehat{p})^2\rangle = (n+\tfrac{1}{2})\hbar m\omega, \tag{8.60}$$

where $n = 0, 1, 2, \ldots$ is the level of the excitation; thus,

$$\langle(\Delta\widehat{x})^2\rangle\langle(\Delta\widehat{p})^2\rangle = \tfrac{1}{4}\hbar^2(1+2n)^2 \geq \tfrac{1}{4}\hbar^2, \tag{8.61}$$

so that both dispersions can decrease simultaneously.

Heisenberg's inequalities are sometimes used to obtain semi-quantitative information about the dynamical behavior of a physical system. For example, consider the case of a particle in a stationary state, subject to a binding potential that depends only on x; (8.52) gives

$$\langle(\Delta\widehat{x})^2\rangle\langle(\Delta\widehat{H})^2\rangle \geq \frac{\hbar^2}{4m^2}\langle\widehat{p}\rangle^2. \tag{8.62}$$

Since $\langle(\Delta H)^2\rangle = 0$ and $\langle(\Delta x)^2\rangle < \infty$, we conclude that the mean value of p is zero (as shown with the examples of Chapter 5). Taking now the commutator

$$[\widehat{p}, \widehat{H}] = i\hbar \widehat{f}, \tag{8.63}$$

where $\widehat{f}(x)$ represents the external force, and using again Eq. (8.52), we obtain

$$\langle(\Delta\widehat{p})^2\rangle\langle(\Delta\widehat{H})^2\rangle \geq \tfrac{1}{4}\hbar^2\langle\widehat{f}\rangle^2. \tag{8.64}$$

Therefore, for any real (i.e., with $\langle(\Delta\widehat{p})^2\rangle$ finite) bound eigenstate of the Hamiltonian, the mean value of the force is zero. This result is to be expected for any stationary state, classical or quantum.

Exercise E8.4. A minimum-dispersion packet

A state for which $\langle(\Delta\widehat{A})^2\rangle\langle(\Delta\widehat{B})^2\rangle = \tfrac{1}{4}\langle\widehat{C}\rangle^2$ is called a minimum-dispersion packet with respect to the variables A and B. Construct a state in x-space, $\psi_{\min}(x)$, for which the Heisenberg inequality becomes an equality (remember that the substitution $\widehat{A} \to \Delta\widehat{A}$ can be made).

Solution. The definition of minimum dispersion requires the function $J(\alpha)$ in Eq. (8.47) to be zero, which only happens if

$$(\alpha\Delta\widehat{A} + i\Delta\widehat{B})\psi_{\min} = 0, \tag{8.65}$$

as follows from Eq. (8.46). In this expression, α must have the value α_m that minimizes J, which is given by Eq. (8.49), $\alpha_m = \langle \widehat{C} \rangle / 2 \langle \widehat{A}^2 \rangle$. The minimum-dispersion packet is therefore a solution of the differential equation

$$\left(\frac{\langle \widehat{C} \rangle}{2 \langle (\Delta \widehat{A})^2 \rangle} \Delta \widehat{A} + i \Delta \widehat{B} \right) \psi_{\min} = 0. \tag{8.66}$$

Taking $\widehat{A} = \widehat{x}$, $\widehat{B} = \widehat{p}$ and moving to the coordinate system with $\langle \widehat{x} \rangle = \langle \widehat{p} \rangle = 0$, $\langle \widehat{C} \rangle = \hbar$, we get

$$\left(\frac{x}{2\sigma^2} + \frac{\partial}{\partial x} \right) \psi_{\min} = 0, \quad \sigma^2 = \langle (\Delta x)^2 \rangle, \tag{8.67}$$

whose solution,

$$\psi_{\min} = \left(\frac{1}{\sqrt{2\pi} \sigma} \right)^{\frac{1}{2}} \exp \left[-\frac{(x - \langle x \rangle)^2}{4\sigma^2} \right], \tag{8.68}$$

is a Gaussian (normal) distribution with variance $\langle (\Delta \widehat{x})^2 \rangle = \sigma^2$ and centered at $\langle \widehat{x} \rangle$.

For a harmonic oscillator, $m\omega \Delta x = \Delta p = \hbar/2$, which means that both quadratures have the same dispersion. As will be shown in Chapter 9, this solution is the coherent, minimal-dispersion packet.[5]

8.5.2 Combining equations of motion and the Heisenberg inequalities

To conclude this section, we will combine Ehrenfest's equation (8.14) with Heisenberg's inequality (8.52). If we set $\widehat{B} = \widehat{H}$ in the latter, we get

$$\langle (\Delta \widehat{A})^2 \rangle \langle (\Delta \widehat{H})^2 \rangle \geq \tfrac{1}{4} \langle \widehat{C} \rangle^2, \tag{8.69}$$

where

$$\langle \widehat{C} \rangle = -i \langle [\widehat{A}, \widehat{H}] \rangle = \hbar \frac{d \langle \widehat{A} \rangle}{dt}, \tag{8.70}$$

as follows from Eq. (8.14). Therefore,

$$\langle (\Delta \widehat{A})^2 \rangle \langle (\Delta \widehat{H})^2 \rangle \geq \tfrac{1}{4} \hbar^2 \left(\frac{d \langle \widehat{A} \rangle}{dt} \right)^2. \tag{8.71}$$

From this it follows, in particular, that when the system is in an eigenstate of either the Hamiltonian or the operator $\widehat{A}$, the average value $\langle \widehat{A} \rangle$ is constant. Equation (8.71) can be written in a slightly different form, which is obtained by dividing it by $(d \langle \widehat{A} \rangle / dt)^2$,

$$\langle (\Delta \widehat{H})^2 \rangle \left\langle \left(\frac{\Delta \widehat{A}}{d \langle \widehat{A} \rangle / dt} \right)^2 \right\rangle \geq \tfrac{1}{4} \hbar^2. \tag{8.72}$$

---

[5] For an oscillator in a *squeezed* state such that $\lambda^{-1} m\omega \Delta x = \lambda \Delta p = \hbar/2$ with $\lambda \neq 1$, the dispersion in one of the variables has increased at the expense of the other; their product remains minimal, but the energy increases. Squeezed states of light are an interesting topic in quantum optics and have been studied extensively experimentally since the 1980s.

This result is known as the *Mandelstam–Tamm*[6,7] *inequality* or *relation* (1945) and can be interpreted as follows. We introduce a time τ_A defined as

$$\tau_A = \sqrt{\langle(\Delta\widehat{A})^2\rangle} \bigg/ \left|\frac{d\langle\widehat{A}\rangle}{dt}\right|, \tag{8.73}$$

representing the mean time required by $\langle\widehat{A}\rangle$ to change by an amount of the order of its standard deviation, which we may call the *evolution time* of $\widehat{A}$. Equation (8.72) can then be written

$$\langle(\Delta\widehat{H})^2\rangle \tau_A^2 \geq \tfrac{1}{4}\hbar^2, \tag{8.74}$$

which shows that the evolution time is not arbitrarily short, but is bounded from below by the standard deviation of the energy,

$$\tau_A(\Delta H) \geq \frac{\hbar}{2}. \tag{8.75}$$

For an eigenstate of $\widehat{H}$, τ_A becomes infinite, which means that the dynamical variables then take an infinitely long time to change, or, in other words, that they do not change. From this point of view, τ_A is a specific measure of the "degree of stationarity" of the system.

8.6 Quantum Canonical Transformations

We are now equipped with the necessary elements to study the evolution of the quantum states using the propagator method introduced in Chapter 7. We then turn to an introduction to more powerful operator procedures that allow us to switch between different representations.

8.6.1 The evolution operator and the propagator

We will study the time evolution of a state $|\psi\rangle$ using Schrödinger's description. For a Hamiltonian $\widehat{H}$ that does not depend on time, the formal solution of the evolution equation (7.140), $i\hbar\frac{d|\psi\rangle}{dt} = \widehat{H}|\psi\rangle$, is

$$|\psi\rangle = e^{-\frac{i}{\hbar}\widehat{H}t}|\psi_0\rangle; \quad |\psi_0\rangle \equiv |\psi(t=0)\rangle. \tag{8.76}$$

This allows us to introduce the *evolution operator*

$$\widehat{S}(t) = e^{-\frac{i}{\hbar}\widehat{H}t}, \tag{8.77}$$

---

[6] Leonid Isaakovich Mandelstam (1879–1944) was a Russian-Soviet physicist co-discoverer of the Raman scattering; he founded one of the two major schools of theoretical physics in the Soviet Union, the other being due to L. Landau.
[7] Igor Yevgenyevich Tamm (1895–1971) was a Russian-Soviet theoretical physicist who received the 1958 Nobel Prize for his co-discovery of the Cherenkov radiation. He also predicted the phonon and together with A. Sakharov proposed the Tokamak system for the (successful) construction of the toroidal magnetic thermonuclear reactor.

so that
$$|\psi(t)\rangle = \widehat{S}(t)|\psi_0\rangle. \tag{8.78}$$

Substituting this in (7.140) and simplifying gives the equation for the evolution operator itself,
$$i\hbar \frac{d\widehat{S}(t)}{dt} = \widehat{H}\widehat{S}(t). \tag{8.79}$$

Equation (8.78) is a formal way of writing an integral expression. To obtain the latter explicitly we multiply it from the left by $\langle x|$ and introduce on the right-hand side an expansion of the unit in terms of the basis; we thus obtain
$$\psi(x,t) = \int K(x,t|x't')|\psi(x',t')dx', \tag{8.80}$$

which is the same as Eq. (7.144), namely $\langle x|\psi(t)\rangle = \int K(x,t|x't')|\psi(x',t')dx'$, with $\psi(x,t) = \langle x|\psi\rangle$ and $t' = 0$. From this comparison it follows that the propagator from t' to $t \geq t'$ is given by
$$K(x,t|x',t') = \langle x|\widehat{S}(t-t')|x'\rangle = \langle x|e^{-\frac{i}{\hbar}\widehat{H}(t-t')}|x'\rangle, \tag{8.81}$$

which shows how the Hamiltonian takes (propagates) the state in x',t' to the state in x at a later time t.

An alternative convenient way to write the propagator in the general form discussed in Chapter 7 is to insert a pair of expansions of the unit and write
$$K(x,t|x',t') = \sum_n \sum_{n'} \langle x|n\rangle \langle n|e^{-\frac{i}{\hbar}\widehat{H}(t-t')}|n'\rangle \langle n'|x'\rangle$$
$$= \sum_n \sum_{n'} e^{-\frac{i}{\hbar}E_{n'}(t-t')}\delta_{nn'}\varphi_n(x)\varphi_{n'}^*(x'), \tag{8.82}$$

which reduces to Eq. (7.145),
$$K(x,t|x',t') = \sum_n \psi_n(x,t)\psi_n^*(x',t'). \tag{8.83}$$

Similarly, by expanding $|\psi_0\rangle$ in terms of the eigenstates $|n\rangle$ of the Hamiltonian
$$|\psi_0\rangle = \sum_n c_n|n\rangle, \tag{8.84}$$

with $\widehat{H}|n\rangle = E_n|n\rangle$, we get
$$|\psi\rangle = e^{-\frac{i}{\hbar}\widehat{H}t}\sum_n c_n|n\rangle = \sum_n c_n e^{-\frac{i}{\hbar}E_n t}|n\rangle, \tag{8.85}$$

where $\widehat{S}(t)$ has been interpreted in terms of its power series,
$$e^{-\frac{i}{\hbar}\widehat{H}t}|n\rangle = \sum_{k=0}^{\infty} \frac{1}{k!}\left(-\frac{i}{\hbar}t\right)^k \widehat{H}^k|n\rangle = \sum_{k=0}^{\infty} \frac{1}{k!}\left(-\frac{i}{\hbar}t\right)^k E_n^k|n\rangle = e^{-\frac{i}{\hbar}E_n t}|n\rangle. \tag{8.86}$$

Rewriting (8.85) in the coordinate representation, we obtain Eq. (7.141). It is common to interpret expressions like $\exp(-i\widehat{H}t/\hbar)$ in terms of their power series expansion.

It may happen that the expansion is not convergent, in which case expressing the initial wave function in terms of the eigenfunctions of $\widehat{H}$ (as was done here) solves the problem.

From (8.77) we note that

$$\widehat{S}^{-1} = e^{\frac{i}{\hbar}\widehat{H}t} = \widehat{S}^{\dagger} \tag{8.87}$$

because $\widehat{H}$ is Hermitian. Therefore, $\widehat{S}^{\dagger} = \widehat{S}^{-1}$, which is the defining property of a unitary operator (see next section). This means that *the time evolution of the system is given by the action of a unitary operator acting on the initial state*. In classical mechanics, the time evolution of a dynamical system is a canonical transformation (i.e., if $f(0)$, $g(0)$ are canonically conjugate, $f(t)$, $g(t)$ are canonically conjugate at any time, see[4]). This general relationship points to the importance of the study of unitary transformations in QM, which will play the role of canonical transformations.

8.6.2 Properties of unitary operators

Let $\widehat{U}$ be a unitary operator,

$$\widehat{U}^{\dagger} = \widehat{U}^{-1}, \tag{8.88}$$

which we use to transform the state vector $|i\rangle$ into $|i'\rangle$ according to the rule

$$|i'\rangle = \widehat{U}|i\rangle, \quad \langle i'| = \langle i|\widehat{U}^{\dagger}. \tag{8.89}$$

The transformation preserves orthonormality because $\langle i'|j'\rangle = \langle i|\widehat{U}^{\dagger}\widehat{U}|j\rangle = \langle i|j\rangle$. This is a first reason for choosing unitary operators to transform state vectors into state vectors. The modified basis is still complete because the closure property is retained,

$$\sum_{i'} |i'\rangle\langle i'| = \sum_{i} \widehat{U}|i\rangle\langle i|\widehat{U}^{\dagger} = \widehat{U}\widehat{U}^{\dagger} = \mathbb{I}. \tag{8.90}$$

In other words, a unitary transformation does not modify the space, it only modifies the basis (like a rotation of the Cartesian coordinate axes in 3D space). The transformation can also be applied to operators, by writing

$$\widehat{F}' = \widehat{U}\widehat{F}\widehat{U}^{\dagger}. \tag{8.91}$$

Note that the matrix elements are invariant under the transformation,

$$F'_{ij} = \langle i'|\widehat{F}'|j'\rangle = \langle i|\widehat{U}^{\dagger}\widehat{F}'\widehat{U}|j\rangle = \langle i|\widehat{U}^{\dagger}\widehat{U}\widehat{F}\widehat{U}^{\dagger}\widehat{U}|j\rangle = \langle i|\widehat{F}|j\rangle = F_{ij}. \tag{8.92}$$

From the unitarity property $\widehat{U}^{\dagger}\widehat{U} = \widehat{U}\widehat{U}^{\dagger} = \mathbb{I}$ it follows that the inverse of the transformation (8.89)–(8.91) is

$$|i\rangle = \widehat{U}^{\dagger}|i'\rangle, \quad \widehat{F} = \widehat{U}^{\dagger}\widehat{F}'\widehat{U}. \tag{8.93}$$

Also the trace of the operator is invariant to a change in representation, because

$$\operatorname{tr} \widehat{F}' = \operatorname{tr} \widehat{U}\widehat{F}\widehat{U}^{-1} = \operatorname{tr} \widehat{F}\widehat{U}^{-1}\widehat{U} = \operatorname{tr} \widehat{F}. \tag{8.94}$$

Throughout the text we have already encountered several quantum unitary operators. Operations that preserve the norm or more generally, the scalar product of real vectors, as well as those that generate rotations, reflections and similar transformations, are performed with orthogonal operators, which satisfy the condition $\hat{O}^{-1} = \hat{O}^T$, so that $\hat{O}\hat{O}^T = \hat{\mathbb{I}}$. The unitary operators are the generalization of the orthogonal operators to complex spaces (of any number of dimensions) $\hat{O} \to \hat{U}$, $\hat{O}^T \to \hat{U}^{T*} \equiv \hat{U}^\dagger$, and thus they satisfy the condition $\hat{U}^{-1} = \hat{U}^\dagger$.

It is clear that the product of two unitary operators is unitary, since each transformation separately preserves the scalar product. From this property it follows that the set of unitary operators (of a given dimension) contains the product of any pair of them; since it also contains the unit $\hat{U} = \hat{\mathbb{I}}$ and the inverse of each operator $\hat{U}^{-1} = \hat{U}^\dagger$, this set forms a continuous group of transformations, called *unitary transformations*, a generalization to the complex case of orthogonal transformations.

8.6.3 Eigenvectors of unitary operators

The eigenvalues of a unitary operator are complex numbers modulo 1, since

$$\langle u_n | \hat{U}^\dagger \hat{U} | u_n \rangle = \langle u_n | u_n \rangle = 1 = u_n^* u_n, \tag{8.95}$$

and we can therefore write

$$u_n = e^{i g_n}, \quad g_n \text{ real}. \tag{8.96}$$

We also have that

$$\langle u_n | u_m \rangle = \langle u_n | \hat{U}^\dagger \hat{U} | u_m \rangle = u_n^* u_m \langle u_n | u_m \rangle = e^{i(g_m - g_n)} \langle u_n | u_m \rangle, \tag{8.97}$$

with solution

$$\langle u_n | u_m \rangle = \delta_{nm}. \tag{8.98}$$

We conclude that the eigenvectors of a unitary operator belonging to different (complex) eigenvalues are orthogonal. Equation (8.96) suggests writing a unitary operator in the form (reminiscent of the time-evolution operator (8.77))

$$\hat{U} = e^{i\hat{G}}. \tag{8.99}$$

From $\hat{U}^{-1} = \hat{U}^\dagger$ we get $e^{-i\hat{G}} = e^{-i\hat{G}^\dagger}$, which shows that $\hat{G}$ must be Hermitian. If $|g_n\rangle$ is an eigenvector of $\hat{G}$ with real eigenvalue g_n, we have

$$\hat{U}|g_n\rangle = e^{i\hat{G}}|g_n\rangle = \sum_k \frac{1}{k!}(i\hat{G})^k|g_n\rangle = \sum_k \frac{1}{k!}(ig_n)^k|g_n\rangle = e^{ig_n}|g_n\rangle = u_n|g_n\rangle, \tag{8.100}$$

which shows that the eigenvectors of $\hat{G}$ are eigenvectors of the unitary operator $e^{i\hat{G}}$.

Exercise E8.5. Invariance of the physical laws under a unitary transformation
Show that the prescribed transformation rules (8.89) and (8.91) keep the form of the physical laws invariant.

Solution. Consider the expression $|\psi\rangle = \hat{F}|\varphi\rangle$; its transform is

$$|\psi'\rangle = \hat{U}|\psi\rangle = \hat{U}\hat{F}|\varphi\rangle = \hat{U}\hat{U}^\dagger \hat{F}' \hat{U}\hat{U}^\dagger |\varphi'\rangle = \hat{F}'|\varphi'\rangle, \tag{8.101}$$

which has exactly the form of the original relation $|\psi\rangle = \widehat{F}|\varphi\rangle$. To show the invariance of the equations of motion against the transformation (8.91), we multiply the equation of motion

$$i\hbar\widehat{\dot{A}} = [\widehat{A},\widehat{H}] \qquad (8.102)$$

by $\widehat{U}$ from the left and by $\widehat{U}^\dagger$ from the right, to obtain

$$i\hbar\widehat{U}\widehat{\dot{A}}\widehat{U}^\dagger = \widehat{U}[\widehat{A},\widehat{H}]\widehat{U}^\dagger = \widehat{U}\widehat{A}\widehat{U}^\dagger\widehat{U}\widehat{H}\widehat{U}^\dagger - \widehat{U}\widehat{H}\widehat{U}^\dagger\widehat{U}\widehat{A}\widehat{U}^\dagger = \widehat{A}'\widehat{H}' - \widehat{H}'\widehat{A}', \qquad (8.103)$$

that is,

$$i\hbar\widehat{\dot{A}}' = [\widehat{A}',\widehat{H}'], \qquad (8.104)$$

which shows that the form of the equation of motion has not changed. From the same proof it follows that the canonical commutation relations also retain their form and, therefore, the quantization rules are invariant under the transformation. In this way it is shown that the unitary transformations performed with the rules (8.89) for the state vectors and (8.91) for the operators, leave the canonical equations of quantum mechanics invariant.

We thus confirm that unitary transformations are in general the quantum equivalent of canonical transformations; in particular, the time evolution of the state vectors, described by Eq. (8.78), is a canonical transformation (in evolution).

8.6.4 Projection operators

In Section 4.6 we saw that an operator of the form

$$\widehat{P}_a = |a\rangle\langle a| \qquad (8.105)$$

is a projection operator. We will now briefly analyze such operators. Applying the operator $\widehat{P}_a$ to an arbitrary state $|Q\rangle$ we get

$$\widehat{P}_a|Q\rangle = |a\rangle\langle a|Q\rangle = c_Q(a)|a\rangle, \qquad (8.106)$$

which is a vector in the $|a\rangle$ direction, with magnitude equal to the component $c_Q(a) = \langle a|Q\rangle$ of $|Q\rangle$ in the $|a\rangle$ direction. The operator $\widehat{P}_a$ thus projects any vector from the Hilbert space $\mathcal{H}$ into the direction (or the subspace) $|a\rangle$. The projection operators are Hermitian and idempotent, as follows from their definition,

$$\widehat{P}_a^\dagger = (|a\rangle\langle a|)^\dagger = |a\rangle\langle a| = \widehat{P}_a, \qquad (8.107)$$

$$\widehat{P}_a^2 = |a\rangle\langle a|a\rangle\langle a| = |a\rangle\langle a| = \widehat{P}_a. \qquad (8.108)$$

This last equation implies that the eigenvalues of a projector are 0 or 1. Conversely, any Hermitian operator $\widehat{P}$ that satisfies the equation $\widehat{P}^2 = \widehat{P}$ is a projector; it projects into the subspace of $\mathcal{H}$ that corresponds to its eigenvalue one. From $\widehat{P} = \widehat{P}^2 = \widehat{P}^n$, n any positive integer, it follows that the most general function $f(\widehat{P}_a)$ of a projector has the form $f(\widehat{P}_a) = A + B\widehat{P}_a$, with A and B independent of $\widehat{P}_a$. Furthermore, since a projector has only one nonzero eigenvalue equal to 1, its trace is 1; note that according to (8.94) this property is independent of the representation.

From the property of completeness it follows that the operator $1 - \widehat{P}$ is orthogonal to $\widehat{P}$; in fact, since $\widehat{P}$ is idempotent, we have that

$$\widehat{P}(1 - \widehat{P}) = \widehat{P} - \widehat{P}^2 = 0. \tag{8.109}$$

This means that the vectors $\widehat{P}|u\rangle$ and $(1 - \widehat{P})|u\rangle$ are orthogonal components for any state $|u\rangle$, and thus it is always possible to decompose a state into two orthogonal states,

$$|u\rangle = \widehat{P}|u\rangle + (1 - \widehat{P})|u\rangle. \tag{8.110}$$

The vector $\widehat{P}|u\rangle$ is an eigenvector of $\widehat{P}$ with eigenvalue 1, since

$$\widehat{P}\left(\widehat{P}|u\rangle\right) = \widehat{P}^2|u\rangle = \widehat{P}|u\rangle. \tag{8.111}$$

In turn, $(1 - \widehat{P})|u\rangle$ is an eigenvector of $\widehat{P}$ with eigenvalue 0,

$$\widehat{P}[(1 - \widehat{P})|u\rangle] = (\widehat{P} - \widehat{P}^2)|u\rangle = 0. \tag{8.112}$$

Note that when an expansion is made using the closure relation, projections are being made. For example,

$$|\psi\rangle = \sum_i |a_i\rangle\langle a_i|\psi\rangle = \sum_i \widehat{P}_i|\psi\rangle = \sum_i |\psi_i\rangle = \sum_i \langle a_i|\psi\rangle |a_i\rangle, \tag{8.113}$$

where $|\psi_i\rangle = \widehat{P}_i|\psi\rangle$ is the projection of $|\psi\rangle$ in the $|a_i\rangle$ direction.

8.6.5 Moving between descriptions

We have seen that unitary transformations in QM keep the normalization of the state functions and the fundamental laws of the theory invariant. Given two orthonormal bases in the same Hilbert space $\mathcal{H}$ we will now find the unitary operator that takes us from one basis to the other. Consider, for example, an operator $\widehat{Q}$ whose eigenfunctions $\{u_\mu\}$ we use to write the generic operator $\widehat{A}$,

$$A_{\mu\nu} = \int u_\mu^* \widehat{A} u_\nu dx. \tag{8.114}$$

Similarly, consider another operator $\widehat{R}$, with eigenfunctions $\{\varphi_m\}$ and write now

$$A_{mn} = \int \varphi_m^* \widehat{A} \varphi_n dx. \tag{8.115}$$

Let $\widehat{U}$ be the operator that takes us from the Q-representation to the R-representation, that is,

$$\varphi = \widehat{U}u, \quad \text{or} \quad \varphi_m = \sum_\lambda U_{\lambda m} u_\lambda. \tag{8.116}$$

From this expression it follows that

$$U_{\mu m} = \int u_\mu^* \varphi_m dx. \tag{8.117}$$

Now we can write by direct substitution

$$A_{mn} = \int \varphi_m^* \widehat{A} \varphi_n dx = \sum_{\lambda\sigma} U_{\lambda m}^* U_{\sigma n} \int dx u_\lambda^* \widehat{A} u_\sigma$$
$$= \sum_{\lambda\sigma} U_{\lambda m}^* U_{\sigma n} A_{\lambda\sigma} = \sum_{\lambda\sigma} \left(\widehat{U}^\dagger\right)_{m\lambda} A_{\lambda\sigma} U_{\sigma n} = \left(\widehat{U}^\dagger \widehat{A} \widehat{U}\right)_{mn}, \tag{8.118}$$

or adding the indexes R and Q as reminders,

$$\hat{A}_R = \hat{U}^\dagger \hat{A}_Q \hat{U}. \tag{8.119}$$

The transition from the Schrödinger description to the Heisenberg description, or vice versa, is also done by a unitary transformation, since it is obtained by the inverse application of the evolution operator (8.77), which "devolves" the state vector. To see this, we apply $\widehat{S}^\dagger = \widehat{S}^{-1}$ from the left to (8.78), $|\psi(t)\rangle = \widehat{S}(t)|\psi_0\rangle$, to get

$$|\psi_H\rangle \equiv \widehat{S}^\dagger|\psi_S\rangle = \widehat{S}^\dagger|\psi(t)\rangle = S^\dagger S|\psi_0\rangle = |\psi_0\rangle. \tag{8.120}$$

$|\psi_S\rangle \equiv |\psi\rangle = |\psi(t)\rangle$ is the state vector in the Schrödinger description, while $|\psi_H\rangle$ represents the state vector in the Heisenberg description. According to (8.91) and (8.120), if $\widehat{F}_S$ represents an operator in the Schrödinger description, its expression in the Heisenberg description is

$$\widehat{F}_H(t) = \widehat{S}^\dagger \widehat{F}_S \widehat{S} = e^{\frac{i}{\hbar}\hat{H}t} \widehat{F}_S e^{-\frac{i}{\hbar}\hat{H}t}, \tag{8.121}$$

from which it follows that, in general, $\widehat{F}_H$ depends on time and evolves (as we have already seen) according to the equation of motion

$$i\hbar \frac{d\widehat{F}_H}{dt} = [\widehat{F}_H, \widehat{H}_H], \tag{8.122}$$

obtained by differentiating (8.121) with respect to time. Note that for the Hamiltonian, $\widehat{H}_H = \widehat{H}_S$ because $\hat{H}$ commutes with itself in any representation.

Exercise E8.6. A detour: "quantum" description of a classical problem
In the context of classical physics, describe the evolution of a system in a Schrödinger version (functions, such as the probability density, evolve as in classical statistical mechanics), or in a Heisenberg version (dynamical variables evolve).

Solution. Consider the generic variable A of a 1D one-particle problem, with $A = A(q, p)$ at $t = 0$; we limit ourselves to the case of analytic functions that can be expressed in the form

$$A(q, p) = \sum_{n,m} \alpha_{nm} q^n p^m. \tag{8.123}$$

At time t we can do the expansion

$$A(q(t), p(t)) = \sum_{n,m} \alpha_{nm} q^n(t) p^m(t), \tag{8.124}$$

or alternatively write it in terms of q and p,

$$A_t(q, p) = \sum_{n,m} \alpha_{nm}(t) q^n p^m. \tag{8.125}$$

In the first case (Heisenberg description) A is expressed as the original function of the new (evolving) variables, while in the second case (Schrödinger description) it involves a new (evolving) function ($\alpha(t) \neq \alpha$ in general) and the original variables. From the equations of motion it follows (now for an arbitrary number of dimensions), with $\hat{L}$ the Liouville operator, that

$$\frac{\partial A_t}{\partial t} = \sum_i \left(\frac{p_i}{m} \frac{\partial}{\partial q_i} + F_i \frac{\partial}{\partial p_i} \right) A_t(q, p) \equiv \hat{L} A_t(q, p). \tag{8.126}$$

Assuming that this equation is integrable, it can be written formally for time-independent forces in the form

$$A_t(q, p) = \left[e^{(t-t_0)\hat{L}} A \right](q, p) = A(q(t), p(t)), \tag{8.127}$$

a result that establishes the connection between the two descriptions.

8.6.6 The interaction description

To deal with physical problems of two interacting subsystems, it is common to decompose the Hamiltonian into a sum of two contributions, one containing the Hamiltonians of the independent parts and the other describing their interaction. We will show that in these cases it is possible to construct an intermediate description between Schrödinger's and Heisenberg's, in which the state vector evolves only due to the action of the interaction Hamiltonian. This is the so-called *interaction description* (or *representation*) introduced in 1934 by Stückelberg,[8] and used successfully for the study of complex problems; in particular it played a fundamental role for the development of quantum electrodynamics.

We start by writing the Hamiltonian of the system in the form

$$\widehat{H} = \widehat{H}_0 + \widehat{V}, \tag{8.128}$$

where $\widehat{H}_0$ is the Hamiltonian of free – that is, noninteracting – fields (or particles), and $\widehat{V}$ represents the Hamiltonian of the interaction between them. We introduce a unitary transformation $\widehat{S}(t)$ defined in terms of the free Hamiltonian $\widehat{H}_0$ only, and with it we move from the original Schrödinger description to the interaction description; that is, we write

$$|\psi_I\rangle = \widehat{S}(t)|\psi_S\rangle, \tag{8.129}$$

where $|\psi_S\rangle$ corresponds to the Schrödinger description for the free Hamiltonian,

$$\widehat{S}(t) = e^{i\widehat{H}_0 t/\hbar}. \tag{8.130}$$

Substituting in the Schrödinger equation

$$i\hbar \frac{\partial |\psi_S\rangle}{\partial t} = (\widehat{H}_0 + \widehat{V})|\psi_S\rangle \tag{8.131}$$

and simplifying, we get (as the student can verify)

$$i\hbar \frac{\partial |\psi_I\rangle}{\partial t} = \widehat{V}_I |\psi_I\rangle, \tag{8.132}$$

---

[8] Ernst Carl Gerlach Stückelberg (1905–1984) was a renowned Swiss mathematician and physicist.

where $\widehat{V}_I$ is the interaction Hamiltonian in the interaction description, that is,

$$\widehat{V}_I = \widehat{S}(t)\widehat{V}\widehat{S}^{-1}(t). \tag{8.133}$$

We see that in this description the state vector $|\psi_I\rangle$ evolves only due to the interaction between the subsystems; if it were zero, we would be making a usual Heisenberg description. On the other hand, if $\widehat{F}_S$ represents a generic operator in the Schrödinger description, then in the interaction description it is written as

$$\widehat{F}_I = \widehat{S}(t)\widehat{F}_S\widehat{S}^\dagger(t). \tag{8.134}$$

From here it follows that the evolution equation for $\widehat{F}_I$ is

$$i\hbar\frac{d\widehat{F}_I}{dt} = [\widehat{F}_I, \widehat{H}_0]. \tag{8.135}$$

This is a really important result; it shows that the time evolution of the operators in the interaction description is generated exclusively by the Hamiltonian of the free subsystems, even when they are in interaction. In other words, in the interaction description, the form of the fundamental dynamical equations is the *same* in the presence or absence of interaction. This, together with Eq. (8.132), makes this description extremely practical for the study of complicated systems, since an essential part of free systems theory, in particular, the set of evolution equations for *free fields*, remains applicable in the presence of interaction. Thus, having solved the problem for $\widehat{H}_0$, it remains only to solve Eq. (8.132) to determine $|\psi_I\rangle$.

8.6.6.1 *Perturbative method based on the interaction description

The perturbative method discussed here is used extensively in QED, including its relativistic version, and in many other problems. The task is to find the solution of the Schrödinger equation

$$i\hbar\frac{\partial\Psi}{\partial t} = \left(\widehat{H}_0 + V\right)\Psi, \tag{8.136}$$

where the interaction Hamiltonian V may depend on time. In the interaction description, the wave function satisfies Eq. (8.132), with

$$V_I(t) = e^{i\widehat{H}_0 t/\hbar} V e^{-i\widehat{H}_0 t/\hbar}, \tag{8.137}$$

and all other operators (except, of course, $\widehat{H}_0$) evolve according to the same rule determined by $\widehat{H}_0$. Indeed, Eq. (8.132) can be formally integrated to give

$$\Psi_I(t) = \widehat{U}(t, t_0)\Psi_I(t_0), \tag{8.138}$$

where the new evolution operator $\widehat{U}(t, t_0)$ evolves due to the perturbation (the old one evolved due to $\widehat{H}_0$),

$$i\hbar\frac{d\widehat{U}(t, t_0)}{dt} = V_I(t)\widehat{U}(t, t_0), \quad \widehat{U}(t_0, t_0) = \mathbb{I}, \tag{8.139}$$

so that

$$\widehat{U}(t, t_0) = \mathbb{I} - \frac{i}{\hbar}\int_{t_0}^{t} \widehat{V}_I'(t')\widehat{U}(t', t_0)dt'. \tag{8.140}$$

This expression is well suited for a perturbative analysis, if the perturbation is sufficiently "small" so that $\hat{U}(t,t_0)$ remains close enough to the initial value $\mathbb{I}$. In such a case we can expand by successive iterations,

$$\hat{U}(t,t_0) = \mathbb{I} - \frac{i}{\hbar}\int_{t_0}^{t} \hat{V}_I(t')\hat{U}(t',t_0)dt'$$

$$= \mathbb{I} - \frac{i}{\hbar}\int_{t_0}^{t} dt' \hat{V}_I(t')\left[\mathbb{I} - \frac{i}{\hbar}\int_{t_0}^{t'} dt'' \hat{V}_I(t'')\hat{U}(t'',t_0)\right]$$

$$= \mathbb{I} - \frac{i}{\hbar}\int_{t_0}^{t} dt' \hat{V}_I(t')\left\{\mathbb{I} - \frac{i}{\hbar}\int_{t_0}^{t'} dt'' \hat{V}_I(t'')\left[\mathbb{I} - \frac{i}{\hbar}\int_{t_0}^{t''} dt''' \hat{V}_I(t''')\hat{U}(t''',t_0)\right]\right\}$$

(8.141)

and so on, so that

$$\hat{U}(t,t_0) = \mathbb{I} + \frac{1}{i\hbar}\int_{t_0}^{t} dt' \hat{V}_I(t') + \left(\frac{1}{i\hbar}\right)^2 \int_{t_0}^{t} dt' \hat{V}_I(t') \int_{t_0}^{t'} dt'' \hat{V}_I(t'') + \cdots \quad (8.142)$$

The term of order n contains n integrals, n factors V and n integration times, so that it is possible to write

$$\hat{U}(t,t_0) = \sum_{n=0}^{\infty} \hat{U}^{(n)}(t,t_0), \quad \hat{U}^{(0)}(t,t_0) = \mathbb{I}, \quad (8.143)$$

$$\hat{U}^{(1)}(t,t_0) = \frac{1}{i\hbar}\int_{t_0}^{t} \hat{V}_I(t')dt', \quad (8.144)$$

$$\hat{U}^{(2)}(t,t_0) = \left(\frac{1}{i\hbar}\right)^2 \int_{t_0}^{t} \hat{V}_I(t')dt' \int_{t_0}^{t'} \hat{V}_I(t'')dt'', \quad (8.145)$$

and so on. The last expression can be conveniently rewritten by changing the order of integration with the help of the Heaviside $\Theta(t)$ function,

$$\int_{t_0}^{t} \hat{V}_I(t')dt' \int_{t_0}^{t'} \hat{V}_I(t'')dt'' = \int_{t_0}^{t} dt' \int_{t_0}^{t} dt'' \Theta(t'-t'')\hat{V}_I(t')\hat{V}_I(t'')$$

$$= \int_{t_0}^{t} dt'' \int_{t_0}^{t} dt' \Theta(t'-t'')\hat{V}_I(t')\hat{V}_I(t'')$$

$$= \int_{t_0}^{t} dt'' \int_{t''}^{t} dt' \hat{V}_I(t')\hat{V}_I(t'')$$

$$= \tfrac{1}{2}\int_{t_0}^{t} dt' \int_{t_0}^{t'} dt'' \hat{V}_I(t')\hat{V}_I(t'')$$

$$+ \tfrac{1}{2}\int_{t_0}^{t} dt' \int_{t'}^{t} dt'' \hat{V}_I(t'')\hat{V}_I(t')$$

$$= \hat{\mathsf{T}}\tfrac{1}{2}\int_{t_0}^{t} dt' \int_{t_0}^{t} dt'' \hat{V}_I(t')\hat{V}_I(t''). \quad (8.146)$$

The last but one equality was obtained by exchanging t' and t''. To write the final result, we introduced the *chronological operator* $\hat{\mathsf{T}}$, which orders the factors to the right in descending order of time; for example,

$$\hat{T}\hat{V}_I(t')\hat{V}_I(t'') = \begin{cases} \hat{V}_I(t')\hat{V}_I(t'') & \text{if } t' \geq t'', \\ \hat{V}_I(t'')\hat{V}_I(t') & \text{if } t'' \geq t'. \end{cases} \qquad (8.147)$$

When reordering the factors chronologically they should be taken as commutative. The multiple integral of order n contains n terms, but using $\hat{T}$ reduces them to a single one extending from t_0 to t, preceded by a factor $1/n!$, just as in the case of $n=2$ above. So we get

$$\hat{U}^{(n)}(t,t_0) = \frac{1}{n!}\left(\frac{-i}{\hbar}\right)^n \hat{T} \int_{t_0}^t \hat{V}_I(t_1)dt_1 \cdots \int_{t_0}^t \hat{V}_I(t_n)dt_n. \qquad (8.148)$$

Summing up these terms gives the final result from (8.142), expressed in synthetic form,

$$\hat{U}(t,t_0) = \hat{T}\exp\left(-\frac{i}{\hbar}\int_{t_0}^t \hat{V}_I(t')dt'\right). \qquad (8.149)$$

An analysis of this result helps to understand the origin of the chronological operator. Deriving with respect to time gives

$$\begin{aligned} i\hbar\frac{d}{dt}\hat{U}(t,t_0) &= i\hbar\frac{d}{dt}\hat{T}\exp\left(-\frac{i}{\hbar}\int_{t_0}^t \hat{V}_I(t')dt'\right) \\ &= \hat{T}\hat{V}_I(t)\exp\left(-\frac{i}{\hbar}\int_{t_0}^t \hat{V}_I(t')dt'\right) \\ &= \hat{V}_I(t)\hat{T}\exp\left(-\frac{i}{\hbar}\int_{t_0}^t \hat{V}_I(t')dt'\right) = \hat{V}_I(t)\hat{U}(t,t_0), \end{aligned} \qquad (8.150)$$

which coincides with Eq. (8.139). Without the chronological operator, the order of the factor $\hat{V}_I(t)$ that comes from the derivative of the exponent would remain imprecise. The $\hat{T}$ sends this factor to the extreme left because its argument has the maximum value t, forcing the correct expression. Taking this into account, the series of equalities in (8.150) is equivalent to a derivation of (8.149).

Exercise E8.7. First-order transition amplitude

Use the preceding method to obtain a formula for the transition amplitude between energy eigenstates, up to first order.

Solution. Consider the eigenstate ψ_n of $\hat{H}_0$; the transition amplitude to another eigenstate ψ_k is

$$\begin{aligned} A_{nk}(t) &\equiv \langle \psi_k | \Psi_{In}(t) \rangle = \langle \psi_k | U(t,t_0) | \psi_n \rangle \\ &= \langle \psi_k | I - \frac{i}{\hbar}\int_{t_0}^t \hat{V}_I(t')dt' + \left(-\frac{i}{\hbar}\right)^2 \int_{t_0}^t \hat{V}_I(t')dt' \int_{t_0}^{t'} \hat{V}_I(t'')dt'' + \cdots | \psi_n \rangle \\ &\quad + \frac{1}{2}\left(-\frac{i}{\hbar}\right)^2 \hat{T}\int_{t_0}^t dt' \int_{t_0}^t dt'' \langle \psi_k | \hat{V}_I(t')\hat{V}_I(t'') | \psi_n \rangle + \ldots \end{aligned} \qquad (8.151)$$

From (8.137) we can write for the first-order transition amplitude

$$A_{nk}^{(1)}(t) = \langle \psi_k | U^{(1)}(t,t_0) | \psi_n \rangle = -\frac{i}{\hbar}\int_{t_0}^t \langle \psi_k | \hat{V}_I | \psi_n \rangle e^{i\omega_{kn}t'}dt'. \qquad (8.152)$$

Thus, at $t \to \infty$ the transition probability is to first order in perturbation theory given by

$$w_{nk}^{(1)} = \left|A_{nk}^{(1)}\right|^2 = \left|\langle\psi_k| U^{(1)}(t, t_0) |\psi_n\rangle\right|^2 = \frac{1}{\hbar^2}\left|\int_{t_0}^{\infty} \langle\psi_k| \hat{V}_I |\psi_n\rangle e^{i\omega_{kn}t} dt\right|^2. \quad (8.153)$$

This result coincides with the formula obtained in Section 17.1 using time-dependent perturbation theory.

8.7 *Integrals of Motion and Symmetries: Noether's Theorem

We have learned from classical mechanics that it is possible to establish a relation between the integrals of motion of a dynamical system and the symmetries of its Hamiltonian (or equivalently, of its Lagrangian or its action function). Successfully developed by Emmy Noether in 1915[9] as a theorem of classical dynamics, this result was subsequently extended to QM. Noether's theorem has become one of the most powerful tools for the study of dynamical systems, both classical and quantum, as will become clear from the brief introduction to the subject in the following sections, or by exploring any advanced text on classical or quantum field theory.

There is a simple connection between the constants of motion and the infinitesimal transformations of the coordinate system that leave the Hamiltonian invariant. For example, the infinitesimal transformation of the generalized coordinate q_i into $q_i + \delta q_i$ transforms H into $H + \delta H$, where $\delta H = (\partial H/\partial q_i)\delta q_i$ (to first order in δq_i); if H is invariant under this infinitesimal transformation, then $\partial H/\partial q_i = 0$ and it follows from $\dot{p}_i = -\partial H/\partial q_i$ that p_i is a constant of the motion. This particular symmetry occurs when H does not depend on q_i (which is then called an ignorable or cyclic coordinate); it then turns out that *the momentum associated with an ignorable coordinate is conserved*. For example, the spatial homogeneity of the free-particle Hamiltonian $H = p_x^2/2m$ leads to the conservation of the linear momentum p_x.

8.7.1 Quantum Noether's theorem

Let $\widehat{G}$ be a Hermitian operator used as the generator to construct the unitary operator $\widehat{U}$

$$\widehat{U} = e^{i\alpha\widehat{G}}, \quad (8.154)$$

where α is a real parameter; taking $\alpha \to \delta\alpha$ infinitesimal for the moment, we can write to first order

$$\widehat{U} = 1 + i\delta\alpha\widehat{G}. \quad (8.155)$$

---

[9] Amalie Emmy Noether (1882–1935) was an outstanding German mathematician and logician who was denied an official position at any university because of her gender. Several scientists, including Albert Einstein, Norbert Wiener, David Hilbert, Pavel Alexandrov, Jean Dieudonné, and Hermann Weyl, consider her as the most important woman in the history of mathematics. In 1915 Hilbert invited her out of step to work in a corner at the University in Göttingen.

$\widehat{G}$ is the *generator of the infinitesimal transformation* $\widehat{U}$ and $\delta\alpha$ is the *parameter* of the transformation. If $\widehat{H}$ is the initial Hamiltonian, the transformed Hamiltonian $\widehat{H}_U$ is given by Eq. (8.91); expanding to first order gives

$$\widehat{H}_U = \widehat{U}\widehat{H}\widehat{U}^{-1} = (1 + i\delta\alpha\widehat{G})\widehat{H}(1 - i\delta\alpha\widehat{G}) = \widehat{H} + i\delta\alpha(\widehat{G}\widehat{H} - \widehat{H}\widehat{G}), \qquad (8.156)$$

namely

$$\widehat{H}_U = \widehat{H} + \delta\widehat{H} = \widehat{H} + i\delta\alpha[\widehat{G},\widehat{H}]. \qquad (8.157)$$

For this transformation to leave the Hamiltonian invariant $\widehat{H}_U = \widehat{H}$, it is required that $\delta H = i\delta\alpha[\widehat{G},\widehat{H}] = 0$ and since $\delta\alpha \neq 0$, it follows that

$$[\widehat{G},\widehat{H}] = 0. \qquad (8.158)$$

This means that $\widehat{G}$ is conserved. Therefore, *the operators representing constants of motion of the system generate unitary infinitesimal transformations that leave the Hamiltonian unchanged*. Or alternatively, if $\widehat{H}$ is invariant under a unitary infinitesimal transformation, then the generator of the transformation is an integral of motion. For example, if $\widehat{H}$ is homogeneous with respect to spatial coordinates, it will be invariant under an infinitesimal translation and momentum will be conserved; in this case, the generator of the transformation is the momentum operator. Similarly, the spatial isotropy of $\widehat{H}$ will guarantee its invariance against infinitesimal rotations generated by the angular momentum operator, which will be conserved, and so on.

The set of operators that commute with the Hamiltonian forms a group, which is called the *group of the Schrödinger equation*. A set of operators forms a group if it contains: (i) the identity operator; (ii) the product operators $\widehat{A}\widehat{B}$ and $\widehat{B}\widehat{A}$ for each pair of operators $\widehat{A}$ and $\widehat{B}$ of the set; (iii) the inverse operator $\widehat{A}^{-1}$ for each operator $\widehat{A}$ of the set, such that $\widehat{A}^{-1}\widehat{A} = \widehat{A}\widehat{A}^{-1} = \widehat{1}$. We can see that if $\widehat{A}$ and $\widehat{B}$ commute with $\widehat{H}$, then the operators $\widehat{A}\widehat{B}$, $\widehat{B}\widehat{A}$, $\widehat{A}^{-1}$, $\widehat{B}^{-1}$ also commute with $\widehat{H}$; furthermore, it is obvious that $\widehat{H}$ commutes with $\widehat{1}$. So the set of operators that commute with the Hamiltonian forms a group.

8.7.2 Finite unitary transformations

It is possible to apply the previous arguments to the case of finite unitary transformations. To see this, let us consider a Hermitian operator $\widehat{G}$ that commutes with the Hamiltonian, so that $\widehat{G}$ corresponds to a conserved quantity. The operator $\widehat{G}$ generates a continuous symmetry transformation $\widehat{U}_\alpha = e^{i\alpha\widehat{G}}$, that leaves $\widehat{H}$ invariant, since from (8.158) it follows that

$$\widehat{H}_\alpha = \widehat{U}_\alpha \widehat{H} \widehat{U^\dagger}_\alpha = \widehat{H}\widehat{U}_\alpha \widehat{U}_\alpha^\dagger = \widehat{H}. \qquad (8.159)$$

With this we verify that every operator that commutes with the Hamiltonian – and therefore corresponds to a conserved quantity – generates a continuous symmetry transformation of the Hamiltonian. Alternatively, we can say that unitary operators whose generators are conserved generate symmetry transformations of the Hamiltonian. Note that the reverse is not necessarily true: the generator of a finite symmetry transformation of the Hamiltonian does not necessarily commute with it.

So far we have only dealt with the changes that are induced in the Hamiltonian, but it is possible to study the general case. Let $\widehat{F}$ be a generic dynamical variable; a unitary infinitesimal transformation changes it to $\widehat{F}' = \widehat{F} + \delta\widehat{F}$ with $\widehat{F}'$ given by (8.91) and $\widehat{U}$ by (8.155), so we get

$$\widehat{F}' = \widehat{F} + \delta\widehat{F} = \widehat{U}\widehat{F}\widehat{U}^\dagger = (1 + i\delta\alpha\widehat{G})\widehat{F}(1 - i\delta\alpha\widehat{G})$$
$$= \widehat{F} + i\delta\alpha(\widehat{G}\widehat{F} - \widehat{F}\widehat{G}), \qquad (8.160)$$

and thus

$$\delta\widehat{F} = i\delta\alpha[\widehat{G}, \widehat{F}], \qquad (8.161)$$

where $\widehat{G}$ is the generator of the infinitesimal transformation and $\delta\alpha$ its parameter,

$$\widehat{U} = 1 + i\delta\alpha\widehat{G}. \qquad (8.162)$$

Equation (8.161) – often written in the form of a functional derivative $\delta\widehat{F}/\delta\alpha = i[\widehat{G}, \widehat{F}]$ – is important, since it establishes the general relationship between the variation $\delta\widehat{F}$ of the dynamical variable $\widehat{F}$ and the generator $\widehat{G}$ of the applied infinitesimal transformation. In particular, if we take $\widehat{G} = \widehat{H}$ and $\delta\alpha = \delta t/\hbar$ (as required by a dimensional analysis), we get the evolution equation

$$i\hbar\delta\widehat{F}/\delta t = i\hbar\dot{\widehat{F}} = [\widehat{F}, \widehat{H}], \qquad (8.163)$$

which again shows the Hamiltonian as the generator of time translations – and thus of the evolution of the system.

8.7.2.1 The rotation operator

We will now use these ideas applied in reverse, to construct the generator of infinitesimal rotations. The Hamiltonian of an isotropic system with central interactions is invariant under infinitesimal rotations $\widehat{T}_{\delta\varphi}$ about an axis in any direction $\widehat{\mathbf{n}}$ passing through the origin. If $\delta\varphi$ is the infinitesimal angle of the rotation, we write

$$\delta\boldsymbol{\varphi} = \widehat{\mathbf{n}}\delta\varphi. \qquad (8.164)$$

Applying this operator to $\psi(\mathbf{r})$ and using the formula $\mathbf{a} \cdot \mathbf{b} \times \mathbf{c} = \mathbf{a} \times \mathbf{b} \cdot \mathbf{c}$ we get

$$\widehat{T}_{\delta\varphi}\psi(\mathbf{r}) = \psi(\mathbf{r} - \delta\boldsymbol{\varphi} \times \mathbf{r}) = \psi(\mathbf{r}) - \delta\boldsymbol{\varphi} \times \mathbf{r} \cdot \nabla\psi = (1 - \delta\boldsymbol{\varphi} \cdot \mathbf{r} \times \nabla)\psi(\mathbf{r}). \qquad (8.165)$$

With $\widehat{\mathbf{L}} = \mathbf{r} \times \widehat{\mathbf{p}}$, $\widehat{T}_{\delta\varphi}$ takes the form

$$\widehat{T}_{\delta\varphi} = 1 - \frac{i}{\hbar}\delta\boldsymbol{\varphi} \cdot \widehat{\mathbf{L}}. \qquad (8.166)$$

Therefore, the *rotation operator* that performs a finite rotation of the coordinate system around the axis $\widehat{\mathbf{n}}$ by the angle φ is

$$\widehat{T}_{\widehat{\mathbf{n}}}(\varphi) = e^{-\frac{i}{\hbar}\varphi\widehat{\mathbf{n}}\cdot\widehat{\mathbf{L}}}. \qquad (8.167)$$

To complete the example, let us apply Eq. (8.161) (noting the change in the sign of $\delta\alpha$) to the operator $\widehat{G} = x\widehat{p}_y - y\widehat{p}_x = \widehat{L}_z$. We have

$$x' = x + i\delta\alpha[\widehat{x}, \widehat{L}_z] = x + i\delta\alpha\left[x(x\widehat{p}_y - y\widehat{p}_x) - (x\widehat{p}_y - y\widehat{p}_x)x\right]$$
$$= x + i\delta\alpha y[-x\widehat{p}_x + \widehat{p}_x x] = x + \hbar\delta\alpha y = x - \epsilon y, \qquad (8.168)$$

where we have put $\epsilon = -\hbar\delta\alpha$. Analogously we get

$$y' = y + \epsilon x, \quad z' = z. \tag{8.169}$$

Similarly,

$$\begin{aligned}
p'_x &= \widehat{p}_x + i\delta\alpha[\widehat{p}_x, \widehat{L}_z] = \widehat{p}_x + i\delta\alpha[\widehat{p}_x x - x\widehat{p}_x]p_y \\
&= \widehat{p}_x + \hbar\delta\alpha\widehat{p}_y = \widehat{p}_x - \epsilon\widehat{p}_y,
\end{aligned} \tag{8.170}$$

$$p'_y = \widehat{p}_y + \epsilon\widehat{p}_x, \quad p'_z = \widehat{p}_z. \tag{8.171}$$

These results show, as expected, that $\widehat{L}_z$ is the generator of rotations about the z-axis.

We take the opportunity to observe that the application of Eq. (8.167) to a spin $\hbar\boldsymbol{\sigma}/2$, with $\boldsymbol{\sigma}$ the vector constructed with Pauli matrices, leads to

$$\widehat{T}_{\widehat{\mathbf{n}}}(\varphi) = e^{-\frac{i}{2}\varphi\widehat{\mathbf{n}}\cdot\widehat{\boldsymbol{\sigma}}}. \tag{8.172}$$

This result has two properties, both peculiar and important. One of them is related to the properties of the Pauli 2×2 matrices, the other to the half rotation angle $\varphi/2$. Together they explain the strange rotation properties of spinors, in particular the fact that they are not invariant with respect to a full rotation by the angle 2π, which leads to a change of sign of the spinor, thus essentially distinguishing spinors from tensors, the latter being, by definition, insensitive to a 2π rotation.

Other important conclusions can be drawn from the preceding results. On the one hand, it is clear that the transformation law $x \to x' = x + (\hbar\delta\alpha)\,y$ due to an infinitesimal rotation around the z-axis, applies to *any* vector operator, which leads us to write more generally that

$$\hat{A}_x \to \hat{A}'_x = \hat{A}_x + (\hbar\delta\alpha)\,\hat{A}_y = \hat{A}_x + i\delta\alpha\left[\hat{A}_x, \hat{L}_z\right], \tag{8.173}$$

whence

$$\left[\hat{L}_z, \hat{A}_x\right] = i\hbar\hat{A}_y. \tag{8.174}$$

Noting that in this expression the indices are in cyclic order, we can write more generally

$$\left[\hat{L}_i, \hat{A}_j\right] = i\hbar\varepsilon_{ijk}\hat{A}_k, \tag{8.175}$$

where ε_{ijk} is the Levi–Civita totally antisymmetric 3D tensor. This law expresses the commutators of *any* vector $\hat{A}$ and the angular momentum operator. In fact, it is often used as the *definition* of a vector operator, namely as the three-component operator that satisfies these commutation rules. In particular, angular momentum is a vector operator, so applying this rule to it with $\hat{A}_i = \hat{L}_i$ gives the commutation rules of the components of the angular momentum operator, namely

$$\left[\hat{L}_i, \hat{L}_j\right] = i\hbar\varepsilon_{ijk}\hat{L}_k. \tag{8.176}$$

In Chapter 11 we will have the opportunity to review and make use of these results.

Note that in the cases analyzed the generator of the transformation is the conjugate canonical variable of the transformed variable. In fact, this relation can be used in general to define the pairs of conjugate canonical variables. It is not difficult to relate this result to the structure of the commutators of these variables, but the development of this observation is left to the student.

8.7.3 Discrete symmetries: Parity

Discrete symmetries are symmetries that describe noncontinuous changes in a system. In geometry, for example, a square has discrete rotational symmetry, since only rotations by multiples of $\pi/2$ preserve the original shape of the square. Here we will discuss one of the most important symmetries in physics, the parity symmetry.

For this purpose we introduce the parity operator $\widehat{P}$ in coordinate space,[10] which produces a mirror reflection of the coordinates x_i or, equivalently, a reflection of the *polar vectors* (such as r and p; see what follows) about the origin, leaving the *axial vectors* (such as L) unchanged. By definition, $\widehat{P}$ satisfies the following relations

$$r_P \equiv \widehat{P} r \widehat{P} = -r, \qquad (8.177)$$

$$p_P \equiv \widehat{P} p \widehat{P} = -p. \qquad (8.178)$$

When $\widehat{P}$ acts on a function of the coordinates, one gets

$$\widehat{P}\psi(r) = a\psi(-r), \qquad (8.179)$$

or $\widehat{P}\psi(r) = \langle r| \widehat{P} |\psi\rangle = \langle -r| \psi\rangle$. Applying the operator $\widehat{P}$ to (8.179) we get

$$\widehat{P}^2 \psi(r) = \widehat{P} a \psi(-r) = a^2 \psi(r) = \psi(r), \qquad (8.180)$$

thus $\widehat{P}^2 = 1$; eigenfunctions of $\widehat{P}$ with eigenvalue $+1$, i.e., $\psi(-r) = \psi(r)$, have even parity, and those with eigenvalue -1, that is, $\psi(-r) = -\psi(r)$, have odd parity. Note that $\widehat{P}$ is a unitary operator, since it is Hermitian with square equal to 1.

When $\widehat{H}(-r) = \widehat{H}(r)$, that is, when the system is invariant with respect to a reflection about the origin, the operator $\widehat{P}$ commutes with the Hamiltonian. This happens in many important cases, including the free particle, 1D square potentials, central potentials, and so on. In these cases, the parity is an integral of motion, so the eigenvalue of $\widehat{P}$ is preserved during the evolution of the system.

Vectors that transform like r or p (i.e., that change the sign of their components against mirror reflection) are called *polar vectors*. Note that since the angular-momentum operator is the product of two (polar) vectors, $\widehat{L} = r \times p$, it is invariant under a mirror reflection of the coordinates. Vectors such as angular momentum, spin, and the magnetic field $B = \nabla \times A$, are examples of *axial vectors* or *pseudo-vectors*; their components do not change sign with the specular reflection generated by $\widehat{P}$.

Just as there are two types of vectors with respect to the parity operator, there are the corresponding *scalars* and *pseudo scalars*. A product $A \cdot B$ with both factors A and B polar or axial vectors is a scalar; when one factor is a polar vector and the other an axial vector, the result is a pseudo-scalar; for example, $r \cdot L$ is pseudo scalar.

Particles may also possess an *intrinsic* parity, expressed in the form of a phase factor that arises as an eigenvalue of the parity operation. In this case too, the operator P is unitary, since $PP^* = P^*P = 1$. Thus

$$\widehat{P}\psi = e^{i\phi}\psi, \qquad (8.181)$$

where the phase factor must be determined experimentally (or, where appropriate, by the corresponding theory). For example, the pions ($\pi^0, \pi^\pm$) have zero spin and

---

[10] Not to be confused with the projection operators $\widehat{P}$ introduced previously.

parity -1, which indicates that they are *pseudo-scalar* particles. As will be seen in Chapter 15, a rotation of 2π does not change the wave function of bosons (integer-spin particles), so it is appropriate to set $a^2 = 1$ for them, giving $a = \pm 1$. In contrast, the wave function of fermions (half-integer-spin particles) is only invariant against rotations of 4π, so $a^4 = 1$ and $\pm 1, \pm i$.

8.7.3.1 Time inversion

The operation usually called *time inversion* – which should more properly be called motion inversion – is defined by the relations ($\hat{\boldsymbol{J}}$ stands for the total angular-momentum operator)

$$\hat{T}\hat{\boldsymbol{r}}\hat{T}^{-1} = \hat{\boldsymbol{r}}, \quad \hat{T}\hat{\boldsymbol{p}}\hat{T}^{-1} = -\hat{\boldsymbol{p}}, \quad \hat{T}\hat{\boldsymbol{J}}\hat{T}^{-1} = -\hat{\boldsymbol{J}}. \tag{8.182}$$

In order for the fundamental commutation rules not to change with time inversion, it is required that $\hat{T}$ be an *antilinear* and *antiunitary* operator.

An operator $\hat{F}$ is antilinear if

$$\hat{F}(a|\varphi\rangle + b|\psi\rangle) = a^*\hat{F}|\varphi\rangle + b^*\hat{F}|\psi\rangle. \tag{8.183}$$

The conjugation operator $\widehat{K}$ is such that $\widehat{K}\varphi = \varphi^*$ for any φ, so it is easy to conclude that $\widehat{K}$ is an antilinear operator. The product of $\widehat{K}$ with a unitary operator is antiunitary; hence, in particular, $\widehat{K}$ is antiunitary. These properties make it difficult to define the action of this operator on the bras and kets and complicate the theory; for that reason we do not insist on this topic here. The point is discussed in more advanced textbooks on QM.

Problems

P8.1 Show that the commutator of two variables that are conserved is also a conserved quantity.

P8.2 Prove the validity of the following expression of the virial theorem for a system of several particles, where α denotes the particles and i, j are their Cartesian coordinates,

$$\sum_\alpha \left\langle -\frac{1}{m^\alpha} \hat{p}_i^\alpha \hat{p}_j^\alpha + x_i^\alpha \frac{\partial V}{\partial x_j^\alpha} \right\rangle = 0. \tag{8.184}$$

P8.3 Show that an attractive potential of the type $V \sim r^n$ only produces stable orbits if $n > -2$.

P8.4 Show that for an electron subject to the one-dimensional potential $V = -Cx$ with $C > 0$, the momentum dispersion Δp is constant.

P8.5 Show that

$$\frac{d\langle(\Delta x)^2\rangle}{dt} = \frac{1}{m}\langle \hat{x}\hat{p} + \hat{p}\hat{x}\rangle - \frac{2}{m}\bar{x}\bar{p}. \tag{8.185}$$

P8.6 In Section 8.4 we proved that if two dynamical variables F, G have simultaneous well-defined values, their respective operators must commute.

Show that this condition is also sufficient, that is, if $\widehat{F}$ and $\widehat{G}$ commute, they have simultaneous well-defined values.

P8.7 Study the motion of a packet of free particles in the Heisenberg description. *Hint:* calculate the dispersions of x and p as functions of time.

P8.8 Show that the transformation $\hat{U}$ given by Eqs. (8.114)–(8.119) is unitary.

P8.9 Use the projection operators to construct an expression for the function $f(\hat{A})$, where $\hat{A}$ is an operator with eigenvectors $|n\rangle$ and eigenvalues a_n.

P8.10 Consider the two-dimensional problem of a particle of mass m moving on the xy-plane under the action of a radial force $f(r)$. Analyze the symmetries of the Hamiltonian and find the associated constants of motion.

Bibliographical Notes

Three highly recommended, very different textbooks are Feynman, Leighton, and Sands (1971), Bohm (1989), and Cohen-Tannoudji et al. (1977).

9 The One-Dimensional Harmonic Oscillator

We are now ready to study in detail one of the most theoretically interesting and useful physical systems: the harmonic oscillator (HO). The HO occurs in the quantum case in a similar way to the classical case, whenever a potential can be approximated by a parabolic curve in the region of a minimum. It is therefore generally applicable to low-energy oscillations around a potential minimum, such as regularly occur in atoms or molecules, for example. Moreover, the frequent use of Fourier analysis to study complex systems in terms of elementary oscillators makes the HO an important problem in theoretical physics, in addition to its methodological interest due to the fact that it can be solved exactly.

The problem we will tackle first, solving the time-dependent Schrödinger equation, is more complicated than determining the oscillator's eigenstates and energy spectrum. However, it will give us a valuable opportunity to appreciate the similarities and differences between the dynamics of the classical and quantum oscillators. As a second step we will study the stationary solutions of the HO, which is one of the most important orthogonal sets of quantum eigenfunctions because of its many applications, properties, and relative simplicity. Further, the use of raising and lowering (or creation and annihilation) operators is shown to be a powerful way of dealing with both material and field oscillators in general. Finally, the problem of two coupled oscillators is used to generalize the Schrödinger equation to the case of more than one particle.

To simplify the treatment, we will concentrate throughout the chapter on the 1D case, which has no degeneracy; the HO in more than one dimension requires angular-momentum theory for a complete treatment, and this naturally leads to degeneracy.

9.1 A Maximally Coherent Oscillating Packet with Minimal Dispersion

We start with the study of a packet of HO, which we choose to be as mathematically simple as possible, since our interest is focused on the physics of the system. Thus, without sacrificing the physics, we focus on a Gaussian packet with both maximum coherence and minimum dispersion, a case of great interest in quantum optics today.

The potential of a 1D quantum HO is given by the usual expression

$$V(x) = \tfrac{1}{2}m\omega^2 x^2, \tag{9.1}$$

where ω is the frequency of oscillation; its Schrödinger equation is

$$i\hbar\frac{\partial\psi}{\partial t} = -\frac{\hbar^2}{2m}\frac{\partial^2\psi}{\partial x^2} + \tfrac{1}{2}m\omega^2 x^2\psi. \tag{9.2}$$

Let us consider a wave packet consisting of oscillators with a *Gaussian distribution* (also known as *normal distribution*) centered at x_0 and allowed to evolve freely from $t = 0$. Due to the linearity of the Schrödinger equation, the packet will retain its Gaussian structure as it evolves, so the time-dependent solution can be written in the general form[1]

$$\psi = A e^{-\alpha(x-\gamma)^2 + f(t) - f(0)}, \quad \alpha(t) > 0, \tag{9.3}$$

where $\alpha = \alpha(t)$ and $\gamma = \gamma(t)$ have initial values

$$\alpha(0) = a, \quad \gamma(0) = x_0, \tag{9.4}$$

so that (9.3) reduces for $t = 0$ to

$$\psi_0(x) \equiv \psi(x,0) = A e^{-a(x-x_0)^2}, \tag{9.5}$$

which corresponds to an initial normal distribution with dispersion $\sigma^2 = 1/4a^2$; $f(t)$ is a function of time that modulates the Gaussian profile and A is the normalization constant.

We now introduce the proposed (anticipated) solution (9.3) in the Schrödinger equation (9.2); by eliminating the common factor ψ, we obtain

$$-i\hbar\dot{\alpha}(x-\gamma)^2 + 2i\hbar\alpha\dot{\gamma}(x-\gamma) + i\hbar\dot{f} = \frac{\hbar^2\alpha}{m} - \frac{2\hbar^2\alpha^2}{m}(x^2 - 2\gamma x + \gamma^2) + \tfrac{1}{2}m\omega^2 x^2. \tag{9.6}$$

This expression contains terms in x^0, x^1 and x^2, which are linearly independent; the respective coefficients must therefore cancel between the right and left sides. The coefficient of x^2 vanishes if

$$-i\hbar\dot{\alpha} + \frac{2\hbar^2}{m}\alpha^2 = \tfrac{1}{2}m\omega^2. \tag{9.7}$$

This equation can be easily solved by writing $\alpha = \lambda \dot{u}/u$ and choosing λ so that the quadratic terms cancel out. The solution,

$$\alpha(t) = -i(m\omega/2\hbar)\cot(\omega t + \beta), \tag{9.8}$$

is a relatively complicated function, since for condition (9.4) to hold with a real, β must be imaginary. The hyperbolic functions that appear due to this expression describe a transient irreversible evolution of the packet, which is added to the periodic motions due to the harmonic terms.

---

[1] We make use of what is called an *ansatz* (a starting assumption). This is a common technique in physics, mathematics and engineering to solve a problem by anticipating the form of the solution, which should then be verified and refined by introducing it into the equation to be solved, taking into account boundary conditions and other details of the problem.

[2] A normal (Gaussian) distribution has the form $Ne^{-(x-\bar{x})^2/2\sigma^2}$, where $\bar{x} = \langle x \rangle$ is the average of x, $\sigma^2 = \langle x^2 \rangle - \langle x \rangle^2$ is the variance of x, and $N = 1/\sqrt{2\pi}\sigma$.

Note, however, that there is a simpler particular solution obtained by setting $\dot{\alpha} = 0$ in (9.7), which gives then $\alpha = m\omega/2\hbar$ (the negative solution is inadmissible). In this case the dispersion of x remains constant and the packet as a whole oscillates *without changing its shape*, that is, coherently, as if it were rigid. For this solution to be realized, the initial dispersion given by (9.4) must coincide with the value calculated from (9.7), $\alpha(0) = \alpha = a$. We will limit ourselves here to this simple and important case, assuming that the initial Gaussian packet has been prepared with just the required dispersion of x,

$$a = \alpha = \frac{m\omega}{2\hbar}. \tag{9.9}$$

Further, the condition that the coefficient of the terms in x vanishes gives a relation between $\dot{\gamma}$ and γ; integrating this differential equation and substituting the value of a given by (9.9) gives

$$\gamma = x_0 e^{-i\omega t}. \tag{9.10}$$

Integrating what remains of Eq. (9.6) after substituting the values of a and γ, we obtain

$$f = -\frac{i}{2}\omega t + \frac{m\omega}{4\hbar} x_0^2 e^{-2i\omega t}. \tag{9.11}$$

Substituting the preceding results into Eq. (9.3) and normalizing in the usual way gives

$$\psi(x,t) = \left(\frac{m\omega}{\pi\hbar}\right)^{\frac{1}{4}} \exp\left[-\frac{m\omega}{2\hbar}(x - x_0 \cos \omega t)^2 \right.$$
$$\left. -i\left(\tfrac{1}{2}\omega t + \frac{m\omega}{\hbar} x_0 x \sin \omega t - \frac{m\omega}{4\hbar} x_0^2 \sin 2\omega t\right)\right]. \tag{9.12}$$

From this wave function we can extract all the available information about the behavior of the packet. In particular, it follows that the particle spatial density is

$$\rho(x,t) = \left(\frac{m\omega}{\pi\hbar}\right)^{\frac{1}{2}} \exp\left(-\frac{m\omega}{\hbar}(x - x_0 \cos \omega t)^2\right), \tag{9.13}$$

which is in fact still Gaussian for all t, with the center oscillating around the origin with frequency ω and amplitude x_0,

$$\langle x \rangle = x_0 \cos \omega t, \tag{9.14}$$

just like a classical oscillator.

The packet (9.12) is a coherent superposition of eigenstates of the Hamiltonian, a particular realization of the usual superposition $\psi(x,t) = \sum_n c_n e^{-iE_n t/\hbar} \varphi_n(x)$ applied to the case at hand. Once we have the eigenstates $\varphi_n(x)$ of the oscillator's Hamiltonian, we can use them to construct this or any other packet of harmonic oscillators of interest.

Exercise E9.1. Minimum dispersion packet of harmonic oscillators

Show that the preceding solution, Eq. (9.12), describes a packet of minimum dispersion.

Solution. From Eq. (9.13) we obtain for the variance of x,

$$\langle(\Delta x)^2\rangle = \frac{1}{4a} = \frac{\hbar}{2m\omega}. \tag{9.15}$$

The calculation of $\langle(\Delta p)^2\rangle$ gives (as the student should check)

$$\langle(\Delta p)^2\rangle = a\hbar^2 = \tfrac{1}{2}m\hbar\omega. \tag{9.16}$$

Multiplying the two previous results we verify that the packet has the minimum possible dispersion,

$$\langle(\Delta x)^2\rangle\langle(\Delta p)^2\rangle = \tfrac{1}{4}\hbar^2. \tag{9.17}$$

In Section 10.2.2 we will obtain the expression for the wave packet, Eq. (8.68), by requiring it to have minimum dispersion, here it was obtained indirectly by requiring the oscillations to be coherent. It thus corresponds to a very special packet, that of minimum dispersion (maximum organization) and coherent oscillations.

Note that the mean energy associated with the packet (9.12),

$$\langle\widehat{E}\rangle = \langle i\hbar\frac{\partial}{\partial t}\rangle = \tfrac{1}{2}\hbar\omega + \tfrac{1}{2}m\omega^2 x_0^2, \tag{9.18}$$

consists of two terms: the second term is the classical oscillation energy $\tfrac{1}{2}m\omega^2 x_0^2$, which may acquire any positive value and goes to zero with the oscillation amplitude x_0. The first term $\tfrac{1}{2}\hbar\omega$ represents the minimum energy of the system. This is the *zero-point energy* (ZPE) of the HO that remains at absolute zero temperature, when the packet is "frozen" (i.e., $x_0 = 0$). The origin and meaning of this term can be understood as follows. Even when the temperature of the system is brought down toward absolute zero, the random motion of the oscillators is maintained by the action of the ZPF. As we will see in the next chapter, the energy associated with these fluctuations is balanced by the ZPE of the field at the oscillation frequency, which is exactly $\tfrac{1}{2}\hbar\omega$.

This inevitable presence of fluctuations (an example of quantum fluctuations) means that we cannot simultaneously build a stationary packet of oscillators with arbitrarily small widths in x and p, since this would imply an arbitrary reduction of the initial energy, while the equilibrium energy is bounded from below at the zero point. In other words, the Heisenberg equality (9.17) is a direct consequence of the fact that the energy of the zero point is nonzero. We can therefore regard Heisenberg's inequalities as a proof of the existence of fluctuations in any quantum system at any temperature, including $T = 0$, precisely the quantum fluctuations that Heisenberg perceived. This shows that notions such as absolute standstill are incompatible with current physical knowledge. In other words, QM is a proof of the existence of irreducible background fluctuations.

To conclude this section, we should add that the ZPE produces observable effects. For example, Brindley and Wood (1929) reported X-ray scattering results from aluminum crystals that could only be explained by the existence of random motions associated with the ZPE. A second example is liquid helium, where the ZPE is so large that it prevents the liquid from solidifying under its own vapor pressure; another effect is the specific heat of the hydrogen molecule, which cannot be understood without the concept of ZPE.

9.2 Eigenfunctions and Eigenvalues of the Hamiltonian

We now turn to the study of the eigenstates of the HO. In Chapter 3 we obtained many of its fundamental properties using the Heisenberg matrix description, which happens to be very well suited for such purposes; here we will study it from the perspective offered by the Schrödinger description, based on the wave function. The stationary 1D Schrödinger equation for the HO is

$$-\frac{\hbar^2}{2m}\frac{d^2\psi}{dx^2} + \tfrac{1}{2}m\omega^2 x^2 \psi = E\psi. \tag{9.19}$$

Before starting the calculations, it is highly recommended to rewrite the equation in terms of dimensionless variables, to give it a canonical form; this will be seen to have several important advantages. Since the equation is homogeneous there is no need to introduce a dimensionless wave function (since it cancels out), so we start by writing

$$x = \alpha_0 \xi, \tag{9.20}$$

with α_0 a constant with dimension of length to be determined, and ξ dimensionless. Substituting in Eq. (9.19) and multiplying by $2m\alpha_0^2/\hbar^2$, we get

$$-\frac{d^2\psi}{d\xi^2} + \frac{m^2\alpha_0^4\omega^2}{\hbar^2}\xi^2\psi = \frac{2m}{\hbar^2}\alpha_0^2 E\psi. \tag{9.21}$$

To simplify this expression we choose α_0 so that the coefficient of $\xi^2\psi$ is unity (this is an arbitrary choice and can be changed if convenient),

$$\alpha_0^2 = \frac{\hbar}{m\omega}. \tag{9.22}$$

As the dimensionless energy variable we take

$$\epsilon = \frac{2m}{\hbar^2}\alpha_0^2 E = \frac{2E}{\hbar\omega}. \tag{9.23}$$

With this choice of parameters, the eigenvalue equation (9.19) becomes (a prime represents a derivation with respect to ξ)

$$-\psi'' + \xi^2\psi = \epsilon\psi. \tag{9.24}$$

Note that the procedure of eliminating physical dimensions not only greatly simplifies the differential equation and makes it more transparent and elegant, but something else is gained, which can often be important. You will never find Eq. (9.19) in a textbook on differential equations, but you will find its formal version (9.24). In addition, Eq. (9.22) tells us that the quantum oscillator has a *natural length scale* with the unit $\sqrt{\hbar/m\omega}$, while Eq. (9.23) shows that the oscillator has the quantity $\hbar\omega/2$ as a natural unit of energy. The appearance of such natural scales gives an immediate idea of the order of magnitude of the dynamical properties of the system. Another advantage of the transformation is that it makes it easy to see how many independent parameters the differential equation has. For example, Eq. (9.19) seems at first glance to suggest the presence of two independent parameters; however, Eq. (9.24) corrects this and shows that there is actually only one, namely ϵ. All these advantages (and possibly others) make it usually advisable to switch to the dimensionless description before starting to solve a new complicated differential problem.

We are interested in solving the eigenvalue problem associated with Eq. (9.24), rather than finding its general (mathematical) solution, which is a different problem. It is therefore convenient to devise a strategy that guarantees in advance that the solutions satisfy the *physical* conditions of the problem, in particular the boundary conditions. The strategy, conventional since its use by Schrödinger himself, consists in studying the behavior of the solutions at the boundaries, selecting those that are suitable for the problem and eliminating the rest. The application of this method to the present case will help to clarify the procedure to be followed in general.

To solve Eq. (9.24) we first study the behavior of the solution at infinity, ψ_∞. When $\xi \to \infty$, we can approximate this equation as $\psi''_\infty - \xi^2 \psi_\infty = 0$. We have omitted the energy term, which becomes negligible with respect to the large terms in the limit. The solution of this equation has the form $\psi_\infty = e^{a\xi^2}$. Since

$$\psi''_\infty = 2a\psi_\infty + 4a^2\xi^2\psi_\infty \approx 4a^2\xi^2\psi_\infty \quad (\text{for } \xi \to \infty), \tag{9.25}$$

with $4a^2 = 1$ the result coincides with the asymptotic equation; therefore $a = \pm\frac{1}{2}$, and the general asymptotic solution reads

$$\psi_\infty = A e^{\frac{1}{2}\xi^2} + B e^{-\frac{1}{2}\xi^2}. \tag{9.26}$$

The condition $\psi \to 0$ as $|\xi| \to \infty$ requires taking $A = 0$; the physically acceptable solutions of (9.24) can therefore be written in the form

$$\psi = e^{-\frac{1}{2}\xi^2} u(\xi), \tag{9.27}$$

where $u(\xi)$ can be at most a polynomial in ξ, since the asymptotic behavior of ψ is already contained in the exponential factor. Indeed, it is easy to see that if u were an infinite power series, it would diverge for $|\xi| \to \infty$ and the asymptotic behavior would pass from $e^{-\frac{1}{2}\xi^2}$ to $e^{\frac{1}{2}\xi^2}$ and ψ would then cease to be square-integrable and finite at all points. Substituting thus (9.27) into (9.24) and simplifying, we obtain the differential equation for the function u,

$$u'' - 2\xi u' + (\epsilon - 1)u = 0. \tag{9.28}$$

This equation and its polynomial solutions are reviewed in Appendix A, Section A.1. It is found that polynomial (and therefore regular) solutions exist if and only if the parameter $\epsilon - 1$ is an even integer,

$$\epsilon - 1 = 2n, \quad n = 0, 1, 2, \ldots \tag{9.29}$$

In this case the equation reduces to the Hermite equation

$$H''_n - 2\xi H'_n + 2n H_n = 0, \tag{9.30}$$

whose solutions are the Hermite polynomials H_n,

$$u = H_n(\xi). \tag{9.31}$$

Therefore, the energy eigenfunctions are

$$\psi_n = C_n e^{-\frac{1}{2}\xi^2} H_n(\xi), \tag{9.32}$$

where H_n is the Hermite polynomial of order n and C_n is the normalization constant, to be determined later. The energy eigenvalues are obtained by solving for ϵ in Eq. (9.29) and using (9.23),

$$E = \hbar\omega \left(n + \tfrac{1}{2}\right). \tag{9.33}$$

This result confirms that the energy of the ground state ($n = 0$) corresponds to the zero point, i.e., the athermal fluctuations. Starting from this minimum energy, the energy eigenvalues increase in uniform steps of value $\hbar\omega$.

The calculation of the normalization constant is greatly simplified by making effective use of the properties of Hermite polynomials, in particular, of the Rodrigues formula (A.6)

$$H_n(\xi) = (-1)^n e^{\xi^2} \frac{d^n e^{-\xi^2}}{d\xi^n}. \tag{9.34}$$

The normalization condition is written, using Eq. (9.32), as

$$\int_{-\infty}^{\infty} \psi_n^* \psi_n dx = 1 = \alpha_0 \int_{-\infty}^{\infty} \psi_n^*(\xi)\psi_n(\xi)d\xi = \alpha_0 |C_n|^2 \int_{-\infty}^{\infty} e^{-\xi^2} H_n^2(\xi) d\xi. \tag{9.35}$$

We introduce (9.34) in this expression to eliminate a factor H_n,

$$\alpha_0 |C_n|^2 (-1)^n \int_{-\infty}^{\infty} H_n(\xi) \frac{d^n e^{-\xi^2}}{d\xi^n} d\xi = 1. \tag{9.36}$$

Integrating by parts n successive times, noting that all integrated terms are zero at the limits $|\xi| = \infty$, we obtain

$$\alpha_0 |C_n|^2 \int_{-\infty}^{\infty} e^{-\xi^2} \frac{d^n H_n(\xi)}{d\xi^n} d\xi = 1. \tag{9.37}$$

Carrying out now the derivatives indicated in (9.34), one obtains the following explicit expression for the Hermite polynomials, Eq. (A.4),

$$H_n(\xi) = (2\xi)^n - \frac{n(n-1)}{1!}(2\xi)^{n-2} + \frac{n(n-1)(n-2)(n-3)}{2!}(2\xi)^{n-4} - \ldots, \tag{9.38}$$

from which it immediately follows that

$$\frac{d^n H_n(\xi)}{d\xi^n} = 2^n n!. \tag{9.39}$$

Substituting in the normalization condition and calculating the remaining integral, which is immediate, we obtain

$$\alpha_0 |C_n|^2 2^n n! \int_{-\infty}^{\infty} e^{-\xi^2} d\xi = \alpha_0 |C_n|^2 2^n n! \sqrt{\pi} = 1; \tag{9.40}$$

whence, taking C_n as real and positive,

$$C_n = \left(\sqrt{\pi} \alpha_0 2^n n!\right)^{-\frac{1}{2}}. \tag{9.41}$$

This completely determines the stationary wave functions of the 1D HO.

The wave functions (9.32) decay exponentially with distance from the origin, so the particles are predominantly in the region of small values of ξ, as shown in Fig. 9.1. However, there is a finite probability that the particles will visit the classically forbidden

9.2 Eigenfunctions and Eigenvalues of the Hamiltonian

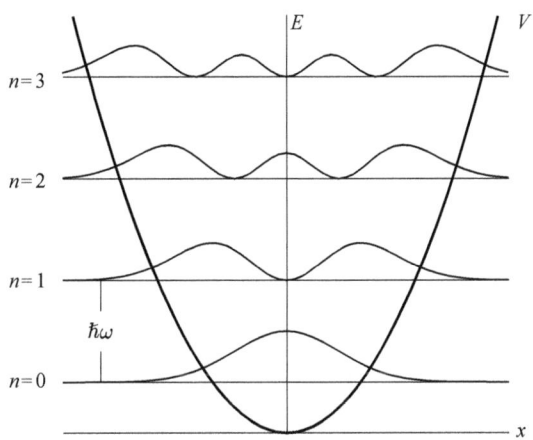

Figure 9.1 Energy levels and probability distribution for the lowest energy states of the harmonic oscillator.

region where the potential exceeds the value of the energy, as can be seen in the same figure. This is a particular example of a general property of quantum systems: the probability of visiting classically forbidden regions may turn out to be very small, but it does not generally vanish. If one adheres strictly to the interpretation of the wave function as a probability amplitude, this is another sign that in no limit does quantum behavior reduce strictly to the classical behavior, as a general rule. For an eigenstate of the Hamiltonian – for example, the ground state – with perfectly fixed energy, in these nonclassical regions the momentum of the particle takes on imaginary values, as happens inside a potential barrier or in any example where tunneling occurs.

The ground state corresponds to a Gaussian distribution centered at the origin and with variance $\langle x^2 \rangle = \frac{1}{2}a_0^2 = \hbar/2m\omega$, cf. Eq. (9.15). Since H_n is a polynomial of order n, the wave function (9.32) has n zeros. This result illustrates in a particularly simple way the general rule that the number of nodes increases with n.

Exercise E9.2. The HO in the momentum representation
Solve the problem of the 1D HO in the momentum representation.

Solution. In the momentum representation we have $x^2 = -\hbar^2 \partial^2/\partial p^2$ and the Schrödinger equation of the oscillator is therefore

$$\frac{p^2}{2m}\varphi(p) - \frac{1}{2}m\omega^2 \hbar^2 \frac{\partial^2 \varphi(p)}{\partial p^2} = E\varphi(p). \tag{9.42}$$

In terms of the dimensionless variable $\xi = p/p_0$, with $p_0 = \sqrt{m\hbar\omega}$, this equation becomes

$$\varphi'' + (\epsilon - \xi^2)\varphi = 0, \tag{9.43}$$

where ϵ is given by Eq. (9.23). Equation (9.43) coincides with (9.24), so the solutions to this eigenvalue problem follow directly from those already obtained with a simple change of notation,

$$\varphi_n(p) = \frac{1}{\sqrt{2^n n! \sqrt{\pi} p_o}} e^{-\frac{1}{2}(p/p_0)^2} H_n(p/p_0), \tag{9.44}$$

$$E_n = \hbar\omega \left(n + \tfrac{1}{2}\right). \tag{9.45}$$

These results show that the statistical behavior of the HO in the momentum space is analogous to that in the configuration space.

Exercise E9.3. Landau[3] levels

Consider an electron moving in the xy-plane, subject to a uniform magnetic field B along the z-axis. Find the possible values of the electron's energy and the corresponding wave functions.

Solution. The Hamiltonian of the particle in a magnetic field is

$$H = \frac{1}{2}m\mathbf{v}^2 = \frac{1}{2m}(\mathbf{p} - e\mathbf{A})^2, \tag{9.46}$$

where $\mathbf{B} = \nabla \times \mathbf{A}$. Since $\mathbf{B} = B\hat{\mathbf{k}}$, we may choose $\mathbf{A} = Bx\hat{\mathbf{j}}$, so that the Hamiltonian reduces to

$$H = \frac{p_x^2}{2m} + \frac{1}{2m}(p_y - eB)^2, \quad B = B_z. \tag{9.47}$$

Note that since the operator p_y commutes with H, it can be replaced by its eigenvalue $p_y = \hbar k_y$. We therefore rewrite H in the form

$$H = \frac{p_x^2}{2m} + \frac{1}{2}m\omega^2 (x - \frac{\hbar k_y}{m\omega})^2, \tag{9.48}$$

where $\omega = |eB/m|$ is the Larmor frequency. This is exactly the Hamiltonian for the quantum harmonic oscillator with the minimum of the potential shifted in the x-direction by $x_0 = \hbar k_y/m\omega$. The total energy is then given by

$$H = \hbar\omega(n + \frac{1}{2}), \quad n \geq 0. \tag{9.49}$$

This result shows that the energies of the electronic motion on the xy-plane are quantized. Each set of wave functions with the same value of n,

$$\psi_n(x, y, z) = e^{ik_y y} \phi_n(x - x_0), \tag{9.50}$$

where ϕ_n are harmonic oscillator eigenstates, is called a *Landau level*. These functions describe the statistical result of the quantized oscillations on the xy-plane.

---

[3] Lev Davidovich Landau (1908–1968) was a Soviet physicist who made fundamental contributions to many areas of theoretical physics, including the foundations of condensed matter physics. He was awarded the Nobel Prize in Physics in 1962 for his development of a theory of superfluidity that explains the properties of liquid helium II at low temperatures.

9.3 Transitions Between States: Selection Rules for the Harmonic Oscillator

As mentioned in Section 3.5 and will be studied in detail in Sections 10.4.3 and 17.2, the interaction of a bound particle (e.g., an atomic electron) with the radiation field induces transitions between stationary states. The most prominent (and simplest) of these transitions are the electric dipole ones, when the interaction is proportional to the dipole moment, er. This means that when the dipole matrix element $\langle n'|er|n\rangle$ is different from zero, electric dipole transitions occur between the states n, n'.

We will use this information to derive the selection rules (i.e., the rules that apply to these transitions) for the 1D HO, which amounts to calculating

$$x_{n'n} = \langle n'|x|n\rangle = \int_{-\infty}^{\infty} \psi_{n'}^* x \psi_n dx. \tag{9.51}$$

The explicit calculation of this integral provides another opportunity to show how an efficient handling of the properties of special functions can greatly simplify the task. Using Eq. (9.32), we first rewrite the integral in the following form

$$x_{n'n} = \alpha_0^2 C_{n'} \int_{-\infty}^{\infty} e^{-\frac{1}{2}\xi^2} \xi H_{n'}(\xi) \psi_n(\xi) d\xi. \tag{9.52}$$

This has been done in order to use the recurrence relation of the Hermite polynomials

$$\xi H_n(\xi) = n H_{n-1}(\xi) + \tfrac{1}{2} H_{n+1}(\xi), \tag{9.53}$$

which allows us to write

$$x_{n'n} = \alpha_0^2 C_{n'} \int_{-\infty}^{\infty} \left(n' H_{n'-1} + \tfrac{1}{2} H_{n'+1}\right) e^{-\frac{1}{2}\xi^2} \psi_n(\xi) d\xi. \tag{9.54}$$

We now express each polynomial in the integrand in terms of the corresponding wave function, again using Eq. (9.32),

$$x_{n'n} = \alpha_0^2 \int_{-\infty}^{\infty} \left(\frac{n' C_{n'}}{C_{n'-1}} \psi_{n'-1}^* + \frac{C_{n'}}{2C_{n'+1}} \psi_{n'+1}^*\right) \psi_n d\xi, \tag{9.55}$$

so that taking into account the orthonormality of the wave functions, we get

$$x_{n'n} = \alpha_0 \left(\frac{n' C_{n'}}{C_{n'-1}} \delta_{n'-1,n} + \frac{C_{n'}}{2C_{n'+1}} \delta_{n'+1,n}\right). \tag{9.56}$$

This expression is different from zero only for $n' - 1 = n$ and $n' + 1 = n$, i.e., for transitions characterized by $|\Delta n| = 1$, which correspond to changes in the energy by the amount $\hbar\omega$. Spontaneous transitions (in the absence of external radiation) occur only for $n' < n$, so only the term with $n' + 1 = n$ contributes when there is no external field,

$$x_{n-1,n} = \frac{\alpha_0 C_{n-1}}{2C_n}. \tag{9.57}$$

Inserting the values of C_n, C_{n-1} given by Eq. (9.41) and taking into account that the matrix x is Hermitian, we get (with $\alpha_0 = \sqrt{\hbar/m\omega}$)

$$x_{n-1,n} = x_{n,n-1} = \alpha_0 \sqrt{\frac{n}{2}}. \tag{9.58}$$

By substituting $n \to n+1$, we obtain the remaining nonzero matrix elements for upward (induced) transitions,

$$x_{n,n+1} = x_{n+1,n} = \alpha_0 \sqrt{\frac{n+1}{2}}. \tag{9.59}$$

Note that the results obtained here by solving the Schrödinger equation are in complete agreement with those obtained in Section 3.5 using the Heisenberg matrix method.

9.4 Raising and Lowering Operators

We now study the problem of the HO from a different perspective, still using the Schrödinger approach. The present procedure is a fundamental part of quantum theory and serves as an introduction to an extremely useful method for dealing with more complicated problems.

From Eq. (9.38) it is easy to show that

$$H'_n = 2n H_{n-1}, \tag{9.60}$$

from where it follows, using (9.32), (9.41), and (9.53), that

$$\frac{1}{\sqrt{2}}\left(\xi + \frac{\partial}{\partial \xi}\right)\psi_n = \sqrt{n}\,\psi_{n-1}, \tag{9.61}$$

$$\frac{1}{\sqrt{2}}\left(\xi - \frac{\partial}{\partial \xi}\right)\psi_n = \sqrt{n+1}\,\psi_{n+1}. \tag{9.62}$$

Introducing the dimensionless operator $\widehat{p}_\xi$

$$\widehat{p}_\xi \equiv -i\frac{\partial}{\partial \xi} = \frac{1}{\sqrt{m\hbar\omega}}\,\widehat{p}_x, \tag{9.63}$$

we define the (*non-Hermitian*) operators

$$\widehat{a} \equiv \tfrac{1}{\sqrt{2}}(\xi + \partial_\xi) = \tfrac{1}{\sqrt{2}}(\xi + i\widehat{p}_\xi) \tag{9.64}$$

$$\widehat{a}^\dagger \equiv \tfrac{1}{\sqrt{2}}(\xi - \partial_\xi) = \tfrac{1}{\sqrt{2}}(\xi - i\widehat{p}_\xi). \tag{9.65}$$

In terms of the canonical variables these operators are

$$\widehat{a} = \frac{1}{\sqrt{2m\hbar\omega}}(m\omega x + i\widehat{p}), \tag{9.66}$$

$$\widehat{a}^\dagger = \frac{1}{\sqrt{2m\hbar\omega}}(m\omega x - i\widehat{p}), \tag{9.67}$$

in coincidence with Eqs. (3.72, 3.73).

Equations (9.61, 9.62) then take the simple form

$$\hat{a}\psi_n = \sqrt{n}\psi_{n-1} \tag{9.68}$$

$$\hat{a}^\dagger \psi_n = \sqrt{n+1}\psi_{n+1}. \tag{9.69}$$

We therefore identify the operator $\hat{a}^\dagger$ as the raising operator and $\hat{a}$ as the lowering operator for the HO introduced in Section 3.5. By successively applying $a^\dagger$ to ψ_0 we get

$$\hat{a}^\dagger \psi_0 = \sqrt{1}\psi_1,$$
$$\hat{a}^\dagger(\hat{a}^\dagger \psi_0) = (\hat{a}^\dagger)^2 \psi_0 = \sqrt{1}\hat{a}^\dagger \psi_1 = \sqrt{1 \cdot 2}\psi_2,$$
$$\hat{a}^\dagger(\hat{a}^\dagger)^2 \psi_0 = (\hat{a}^\dagger)^3 \psi_0 = \sqrt{1 \cdot 2}\hat{a}^\dagger \psi_2 = \sqrt{1 \cdot 2 \cdot 3}\psi_3, \tag{9.70}$$

and so on. From this we see that by applying $a^\dagger$ to ψ_0 n times, we obtain an important expression for the normalized wave function of state n in terms of the normalized wave function of the ground state,

$$\psi_n = \frac{1}{\sqrt{n!}}(\hat{a}^\dagger)^n \psi_0. \tag{9.71}$$

Taking now $n = 0$ in (9.68), we obtain

$$\hat{a}\psi_0 = 0, \tag{9.72}$$

which, with the introduction of the differential expression of the operator $\hat{a}$, Eq. (9.64), gives a *first-order* differential equation for the ground state,

$$\left(\xi + \frac{\partial}{\partial \xi}\right)\psi_0 = 0. \tag{9.73}$$

The normalized solution of this equation is

$$\psi_0 = C_0 e^{-\frac{1}{2}\xi^2}, \tag{9.74}$$

where C_0 is given by Eq. (9.41) with $n = 0$. From this ψ_0 we obtain ψ_n by simple successive algebraic and differentiation processes using (9.71). Note in particular that the equation $\hat{a}\psi_0 = 0$, which *defines* the ground state, *already contains* the correct boundary condition $\psi_0(\xi) \to 0$ for $|\xi| \to \infty$, which guarantees, via Eq. (9.71), that all ψ_n generated with this formalism are square integrable. How this was possible can be seen by proceeding as follows. The Schrödinger equation (9.24) for the HO can be factored interchangeably in the following two forms,

$$\tfrac{1}{2}\hbar\omega \left(\xi + \frac{\partial}{\partial \xi}\right)\left(\xi - \frac{\partial}{\partial \xi}\right)\psi = \left(E + \tfrac{1}{2}\hbar\omega\right)\psi, \tag{9.75}$$

$$\tfrac{1}{2}\hbar\omega \left(\xi - \frac{\partial}{\partial \xi}\right)\left(\xi + \frac{\partial}{\partial \xi}\right)\psi = \left(E - \tfrac{1}{2}\hbar\omega\right)\psi. \tag{9.76}$$

The second of these expressions vanishes for the ground state with $E = \hbar\omega/2$; then we can *define* the ground state through the condition (9.73) whose solution (9.74) automatically satisfies the boundary condition $\psi_0 \underset{\xi\to\infty}{\to} 0$. The alternative possibility that follows from Eq. (9.75),

$$\left(\xi - \frac{\partial}{\partial \xi}\right)\psi = 0, \tag{9.77}$$

leads to a function $C_0 e^{+\frac{1}{2}\xi^2}$ that is not square integrable.[4] Thus we see that the definition of the ground state given by Eq. (9.72) is a simple way to guarantee the correct boundary condition at infinity and an energy spectrum that is bounded from below. Incidentally, these results are at the root of method of quantization by factorization, which is used in many applications.

9.4.1 Eigenvalue equations for ψ_n

By applying $\widehat{a}^\dagger$ to Eq. (9.68) from the left and using (9.69), we get $\widehat{a}^\dagger \widehat{a} \psi_n = \sqrt{n} \widehat{a}^\dagger \psi_{n-1} = n \psi_n$. The result,

$$\widehat{a}^\dagger \widehat{a} \psi_n = n \psi_n, \tag{9.78}$$

shows that ψ_n is an eigenfunction of the operator $\widehat{a}^\dagger \widehat{a}$ with eigenvalue n. If we reverse the order of the operations, we get similarly

$$\widehat{a} \widehat{a}^\dagger \psi_n = (n+1) \psi_n, \tag{9.79}$$

i.e., ψ_n is also an eigenfunction of the operator $\widehat{a} \widehat{a}^\dagger$, but with eigenvalue $n+1$. Adding and subtracting these two results gives

$$(\widehat{a} \widehat{a}^\dagger + \widehat{a}^\dagger \widehat{a}) \psi_n = (2n+1) \psi_n \tag{9.80}$$

and

$$(\widehat{a} \widehat{a}^\dagger - \widehat{a}^\dagger \widehat{a}) \psi_n = \psi_n. \tag{9.81}$$

These expressions are valid for all ψ_n, so the latter is equivalent to the commutation rule obtained in Section 3.5,

$$[\widehat{a}, \widehat{a}^\dagger] = 1. \tag{9.82}$$

In turn, (9.80) suggests that the operator $\widehat{a} \widehat{a}^\dagger + \widehat{a}^\dagger \widehat{a}$ is proportional to the Hamiltonian. Indeed, in terms of the dimensionless variables ξ and p_ξ the Hamiltonian is

$$\widehat{H} = \frac{\widehat{p}^2}{2m} + \tfrac{1}{2} m \omega^2 x^2 = \tfrac{1}{2} \hbar \omega (\widehat{p}_\xi^2 + \xi^2). \tag{9.83}$$

For its part, from the definitions (9.64, 9.65, 9.66, 9.67) it follows that

$$\widehat{a} \widehat{a}^\dagger + \widehat{a}^\dagger \widehat{a} = \tfrac{1}{2}(\xi + i \widehat{p}_\xi)(\xi - i \widehat{p}_\xi) + \tfrac{1}{2}(\xi - i \widehat{p}_\xi)(\xi + i \widehat{p}_\xi) = \widehat{p}_\xi^2 + \xi^2, \tag{9.84}$$

so that

$$\widehat{H} = \tfrac{1}{2} \hbar \omega (\widehat{a} \widehat{a}^\dagger + \widehat{a}^\dagger \widehat{a}). \tag{9.85}$$

Eliminating from here the product $\widehat{a} \widehat{a}^\dagger$ by means of the commutation relation between $\widehat{a}$ and $\widehat{a}^\dagger$, we obtain a very important alternative expression for the Hamiltonian of the HO,

$$\widehat{H} = \hbar \omega \left(\widehat{a}^\dagger \widehat{a} + \tfrac{1}{2} \right). \tag{9.86}$$

---

[4] In this alternative ($E = -\hbar \omega/2$) the energy of the "ground" state is negative (which cannot be the case for a harmonic oscillator); since all the energy values obtained from this state by subtracting the quantity $k \hbar \omega$ are negative, there is no lower bound. This would mean that the system could spontaneously decay to an energy $-\infty$, which makes the solution physically inadmissible.

The eigenvalues of the Hamiltonian now follow from Eqs.(9.85) and (9.80), or from (9.86) and (9.78),

$$E = \langle n|\widehat{H}|n\rangle = \hbar\omega\left(n + \tfrac{1}{2}\right). \tag{9.87}$$

These results show that using the operators $\widehat{a}$ and $\widehat{a}^\dagger$ one can study the HO problem with advantage. In this alternate description, an oscillator of frequency ω is simply a system *defined* by the trio of equations

$$\widehat{H} = \hbar\omega\widehat{a}^\dagger\widehat{a} + \tfrac{1}{2}\hbar\omega, \qquad [\widehat{a},\widehat{a}^\dagger] = 1, \qquad \widehat{a}|0\rangle = 0, \tag{9.88}$$

$|0\rangle$ being its ground (or vacuum) state.

These equations coincide with Eqs. (3.74) and (3.75) of Chapter 3 and, as mentioned there, they are applicable to any quantum oscillator. In particular, they can be used to describe radiation field oscillators, see Section 17.5. When applied to quantized fields, the method is conventionally called *second quantization*, and $\widehat{a}^\dagger$, $\widehat{a}$ are then called *creation* and *annihilation* operators, respectively, as they increase and decrease the number of field quanta (see Section 17.6). When applied to QM, as we have just done, it is often called the *factorization method*, referring to the fact that the Hamiltonian of the excitations $\widehat{H}_{\text{eff}} \equiv \widehat{H} - \tfrac{1}{2}\hbar\omega$ is written as $\hbar\omega\widehat{a}^\dagger\widehat{a}$; in this case $\widehat{a}^\dagger, \widehat{a}$ are called raising and lowering operators, for the reasons explained above.

9.4.2 Algebraic construction of eigenstates and eigenvalues

The formalism introduced in the previous section lends itself to an algebraic procedure for the construction of the eigenstates and eigenvalues of the Hamiltonian, by using Eqs. (9.88). From these equations we can obtain the commutators of $\widehat{a}$ and $\widehat{a}^\dagger$ with $\widehat{H}$ (and therefore with the number operator $\widehat{a}^\dagger\widehat{a}$). For example,

$$[\widehat{H},\widehat{a}] = \hbar\omega\left[\widehat{a}^\dagger\widehat{a},\widehat{a}^\dagger\right] = \hbar\omega\left[\widehat{a}^\dagger\widehat{a}\widehat{a}^\dagger - \widehat{a}^{\dagger 2}\widehat{a}\right] = \hbar\omega\widehat{a}^\dagger\left[\widehat{a}\widehat{a}^\dagger - \widehat{a}^\dagger\widehat{a}\right]. \tag{9.89}$$

Thus,

$$\left[\widehat{H},\widehat{a}^\dagger\right] = \hbar\omega\widehat{a}^\dagger, \tag{9.90}$$

and similarly,

$$[\widehat{H},\widehat{a}] = -\hbar\omega\widehat{a}. \tag{9.91}$$

Now let $|E\rangle$ be an eigenstate of $\widehat{H}$ with eigenvalue E, $\widehat{H}|E\rangle = E|E\rangle$. Using (9.91) we get

$$\widehat{H}\widehat{a}|E\rangle = (\widehat{a}\widehat{H} - \hbar\omega\widehat{a})|E\rangle = (E - \hbar\omega)\widehat{a}|E\rangle, \tag{9.92}$$

whence $\widehat{a}|E\rangle$ is also an eigenstate of $\widehat{H}$, with eigenvalue $E - \hbar\omega$. Successively applying the operator $\widehat{a}$ in the same way, we conclude that $\widehat{H}$ has the eigenvalues E, $E - \hbar\omega$, $E - 2\hbar\omega, \ldots, E - n\hbar\omega, \ldots$ On the other hand, we can write for the energy expectation value (recall that $\langle \widehat{F}^\dagger \widehat{F}\rangle \geq 0$ for any $\widehat{F}$)

$$E = \langle \widehat{H}\rangle = \hbar\omega\left(\langle \widehat{a}^\dagger\widehat{a}\rangle + \tfrac{1}{2}\right) \geq \tfrac{1}{2}\hbar\omega, \tag{9.93}$$

so that the minimum energy satisfies

$$E_{\min} \equiv E_0 \geq \tfrac{1}{2}\hbar\omega > 0. \tag{9.94}$$

The process of lowering the energy with the $\hat{a}$ operator, illustrated in Eq. (9.92), must come to an end, since $E_{min} > 0$. This is achieved only by requiring that

$$\hat{a}|E_0\rangle = 0, \tag{9.95}$$

so that states of lower energy can no longer be created by applying the operator $\hat{a}$. Applying $\hat{H}$ given to the ground state $|E_0\rangle$ then gives

$$\hat{H}|E_0\rangle = \hbar\omega\left(\hat{a}^\dagger\hat{a} + \tfrac{1}{2}\right)|E_0\rangle = \tfrac{1}{2}\hbar\omega|E_0\rangle = E_0|E_0\rangle = E_{min}|E_0\rangle. \tag{9.96}$$

Since this minimum energy is reached from E after n steps of value $\hbar\omega$, it follows that

$$E_n = \tfrac{1}{2}\hbar\omega + n\hbar\omega. \tag{9.97}$$

By iterating, it is not difficult to show that

$$|E_n\rangle = \frac{(\hat{a}^\dagger)^n}{\sqrt{n!}}|E_0\rangle. \tag{9.98}$$

We thus verify that Eqs. (9.88) completely define a quantum HO of frequency ω.

9.4.3 The displacement operator

A displacement operator in QM effects displacements in phase space: it can change the position or the momentum or both. This operator is particularly useful in quantum optics for describing the state of radiation field modes, which are treated as quantum HOs. Similar to how the creation operator $\hat{a}^\dagger$ generates (by successive application) the occupation number state $|n\rangle$, the displacement operator $\hat{D}(\alpha)$ generates the *coherent state* $|\alpha\rangle$,

$$|\alpha\rangle = \hat{D}(\alpha)|0\rangle. \tag{9.99}$$

Alternative expressions for $\hat{D}(\alpha)$ are

$$\hat{D}(\alpha) = e^{\alpha\hat{a}^\dagger - \alpha^*\hat{a}} \tag{9.100}$$

and

$$\hat{D}(\alpha) = e^{-\tfrac{1}{2}|\alpha|^2} e^{\alpha\hat{a}^\dagger} e^{-\alpha^*\hat{a}}, \tag{9.101}$$

where α is a complex number. By applying this operator to the vacuum state, one obtains after some algebra the following expression for a coherent state in terms of number states,

$$|\alpha\rangle = e^{-\tfrac{1}{2}|\alpha|^2} \sum_{n=0}^{\infty} \frac{\alpha^n}{\sqrt{n!}}|n\rangle. \tag{9.102}$$

In quantum optics, coherent states are produced by a laser, which is the most coherent source of oscillations currently available (thermal light, by contrast, can be described as a statistical mixture of coherent states). Coherent states more generally have become a major topic in mathematical physics and in applied mathematics, with applications ranging from quantization to signal processing and image processing.

Exercise E9.4. The displacement operator applied to the HO

Using (9.101) for $\widehat{D}(\alpha)$, show that this operator displaces both the $\widehat{a}$ and $\widehat{a}^\dagger$ operators. Further, show that the coherent state is an eigenstate of the annihilation operator.

Solution. To show that $\widehat{D}(\alpha)$ given by (9.101) is a displacement operator of $\widehat{a}$, we write it in the form

$$\widehat{D}(\alpha) = e^{\alpha \widehat{a}^\dagger} e^{-\frac{1}{2}|\alpha|^2 - \alpha^* \widehat{a}} \equiv e^{\alpha \widehat{a}^\dagger} F_0(\widehat{a}). \tag{9.103}$$

Using the property

$$[\widehat{a}, (\widehat{a}^\dagger)^n] = n(\widehat{a}^\dagger)^{n-1} \tag{9.104}$$

allows us to write

$$\widehat{a}\widehat{D}(\alpha) = \widehat{a} \sum \frac{\alpha^n}{n!} (\widehat{a}^\dagger)^n F_0(\widehat{a}) = \sum \frac{\alpha^n}{n!} [(\widehat{a}^\dagger)^n \widehat{a} + n(\widehat{a}^\dagger)^{n-1}] F_0(\widehat{a}). \tag{9.105}$$

Since $F_0(\widehat{a})$ and $\widehat{a}$ commute, we can write

$$\widehat{a}\widehat{D}(\alpha) = \widehat{D}(\alpha)\widehat{a} + \alpha \sum \frac{\alpha^{n-1}}{(n-1)!} (\widehat{a}^\dagger)^{n-1} F_0(\widehat{a}) = \widehat{D}(\alpha)\widehat{a} + \alpha \widehat{D}(\alpha), \tag{9.106}$$

that is,

$$\widehat{a}\widehat{D}(\alpha) = \widehat{D}(\alpha)(\widehat{a} + \alpha). \tag{9.107}$$

Multiplying by D^{-1} from the left gives

$$\widehat{D}^{-1}(\alpha)\widehat{a}\widehat{D}(\alpha) = \widehat{a} + \alpha, \tag{9.108}$$

which is the requested result. Similarly,

$$\widehat{D}^{-1}(\alpha)\widehat{a}^\dagger \widehat{D}(\alpha) = \widehat{a}^\dagger + \alpha^*. \tag{9.109}$$

This is why the $\widehat{D}(\alpha)$ operator is called the (Glauber)[5] displacement operator.

Applying Eq. (9.107) to the vacuum state $|0\rangle$ and using (9.99), we get

$$\widehat{a}|\alpha\rangle = \alpha|\alpha\rangle, \tag{9.110}$$

which shows that the coherent state $|\alpha\rangle$ is an eigenstate of the annihilation operator, with eigenvalue α. Alternatively, one can use Eqs. (9.68) and (9.102) for the same purpose. In fact, a coherent state is *defined* as an eigenstate of the annihilation operator, as in Eq. ((9.110)).

To conclude, we note that the operator $\widehat{D}(\alpha)$ can be expressed in terms of the variables $\widehat{x}$ and $\widehat{p}$, using (9.100). With $\widehat{a} = \frac{1}{\sqrt{2}}(\frac{\widehat{x}}{\alpha_0} + i\frac{\alpha_0}{\hbar}\widehat{p}), \widehat{a}^\dagger = \frac{1}{\sqrt{2}}(\frac{\widehat{x}}{\alpha_0} - i\frac{\alpha_0}{\hbar}\widehat{p}), \alpha_0 = \sqrt{\hbar/m_0\omega}$, we get

---

[5] Roy Jay Glauber (1925)–(2018) was a US-born theoretical physicist, best known for his work on optical coherence and his model of photodetection. He was awarded the Nobel Prize in 2005 for his pioneering contributions to quantum optics.

$$\alpha \widehat{a}^\dagger - \alpha^* \widehat{a} = i(k\widehat{x} - r\widehat{p}),$$

$$k = \frac{\sqrt{2}}{\alpha_0} \operatorname{Im}\alpha, \quad r = \sqrt{2}\alpha_0 \operatorname{Re}\alpha, \qquad (9.111)$$

whence

$$\widehat{D}(\alpha) = e^{i(k\widehat{x} - r\widehat{p})}. \qquad (9.112)$$

The fact that this operator depends on both the $\widehat{x}$ and $\widehat{p}$ operators can be used to construct a specific quantum description in phase space.

9.4.4 Time dependence of elementary operators

Applying the Heisenberg equation of motion $i\hbar \dot{\widehat{F}} = [\widehat{F}, \widehat{H}]$ to the operator $\widehat{a}$ and using (9.91), we get

$$\dot{\widehat{a}} = -i\omega \widehat{a}, \qquad (9.113)$$

with solution

$$\widehat{a} = \widehat{a}_0 e^{-i\omega t}, \quad \widehat{a}_0 = \widehat{a}(0), \qquad (9.114)$$

and similarly,

$$\widehat{a}^\dagger = \widehat{a}_0^\dagger e^{i\omega t}. \qquad (9.115)$$

Reversing Eqs. (9.64, 9.65) and using the preceding results, we obtain the expressions that determine the time evolution of $\widehat{x}$ and $\widehat{p}$,

$$\widehat{x}(t) = \frac{\alpha_0}{\sqrt{2}}(\widehat{a} + \widehat{a}^\dagger) = \sqrt{\frac{\hbar}{2m\omega}}(\widehat{a}_0 e^{-i\omega t} + \widehat{a}_0^\dagger e^{i\omega t}) \qquad (9.116)$$

$$\widehat{p}(t) = \frac{\hbar}{i\sqrt{2}\alpha_0}(\widehat{a} - \widehat{a}^\dagger) = -i\sqrt{\frac{m\omega\hbar}{2}}(\widehat{a}_0 e^{-i\omega t} - \widehat{a}_0^\dagger e^{i\omega t}), \qquad (9.117)$$

or in terms of $\widehat{x}_0$ and $\widehat{p}_0$,

$$\widehat{x} = \widehat{x}_0 \cos \omega t + \frac{\widehat{p}_0}{m\omega} \sin \omega t, \qquad (9.118)$$

$$\widehat{p} = \widehat{p}_0 \cos \omega t - m\omega \widehat{x}_0 \sin \omega t, \qquad (9.119)$$

which correspond to the elementary oscillators obtained in Chapter 3. Note that the quantum and classical expressions are formally identical.

Exercise E9.5. Time-dependent commutators

As is clear from Eqs. (9.118, 9.119), operators that commute at equal times do not necessarily commute at different times. Calculate the commutation relations $[\widehat{x}(t), \widehat{x}_0]$, $[\widehat{x}(t), \widehat{p}_0]$, and the corresponding Heisenberg inequalities for the HO, for arbitrary t.

Solution. Using Eqs. (9.118, 9.119) and the basic quantum commutators, we get

$$[\widehat{x}(t), \widehat{x}_0] = \frac{1}{m\omega}[\widehat{p}_0, \widehat{x}_0] \sin \omega t = \frac{-i\hbar}{m\omega} \sin \omega t,$$

$$[\widehat{x}(t), \widehat{p}_0] = [\widehat{x}_o, \widehat{p}_0] \cos \omega t = i\hbar \cos \omega t$$

$$\Delta \widehat{x}(t)\Delta \widehat{x_0} \geq \frac{\hbar}{2m\omega}|\sin \omega t|,$$

$$\Delta \widehat{x}(t)\Delta \widehat{p_0} \geq \tfrac{1}{2}\hbar|\cos \omega t|. \qquad (9.120)$$

Note that for the free particle ($\omega \to 0$), the first commutator becomes $[\widehat{x}(t),\widehat{x_0}] = -i\hbar t/m$; the corresponding Heisenberg inequality indicates that in this case the dispersion of x increases linearly with time for not too long times.

9.5 Two Coupled Harmonic Oscillators

The coupled HO problem is interesting from both a physical and a mathematical point of view. Physically, because it allows modeling the small-amplitude oscillations of composite systems, such as the internal oscillations of polyatomic molecules. Mathematically, because it raises a new problem: the need to generalize the Schrödinger equation to the case of more than one particle. We will deal with the latter here specifically for the HO, with more detailed discussions in subsequent chapters.

We write the stationary Schrödinger equation in the form

$$E\psi = \widehat{H}\psi, \qquad (9.121)$$

and proceed to the case of more than one particle by writing the total Hamiltonian as the sum of the contribution of each individual Hamiltonian H_i, plus the contributions due to their mutual interactions, that is,[6]

$$\widehat{H} = \sum_i \widehat{H}_i + \sum_{i>j} \widehat{H}_{ij}, \qquad (9.122)$$

with $i, j = 1, 2, \ldots, N$,

$$\widehat{H}_i = \frac{\widehat{p}_i^2}{2m_i} + V(r_i) = -\frac{\hbar^2}{2m_i}\nabla_i^2 + V(r_i), \qquad (9.123)$$

the Hamiltonian of the particle i, and

$$\widehat{H}_{ij} = V(r_i, r_j) \qquad (9.124)$$

the interaction potential between particles i and j. For example, for two harmonically coupled collinear 1D harmonic oscillators of frequencies ω_1 and ω_2, and equal mass m, we have, assuming a bilinear interaction,

$$\widehat{H} = -\frac{\hbar^2}{2m}\left(\frac{\partial^2}{\partial x_1^2} + \frac{\partial^2}{\partial x_2^2}\right) + \tfrac{1}{2}m\omega_1^2 x_1^2 + \tfrac{1}{2}m\omega_2^2 x_2^2 - m\omega_{12}^2 x_1 x_2. \qquad (9.125)$$

Let us solve this problem as an example; the ideas we will use are directly applicable to any number of coupled oscillators, in any number of dimensions.

---

[6] This procedure is common and, in a sense, shared with classical physics. In neither of these theories are there rules that specify the Hamiltonian; it has to be constructed from empirical principles. In QM, the most common recipe is to start with the corresponding classical Hamiltonian, as is done here, and give it the appropriate quantum avail for the particular case.

In the corresponding classical problem, the solution is obtained by moving to a new coordinate system y related to the original system x by means of an orthogonal transformation that diagonalizes the matrix of the coupling constants (of the frequencies, in the present notation). In the y-system the variables separate, reducing the problem to that of two decoupled oscillators, each oscillating at its own frequency. For this reason, the y_i are called *normal* coordinates and the characteristic frequencies are the normal modes of oscillation.

We will show that the same separation can be used in the quantum case. The task is simplified by introducing the column matrix x of two components

$$x = \begin{pmatrix} x_1 \\ x_2 \end{pmatrix} \tag{9.126}$$

and the new variables $y_i = \sum_j a_{ij} x_j$ $(i, j = 1, 2)$ also in matrix form,

$$y = ax, \tag{9.127}$$

where y is a column matrix with components y_1, y_2, and a is a 2×2 matrix. Let b correspond to the inverse of a,

$$x = by, \quad b = a^{-1}. \tag{9.128}$$

From (9.127) we have

$$\frac{\partial}{\partial x_i} = \sum_j \frac{\partial y_j}{\partial x_i} \frac{\partial}{\partial y_j} = \sum_j a_{ji} \frac{\partial}{\partial y_j} = \sum_j a^T_{ij} \frac{\partial}{\partial y_j}, \tag{9.129}$$

where a^T is the transpose of a. From this it follows that if $\widehat{p}_x$ is the matrix of the momenta in terms of the variables x, and $\widehat{p}_y$ the corresponding matrix expressed in terms of the variables y, then

$$\widehat{p}_x = a^T \widehat{p}_y. \tag{9.130}$$

We can now write the kinetic energy operator in the form

$$\widehat{T} \equiv \frac{1}{2m} \widehat{p}_x^T \widehat{p}_x = \frac{1}{2m} \widehat{p}_y^T a a^T \widehat{p}_y. \tag{9.131}$$

For the potential energy we get

$$V \equiv \tfrac{1}{2} m x^T \omega^2 x = \frac{m}{2} y^T b^T \omega^2, \quad \text{with } \omega^2 = \begin{pmatrix} \omega_1^2 & -\omega_{12}^2 \\ -\omega_{12}^2 & \omega_2^2 \end{pmatrix}. \tag{9.132}$$

For (9.131) to be reduced to a quadratic form in the variables p_y, we take

$$aa^T = 1, \tag{9.133}$$

which is true if the matrix a is orthogonal, i.e., if

$$a^T = a^{-1} = b. \tag{9.134}$$

On the other hand, for the potential energy to be reduced to a quadratic form in the variables y and for the problem to become separable, it is required that

$$b^T\omega^2 b = \Omega^2, \tag{9.135}$$

where Ω^2 is a diagonal matrix. With (9.134), this condition takes the form

$$a\omega^2 a^{-1} = \Omega^2, \tag{9.136}$$

that is, a is the matrix that diagonalizes the matrix ω^2. The procedure leads to the normal modes of the system, which are the independent modes. As already mentioned, these results are the same as the classical ones.

Suppose we have made the transition to the normal coordinate system y; the Hamiltonian is now

$$\widehat{H} = \widehat{H}_1 + \widehat{H}_2, \tag{9.137}$$

where

$$\widehat{H}_1 = -\frac{\hbar^2}{2m}\frac{\partial^2}{\partial y_1^2} + \tfrac{1}{2}m\Omega_1^2 y_1^2, \tag{9.138}$$

and similarly for $\widehat{H}_2$; Ω_1^2 and Ω_2^2 are the eigenvalues of the matrix Ω^2, that is, the solutions of Eq. (9.136). It is left to the student as an exercise to find these solutions. Since $\widehat{H}$ is now the sum of two independent terms, the wave function is factorizable in terms of the respective eigenfunctions,

$$\psi(y_1, y_2) = \psi_{n_1}(y_1)\psi_{n_2}(y_2), \tag{9.139}$$

as can be verified by substituting in the Schrödinger equation,

$$\begin{aligned}\widehat{H}\psi &= (\widehat{H}_1 + \widehat{H}_2)\psi_1(1)\psi_2(2) \\ &= (\widehat{H}_1\psi_1)\psi_2 + \psi_1(\widehat{H}_2\psi_2) = E_1\psi_1\psi_2 + E_2\psi_1\psi_2 = (E_1 + E_2)\psi,\end{aligned} \tag{9.140}$$

and the total energy is the sum of the respective energy eigenvalues,

$$E = E_1 + E_2. \tag{9.141}$$

Since $\psi_{n_1}(y_1)$ and $\psi_{n_2}(y_2)$ are the eigenfunctions of the HO in states n_1 and n_2, respectively, the total energy E is

$$E = \hbar\Omega_1\left(n_1 + \tfrac{1}{2}\right) + \hbar\Omega_2\left(n_2 + \tfrac{1}{2}\right). \tag{9.142}$$

The product wave function (9.139) indicates that the oscillators y_i are uncorrelated; this is how the orthogonality of the normal coordinates manifests itself in the quantum case. With respect to the original variables x_i, the oscillators are correlated and the corresponding wave function is not factorizable, since

$$\psi(x_1, x_2) = \psi_{n_1}(a_{11}x_1 + a_{12}x_2)\psi_{n_2}(a_{21}x_1 + a_{22}x_2). \tag{9.143}$$

The total energy, Eq. (9.142), is obtained by solving Eq. (9.136); the fact that this is a quadratic equation for Ω indicates a splitting of the energy levels, due to the coupling of the oscillators. In Section 13.3.2 this problem is revisited to illustrate the application of perturbation theory.

Problems

P9.1 Derive explicitly Eqs. (9.15) and (9.16) for the minimum-dispersion packet.

P9.2 Show that the stationary states of the harmonic oscillator satisfy the following relations:

$$E_n = m\omega^2 \langle n|x^2|n\rangle; \quad \langle n|\widehat{T}|n\rangle = \langle n|\widehat{V}|n\rangle = \tfrac{1}{2}E_n. \tag{9.144}$$

Discuss these results from the point of view of the virial theorem.

P9.3 Determine the energy eigenvalues for the 1D Schrödinger equation with the potential

$$V(x) = \begin{cases} \tfrac{1}{2}m\omega^2 x^2, & x > 0; \\ +\infty, & x < 0. \end{cases} \tag{9.145}$$

P9.4 Find the probability that the particles lie within the classically allowed range for the eigenstates of the HO, and determine the value of this probability for the ground state. *Note*: erf(1) = 0.8427...

P9.5 Small-amplitude vibrations in diatomic molecules can be studied using as a model for molecular vibrational potential the potential of the HO, $V(x) = \tfrac{1}{2}k_f x^2$, with k_f the force constant. For a typical diatomic molecule, $k_f \simeq 1 \times 10^3$ J·m^{-2}.

a) Use this data to estimate the value of the zero-point vibrational energy.
b) Estimate the spacing, in energy, between two successive states.

P9.6 Solve the 3D HO problem in Cartesian coordinates. Discuss the degeneracy for the isotropic case.

P9.7 A one-dimensional oscillator moves in the external electric field generated by the potential $-e\mathcal{E}x(t)$.

a) State the equations of motion for $\hat{x}(t)$ and $\hat{p}(t)$ and show that they have precisely the form of the corresponding classical equations.
b) Solve for $\hat{x}(t)$ and $\hat{p}(t)$ as a function of $\hat{x}(0)$ and $\hat{p}(0)$.
c) Determine the commutation relations $[x(t_1), x(t_2)]$ for arbitrary $t_1 - t_2$.

P9.8 Consider a HO of frequency ω in its ground state. At time $t = 0$, the oscillation frequency is abruptly reduced to the value $\omega' = \omega/k$, with $1 < k < \infty$. Calculate $\psi(x,t)$ for $t > 0$. Determine:

a) the probability that the system is in the eigenstate with energy E_n at time t;
b) the value of n for which this probability reaches its maximum.

Discuss the results.

P9.9 Derive the evolution equations for $\langle \hat{x}\rangle$ and $\langle \hat{p}\rangle$ for systems with Hamiltonians

$$\widehat{H} = \frac{\hat{p}^2}{2m} + \tfrac{1}{2}m\left(\omega^2 x^2 + \beta x + \mathcal{E}\right), \tag{9.146}$$

$$\widehat{H} = \frac{\hat{p}^2}{2m} + \tfrac{1}{2}m\omega^2 x^2 - \frac{A}{x^2}. \tag{9.147}$$

Solve the equations for the first case.

P9.10 Derive the creation and annihilation operators for the HO, based on the description of the system in momentum space.

P9.11 Calculate the matrix elements x_{nm} and x^2_{nm} for the HO, using $x = \sqrt{\frac{\hbar}{2m\omega}}(\hat{a}+\hat{a}^\dagger)$ and Eqs. (9.68, 9.69).

P9.12 Show that

$$e^{\lambda\hat{a}}\hat{a}^\dagger e^{-\lambda\hat{a}} = \hat{a}^\dagger + \lambda, \tag{9.148}$$

$$e^{\lambda\hat{a}^\dagger}\hat{a}e^{-\lambda\hat{a}^\dagger} = \hat{a} - \lambda, \tag{9.149}$$

$$e^{\lambda\hat{a}}f(\hat{a},\hat{a}^\dagger)e^{-\lambda\hat{a}} = f(\hat{a},\hat{a}^\dagger + \lambda), \quad \text{and so on.} \tag{9.150}$$

where f represents a function admitting a power series expansion of $\hat{a}$ and $\hat{a}^\dagger$.

P9.13 Determine the eigenvalues of the Hamiltonian

$$\hat{H} = h_0\hat{a}^\dagger\hat{a} + h_1(\hat{a}+\hat{a}^\dagger). \tag{9.151}$$

Hint: Introduce a new pair of creation and annihilation operators that diagonalize $\hat{H}$.

P9.14 Show that if $\hat{A} = \hat{A}(t=0)$ is an operator at time $t = 0$, then $\hat{A}(-t)\psi(x,t) = A\psi(x,t)$, where $\psi(x,0)$ is the eigenstate of $\hat{A}$. Use this result to build the Feynman propagator of the harmonic oscillator. Verify that it approaches the free-particle propagator as $\omega \to 0$. Why does the same happen for ω fixed ($< \infty$) when $t - t' \to 0$?

P9.15 Prove that if $f(\hat{a}^\dagger)$ is a polynomial in $\hat{a}^\dagger$, then

$$\hat{a}f(\hat{a}^\dagger)|0\rangle = \frac{df(\hat{a}^\dagger)}{d\hat{a}^\dagger}|0\rangle, \tag{9.152}$$

$$e^{\lambda\hat{a}}f(\hat{a}^\dagger)|0\rangle = f(\hat{a}^\dagger + \lambda)|0\rangle. \tag{9.153}$$

P9.16 Prove that

$$e^{\alpha\hat{a}+\beta\hat{a}^\dagger} = e^{\alpha\hat{a}}e^{\beta\hat{a}^\dagger}e^{-\alpha\beta\hbar/2}. \tag{9.154}$$

P9.17 Calculate the normal frequencies Ω_1 and Ω_2 of the Hamiltonian given by Eq. (9.125).

P9.18 Two particles with masses m_1 and m_2 moving along a straight line interact through the potential $V(x_1, x_2) = a(x_1 - x_2) + b(x_1 - x_2)^2$, $b > 0$. Determine the eigenenergies and the corresponding wave functions.

Bibliographical Notes

The techniques developed in this chapter are widely used in quantum optics. Standard references are Mandel and Wolf (1995), Cohen-Tannoudji et al. (1977), and Scully and Zubairy (1997).

At a more accessible level, chapter 3 of Gerry and Knight (2012) focuses on coherent states in quantum optics.

10 The Physics Underlying Quantum Phenomena

At this point, after having studied the previous chapters, you should be sufficiently familiar with quantum formalism to be ready to pause for a moment and delve into the underlying physics. It is likely that along the way you have been puzzled by some strange or counter-intuitive quantum features and have kept your puzzlement to yourself. This chapter is devoted to explaining some of the basic physics behind the theory, and in this respect it differs from most textbooks. Although you can skip this material if you are only interested in a more formal or conventional course in QM, we strongly recommend that you take the opportunity to join us in this exciting and timely endeavor. The aim is that you come away with a clearer picture of quantum behavior and its causes, which will also hopefully make it easier for you to understand the further reading.

We have learned from previous chapters that randomness enters the description in the form of "quantum fluctuations," considered to be a signature of quantumness. A number of different stochastic theories have been developed over the years to shed light on the nature of quantum stochasticity. In particular, *stochastic quantum mechanics* (SQM) and *stochastic electrodynamics* (SED) provide valuable and complementary insights, as we shall see.

The phenomenological approach of SQM, which is schematically presented in the next section, uses simple and straightforward mathematics with a clear physical meaning. However, like any phenomenological theory, it has the drawback that the source of the stochasticity remains unidentified.

We recall that already in Section 2.9.1 the interaction with the fluctuating electromagnetic vacuum or zero-point field (ZPF) was pointed out as a natural cause for both, the quantum stochasticity and the presence of wave-like phenomena; this is the starting point for SED. In this chapter we will show that it is indeed possible to clarify central aspects of QM by considering the coupling of matter with the radiation field, including the zero-point component. In particular, the intriguing operator formalism finds a first-principles explanation that gives substance to the basic postulates introduced by Heisenberg a century ago.

Finally, Section 5 introduces the hydrodynamic quantum analogue (HQA) of a walking droplet, a striking macroscopic example of a particle-wave system capable of expressing quantum-like features under appropriate conditions. The discussion of characteristic wave-like manifestations arising from the particle–field interaction in the case of quantum systems is left to future chapters.

10.1 Quantum Mechanics as a Stochastic Theory. The Nelson Theory

The best known description of QM as a stochastic process, called *stochastic mechanics*, was put forward by Nelson[1] (1966), who considered the underlying process as classical and Brownian in nature.[2] Quantum mechanics and Brownian motion do indeed lend themselves to a stochastic description in x-space, with notable features in common, the most important being that they both can be treated as Markov (memoryless) processes. Put simply, in a Markov process the probability of each event depends only on the state reached by the system in just the last previous event, which means that the system has no memory of its past and predictions about future outcomes depend only on its current state.[3]

However, Brownian motion and QM are two *different* stochastic processes; in particular, while the latter is time-reversible and conservative, the former is irreversible and dissipative. The need to distinguish between the two led the present authors to develop an alternative version of the theory called *stochastic quantum mechanics* (SQM), which is briefly presented here. The stochastic nature of the dynamics requires a statistical treatment, which is carried out in x-space; the description is general enough to encompass a range of physical phenomena in which the stochastic process is Markovian.

The basic kinematic elements are obtained by averaging over the ensemble of particles that cross a given point x at time t. By taking a time interval Δt around t that is small but *finite*, and expanding in a Taylor series up to the second order, two different velocities are obtained, namely the *flow* or *systematic velocity* v, which is a local mean measure of the nearly regular flow,

$$v(x,t) = \overline{\frac{x(t+\Delta t) - x(t-\Delta t)}{2\Delta t}} = \hat{\mathcal{D}}_c x, \qquad (10.1)$$

where

$$\hat{\mathcal{D}}_c = \frac{\partial}{\partial t} + v \cdot \nabla, \qquad (10.2)$$

---

[1] Edward Nelson (1932–2014) was a US mathematician, well known for his original work on mathematical physics and mathematical logic.

[2] The term *Brownian motion* originally referred to the random movement of pollen particles suspended in water, observed by the British botanist Robert Brown in 1827. In 1905, Einstein developed a theory that explained Brown's observations by recognizing that the (unobservable) impact of the water molecules on the particle would produce irregular motions that could be directly observed under the microscope. Einstein related the x–diffusion coefficient of such motions to the mean square displacement of the particle, as in Eq. (10.5), and to measurable physical quantities, thus laying the theoretical foundations for precision measurements that could reveal the reality of atoms, which was the leitmotif of this Einsteinian investigation.

[3] As we will see in Section 3, the detailed dynamics is not strictly memoryless, but the current (coarse-grained) quantum description is.

and a *diffusive velocity* **u** resulting from the dispersion,

$$u(x,t) = \frac{\overline{x(t+\Delta t) + x(t-\Delta t) - 2x(t)}}{2\Delta t} = \hat{\mathcal{D}}_s x, \tag{10.3}$$

where

$$\hat{\mathcal{D}}_s = u \cdot \nabla + D\nabla^2. \tag{10.4}$$

The symbol $\overline{(\cdot)}$ denotes the ensemble averaging over the finite time interval Δt; it is thus a *coarse-grained* average. Note that $u(x,t)$ is only different from zero if the mean differences $\overline{x(t+\Delta t) - x(t)}$ and $\overline{x(t) - x(t-\Delta t)}$ are different, which for small Δt only occurs due to fluctuations in the positions $x(t-\Delta t)$ and $x(t+\Delta t)$ among the particles of the ensemble.

The diffusion coefficient,

$$D = \frac{\overline{(\Delta x)^2}}{2\Delta t} = \frac{\sigma_x^2}{2\Delta t}, \tag{10.5}$$

is taken as a constant scalar. The Taylor series expansion has been truncated to make the definitions of the mean velocities v and u consistent with the Markovian assumption: the position $x(t)$ does not depend on the positions x at times prior to $t - \Delta t$.

From the generality of the preceding procedure it follows that, given a dynamical system containing a stochastic source, *both velocities v and u are required* to describe it. For this reason, it is common in the theory of Markov processes to refer to the diffusive velocity (associated with the *osmotic* force, which will be discussed later) in addition to the flow velocity.

10.1.1 Quantum mechanics as a Markov process

The two time-derivative operators (10.2) and (10.4), applied to the velocities (10.1) and (10.3), give four different accelerations, which in turn lead to a pair of equations of motion: the first is invariant with respect to time reversal, whereas the second changes sign with time reversal,

$$m\left(\hat{\mathcal{D}}_c v - \lambda \hat{\mathcal{D}}_s u\right) = f, \tag{10.6}$$

$$m\left(\hat{\mathcal{D}}_c u + \hat{\mathcal{D}}_s v\right) = 0, \tag{10.7}$$

where λ is a real parameter and f is the external force acting on the particle. If the force contains a component f_- that changes sign on time reversal, this must be added to the right-hand side of Eq. (10.7).

Note that Eq. (10.6) is a stochastic generalization of Newton's second law; Eq. (10.7), on the other hand, is the gradient of the continuity equation expressing the local conservation of particles,

$$\frac{\partial \rho}{\partial t} + \nabla \cdot (\rho v) = 0, \tag{10.8}$$

with the diffusive velocity $u(x,t)$ given by

$$u(x,t) = D\frac{\nabla\rho}{\rho}, \qquad (10.9)$$

as is confirmed in what follows. Equations (10.6) and (10.7) can be combined into a single one by multiplying the second equation by $\sqrt{-\lambda}$; this gives

$$\hat{\mathcal{D}}_\lambda p_\lambda = f, \qquad (10.10)$$

where

$$p_\lambda = m(v + \sqrt{-\lambda}u), \qquad (10.11)$$

and

$$\hat{\mathcal{D}}_\lambda = \hat{\mathcal{D}}_c + \sqrt{-\lambda}\hat{\mathcal{D}}_s. \qquad (10.12)$$

These equations show that the specific dynamical properties of the system depend strongly on the sign of λ. Since the magnitude of λ can be absorbed in the value of D, we take $\lambda = \pm 1$. The value $\lambda = -1$ describes a classical process of the Brownian type. However, as Brownian motion is dissipative, this method is not the most suitable for a simple description, since the inclusion of the required (velocity-dependent) dissipative term leads to a nonlinear, parabolic differential equation after integrating Eq. (10.10).

In contrast, with $\lambda = 1$ and setting

$$D = \frac{\hbar}{2m}, \qquad (10.13)$$

we obtain the *Schrödinger* equation

$$-\frac{\hbar^2}{2m}\nabla^2\psi(x,t) + V(x)\psi(x,t) = i\hbar\frac{\partial\psi(x,t)}{\partial t} \qquad (10.14)$$

and its complex conjugate, where $\psi(x,t)$ is a complex function such that

$$\rho(x,t) = |\psi(x,t)|^2, \qquad (10.15)$$

$$v = i\frac{\hbar}{2m}\left(\frac{\nabla\psi^*}{\psi^*} - \frac{\nabla\psi}{\psi}\right), \qquad (10.16)$$

and

$$u = \frac{\hbar}{2m}\left(\frac{\nabla\psi^*}{\psi^*} + \frac{\nabla\psi}{\psi}\right) = \frac{\hbar}{2m}\frac{\nabla\rho}{\rho}. \qquad (10.17)$$

From these equations we obtain

$$-i\hbar\nabla\psi = \hat{p}\psi = m(v - iu)\psi, \qquad (10.18)$$

which shows that the diffusive velocity u appears naturally (and more often than you might suppose) in the quantum description, although it is usually overlooked; see Exercise E10.1 and Section 4.4.

The value of the diffusion coefficient D has been set by hand to satisfy the needs of the Schrödinger equation. To *derive* Eq. (10.13) we need to lay the theory on a physical foundation, i.e., to specify the cause of stochasticity. This is where SED comes in, as we will see below.

Exercise E10.1. Role of the diffusive velocity u in QM

Compare the SQM and QM formulas obtained for the variance of the momentum and the mean kinetic energy, and discuss the results.

Solution. Writing the wave function in polar form,

$$\psi(x,t) = \sqrt{\rho(x,t)}\, e^{iS(x,t)}, \tag{10.19}$$

where S is a real function of x, t, we obtain from Eq. (10.16)

$$v = \frac{\hbar}{m}\nabla S. \tag{10.20}$$

This equation together with Eq. (10.17) give the correct quantum result when introduced in Eq. (10.18),

$$\langle \hat{p} \rangle = \hbar \langle \nabla S \rangle = m \langle v \rangle; \quad \langle u \rangle = 0. \tag{10.21}$$

On the other hand, the quantum expectation value of $\hat{p}^2$ gives

$$\langle \hat{p}^2 \rangle = -\hbar^2 \int \psi^* \nabla^2 \psi\, dx$$

$$= -\hbar^2 \int \left[\sqrt{\rho}\nabla^2 \sqrt{\rho} - \rho(\nabla S)^2 + 2i\sqrt{\rho}\nabla S \cdot \nabla\sqrt{\rho} + i\rho\nabla^2 S\right] dx$$

$$= -\hbar^2 \int \left[\sqrt{\rho}\nabla^2\sqrt{\rho} - \rho(\nabla S)^2\right] dx = \hbar^2 \langle (\nabla S)^2 \rangle + \tfrac{1}{4}\hbar^2 \left\langle \left(\frac{\nabla \rho}{\rho}\right)^2 \right\rangle, \tag{10.22}$$

since the imaginary contribution cancels out by an integration by parts, and

$$\int \sqrt{\rho}\nabla^2\sqrt{\rho}\, dx = \left\langle \frac{\nabla^2 \sqrt{\rho}}{\sqrt{\rho}} \right\rangle = -\int \rho \left(\frac{\nabla\sqrt{\rho}}{\sqrt{\rho}}\right)^2 dx = -\tfrac{1}{4}\left\langle \left(\frac{\nabla\rho}{\rho}\right)^2 \right\rangle. \tag{10.23}$$

Therefore,

$$\langle \hat{p}^2 \rangle = \langle (mv)^2 \rangle + \tfrac{1}{4}\hbar^2 \left\langle \left(\frac{\nabla\rho}{\rho}\right)^2 \right\rangle. \tag{10.24}$$

With u given by Eq. (10.17), we obtain two contributions to the kinetic energy,

$$\langle T \rangle = \tfrac{1}{2} m \langle v^2 + u^2 \rangle. \tag{10.25}$$

Similarly, the resulting variance of the momentum,

$$(\Delta p)^2 = \langle (\Delta \hat{p})^2 \rangle = m^2 \langle (\Delta v)^2 \rangle + m^2 \langle (\Delta u)^2 \rangle, \tag{10.26}$$

consists of an extrinsic part due to the dispersion of the flow velocity (e.g., produced by a distribution of the initial velocities) and an intrinsic part due to the unavoidable dispersion of the diffusive velocity. These results show that the velocity u plays a decisive role in QM. Among other things, the dispersion of u is found to be responsible for both the nonzero ground-state energy and the nonzero minimum value of the Heisenberg inequality.

10.1.2 Evidence of diffusion in quantum mechanics nonlocality of the description

Another distinctive and persistent manifestation of the diffusive velocity $\boldsymbol{u}$ in QM is the so-called *quantum potential*, which is derived from the time-dependent Schrödinger equation (see Section 7.2),

$$i\hbar\frac{\partial \psi}{\partial t} = -\frac{\hbar^2}{2m}\nabla^2 \psi + V\psi, \tag{10.27}$$

by multiplying it by ψ^* and separating it into real and imaginary parts. Using Eq. (10.19), the real part gives the continuity equation, $\frac{\partial \rho}{\partial t} + \nabla \cdot (\boldsymbol{v}\rho) = 0$, and the imaginary part gives the dynamical equation

$$\hbar\frac{\partial S}{\partial t} + \frac{\hbar^2}{2m}(\nabla S)^2 + V - \frac{\hbar^2}{2m}\frac{\nabla^2 \sqrt{\rho}}{\sqrt{\rho}} = 0, \tag{10.28}$$

which can be rewritten as

$$\hbar\frac{\partial S}{\partial t} + \frac{1}{2}m\boldsymbol{v}^2 + V + V_Q = 0, \tag{10.29}$$

where

$$V_Q = -\frac{\hbar^2}{2m}\left(\frac{\nabla^2 \sqrt{\rho}}{\sqrt{\rho}}\right). \tag{10.30}$$

A formal analogy to classical mechanics suggests to interpret Eq. (10.29) as a Hamilton–Jacobi equation with an additional potential V_Q given by Eq. (10.30). This term, known as the quantum potential, is the basis of Bohm's causal theory of QM (Bohm 1951). In this interpretation, $\boldsymbol{v}$ is taken to be the velocity field of a *single* particle located at $\boldsymbol{x}$ at time t and describing a real trajectory,

$$\frac{d\boldsymbol{x}}{dt} = \frac{\hbar}{m}\nabla S(\boldsymbol{x},t)|_{\boldsymbol{x}=\boldsymbol{x}(t)}, \tag{10.31}$$

where $S(\boldsymbol{x},t)$ is the action function of the problem, and the wave function is taken to represent a real field that *guides* the particle, known thus as the *pilot wave*, following a suggestion by de Broglie. Trajectories are sensitive to the initial conditions and therefore produce patterns such as the one illustrated in Fig. 10.1 for a uniformly distributed packet scattered by two Gaussian slits.

The pilot-wave equation (10.31) was introduced in 1925–1928 by de Broglie in his efforts to unify the laws of mechanics and optics, following his fundamental proposal on the wave behavior of particles in motion. For Bohm, on the other hand, the starting point is Eq. (10.28), which contains the quantum potential. What is known as Bohmian mechanics is a mixture of the two theories and is recognized for its success in restoring realism, objectivity and determinism to QM.

However, this success comes at the cost of the nonlocality embodied in the theory. Formally, the only difference between a classical problem and the corresponding quantum problem is the presence of V_Q in the latter; but unlike the local classical potential V, V_Q depends on the spatial distribution $\rho(\boldsymbol{x},t)$ (see Eq. (10.30)), which in turn is influenced by V_Q, in a kind of feedback loop. The dependence of V_Q on $\rho(\boldsymbol{x},t)$ gives

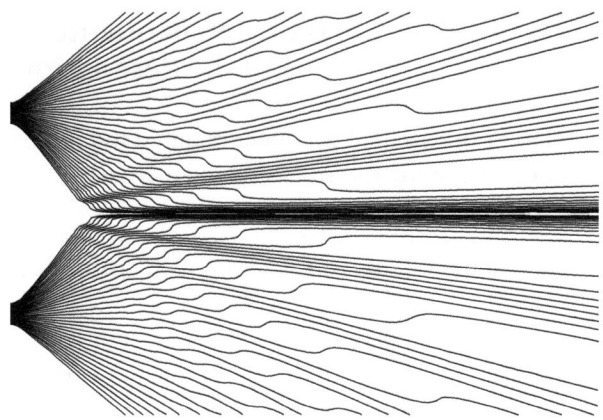

Figure 10.1 The trajectories of an initially uniform distribution of particles scattered by two Gaussian slits, according to Bohm's theory. Adapted from Philippidis et al. (1979).

(10.29) a nonlocal and statistical content. This nonlocality becomes manifest as soon as V_Q comes into play.

By introducing Eq. (10.17) into (10.30), we get

$$V_Q = -\frac{1}{2}\left(m\boldsymbol{u}^2 + \hbar \boldsymbol{\nabla} \cdot \boldsymbol{u}\right), \tag{10.32}$$

which suggests reading the "quantum potential" as a nonlocal kinematic term associated with the diffusive nature of the dynamics, i.e., as an energy contribution due to "quantum fluctuations." Whichever view is adopted, the term is clearly of paramount importance in determining much of the quantum behavior.

Exercise E10.2. Estimating the time resolution of the quantum description

One of the important lessons to be learned from the stochastic formulation of QM is the confirmation that the quantum description has a time resolution limit. Using the formulas derived in this section, obtain an estimate for Δt.

Solution. Indeed, the development of the stochastic theory required the introduction of a time interval Δt, which by definition is very short but not zero; no better estimate of it is proposed. This Δt represents a minimum value for the precision of the description of the temporal behavior. Thus, any phenomenon or detail shorter in time than Δt remains outside the possibilities of (nonrelativistic) quantum theory.

Let us consider the fluctuations of the position variable x; according to Eq. (10.5) the variance of x is given by

$$\overline{(\Delta x)^2} = 2D\Delta t, \tag{10.33}$$

where the diffusion coefficient of the position variable D is phenomenologically ascribed the value $D = \hbar/2m$ in the quantum case, according to Eq. (10.13). Combining this with the above equation gives

$$\overline{(\Delta x)^2} = (\hbar/m)\,\Delta t. \tag{10.34}$$

So all we need to get an estimate for Δt is to assign a minimum characteristic value to the mean velocity $\bar{v} = \sqrt{\overline{(\Delta x)^2}/\Delta t}$. To do this, consider an orbiting atomic electron; its kinetic energy is of the order $\alpha^2 mc^2$, where as usual, α is the fine structure constant (see Section 12.5); hence

$$\overline{(\Delta x)^2}/(\Delta t)^2 = \frac{\hbar}{m\Delta t} \simeq \alpha^2 c^2. \tag{10.35}$$

It follows that

$$\Delta t \simeq \frac{\hbar}{m\alpha^2 c^2} = \frac{1}{\alpha^2 \omega_C} = \frac{1}{2\pi\alpha^2}\tau_C, \tag{10.36}$$

where τ_C is the Compton time of the electron. With the above estimates, Δt is about 3×10^3 times the Compton time for the electron, i.e., $\Delta t \sim 10^{-17}$ s.

Note further that if we assume the dispersion of p to be of the order of $m\bar{v}$, then by combining the above equations we get

$$\overline{(\Delta x)^2}\,\overline{(\Delta p)^2} \simeq \hbar^2, \tag{10.37}$$

implying that the limited time resolution results in a limited resolution in x and p.

10.2 Stochastic Electrodynamics. Basic Properties of the ZPF

Stochastic electrodynamics has been developed on the basis of the existence of the ZPF as a real, omnipresent, fluctuating Maxwellian field in permanent interaction with matter. A number of authors have contributed to SED, with different approaches and views on its relation to classical mechanics, QM, SQM and even QED (for a few examples see the bibliographical notes at the end of the chapter). Over the years, important results have been obtained, some of which will be presented in this and following chapters, using procedures that are generally more transparent and direct than the corresponding quantum-mechanical and QED ones.

The vacuum or zero-point radiation field, i.e., the nonthermal field proposed by Planck – corresponding to "zero photons" in the language of quantum theory – has an average energy per normal mode of value $\hbar\omega/2$, which implies a spectral energy density of value

$$\rho_0(\omega) = \frac{\hbar\omega^3}{2\pi^2 c^3}, \tag{10.38}$$

as explained in Section 2.9 of Chapter 2. The spectral energy density $\rho \sim \omega^3$ is the *only* one consistent with relativity, as was shown by Einstein and Hopf (1910). The argument is as follows.

When a charged or polarizable body moving at a constant velocity v interacts with an electromagnetic field of spectral energy density $\rho(\omega)$, the field exerts on it a force F due to the Doppler effect, given by

$$\boldsymbol{F} = A\left(\rho - \tfrac{1}{3}\omega\frac{\partial\rho}{\partial\omega}\right)\boldsymbol{v}, \tag{10.39}$$

where the constant A depends on the geometry of the body. But the vacuum field must not exert any force on a body in uniform motion; this is true for any velocity v only if $\rho \sim \omega^3$.

From Eq. (10.38) it follows that the total energy of the ZPF is infinite, since the field contains in principle modes of frequencies ω covering the entire spectrum (0 to ∞). Much has been written about this problem, as it is part of contemporary cosmology and quantum theory in general, which deals with a variety of vacuum fields in addition to the electromagnetic one. Several plausible solutions have been proposed, including the possibility that some (or all) of the dark energy in the universe is associated with the vacua, whose energy is positive in some cases and negative in others. Without getting into unresolved cosmological arguments, it is a fact that atomic particles do not respond to extremely high frequencies; atoms are known not to radiate or absorb at frequencies above a cut-off frequency, that can be set around the Compton frequency $\omega_c = mc^2/\hbar$ of the electron ($\sim 10^{21}\,\text{s}^{-1}$). This sets a natural limit to the energy content of the ZPF, which is large but not infinite. In Chapter 17 we will find further justification for the introduction of this frequency cut-off.

To describe the basic statistical properties of the ZPF, we introduce the Fourier transform of a (Cartesian) electric field component $E_i(t)$, $i = 1, 2, 3$,

$$E_i(\omega) = \frac{1}{2\pi} \int_{-\infty}^{\infty} E_i(t)\, e^{-i\omega t}\, dt, \tag{10.40}$$

$$E_i(t) = \int_{-\infty}^{\infty} E_i(\omega)\, e^{i\omega t}\, d\omega, \tag{10.41}$$

$$E_i^*(-\omega) = E_i(\omega). \tag{10.42}$$

$E_i(\omega)$ is assumed to have a Gaussian (normal) distribution; this is the simplest and most natural proposal, given the vast number of sources contributing to it. The average over all its possible realizations is of course zero; denoting this ensemble average by $\langle\,\rangle_0$, and considering (10.40),

$$\langle E_i(t)\rangle_0 = 0, \quad \langle E_i(\omega)\rangle_0 = 0. \tag{10.43}$$

To fully characterize the Gaussian distribution, it is sufficient to specify some other statistical property, such as the autocorrelation of the field. This can be done using Eq. (10.38) for the spectral energy density $\rho_0(\omega)$. The total energy density can be written in one of the following alternative forms,

$$u = \frac{\langle \mathbf{E}^2(t)\rangle_0}{4\pi} = \frac{3}{4\pi} \int_{-\infty}^{\infty} d\omega \int_{-\infty}^{\infty} d\omega' \langle E_i(\omega) E_i^*(\omega')\rangle_0 e^{i(\omega-\omega')t}$$

$$= \int_0^{\infty} \rho_0(\omega)\, d\omega = \tfrac{1}{2} \int_{-\infty}^{\infty} \rho_0(\omega)\, d\omega, \tag{10.44}$$

where $\rho(-\omega) = \rho(\omega)$, $\omega > 0$. For the second and last equalities to hold, it is required that

$$\langle E_i(\omega) E_j^*(\omega')\rangle_0 = \frac{2\pi}{3} \rho_0(\omega) \delta(\omega - \omega') \delta_{ij}. \tag{10.45}$$

This expression means that the different modes of the field are uncorrelated, and shows that the spectral energy density is the Fourier transform of the autocorrelation function of the field up to a numerical factor,

$$\langle E_i(t')E_j(t)\rangle_0 = \varphi(t'-t)\delta_{ij} = \frac{2\pi}{3}\delta_{ij}\int \rho_0(\omega)e^{i\omega(t'-t)}d\omega. \tag{10.46}$$

The integration results in an autocorrelation that decreases with time as t^{-4}; for a white (ω-independent) noise one gets instead the much slower decrease t^{-1}. The ω^3-dependence of the spectral energy density means that the particles are subject to a highly colored noise.

10.2.1 The Casimir forces, direct manifestation of the ZPF

In 1948, Casimir[4] predicted that the presence of two parallel, neutral conducting plates would modify the distribution of vacuum field modes in the vicinity of the plates, particularly in the space between them, resulting in an effective force of attraction given by

$$F = -\frac{\pi^2 \hbar c A}{240 R^4}, \tag{10.47}$$

where A is the area of the plates and R is the distance separating them. Half a century later the effect was experimentally confirmed. Casimir's analysis of idealized metal plates has been extended to arbitrary dielectrics and realistic boundaries, giving rise to a new branch of physics. In most of the literature on the subject, the calculations are carried out in the context of quantum field theory, where the vacuum is taken to be the ground state of the quantized electromagnetic field, conventionally called *vacuum fluctuations* (see Section 17.3).

The SED approach successfully reproduces the quantum predictions under a wide range of conditions, with the advantage of greater simplicity and physical transparency. The following exercise illustrates this for the Casimir force between two parallel metal plates. These calculations are so convincing that they are often taken as a demonstration of the reality of the ZPF and its action on quantum matter.

Exercise E10.3. Casimir force due to the ZPF
Calculate the Casimir force between two parallel metal plates, taking into account the modification of the ZPF by the presence of the plates.

Solution. Let us take two square plates of side L, separated by a distance R, which is considered small compared to L. To calculate the effect of the plates on the field, we place a vertical perfectly conducting wall at a large distance D from the left plate ($D \to \infty$ at the end of the calculation), and enclose the array within conducting walls as shown in Fig. 10.2.

---

[4] Hendrik Brugt Gerhard Casimir (1909–2000) was a Dutch physicist who made significant contributions to the field of QM and QED. He is best known for his work on the Casimir effect and for his research on the two-fluid model of superconductors. For 30 years from 1942, Casimir worked and (co-)directed the Philips Physics Laboratory in Eindhoven, the Netherlands, where many fundamental steps in the development and applications of electronics were made.

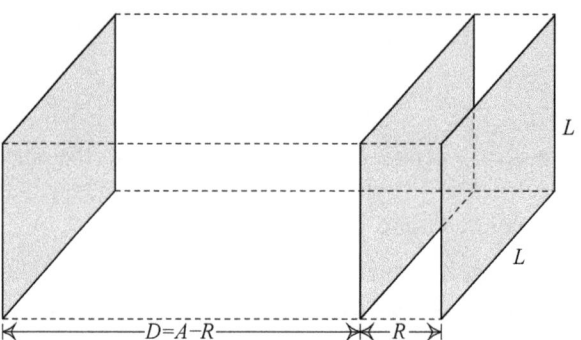

Figure 10.2 A simple model for calculating the Casimir force between two plates separated by a distance R. To complete the calculation, a conducting wall is placed at a large distance D from the left plate and the array is surrounded by conducting walls to form a parallelopiped.

The field trapped between the plates can be expanded in normal modes as usual. Given the boundary conditions on the electric and magnetic components, it is easy to see that the allowed frequencies between the plates are given by the formula

$$\omega_{lmn} = ck_{lmn} = c\pi \left(\frac{l^2 + m^2}{L^2} + \frac{n^2}{R^2}\right)^{\frac{1}{2}}, \quad (10.48)$$

where l, m, n are positive integers or zero. The energy of this field is given by the triple sum over modes,

$$E(R) = 2\sum_{lmn}{}' \frac{\hbar \omega_{lmn}}{2} = \pi \hbar c \sum_{lmn}{}' \left(\frac{l^2 + m^2}{L^2} + \frac{n^2}{R^2}\right)^{\frac{1}{2}}. \quad (10.49)$$

The factor 2 arises from the two independent polarizations for $l, m, n \neq 0$. Only one of these numbers can be zero at a time. When this happens a factor $1/2$ must be inserted because there is only one polarization; this is the meaning of the prime in the summation sign. For large values of L, the double sum over l and m transforms into a double integral over k_x and k_y, respectively,

$$\sum_{lm} \longrightarrow_{L\to\infty} \left(\frac{L}{2\pi}\right)^2 \int_{-\infty}^{\infty} dk_x \int_{-\infty}^{\infty} dk_y, \quad (10.50)$$

whence

$$E(R) = \frac{\hbar c L^2}{\pi^2} \int_{-\infty}^{\infty} dk_x \int_{-\infty}^{\infty} dk_y \sum_{n}{}' \left(k_x^2 + k_y^2 + \frac{\pi^2 n^2}{R^2}\right)^{\frac{1}{2}}. \quad (10.51)$$

Setting $k_x^2 + k_y^2 = k_\rho^2$ we perform the angular integration on the xy-plane, to get

$$E(R) = \frac{\hbar c L^2}{4\pi} \int_0^\infty dk_\rho^2 \sum_{n}{}' k_n, \quad (10.52)$$

with $k_n = \left(k_\rho^2 + \frac{\pi^2 n^2}{R^2}\right)^{\frac{1}{2}}$. Similarly, the energy of the ZPF between the conducting wall and the left plate is given by $E(D)$, so the energy of the whole configuration is $E(R) + E(D)$.

The Casimir energy $E_C(R)$, i.e., the energy change produced by the central plate, is obtained by subtracting the energy in the absence of this plate and making the distance D of the wall tend to infinity,

$$E_C(R) = \lim_{D \to \infty} E(R) + E(D) - E(D+R). \tag{10.53}$$

This expression represents the work performed to bring the central plate adiabatically to its position at a distance R from the right plate.

From Eq. (10.52) it is clear that $E(R)$ is infinite and the same applies to the remaining terms in (10.53). However, since even well conducting materials become poor conductors at very short wavelengths, a cut-off in the wavenumber k_n should be introduced to obtain a finite value for the integrals. A convenient cut-off function is $\exp(-k_n/k_c)$, where k_c is assumed to be very large. Equation (10.52) gives after a change of variable

$$E(R;k_c) = \frac{\hbar c L^2 k_c^3}{2\pi} {\sum_n}' \int_{n/x}^{\infty} ds\, s^2 e^{-s}, \tag{10.54}$$

with $s = k_n/k_c$ and $x = R k_c/\pi$. Integrating gives

$$E(R;k_c) = \frac{\hbar c L^2 k_c^3}{2\pi} \frac{d^2}{dx^2} x^2 {\sum_n}' e^{-n/x} = \frac{\hbar c L^2 k_c^3}{2\pi}\left(-1 + 2\Delta + \frac{2\varsigma \Delta^2}{x} - \frac{\varsigma \Delta^2}{x^2} + \frac{2\varsigma \Delta^3}{x^2}\right), \tag{10.55}$$

with $\varsigma = \exp(-1/x)$ and $\Delta = (1-\varsigma)^{-1}$. By introducing this expression into (10.53), we obtain after some algebra the exact result

$$E_C(R;k_c) = \frac{\hbar c L^2 k_c^3}{2\pi}\left(-1 - 6x + 2\Delta + \frac{2\varsigma \Delta^2}{x} - \frac{\varsigma \Delta^2}{x^2} + \frac{2\varsigma \Delta^3}{x^2}\right). \tag{10.56}$$

For macroscopic objects, $x \gg 1$ and the above expression takes the simple form (with $L^2 = A$)

$$E_C(R;k_c) = -\frac{\pi^2 \hbar c A}{720 R^3}, \tag{10.57}$$

which gives

$$F = -\frac{\partial E_C}{\partial R} = -\frac{\pi^2 \hbar c A}{240 R^4}, \tag{10.58}$$

in agreement with Casimir's formula.

Note that this result is independent of k_c for large values of the cut-off frequency, i.e., for good metallic conductors. In the case of dielectric materials or very small objects, the Casimir energy E_C has a more complicated behavior as a function of distance, as can be seen from Eq. (10.56).

10.2.2 Initial SED analysis of the harmonic oscillator

The early SED treatment (around 1960) of the HO already showed that the ZPF is somehow responsible for its quantum behavior. The following presentation is necessarily schematic; the interested reader is invited to consult the bibliography at the end of the chapter for details.

10.2.2.1 Dynamics of the linear oscillator

The usual starting point is the equation of motion for the nonrelativistic 1D HO in the long-wavelength or electric-dipole approximation, Eq. ((2.3)),

$$\ddot{x} + \omega_0^2 x - \tau \dddot{x} = \frac{e}{m} E(t), \tag{10.59}$$

where

$$\tau = \frac{2e^2}{3mc^3}. \tag{10.60}$$

We recall that the term $m\tau \dddot{x}$ is the self-force due to the radiation reaction; for an electron, $\tau \approx 10^{-23}$ s.

$E(t)$ is an electric component of the random ZPF, which means that (10.59) is a stochastic differential equation, i.e., each realization of $E(t)$ leads to a different trajectory, even under the same initial conditions.[5] Since one cannot know in advance which $E(t)$ will be realized in a particular instance, one can only make statistical predictions about trajectories from the statistical properties of the field E.

The stationary solution of Eq. (10.59) is obtained by taking the Fourier transform of $x(t)$ and computing the inverse transform. Since our interest centers on the *stationary motions generated and maintained by the field*, we neglect the transient part of the solution that decays exponentially with time, at a rate given by $\tau \omega_0^2$. This gives

$$x(t) = \frac{e}{m} \int_{-\infty}^{\infty} d\omega \frac{E(\omega)e^{i\omega t}}{\omega_0^2 - \omega^2 + i\tau \omega^3} + \text{c. c.}, \tag{10.61}$$

which means that for general initial conditions, the system will reach an equilibrium state – which, as we shall see, corresponds to the ground state of QM. It is important to note that Eq. (10.61) expresses the effect of the direct action of the ZPF on the particle in a stationary state, suggesting that a similar procedure should be applied to the study of stationary states more generally. This observation is the starting point for the SED approach to be presented in later sections.

10.2.2.2 Minimal fluctuations and minimal energy of the oscillator

By averaging (10.61) over the ensemble using Eq. (10.43), it becomes clear that at equilibrium, the oscillators remain on average centered at the origin,

$$\langle x(t) \rangle_0 = 0. \tag{10.62}$$

---

[5] Equations of motion of this type are given the generic name *Langevin equation*; more specifically, the Langevin equation of SED is often called the *Braffort–Marshall equation*.

Further, Eqs. (10.61) and (10.45) give for the average of x^2

$$\langle x^2 \rangle_0 = \frac{2\pi c^3 \tau}{m} \int_0^\infty d\omega \frac{\rho_0(\omega)}{\Delta(\omega;\omega_0)}, \qquad (10.63)$$

where

$$\Delta = (\omega_0^2 - \omega^2)^2 + \tau^2 \omega^6. \qquad (10.64)$$

To carry out the integration we take into account that for the relevant (atomic) frequencies $\omega_0 \leq 10^{15}\, s^{-1}$,

$$\tau\omega_0 \leq 10^{-8} \ll 1, \qquad (10.65)$$

and the denominator Δ becomes very small for $\omega \sim \omega_0$. Introducing $y = \omega - \omega_0$, noting that the main contribution to the integral comes from y close to zero, and using

$$\lim_{\gamma \to 0} \frac{\gamma}{\gamma^2 + y^2} = \pi \delta(y), \qquad (10.66)$$

we obtain

$$\langle x^2 \rangle_0 = \frac{\pi^2 c^3}{m\omega_0^4} \rho_0(\omega_0) = \frac{\hbar}{2m\omega_0}, \qquad (10.67)$$

and the dispersion of x is given by

$$\sigma_x^2 \equiv \langle (\Delta x)^2 \rangle_0 = \langle x^2(t) \rangle_0 - \langle x(t) \rangle_0^2 = \frac{\hbar}{2m\omega_0}. \qquad (10.68)$$

Similarly, the higher moments $\langle x^n(t) \rangle_0 \geq 2$ can be shown to correspond to a Gaussian distribution with zero mean value and variance σ_x^2. Therefore, the spatial density normalized to unity is

$$\rho(x) = \frac{1}{\sqrt{2\pi\sigma_x^2}} e^{-x^2/2\sigma_x^2} = \sqrt{\frac{m\omega_0}{\pi\hbar}} e^{-m\omega_0 x^2/\hbar}, \qquad (10.69)$$

in agreement with the quantum result.

The similar calculation of the moments of p gives

$$\langle p(t) \rangle_0 = 0, \qquad (10.70)$$

$$\sigma_p^2 \equiv \langle (\Delta p)^2 \rangle_0 = \langle p^2 \rangle_0 - \langle p \rangle_0^2 = \tfrac{1}{2} m\hbar\omega_0, \qquad (10.71)$$

and the mean energy of the ground state is therefore

$$E_0 = \langle \widehat{H} \rangle_0 = \left\langle \frac{p^2}{2m} + \tfrac{1}{2} m\omega_0^2 x^2 \right\rangle_0 = \frac{\sigma_p^2}{2m} + \tfrac{1}{2} m\omega_0^2 \sigma_x^2, \qquad (10.72)$$

i.e.,

$$E_0 = \tfrac{1}{2}\hbar\omega_0, \qquad (10.73)$$

again in accord with QM. Note that this mechanical energy *is due to the fluctuations induced by the ZPF on the particle.* The equality $\sigma_x^2 \sigma_p^2 = \tfrac{1}{4}\hbar^2$ – the minimum possible value of the Heisenberg inequality $\sigma_x^2 \sigma_p^2 \geq \tfrac{1}{4}\hbar^2$ – is an indication of the irreducible nature of the fluctuations of the oscillator: they will linger there as long as the electron interacts with the ZPF.

10.2.2.3 Canonical ensemble of harmonic oscillators

The previous procedure can be extended to construct the spectral energy density of material oscillators in equilibrium with the field oscillators at a temperature $T > 0$, given by the complete Planck distribution (including the zero-point energy). In terms of the dimensionless variable $\beta E_0 = \hbar\omega/2kT$ it is written as follows (see Eq. (2.83) of Chapter 2),

$$\rho_T(\omega) = \rho_0(\omega)\frac{e^{2\beta E_0} + 1}{e^{2\beta E_0} - 1} = \rho_0(\omega)\coth\beta E_0. \tag{10.74}$$

This equilibrium distribution of the mechanical oscillators in configuration space is again a Gaussian function, with $\langle x^2\rangle_T$ calculated as in Eq. (10.63), with $\rho_0(\omega)$ replaced by $\rho_T(\omega)$, whence

$$\rho(x;T) = \sqrt{\frac{m\omega_0}{\pi\hbar\coth\beta E_0}}e^{-\frac{m\omega_0}{\hbar}x^2\tanh\beta E_0}, \tag{10.75}$$

and

$$E = E_0\coth\beta E_0. \tag{10.76}$$

By expanding the results in a power series with respect to the variable $e^{-2\beta E_0}$, we obtain

$$E = E_0\frac{1 + e^{-2\beta E_0}}{1 - e^{-2\beta E_0}} = \frac{1}{Z}\sum_n E_n e^{-2\beta E_0 n}, \tag{10.77}$$

with the partition function Z given by

$$Z = \sum_n e^{-2\beta E_0 n} = \frac{1}{1 - e^{-2\beta E_0}} \tag{10.78}$$

and $E_n = E_0(2n + 1)$. By expanding (10.75) with respect to the same variable,

$$\rho(x, T) = \frac{1}{Z}\sum_n \rho_n(x)e^{-(2n+1)\beta E_0}, \tag{10.79}$$

the quantum expression involving the Hermite polynomial $H_n(x)$

$$\rho_n(x) = |\psi_n(x)|^2 = c_n^2 e^{-x^2/2\sigma_x^2}H_n^2(x) \tag{10.80}$$

for the spatial density of state n is obtained as in Section 9.2 – but without the need to use the Schrödinger equation.

The early SED results, particularly those on the HO, were instrumental in opening up SED as a legitimate field of research. Still over time it became clear that the HO could be deceptive; what you learn from it could not automatically be extended to other systems. Something fundamental was not taken into account, namely that the action of the ZPF on the particle is not just a perturbation of classical Newtonian motion, but leads to a *qualitative* change in its behavior, which is even reflected in the kinematics, as will be seen in Section 4.

In preparation, we will show how the general theory of stochastic processes provides interesting insights into the statistical content of QM.

10.3 Statistical Approach: The Fokker–Planck Equation

In the equation of motion (10.59)

$$m\ddot{x} - f(x) - m\tau\dddot{x} = eE(t), \qquad (10.81)$$

the right-hand side is the electric force due to the random ZPF, and therefore its solution requires a statistical treatment. The general procedure, well known from the theory of stochastic processes (see the bibliography at the end of the chapter), goes as follows. As a first step, starting from Eq. (10.81), one derives an equation for the probability density in the whole (particle+field) phase space. Then, through a smoothing process, one gets rid of the field variables and obtains a generalized Fokker-Planck equation (GFPE) for the particle phase-space probability density $Q(x, p, t)$. This is an integro-differential equation, or equivalently, a differential equation with an infinite number of time-dependent terms, which in principle gives a statistical description of the particle dynamics for all times since the initial time when the particle and field begin to interact. It is a highly complicated differential equation that is impossible to solve exactly; however, some general observations can be drawn from it.

At the beginning, the system is far from equilibrium; in this regime, the main effect of the zpf on the particle is due to the modes of very high frequency, close to 10^{21}s^{-1}, which produce violent accelerations and randomize the motion. Eventually, the interplay between the electric field force and radiation reaction drives the system to equilibrium; in this regime, the initial conditions have become irrelevant and the Markovian approximation applies. The time averaging implicit in the Markov description which, as discussed in Section 1 above, is estimated to be of the order of 10^{-17}s, prevents the quantum description from capturing the finer details of the motion.

In the Markov approximation, describing the slow dynamics, the GFPE is simplified by retaining terms up to second order. The ensuing Fokker-Planck equation (x_i, p_i are the Cartesian coordinates and corresponding momenta),

$$\frac{\partial Q}{\partial t} + \frac{1}{m}\frac{\partial}{\partial x_i}p_i Q + \frac{\partial}{\partial p_i}f_i Q = -m\tau \frac{\partial}{\partial p_i}\dddot{x}_i Q + D_{ij}^{px}\frac{\partial^2 Q}{\partial p_i \partial x_j} + D_{ij}^{pp}\frac{\partial^2 Q}{\partial p_i \partial p_j}, \qquad (10.82)$$

contains remnants of the memory build-up, expressed through the diffusion tensors

$$D_{ij}^{px}(t) = e\langle x_i E_j\rangle_0 = e^2 \int_{-\infty}^{t} ds\, \frac{\partial x_i(t)}{\partial p_k(s)}\bigg|_{x^{(0)}} \langle E_k(s)E_j(t)\rangle_0, \qquad (10.83)$$

$$D_{ij}^{pp}(t) = e\langle p_i E_j\rangle_0 = e^2 \int_{-\infty}^{t} ds\, \frac{\partial p_i(t)}{\partial p_k(s)}\bigg|_{x^{(0)}} \langle E_k(s)E_j(t)\rangle_0, \qquad (10.84)$$

where the derivatives are calculated to zero order in the coupling constant e, and the field correlation $\langle E_k(s)E_j(t)\rangle_0$ is given by Eq. (10.46),

$$\langle E_k(s)E_j(t)\rangle_0 = \varphi(s-t)\delta_{kj}. \qquad (10.85)$$

Equation (10.82) provides key statistical information about the evolution of the system in the quantum regime. Multiplying it from the left by a dynamical function

$\mathcal{G}(\boldsymbol{x}, \boldsymbol{p}, t)$ and integrating over the phase space (assuming in the integration by parts that the system is bounded so that Q vanishes at infinity), one gets

$$\frac{d}{dt}\langle \mathcal{G}\rangle = \left\langle\frac{d\mathcal{G}}{dt}\right\rangle_{\mathrm{nr}} + m\tau\left\langle\dddot{x}_i\frac{\partial\mathcal{G}}{\partial p_i}\right\rangle - e^2\left\langle\frac{\partial\mathcal{G}}{\partial p_i}\hat{D}_i\right\rangle, \qquad (10.86)$$

where

$$\left\langle\frac{d\mathcal{G}}{dt}\right\rangle_{\mathrm{nr}} = \left\langle\frac{\partial\mathcal{G}}{\partial t}\right\rangle + \left\langle\dot{x}_i\frac{\partial\mathcal{G}}{\partial x_i} + f_i\frac{\partial\mathcal{G}}{\partial p_i}\right\rangle \qquad (10.87)$$

corresponds to the (Liouvillian) nonradiative contribution to $\langle d\mathcal{G}/dt\rangle$. This equation describes the evolution of the mean value of $\mathcal{G}$ in line with quantum mechanics, i.e., in the radiationless approximation, as will be shown below.

The remaining terms in Eq. (10.86), with the diffusion operator $\hat{D}_i$ given by

$$e^2\hat{D}_i = D_{ij}^{px}\frac{\partial}{\partial x_j} + D_{ij}^{pp}\frac{\partial}{\partial p_j}, \qquad (10.88)$$

represent the radiative corrections; they contain information that is not part of QM and pertains to the realm of (nonrelativistic) QED. In Chapter 17 this information will be used to obtain the general formulas for the atomic transition probabilities and the Lamb shift.

Note that for $\mathcal{G} = p_i$ in Eq. (10.86), one gets

$$m\langle\ddot{x}_i\rangle = \langle f_i\rangle + m\tau\langle\dddot{x}_i\rangle - e\langle\hat{D}_i\rangle, \qquad (10.89)$$

which is the Ehrenfest version of Newton's second law including the radiative terms: $m\tau\langle\dddot{x}_i\rangle$ is the average radiation reaction, and $-e\langle\hat{D}_i\rangle$ represents an effective force originating from the electric component of the ZPF; this *diffusive force* plays a role analogous to that played by the osmotic force in Brownian-motion theory.

Exercise E10.4. Time inversion of the evolution equation
Analyze the behavior of the terms appearing in Eq. (10.86) with respect to the time inversion operator, for a function $\mathcal{G}$ that is invariant under time inversion. Assume that enough time has elapsed so that the time integrals in Eqs. (10.83) and (10.84) can be extended to $t \to \infty$.

Solution. The Liouvillian contribution to Eq. (10.86), given by (10.87), contains only terms that change sign under time inversion. In contrast, the two remaining terms in (10.86), which represent the radiative contribution, are invariant under time inversion. Therefore, for long times, assuming that the system has reached a stationary state, $\frac{d}{dt}\langle\mathcal{G}\rangle = 0$, the equation can be separated into two equations,

$$\left\langle\frac{d\mathcal{G}}{dt}\right\rangle_{\mathrm{nr}} = 0, \quad m\tau\left\langle\dddot{x}_i\frac{\partial\mathcal{G}}{\partial p_i}\right\rangle - e^2\left\langle\frac{\partial\mathcal{G}}{\partial p_i}\hat{D}_i\right\rangle = 0. \qquad (10.90)$$

10.3.1 Mean evolution of a constant of motion: A fluctuation-dissipation relation

When $\mathcal{G} = \xi(\mathbf{x}, \mathbf{p})$ represents an integral of motion, or more generally a function of integrals of motion, (10.87) gives

$$\left\langle \dot{x}_i \frac{\partial \xi}{\partial x_i} + f_i \frac{\partial \xi}{\partial p_i} \right\rangle = 0. \tag{10.91}$$

as the student may verify using Hamilton's equations. Equation (10.86) reduces then to

$$\frac{d}{dt}\langle \xi \rangle = m\tau \left\langle \dddot{x}_i \frac{\partial \xi}{\partial p_i} \right\rangle - e^2 \left\langle \frac{\partial \xi}{\partial p_i} \hat{D}_i \right\rangle, \tag{10.92}$$

which shows that only the radiative terms contribute to the evolution of $\langle \xi \rangle$. The system reaches a state of equilibrium when

$$m\tau \left\langle \dddot{x}_i \frac{\partial \xi}{\partial p_i} \right\rangle = e^2 \left\langle \frac{\partial \xi}{\partial p_i} \hat{D}_i \right\rangle. \tag{10.93}$$

This equation is a general diffusion- (or fluctuation-) dissipation relation: the left side represents the average loss of ξ per unit time due to the radiation reaction and the right term represents the average exchange of ξ per unit time with the background field. In particular, when ξ is the Hamiltonian, the condition (10.93) implies that the system has reached a stationary state (an energy eigenstate). It is precisely the combined effect of radiation (dissipation) and diffusion that allows the system to achieve the equilibrium necessary to remain in a stationary quantum state. In Chapter 17, the above equations will be used to calculate the probability of transition between stationary states when the equilibrium is perturbed.

10.4 Transition to the Quantum Description in Terms of Operators

This section is devoted to discussing the change of role of the dynamical variables in the passage to the quantum regime and to explaining the emergence of the corresponding matrix formalism.

We start by recalling the HO case presented in Section 10.2.2, which showed that in the stationary state, the particle responds to the electric component of the radiation field, which acts as a driving force. As shown in Eqs. (10.63)–(10.66), the oscillator's response is a sharp resonance at the frequency ω_0. This is confirmed by the selection rules for dipolar transitions obtained in Section 9.3. A similar situation occurs in the more general case of a conservative (binding) force, when the system can respond resonantly to field modes of more than one frequency. Our purpose in the following is to study the response of the system in such case.

A central observation is that in the transition to the quantum description, the Hamiltonian (symplectic) structure of classical physics is preserved. Therefore, the approach will be Hamiltonian.

10.4.1 Hamiltonian dynamics

Consider an (otherwise) classical charged particle subject to an electromagnetic radiation field. Due to the electrodynamic nature of the problem, we start with the Hamiltonian of the complete system (particle+field),

$$H = \frac{1}{2m}\left(\boldsymbol{p} - \frac{e}{c}\boldsymbol{A}\right)^2 + V(\boldsymbol{r}) + H_\mathrm{R}, \tag{10.94}$$

where $\boldsymbol{A}$ is the vector potential of the radiation field—the ZPF, at a minimum—and H_R is its Hamiltonian. The interaction Hamiltonian H_I has been included in this equation by means of the *minimal coupling*,[6]

$$H_\mathrm{I} = -\frac{e}{mc}\boldsymbol{A}\cdot\boldsymbol{p} + \frac{e^2}{2mc^2}\boldsymbol{A}^2. \tag{10.95}$$

(By choosing the Coulomb gauge $\boldsymbol{\nabla}\cdot\boldsymbol{A} = 0$, one avoids commutation problems that can arise in the quantum case, since $\hat{\boldsymbol{A}}$ and $\hat{\boldsymbol{p}}$ then commute.) The radiation field is expressed as usual in terms of Fourier plane waves,

$$\boldsymbol{A} = \sum_{n,\sigma}\left(\frac{\pi\hbar c^2}{V\omega_n}\right)^{1/2}\boldsymbol{\epsilon}_{n\sigma}a_{n\sigma}\exp[-i(\omega_n t - \boldsymbol{k}_n\cdot\boldsymbol{r})] + \mathrm{c.c.} \tag{10.96}$$

assuming that the system is confined in a (normalization) volume V, usually a parallelopiped, which must be made to tend to infinity at the end of the calculations. The field component $(\boldsymbol{n},\sigma)$ propagates in the direction $\boldsymbol{k}_n$ with $\mathbf{n} = (n_1,n_2,n_3)$, n_i integer; $\omega_n = ck_n$ is its frequency and $\boldsymbol{\epsilon}_{n\sigma}$ ($\sigma = 1,2$)[7] are the two polarization vectors, orthogonal to each other and to $\boldsymbol{k}_n$,

$$\boldsymbol{\epsilon}_{n\sigma}\cdot\boldsymbol{\epsilon}_{n\sigma'} = \delta_{\sigma\sigma'}, \quad \boldsymbol{k}_n\cdot\boldsymbol{\epsilon}_{n\sigma} = 0, \tag{10.97}$$

$$\sum_\sigma \epsilon_{n\sigma i}\epsilon_{n\sigma j} + (k_{ni}k_{nj}/k_n^2) = \delta_{ij}, \tag{10.98}$$

where $\epsilon_{n\sigma i}$ is the i-Cartesian component.

The mode amplitudes $a_{n\sigma}$ are complex random variables, such that $\boldsymbol{A}$ has the correct statistical properties. With the coefficients introduced in the expansion (10.96), the $a_{n\sigma}$ are dimensionless and have the following properties, where $\langle g\rangle_0$ as before denotes the average of g,

$$\langle a_{n\sigma}\rangle_0 = 0, \quad \langle a^*_{n\sigma}\rangle_0 = 0, \tag{10.99}$$

$$\langle a_{n\sigma}a_{n'\sigma'}\rangle_0 = 0, \quad \langle a^*_{n\sigma}a^*_{n'\sigma'}\rangle_0 = 0, \tag{10.100}$$

$$\langle a_{n\sigma}a^*_{n'\sigma'}\rangle_0 = \delta_{nn'}\delta_{\sigma\sigma'}, \tag{10.101}$$

which implies that the field amplitudes corresponding to different modes are statistically independent, i.e., uncorrelated.

---

[6] It is considered in general that the minimal coupling of matter to the electromagnetic field is sufficient and that any further terms required should emerge from the minimal theory; see Section 12.7.1

[7] In principle, a vector field has three degrees of polarization. However, the longitudinal polarization is lost in this case because the electromagnetic field is massless.

For the Hamiltonian H_R of the ZPF one obtains, using (10.96),

$$H_\mathrm{R} = \frac{1}{8\pi}\int \left(E^2 + B^2\right) d^3x = \tfrac{1}{2}\sum_{n,\sigma} \hbar\omega_n a^*_{n\sigma} a_{n\sigma}, \qquad (10.102)$$

so that the average energy per mode is

$$\langle H_\mathrm{R}(\omega_n,\sigma)\rangle_0 = \tfrac{1}{2}\hbar\omega_n \langle a^*_{n\sigma} a_{n\sigma}\rangle = \tfrac{1}{2}\hbar\omega_n. \qquad (10.103)$$

Using the Hamiltonian (10.94) one can derive the Hamilton equations for the particle in the usual way,

$$\dot{x}_i = \{x_i, H\}, \qquad (10.104)$$
$$\dot{p}_i = \{p_i, H\}, \qquad (10.105)$$

where the Poisson brackets are calculated not only with respect to the canonical variables of the particle, as in a mechanical system, but with respect to the entire set of canonical variables of both the particle and the field.[8] With H given by (10.94), the Hamilton equations lead to an equation of motion containing a time integral; if this is expanded as a series of time derivatives and truncated to the third order as usual, Eq. (10.81) is obtained in the dipole approximation.

10.4.2 The new kinematics

To analyze the change in the kinematics, we start from the Poisson bracket of the particle's canonical variables,

$$\{x_j, p_i\}_{xp} = \sum_k \left(\frac{\partial x_j}{\partial x_k}\frac{\partial p_i}{\partial p_k} - \frac{\partial p_i}{\partial x_k}\frac{\partial x_j}{\partial p_k}\right), \qquad (10.106)$$

which satisfies

$$\{x_j, p_i\}_{xp} = \delta_{ij}, \qquad (10.107)$$

where $i,j,k = 1,2,3$, and the variables and the derivatives are taken at the same time t.

The *full* set of canonical variables at any time comprises both those of the particle, $\{x_i; p_i\}$, and those of the field modes, $\{\mathrm{q}_\alpha; \mathrm{p}_\alpha\}$ (roman typography is used for the field's canonical variables to distinguish them from those of the particle, and a discrete set of field modes is considered for reasons that will become clear later), i.e.,

$$\{q; p\} = \{x_i, \mathrm{q}_\alpha; p_i, \mathrm{p}_\alpha\}. \qquad (10.108)$$

---

[8] In classical theory, the phase space of a physical system consists of all the possible values of the (independent) position and momentum variables allowed by the system. It is naturally endowed with a Poisson bracket or symplectic form (see (10.107)), which allows one to formulate the Hamilton equations and describe the dynamics of the system through phase space in time. If the complete set of canonical variables is $\{q;p\}$, the Poisson bracket of any pair of functions $f(q,p,t), g(q,p,t)$ is

$$\{f,g\} = \sum_{q,p}\left[(\partial f/\partial q)(\partial g/\partial p) - (\partial f/\partial p)(\partial g/\partial q)\right]$$

where the sum extends over all possible values of (q, p). See also footnote 4.

Note that a semicolon is used to distinguish a set of variables from a Poisson bracket. At the initial time t_o, when particle and field begin to interact, the canonical variables are

$$\{q_o; p_o\} = \{x_{io}, q_{\alpha o}; p_{io}, p_{\alpha o}\}, \tag{10.109}$$

where x_{io}, p_{io} are the initial values of the particle's variables, and $q_{\alpha o}, p_{\alpha o}$ are those of the field. Since the whole system is Hamiltonian, the variables at times t_0 and t are related by a canonical transformation, and the particle's Poisson bracket at time t can be taken with respect to either set of variables,

$$\{x, p\}_{xp} = \{x, p\}_{x_o p_o} + \{x, p\}_{q_{\alpha o} p_{\alpha o}}. \tag{10.110}$$

Therefore, according to Eq. (10.107),

$$\{x_i(t), p_j(t)\}_{x_o p_o} + \{x_i(t), p_j(t)\}_{q_{\alpha o} p_{\alpha o}} = \delta_{ij}. \tag{10.111}$$

Now according to the discussion in Section 10.3, when in a stationary state, the particle has lost track of its initial conditions $\boldsymbol{x}_o, \boldsymbol{p}_o$ and responds resonantly to the action of the field. Then the first term in Eq. (10.111) vanishes,

$$\{x_i(t), p_j(t)\}_{q_o p_o} \to \{x_i(t), p_j(t)\}_{q_{\alpha o} p_{\alpha o}}, \tag{10.112}$$

and the particle's Poisson bracket becomes defined by the canonical variables of the field modes α with which it interacts,

$$\{x_i(t), p_j(t)\}_{q_{\alpha o} p_{\alpha o}} = \delta_{ij}. \tag{10.113}$$

Using the normal field amplitudes introduced in Eq. (10.96), related to $q_{\alpha o}, p_{\alpha o}$ by the transformation

$$\omega_\alpha q_{\alpha o} = \sqrt{\hbar \omega_\alpha / 2}(a_\alpha + a_\alpha^*), \quad p_{\alpha o} = -i\sqrt{\hbar \omega_\alpha / 2}(a_\alpha - a_\alpha^*), \tag{10.114}$$

we write the Poisson bracket of two functions f, g with respect to a_α, a_α^*,

$$\{f, g\}_{aa^*} = \sum_\alpha \left(\frac{\partial f}{\partial a_\alpha} \frac{\partial g}{\partial a_\alpha^*} - \frac{\partial g}{\partial a_\alpha} \frac{\partial f}{\partial a_\alpha^*} \right) = i\hbar \{f, g\}_{q_{\alpha o} p_{\alpha o}}. \tag{10.115}$$

Applied to the particle's canonical variables, (10.115) reads

$$i\hbar \{x_i, p_j\}_{q_{\alpha o} p_{\alpha o}} = \{x_i, p_j\}_{aa^*}, \tag{10.116}$$

and therefore, according to Eq. (10.113), for the particle in a stationary state we get

$$\{x_i, p_j\}_{aa^*} = i\hbar \delta_{ij}. \tag{10.117}$$

This result indicates that the relation between the particle variables x_i and p_j becomes determined by their functional dependence on the normal field amplitudes $\{a_\alpha; a_\alpha^*\}$, with the scale given by Planck's constant. The symplectic structure is inherited from classical mechanics, but the meaning of the quantities x_i, p_j has changed, as we will see below.

10.4.3 From dynamical variables to quantum operators

When the particle is subject solely to the ZPF, the only possible stationary state is the ground state $n = 0$; when there is an additional (external) field, e.g., a thermal field, the particle can reach an excited state $n > 0$. Given the general character of (10.117), it can be applied to any stationary state n. We therefore label the variables x_i, p_j with the subindex n, and write using (10.115) (limiting the discussion to the 1D case, for simplicity, and henceforth omitting the subindices aa^* from the Poisson bracket),

$$\{x, p\}_{nn} = \sum_\alpha \left(\frac{\partial x_n}{\partial a_\alpha} \frac{\partial p_n}{\partial a_\alpha^*} - \frac{\partial p_n}{\partial a_\alpha} \frac{\partial x_n}{\partial a_\alpha^*} \right) = i\hbar. \tag{10.118}$$

The constant value $i\hbar$ of this bilinear form implies that x_n and $p_n = m\dot{x}_n$ are linear functions of the set $\{a_\alpha; a_\alpha^*\}$. This means that the (electric component of the) field has taken control of the particle's behavior. The field modes involved are those to which the particle responds resonantly, i.e., those that can take it from state n to another state, say k, i.e., $\{a_\alpha; a_\alpha^*\} \to \{a_{nk}; a_{nk}^*\}$,

$$x_n(t) = \sum_k x_{nk} a_{nk} e^{-i\omega_{kn} t} + \text{c.c.}, \quad p_n(t) = \sum_k p_{nk} a_{nk} e^{-i\omega_{kn} t} + \text{c.c.}, \tag{10.119}$$

where a_{nk} is associated with the field mode connecting state n with state k, and x_{nk}, $p_{nk} = -im\omega_{kn} x_{nk}$ are the respective dipolar response coefficients. Substituting these expressions into Eq. (10.118) gives

$$\{x, p\}_{nn} = 2im \sum_k \omega_{kn} |x_{nk}|^2 = i\hbar \tag{10.120}$$

for any n. Further, since the $a_{nk}, a_{n'k}$ connecting different states n, n' with state k are independent random variables, as mentioned in Section 10.4.1, by combining Eqs. (10.115) and (10.119) one obtains

$$\{x, p\}_{nn'} = i\hbar \delta_{nn'}. \tag{10.121}$$

The quantities x_{nk} and a_{nk} refer to the transition $n \to k$, while x_{kn} and a_{kn} refer to the inverse transition, with $\omega_{nk} = -\omega_{kn}$; therefore, from (10.119), $x_{nk}^*(\omega_{nk}) = x_{kn}(\omega_{kn})$, $p_{nk}^*(\omega_{nk}) = p_{kn}(\omega_{kn})$, $a_{nk}^*(\omega_{nk}) = a_{kn}(\omega_{kn})$, whence (10.121) takes the form

$$\sum_k (x_{nk} p_{kn'} - p_{n'k} x_{kn}) = i\hbar \delta_{nn'}. \tag{10.122}$$

The coefficients x_{nk} and p_{nk} can thus be identified as the elements of two matrices $\hat{x}$ and $\hat{p}$, respectively, with as many rows and columns as there are different states, and the Poisson bracket (10.121) transforms into the commutator

$$\left[\hat{x}, \hat{p}\right]_{nn'} = i\hbar \delta_{nn'}, \tag{10.123}$$

i.e., $\left[\hat{x}, \hat{p}\right] = i\hbar \mathbb{I}$, where $\mathbb{I}$ is the $n \times n$ unit matrix, or

$$\left[\hat{x}, \hat{p}\right] = i\hbar \tag{10.124}$$

for short. Equation (9.3) is thus identified with the Thomas-Reiche-Kuhn sum rule, Eq. (3.82).

In conclusion, the classical phase-space variables x, p are replaced by the corresponding operators $\hat{x}, \hat{p}$, and the Poisson brackets are replaced by commutators,

$$\{x_i, p_j\} \Rightarrow \frac{1}{i\hbar}\left[\hat{x}_i, \hat{p}_j\right]. \tag{10.125}$$

The relationship between Poisson brackets and commutators, formally established by Dirac as a matter of fact, finds its physical basis here. The commutator (10.124) is no longer a postulate, as presented in Chapter 3, but an expression of the response of the particle to the field when the stationary regime is reached. Although the ZPF has explicitly disappeared from the picture – from the description of QM! – it has left Planck's constant $\hbar$ as an indelible mark. The vanishing of the ZPF means that although we are dealing with a causal phenomenon, the theory becomes non-causal, with unexplained quantum fluctuations. This is the origin of some of the QM conceptual problems mentioned in Chapter 1.

The matrix element x_{ln} represents the amplitude (or coefficient) of the *linear resonant response* of the particle to the field mode (ln) that takes it from state n to state l. This result is consistent with the quantum description and, as will be shown in Sections 12.2 and 17.2, is expressed in the selection rules for atomic dipole transitions, which depend directly on the coefficients x_{ln}.

It is important to note that while there is a formal one-to-one correspondence between variables and operators, they differ substantially in nature *and* physical meaning. Since $\hat{x}, \hat{p}$ do not describe particle trajectories, there is no phase space associated with them. Of course, this does not mean that trajectories have ceased to exist; they are simply not part of the quantum description. At the cost of a detailed dynamics in space-time, the Hilbert space formalism provides a compact and elegant description in terms of the elements connecting the quantum states.

10.4.4 Applying the quantum formalism

Equation (10.124) is used as the basis for the general transformation of dynamical variables into operators. In particular, it allows one to formulate the Heisenberg equations that govern the quantum dynamics in Hilbert space. Once x and p become operators, all dynamical variables $G(x, p; t)$ become operators $\widehat{G}(\hat{x}, \hat{p}; t)$ that act on the states, which are represented by column matrices.

Therefore, when applying Eq. (10.86) and the following to study the dynamics in the quantum regime, one must express them in terms of operators instead of c-variables, and replace the averages by expectation values. In particular, this means that the Liouvillian (radiationless) equation (10.87), which can be written as the average of Hamilton's equation (with $H = (\boldsymbol{p}^2/2m) + V$),

$$\langle \dot{G} \rangle = \left\langle \frac{\partial G}{\partial t} \right\rangle + \langle \{G, H\} \rangle, \tag{10.126}$$

takes the form

$$\langle \dot{\hat{G}} \rangle = \left\langle \frac{\partial \hat{G}}{\partial t} \right\rangle + i\hbar \left\langle \left[\hat{G}, \hat{H}\right] \right\rangle, \tag{10.127}$$

where the angular bracket is to be interpreted as the expectation value of the Heisenberg equation. Experience shows that this replacement is usually straightforward, although occasionally some additional analysis is required to take account of the order of the (noncommuting) operators. A representative example is the calculation of the hydrogen Lamb shift in Section 17.4.4.

To recapitulate, quantum mechanics – whether in the Heisenberg or any other representation – is obtained by taking the radiationless approximation of the SED statistical equations in the Markovian limit, which implies a coarse graining over time. For very short times there is no guarantee that the quantum equations hold. The stationary quantum regime is ensured by a general fluctuation-dissipation relation, which expresses the balanced effect of the radiation reaction and the zero-point-field-induced stochasticity. In this regime the operators $\hat{x}, \hat{p}$ represent the resonant response of the particle to the field, which induces transitions between states, and all dynamical variables become functions of these operators. The classical Hamiltonian structure is preserved by the quantum formalism in terms of operators. The radiative terms in the SED equations contain valuable additional information that takes us beyond QM.

10.5 Hydrodynamic Quantum Analogs

We have seen in Section 10.1 that SQM leads to a differential equation that is parabolic (and dissipative) in the classical case of Brownian motion and hyperbolic (wave-like and conservative) in the quantum case. Furthermore, we have just seen that quantization can be understood as the result of the balancing action of the background field, which compensates for the particle's radiation. The fact that this underlying physical field, which acts on the particles and mediates between them, is hidden from the quantum description introduces *apparent* nonlocalities that permeate the whole of quantum theory.

It turns out that a similar transition from a classical dynamics to one that reproduces characteristic quantum features, has been observed on a macroscopic scale in experiments with oil droplets bouncing on a vibrating bath of the same oil. Under certain conditions the droplet can self-propel due to its interaction with the traveling waves it generates. As the walking droplet bounces repeatedly off the surface of the oil bath, the sequence of waves generated is superimposed. In the high-memory regime (close to the Faraday threshold, beyond which standing waves form even in the absence of the drop), the resulting waves decay very slowly in time, so the trajectory of the droplet is significantly influenced by waves accumulated since the distant past, resulting in memory-induced wave-like particle statistics. From the perspective of SED, the system has a suggestive quantum similarity, with the advantage that the droplet trajectories can be visualized, and the origin of the apparent nonlocalities can be understood as arising from the (local) wave-particle interaction.

This topic has given rise to an attractive and vibrant new field of research known as *hydrodynamic quantum analogs* (HQA). Some relevant references are given at the end of the chapter. To illustrate the similarities, below is a brief description of an experiment with circularly orbiting droplets carried out by Fort et al. (2010).

In this experiment the oil is contained in a small cylindrical cell (150 mm diameter) which is vibrated vertically and rotated about a central vertical axis at an angular velocity Ω. When the cell is set in rotation, the walker experiences a Coriolis force. Seen from the rotating frame, the droplet trajectory is a circular orbit. For long-term memory, as Ω is increased, the radius of the orbit undergoes transitions between successive values and a slight increase in velocity is observed as the orbits become smaller. The result is that each droplet describes a circular trajectory with a radius that can have any of the values given by

$$R_n \simeq \frac{1}{2}\left(n + \frac{1}{2}\right)\lambda_F, \qquad (10.128)$$

where $n = 0, 1, 2, 3, \ldots$ and λ_F corresponds to the Faraday wavelength of the oil bath.

In QM, the motion of a charge moving in a plane and subject to a homogeneous magnetic field perpendicular to the plane is quantified in Landau levels (see Exercise 9.33). In the case of circular orbits, the charge can orbit in equidistant rings where the radius takes discrete values given by

$$r_n = \frac{1}{\pi}\left(n + \frac{1}{2}\right)\lambda, \qquad (10.129)$$

and λ is the de Broglie wavelength.

While the classical Faraday and the quantum de Broglie pilot-wave systems are of completely different scales, their theoretical descriptions differ only in minor details and the mathematical formulae (10.128), (10.129) have similar structure and the same meaning.

How far analogies like this one can be taken remains to be seen. However, to the more or less "obvious" analogy we must add a profound fundamental kinship. As discussed in several places in this book, with the introduction of the ZPF, QM becomes a causal local theory of nature; it becomes a true matter-field theory, just like the theory of the walking droplets, where quantum – or "apparent" quantum – behavior arises through the action of the immediate field on the particle, as modulated by the particle's own motion.

Problems

P10.1 Two very thin metal plates, 10 Å on each side and initially neutral, are 20 Å apart. Calculate the number of electrons that need to be added to each plate for the Coulomb repulsion to counteract the Casimir attraction.

P10.2 Construct the general solution $x(t)$ for the HO immersed in the ZPF with arbitrary initial conditions, and show that the stationary solution (for $t \to \infty$) is given approximately by Eq. (10.56).

P10.3 Derive in detail the autocorrelation function and the variance of $p(t)$ for the stationary HO.

P10.4 Show that the definitions used in early SED give for the dispersion of the ground state energy of the HO $\sigma_H^2 = E_0^2$.

P10.5 Stationary white noise is so named because its spectral energy density is a constant, the same for all frequencies. Show that the autocorrelation function of a white noise is $\langle E(t') E(t) \rangle = A\delta(t' - t)$ and express the value of the constant A in terms of the power density.

P10.6 Consider the 1D infinite square well of width a. a) Using the eigenfunctions obtained in Exercise 4.4, determine the transition frequencies ω_{nk} and the corresponding response coefficients (matrix elements) x_{kn} for $n = 1$. b) Carry out the same calculations for $n = 2$. c) Make a drawing of the spectral absorption bands obtained for the two cases.

Bibliographical Notes

The standard reference on SQM is Nelson (1966). See also de la Peña (1969) and Nelson (2012).

For the distinction between Brownian motion and QM as two stochastic processes, see de la Peña and Cetto (1975, 1982).

For a simulation of Brownian motion, see https://en.wikipedia.org/wiki/Brownian_motion#/media/File:Brownian_motion_large.gif.

For the introduction of the quantum potential see Bohm (1951).

For a detailed discussion of causality and nonlocality in the context of Bohm's theory, see de la Peña, Cetto, and Valdés-Hernández (2015), Chapter 8.

To familiarize the reader with the theory of stochastic processes, in particular with the Fokker-Planck equation, we recommend van Kampen (1981) and Risken (1996).

To learn about classical electrodynamics, we recommend Zangwill (2012), a comprehensive textbook that emphasizes physics without sacrificing mathematical rigor.

For pioneering work on SED, see Marshall (1963, 1965), Santos (1968), and Boyer (1979). For a more complete exposition, which includes some aspects of stochastic optics, see de la Peña, and Cetto (1996).

On more recent developments in SED and its connection with SQM and QM, see de la Peña, Cetto, and Valdés-Hernández (2015). In Cetto, de la Peña, and Valdés-Hernández (2021) and de la Peña and Cetto (2024) the transition to quantum operators is explained in detail.

For the hydrodynamic quantum analogs and the connection with de Broglie's pilot wave, see Bush (2015) and Bush and Oza (2020). For the experiment described in Section 5, see Fort et al. (2010).

11 Angular Momentum Theory

In this chapter we will develop the quantum theory of angular momentum. As in classical mechanics, working with angular momentum requires that we leave the one-dimensional space in which we have been working so far, with all the complications that this entails.

For two- or three-dimensional problems, the notion of angular momentum is not only very useful, but to a large extent necessary. In QM, this importance is accentuated by at least three facts: (i) central problems are of particular interest (the problem of the H atom is paradigmatic); (ii) many particles have an intrinsic angular momentum (the spin); and (iii) the physics of angular momentum involves features such as the so-called *space quantization*, that have no classical analog but are essential for describing the behavior of quantum systems, even the simplest atom. Moreover, since angular momentum is described by an operator, studying its properties gives us another opportunity to learn how to deal with the quantum formalism.

11.1 Orbital Angular Momentum. Spherical Harmonics

We saw in Section 8.7.2 that the orbital angular momentum operator $\widehat{L}$ can be defined in QM precisely as in classical theory,

$$\widehat{L} = \widehat{r} \times \widehat{p}. \tag{11.1}$$

If you have not read that section, take Eq. (11.1) as the definition. Note that $\widehat{r} \times \widehat{p} = -\widehat{p} \times \widehat{r}$ in classical physics, so there is no problem with the order of the factors when moving from the classical to the quantum case. In Cartesian coordinates, the components of $\widehat{L}$ are

$$\widehat{L}_i = \varepsilon_{ijk} j \colon x_j \widehat{x} \widehat{p}_k = -i\hbar \varepsilon_{ijk} x_j \frac{\partial}{\partial x_k} \tag{11.2}$$

(summing over repeated indices is understood; ε_{ijk} is the totally anti-symmetric 3-D Levi–Civita tensor). In spherical coordinates we have

$$\widehat{L} = -i\hbar \left(\widehat{a}_\varphi \frac{\partial}{\partial \theta} - \widehat{a}_\theta \frac{1}{\sin\theta} \frac{\partial}{\partial \varphi} \right), \tag{11.3}$$

whence $\widehat{L}$ commutes with functions of r and of $\partial/\partial r$, that is,

$$\left[\widehat{L}, f\left(r, \frac{\partial}{\partial r}\right) \right] = 0. \tag{11.4}$$

This feature will prove very useful.

We will now study the commutation properties of $\widehat{\boldsymbol{L}}$ with different dynamical variables. We start with the following equations in Cartesian coordinates,

$$[x_i, P(\widehat{p})] = i\hbar \frac{\partial P}{\partial \widehat{p}_i}, \quad [\widehat{p}_i, Q(\widehat{x})] = -i\hbar \frac{\partial Q}{\partial \widehat{x}_i}. \tag{11.5}$$

From the first expression it follows that

$$[x_i \widehat{p}_j, P(\widehat{p})] = x_i \widehat{p}_j P(\widehat{p}) - P(\widehat{p}) x_i \widehat{p}_j = (x_i P - P x_i) \widehat{p}_j = i\hbar \frac{\partial P}{\partial \widehat{p}_i} \widehat{p}_j. \tag{11.6}$$

Interchanging the indices i and j and taking the difference between the two results gives

$$[\widehat{L}_k, P(\widehat{p})] = [x_i \widehat{p}_j - x_j \widehat{p}_i, P(\widehat{p})] = i\hbar \left(\frac{\partial P}{\partial p_i} \widehat{p}_j - \frac{\partial P}{\partial p_j} \widehat{p}_i \right). \tag{11.7}$$

Similarly, from Eq. (11.5) it follows that

$$[x_i \widehat{p}_j, Q(x)] = x_i \widehat{p}_j Q(x) - x_i Q(x) \widehat{p}_j = x_i [\widehat{p}_j, Q(x)] = -i\hbar x_i \frac{\partial Q}{\partial x_j}, \tag{11.8}$$

which gives

$$[\widehat{L}_k, Q(x)] = [x_i \widehat{p}_j - x_j \widehat{p}_i, Q(x)] = -i\hbar \left(x_i \frac{\partial Q}{\partial x_j} - x_j \frac{\partial Q}{\partial x_i} \right). \tag{11.9}$$

In Eqs. (11.7) and (11.9) the indices i, j, k are in cyclic order. As a simple example of application of these results, we calculate the commutators $[\widehat{L}_i, x_j]$ and $[\widehat{L}_i, \widehat{p}_j]$,[1]

$$[\widehat{L}_x, x] = -i\hbar \left(y \frac{\partial x}{\partial z} - z \frac{\partial x}{\partial y} \right) = 0,$$

$$[\widehat{L}_x, y] = -i\hbar \left(y \frac{\partial y}{\partial z} - z \frac{\partial y}{\partial y} \right) = i\hbar z,$$

$$[\widehat{L}_x, z] = -i\hbar \left(y \frac{\partial z}{\partial z} - z \frac{\partial z}{\partial y} \right) = -i\hbar y; \tag{11.10}$$

$$[\widehat{L}_x, \widehat{p}_x] = -i\hbar \left(p_y \frac{\partial p_x}{\partial p_z} - p_z \frac{\partial p_x}{\partial p_y} \right) = 0,$$

$$[\widehat{L}_x, \widehat{p}_y] = i\hbar \widehat{p}_z, \quad [\widehat{L}_x, \widehat{p}_z] = -i\hbar \widehat{p}_y. \tag{11.11}$$

Of particular importance are the commutators of the components of $\widehat{\boldsymbol{L}}$, which we proceed to determine. Since the $\widehat{L}_i$ are linear combinations of components of $\boldsymbol{r}$ and $\widehat{\boldsymbol{p}}$, it follows that they do not commute with each other (we give the Cartesian index the values 1, 2, 3 or x, y, z, as convenient). Specifically,

$$[\widehat{L}_x, \widehat{L}_y] = [\widehat{L}_x, z\widehat{p}_x] - [\widehat{L}_x, x\widehat{p}_z]$$
$$= [\widehat{L}_x, z]\widehat{p}_x - x[\widehat{L}_x, \widehat{p}_z] = -i\hbar y \widehat{p}_x + i\hbar x \widehat{p}_y = i\hbar L_z, \tag{11.12}$$

where we have used (11.10) and (11.11). To the commutator

$$[\widehat{L}_x, \widehat{L}_y] = i\hbar \widehat{L}_z \tag{11.13}$$

---

[1] Note that Eqs. (11.10) and (11.11) are particular cases of the general law $\left[\hat{L}_i, \hat{A}_j \right] = i\hbar \varepsilon_{ijk} \hat{A}_k$ derived in Section 8.7.2.

we may add the cyclic permutations,

$$[\hat{L}_y, \hat{L}_z] = i\hbar \hat{L}_x, \quad [\hat{L}_z, \hat{L}_x] = i\hbar \hat{L}_y, \tag{11.14}$$

and write the results formally and succinctly as

$$\hat{\mathbf{L}} \times \hat{\mathbf{L}} = i\hbar \hat{\mathbf{L}}. \tag{11.15}$$

(Note that the cross product of a vector by itself is not necessarily zero, if the vector is an operator.)

To calculate the commutator of a component $\hat{L}_i$ with the operator $\hat{\mathbf{L}}^2$, we write

$$\hat{\mathbf{L}}^2 = \hat{L}_x^2 + \hat{L}_y^2 + \hat{L}_z^2, \tag{11.16}$$

so we get, for example,

$$[\hat{\mathbf{L}}^2, \hat{L}_z] = [\hat{L}_x^2 + \hat{L}_y^2, \hat{L}_z]$$
$$= \hat{L}_x[\hat{L}_x, \hat{L}_z] + [\hat{L}_x, \hat{L}_z]\hat{L}_x + \hat{L}_y[\hat{L}_y, \hat{L}_z] + [\hat{L}_y, \hat{L}_z]\hat{L}_y = 0, \tag{11.17}$$

i.e.,

$$[\hat{\mathbf{L}}^2, \hat{L}_i] = 0, \quad i = 1, 2, 3. \tag{11.18}$$

The result just obtained was expected, since $\hat{\mathbf{L}}^2$ is a scalar operator, and thus invariant to rotations of the coordinate system, and the $\hat{L}_k$ are the generators of such rotations. Since $\hat{\mathbf{L}}^2$ commutes with each component $\hat{L}_i$, it looks as if simultaneous eigenstates of $\hat{\mathbf{L}}^2$ and the three $\hat{L}_i$ exist. However, Eqs. (11.13) and (11.14) show that the $\hat{L}_i$ do not commute with each other, so at most one of them can be fixed in a given state. This means that although all three components are integrals of motion, only one of them can have a well-defined value. Taking, in a purely conventional way, L_z as the variable with a well-defined value, we can construct the simultaneous eigenstate of $\hat{\mathbf{L}}^2$ and $\hat{L}_z$, namely $|L\, L_z\rangle$, which in general is not an eigenstate of $\hat{L}_x$ and $\hat{L}_y$. In fact, by applying Heisenberg's inequality to Eq. (11.13), we obtain

$$\langle (\Delta \hat{L}_x)^2 \rangle \langle (\Delta \hat{L}_y)^2 \rangle \geq \tfrac{1}{4}\hbar^2 |\langle \hat{L}_z \rangle|^2, \tag{11.19}$$

which shows that when $\langle \hat{L}_z \rangle \neq 0$ (its value is well defined), the components $\hat{L}_x$ and $\hat{L}_y$ are necessarily distributed. However, if $\langle \hat{L}_z \rangle = 0$, all projections of $\hat{\mathbf{L}}$ can be well defined, with $\langle \hat{L}_x \rangle = \langle \hat{L}_y \rangle = \langle \hat{L}_z \rangle = 0$. Let us look at this last case more closely. If $\langle \hat{\mathbf{L}} \rangle = 0$ or, what is equivalent,

$$\hat{\mathbf{L}}|\psi_0\rangle = 0, \tag{11.20}$$

where $|\psi_0\rangle$ corresponds to the zero eigenvalue of L_i, the three Cartesian components of $\hat{\mathbf{L}}$ are well defined (all three eigenvalues are 0). Writing the wave function as $\psi_0(r,\theta,\varphi) = \langle r,\theta,\varphi|\psi_0\rangle$, we see from (11.3) and (11.20) that the following differential equation must hold

$$\left(\hat{\mathbf{a}}_\varphi \frac{\partial}{\partial \theta} - \hat{\mathbf{a}}_\theta \frac{1}{\sin\theta} \frac{\partial}{\partial \varphi} \right) \psi_0 = 0, \tag{11.21}$$

whose general solution

$$\psi_0 = \psi(r), \tag{11.22}$$

does not depend on the angular variables, that is, it is a state with spherical symmetry. These are the so-called s-states, with zero orbital angular momentum. For any other state, assuming that $\widehat{L}_z$ is well defined, the variables $\widehat{L}_x$ and $\widehat{L}_y$ are dispersive.

11.1.1 Dynamics of angular momentum

From Eq. (11.7) with $P(\hat{p}) = \widehat{\boldsymbol{p}}^2/2m$, we see that the three components of the angular momentum commute with the kinetic-energy operator,

$$\left[\widehat{L}_x, \frac{\widehat{\boldsymbol{p}}^2}{2m}\right] = \frac{i\hbar}{2m}\left(\frac{\partial \widehat{\boldsymbol{p}}^2}{\partial p_y}\widehat{p}_z - \frac{\partial \widehat{\boldsymbol{p}}^2}{\partial p_z}\widehat{p}_y\right) = \frac{i\hbar}{m}(\widehat{p}_y\widehat{p}_z - \widehat{p}_z\widehat{p}_y) = 0, \qquad (11.23)$$

or in vector form,

$$\left[\widehat{\boldsymbol{L}}, \frac{\widehat{\boldsymbol{p}}^2}{2m}\right] = 0. \qquad (11.24)$$

On the other hand, applying Eq. (11.9) to the potential energy, we get

$$[\widehat{L}_x, V(\boldsymbol{r})] = -i\hbar\left(y\frac{\partial V}{\partial z} - z\frac{\partial V}{\partial y}\right) = i\hbar(yF_z - zF_y), \qquad (11.25)$$

where $F_i = -\partial V/\partial x_i$ is the external force. Since $yF_z - zF_y$ is the x-component of the momentum of the external force (or torque) $\widehat{\boldsymbol{M}}$,

$$\widehat{\boldsymbol{M}} = \widehat{\boldsymbol{r}} \times \widehat{\boldsymbol{F}}, \qquad (11.26)$$

we can write the preceding result in vector form,

$$[\widehat{\boldsymbol{L}}, V(\boldsymbol{r})] = i\hbar\widehat{\boldsymbol{M}}. \qquad (11.27)$$

Adding Eqs. (11.24) and (11.27) and introducing the Hamiltonian $\widehat{H} = \widehat{\boldsymbol{p}}^2/2m + V(\boldsymbol{r})$ gives

$$[\widehat{\boldsymbol{L}}, \widehat{H}] = i\hbar\widehat{\boldsymbol{M}}. \qquad (11.28)$$

This equation shows that if the external force exerts a torque $\boldsymbol{M}$ on the particle, $\widehat{\boldsymbol{L}}$ does not commute with the Hamiltonian. Combining Eq. (11.28) with the Heisenberg equation of motion applied to $\widehat{\boldsymbol{L}}$ (writing $\dot{\widehat{\boldsymbol{L}}} = \widehat{\dot{\boldsymbol{L}}}$),

$$i\hbar\dot{\widehat{\boldsymbol{L}}} = [\widehat{\boldsymbol{L}}, \widehat{H}], \qquad (11.29)$$

we obtain the law of evolution of the angular momentum

$$\frac{d\widehat{\boldsymbol{L}}}{dt} = \widehat{\boldsymbol{M}}, \qquad (11.30)$$

which has the same form as the corresponding classical law. We see that $\widehat{\boldsymbol{M}}$ is the cause of the evolution of $\widehat{\boldsymbol{L}}$.

For radial or central forces (for which $\widehat{\boldsymbol{M}} = 0$), $\widehat{H}$ and $\widehat{\boldsymbol{L}}$ commute and therefore have common eigenvectors. The reason for this result is that in this case the Hamiltonian is invariant with respect to rotations of the coordinate system and therefore invariant with respect to the generator of rotations, $\widehat{\boldsymbol{L}}$.

When $\langle\widehat{M}\rangle \neq 0$, we can use Eq. (8.73) to determine a characteristic time of evolution τ_L of the angular momentum. If $\langle\Delta L\rangle$ represents the standard deviation of $\widehat{L}$ in the given state, we have

$$\tau_L = \frac{\langle\Delta L\rangle}{|\langle\widehat{M}\rangle|}. \tag{11.31}$$

Therefore, according to the Mandelstam–Tamm inequality (8.75) applied to L,

$$\langle(\Delta\widehat{H})^2\rangle \left\langle \left(\frac{\Delta\widehat{L}}{d\langle\widehat{L}\rangle/dt}\right)^2 \right\rangle \geq \tfrac{1}{4}\hbar^2, \tag{11.32}$$

the evolution time τ_L is not arbitrarily short but is bounded from below by the standard deviation of the energy.

11.1.2 Spherical harmonics and their eigenvalues

We will now establish the differential equations for the eigenfunctions and eigenvalues of the orbital angular-momentum operators. It is convenient to use spherical coordinates for this purpose, since the angular-momentum operator commutes with the variable r and acts only on the angular variables θ and φ, so it is a function of only these two variables.

We look for simultaneous eigenfunctions of $\widehat{L}^2$ and one of its components, say, $\widehat{L}_z$, so the equations to solve are

$$\widehat{L}^2 Y(\theta,\varphi) = \hbar^2 \lambda Y(\theta,\varphi), \tag{11.33}$$

$$\widehat{L}_z Y(\theta,\varphi) = \hbar m Y(\theta,\varphi), \tag{11.34}$$

where $Y(\theta,\varphi)$ are the eigenfunctions, and the eigenvalues λ and m are dimensionless parameters to be determined. From (11.3) squared, we get

$$\widehat{L}^2 = -\hbar^2 \left(\widehat{a}_\varphi \partial_\theta - \frac{\widehat{a}_\theta}{\sin\theta}\partial_\varphi\right)^2$$
$$= -\hbar^2 \left[\widehat{a}_\varphi \cdot \partial_\theta(\widehat{a}_\varphi \partial_\theta) - \widehat{a}_\varphi \cdot \partial_\theta\left(\frac{\widehat{a}_\theta}{\sin\theta}\partial_\varphi\right)\right.$$
$$\left. - \frac{\widehat{a}_\theta}{\sin\theta}\cdot\partial_\varphi(\widehat{a}_\varphi \partial_\theta) + \frac{\widehat{a}_\theta}{\sin\theta}\cdot\partial_\varphi\left(\frac{\widehat{a}_\theta}{\sin\theta}\partial_\varphi\right)\right]. \tag{11.35}$$

From the expressions for $\widehat{a}_r, \widehat{a}_\theta$ and $\widehat{a}_\varphi$ in terms of the Cartesian unit vectors $\widehat{a}_x, \widehat{a}_y, \widehat{a}_z$,

$$\widehat{a}_r = \widehat{a}_x \sin\theta\cos\varphi + \widehat{a}_y \sin\theta\sin\varphi + \widehat{a}_z \cos\theta, \tag{11.36}$$
$$\widehat{a}_\theta = \widehat{a}_x \cos\theta\cos\varphi + \widehat{a}_y \cos\theta\sin\varphi - \widehat{a}_z \sin\theta, \tag{11.37}$$
$$\widehat{a}_\varphi = -\widehat{a}_x \sin\varphi + \widehat{a}_y \cos\varphi, \tag{11.38}$$

it follows that

$$\partial_\theta \widehat{a}_\varphi = 0, \quad \partial_\theta \widehat{a}_\theta = -\widehat{a}_r,$$
$$\partial_\varphi \widehat{a}_\theta = \widehat{a}_\varphi \cos\theta, \quad \partial_\varphi \widehat{a}_\varphi = -\widehat{a}_r \sin\theta - \widehat{a}_\theta \cos\theta. \tag{11.39}$$

Plugging these results into Eq. (11.35) we get

$$\widehat{L}^2 = -\hbar^2 \left[\partial_\theta^2 + \frac{\cos\theta}{\sin\theta} \partial_\theta + \frac{1}{\sin^2\theta} \partial_\varphi^2 \right]$$
$$= -\hbar^2 \left[\frac{1}{\sin^2\theta} \partial_\varphi^2 + \frac{1}{\sin\theta} \partial_\theta (\sin\theta \, \partial_\theta) \right]. \quad (11.40)$$

To construct the operator $\widehat{L}_z$ we project the $\widehat{\mathbf{L}}$ given by Eq. (11.3) in the z-direction,

$$\widehat{L}_z = -i\hbar \widehat{\mathbf{a}}_z \cdot \left(\widehat{\mathbf{a}}_\varphi \partial_\theta - \frac{\widehat{\mathbf{a}}_\theta}{\sin\theta} \partial_\varphi \right). \quad (11.41)$$

Since from Eqs. (11.37) and (11.38) it follows that $\widehat{\mathbf{a}}_z \cdot \widehat{\mathbf{a}}_\varphi = 0$, $\widehat{\mathbf{a}}_z \cdot \widehat{\mathbf{a}}_\theta = -\sin\theta$, the result is

$$\widehat{L}_z = -i\hbar \partial_\varphi. \quad (11.42)$$

With (11.40) and (11.42), the eigenvalue equations (11.33) and (11.34) take the form

$$\frac{\partial Y}{\partial \varphi} = imY, \quad (11.43)$$

$$\frac{1}{\sin\theta} \frac{\partial}{\partial \theta} \left(\sin\theta \frac{\partial Y}{\partial \theta} \right) + \frac{1}{\sin^2\theta} \frac{\partial^2 Y}{\partial \varphi^2} = -\lambda Y. \quad (11.44)$$

The solution of this pair of equations is simplified by first solving Eq. (11.43) for $\widehat{L}_z$,

$$Y(\theta, \varphi) = \Theta(\theta) e^{im\varphi}. \quad (11.45)$$

For this to be a physical (i.e., single-valued) solution, its value must be the same for φ and $\varphi + 2\pi m$, where

$$|m| = 0, 1, 2, \ldots \quad (11.46)$$

Therefore the eigenvalues $\hbar m$ of $\widehat{L}_z$ are *integer* multiples, positive or negative, of $\hbar$: $0, \pm\hbar, \pm 2\hbar, \ldots$ Since the z-direction is arbitrary, the result must be understood in the sense that the projection of the orbital momentum in any given direction is quantized, so that the eigenvalues of the operator $\mathbf{n} \cdot \widehat{\mathbf{L}}$, where $\mathbf{n}$ is a unit vector in that direction, are $0, \pm\hbar, \pm 2\hbar, \ldots$ This is the so-called *space quantization*,[2] perhaps one of the most counterintuitive concepts in QM. For historical reasons (related to its important applications, see Section 12.7), the quantum number m is called the *magnetic quantum number*.

Substituting (11.45) in Eq. (11.44), we get

$$\frac{1}{\sin\theta} \frac{\partial}{\partial \theta} \left(\sin\theta \frac{\partial}{\partial \theta} \right) \Theta - \frac{m^2}{\sin^2\theta} \Theta + \lambda \Theta = 0. \quad (11.47)$$

We recognize here the *Legendre differential equation*, which has a solution for all values of the parameter λ. In general, however, such solutions diverge for some value of the variable θ within the interval $(0, \pi)$, so they are physically inadmissible. Equation (11.47) has regular solutions in $(0, \pi)$ for $m \geq 0$ if and only if

$$\lambda = l(l+1), \quad l \geq |m| \quad (11.48)$$

---

[2] This term does not refer to the quantization of space of course, but to the possible values of the angular-momentum projections in a given direction.

with l an integer, $l = 0, 1, 2, \ldots$ This means that also the magnitude of the orbital angular momentum is quantized.[3]

The appropriate solutions of Eq. (11.47) are the associated Legendre polynomials $\Theta(\theta) = P_l^m(x)$, with $x = \cos\theta$, which can be obtained using the Rodrigues formula

$$P_l^m(x) = \frac{1}{2^l l!}(1-x^2)^{m/2}\frac{d^{l+m}}{dx^{l+m}}(x^2-1)^l. \tag{11.49}$$

These functions are orthogonal with respect to the index l,

$$\int_{-1}^{1} P_l^m(x)P_{l'}^m(x)dx = \frac{2}{2l+1}\frac{(l+m)!}{(l-m)!}\delta_{ll'}, \tag{11.50}$$

while the factor $e^{im\varphi}$ in Eq. (11.45) takes care of the orthogonality with respect to the index m. It is thus established that the simultaneous eigenfunctions of $\boldsymbol{L}^2$ and $\widehat{L}_z$ are the *spherical harmonics* $Y_l^m(\theta,\varphi) \sim P_l^m(\cos\theta)e^{im\varphi}$. Using formula (11.50) it is not difficult to show that the Y's normalized over the entire solid angle are

$$Y_l^m(\theta,\varphi) = \sqrt{\frac{(2l+1)(l-m)!}{4\pi(l+m)!}}(-)^m P_l^m(\cos\theta)e^{im\varphi}, \tag{11.51}$$

for $m \geq 0$.[4]

For $m < 0$ we write

$$Y_l^m = (-)^m Y_l^{-m*}. \tag{11.52}$$

The functions Y_l^m thus defined form an orthonormal set, so for $-l \leq m \leq l$ they give

$$\int_0^{2\pi}\int_0^{\pi} Y_l^{m*}(\theta,\varphi)Y_{l'}^{m'}(\theta,\varphi)\sin\theta d\theta d\varphi = \delta_{ll'}\delta_{mm'}. \tag{11.53}$$

Spherical harmonics have a well-defined parity. To determine it, we observe from Fig. 11.1 that when the coordinates are reflected with respect to the origin, the angles θ and φ change to

$$\theta \to \theta' = \pi - \theta; \quad \varphi \to \varphi' = \varphi + \pi, \tag{11.54}$$

$$x = \cos\theta \to \cos(\pi-\theta) = -\cos\theta = -x, \tag{11.55}$$

which gives a factor $(-1)^{l+m}$; the function Y is then affected by the factor $(-1)^{l+2m} = (-1)^l$. We write this as an eigenvalue equation for the parity operator,

$$\widehat{P}Y_l^m(\theta,\varphi) = (-)^l Y_l^m(\theta,\varphi), \tag{11.56}$$

to show explicitly that the parity of the spherical harmonics is that of the angular momentum l,

$$P = (-1)^l. \tag{11.57}$$

---

[3] It is customary to speak of the magnitude of the angular momentum in units of $\hbar$.
[4] Different conventions are used in the literature for defining both the normalization of spherical harmonics and their normalization condition. For example, the normalization factor of Eq. (11.51) may vary from one text to another. The definition of spherical harmonics for negative m can also vary between different authors. The present one is used in the book by Condon and Shortley (1935), which is a very common reference.

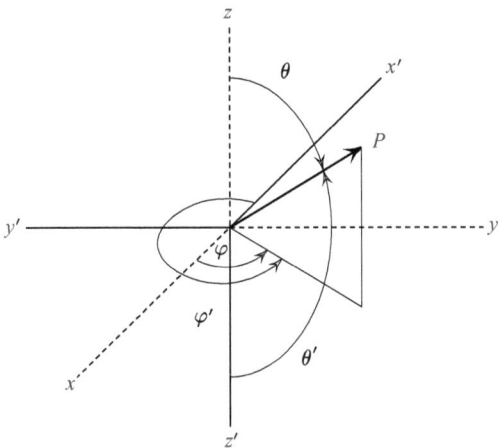

Figure 11.1 Upon reflection of the coordinates with respect to the origin, the angles θ and φ change to $\pi - \theta$ and $\varphi + \pi$, respectively.

11.1.3 Angular momentum dispersion

From the Heisenberg inequalities and the commutation relations (11.10) it follows that

$$\langle(\Delta\widehat{L}_z)^2\rangle\langle(\Delta y)^2\rangle \geq \tfrac{1}{4}\hbar^2\langle x\rangle^2, \tag{11.58}$$

$$\langle(\Delta\widehat{L}_x)^2\rangle\langle(\Delta y)^2\rangle \geq \tfrac{1}{4}\hbar^2\langle z\rangle^2, \tag{11.59}$$

and so on. From the first inequality we see that the eigenstates of $\widehat{L}_z$ necessarily have $\langle x\rangle = 0$ and $\langle y\rangle = 0$ (this second condition follows from the first by symmetry). The second inequality tells us that in such a case $\langle z\rangle \neq 0$, as expected for $\langle L_z\rangle \neq 0$. In other words, there is rotation about the z-axis. Therefore, an eigenstate of $\langle L_z\rangle \neq 0$ means precession around z.

This is confirmed by Eq. (11.48), the meaning of which can be understood as follows. The maximum value of the projection of $\widehat{L}$ in any given direction is $\hbar m_{\max} = \hbar l$, according to (11.48); from $|\langle\widehat{L}\rangle|_{\max} = \langle\widehat{L}_z\rangle_{\max} = \hbar l$ it follows that $\langle(\Delta\widehat{L})^2\rangle_{\min} = \langle\widehat{L}^2\rangle - (\langle\widehat{L}\rangle_{\max})^2 = \hbar^2 l(l+1) - \hbar^2 l^2 = \hbar^2 l$. In other words, if $l \neq 0$, the angular momentum *always* precesses around some axis. It is this inevitable precession that causes the x- and y-components to change continuously, albeit averaging to zero, as can be seen in Fig. 11.2. From Eq. (11.19) and the figure, we see that when $\langle L_z\rangle$ is maximum, the dispersions on the xy-plane are minimum, which allows us to write

$$\left[\langle(\Delta\widehat{L}_x)^2\rangle\langle(\Delta\widehat{L}_y)^2\rangle\right]_{\min} = \tfrac{1}{4}\hbar^2\langle\widehat{L}_z\rangle_{\max}^2 = \tfrac{1}{4}\hbar^4 l^2. \tag{11.60}$$

Since $\langle\widehat{L}_x\rangle = \langle\widehat{L}_y\rangle = 0$, $\langle(\Delta\widehat{L}_x)^2\rangle_{\min} = \langle\widehat{L}_x^2\rangle_{\min} = \tfrac{1}{2}\hbar^2 l$, and therefore,

$$\langle\widehat{L}_x^2 + \widehat{L}_y^2\rangle_{\min} = \hbar^2 l. \tag{11.61}$$

Consequently,

$$\langle\widehat{L}^2\rangle = \langle\widehat{L}_x^2 + \widehat{L}_y^2\rangle_{\min} + \langle\widehat{L}_z^2\rangle_{\max} = \hbar^2 l + \hbar^2 l^2 = \hbar^2 l(l+1), \tag{11.62}$$

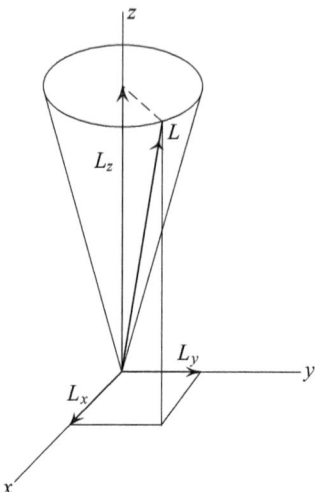

Figure 11.2 For eigenstates of L_z, the angular-momentum vector precesses around the z-axis, with its x- and y-components oscillating periodically around the origin.

which confirms that the term $\hbar^2 l$ in the eigenvalue of $\widehat{L}^2$ comes from the precession of the angular momentum around the quantization axis, that is, from the variances $\langle (\Delta \widehat{L}_x)^2 \rangle$ and $\langle (\Delta \widehat{L}_y)^2 \rangle$.

In this analysis we have assumed that the averages over the ensemble coincide with the averages over time, obtained by following a single particle. For example, in Fig. 11.2 we can see that as the particle describes an orbit, $\widehat{L}_x$ takes on different values, positive and negative, so that the average over one revolution (one period) is zero. But we can interpret the drawing differently, by considering that it corresponds to a particle at a given moment, and that at that *same moment* other particles of the ensemble would be in other possible positions of the orbit. In this case, the average of $\widehat{L}_x$ over the ensemble is zero. If we stick to this second view, the whole preceding description will be consistent, but we will have to reinterpret Fig. 11.2, understanding that there are infinitely many vectors L coexisting in the ensemble, uniformly distributed around the circumference. If we prefer the temporal interpretation, then we will have to propose that $\langle \widehat{L}_x \rangle = \overline{L_x}^t$, where the time average refers to a single particle. As mentioned earlier, this equality means that the system has ergodic properties (equivalence of the ensemble average and the individual time average).

It is customary to refer to states with $l = 0$ as s states, with $l = 1$ as p states, and so on, the order of the letters s, p, d, f, g, corresponding to $0, 1, 2, 3, 4, 5$; hereafter the order is alphabetical. The current use of the letters s (sharp), p (principal), d (diffusive), f (fundamental)... is nothing more than a legacy of the prequantum spectroscopic tradition.

Exercise E11.1. An action-angle Heisenberg inequality
In the Heisenberg–Robertson inequality (8.54) one obtains by taking $\hat{A} = \varphi$, $\hat{B} = L_z = -i\hbar(\partial/\partial\varphi)$

$$\Delta_\psi \varphi \Delta_\psi L_z \geq \frac{1}{2}\hbar. \tag{11.63}$$

Check the validity of this expression.

Solution. Since $\Delta\varphi \leq 2\pi$, for (11.63) to be valid it is required that $\Delta_\psi \hat{L}_z \geq \hbar/4\pi$, which is violated if ψ is chosen as one of the eigenfunctions $Y_l^m(\theta,\varphi)$ of L_z. This contradiction shows that the inequality (11.63) is only valid under restricted conditions. The difficulty arises from the fact that the inequalities (8.53) and (8.54) hold only if the two operators $\hat{A}$ and $\hat{B}$ are Hermitian; but $\hat{L}_z$ is only Hermitian in the space of periodic functions of φ with period 2π. Indeed, it follows from

$$\int \psi_1^* \hat{L}_z \psi_2 d\varphi = -i\hbar \int \psi_1^* \frac{\partial \psi_2}{\partial \varphi} d\varphi = -i\hbar \psi_1^* \psi_2 \big|_0^{2\pi} + i\hbar \int \psi_2 \frac{\partial \psi_1^*}{\partial \varphi} d\varphi \tag{11.64}$$

that the Hermiticity condition holds if and only if the integrated term vanishes, which in general only happens if ψ_1 and ψ_2 have period 2π. In the previous case, this condition is not met, since the function $\varphi Y_{lm}(\theta,\varphi)$ is not periodic in φ.

To obtain a correct inequality that replaces (11.63), we can proceed in several ways; one of them is the following. Let us take a periodic function A of φ with period 2π, otherwise arbitrary; then $A(\varphi) = A(\varphi + 2\pi)$. Since $[\hat{A},\hat{L}_z] = i\hbar A'$, from the inequality (8.53) we get

$$\Delta_\psi A \Delta \hat{L}_z \geq \frac{1}{2}\hbar \,|\langle A'\rangle|, \tag{11.65}$$

that is,

$$\Delta_A \varphi \Delta \hat{L}_z \geq \frac{1}{2}\hbar, \tag{11.66}$$

where

$$\Delta_A \varphi = \sqrt{\frac{\langle (\Delta A)^2\rangle}{|\langle dA/d\varphi\rangle|}}. \tag{11.67}$$

We conclude that an inequality of the Mandelstam–Tamm type is the correct form of the Heisenberg inequality for angle variables. Alternatively, we can give a correct form to Eq. (11.63) by replacing φ with the periodic function $A = \varphi$ mod 2π. A Fourier expansion gives

$$A = \pi + i \sum_{-\infty}^{\infty} \frac{1}{k} e^{ik\varphi}, \quad (k \neq 0), \tag{11.68}$$

and therefore, $A' = -\sum e^{ik\varphi}$ ($k \neq 0$). Replacing A and A' in Eq. (11.65) gives the desired inequality. Note that A' is highly singular, since for $\varphi \neq 2\pi n, n = 0,1,2,\ldots$, it is zero, but for $\varphi = 2\pi n$ it is equal to $-\infty$. These singularities are the ones that cancel $\langle A'\rangle$ when calculating the expectation value of eigenstates of $\hat{L}_z$. Conversely, in the initial calculation with $A = \varphi$ we have $\langle A'\rangle = 1$, and the inconsistency is produced.

11.2 Radial Hamiltonian for Central Forces

We have seen that when the external force is central (or radial), the Hamiltonian commutes with $\widehat{L}^2$ and with $\widehat{L}_z$, so that there are simultaneous eigenfunctions of these three operators. In this section we will find the general form of these eigenfunctions, which are of great theoretical importance.

In the spherical-coordinate system, the Laplacian operator is

$$\nabla^2 = \frac{1}{r^2}\frac{\partial}{\partial r}\left(r^2\frac{\partial}{\partial r}\right) + \frac{1}{r^2 \sin\theta}\frac{\partial}{\partial\theta}\left(\sin\theta\frac{\partial}{\partial\theta}\right) + \frac{1}{r^2 \sin^2\theta}\frac{\partial^2}{\partial\varphi^2}. \tag{11.69}$$

Using Eq. (11.40), we can rewrite this operator in the form

$$\widehat{p}^2 = -\hbar^2 \nabla^2 = -\frac{\hbar^2}{r^2}\frac{\partial}{\partial r}\left(r^2\frac{\partial}{\partial r}\right) + \frac{1}{r^2}\widehat{L}^2. \tag{11.70}$$

This suggests writing

$$\widehat{p}^2 = \widehat{p}_r^2 + \frac{\widehat{L}^2}{r^2}, \tag{11.71}$$

where $\widehat{p}_r$ represents the *canonical momentum conjugate to r*. Since

$$\boldsymbol{r}\cdot\widehat{\boldsymbol{p}} = -i\hbar r\frac{\partial}{\partial r}, \tag{11.72}$$

using (11.70), we can write the square of $\widehat{p}_r$ in the form

$$\widehat{p}_r^2 = -\frac{\hbar^2}{r^2}\frac{\partial}{\partial r}\left(r^2\frac{\partial}{\partial r}\right) = -\frac{\hbar^2}{r^2}\left[\left(r\frac{\partial}{\partial r}\right)\left(r\frac{\partial}{\partial r}\right) + r\frac{\partial}{\partial r}\right], \tag{11.73}$$

or alternatively,

$$\widehat{p}_r^2 = \frac{1}{r^2}[(\boldsymbol{r}\cdot\widehat{\boldsymbol{p}})^2 - i\hbar(\boldsymbol{r}\cdot\widehat{\boldsymbol{p}})]. \tag{11.74}$$

From this it follows that

$$\widehat{p}_r = \frac{1}{r}(\boldsymbol{r}\cdot\widehat{\boldsymbol{p}} - i\hbar) = -i\hbar\left(\frac{\partial}{\partial r} + \frac{1}{r}\right). \tag{11.75}$$

This is the canonical conjugate momentum of the variable r (see Problem 11.7).

Exercise E11.2. Conjugate momentum of the radial variable r

Guided by the correspondence with the classical variable p_r, derive the correct quantum expression for the momentum operator conjugate to the variable r.

Solution. The canonical conjugate momentum of r in classical mechanics is $p_r = \frac{1}{r}\boldsymbol{p}\cdot\boldsymbol{r}$. Taking into account that $\widehat{\boldsymbol{p}}$ and $\widehat{\boldsymbol{r}}$ do not commute, we propose a linear combination of the form

$$\widehat{p}_r = \lambda\widehat{\boldsymbol{p}}\cdot\frac{\widehat{\boldsymbol{r}}}{r} + (1-\lambda)\frac{\widehat{\boldsymbol{r}}}{r}\cdot\widehat{\boldsymbol{p}}, \tag{11.76}$$

with λ to be determined, to correspond to the preceding formula. It is easy to see that this $\hat{p}_r$ is Hermitian only for $\lambda = 1/2$, which gives the symmetric combination

$$\hat{p}_r = \tfrac{1}{2}\left(\hat{\boldsymbol{p}}\cdot\frac{\hat{\boldsymbol{r}}}{r} + \frac{\hat{\boldsymbol{r}}}{r}\cdot\hat{\boldsymbol{p}}\right). \tag{11.77}$$

A little algebra using the basic commutator $[\hat{x}_i, \hat{p}_j] = i\hbar\delta_{ij}$ (or alternatively, Eq. (11.70)) shows that this expression is equivalent to the one given in (11.75). Since we want the eigenvalues of $\hat{p}_r$ to be real, the final result is unique.

11.2.1 The radial Schrödinger equation

Using Eq. (11.71), we can write the Hamiltonian for a central problem in the form

$$\hat{H} = \frac{\hat{p}_r^2}{2m} + \frac{\hat{L}^2}{2mr^2} + V(r). \tag{11.78}$$

Therefore, there are simultaneous eigenstates $|nlm\rangle$ of the Hamiltonian $\hat{H}$ (with principal quantum number n), $\hat{L}^2$ (with orbital angular-momentum quantum number l) and $\hat{L}_z$ (with magnetic quantum number m). Furthermore, since the angular dependence of the Hamiltonian is restricted to the term containing $\hat{L}^2$, the wave function $\psi_{nlm}(r,\theta,\varphi) \equiv \langle r\theta\varphi|nlm\rangle$, that is, the coordinate representation of the vector $|nlm\rangle$, can be factored as the product of a radial function $R_n(r)$ and the angular function, that is, the spherical harmonic $Y_l^m(\theta,\varphi)$. In fact, if we write

$$\psi_{nlm}(r,\theta,\varphi) = R_{nl}(r)Y_l^m(\theta,\varphi) \tag{11.79}$$

and substitute in the Schrödinger equation, using

$$\hat{L}^2 Y_l^m = \hbar^2 l(l+1) Y_l^m, \tag{11.80}$$

we get

$$\hat{H}\psi = \left[\frac{\hat{p}_r^2}{2m} + \frac{\hbar^2 l(l+1)}{2mr^2} + V(r)\right] R(r)Y_l^m = E R(r)Y_l^m. \tag{11.81}$$

Omitting the common factor Y_l^m, which is now irrelevant, this equation reduces to the *radial Schrödinger equation*,

$$\left[\frac{\hat{p}_r^2}{2m} + \frac{\hbar^2 l(l+1)}{2mr^2} + V(r)\right] R(r) = E R(r). \tag{11.82}$$

Three observations are worth making. First, the factorization just done is independent of $V(r)$, so the spherical harmonics are the angular wave functions of *any* central problem. Second, the effective potential that appears in the radial equation is the sum of $V(r)$ and the term $\hbar^2 l(l+1)/2mr^2$, which represents the so-called *centrifugal barrier* that prevents the particles from approaching the potential-generating center (the nucleus of an atom, for example) without limit. This centrifugal repulsion is only canceled out when the angular momentum is zero. Finally, the eigenstates of $\hat{H}$ are degenerate with respect to m, since the eigenvalues of the energy depend in general on l, but not on m, which does not appear in Eq. (11.82) This allowed us to write $R(r)$

in the explicit form $R_{nl}(r)$ in Eq. (11.79). In the next chapter we will use the radial equation (11.82) to solve several important problems, including that of the H atom.

Exercise E11.3. Particle flow for a central-force problem

Show that if the particles move under the action of a central potential, in an energy and angular momentum eigenstate, the flow is purely equatorial.

Solution. In spherical coordinates, the components of the particle flow vector are (we write M for the particle mass)

$$j_r = \frac{i\hbar}{2M}\left(\psi\frac{\partial\psi^*}{\partial r} - \psi^*\frac{\partial\psi}{\partial r}\right), \qquad j_\theta = \frac{i\hbar}{2Mr}\left(\psi\frac{\partial\psi^*}{\partial\theta} - \psi^*\frac{\partial\psi}{\partial\theta}\right), \qquad (11.83)$$

$$j_\varphi = \frac{i\hbar}{2Mr\sin\theta}\left(\psi\frac{\partial\psi^*}{\partial\varphi} - \psi^*\frac{\partial\psi}{\partial\varphi}\right). \qquad (11.84)$$

From Eqs. (11.79) and (11.45) it follows that the energy and angular momentum eigenfunction for a central problem can be written in the form

$$\psi = R(r)\Theta(\theta)e^{im\varphi}, \qquad (11.85)$$

where $R(r)$ and $\Theta(\theta)$ are real; therefore, the only nonzero component of the flow is

$$j_\varphi = \frac{\hbar}{M}\frac{R^2(r)\Theta^2(\theta)}{r\sin\theta}m = \frac{\hbar}{M}\frac{\rho(r,\theta)}{r\sin\theta}m. \qquad (11.86)$$

This result represents a net average circulation around the polar axis of the particles with angular momentum projection $\hbar m$.

Exercise E11.4. Schrödinger equation in curvilinear coordinates

According to Eqs. (11.73) and (11.75), the operator representing p_r is not simply $-i\hbar(\partial/\partial r)$, so that the contribution of the radial component to the energy is not simply $(-\hbar^2/2m)(\partial^2/\partial r^2)$. This shows that care must be taken when defining operators in curvilinear coordinates. Derive the Schrödinger equation in an arbitrary curvilinear coordinate system.

Solution. Consider the transition from the Cartesian coordinate system to an arbitrary curvilinear one $\{u_i\}$. Suppose that the coordinate transformation is $x_k = f_k(u_i)$; it follows that

$$\dot{x}_k = \sum_i \frac{\partial f_k}{\partial u_i}\dot{u}_i \qquad (11.87)$$

and the kinetic energy becomes (various particles are considered, whose masses can be different)

$$T = \frac{1}{2}\sum_k m_k \dot{x}_k^2 = \frac{1}{2}\sum_{i,j,k} m_k \frac{\partial f_k}{\partial u_i}\frac{\partial f_k}{\partial u_j}\dot{u}_i\dot{u}_j = \frac{1}{2}\sum_{i,j} g_{ij}\dot{u}_i\dot{u}_j, \qquad (11.88)$$

where the metric tensor of the transformation is given by $g_{ij}(u_k) = \sum_k m_k \frac{\partial f_k}{\partial u_i}\frac{\partial f_k}{\partial u_j}$.

The coefficients of the velocities in the last expression of (11.88), which play the role of the effective masses, are now a function of the coordinates. If you want to quantize this expression, you have the problem of defining the order in which, for example, the powers of the u_i and the factor $\dot{u}_i$ in (11.87) should appear. The easiest way to proceed is to express the problem in Cartesian coordinates, quantize, and then return to the curvilinear coordinate system. To write the Schrödinger equation correctly, however, it is sufficient to write the kinetic energy precisely in the form (11.88), where the components of the metric tensor are given by the structure of the square of the line element in the new coordinate system,

$$ds^2 = \sum_{i,j} g_{ij} du_i du_j, \qquad T = \frac{1}{2}m \left(\frac{ds}{dt}\right)^2 = \frac{1}{2} \sum_{i,j} g_{ij} \dot{u}_i \dot{u}_j. \qquad (11.89)$$

If g is the determinant of the matrix $g_{ij}(u_s)$, and the matrix inverse is $(g_{ij})^{-1} = g^{ij}$, then the Laplacian in the u_k coordinate system is (in any number of dimensions; if there are N equal particles, the sum goes up to $3N$)

$$\nabla^2 = \frac{1}{\sqrt{g}} \sum_{i,j} \frac{\partial}{\partial u_i} \left(\sqrt{g} g^{ij} \frac{\partial}{\partial u_j}\right). \qquad (11.90)$$

This is precisely the method used earlier to write the Hamiltonian in spherical coordinates. In fact, in this case

$$ds^2 = dr^2 + r^2 d\theta^2 + r^2 \sin^2\theta d\varphi^2 = \sum_{i,j} g_{ij} du_i du_j, \qquad (11.91)$$

from which it follows that the matrix g of the metric tensor is diagonal, with elements

$$g_{11} = 1, \qquad g_{22} = r^2, \qquad g_{33} = r^2 \sin^2\theta, \qquad g_{12} = g_{13} = g_{23} = 0, \qquad (11.92)$$

and therefore

$$g^{11} = 1, \quad g^{22} = r^{-2}, \quad g^{33} = (r\sin\theta)^{-2}, \quad g^{12} = g^{13} = g^{23} = 0, \quad g = r^4 \sin^2\theta. \qquad (11.93)$$

Substituting these values into the expression for the Laplacian leads directly to Eq. (11.69). The Schrödinger equation for particles of equal mass in the curvilinear u_k coordinate system is then

$$i\hbar \frac{\partial \psi}{\partial t} = -\frac{\hbar^2}{2m} \frac{1}{\sqrt{g}} \sum_{i,j} \frac{\partial}{\partial u_i} \left(\sqrt{g} g^{ij} \frac{\partial}{\partial u_j}\right) \psi + V(u_i)\psi. \qquad (11.94)$$

11.3 Matrix Representation of the Angular Momentum. The Pauli Matrices

So far we have concentrated on the study of the angular momentum associated with the *orbital* motion of particles. However, it is a fact that elementary particles (electron, proton, neutron, etc.) have, in addition to mass and charge, other properties associated with quantum numbers, among which the *spin* angular momentum stands out.

Particles in general can have an "internal" angular momentum and an associated quantum number. While the orbital angular momentum can only be an integer multiple of $\hbar$, the spin can be an integer or half-integer of $\hbar$. For example, electrons and nucleons (protons and neutrons) have spin $\hbar/2$, pions have spin 0, photons have spin $\hbar$, and so on. The coordinate representation we have used so far turns out to be too restrictive, since it excludes from the outset both the possibility of an intrinsic angular momentum and of half-integer values for it. The theory must therefore be generalized to include spin.

Following the preceding arguments, the total angular momentum $\widehat{\boldsymbol{J}}$ (which we will simply call angular momentum when there is no confusion) is generally defined as the sum of orbital and spinorial contributions,

$$\widehat{\boldsymbol{J}} = \widehat{\boldsymbol{L}} + \widehat{\boldsymbol{S}}. \tag{11.95}$$

The definition $\boldsymbol{r} \times \widehat{\boldsymbol{p}}$ only applies to the orbital part of $\widehat{\boldsymbol{J}}$, and the commutation properties of $\widehat{L}_i$ with $\widehat{L}_j$ and with $\widehat{\boldsymbol{L}}^2$ were obtained from the definition of $\widehat{\boldsymbol{L}}$. We will now use the fact that the commutation relations are independent of the representation. Therefore, we define the angular-momentum operator as the vector operator whose components satisfy the following commutation relations,

$$[\widehat{J}_i, \widehat{J}_j] = i\hbar\varepsilon_{ijk}\widehat{J}_k, \tag{11.96}$$

$$[\widehat{\boldsymbol{J}}^2, \widehat{J}_i] = 0. \quad i = 1, 2, 3, \tag{11.97}$$

write a common eigenstate of $\widehat{\boldsymbol{J}}^2$ and $\widehat{J}_z$ – which, as we know from (11.96), is neither an eigenstate of $\widehat{J}_x$ nor of $\widehat{J}_y$ – as $|jm\rangle$, and call j the (total) angular momentum quantum number and m, the quantum number corresponding to the z-projection of the angular momentum. The eigenvalue equations now take the form

$$\widehat{\boldsymbol{J}}^2|jm\rangle = \hbar^2\lambda|jm\rangle, \tag{11.98}$$

$$\widehat{J}_z|jm\rangle = \hbar m|jm\rangle, \tag{11.99}$$

where the eigenvalue λ will obviously be a function of j, which must reduce to $l(l+1)$ when $\widehat{\boldsymbol{J}}$ is reduced to $\widehat{\boldsymbol{L}}$ (i.e., when the spin is zero).

We now determine the possible values of j and m and construct a matrix representation of the angular-momentum operators in the basis $|jm\rangle$. The calculation is an excellent, simple example of the pure matrix methods of QM, in the style of those developed with the advent of matrix mechanics.

11.3.1 Matrix elements of angular-momentum components

The matrix elements of $\widehat{\boldsymbol{J}}^2$ and $\widehat{J}_z$ are immediately obtained, since according to (11.98) and (11.99), these matrices are diagonal,

$$\langle j'm'|\widehat{\boldsymbol{J}}^2|jm\rangle = \hbar^2\lambda\langle j'm'|jm\rangle = \hbar^2\lambda\delta_{jj'}\delta_{mm'}, \tag{11.100}$$

$$\langle j'm'|\widehat{J}_z|jm\rangle = \hbar m\langle j'm'|jm\rangle = \hbar m\delta_{jj'}\delta_{mm'}. \tag{11.101}$$

To calculate the matrix elements of $\widehat{J}_x$ and $\widehat{J}_y$, it is useful to introduce the linear combinations $\widehat{J}_+$ and $\widehat{J}_-$, defined as follows,

$$\widehat{J}_\pm \equiv \tfrac{1}{\sqrt{2}}(\widehat{J}_x \pm i\widehat{J}_y). \tag{11.102}$$

These are known as raising and lowering operators, for reasons that will become clear in Section 3.2.

Since the $\widehat{J}_i$ are Hermitian, it follows that $\widehat{J}_+^\dagger = \widehat{J}_-$ and $\widehat{J}_-^\dagger = \widehat{J}_+$. Here we have another important example of very commonly used operators that are not Hermitian, but one is the adjoint of the other, as happened with the $\hat{a}^\dagger, \hat{a}$ operators of the HO. Using (11.96) and (11.97) it is possible to prove the following commutation properties

$$\left[\widehat{\boldsymbol{J}}^2, \widehat{J}_\pm\right] = 0, \tag{11.103}$$

$$\left[\widehat{J}_z, \widehat{J}_\pm\right] = \pm\hbar\widehat{J}_\pm, \tag{11.104}$$

$$\left[\widehat{J}_+, \widehat{J}_-\right] = \hbar\widehat{J}_z. \tag{11.105}$$

In addition, we have the identities

$$\widehat{J}_+\widehat{J}_- = \tfrac{1}{2}\left(\widehat{\boldsymbol{J}}^2 - \widehat{J}_z(\widehat{J}_z - \hbar)\right), \tag{11.106}$$

$$\widehat{J}_-\widehat{J}_+ = \tfrac{1}{2}\left(\widehat{\boldsymbol{J}}^2 - \widehat{J}_z(\widehat{J}_z + \hbar)\right). \tag{11.107}$$

Note that these two equations can be rewritten in the following useful form (keep them handy),

$$\widehat{\boldsymbol{J}}^2 = 2\widehat{J}_+\widehat{J}_- + \widehat{J}_z^2 - \hbar\widehat{J}_z, \tag{11.108}$$

$$\widehat{\boldsymbol{J}}^2 = 2\widehat{J}_-\widehat{J}_+ + \widehat{J}_z^2 + \hbar\widehat{J}_z. \tag{11.109}$$

The difference of these two expressions gives Eq. (11.105). From (11.98), (11.99), and (11.106), it follows that

$$\widehat{J}_+\widehat{J}_-|jm\rangle = \tfrac{1}{2}\hbar^2[\lambda - m(m-1)]|jm\rangle, \tag{11.110}$$

while (11.107) gives

$$\widehat{J}_-\widehat{J}_+|jm\rangle = \tfrac{1}{2}\hbar^2[\lambda - m(m+1)]|jm\rangle, \tag{11.111}$$

from which it follows that

$$\langle jm|\widehat{J}_+^\dagger \widehat{J}_+|jm\rangle = \langle jm|\widehat{J}_-\widehat{J}_+|jm\rangle = \tfrac{1}{2}\hbar^2[\lambda - m(m+1)] \geq 0$$

(the inequality follows from Eq. (8.43)), whence

$$\lambda - m(m+1) \geq 0. \tag{11.112}$$

Analogously, from (11.110) we obtain

$$\lambda - m(m-1) \geq 0. \tag{11.113}$$

Rewriting the first inequality in the form $\lambda \geq m(m+1)$ and adding 1/4 to both sides, we get

$$\lambda + \tfrac{1}{4} \geq m^2 + m + \tfrac{1}{4} = \left(m + \tfrac{1}{2}\right)^2. \tag{11.114}$$

We now define a number j by the formula

$$j + \tfrac{1}{2} = \sqrt{\lambda + \tfrac{1}{4}}. \tag{11.115}$$

To investigate the meaning of j we get λ from this equation,

$$\lambda = \left(j + \tfrac{1}{2}\right)^2 - \tfrac{1}{4} = j(j+1), \tag{11.116}$$

which is just the generalization of Eq. (11.48) to the angular momentum j, so that we identify the parameter j in (11.115) with the quantum number j and the eigenvalues of $\boldsymbol{J}^2$ with $\hbar^2 j(j+1)$. Using (11.115), Eq. (11.114) becomes $\left(j+\tfrac{1}{2}\right)^2 \geq \left(m+\tfrac{1}{2}\right)^2$, that is,

$$j + \tfrac{1}{2} \geq |m + \tfrac{1}{2}|. \tag{11.117}$$

If $m \geq -\tfrac{1}{2}$, then $j \geq m$; if $m < -\tfrac{1}{2}$, then $m \geq -(j+1)$. Combining both results, we get

$$-j - 1 \leq m \leq j. \tag{11.118}$$

Analogously, (11.113) can be analyzed to conclude that

$$-j \leq m \leq j + 1. \tag{11.119}$$

Both expressions (11.118) and (11.119) must be satisfied, so that

$$-j \leq m \leq j. \tag{11.120}$$

Note that, as before, $m_{\max} = j \leq \sqrt{j(j+1)}$, so that the notion of precession must be extended to any angular momentum $j \neq 0$. Later we will determine the values that j and m can take.

We now turn to the construction of the matrix elements of $\widehat{J}_x$ and $\widehat{J}_y$. Since $\widehat{J}_+$ commutes with $\widehat{\boldsymbol{J}}^2$, every matrix element of $\left[\widehat{J}_+, \widehat{\boldsymbol{J}}^2\right]$ is zero, that is,

$$\langle j'm'|\widehat{\boldsymbol{J}}^2 J_+ - \widehat{J}_+ \widehat{\boldsymbol{J}}^2|jm\rangle = \hbar^2(\lambda' - \lambda)\langle j'm'|\widehat{J}_+|jm\rangle = 0, \tag{11.121}$$

where $\lambda = j(j+1), \lambda' = j'(j'+1)$. An analogous result is obtained for $\widehat{J}_-$ by taking the adjoint of the previous expression. We conclude that if $j \neq j'$, the matrix elements of $\widehat{J}_+$ and $\widehat{J}_-$ are zero. Using (11.102) and (11.99) we can synthetically write

$$\langle j'm'|\widehat{J}_s|jm\rangle \sim \delta_{jj'}, \quad \text{for } s = x, y, z, +, -, \tag{11.122}$$

which tells us that all matrix elements of $\widehat{J}_s$ between states corresponding to different values of j are zero, as is the case with $\widehat{\boldsymbol{J}}^2$. Therefore, the Hilbert space of angular momentum is partitioned into rotation-invariant orthogonal subspaces, each with a well-defined value of j. The fact that the subspaces of a given j are invariant with respect to rotation means that when a rotation is performed, only the vectors belonging to one such subspace are mixed with each other. Therefore, in the following we will consider matrix elements between states with the same j; only the projections m, m' may be different.

11.3 Matrix Representation of the Angular Momentum. The Pauli Matrices

Using the commutator between $\widehat{J}_z$ and $\widehat{J}_+$, (11.104), we now write $\hbar \langle jm'|\widehat{J}_+|jm\rangle = \langle jm'|\widehat{J}_z\widehat{J}_+ - \widehat{J}_+\widehat{J}_z|jm\rangle = \hbar(m'-m)\langle jm'|\widehat{J}_+|jm\rangle$, that is, $(m'-m-1)\langle jm'|\widehat{J}_+|jm\rangle = 0$. The general solution of this equation is

$$\langle jm'|\widehat{J}_+|jm\rangle = \hbar C_{jm}\delta_{m',m+1}, \tag{11.123}$$

where the coefficient C_{jm} remains to be determined. Taking the adjoint we get

$$\langle jm|\widehat{J}_-|jm'\rangle = \hbar C^*_{jm}\delta_{m',m+1}, \tag{11.124}$$

or, interchanging the indices m and m',

$$\langle jm'|\widehat{J}_-|jm\rangle = \hbar C^*_{jm-1}\delta_{m'+1,m}. \tag{11.125}$$

To determine C_{jm} we turn to the matrix element of $\widehat{J}_-\widehat{J}_+$, obtained from (11.111) with λ given by (11.116). Inserting an expansion of the unit in terms of the projectors $|jm'\rangle\langle jm'|$, and using (11.123) and (11.124), we obtain

$$\langle jm|\widehat{J}_-\widehat{J}_+|jm\rangle = \tfrac{1}{2}\hbar^2[j(j+1) - m(m+1)]$$
$$= \sum_{j'm'}\langle jm|\widehat{J}_-|j'm'\rangle\langle j'm'|\widehat{J}_+|jm\rangle = \sum_{m'}\langle jm|\widehat{J}_-|jm'\rangle\langle jm'|\widehat{J}_+|jm\rangle$$
$$= \hbar^2\sum_{m'} C^*_{jm}\delta_{m',m+1}C_{jm}\delta_{m',m+1} = \hbar^2|C_{jm}|^2, \tag{11.126}$$

that is,

$$|C_{jm}|^2 = \tfrac{1}{2}[j(j+1) - m(m+1)]$$
$$= \tfrac{1}{2}[j(j+1) + jm - m(m+1) - mj] = \tfrac{1}{2}(j+m+1)(j-m). \tag{11.127}$$

The phase is not determined by the above relations, so it is chosen arbitrarily; the most common convention, which we will adopt here, is to take C_{jm} to be real and positive, that is,

$$C_{jm} = \sqrt{\tfrac{1}{2}(j+m+1)(j-m)}. \tag{11.128}$$

Substitution of this value in (11.123) and (11.125) gives the full matrix elements of $\widehat{J}_+$ and $\widehat{J}_-$,

$$\langle j,m+1|\widehat{J}_+|j,m\rangle = \hbar\sqrt{\tfrac{1}{2}(j+m+1)(j-m)},$$
$$= \hbar\sqrt{\tfrac{1}{2}\left[(j+1)j - (m+1)m\right]}, \tag{11.129}$$

$$\langle j,m-1|\widehat{J}_-|j,m\rangle = \hbar\sqrt{\tfrac{1}{2}(j+m)(j-m+1)},$$
$$= \hbar\sqrt{\tfrac{1}{2}\left[(j+1)j - (m-1)m\right]}. \tag{11.130}$$

All other matrix elements of these operators are zero. From (11.129) and the definitions of $\widehat{J}_+$ and $\widehat{J}_-$, (11.102), we get the nonzero matrix elements of $\widehat{J}_x$ and $\widehat{J}_y$. Collecting results,

$$\langle j,m\pm 1|\widehat{J}_x|j,m\rangle = \tfrac{1}{2}\hbar\sqrt{(j\mp m)(j\pm m+1)}, \tag{11.131}$$

$$\langle j,m\pm 1|\widehat{J}_y|j,m\rangle = \mp i\tfrac{1}{2}\hbar\sqrt{(j\mp m)(j\pm m+1)}, \tag{11.132}$$

$$\langle jm'|\widehat{J}_z|jm\rangle = \hbar m \delta_{mm'}, \tag{11.133}$$

$$\langle jm'|\widehat{\mathbf{J}}^2|jm\rangle = \hbar^2 j(j+1)\delta_{mm'}. \tag{11.134}$$

This completes the construction of the matrix representation of the angular-momentum operators.

11.3.2 Angular-momentum raising and lowering operators

We now show that the $\widehat{J}_\pm$ are raising and lowering operators acting on m, which in turn allows us to determine the possible values of j and m. To do this, we express the effect of $\widehat{J}_+$ on $|jm\rangle$ in the general form

$$\widehat{J}_+|jm\rangle = \sum_{m'} B_{jm'}|jm'\rangle. \tag{11.135}$$

From this and (11.123) it follows that

$$\langle jm''|\widehat{J}_+|jm\rangle = \sum_{m'} B_{jm'}\delta_{m'm''} = B_{jm''} = \hbar C_{jm}\delta_{m'',m+1}. \tag{11.136}$$

Substituting into the previous expression the value of the coefficients $B_{jm'}$ just obtained, we get

$$\widehat{J}_+|jm\rangle = \sum_{m'} \hbar C_{jm}\delta_{m',m+1}|jm'\rangle,$$

$$\widehat{J}_+|jm\rangle = \hbar C_{jm}|j,m+1\rangle, \tag{11.137}$$

so that the operator $\widehat{J}_+$ effectively increases the value of the projection m by one. Similarly, we can see that $\widehat{J}_-$ decreases the value of m by one,

$$\widehat{J}_-|jm\rangle = \hbar C_{j,m-1}|j,m-1\rangle. \tag{11.138}$$

The states with different m can be generated from the state with the minimum m (which is $-j$) by repeated application of the raising operator. Since each application of $\widehat{J}_+$ increases the value of m by one, m can only vary in unit steps, taking the values $-j, -j+1, -j+2, \ldots, j-1, j$. The number of steps needed to cover all possible values of m, from $m = -j$ to $m = j$, is $2j$, and we get

$$\widehat{J}_+|j,-j\rangle = \hbar C_{j,-j}|j,-j+1\rangle,$$

$$(\widehat{J}_+)^2|j,-j\rangle = \hbar^2 C_{j,-j} C_{j,-j+1}|j,-j+2\rangle,$$

$$\cdots$$

$$(\widehat{J}_+)^{2j}|j,-j\rangle = \hbar^{2j} C_{j,-j} C_{j,-j+1} \cdots C_{j,j}|jj\rangle. \tag{11.139}$$

Since the total number of steps is an integer, j can take integer values $0, 1, 2, \ldots$ or half-integer values $1/2, 3/2, \ldots$ In this way, we verify that the theory covers all the cases of interest, allowing both the integer orbital momentum and the integer or half-integer spinor momentum to be taken into account. We have thus fully determined the angular-momentum operators and their eigenvalues.

For a given j, the matrices must contain the $2j+1$ possible values of m from $-j$ to $+j$ in unit steps, so they must be square matrices with $2j+1$ rows and columns. The corresponding $2j+1$ basis vectors, $|j,-j\rangle, |j,-j+1\rangle, \ldots, |j, j-1\rangle, |j, j\rangle$, can be selected as

$$|j,j\rangle = \begin{pmatrix} 1 \\ 0 \\ 0 \\ \vdots \end{pmatrix}; \quad |j,j-1\rangle = \begin{pmatrix} 0 \\ 1 \\ 0 \\ \vdots \end{pmatrix}; \quad |j,j-2\rangle = \begin{pmatrix} 0 \\ 0 \\ 1 \\ \vdots \end{pmatrix}, \quad \text{and so on.} \quad (11.140)$$

The matrices $\widehat{J}_z$ and $\widehat{J}^2$ take the diagonal form

$$\widehat{J}_z = \hbar \begin{pmatrix} j & 0 & 0 & \cdots \\ 0 & j-1 & 0 & \cdots \\ \vdots & \vdots & \ddots & \cdots \\ 0 & 0 & 0 & -j \end{pmatrix}, \quad \widehat{J}^2 = \hbar^2 j(j+1) \begin{pmatrix} 1 & 0 & 0 & \cdots \\ 0 & 1 & 0 & \cdots \\ 0 & 0 & 1 & \cdots \\ \vdots & \vdots & \vdots & \ddots \end{pmatrix}, \quad (11.141)$$

while the matrices representing $\widehat{J}_x$ and $\widehat{J}_y$ have their nonzero elements shifted by one unit to either side of the main diagonal. If we denote the nonzero elements schematically by C_s, the structure of $\widehat{J}_x$, for example, is

$$\widehat{J}_x = \frac{\hbar}{2} \begin{pmatrix} 0 & C_a & 0 & 0 & \cdots \\ C_a & 0 & C_b & 0 & \cdots \\ 0 & C_b & 0 & C_c & \cdots \\ 0 & 0 & C_c & 0 & \cdots \\ \vdots & \vdots & \vdots & \vdots & \ddots \end{pmatrix} \quad (11.142)$$

and similarly for $\widehat{J}_y$, but with imaginary elements.

11.3.3 The spin angular momentum 1/2: Pauli matrices

We will use the above results to construct the angular-momentum operators for spin 1/2. The results obtained here will be fundamental for studying important problems in which the electron spin in particular plays a relevant role, but they also apply to other particles with spin 1/2.

To study spin 1/2, we take $l = 0$ and $\widehat{J} = \widehat{S}$, with $j = s = 1/2$. The projection m can only take the values $m = \pm 1/2$ so the spinor space is two-dimensional. In the matrix representation the basis vectors are

$$|+\rangle \equiv |\tfrac{1}{2}, \tfrac{1}{2}\rangle \to \begin{pmatrix} 1 \\ 0 \end{pmatrix}; \quad |-\rangle \equiv |\tfrac{1}{2}, -\tfrac{1}{2}\rangle \to \begin{pmatrix} 0 \\ 1 \end{pmatrix}. \quad (11.143)$$

Thus, the wave function of an electron (or any particle with spin 1/2), called a spinor, has two components, each corresponding to one of the possible projections of the spin in spin space.

The general state consists of a superposition of states ψ_+ with spin up $|+\rangle$ and states ψ_- with spin down $|-\rangle$, which we write as

$$|\psi\rangle = \psi_+|+\rangle + \psi_-|-\rangle = \begin{pmatrix} 1 \\ 0 \end{pmatrix} \psi_+ + \begin{pmatrix} 0 \\ 1 \end{pmatrix} \psi_- = \begin{pmatrix} \psi_+ \\ \psi_- \end{pmatrix}. \quad (11.144)$$

By taking another spinor $|\varphi\rangle$ with components φ_+ and φ_-, we can construct the scalar product (the result is a pseudo-scalar, not a scalar)

$$\langle\varphi|\psi\rangle = \begin{pmatrix} \varphi_+^* & \varphi_-^* \end{pmatrix} \begin{pmatrix} \psi_+ \\ \psi_- \end{pmatrix} = \varphi_+^* \psi_+ + \varphi_-^* \psi_-. \qquad (11.145)$$

In particular, with $\psi = \varphi$ we get

$$\langle\psi|\psi\rangle = |\psi_+|^2 + |\psi_-|^2 = \rho_+ + \rho_-. \qquad (11.146)$$

Suppose that ψ_+, ψ_- are in the coordinate representation; then the particle density in the configuration space is given by the sum of the densities corresponding to spin up and spin down, with no interference between them. Note that $\langle\psi|\psi\rangle$ is a scalar in spin space, but *not* necessarily the square of the norm; it is the density associated with ψ, since the vectors in this section refer exclusively to spin space, and their coefficients are wave functions in other spaces, such as $\psi_\pm = \psi_\pm(\boldsymbol{x}, t)$.

According to Eqs. (11.141) and (11.142), the matrices $\hat{S}_x, \hat{S}_y$ and $\hat{S}_z$ have the form

$$\hat{S}_x = \tfrac{1}{2}\hbar \begin{pmatrix} 0 & C_1 \\ C_1 & 0 \end{pmatrix}, \quad \hat{S}_y = \tfrac{1}{2}\hbar \begin{pmatrix} 0 & iC_2 \\ -iC_2 & 0 \end{pmatrix}, \quad \hat{S}_z = \tfrac{1}{2}\hbar \begin{pmatrix} 1 & 0 \\ 0 & -1 \end{pmatrix}, \qquad (11.147)$$

with the coefficients C_1, C_2 obtained from Eqs. (11.131) and (11.132). For example, if C_1 corresponds to $m = -1/2$ (upper sign in Eq. (11.131)), then $C_1 = 1$. Completing the calculations gives

$$\hat{S}_x = \tfrac{1}{2}\hbar \begin{pmatrix} 0 & 1 \\ 1 & 0 \end{pmatrix}, \quad \hat{S}_y = \tfrac{1}{2}\hbar \begin{pmatrix} 0 & -i \\ i & 0 \end{pmatrix}, \quad \hat{S}_z = \tfrac{1}{2}\hbar \begin{pmatrix} 1 & 0 \\ 0 & -1 \end{pmatrix}. \qquad (11.148)$$

It is usual to write these results in terms of the Hermitian *Pauli matrices* $\hat{\sigma}_i$, which were introduced in Section 3.7, that is,

$$\hat{\boldsymbol{S}} = \tfrac{1}{2}\hbar \hat{\boldsymbol{\sigma}}, \qquad (11.149)$$

where

$$\hat{\sigma}_1 = \begin{pmatrix} 0 & 1 \\ 1 & 0 \end{pmatrix}, \quad \hat{\sigma}_2 = \begin{pmatrix} 0 & -i \\ i & 0 \end{pmatrix}, \quad \hat{\sigma}_3 = \begin{pmatrix} 1 & 0 \\ 0 & -1 \end{pmatrix}. \qquad (11.150)$$

These matrices together with the identity matrix form a complete set of linearly independent 2×2 operators in the 2D Hilbert space, and any operator in this space can be written as a linear combination of them. Using (11.148) and (11.150) in combination with the fundamental commutation rules of angular-momentum operators (11.96), we obtain the commutation rules

$$[\hat{\sigma}_i, \hat{\sigma}_j] = 2i\varepsilon_{ijk}\hat{\sigma}_k. \qquad (11.151)$$

Furthermore, from the definition (11.96) we can immediately verify that the square of each Pauli matrix is the unit,

$$\hat{\sigma}_i^2 = \mathbb{I}. \qquad (11.152)$$

These two properties can be summarized in the following expression, which is *specific* of Pauli matrices,

$$\hat{\sigma}_i \hat{\sigma}_j = \mathbb{I}\delta_{ij} + i\varepsilon_{ijk}\hat{\sigma}_k. \qquad (11.153)$$

The raising and lowering operators for spin $\frac{1}{2}$ are

$$\hat{\sigma}_+ = \tfrac{1}{\sqrt{2}}(\hat{\sigma}_1 + i\hat{\sigma}_2) = \sqrt{2}\begin{pmatrix} 0 & 1 \\ 0 & 0 \end{pmatrix}, \quad \hat{\sigma}_- = \sqrt{2}\begin{pmatrix} 0 & 0 \\ 1 & 0 \end{pmatrix}, \tag{11.154}$$

and the projection operators are

$$\hat{P}^+ = |+\rangle\langle+| = \begin{pmatrix} 1 & 0 \\ 0 & 0 \end{pmatrix}, \quad \hat{P}^- = |-\rangle\langle-| = \begin{pmatrix} 0 & 0 \\ 0 & 1 \end{pmatrix}. \tag{11.155}$$

For further properties of Pauli matrices, see Section 3.7 and Exercise E4.7.
For reference, the angular momentum matrices for $j = 1$ are (see Problem 11.10):

$$\hat{J}_x = \frac{\hbar}{\sqrt{2}}\begin{pmatrix} 0 & 1 & 0 \\ 1 & 0 & 1 \\ 0 & 1 & 0 \end{pmatrix}, \quad \hat{J}_y = \frac{\hbar}{\sqrt{2}}\begin{pmatrix} 0 & -i & 0 \\ i & 0 & -i \\ 0 & i & 0 \end{pmatrix}, \quad \hat{J}_z = \hbar\begin{pmatrix} 1 & 0 & 0 \\ 0 & 0 & 0 \\ 0 & 0 & -1 \end{pmatrix}. \tag{11.156}$$

In the coordinate representation, the three angular components of the wave function for $l = 1$, corresponding to $m = 1, 0, -1$, are the spherical harmonics Y_1^1, Y_1^0 and Y_1^{-1}, respectively.

11.4 Addition of Two Angular Momenta

There are quantum systems whose components can eventually stop interacting. For example, an atomic electron can be removed by a photon, or an atomic nucleus can decay into different parts that no longer interact, and so on. In some cases the elements into which the system is split do not even have an existence of their own before the decay, but are created during the very process that produces the decay; this happens, for example, when photons in an external field give rise to an electron–positron pair.

Here we study the case where the quantum system can be considered to be effectively composed of parts, each of which has a well-defined angular momentum. An example would be an atom composed of electrons and the nucleus. We then ask for the possible values of the angular momentum of the whole system. It is clear that we can pose the problem in the reverse sense: given the angular-momentum state of the composite system, we ask about the possible angular momentum states of its constituent parts. Due to the mathematical complexity of the problem, we will restrict ourselves to the case of only two parts; the analysis of systems with three or more constituents can be found in the literature.

From a physical point of view, the problem is interesting for several reasons. To get a qualitative idea of the situation, we note that the results of the previous section are general, that is, they apply to any quantum system, simple or complex. In particular, this means that the angular momenta of both the parts and the whole are constrained to be integers or half-integers. When two arbitrary angular momenta $\widehat{\boldsymbol{J}}_1$ and $\widehat{\boldsymbol{J}}_2$ are coupled, the resulting angular momentum $\widehat{\boldsymbol{J}} = \widehat{\boldsymbol{J}}_1 + \widehat{\boldsymbol{J}}_2$ must be an integer or half-integer. This indicates the existence of strong "interaction" phenomena between the different angular momenta of the system, so that the eigenstates that finally emerge as the product of all the interactions obey laws as singular as that of the spatial quantization of the angular momenta (total or partial). For example, we will see that in the stationary state of an atomic system, two electrons are coupled in such a way that their total spin is only 0 or 1. From the mathematical point of view, the problem of the coupling of two angular momenta J_1 and J_2 corresponds to the construction of the Hilbert-space product $\mathcal{H}_{J_1} \otimes \mathcal{H}_{J_2}$ and its decomposition into rotation-invariant subspaces.

11.4.1 Clebsch–Gordan coefficients

Our purpose is to express the eigenstates of the total angular momentum in terms of the eigenstates of its components. First, we note that the angular-momentum operators of each particle operate on different variables – that is, they act in different spaces – so they commute with each other,

$$[\widehat{J}_{1k}, \widehat{J}_{2l}] = 0 \quad (k, l = 1, 2, 3, +, -). \tag{11.157}$$

For the rest, the commutation properties of $\widehat{J}_{1i}$ and $\widehat{J}_{2i}$ are the usual ones, Eqs. (11.96) and (11.97). We will consider that each subsystem is in an angular-momentum eigenstate, with eigenvalues (j_1, m_1) and (j_2, m_2), respectively. Thus,

$$\widehat{\boldsymbol{J}}_i^2 |j_i m_i\rangle = \hbar^2 j_i(j_i + 1) |j_i m_i\rangle, \quad i = 1, 2. \tag{11.158}$$

The angular-momentum operator of the system is

$$\widehat{\boldsymbol{J}} = \widehat{\boldsymbol{J}}_1 + \widehat{\boldsymbol{J}}_2, \tag{11.159}$$

with projection on the z-axis given by

$$\widehat{J}_z = \widehat{J}_{1z} + \widehat{J}_{2z}, \tag{11.160}$$

and square angular momentum

$$\widehat{\boldsymbol{J}}^2 = \widehat{\boldsymbol{J}}_1^2 + \widehat{\boldsymbol{J}}_2^2 + 2\widehat{\boldsymbol{J}}_1 \cdot \widehat{\boldsymbol{J}}_2. \tag{11.161}$$

Since $\widehat{J}_z$ commutes with $\widehat{\boldsymbol{J}}_1^2$ and with $\widehat{\boldsymbol{J}}_2^2$, there are simultaneous eigenfunctions of the four operators $\widehat{\boldsymbol{J}}_1^2, \widehat{\boldsymbol{J}}_2^2, \widehat{\boldsymbol{J}}^2, \widehat{J}_z$. To construct them we proceed as follows. We call $|j_1 j_2 j m\rangle$ the state vector with the quantum numbers j_1, j_2, j, m corresponding to well-defined values of $\widehat{\boldsymbol{J}}_1^2, \widehat{\boldsymbol{J}}_2^2, \widehat{\boldsymbol{J}}^2, \widehat{J}_z$, respectively. We can then write, introducing an expansion in terms of the basis $|j_1 m_1\rangle |j_2 m_2\rangle$ for given j_1 and j_2 (remember that $|j_1 m_1\rangle$ and $|j_2 m_2\rangle$ are vectors in different spaces),

$$\begin{aligned} |j_1 j_2 j m\rangle &= \sum_{m_1 m_2} |j_1 m_1\rangle |j_2 m_2\rangle \langle j_1 m_1| \langle j_2 m_2 | j_1 j_2 j m\rangle \\ &= \sum_{m_1 m_2} (j_1 j_2 m_1 m_2 | j m) |j_1 m_1\rangle |j_2 m_2\rangle. \end{aligned} \tag{11.162}$$

The coefficients of the expansion

$$(j_1 j_2 m_1 m_2 | j m) \equiv \langle j_1 m_1 | \langle j_2 m_2 | j_1 j_2 j m \rangle \tag{11.163}$$

are called *Clebsch–Gordan coefficients*[5] (hereafter C-G coefficients).[6] These coefficients are the elements of a unitary matrix that transforms the representation in which the angular momentum of each subsystem and its z-component are fixed, into another in which the angular momentum of the composite system and its projection m are fixed.

Because of (11.160), the addition over m_1 and m_2 in the preceding expression can only be performed over the values that satisfy the condition

$$m = m_1 + m_2. \tag{11.165}$$

To determine the values of j that can be realized with the given values of j_1 and j_2, first note that the maximum possible m is

$$m_{\max} = (m_1 + m_2)_{\max} = j_1 + j_2, \tag{11.166}$$

which is only realized with the state $|j_1, j_1\rangle|j_2, j_2\rangle$ and corresponds to $j = j_1 + j_2$ with $m = j$. The next value of m,

$$m_{\max} - 1 = j_1 + j_2 - 1, \tag{11.167}$$

can be obtained with the states $|j_1, j_1 - 1\rangle|j_2, j_2\rangle$ and $|j_1, j_1\rangle|j_2, j_2 - 1\rangle$, whose linear combinations allow us to construct states with $j = j_1 + j_2$ and with $j = j_1 + j_2 - 1$; we conclude that this last value of j is also realized. Repeating this procedure, we can continue to reduce the value of j, while maintaining the possibility of constructing states with the given basis, until we reach the value $j = |j_1 - j_2|$, after which it is not possible to continue. Therefore, the possible values of j generated by the pair j_1, j_2 are the integers or semi-integers that differ by unit steps and satisfy the triangle relation,

$$|j_1 - j_2| \leq j \leq j_1 + j_2. \tag{11.168}$$

11.4.1.1 Some properties of the C–G coefficients

The usual phase convention is

$$\langle j_1, j_2, j_1, j - j_1 | jj \rangle \quad \text{real and positive.} \tag{11.169}$$

With this convention, all C-G coefficients are real, so the transformation matrix is not only unitary but also orthogonal ($U^T = U^{-1}$). The orthogonality condition implies the two equalities

$$\sum_{jm} (j_1 j_2 m_1 m_2 | jm)(j_1 j_2 m'_1 m'_2 | jm) = \delta_{m_1 m'_1} \delta_{m_2 m'_2} \tag{11.170}$$

$$\sum_{m_1 m_2} (j_1 j_2 m_1 m_2 | jm)(j_1 j_2 m_1 m_2 | j'm') = \delta_{jj'} \delta_{mm'}. \tag{11.171}$$

---

[5] Alfred Clebsch (1833–1872) was a Prussian mathematician; Paul Gordan (1837–1912) was a German mathematician, Emmy Noether's doctoral supervisor.

[6] These coefficients have different names in the literature; they are also called also Wigner coefficients or simply vector addition coefficients. Sometimes the definitions are slightly modified and, more often, the conventions regarding their arbitrary phases are changed. Certain coefficients proportional to C-G are often used because they are more symmetric; these are the *3-j symbols*, defined as

$$\begin{pmatrix} j_1 & j_2 & j_3 \\ m_1 & m_2 & m_3 \end{pmatrix} = \frac{(-1)^{j_1 - j_2 - m_3}}{(2j_3 + 1)^{1/2}} (j_1 j_2 m_1 m_2 | j_3, -m_3). \tag{11.164}$$

Using these properties, we can invert (11.162) to obtain

$$|j_1 m_1\rangle|j_2 m_2\rangle = \sum_{j,m} (j_1 j_2 m_1 m_2 | jm)|j_1 j_2 jm\rangle. \tag{11.172}$$

It is possible, although tedious, to determine the C-G coefficients using general procedures. For example, from the expression whose validity is directly derived from (11.162),

$$\widehat{J}_\pm |j_1 j_2 jm\rangle = \sum_{m_1 m_2} (j_1 j_2 m_1 m_2 | jm)(\widehat{J}_{1\pm} + \widehat{J}_{2\pm})|j_1 m_1\rangle|j_2 m_2\rangle, \tag{11.173}$$

one obtains with the help of Eqs. (11.137), (11.138) and (11.172)

$$\sqrt{(j \pm m + 1)(j \mp m)}(j_1 j_2 m_1 m_2 | j, m \pm 1)$$
$$= \sqrt{(j_1 \mp m_1 + 1)(j_1 \pm m_1)}(j_1 j_2, m_1 \pm 1, m_2 | jm)$$
$$+ \sqrt{(j_2 \mp m_2 + 1)(j_2 \pm m_2)}(j_1, j_2, m_1, m_2 \pm 1 | jm). \tag{11.174}$$

For $m = j$ and the upper sign, the left side vanishes, which gives a series of relations between the coefficients $(j_1 j_2 m_1 m_2 | jj)$. Using the coefficient given in (11.169) as a reference, all the others are obtained from these relations and are in fact real. If we now take the lower sign, we obtain successively the coefficients for $m = j - 1, j - 2$, and so on. This method was used by G. Racah[7] to obtain an explicit expression for the C-G coefficients, which is extremely complicated, so we will omit it. Note that Eq. (11.174) is sufficient to obtain all the C-G coefficients. Its detailed derivation and use can be studied in one of the specialized texts (references at the end of the chapter).

11.4.1.2 Calculation of some C-G coefficients

With the twofold purpose of gaining some experience with the algebra involved – without going into its computational difficulties – and of obtaining results that will be very useful to us later, we will calculate the coupling coefficients for two angular momenta 1/2 from first principles (in particular we will not use equations like (11.174)).

As follows from the triangle relation (11.168), if $j_1 = j_2 = \frac{1}{2}$ the possible values of j are 0 and 1 (antiparallel and parallel spins, respectively), giving rise to four states that are grouped naturally into a singlet with $j = 0$ and $m = 0$, and a triplet with $j = 1$ and $m = 1, 0, -1$.[8] These are the invariant subspaces of dimensions 1 and 3 respectively of the product space $2 \otimes 2 = 3 \oplus 1$.[9] The three states belonging to $j = 1$ are symmetric with respect to the exchange of particles 1 and 2, Eqs. (11.175, 11.177, 11.181), while the (singlet) state belonging to $j = 0$ is antisymmetric, Eq. (11.182). Since rotation does not change these symmetry properties, the decomposition $3 \oplus 1$ yields rotation-invariant subspaces; in other words, the components of the triplet mix with each other under rotation, but not with that of the singlet and vice versa.

---

[7] Giulio Racah (1909–1965) was an Italian–Israeli physicist and mathematician.
[8] The terms singlet and triplet have a spectroscopic origin: while the singlet state gives rise to a single spectral line, the triplet state gives rise to a triplet splitting of the spectral lines when the system is subjected to a magnetic field.
[9] The notation used refers to summation and multiplication, respectively, in Hilbert space.

11.4 Addition of Two Angular Momenta

We will denote the vectors of the composite states as $|c;jm\rangle \equiv |j_1 j_2 jm\rangle$, where the c reminds us of the composite system; when there is no danger of confusion, we will omit the c. Let us first analyze the triplet. The state $|jm\rangle = |11\rangle$ can only be obtained with the values $m_1 = m_2 = \frac{1}{2}$, so we have (in products of kets, the first ket refers to particle 1 and the second to particle 2)

$$|11\rangle = |+\rangle|+\rangle. \tag{11.175}$$

In more explicit notation we would write this equation in the form

$$|c;11\rangle = \left|\tfrac{1}{2}\tfrac{1}{2}11\right\rangle = \left|\tfrac{1}{2}\tfrac{1}{2}\right\rangle\left|\tfrac{1}{2}\tfrac{1}{2}\right\rangle. \tag{11.176}$$

Similarly, the state with $j=1, m=-1$ can only be obtained from the projections $-\frac{1}{2}$ of j_1 and j_2,

$$|1,-1\rangle = |-\rangle|-\rangle. \tag{11.177}$$

The remaining state of the triplet, $|10\rangle$, has two components, the states with $m_1 = \frac{1}{2}$, $m_2 = -\frac{1}{2}$, and with $m_1 = -\frac{1}{2}$, $m_2 = \frac{1}{2}$. We can obtain it by applying $\hat{J}_- \equiv \hat{J}_{1-} + \hat{J}_{2-}$ to $|11\rangle$ or $\hat{J}_+ \equiv \hat{J}_{1+} + \hat{J}_{2+}$ to $|1,-1\rangle$. For example,

$$\hat{J}_+|1,-1\rangle = C_{1,-1}|10\rangle = \left(\hat{J}_{1+} + \hat{J}_{2+}\right)|-\rangle|-\rangle = \left(\hat{J}_{1+}|-\rangle\right)|-\rangle + |-\rangle\left(\hat{J}_{2+}|-\rangle\right). \tag{11.178}$$

Using (11.137) we get

$$C_{1,-1}|10\rangle = C_{\frac{1}{2},-\frac{1}{2}}|+\rangle|-\rangle + C_{\frac{1}{2},-\frac{1}{2}}|-\rangle|+\rangle. \tag{11.179}$$

We calculate the coefficients using (11.128),

$$\frac{C_{\frac{1}{2},-\frac{1}{2}}}{C_{1,-1}} = \sqrt{\frac{\left(\frac{1}{2} - \frac{1}{2} + 1\right)\left(\frac{1}{2} + \frac{1}{2}\right)}{(1-1+1)(1+1)}} = \frac{1}{\sqrt{2}}, \tag{11.180}$$

and finally we get

$$|10\rangle = \tfrac{1}{\sqrt{2}}\Big(|+\rangle|-\rangle + |-\rangle|+\rangle\Big). \tag{11.181}$$

To construct the singlet state $|00\rangle$, we note that it is a linear combination of the same states that make up $|10\rangle$, and that it must be orthogonal to the latter. It is a simple exercise to show that

$$|00\rangle = \tfrac{1}{\sqrt{2}}\Big(|+\rangle|-\rangle - |-\rangle|+\rangle\Big). \tag{11.182}$$

With this, all coupling coefficients for $j_1 = j_2 = \frac{1}{2}$ have been determined. Later, when we deal with two-electron problems (such as the He atom, or Bell states), we will have the opportunity to use these important results.

We will solve a second example that generalizes the previous one and gives us results that will also be used later. This is the coupling between an arbitrary angular momentum j_1 and the spin angular momentum $j_2 = s = 1/2$. The calculation is greatly simplified by using the following equality, which can be derived from the given definitions,

$$\hat{J}^2 = (\hat{J}_1 + \hat{S})^2 = \hat{J}_1^2 + \hat{S}^2 + 2\hat{J}_{1z}\hat{S}_z + 2\hat{J}_{1+}\hat{S}_- + 2\hat{J}_{1-}\hat{S}_+. \tag{11.183}$$

Note that this equality contains the following,

$$\hat{J}_1 \cdot \hat{S} = \hat{J}_{1z}\hat{S}_z + \hat{J}_{1+}\hat{S}_- + \hat{J}_{1-}\hat{S}_+. \tag{11.184}$$

More generally, for two arbitrary angular momenta $\hat{J}_1$ and $\hat{J}_2$,

$$\hat{J}_1 \cdot \hat{J}_2 = \hat{J}_{1z}\hat{J}_{2z} + \hat{J}_{1+}\hat{J}_{2-} + \hat{J}_{1-}\hat{J}_{2+}. \tag{11.185}$$

These expressions are very useful in practice and should be kept in mind when performing calculations involving two angular momenta.

Returning to the previous problem, we suggest writing

$$|c; j, m_1 + \tfrac{1}{2}\rangle = a|j_1, m_1\rangle|+\rangle + b|j_1, m_1 + 1\rangle|-\rangle, \tag{11.186}$$

and a similar relationship for composite states with $m = m_1 - \tfrac{1}{2}$. The possible values of j are $j_1 + \tfrac{1}{2}$ and $j_1 - \tfrac{1}{2}$, according to the triangle law. We now apply the operator $\hat{J}^2$ written in the form (11.183) to the last expression, to obtain, using $\langle\hat{J}^2\rangle = \hbar^2 S(S+1) = \hbar^2 \tfrac{1}{2}(\tfrac{3}{2}) = \tfrac{3}{4}\hbar^2$,

$$\hat{J}^2|j, m_1 + \tfrac{1}{2}\rangle = \hbar^2 j(j+1)\left|j, m_1 + \tfrac{1}{2}\right\rangle$$

$$= \hbar^2 j(j+1)\left[a|j_1, m_1\rangle|+\rangle + b|j_1, m_1 + 1\rangle|-\rangle\right]$$

$$= \hbar^2 a\left[\left(j_1(j_1+1) + \tfrac{3}{4} + 2m_1 \cdot \tfrac{1}{2}\right)|j_1 m_1\rangle|+\rangle\right.$$

$$\left. + 2C_{j_1 m_1} C_{\tfrac{1}{2},-\tfrac{1}{2}}|j_1, m_1 + 1\rangle|-\rangle\right]$$

$$+ \hbar^2 b\left[\left(j_1(j_1+1) + \tfrac{3}{4} + 2(m_1+1)(-\tfrac{1}{2})\right)|j_1, m_1+1\rangle|-\rangle\right.$$

$$\left. + 2C_{j_1 m_1} C_{\tfrac{1}{2},-\tfrac{1}{2}}|j_1 m_1\rangle|+\rangle\right]. \tag{11.187}$$

We regroup the terms containing $|j_1 m_1\rangle|+\rangle$ and those containing $|j_1 m_1+1\rangle|-\rangle$, which must be cancelled separately. This condition leads to the system of equations

$$\left[j(j+1) - j_1(j_1+1) - \tfrac{3}{4} - m_1\right]a - 2C_{\tfrac{1}{2},-\tfrac{1}{2}}C_{j_1 m_1}b = 0, \tag{11.188}$$

$$-2C_{\tfrac{1}{2},-\tfrac{1}{2}}C_{j_1 m_1}a + \left[j(j+1) - j_1(j_1+1) - \tfrac{3}{4} + m_1 + 1\right]b = 0. \tag{11.189}$$

It is easy to show that the determinant of this system is zero if and only if j takes the values $j_1 + \tfrac{1}{2}$ or $j_1 - \tfrac{1}{2}$, as is expected from the triangle relation. If we take $j = j_1 + \tfrac{1}{2}$ and substitute it into Eq. (11.188), we obtain the quotient a/b. If we now impose the normalization condition $a^2 + b^2 = 1$, and taking a positive, we get

$$a = \sqrt{\frac{j_1 + m_1 + 1}{2j_1 + 1}}, \quad b = \sqrt{\frac{j_1 - m_1}{2j_1 + 1}}. \tag{11.190}$$

This process is repeated until the four possible states are constructed. The final results are shown in Table 11.1. If we take the case $j_1 = 1/2$ in the table, we recover the C-G coefficients for the coupling studied in the previous example.

Table 11.1 Clebsch–Gordan coefficients for the coupling of an arbitrary angular momentum j to a spin 1/2 angular momentum.

	$\left(j_1 \tfrac{1}{2} m_1 m_2 \mid j m\right)$	
J	$m_2 = \tfrac{1}{2}$	$m_2 = -\tfrac{1}{2}$
$j_1 + \tfrac{1}{2}$	$\left(\dfrac{j_1+m+\tfrac{1}{2}}{2j_1+1}\right)^{1/2}$	$\left(\dfrac{j_1-m+\tfrac{1}{2}}{2j_1+1}\right)^{1/2}$
$j_1 - \tfrac{1}{2}$	$-\left(\dfrac{j_1-m+\tfrac{1}{2}}{2j_1+1}\right)^{1/2}$	$\left(\dfrac{j_1+m+\tfrac{1}{2}}{2j_1+1}\right)^{1/2}$

Exercise E11.5. Scalar nature of the singlet spin state

Show that the entangled state of two spin-1/2 particles,

$$|\Psi\rangle_\lambda = \frac{1}{\left(1+|\lambda^2|\right)^{1/2}} [|+\rangle|-\rangle + \lambda |-\rangle|+\rangle] \qquad (11.191)$$

is a scalar (i.e., invariant against rotation) if and only if $\lambda = -1$, which is the antisymmetric combination upon particle exchange. The importance of entangled states will become clear in Section 14.5.

Solution. Let us consider a new basis $|\alpha\rangle$ and $|\beta\rangle$, whose axes are oriented at angles θ, ϕ with respect to the original basis $|+\rangle, |-\rangle$. The matrix describing this rotation has the form given by Eq. (8.167) (with the substitution $L \to S = \hbar\sigma/2$); we can write it in the form

$$\hat{T} = \begin{pmatrix} a & b \\ b^* & -a \end{pmatrix}, \qquad a = \cos\theta^2, \qquad b = \frac{\sin\theta}{2} e^{i\phi/2}. \qquad (11.192)$$

Therefore, $|\alpha\rangle = \hat{T}|+\rangle = a|+\rangle + b^*|-\rangle$, $|\beta\rangle = \hat{T}|-\rangle = b|+\rangle - a|-\rangle$, or inverting, $|+\rangle = a|\alpha\rangle + b^*|\beta\rangle$, $|-\rangle = b|\alpha\rangle - a|\beta\rangle$.

We substitute these results into the expression for $|\Psi\rangle_\lambda$ with $N_\lambda = \left(1+|\lambda|^2\right)^{-1/2}$ and rearrange to obtain

$$|\Psi\rangle_\lambda = N_\lambda [|+\rangle|-\rangle + \lambda|-\rangle|+\rangle] = N(1+\lambda)\left[ab|\alpha\rangle|\alpha\rangle - ab^*|\beta\rangle|\beta\rangle\right]$$
$$+ N\left[\left(|b|^2 - \lambda a^2\right)|\beta\rangle|\alpha\rangle - \left(a^2 - \lambda|b|^2\right)|\alpha\rangle|\beta\rangle\right]. \qquad (11.193)$$

For $\lambda = -1$ this expression reduces to ($a^2 + |b|^2 = 1$):

$$|\Psi\rangle_{-1} = \frac{1}{\sqrt{2}}[|\beta\rangle|\alpha\rangle - |\alpha\rangle|\beta\rangle], \qquad (11.194)$$

which has exactly the initial form of $|\Psi\rangle$ with $\lambda = -1$. This invariance shows that in (11.191) with $\lambda = -1$ (but only with this value), the orientation of the axes defining the states $|+\rangle$ and $|-\rangle$ is arbitrary. In other words, the vector $|\Psi\rangle_{-1}$ in Eq. (11.194) contains no information about a preferred direction. This means that $|\Psi\rangle_{-1}$ is a scalar with respect to rotation.

Problems

P11.1 Determine the relationship between the angular momentum defined in two different inertial reference systems when:

a) the systems are at relative rest, with origins separated by a distance a,
b) the systems move with constant relative velocity v.

P11.2 Prove that

$$[\hat{\mathbf{n}} \cdot \hat{\mathbf{L}}, \mathbf{r}] = -i\hbar(\hat{\mathbf{n}} \times \mathbf{r}), \quad [\hat{\mathbf{n}} \cdot \hat{\mathbf{L}}, \hat{\mathbf{p}}] = -i\hbar(\hat{\mathbf{n}} \times \hat{\mathbf{p}}), \quad [\hat{\mathbf{n}} \cdot \hat{\mathbf{L}}, \hat{\mathbf{L}}] = -i\hbar(\hat{\mathbf{n}} \times \hat{\mathbf{L}}). \tag{11.195}$$

In the preceding expressions, $\hat{\mathbf{n}}$ is a constant unit vector in any direction.

P11.3 Show that the total orbital momentum of a two-particle system can be expressed in terms of the center-of-mass and relative coordinates in the form

$$\hat{\mathbf{L}} = \mathbf{R} \times \hat{\mathbf{P}} + \mathbf{r} \times \hat{\mathbf{p}}, \tag{11.196}$$

where

$$\mathbf{r} = \mathbf{r}_1 - \mathbf{r}_2, \quad \hat{\mathbf{p}} = m\left(\frac{\hat{\mathbf{p}}_1}{m_1} - \frac{\hat{\mathbf{p}}_2}{m_2}\right), \quad m = \frac{m_1 m_2}{m_1 + m_2},$$

$$\mathbf{R} = \frac{m_1 \mathbf{r}_1 + m_2 \mathbf{r}_2}{M}, \quad \hat{\mathbf{P}} = \hat{\mathbf{p}}_1 + \hat{\mathbf{p}}_2, \quad M = m_1 + m_2. \tag{11.197}$$

P11.4 Show that any function that depends only on r is an eigenfunction of $\hat{L}_z$ and $\hat{L}^2$, with both eigenvalues equal to zero, and that any function $g(z, r)$ is an eigenfunction of the z component with zero eigenvalue. Determine the eigenfunction of $\hat{L}_x$ corresponding to zero eigenvalue and angular momentum 1.

P11.5 Calculate the expectation value of the operator $\frac{1}{2}(\hat{L}_x \hat{L}_y + \hat{L}_y \hat{L}_x)$ and its square, in an eigenstate of $\hat{L}^2$ and $\hat{L}_z$.

P11.6 Determine the expression for the eigenfunctions of $\hat{L}^2$ and $\hat{L}_z$ in momentum space.

P11.7 Prove that there is no solution to the eigenvalue problem $\hat{p}_r R(r) = \alpha R(r)$ that satisfies the condition $\lim_{r \to 0} r R(r) = 0$. What can be concluded from here?

P11.8 Consider an eigenfunction $\psi_{nlm}(\mathbf{r})$ of a radial Hamiltonian and show that

$$\psi' = e^{i\alpha \hat{\mathbf{n}} \cdot \hat{\mathbf{L}}/\hbar} \psi_{nlm}(\mathbf{r}) \tag{11.198}$$

is an eigenfunction of the Hamiltonian with the same angular momentum l, independent of the value of α and the orientation of the $\hat{\mathbf{n}}$ axis. Is it also an eigenstate of $\hat{L}_z$?

P11.9 Construct the matrices representing the angular-momentum operators for $j = 1$ and $j = 3/2$.

P11.10 Since $\hat{S}_i$ are the components of a vector, they are transformed under rotation according to the laws of vectors. For example, under rotation around the z-axis by an angle θ, which performs the transformation $\hat{\mathbf{S}} \to \hat{\mathbf{S}}'$,

$$\hat{S}'_x = \hat{S}_x \cos\theta + \hat{S}_y \sin\theta, \quad \hat{S}'_y = -\hat{S}_x \sin\theta + \hat{S}_y \cos\theta, \quad \hat{S}'_z = \hat{S}_z. \tag{11.199}$$

Use the commutation rules for $\hat{S}_x$, $\hat{S}_y$, and $\hat{S}_z$ to show that the transformed components $\hat{S}'_x$, $\hat{S}'_y$ and $\hat{S}'_z$ satisfy exactly the same rules.

P11.11 If $\hat{n}$ is a constant unit vector with direction cosines l, m, n with respect to the x-, y- and z-axes, respectively, show that

$$\hat{\sigma}_n = \hat{n}\cdot\hat{\sigma} = \begin{pmatrix} n & l-im \\ l+im & -n \end{pmatrix}, \quad (\hat{n}\cdot\hat{\sigma})^2 = \mathbb{I}. \quad (11.200)$$

Confirm that the eigenvalues of $\hat{\sigma}_n$ are ± 1. What does this mean?

P11.12 The spin projection of an electron on a certain z-axis is $\hbar/2$. Determine the probability that the projection of this spin on a certain z'-axis have the value $\hbar/2$ or $-\hbar/2$, as well as the average value of this projection.

P11.13 Consider an ion with spin 1, characterized by the Hamiltonian

$$\hat{H} = D\hat{S}_z^2 + E\left(\hat{S}_x^2 + \hat{S}_y^2\right), \quad (11.201)$$

with D and E constant, $D \gg E$. Determine the energy levels.

P11.14 Consider a system of angular momentum 1 represented by the state vector

$$\psi = \frac{1}{\sqrt{26}} \begin{pmatrix} 1 \\ 4 \\ -3 \end{pmatrix}. \quad (11.202)$$

What is the probability that a measurement of $\hat{L}_x$ returns the value zero?

P11.15 Calculate directly the Clebsch–Gordan coefficients for the coupling of angular momenta 1/2 and 1. Compare your results with those in Table 11.1.

P11.16 An orbital angular momentum $\hat{L}$ couples with a spin $\hat{S}$ to produce a state of total angular momentum $\hat{J}$. Find out what angles between the vectors $\hat{L}$ and $\hat{S}$ are allowed by quantum rules.

P11.17 Is it possible for one photon to spontaneously decay into two photons? And into three photons? Explain your answers.

P11.18 The deuteron has spin 1. What are the possible states of spin and total angular momentum of a system of two deuterons when their total orbital angular momentum is L?

P11.19 Consider a spin 1/2 particle confined by a central potential.

a) Determine the wave functions that are simultaneously eigenfunctions of the operators $\hat{L}^2$, $\hat{J}^2$ and $\hat{J}_z = \hat{L}_z + \hat{S}_z$.

b) When an interaction term of the form $\gamma \hat{L}\cdot\hat{S}$ with γ small is added to the Hamiltonian, what are the eigenfunctions of the system?

Bibliographical Notes

A highly recommended reference is the classic book by Condon and Shortley (1964). Specific references on angular momentum are Brink and Satchler (1968), Edmonds (1957), Khersonskii, Moskalev, and Varshalovich (1988), and Hassani (2013).

12 Central Potentials. The Hydrogen Atom

Central symmetry plays an important role in many physical systems, such as the rotor, the isotropic harmonic oscillator, the van der Waals potential, the diatomic molecule or the hydrogen atom. In these cases, the system typically consists of two particles interacting via a central potential. It is therefore convenient to begin this chapter by transforming the Hamiltonian of the two-particle system into one that separates the center-of-mass motion from the relative motion. When the potential is spherically symmetric, the relative motion is conveniently described in spherical coordinates to take advantage of the conservation of angular momentum and the solution of the corresponding eigenvalue problem as discussed in the previous chapter. We therefore begin this chapter by reducing the 6D two-body problem to the 3D problem of relative motion under a central potential.

12.1 Reduction of the Two-Body Problem

Consider a physical system consisting of two interacting parts that can be thought of as a pair of particles. The Hamiltonian consists of the Hamiltonians of the two particles plus the interaction potential. When the interaction depends only on the relative position of the particles, we write

$$\widehat{H}(\widehat{\boldsymbol{p}}_1, \widehat{\boldsymbol{p}}_2, \boldsymbol{r}_1, \boldsymbol{r}_2) = \widehat{H}_1(\widehat{\boldsymbol{p}}_1, \boldsymbol{r}_1) + \widehat{H}_2(\widehat{\boldsymbol{p}}_2, \boldsymbol{r}_2) + V(\boldsymbol{r}_1 - \boldsymbol{r}_2), \tag{12.1}$$

with the single-particle Hamiltonians given by

$$\widehat{H}_i = \frac{\widehat{\boldsymbol{p}}_i^2}{2m_i} + V_i(\boldsymbol{r}_i) = -\frac{\hbar^2}{2m_i}\nabla_i^2 + V_i(\boldsymbol{r}_i), \quad i = 1, 2. \tag{12.2}$$

The Schrödinger equation is then (here we will analyze only the stationary case)

$$E\Psi = \widehat{H}\Psi = \left[-\frac{\hbar^2}{2m_1}\nabla_1^2 - \frac{\hbar^2}{2m_2}\nabla_2^2 + V_1(\boldsymbol{r}_1) + V_2(\boldsymbol{r}_2) + V(\boldsymbol{r}_1 - \boldsymbol{r}_2)\right]\Psi. \tag{12.3}$$

Before proceeding, it is worth noting the following. In the preceding equation, $\Psi = \Psi(\boldsymbol{r}_1, \boldsymbol{r}_2)$ is a function of six variables, three for each of the particles; more generally, the wave function will be a function of $3N$ variables, where N is the number of (point-like) particles. This means that it is not possible to interpret Ψ as a physical wave in the conventional sense of the term, because if it were, we could only describe it in terms of the three variables that specify the point of physical space in which it is being considered. An electromagnetic wave, in contrast, satisfies the wave equation in 3D space (apart from time), regardless of the number of sources that produce it. Nevertheless, $\Psi(\mathbf{r}_1, \mathbf{r}_2)$ can still be interpreted as a probability (density) amplitude, since a probability density can depend on any number of variables.

For simplicity, we restrict ourselves to the case of an isolated system, where the potentials V_1 and V_2 produced by external fields are zero, and only the (internal) interaction between the particles remains. Under this condition, Eq. (12.3) takes the form

$$E\Psi = \left[-\frac{\hbar^2}{2m_1}\nabla_1^2 - \frac{\hbar^2}{2m_2}\nabla_2^2 + V(\mathbf{r}_1 - \mathbf{r}_2) \right]\Psi. \quad (12.4)$$

The structure of this equation suggests going from the variables $\mathbf{r}_1, \mathbf{r}_2$ to the relative and CM variables,[1]

$$\mathbf{r} = \mathbf{r}_1 - \mathbf{r}_2, \quad \mathbf{R} = \frac{m_1 \mathbf{r}_1 + m_2 \mathbf{r}_2}{M}, \quad (12.5)$$

where $M = m_1 + m_2$. From (12.5) it follows that

$$\nabla_1 = \nabla_r + \frac{m_1}{M}\nabla_R, \quad (12.6)$$

$$\nabla_2 = -\nabla_r + \frac{m_2}{M}\nabla_R. \quad (12.7)$$

Taking the square of these operators we get

$$\frac{\nabla_1^2}{m_1} + \frac{\nabla_2^2}{m_2} = \frac{\nabla_r^2}{m} + \frac{\nabla_R^2}{M}, \quad (12.8)$$

where m is the reduced mass of the system (smaller than m_1, m_2),

$$m = \frac{m_1 m_2}{m_1 + m_2} = m_1 \frac{1}{1 + \frac{m_1}{m_2}} = m_2 \frac{1}{1 + \frac{m_2}{m_1}}. \quad (12.9)$$

Substituting this result into (12.4), we get

$$E\Psi = \left[-\frac{\hbar^2}{2M}\nabla_R^2 - \frac{\hbar^2}{2m}\nabla_r^2 + V(\mathbf{r}) \right]\Psi. \quad (12.10)$$

This equation is solved using the method of separation of variables. We write Ψ in the form

$$\Psi(\mathbf{R}, \mathbf{r}) = \Phi(\mathbf{R})\psi(\mathbf{r}) \quad (12.11)$$

and split the energy into two parts,

$$E = E_R + E_r, \quad (12.12)$$

so that we can write

$$(E_R + E_r)\Phi\psi = -\frac{\hbar^2}{2M}\psi\nabla_R^2\Phi + \Phi\left[-\frac{\hbar^2}{2m}\nabla_r^2 + V(\mathbf{r}) \right]\psi. \quad (12.13)$$

which can be broken down into the pair of equations

$$E_R \Phi(\mathbf{R}) = -\frac{\hbar^2}{2M}\nabla_R^2 \Phi(\mathbf{R}), \quad (12.14)$$

---

[1] Section 18.1 deals with the problem of removing CM coordinates in a slightly more general way.

$$E_r\psi(r) = -\frac{\hbar^2}{2m}\nabla_r^2\psi(r) + V\psi(r). \tag{12.15}$$

Equation (12.14) describes the free motion of a particle of total mass M (such purely formal "particles" without a real counterpart are called *quasiparticles*); it is the equation for the CM moving freely with energy E_R. This motion is irrelevant for our purposes and it is better to switch to a coordinate system in which the CM is at rest (CM system), so that we can take the trivial solution $E_R = 0$ and $\Phi = $ const.

Equation (12.15), on the other hand, describes the relative motion of a quasiparticle of reduced mass m acted upon by the interaction potential $V(r)$. This is the Schrödinger equation for the two-particle central problem, reduced to a one-quasiparticle problem, the equation of real interest for us. It must be remembered that in (12.15) r is not the coordinate of one of the two particles and m is the reduced mass of the system. In the case of the H atom, for example, the reduced mass of the electron-proton system differs very little from the mass of the electron, because the mass of the proton is almost 2,000 times that of the electron, but the difference is noticeable to spectroscopists. The proton itself, due to its large relative mass, remains almost motionless very close to the origin of the coordinates.

12.1.1 Canonical variables for the central problem

From the Heisenberg equations of motion applied to the variables r and R it follows that the relative and CM momenta are given by

$$\widehat{p} = m\left(\frac{\widehat{p}_1}{m_1} - \frac{\widehat{p}_2}{m_2}\right), \quad \widehat{P} = \widehat{p}_1 + \widehat{p}_2. \tag{12.16}$$

These variables obey the quantum laws of independent particles, since their commutation relations are

$$[\widehat{r}_i, \widehat{p}_j] = i\hbar\delta_{ij}, \quad [\widehat{R}_i, \widehat{P}_j] = i\hbar\delta_{ij}, \quad [\widehat{r}_i, \widehat{P}_j] = [\widehat{R}_i, \widehat{p}_j] = 0, \tag{12.17}$$

which shows that the change of variables from r_1, r_2 to R, r and from $\widehat{p}_1, \widehat{p}_2$ to $\widehat{P}, \widehat{p}$ is a canonical transformation; this observation could have been used to perform the separation of variables. It can also be shown that

$$L_1 + L_2 = l + L, \tag{12.18}$$

where

$$\widehat{L}_{1,2} = r_{1,2} \times \widehat{p}_{1,2}, \quad \widehat{l} = r \times \widehat{p}, \quad \widehat{L} = R \times \widehat{P}, \tag{12.19}$$

where $\widehat{L}_1$ and $\widehat{L}_2$ the angular momenta of particles 1 and 2, respectively, $\widehat{l}$ is the angular momentum of the relative motion and $\widehat{L}$ is that of the CM.

Equation (12.15) can be solved using the techniques of Chapter 11; in particular, we concluded there that the solution of the Schrödinger equation for a central problem is the product of a spherical harmonic $Y_l^m(\theta, \varphi)$ and the solution of the radial equation, that is,

$$\psi_{nlm} = R_{nl}(r)Y_l^m(\theta, \varphi). \tag{12.20}$$

12.1 Reduction of the Two-Body Problem

The radial equation for $R(r)$ is given by Eq. (11.82),

$$\left[\frac{\widehat{p}_r^2}{2m} + \frac{\hbar^2 l(l+1)}{2mr^2} + V(r)\right] R(r) = E R(r), \tag{12.21}$$

where the index r has been omitted from the energy eigenvalue, $E = E_r$. The operator

$$\widehat{p}_r = -i\hbar \frac{1}{r}\frac{\partial}{\partial r} r = -i\hbar \left(\frac{\partial}{\partial r} + \frac{1}{r}\right) \tag{12.22}$$

is the canonical conjugate momentum of r, and $\widehat{p}_r^2$ is

$$\widehat{p}_r^2 = -\frac{\hbar^2}{r^2}\frac{\partial}{\partial r}\left(r^2 \frac{\partial}{\partial r}\right) = -\frac{\hbar^2}{r}\frac{\partial^2}{\partial r^2} r. \tag{12.23}$$

Due to the form of $\widehat{p}_r^2$, (12.21) is not strictly a 1D Schrödinger equation; but by introducing a new radial function $u(r)$ defined as

$$u = rR, \tag{12.24}$$

it becomes one,

$$\left[-\frac{\hbar^2}{2m}\frac{d^2}{dr^2} + \frac{\hbar^2 l(l+1)}{2mr^2} + V\right] u = Eu. \tag{12.25}$$

Thus the 6D central two-body problem has been formally reduced to a 1D problem for a quasiparticle associated with the relative radial motion and driven by the effective potential $V + \hbar^2 l(l+1)/2mr^2$ in the region $r \geq 0$.

Due to the factors $1/r$, both $\widehat{p}_r$ and Eq. (12.25) remain undefined for $r=0$; therefore it is necessary to specify the boundary condition that must be satisfied at the origin. To this aim, we observe that if $\widehat{p}_r$ is Hermitian with respect to the (square-integrable) eigenfunctions of the Hamiltonian, then the Hamiltonian will also be Hermitian with respect to its eigenfunctions. From an integration by parts it follows that

$$\int_0^\infty R_2^*(\widehat{p}_r R_1) r^2 \, dr = -i\hbar r^2 R_2^* R_1 \Big|_0^\infty + \int_0^\infty R_1 (\widehat{p}_r^* R_2^*) r^2 \, dr, \tag{12.26}$$

where R_1 and R_2 are two arbitrary radial functions. Therefore we see that if $\lim_{r\to\infty} (rR_1) = 0$ (and an analogous behavior for R_2), $\widehat{p}_r$ is Hermitian if $\lim_{r\to 0} r^2 R_1 R_2^* = 0$. This is true in particular when the condition

$$u(0) = 0 \tag{12.27}$$

is imposed. For all nonrelativistic problems, this boundary condition is sufficient to select the acceptable radial solutions at the origin; it is equivalent to requiring that the radial function $R(r) = u(r)/r$ remains finite at the origin, or, alternatively, to considering that in (12.25) the potential V is infinite for $r < 0$. In other words, the acceptable solutions are the *odd* solutions (which vanish at the origin) of the 1D Schrödinger problem with potential $V(|x|) + \hbar^2 l(l+1)/2mx^2$.

Two additional remarks are useful here, both arising from the fact that the differential equation for u is precisely a 1D Schrödinger equation. The first is that the u's obtained for any potential form a complete set in $[0, \infty)$. The second is that the WKB method can be applied directly to the radial equation for u; for example, this method has been used successfully to investigate positive-energy solutions for the Coulomb potential.

Exercise E12.1. Central problem in two and three dimensions

Show that the discrete energy spectrum of a central problem is different when the system is treated in two or three dimensions.

Solution. That this must be the case in general can be seen by referring to the example of the isotropic harmonic oscillator. Treating the problem in a Cartesian coordinate system we obtain for the eigenvalues of the Hamiltonian the formula $E_n = \hbar\omega \sum_{i=1}^{2} (n_i + \frac{1}{2}) = \hbar\omega(n+1)$ where n is an integer in the 2D case, while in the 3D case we get $E_n = \hbar\omega \sum_{i=1}^{3} (n_i + \frac{1}{2}) = \hbar\omega(n + \frac{3}{2})$ with n integer.

In the general 3D case, the radial Schrödinger equation is (12.25),

$$\left[-\frac{\hbar^2}{2m} \frac{d^2}{dr^2} + l(l+1) \frac{\hbar^2}{2mr^2} + V(r) \right] u(r) = E u(r). \tag{12.28}$$

To treat the 2D case we use a polar plane coordinate system and write the wave function in the form

$$\psi = \frac{u(r)}{\sqrt{r}} e^{ik\phi}. \tag{12.29}$$

The usual requirement that ψ be single-valued demands that $|k|$ be an integer. Proceeding as in the derivation of Eq. (12.28), we get

$$\left[-\frac{\hbar^2}{2m} \frac{d^2}{dr^2} + \left(k - \frac{1}{2}\right)\left(k + \frac{1}{2}\right) \frac{\hbar^2}{2mr^2} + V(r) \right] u(r) = E u(r). \tag{12.30}$$

The comparison of Eqs. (12.28) and (12.30) shows that the eigenvalues of H of the 2D case are obtained from the 3D case by substituting the integer l for the half-integer $k - 1/2$ (obviously E does not depend on the sign of k).

Exercise E12.2. Recurrence relation between expectation values of powers of r

Use the formula for the canonical momentum operator $\hat{p}_r$ to obtain a recurrence relation between the expectation values of powers of r, both for the hydrogen-like atom and for the isotropic 3D harmonic oscillator.

Solution. In a bound stationary system, the virial theorem states that $\langle \dot{\hat{A}} \rangle = 0$ for any $\hat{A}$, $\langle [\hat{A}, \hat{H}] \rangle = 0$. Let us apply this result to the operator $\hat{A} = \hat{p}_r r^{s+1} = (\hat{\boldsymbol{p}} \cdot \boldsymbol{r}) r^s$; it is obvious that for s = 0, the usual virial theorem holds. By taking the time derivative of $\hat{A}$ we get

$$\frac{d}{dt} \langle \hat{p}_r r^{s+1} \rangle = \langle \dot{\hat{p}}_r r^{s+1} \rangle + \langle \hat{p}_r \dot{r} r^s \rangle + \cdots + \langle \hat{p}_r r^s \dot{r} \rangle = 0. \tag{12.31}$$

To simplify this expression we write $\dot{r} = \hat{p}_r/m$ and pass all the $\hat{p}_r$'s to the left using the commutation relation

$$[r^n, \hat{p}_r] = i\hbar n r^{n-1}. \tag{12.32}$$

The result is

$$\langle \dot{\hat{p}}_r r^{s+1} \rangle + \frac{s+1}{m} \langle \hat{p}_r^2 r^s \rangle + \frac{is(s+1)}{2} \frac{\hbar}{m} \langle \hat{p}_r r^{s-1} \rangle = 0. \tag{12.33}$$

We now simplify each term separately. The last term can be simplified by noting that the generalized virial theorem itself implies that $\frac{d}{dt}\langle r^s \rangle = 0$; proceeding with this expression as before, we get

$$\langle \hat{p}_r r^{s-1} \rangle = -i(s-1)\frac{\hbar}{2}\langle r^{s-2} \rangle. \tag{12.34}$$

To simplify the first two terms, we must use the Hamiltonian of the system. We first take the case of the hydrogen-like atom, and write $\hat{H}$ in the form

$$\hat{H} = \frac{\hat{p}_r^2}{2m} + \frac{\hbar^2 l(l+1)}{2mr^2} - \frac{C}{r}. \tag{12.35}$$

This gives an expression for $\dot{p}_r$ in terms of r,

$$\dot{p}_r = \frac{-i}{\hbar}[\hat{p}_r, \hat{H}] = \frac{\hbar^2 l(l+1)}{mr^3} - \frac{C}{r^2}. \tag{12.36}$$

Substituting E for $\hat{H}$, we get an expression for p_r^2 in terms of r,

$$\hat{p}_r^2 = 2m\left[E - \frac{\hbar^2 l(l+1)}{2mr^2} + \frac{C}{r}\right]. \tag{12.37}$$

Substitution of the three previous results in (12.33) gives the following recurrence relation for the hydrogen-like atom, *known as the* Kramers formula,

$$2E(s+1)\langle r^s \rangle + C(2s+1)\langle r^{s-1} \rangle + \frac{\hbar^2}{m}s\left(\frac{s^2-1}{4} - l(l+1)\right)\langle r^{s-2} \rangle = 0. \tag{12.38}$$

Following the same procedure with the Hamiltonian of the isotropic oscillator with potential $V = \frac{1}{2}m\omega^2 r^2$, we obtain the recurrence relation

$$2E(s+1)\langle r^s \rangle - \frac{1}{2}m\omega^2(2s+4)\langle r^{s+2} \rangle + \frac{\hbar^2}{m}s\left(\frac{s^2-1}{4} - l(l+1)\right)\langle r^{s-2} \rangle = 0. \tag{12.39}$$

Note that by taking $\hbar = 0$, the generalization of the classical virial theorem is obtained in both cases.

An interesting exercise is to prove that for any central potential the following recurrence relation holds,

$$2E(s+1)\langle r^s \rangle - 2(s+1)\langle r^s V \rangle - \langle r^{s+1} V' \rangle + \frac{\hbar^2}{m}s\left(\frac{s^2-1}{4} - l(l+1)\right)\langle r^{s-2} \rangle = 0. \tag{12.40}$$

12.2 Angular Momentum Selection Rules for Dipole Transitions

The results obtained earlier can be applied to determine the angular-momentum dependence of the allowed radiative dipole transitions between quantum states for any central problem. For this purpose we recall from Section 9.3 that a dipole transition between two states $|nlm\rangle$ and $|n'l'm'\rangle$ can occur if the matrix element $\langle n'l'm'|\mathbf{r}|nlm\rangle$ is different from zero. We write $\mathbf{r} = \hat{\mathbf{a}}_r r$ to factor out the angular and radial integrals as follows, noting that the state vector $|nlm\rangle$ is a shorthand for the product vector $|nl\rangle|lm\rangle$. We can easily see this by writing the wave function (12.20) in Dirac notation:

$$\psi_{nlm}(r,\theta,\varphi) = R_{nl}(r)Y_l^m(\theta,\varphi) = \langle r,\theta,\varphi|nlm\rangle = \langle r|nl\rangle\langle\theta\varphi|lm\rangle. \quad (12.41)$$

Hence it follows by comparison that

$$R_{nl}(r) = \langle r|nl\rangle, \quad Y_l^m(\theta,\varphi) = \langle\theta\varphi|lm\rangle, \quad (12.42)$$

and we can write

$$\langle n'l'm'|\mathbf{r}|nlm\rangle = \langle n'l'|r|nl\rangle\langle l'm'|\widehat{\mathbf{a}}_r|lm\rangle, \quad (12.43)$$

where $\widehat{\mathbf{a}}_r$ is the unit vector

$$\widehat{\mathbf{a}}_r = \widehat{\mathbf{a}}_x \sin\theta\cos\varphi + \widehat{\mathbf{a}}_y \sin\theta\sin\varphi + \widehat{\mathbf{a}}_z \cos\theta$$
$$= \widehat{\mathbf{a}}_+ \sin\theta e^{i\varphi} + \widehat{\mathbf{a}}_- \sin\theta e^{-i\varphi} + \widehat{\mathbf{a}}_z \cos\theta \equiv \widehat{\mathbf{a}}_+ \eta_+ + \widehat{\mathbf{a}}_- \eta_- + \widehat{\mathbf{a}}_z \eta_z, \quad (12.44)$$

$$\widehat{\mathbf{a}}_\pm = \tfrac{1}{2}(\widehat{\mathbf{a}}_x \mp i\widehat{\mathbf{a}}_y),$$
$$\eta_\pm = \sin\theta e^{\pm i\varphi}, \quad \eta_z = \cos\theta. \quad (12.45)$$

The change of $\widehat{\mathbf{a}}_r$ is described in terms of a clockwise or a counter-clockwise rotation around the z-axis plus an oscillation about the z-axis. The matrix elements of these components can be written in the form

$$\langle l'm'|\eta_+|lm\rangle = \int Y_{l'}^{m'*} \sin\theta e^{i\varphi} Y_l^m \, d\Omega = [A_{lm}^+ \delta_{l',l+1} + B_{lm}^+ \delta_{l',l-1}]\delta_{m',m+1}, \quad (12.46)$$

$$\langle l'm'|\eta_-|lm\rangle = \int Y_{l'}^{m'*} \sin\theta e^{-i\varphi} Y_l^m \, d\Omega = [A_{lm}^- \delta_{l',l+1} + B_{lm}^- \delta_{l',l-1}]\delta_{m',m-1}, \quad (12.47)$$

$$\langle l'm'|\eta_z|lm\rangle = \int Y_{l'}^{m'*} \cos\theta Y_l^m \, d\Omega = [A_{lm} \delta_{l',l+1} + B_{lm} \delta_{l',l-1}]\delta_{m',m}. \quad (12.48)$$

The value of the coefficients A_{lm} and B_{lm} in these expressions is obtained by using the recurrence relations of the spherical harmonics and their orthogonality properties; see Section A.1. Substituting Eqs. (12.46–12.48) into (12.43) we see that the selection rules for all three motions demand that l change by unity. The physical reason for this result is that the radiated or absorbed photon has spin 1 and the total angular momentum (atom + radiation) is conserved during the transition. The projection L_z can be preserved during the transition (for oscillations about the z-axis) or change by one (for rotations around the z-axis). In short, the angular-momentum selection rules for any central problem are

$$|\Delta l| = 1, \quad |\Delta m| = 0, 1. \quad (12.49)$$

12.3 The 3D Rigid Rotor

The simplest problem that we can treat with the previous results is that of the rotor, that is, the free movement of a particle on the surface of a sphere, or else, the rigid rotation of a diatomic molecule such as CO, HCl, etc., seen from the CM system. If we assume that the molecule rotates around its CM without changing the interatomic distance a, the potential is reduced to a constant, which we can take to be zero (the reference level of the potential energy is arbitrary), so that the Schrödinger equation (12.25) becomes (m stands now for the reduced mass of the molecule)

$$\left[-\frac{\hbar^2}{2m}\frac{d^2}{dr^2} + \frac{\hbar^2 l(l+1)}{2ma^2}\right] u(r) = Eu(r), \quad r = a. \tag{12.50}$$

Since $u(a)$ is a constant, it immediately follows that the energy is given by

$$E_l = \frac{\hbar^2 l(l+1)}{2ma^2} = \frac{\hbar^2 l(l+1)}{2I}, \tag{12.51}$$

where $I = ma^2$ is the moment of inertia of the molecule. The wave functions $Y_l^m(\theta, \varphi)$ with $|m| \leq l$ are associated with the eigenvalue E_l, so the level l has a degeneracy of order $2l + 1$. The origin of this degeneracy lies in the spherical symmetry of the Hamiltonian, which implies that all directions are equivalent. In the s-state ($l = 0$) the spatial probability density is

$$\rho_{00} = |Y_0^0|^2 = \frac{1}{4\pi}, \tag{12.52}$$

which shows that the direction of the interatomic axis is uniformly distributed over the sphere. This solution therefore corresponds to the case where all orientations of the molecule are equally probable. In states with $l \neq 0$, favored molecular orientations appear on average; for example, for $l = 1$ the probability densities depend on m, resulting in

$$\rho_{11} = \rho_{1-1} = |Y_1^1|^2 = |Y_1^{-1}|^2 = \frac{3}{8\pi}\sin^2\theta, \tag{12.53}$$

$$\rho_{10} = |Y_1^0|^2 = \frac{3}{4\pi}\cos^2\theta. \tag{12.54}$$

These expressions show that for $m = \pm 1$, the most probable orbits are in the xy-plane, the direction of rotation being clockwise for $m = 1$ and the opposite for $m = -1$. Conversely, for $m = 0$, the most probable orbits are in planes containing the z-axis.

A word of caution is needed when interpreting the above results. If we interpret E_l as the energy of the rotor in state l minus the energy of the zero point (which we omit by arbitrarily taking $V(a)$ to be zero), we see that what Eq. (12.51) predicts is the separation between the rotor energy levels, not the absolute value of the energies.

To study the rigid rotation of bodies more complicated than the diatomic molecule, it is necessary to construct the eigenfunctions of the angular-momentum operator for any j (integer or half-integer), expressed in terms of the Euler angles; these eigenfunctions are the Wigner functions $D_{mm'}^{(j)}(\alpha, \beta, \gamma)$. The interested reader will find full discussions of such problems in texts on angular momentum.

The selection rules obtained in the previous section allow us to determine the rotational spectrum of diatomic molecules. In the case of the rigid rotor, $\mathbf{r} = a\hat{\mathbf{a}}_r$, the states are described by just the two angular-momentum quantum numbers l, m. The probability of transition per unit time between the states $|lm\rangle$ and $|l'm'\rangle$ is given by

$$A_{lm,l'm'} = \frac{4e^2 a^2 \omega_{ll'}^3}{3\hbar c^3}|\langle l'm'|\hat{\mathbf{a}}_r|lm\rangle|^2, \tag{12.55}$$

and the selection rules (12.49) apply. According to Eq. (12.51) the transition frequencies are

$$\omega_{l,l-1} = \frac{E_l - E_{l-1}}{\hbar} = \frac{\hbar}{2I}[l(l+1) - (l-1)l] = \frac{\hbar}{I}l = l\omega_{10}. \tag{12.56}$$

Consequently, the spectrum of a diatomic molecule is composed of a series of equidistant lines separated from each other by $\omega_{10} = \hbar/I$. The rotational spectrum of a diatomic molecule normally falls in the far infrared, with wavelengths of the order of 10^{-2} cm, and its study is extremely useful, as it provides information about the moment of inertia of the molecule and hence its dimensions.

Exercise E12.3. Time-dependent approach to the rigid rotor

Obtain the general solution of the time-dependent Schrödinger equation for a rigid rotor and check explicitly that the mean values of the operators $\widehat{H}$, $\widehat{L}_z$, and $\widehat{L}_x$ are conserved. What general conclusions can be drawn from these results?

Solution. The general solution for a rigid rotor can be written in the form

$$\psi(t) = \sum_{l,m} C_{lm} Y_l^m(\theta, \varphi) e^{-iE_l t/\hbar}. \tag{12.57}$$

The calculation of the required expectation values is immediate and gives

$$\langle \widehat{H} \rangle = \sum_{l,m} E_l |C_{lm}|^2, \qquad \langle \widehat{L}_z \rangle = \hbar \sum_{l,m} m |C_{lm}|^2, \tag{12.58}$$

$$\langle \widehat{L}_x \rangle = -\hbar \sum_{l,m} \sqrt{(l \pm m + 1)(l \mp m)} C^*_{l,m \pm 1} C_{lm}. \tag{12.59}$$

Since $\widehat{H} = \widehat{L}^2/2I$ does not explicitly depend on time, the constancy of $\langle \widehat{H} \rangle$ is simply a result of energy conservation. Since $\widehat{L}_z$, $\widehat{L}_+$, and $\widehat{L}_-$ commute with $\widehat{H}$ (which is invariant against arbitrary rotations), they are constant, even though they do not commute with each other. The mean value of $\widehat{L}_z$ turns out to be constant simply because its eigenfunctions were used as the basis.

It is more interesting to analyze how the constancy of $\langle \widehat{L}_x \rangle$ is produced. In this case, the expectation value is not a sum of terms proportional to the squares of the modules of the coefficients $C_{lm} e^{-iE_l t/\hbar}$ of the expansion, as is the case with $\langle \widehat{H} \rangle$ and $\langle \widehat{L}_z \rangle$, but a sum of bilinear combinations of these coefficients, belonging to the same value of l and different values of m.

The temporal independence of these terms is due to the fact that the energy of the state $|lm\rangle$ depends only on l and not on m, $E_{lm} = E_l$. Consequently, we can conclude that if a system has two or more operators that commute with the Hamiltonian, but not with each other, degeneracy occurs.

12.4 The Free Particle in 3D

The solution of the free-particle problem can be expressed in terms of the common eigenfunction of $\widehat{\boldsymbol{p}}$ and $\widehat{H}$, which is $\psi = Ae^{i\boldsymbol{k}\cdot\boldsymbol{r}}$, where $\boldsymbol{k} = \boldsymbol{p}/\hbar$. This is the extension to

3D of the free-particle solution found in Section 7.1, and represents a plane wave traveling in the direction of the momentum. On the other hand, the free-particle problem is a special case of a central problem, so it should be possible to express the solution in terms of the angular-momentum eigenfunctions $Y_l^m(\theta, \varphi)$, multiplied by the solution R of the radial equation for the free particle, Eq. (12.21) with $V = 0$,

$$\left[\frac{\hat{p}_r^2}{2m} + \frac{\hbar^2 l(l+1)}{2mr^2}\right] R(r) = E R(r). \tag{12.60}$$

As shown in Section A.1 of the Mathematical Appendix, the solutions of this equation are the spherical Bessel functions $j_l(kr)$, which form a complete basis. Therefore we can use the functions

$$\psi_{klm} = A j_l(kr) Y_l^m(\theta, \varphi)$$

as a basis to represent the plane wave in three-dimensional space. This decomposition of a plane wave into spherical waves is very useful in dispersion theory, as will be seen in Chapter 18; we derive it here as a further exercise in dealing with orthogonal functions.

To begin the analysis, we note that the function $\psi = A e^{i\mathbf{k}\cdot\mathbf{r}}$ with $\mathbf{k}$ in any direction is not an eigenfunction of $\hat{L}_z$, that is, it has no well-defined angular momentum in the z-direction. Therefore, the required superposition must include all possible values of m; below (Eq. 12.70) we will see that this is the case. However, we can simplify things considerably by orienting the z-axis along $\mathbf{k}$; then the wave function is an eigenfunction of $\hat{L}_z$, since $\mathbf{k}\cdot\mathbf{r} = kz = kr\cos\theta$, and

$$\hat{L}_z e^{ikz} = -i\hbar \frac{\partial}{\partial \varphi} e^{ikr\cos\theta} = 0, \tag{12.61}$$

that is, the eigenvalue of $\hat{L}_z$ is zero. This means that the expansion of a plane wave with momentum in the z-direction contains only spherical waves with $m = 0$. This result is obvious from the geometrical point of view, since if the z-axis is aligned with $\mathbf{p}$, then $\mathbf{L} = \mathbf{r} \times \mathbf{p}$ lies in the xy-plane.

Since for $m = 0$ the spherical harmonics reduce to the Legendre polynomials $P_l(\cos\theta)$, we write

$$e^{ikz} = e^{ikr\cos\theta} = \sum_{l=0}^{\infty} A_l j_l(kr) P_l(\cos\theta). \tag{12.62}$$

To determine the expansion coefficients A_l, we multiply by $P_{l'}(\cos\theta)$, integrate from 0 to π and use the orthogonality relations of the Legendre polynomials, which in terms of the variable $s = \cos\theta$ are (see Eq. (A.20)),

$$\int_{-1}^{1} P_l(s) P_{l'}(s) \, ds = \frac{2}{2l+1} \delta_{ll'}, \tag{12.63}$$

so we get

$$A_l j_l(kr) = \frac{2l+1}{2} \int_{-1}^{1} e^{ikrs} P_l(s) \, ds. \tag{12.64}$$

Using the Rodrigues formula, Eq. (A.19),

$$P_l(s) = \frac{1}{2^l l!} \frac{d^l}{ds^l} (s^2 - 1)^l \tag{12.65}$$

and integrating by parts l successive times, we obtain

$$\begin{aligned} A_l j_l(kr) &= \frac{2l+1}{2} \frac{(-1)^l}{2^l l!} \int_{-1}^{1} (s^2-1)^l \frac{d^l}{ds^l} e^{ikrs} \, ds \\ &= \frac{2l+1}{2} \frac{(ikr)^l}{2^l l!} \int_{-1}^{1} (1-s^2)^l e^{ikrs} ds. \end{aligned} \tag{12.66}$$

On the other hand, from the theory of Bessel functions we have the following integral formula

$$\int_{-1}^{1} (1-x^2)^l e^{i\alpha x} \, dx = \sqrt{\frac{2\pi}{\alpha}} l! \left(\frac{2}{\alpha}\right)^l J_{l+1/2}(\alpha) = 2l! \left(\frac{2}{\alpha}\right)^l j_l(\alpha), \tag{12.67}$$

where the second equality follows from defining the spherical Bessel functions j_l in terms of the cylindrical Bessel functions $J_{l+1/2}$, Eq. (A.62),

$$j_l(\rho) = \sqrt{\frac{\pi}{2\rho}} J_{l+1/2}(\rho). \tag{12.68}$$

Substituting in the previous expression and canceling the common factor $j_l(kr)$, we obtain $A_l = i^l(2l+1)$. Using this result in Eq. (12.62), we obtain the desired expansion,

$$e^{ikz} = \sum_{l=0}^{\infty} i^l (2l+1) j_l(kr) P_l(\cos\theta). \tag{12.69}$$

The expression for any k can be obtained from here by interpreting θ as the angle between k and r and passing to an arbitrary coordinate system; since this is not of direct interest to us, we give only the final result:

$$e^{i\mathbf{k}\cdot\mathbf{r}} = 4\pi \sum_{l,m} i^l j_l(kr) Y_l^m(\theta,\varphi) Y_l^{m*}(\theta_k,\varphi_k), \tag{12.70}$$

where θ_k, φ_k correspond to the vector $\mathbf{k}$, and θ, φ correspond to the vector $\mathbf{r}$.

12.5 The Hydrogen-Like Atom

The ground is finally prepared for the study of the central problem of atomic physics: the hydrogen atom. The system simply consists of a single electron around the nucleus. More generally, we can consider a hydrogen-like atom consisting of an electron and a nucleus of arbitrary positive charge Ze, where Z is the number of protons in the nucleus (atomic number). The problem, as simple as it may sound, is not trivial, and a large number of results had to be accumulated before it was possible to approach it using Schrödinger's theory. The results are worth the effort because, as we shall see, the theory allows it to be studied in detail and gives satisfactory results in all cases.

We start by analyzing the asymptotic behavior of the radial function $u(r)$ when $r \to 0$ and when $r \to \infty$ for potentials $V(r)$ satisfying two conditions: (a) $\lim_{r \to 0} r^2 V(r) = 0$, and (b) $\lim_{r \to \infty} V(r) = 0$. From Eq. (12.25) and condition (a) it follows that sufficiently close to the origin the equation for u reduces to

$$\frac{d^2 u}{dr^2} - \frac{l(l+1)}{r^2} u = 0, \tag{12.71}$$

which has a solution of the form r^s if $s(s-1) - l(l+1) = 0$, that is, if

$$s = \tfrac{1}{2} \pm \left(l + \tfrac{1}{2}\right) = l+1, -l. \tag{12.72}$$

The solution with s negative or zero, $s = -l$, is not acceptable, because it does not satisfy the condition (12.27), $u(0) = 0$. Hence the behavior of u near the origin is

$$u \sim r^{l+1}. \tag{12.73}$$

In the limit of very large r, the effective potential vanishes by property (b) and Eq. (12.25) reduces to

$$-\frac{\hbar^2}{2m}\frac{d^2 u}{dr^2} = Eu, \tag{12.74}$$

whose solution is (omitting the normalization factor)

$$u = e^{\beta r}, \tag{12.75}$$

with the constant β given by

$$\beta^2 = -\frac{2mE}{\hbar^2}. \tag{12.76}$$

We immediately see that there are two possible cases with completely different asymptotic behavior of u at infinity. If the energy E is negative, β is real, with a positive and a negative value of equal magnitude. In this case, the wave function increases or decreases exponentially at infinity; the integrability condition of the square of the wave function eliminates the solution with positive β, so that for negative energies we obtain

$$u \sim e^{-\beta r}, \quad \beta = \frac{\sqrt{2m|E|}}{\hbar} \quad (r \to \infty). \tag{12.77}$$

On the other hand, if the energy is positive, β takes on imaginary values, the wave function is oscillatory at infinity, and the required combination of the two solutions is determined by the demand that u correctly connects with the solution coming from the origin.[2]

Physically, these results are easy to understand. Bound states form when $E < 0$, as shown in Fig. 12.1; this case classically corresponds to the appearance of closed orbits (elliptical in the Coulomb case), and naturally the electron density must decrease rapidly as we move away from the center of attraction, giving rise to the exponential decrease in u. On the other hand, $E > 0$ if, for example, electrons are thrown against the center of attraction, which cannot capture them due to the excess energy but deflects them from their initial trajectory. This case classically corresponds to open orbits (hyperbolic in the Coulomb case); the electrons can escape to arbitrary distances from the scattering center, and the wave function remains finite as r increases. In line with the general discussion in Chapter 4, we conclude that the bound state corresponds to negative energies with a discrete spectrum, while the scattering states correspond to positive energies with a continuous spectrum.

Specifically for the hydrogen-like atom, the Coulomb potential is

$$V = -\frac{Ze^2}{r} \tag{12.78}$$

---

[2] Note that when $E > 0$, the wave function at infinity can be written as the superposition of an outgoing spherical wave (propagating from the center to the periphery) $R \sim e^{ikr}/r$, and an incoming wave $R \sim e^{-ikr}/r$, with $k = \sqrt{2mE}/\hbar$. This is the form we will have to take for the asymptotic wave function when we study the problem of particle scattering by potentials in Chapter 18.

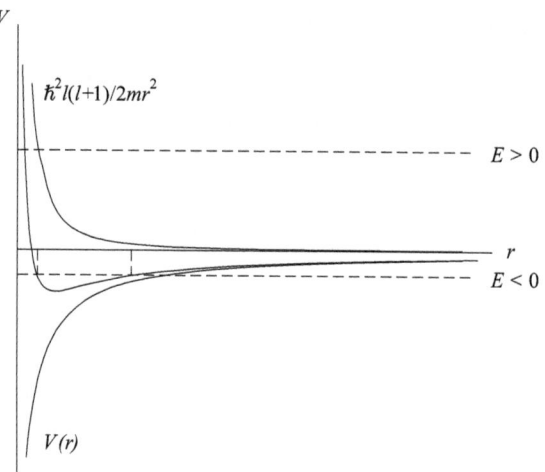

Figure 12.1 Effective potential, constructed as the sum of an attractive potential (a Coulomb potential in the case of the hydrogen-like atom) and the centrifugal barrier potential. Negative energies correspond to bound states, while positive energies correspond to Rutherford scattering.

and the radial equation becomes

$$-\frac{\hbar^2}{2m}\frac{d^2u}{dr^2} + \left(\frac{\hbar^2 l(l+1)}{2mr^2} - \frac{Ze^2}{r}\right)u = Eu. \quad (12.79)$$

In this chapter we will mainly study the solutions corresponding to bound states. The scattering of particles by a uniform spherical potential is discussed in Chapter 18. It is convenient to rewrite Eq. (12.79) in dimensionless form, which we do by introducing the variable ρ defined as

$$\rho = 2\beta r, \quad (12.80)$$

where β is given by (12.77) (the value 2 is chosen for convenience; other authors prefer 1). With this change of variable, we get ($u' = du/d\rho$)

$$u'' + \left[-\frac{1}{4} + \frac{Z/\beta a_0}{\rho} - \frac{l(l+1)}{\rho^2}\right]u = 0, \quad (12.81)$$

where

$$a_0 = \frac{\hbar^2}{me^2}. \quad (12.82)$$

In order to guarantee the correct asymptotic behavior of u we use previous results to write the radial function in the form

$$u = \rho^{l+1}e^{-\rho/2}Q(\rho), \quad (12.83)$$

where $Q(\rho)$ is to be determined by solving the differential equation obtained by substituting (12.83) in (12.81), that is,

$$Q'' + \left(\frac{2(l+1)}{\rho} - 1\right)Q' + \frac{(Z/\beta a_0) - l - 1}{\rho}Q = 0. \quad (12.84)$$

To find out what boundary conditions we need to impose on $Q(\rho)$, we proceed as follows. At the origin, (12.73) is satisfied with $u \approx \rho^{l+1} Q(0)$. As $r \to \infty$, if Q is expressed as a power series of ρ, this series must reduce to a polynomial to grow slower than an exponential and be physically acceptable. In summary, the solutions of (12.84) that we are interested in are polynomials. A study of the differential equation shows that such solutions exist if and only if the numerical coefficient of the last term is a nonnegative integer, that is, if

$$\frac{Z}{\beta a_0} - l - 1 = k, \quad k = 0, 1, 2, \ldots \tag{12.85}$$

As shown in Section A.1 of the Appendix A, the polynomial solutions of Eq. (12.84) are the associated (or generalized) Laguerre polynomials,

$$Q = L_k^{2l+1}(\rho) = \frac{1}{k!} e^\rho \rho^{-(2l+1)} \frac{d^k}{d\rho^k} e^{-\rho} \rho^{k+2l+1}. \tag{12.86}$$

It is convenient to write the quantization condition of the energy, (12.85), in the form

$$\frac{Z}{\beta a_0} = k + l + 1 \equiv n, \tag{12.87}$$

introducing the *principal quantum number* n, which can take the values $1, 2, 3, \ldots$, while the angular momentum $l = n - 1 - k$ is restricted to integers satisfying the condition

$$l \leq n - 1. \tag{12.88}$$

Substituting the value of β given by Equation (12.87) into (12.76) and using (12.82), we obtain Bohr's formula for the energy (see Section 2.5),

$$E_n = -\frac{Z^2 e^2}{2 a_0 n^2} = -\frac{Z^2 m e^4}{2 \hbar^2 n^2}. \tag{12.89}$$

From this we see that the wavelength associated with an atomic transition with n and Δn of the order of unity is of the order of $2\pi \hbar c / E \sim a_0/\alpha \simeq 137 a_0$ ($\alpha = e^2/\hbar c \simeq 1/137$ is the fine structure constant[3] and $Z = 1$ has been set), that is, two orders of magnitude larger than the atomic radius (as we will see later, the radius of hydrogen in its ground state is of the order of a_0, called *Bohr radius*). The Rydberg constant predicted by QM coincides with that previously obtained by Bohr, since by substituting $E_n = -\hbar c R / n^2$ we obtain

$$R = \frac{m e^4}{2 \hbar^3 c}. \tag{12.90}$$

We will have an opportunity to comment on these formulas in Section 12.6.

This seems a suitable place for an important observation. Quantum formalism says that the angular momentum for the s-states of a central problem (including the H atom) is zero. However, to interpret this to mean that the electron is at rest would contradict basic laws of physics; in fact, the electron would collide with the nucleus under the

---

[3] The (dimensionless) fine-structure constant α was introduced into physics in 1916 by Arnold Sommerfeld to account for the relativistic splitting of atomic spectral lines, hence its name. It was later found to be a fairly ubiquitous constant, appearing frequently in the physics of atomic-scale systems and QED as the measure of the strength of the electromagnetic force. Its currently accepted value is 1/137.035999206, to an accuracy of 11 digits.

action of the Coulomb force. In seeking a way out of this difficulty, it will be important to remember that the preceding formulas do not refer to individual electrons but only provide statistical information.

12.5.1 Energy eigenfunctions

The radial functions of the H atom are obtained by inserting Eq. (12.86) into (12.83) and remembering that $R = u/r$. This gives[4]

$$R_{nl}(\rho) = c_{nl}\rho^l e^{-\rho/2} L_{n-l-1}^{2l+1}(\rho). \tag{12.91}$$

The normalization coefficient is obtained from the condition $\int_0^\infty r^2 R_{nl}^2(r)dr = 1$ and gives (see Problem 12.3)

$$c_{nl} = \frac{2}{n^2}\sqrt{\frac{Z^3}{a_0^3}\frac{(n-l-1)!}{(n+l)!}}. \tag{12.92}$$

Using (12.20), the complete energy eigenfunctions for the hydrogen-like atom are therefore given by

$$\psi_{nlm} = R_{nl}(\rho)Y_l^m(\theta,\varphi), \tag{12.93}$$

where $R_{nl}(\rho)$ is given by (12.91) in terms of the associated Laguerre polynomials (12.86), and $\rho = (2\sqrt{2m|E|}/\hbar)r$.

Since the energy depends only on the principal quantum number, for $n > 1$ there are several wave functions associated with the same energy, that is, the states are generally degenerate. The order of degeneracy is determined by considering that for each value of l there are $2l+1$ values of m and for each n there are n possible values of l $(0, 1, 2, \ldots, n-1)$; then the total number of linearly independent wave functions for a given n is

$$\sum_{l=0}^{n-1}(2l+1) = n^2. \tag{12.94}$$

This degeneracy has two causes. On the one hand, the absence of a preferred spatial direction introduces the degeneracy with respect to m (by adding an external field that fixes a spatial direction, this degeneracy is broken, as we will see in Section 12.6). On the other hand, the shape of the Coulomb potential $\sim r^{-1}$ determines the degeneracy with respect to l (it is enough to introduce any additional potential, even a central one, that changes this shape to make this degeneracy disappear, as shown in Problem P12.9).

The energy of the ground state, corresponding to $n = 1$, $l = 0$, is

$$E_1 = -\frac{Z^2 e^2}{2a_0} \simeq -13.6 Z^2 \text{ eV}. \tag{12.95}$$

This means that in order to extract the orbital electron from an H atom ($Z = 1$) in its ground state, at least 13.6 eV of work needs to be invested; this is the theoretical

---

[4] The notation and normalization of the associated Laguerre polynomials varies widely in the literature. For example, what we call L_{n-l-1}^{2l+1} is called L_{n+l}^{2l+1} by other authors. This follows from the fact that we have defined the associated Laguerre polynomials $L_n^k(x)$ in terms of the Laguerre polynomials $L_n^0(x) = L_n(x)$ by the expression $L_{q-p}^p(x) = (-1)^p \left(d^p/dx^p\right) L_q(x)$. This is a common convention, but not the only one; other authors prefer to write $L_q^p(x) = (-1)^p \left(d^p/dx^p\right) L_q(x)$, which is equivalent to L_{q-p}^p.

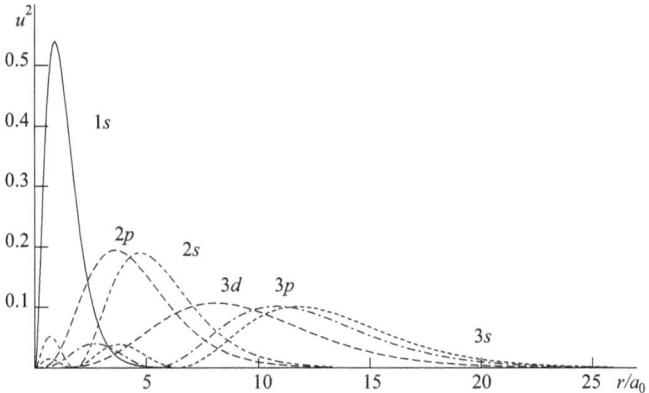

Figure 12.2 Behavior of the radial probability density $u^2(r) = r^2 R^2(r)$ for the H atom, for small quantum numbers. Note that for $l = n - 1$, $u^2(r)$ has no nodes.

value of the ionization energy of H, which agrees well with the experimental value. The first excited state corresponds to $n = 2$ and has a degeneracy of 4, with one state corresponding to $l = 0$ and the other three to $l = 1$, and so on.

From (12.91) it is clear that the radial function $R(r)$ varies in general in a complicated way with r, and can have several maxima and nodes; but in the particular case where l reaches its maximum value $n - 1$, the associated Laguerre polynomial reduces to $L_{n-l-1}^{2l+1}(\rho) = L_0^{2l+1} = 1$, and the function $r^2 R_{n,n-1}^2$ then has a single maximum, located at

$$r_{\max} = \frac{a_0}{Z} n^2. \tag{12.96}$$

In the ground state ($n = 1$), $r_{\max} = a_0 = \hbar^2/me^2$; this is the so-called *Bohr radius* of H and has a value of 0.53×10^{-8} cm, just over half an angstrom. For $l < n-1$, the function $r^2 R^2$ has more than one maximum. If the case $l = n-1$ is considered as a circular orbit, in analogy to the classical case, the lower values of l correspond to elliptical orbits. It follows that the quantum number $k = n - l - 1$ measures the eccentricity of the orbit, although it is an indirect measure. Fig. 12.2 shows the behavior of $u^2 = r^2 R^2(r)$ for some small values of nl.

12.6 The Hydrogen Spectrum

To determine the transition (selection) rules for the H atom we write again $\mathbf{r} = \hat{\mathbf{a}}_r r$ and factor out the angular and radial integrals, $\langle n'l'm'|\mathbf{r}|nlm\rangle = \langle n'l'|r|nl\rangle\langle l'm'|\hat{\mathbf{a}}_r|lm\rangle$. Direct calculation shows that the integral $\langle n'l'|r|nl\rangle = \int R_{n'l'}(r)R_{nl}(r)r^3 dr$ is not zero for any combination nn'. Thus the selection rules for H are those given previously for the general central problem, $\Delta l = \pm 1, \Delta m = 0, \pm 1$, to which we add the radial rule, namely arbitrary integer.

Spontaneous transitions to the n' level occur from states with $n > n'$ and produce photons of frequency

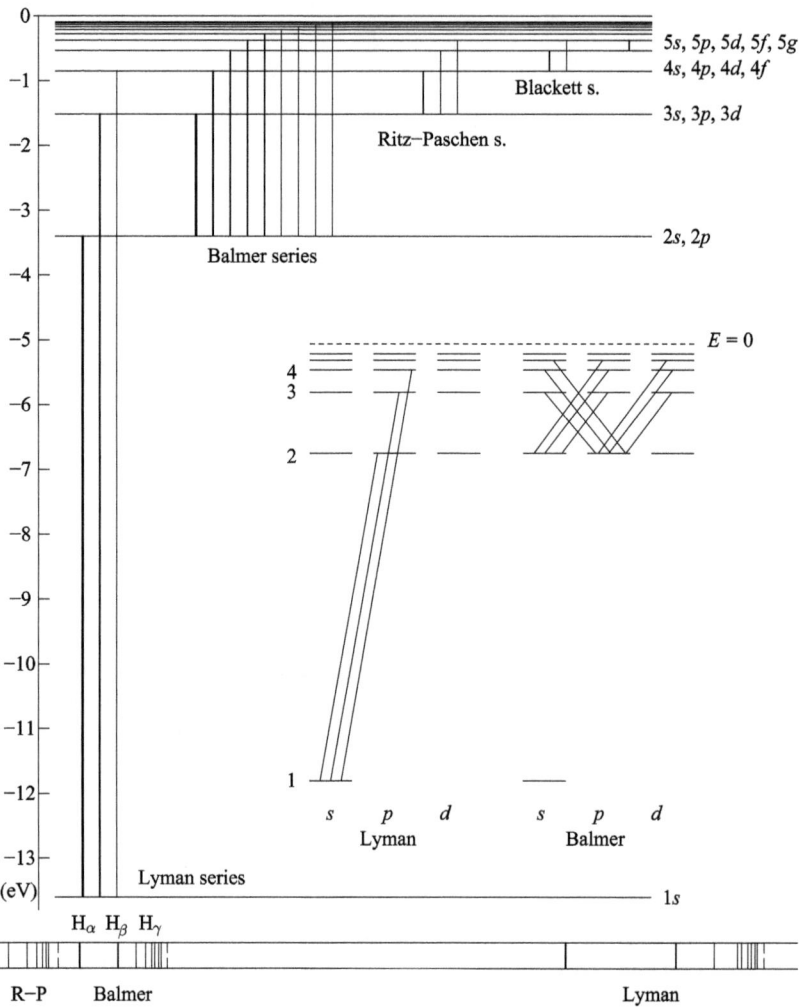

Figure 12.3 Diagram of the energy levels of the H atom, showing the allowed transitions and the corresponding frequency spectrum. The insets show the transitions that give rise to the Lyman and Balmer series.

$$\omega_{nn'} = \frac{E_n - E_{n'}}{\hbar} = Z^2 Rc \left(\frac{1}{n'^2} - \frac{1}{n^2} \right). \tag{12.97}$$

In the following we will restrict ourselves to the case of the H atom, $Z = 1$. Formula (12.97) predicts the existence of several spectral series, each characterized by a fixed value of n' and a variable value of n, with the only restriction being $n > n'$. Transitions to the ground state $1s$ ($l = 0$) occur from all np states ($l = 1$) with $n > 1$, giving rise to the Lyman series (see Fig. 12.3), whose lines have the frequencies

$$\omega = Rc \left(\frac{1}{1^2} - \frac{1}{n^2} \right), \quad n = 2, 3, 4, \ldots \quad \text{(Lyman series)}.$$

The large energy difference between the ground state and the excited states causes all lines in this series to fall in the ultraviolet.

A second series is generated with the transitions to the first excited level consisting of the states $2s$ and $2p$; to the state $2s$ there can be transitions from states $np, n > 2$, while to $2p$ ($l = 1$) there are transitions from states ns ($l = 0$) and nd ($l = 2$) with $n > 2$ (see Fig. 12.3). Since the energy does not depend on l, these spectral lines coincide and their frequencies are

$$\omega = Rc \left(\frac{1}{2^2} - \frac{1}{n^2} \right), \quad n = 3, 4, 5, \ldots \quad \text{(Balmer series)}. \tag{12.98}$$

This is the well-known Balmer series, whose lines H_α, H_β, H_γ, H_δ, ... correspond to the visible part of the H spectrum.

The Ritz–Paschen series follows, with frequencies

$$\omega = Rc \left(\frac{1}{3^2} - \frac{1}{n^2} \right), \quad n = 4, 5, 6, \ldots \quad \text{(Ritz–Paschen series)}, \tag{12.99}$$

followed by the Blackett[5] series, and so on, all of them in the infrared. Note that each series has a minimum and maximum frequency, that is, it occupies a well-defined region of the spectrum. For example, the Balmer series lies between the frequencies $\omega_{\min}^B = (5/36)Rc$ and $\omega_{\max}^B = Rc/4$; this interval contains an infinite number of lines, which become more and more concentrated as they approach the higher frequency extreme, while their intensity decreases; the variable thickness of the lines in the lower part of Fig. 12.3 is intended to show this phenomenon. The formulas for the frequencies of the different spectral series are in perfect agreement with those obtained using Bohr's atomic model, but such a simple model was not able to predict the relative intensity of the spectral lines. This will be the subject of the next section.

12.6.1 Lifetimes of the excited H states

As mentioned in Chapter 10, in the presence of an external radiation field, transitions between the stationary states can take place. In the absence of external radiation, the spectral lines are due only to spontaneous emission from higher excited states to lower excited states or to the ground state. The relative intensity of the spectral lines, such as those shown in Fig. 12.3 for the H atom, is determined by the respective probability per unit time of spontaneous emission, which is inversely proportional to the average time the atom remains in the excited state before making the transition. To calculate the transition probability we use Einstein's formula for the spontaneous transition coefficient, Eq. (17.134) and proceed as follows.

First we consider only those transitions where the angular momentum increases by one, that is, $(nlm) \to (n', l+1, m')$. From Eqs. (12.44)–(12.48) it follows that

$$\langle l+1, m' | \widehat{\mathbf{a}}_r | lm \rangle = \widehat{\mathbf{a}}_+ A_{lm}^+ \delta_{m',m+1} + \widehat{\mathbf{a}}_- A_{lm}^- \delta_{m',m-1} + \widehat{\mathbf{a}}_z A_{lm} \delta_{m',m}. \tag{12.100}$$

---

[5] Patrick Maynart Stuart Blackett (Lord Blackett) (1897–1974) FRS was a British experimental physicist known for his work on cloud chambers and cosmic rays; he was awarded the 1948 Nobel Prize for his investigations on cosmic rays; Blackett was the first person that intentionally produced and demonstrated the transmutation of an element into another ($N + \alpha \to F$).

Each of the three components of this result contributes to transitions with different m'; therefore,

$$|\langle l+1, m+1|\hat{a}_r|lm\rangle + \langle l+1, m-1|\hat{a}_r|lm\rangle + \langle l+1, m|\hat{a}_r|lm\rangle|^2$$
$$= \tfrac{1}{2}(A_{lm}^+)^2 + \tfrac{1}{2}(A_{lm}^-)^2 + A_{lm}^2 = \frac{l+1}{2l+1}. \tag{12.101}$$

The final result was obtained by calculating the different coefficients A_{nl} using the recurrence relations of the spherical harmonics. Therefore, the probability of emission from state n, l, m, to state $n', l+1$, all allowed m' taken together, is

$$A_{nlm \to n', l+1, \text{all } m'} = \frac{4e^2\omega_{nn'}^3}{3\hbar c^3} \frac{l+1}{2l+1} |\langle n'l+1|r|nl\rangle|^2. \tag{12.102}$$

Similarly, for transitions where the orbital momentum decreases by one we obtain

$$A_{nlm \to n', l-1, \text{all } m'} = \frac{4e^2\omega_{nn'}^3}{3\hbar c^3} \frac{l}{2l-1} |\langle n'l-1|r|nl\rangle|^2. \tag{12.103}$$

Exercise E12.4. Lifetime of the lowest excited state

Use the preceding results to calculate the lifetime of the lowest excited state, the $2p$ state of H.

Solution. From this state the atom can only decay to the $1s$ state – producing the so-called Lyman α line, L_α – so this single transition determines the lifetime of the $2p$ state. Using Eq. (12.103), we get

$$\frac{1}{\tau_{2p}} = A_{2p \to 1s} = \frac{4e^2\omega_{21}^3}{3\hbar c^3} |\langle 10|r|21\rangle|^2. \tag{12.104}$$

The radial functions needed are ($\rho = 2r/a_0 n$)

$$R_{10} = \frac{2}{a_0^{3/2}} e^{-r/a_0}, \quad R_{21} = \frac{1}{\sqrt{24}a_0^{3/2}} \frac{r}{a_0} e^{-r/2a_0}, \tag{12.105}$$

and the required integral gives

$$\int_0^\infty R_{10}(r) R_{21}(r) r^3 \, dr = 4(2/3)^5 \sqrt{6} a_0. \tag{12.106}$$

On the other hand, we have from (12.97) that $\omega_{21} = 3Rc/4$ and $a_0 R = \alpha/2$, where $\alpha = e^2/\hbar c \approx 1/137$, so that

$$\frac{e^2\omega_{21}^3}{\hbar c^3} a_0^2 = 2(3\alpha/8)^3 Rc. \tag{12.107}$$

Substituting these results in (12.104), we get

$$\frac{1}{\tau_{2p}} = \tfrac{1}{3}(8/9)^3 \alpha^3 Rc. \tag{12.108}$$

Given the approximate value for the Rydberg constant, $Rc \approx 3.3 \times 10^{15}$ s^{-1}, the result obtained for the lifetime of the $2p$ state is

$$\tau_{2p} \approx 2.1 \times 10^{-9} \text{ s}, \tag{12.109}$$

which seems a very short time indeed. However, a semiclassical estimate of the orbital period of the $2p$ state (equating the mean kinetic energy written as $\frac{1}{2}ma_0^2\Omega^2$ with $-E_2$) gives $T_{2p} = 2\pi/\Omega \approx 2\pi/Rc$, from which $\tau_{2p}/T_{2p} \approx \alpha^{-3} \simeq 2.5 \times 10^6$; this means that the electron has time to complete its orbit several million times before it decays. We can therefore justifiably speak of a quasi-stationary state.

12.6.2 Significance of the reduced mass

A few remarks on the scope of quantum analysis are in order. In the preceding formulas, m represents the reduced mass of the orbital electron; therefore, if M is the mass of the atomic nucleus and m_e that of the electron, the reduced mass is approximately given by

$$m = \frac{m_e M}{m_e + M} \approx m_e \left(1 - \frac{m_e}{M}\right). \tag{12.110}$$

This result shows that the spectrum depends slightly on the mass of the nucleus through the Rydberg constant $R \propto m$ (see (12.90)). The dependence of R on m, due to the motion of the nucleus with respect to the CM, produces important and interesting effects. If we call R_∞ the Rydberg constant for a nucleus of infinite mass, we have that the Balmer formula for a nucleus of mass M is given by

$$\omega_{nn'} = Z^2 R_\infty c \left(1 - \frac{m_e}{M}\right)\left(\frac{1}{n'^2} - \frac{1}{n^2}\right). \tag{12.111}$$

The different isotopes of H (protium, deuterium, and tritium) have different values of nuclear mass.[6] The emission spectrum of natural H is therefore the superposition of three similar spectra, slightly shifted on the frequency axis and with relative intensities essentially proportional to the concentration of each isotope.

On the other hand, if we apply Eq. (12.111) to He$^+$ for which $Z = 2$, we obtain

$$\omega_{nn'} = 4R_\infty c \left(1 - \frac{m_e}{M_{He}}\right)\left(\frac{1}{n'^2} - \frac{1}{n^2}\right) = R_\infty c \left(1 - \frac{m_e}{M_{He}}\right)\left(\frac{1}{(n'/2)^2} - \frac{1}{(n/2)^2}\right). \tag{12.112}$$

In particular, for $n' = 4$ and $n = 5, 6, 7, 8, \ldots$, we get

$$\omega_{n4} = R_\infty c \left(1 - \frac{m_e}{m_{He}}\right)\left(\frac{1}{2^2} - \frac{1}{(n/2)^2}\right); \quad \frac{n}{2} = \frac{5}{2}, 3, \frac{7}{2}, 4, \ldots \tag{12.113}$$

If we ignore the correction due to nuclear motion, this looks like the Balmer series of H $\left(\frac{n}{2} = 3, 4, 5, \ldots\right)$ plus an anomalous series of intermediate lines corresponding to half-integer "quantum numbers" $\left(\frac{n}{2} = \frac{5}{2}, \frac{7}{2} \ldots\right)$. The phenomenon was effectively observed in stellar spectra by Pickering and theoretically explained by Bohr as a regular spectrum, with integer quantum numbers, but due to He. This was confirmed by a careful analysis of the position of the spectral lines, which showed that the corresponding Rydberg constant was indeed $R_{He} = R_\infty \left(1 - \frac{m_e}{M_{He}}\right)$ and not $R_H = R_\infty \left(1 - \frac{m_e}{M_H}\right)$.

---

[6] The nucleus of protium has no neutrons, that of D has one neutron, and that of T, has two. The mass of a neutron is essentially the same as that of a proton.

The results thus validate Schrödinger's description of the two-body problem, although at the time they were "smoking guns" in favor of half-integer quantum numbers.

12.7 The Atom in an Electromagnetic Field. The Zeeman Effect

In 1896, after several decades of research initiated by Faraday and long before the advent of QM, Pieter Zeeman[7] – following a suggestion by Lorentz – demonstrated that the spectral lines of sodium are modified by the action of an external magnetic field. In the course of a few years, it became clear that the modification essentially consisted of a splitting of the emission lines into multiplets (doublets, triplets, etc.) characteristic of the element and dependent on the intensity of the external field. This phenomenon, now known as the Zeeman effect, played a key role in the study of the magnetic properties of the atom, and, in particular, contributed to the discovery of the spin of the electron and its associated magnetic moment.

If the atom has a nonzero electronic spin, the spectral lines decompose in a very complicated way, giving rise to the so-called *anomalous Zeeman effect*, a problem that will be treated in Section 14.3, based on the Pauli electron spin theory. Since we will not consider the effects due to electronic spin here, the results will only apply to atoms whose total electronic spin is zero (this automatically excludes H). This case is known as the *normal Zeeman effect* or the *Paschen–Back effect*.

12.7.1 The minimal coupling

To study the normal Zeeman effect, we need to write the Schrödinger equation for a (spinless) atom immersed in a magnetic field. In general, a body with electromagnetic properties can have an electric charge and a range of electric and magnetic moments, either intrinsic or induced by external fields. When interacting with the field, the charge is coupled to the potentials Φ and A of the electromagnetic field, while the higher moments are coupled to the field through the derivatives of the potentials.

The interaction energy of a charge distribution with the external field Φ can be expressed in terms of electric multipoles by means of a Taylor expansion of the potential around the origin of the charge distribution $\rho(r)$,

$$\int \rho(r)\Phi(r)d^3x = \int \rho(r)\left[\Phi(0) + \sum x_i \partial_i \phi(0) + 1/2 \sum x_i x_j \partial_i \partial_j \Phi(0) + \dots \right]d^3x. \quad (12.114)$$

Identifying $\int \rho(r)d^3x$ with the total charge q, $\int x_i \rho(r)d^3x$ with the i-component of the electric dipole moment d, $\int x_i x_j \rho(r)d^3x$ with the ij-component of the electric quadrupole moment $\overleftrightarrow{Q}$, and so on, the interaction energy can be written as

$$\int \rho(r)\Phi(r)d^3x = q\Phi(0) + \sum d_i \partial_i \Phi(0) + 1/2 \sum Q_{ij}\partial_i \partial_j \Phi(0) + \dots \quad (12.115)$$

---

[7] Pieter Zeeman (1865–1943) was a Dutch physicist who shared the 1902 Nobel Prize with his mentor H. A. Lorentz for their discovery of the Zeeman effect.

This expression shows that while q is directly coupled to the potential Φ, the higher moments d_i, Q_{ij}, etc., are coupled to the electromagnetic field through the derivatives of the potential. When the coupling between the particle and the field is *only* through the potentials, it is said to be *minimal*.[8] It is an empirical fact, usually raised to the status of a principle (there is no conclusive theoretical reason why this should be so), that elementary particles (in particular, the electron) are minimally coupled to the field. Once this principle is applied, no additional couplings are required. We will use this principle to extend the Schrödinger equation to the electromagnetic case.

As in classical theory, when there is an interaction with the field $\boldsymbol{A}$, the canonical momentum of the quantum particle is

$$\boldsymbol{p} = m\boldsymbol{v} + \frac{e}{c}\boldsymbol{A}. \tag{12.116}$$

By eliminating the *mechanical* or *kinetic* momentum $m\boldsymbol{v}$, the Hamiltonian is properly expressed in terms of the canonical momentum $\boldsymbol{p}$,[9]

$$H = \frac{1}{2m}\left(\boldsymbol{p} - \frac{e}{c}\boldsymbol{A}\right)^2 + e\Phi, \tag{12.117}$$

and the stationary Schrödinger equation becomes

$$E\psi = \frac{1}{2m}\left(\hat{\boldsymbol{p}} - \frac{e}{c}\boldsymbol{A}\right)^2 \psi + e\Phi\psi. \tag{12.118}$$

The minimal-coupling principle allows us to postulate this as the Schrödinger equation when there is an external electromagnetic field. The validity of this hypothesis is confirmed by the predictions of the theory so constructed, as we shall have opportunity to see.

Expanding the term in parentheses in (12.118) gives the term

$$\hat{\boldsymbol{p}} \cdot \boldsymbol{A}\psi = \boldsymbol{A} \cdot \hat{\boldsymbol{p}}\psi - i\hbar(\nabla \cdot \boldsymbol{A})\psi. \tag{12.119}$$

In the Coulomb gauge, for which $\nabla \cdot \boldsymbol{A} = 0$, the operators $\hat{\boldsymbol{p}}$ and $\widehat{\boldsymbol{A}}$ commute and (12.118) reduces to

$$E\psi = \frac{\hat{\boldsymbol{p}}^2}{2m}\psi + e\Phi\psi - \frac{e}{mc}\boldsymbol{A} \cdot \hat{\boldsymbol{p}}\psi + \frac{e^2}{2mc^2}\boldsymbol{A}^2\psi. \tag{12.120}$$

With $\widehat{H}_0$ being the Hamiltonian of the problem in the absence of a magnetic field,

$$\widehat{H}_0 = \frac{\hat{\boldsymbol{p}}^2}{2m} + e\Phi, \tag{12.121}$$

we have

$$E\psi = \widehat{H}_0\psi + \left[-\frac{e}{mc}\boldsymbol{A} \cdot \hat{\boldsymbol{p}} + \frac{e^2}{2mc^2}\boldsymbol{A}^2\right]\psi. \tag{12.122}$$

Thus the interaction Hamiltonian consists of a linear term, proportional to $\boldsymbol{A} \cdot \hat{\boldsymbol{p}}$, and a quadratic term, proportional to $\boldsymbol{A}^2$, which depends only on the field. If in this equation we make the substitution $E\psi \to i\hbar\frac{\partial \psi}{\partial t}$ to go to the time-dependent case and apply to

---

[8] We prefer to use the term *minimal*, since minimalist refers to a theory built with a minimum of elements, i.e. *structurally minimal*.

[9] Compare with the Hamiltonian used in Section 10.4.1 for the SED derivation of the operator formalism; here we omit the Hamiltonian of the radiation field because it is irrelevant to the present calculation.

the resulting equation a procedure analogous to that used in Section 7.2, we obtain for the flow density

$$\hat{j} = \frac{1}{2m}\left[\psi^*(\hat{p} - \frac{e}{c}A)\psi - \psi(\hat{p} - \frac{e}{c}A)\psi^*\right]. \tag{12.123}$$

12.7.2 The normal Zeeman effect

Except when the external field is extraordinarily strong, the quadratic Zeeman effect is very weak compared to the linear one, so we can limit ourselves to studying the linear effect. We will assume that the magnetic field, with intensity B, is constant and homogeneous. Since $\boldsymbol{B} = \nabla \times \boldsymbol{A}$, for a homogeneous field we can write

$$\boldsymbol{A} = \tfrac{1}{2}\boldsymbol{B} \times \boldsymbol{r}, \tag{12.124}$$

as follows from

$$\nabla \times \boldsymbol{A} = \tfrac{1}{2}\nabla \times (\boldsymbol{B} \times \boldsymbol{r}) = \tfrac{1}{2}[(\nabla \cdot \boldsymbol{r})\boldsymbol{B} - (\boldsymbol{B} \cdot \nabla)\boldsymbol{r}] = \boldsymbol{B},$$
$$\nabla \cdot \boldsymbol{A} = \tfrac{1}{2}\nabla \cdot (\boldsymbol{B} \times \boldsymbol{r}) = \tfrac{1}{2}[\boldsymbol{r} \cdot \nabla \times \boldsymbol{B} - \boldsymbol{B} \cdot \nabla \times \boldsymbol{r}] = 0. \tag{12.125}$$

Therefore,

$$\boldsymbol{A} \cdot \hat{\boldsymbol{p}} = \tfrac{1}{2}(\boldsymbol{B} \times \boldsymbol{r}) \cdot \hat{\boldsymbol{p}} = \tfrac{1}{2}\boldsymbol{B} \cdot (\boldsymbol{r} \times \hat{\boldsymbol{p}}) = \tfrac{1}{2}\boldsymbol{B} \cdot \hat{\boldsymbol{L}}, \tag{12.126}$$

where $\hat{\boldsymbol{L}}$ is the orbital angular momentum operator, and the Schrödinger equation is written, in the linear approximation, as

$$E\psi = \hat{H}_0\psi - \frac{e}{2mc}\boldsymbol{B} \cdot \hat{\boldsymbol{L}}\psi. \tag{12.127}$$

The last term in this equation can be interpreted as due to an orbital magnetic moment $\hat{\boldsymbol{\mu}}$ associated with the orbital angular momentum $\hat{\boldsymbol{L}}$ which, when coupled to the magnetic field $\boldsymbol{B}$, produces the mutual energy

$$\hat{H}_{\text{mag}} = -\hat{\boldsymbol{\mu}} \cdot \boldsymbol{B}. \tag{12.128}$$

Comparing this expression with (12.127), we see that the orbital magnetic moment operator is

$$\hat{\boldsymbol{\mu}} = \frac{e}{2mc}\hat{\boldsymbol{L}}. \tag{12.129}$$

The coefficient of proportionality between the magnetic moment of a charge and the angular momentum that generates it is known as the *gyromagnetic ratio*; in the present case its value (omitting the sign) is $e_0/2mc$. The z-component of this magnetic moment has the eigenvalues (to avoid confusion, we write the mass of the electron as m_e)

$$\mu_z = \frac{e\hbar}{2m_ec}m, \tag{12.130}$$

where m is the projection of the orbital angular momentum on the z-axis. This result, among others, justifies the name *magnetic quantum number* given to m by Sommerfeld. To simplify the writing, it is customary to introduce the *Bohr magneton* μ_0, defined as

$$\mu_0 = \frac{e_0\hbar}{2m_ec} = 9.27 \times 10^{-21} \text{ erg gauss}^{-1}, \tag{12.131}$$

where e_0 is the absolute value of the charge of the electron. Then $\mu_z = -\mu_0 m$, and the Schrödinger equation becomes

$$E\psi = \widehat{H}_0\psi + \frac{\mu_0}{\hbar}\mathbf{B} \cdot \widehat{\mathbf{L}}\psi. \tag{12.132}$$

The simplest way to solve this equation is to choose the z-axis in the direction of the magnetic field, so that

$$E\psi = \widehat{H}_0\psi + \frac{\mu_0}{\hbar}B\widehat{L}_z\psi \equiv \widehat{H}\psi. \tag{12.133}$$

When $\widehat{H}_0$ corresponds to a central potential (not necessarily that of the H atom), $[\widehat{L}_z, \widehat{H}] = 0$ and the wave functions ψ can be taken as simultaneous eigenfunctions of $\widehat{H}$ with eigenvalue E and of $\widehat{L}_z$ with eigenvalue $\hbar m$. In this subspace the preceding equation reduces to

$$E\psi = \widehat{H}_0\psi + \mu_0 B m \psi = \left(\widehat{H}_0 + \mu_0 B m\right)\psi, \tag{12.134}$$

from which it follows that the eigenfunctions of $\widehat{H}$ and $\widehat{H}_0$ coincide, but not their eigenvalues, which are related by the equation

$$E = E_0 + \mu_0 B m, \tag{12.135}$$

where E_0 is an eigenvalue of $\widehat{H}_0$.

Equation (12.135) allows us to derive the frequencies of the emission lines in the presence of the magnetic field **B**. For the transitions between two levels 1 and 2, we get

$$\omega = \frac{E_1 - E_2}{\hbar} = \frac{E_{01} - E_{02}}{\hbar} + \frac{\mu_0 B}{\hbar}(m_1 - m_2). \tag{12.136}$$

If we call ω_0 the frequency of the corresponding spectral line in the absence of field B, we finally get

$$\omega = \omega_0 + \frac{\mu_0 B}{\hbar}\Delta m. \tag{12.137}$$

From the selection rules for the central problem we know that $\Delta m = 0, \pm 1$; therefore, the field transforms each line of the original spectrum into a triplet with frequencies ω_0, $\omega_0 \pm (\mu_0 B/\hbar)$, that is, the original frequency plus a pair of lines symmetrically shifted with respect to the first one by the amount $\mu_0 B/\hbar$. It is precisely the m selection rule that ensures that, although each level is split into $2l + 1$ contiguous levels, only triplets appear in the spectrum, as shown in Fig. 12.4. These results confirm that it is sufficient to break the spherical symmetry of the Hamiltonian for the degeneracy in the magnetic quantum number m to disappear.

We can express the preceding result in another interesting way, by noting from the definition of the Bohr magneton that $\mu_0 B/\hbar$ does not depend on $\hbar$; this is in fact the Larmor frequency Ω_L, that is, the precession frequency of a magnetic moment in an external field,

$$\Omega_L = \frac{e_0 B}{2mc} = \frac{\mu_0 B}{\hbar}. \tag{12.138}$$

Consequently, Eq. (12.137) says that the triplet of frequencies are the original and the original plus or minus the Larmor frequency. Since this frequency does not depend on

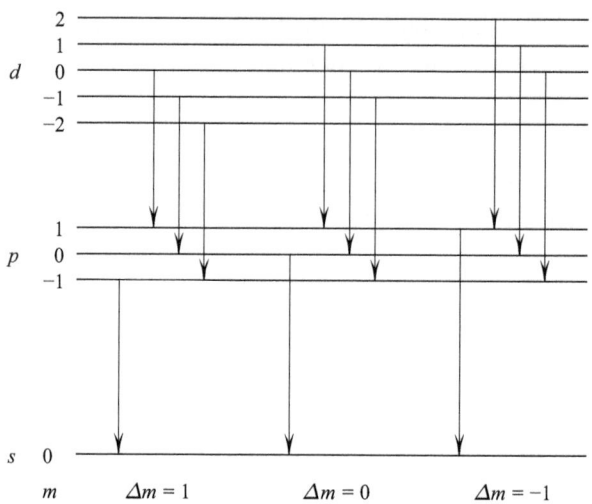

Figure 12.4 Transitions allowed by the selection rule $\Delta m = 0, \pm 1$, between atomic states s, p, and d. Due to the equidistance between levels with successive values of m, the three transitions between states p and d with a fixed value of Δm produce the same spectral line, so the spectrum consists of a series of triplets.

$\hbar$, it was possible to derive it from the classical Lorentz model of the atom. In fact, we have seen that Eq. (12.130) predicts for the orbital motion the gyromagnetic ratio

$$\frac{\mu_z}{L_z} = -\frac{\mu_0}{\hbar} = -\frac{e_0}{2mc}. \qquad (12.139)$$

The agreement of this expression with the corresponding classical one led to qualifying this part of the Zeeman effect as "normal."

12.7.3 Gauge transformations of the electromagnetic field

From electrodynamics we know that an electromagnetic field, described by its electric and magnetic components E and B, is invariant under a local transformation

$$A \rightarrow A' = A + \nabla \chi, \quad \Phi \rightarrow \Phi' = \Phi - \frac{1}{c}\frac{\partial \chi}{\partial t}, \qquad (12.140)$$

where $\chi = \chi(x, t)$ is an arbitrary scalar function. This *gauge transformation* in no way affects the physical results in classical physics, which allows us to conclude that the potentials A, Φ are mathematical entities with no direct correspondence to the elements of the physical system being described. We will now see that something similar occurs in Schrödinger's theory, which manifests itself in the fact that, under a gauge transformation, the wave function is transformed into another physically equivalent one that differs from the first one by a local phase.

Using the appropriate Hamiltonian in the presence of an electromagnetic field, Eq. (12.117), the Schrödinger equation reads

$$i\hbar\frac{\partial \psi}{\partial t} = \frac{1}{2m}\left(-i\hbar\nabla - \frac{e}{c}A\right)^2 \psi + (V + e\Phi)\psi. \qquad (12.141)$$

The field gauge transformation modifies this equation, and hence the wave function. Since the density $\rho(\boldsymbol{x}, t) = |\psi(\boldsymbol{x},t)|^2$ must not be changed by this transformation, the only possibility is a phase shift in ψ. Therefore, the *quantum gauge transformation* must be defined as in (12.140), with

$$\psi \to \psi' = \psi e^{i(e/\hbar c)\chi}. \tag{12.142}$$

The transformed Schrödinger equation is

$$i\hbar \frac{\partial \psi'}{\partial t} = \frac{1}{2m} \left(-i\hbar \nabla - \frac{e}{c} \boldsymbol{A}' \right)^2 \psi' + \left(V + e\Phi' \right) \psi'. \tag{12.143}$$

To show that this equation coincides with (12.141) we rewrite it in the form

$$i\hbar e^{-i(e/\hbar c)\chi} \frac{\partial \psi'}{\partial t} = \frac{e^{-i(e/\hbar c)\chi}}{2m} \left(-i\hbar \nabla - \frac{e}{c} \boldsymbol{A}' \right)^2 \psi' + \left(V + e\Phi' \right) \psi. \tag{12.144}$$

Since

$$i\hbar e^{-i(e/\hbar c)\chi} \frac{\partial \psi'}{\partial t} = i\hbar e^{-i(e/\hbar c)\chi} \frac{\partial}{\partial t} \left[\psi e^{i(e/\hbar c)\chi} \right]$$
$$= i\hbar \frac{\partial \psi}{\partial t} - \frac{e}{c} \frac{\partial \chi}{\partial t} \psi = i\hbar \frac{\partial \psi}{\partial t} + \left[e\Phi' - e\Phi \right] \psi, \tag{12.145}$$

Eq. (12.144) reduces to

$$i\hbar \frac{\partial \psi}{\partial t} = \frac{1}{2m} e^{-i(e/\hbar c)\chi} \left(-i\hbar \nabla - \frac{e}{c} \boldsymbol{A}' \right)^2 \psi' + (V + e\Phi) \psi. \tag{12.146}$$

Now we observe that

$$e^{-i(e/\hbar c)\chi} \left(-i\hbar \nabla - \frac{e}{c} \boldsymbol{A}' \right) \psi e^{i(e/\hbar c)\chi}$$
$$= e^{-i(e/\hbar c)\chi} \left[-i\hbar e^{i(e/\hbar c)\chi} (\nabla \psi + i\psi \,(e/\hbar c)\, \nabla \chi) - \frac{e}{c} (\boldsymbol{A} + \nabla \chi) \psi e^{i(e/\hbar c)\chi} \right]$$
$$= -i\hbar (\nabla \psi + i\,(e/\hbar c)\, \psi \nabla \chi) - \frac{e}{c} (\boldsymbol{A} + \nabla \chi) \psi = \left(-i\hbar \nabla - \frac{e}{c} \boldsymbol{A} \right) \psi. \tag{12.147}$$

With this result, the transformed equation is effectively equal to the original.

We now need to look at what happens to important expectation values, such as momentum or velocity. From Eq. (12.147) it follows that

$$\psi'^* \left(-i\hbar \nabla - \frac{e}{c} \boldsymbol{A}' \right) \psi' = \psi^* \left(-i\hbar \nabla - \frac{e}{c} \boldsymbol{A} \right) \psi, \tag{12.148}$$

which shows that the expectation value of the velocity $\boldsymbol{v} = \left(\boldsymbol{p} - \frac{e}{c} \boldsymbol{A} \right) / m$ is gauge invariant, but the expectation value of the mechanical momentum $\boldsymbol{p}$ is not. This result confirms that in the presence of an electromagnetic field it is preferable to work with the canonical momentum rather than the mechanical momentum. Note that although the expression $-i\hbar \nabla - \frac{e}{c} \boldsymbol{A}$ explicitly depends on the norm, its expectation and eigenvalues are gauge invariant (the change in phase of the wave function compensates for the change in potential). It follows from Eq. (12.148) that the current density (12.123) is also gauge invariant.

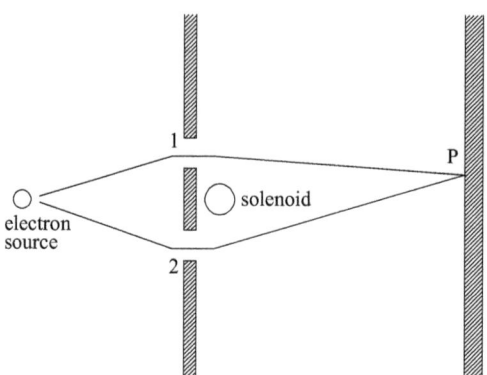

Figure 12.5 Schematic of an Aharonov–Bohm interference experiment. Even though outside of the solenoid $B = 0$, the current through the solenoid causes a shift of the interference pattern on the screen.

12.7.3.1 The Aharonov–Bohm effect

As mentioned earlier, in classical electrodynamics the vector and scalar potentials A, Φ play an auxiliary role; they are introduced as mathematical tools such that the electric and magnetic fields are correctly given by

$$E = -\nabla\Phi - \frac{1}{c}\frac{\partial A}{\partial t}, \quad B = \nabla \times A. \tag{12.149}$$

This allows different sets of potentials A', Φ' to represent the same fields, provided they are related to the original set by the gauge transformation (12.140).

As shown earlier, the Schrödinger equation is also gauge invariant, as are the average values (and eigenvalues) of the dynamical variables. However, in contrast to the classical case where the potentials have no direct physical meaning, there are quantum situations where the electromagnetic potentials do play a role. One case that has gained great importance is the *Aharonov–Bohm effect,* predicted by Aharonov and Bohm in 1959.[10] The effect was experimentally verified by G. Möllenstedt and W. Bayh in 1962 and conclusively demonstrated by A. Tonomura in 2006, using electron-interference experiments.

The apparatus for producing the Aharonov–Bohm effect is shown schematically in Fig. 12.5. As can be seen, it is a variant of the two-slit electron diffraction experiment, similar to that shown in Figure 4.1, with the addition of a very thin and long solenoid placed immediately behind the slits. Because of its size and position, the solenoid is located in a region of space inaccessible to the electrons. When a constant electric current is passed through it, a magnetic field is produced inside it, but the field outside the solenoid is zero, even though the vector potential itself is not zero in this

---

[10] The effect had been predicted by W. Ehrenberg and R. E. Siday in 1949. When Aharonov and Bohm published their work in 1959 they were unaware of Ehrenberg and Siday's work, but they credited it in a subsequent paper in 1961.

outer region: $\nabla \times A_{\text{ext}}(x) = 0$, but $A_{\text{ext}}(x) \neq 0$. The point is that the presence of the potential is sufficient to produce a *shift* in the diffraction pattern observed on the detection screen.[11]

The magnetic flux inside the solenoid is

$$\Phi = \iint B \cdot dS = \iint (\nabla \times A(x)) \cdot dS = \oint A \cdot dx, \qquad (12.150)$$

as follows from the application of Stokes' theorem. Outside the solenoid, however, $\nabla \times A_{\text{ext}} = 0$ and A_{ext} can be expressed as the gradient of a scalar χ, $A_{\text{ext}}(x) = \nabla \chi(x)$. On the other hand, as mentioned earlier, the presence of A_{ext} is counteracted by the gauge transformation $\psi = \psi^0 e^{i(e/\hbar c)\chi}$, where ψ^0 is the solution in the absence of the potential. Applying this idea to the paths on either side of the solenoid, numbered $i = 1, 2$, we can write $\psi_1 = \psi_1^0 e^{i(e/\hbar c)\chi_1}$, $\psi_2 = \psi_2^0 e^{i(e/\hbar c)\chi_2}$, where $\chi_i = \int_i A \cdot dx$. Since A is irrotational in each of the two regions, these integrals depend only on the endpoints of the trajectories. We can now determine the wave function at a point P on the screen *that is common to both trajectories*,

$$\begin{aligned}\psi_P &= \psi_1^0 e^{i(e/\hbar c)\chi_1} + \psi_2^0 e^{i(e/\hbar c)\chi_2} \\ &= e^{i(e/\hbar c)\chi_1}\left(\psi_1^0 + \psi_2^0 e^{i(e/\hbar c)(\chi_2-\chi_1)}\right) \\ &= e^{i(e/\hbar c)\chi_1}\left(\psi_1^0 + \psi_2^0 e^{i(e/\hbar c)\oint_S A\cdot dx}\right),\end{aligned} \qquad (12.151)$$

that is,

$$\psi_P = \psi_1^0 + \psi_2^0 e^{i(e/\hbar c)\Phi}. \qquad (12.152)$$

In the final result, a global phase has been omitted, as it has no observable effect. The fact that the phase difference depends on the magnetic flux in the solenoid, makes the relative position of the observed diffraction pattern dependent on this flux, even though the particles travel through regions of space where there is no magnetic field and therefore do not experience any of the usual forces that would deflect them.

This observation is important because it suggests a more general view of the problem: an effect can occur when the action over closed trajectories is modified by an external field, be it electromagnetic, gravitational, due to spin effects, and so on. This occurs naturally in quantum particle interferometry, and the effects appear to be due to topological phases generated by the presence of the fields. More recent analyses suggest that the phase shift can be viewed as being produced by the vector potential of a solenoid acting on the electron, or the vector potential of the electron acting on the solenoid, or the electron and solenoid currents acting on the quantized vector potential. In any case, there is still no general consensus on the ultimate physical significance of the Aharonov–Bohm effect.

---

[11] As shown in figure 7 of Möllenstedt and Bayh (1962), it is only the diffraction pattern that is shifted; the envelope in which it is contained does not suffer any deviation.

Problems

P12.1 Consider a diatomic molecule made up of ions with charge $\pm q$ and masses m_1 and m_2, respectively. Show that:

(a) The external gravitational field (near the earth's surface) produces an effect on the movement of the CM of the molecule, but not on its internal (relative) motion;

(b) An external uniform electric field $\mathcal{E}$ along the z-axis does not affect the CM motion, but it does affect the internal one, since it induces a dipole electric moment given by $\frac{\partial}{\partial \mathcal{E}}(q\mathcal{E} \cdot r) = qz$. Hint: study the general case corresponding to the potential $V(r) = f_1 z_1 + f_2 z_2$ and take the particular case.

P12.2 Two oscillators interact with each other linearly, in such a way that the Hamiltonian of the system is

$$H = \frac{p_1^2}{2m_1} + \frac{p_2^2}{2m_2} + \tfrac{1}{2}m_1\omega^2 x_1^2 + \tfrac{1}{2}m_2\omega^2 x_2^2 + \tfrac{1}{2}m\omega^2 \beta(x_1 - x_2)^2. \quad (12.153)$$

Show that the relative and CM coordinates coincide with the normal coordinates, separate these variables, and determine the eigenvalues of the energy. Study the energy spectrum for the attractive ($\beta > 0$) and repulsive ($\beta < 0$) cases, in the limits $|\beta| \ll 1$ and $|\beta| \gg 1$.

P12.3 Determine the normalization coefficient of the radial wave function of the hydrogen-like atom.

P12.4 By comparing the corresponding differential equations, show that the Laguerre polynomials can be written as a confluent hypergeometric function in the form

$$Q_k^{2l+1}(\rho) = {}_1F_1(-k, 2(l+1); \rho). \quad (12.154)$$

Use this result to obtain the physically acceptable values of the quantum number k.

P12.5 Show that the eccentricity of the hydrogen-like orbits can be taken as $\epsilon = \sqrt{1 - l(l+1)/n^2}$. Note from here that the minimum eccentricity (closest to circular orbits) corresponds to $l = n-1$ and is $\epsilon_{\min} = 1/\sqrt{n}$, which tends to zero as $n \to \infty$.

P12.6 Derive the following results for the expectation value of r^n for the H atom, where $\rho_{at} = r/a_0$ is measured in atomic units ($m = \hbar = c = 1$), that is, $\rho_{at} = (2n/Z)\rho$:

$$\langle \rho_{at} \rangle = \frac{1}{2Z}\left[3n^2 - l(l+1)\right]; \quad \langle \rho_{at}^2 \rangle = \frac{n^2}{2Z^2}\left[5n^2 + 1 - 3l(l+1)\right];$$

$$\left\langle \frac{1}{\rho_{at}} \right\rangle = \frac{Z}{n^2}; \quad \left\langle \frac{1}{\rho_{at}^2} \right\rangle = \frac{Z^2}{n^3(l+\tfrac{1}{2})}; \quad \left\langle \frac{1}{\rho_{at}^3} \right\rangle = \frac{Z^3}{n^3 l(l+1)(l+\tfrac{1}{2})}. \quad (12.155)$$

P12.7 Derive the recurrence relation for $\langle r^s \rangle$ for the potential $V = qr^n$. Verify that the results of Exercise E12.2 are particular cases of this expression.

P12.8 Show that in the ground state of the H atom the expectation value of r^n is

$$\langle 100|r^n|100\rangle = \frac{1}{2}\left(\frac{a_0}{2Z}\right)^n (n+2)! \tag{12.156}$$

P12.9 Solve the H atom problem with an additional potential γ/r^2 and show that for any value of the parameter $\gamma \neq 0$, the degeneracy with respect to l is broken.

P12.10 Find the probability that in a hydrogen atom in its ground state, the electron and the proton separate beyond the value allowed by classical mechanics at the same energy.

P12.11 An H atom is in the state $n = 2, l = 1, m = 0$. Express the corresponding electronic wave function in momentum space.

P12.12 In classical electrodynamics the magnetic moment μ_z produced by an electric current density $\boldsymbol{J}_{\text{el}}$ is given by

$$\mu_z = \frac{1}{2c} \int (\boldsymbol{r} \times \boldsymbol{J}_{\text{el}})_z d^3x. \tag{12.157}$$

Show that for discrete charges q this expression reduces to $\mu_z = \frac{q}{2m_0 c} L_z$ and, transferring this result to the quantum case, show that

$$\langle \hat{\mu}_z \rangle = e\frac{\langle \hat{L}_z \rangle}{2m_e c} = \frac{e\hbar}{2m_e c} m \tag{12.158}$$

if the wave function has the form $\psi = \Phi(r,\theta)e^{im\varphi}$, with Φ a real function.

P12.13 Show that when the quadratic effects of the magnetic field are taken into account, the magnetic moment of an atom is

$$\mu = -\mu_0 m - \frac{e^2 B}{6 m_e c^2} \langle r^2 \rangle. \tag{12.159}$$

Notes: The first term represents a permanent magnetic moment (independent of the external field) and can have any sign; this is the *paramagnetic component* of the magnetic moment. The second term (which, because it comes from a quadratic effect, is generally very small for weak fields) represents an induced magnetic moment (it vanishes when $B = 0$), it is always negative and exists for all atoms; this is the *diamagnetic moment*.

P12.14 Using (17.134) for the Einstein A coefficient, derive in detail Eqs. (12.102) and (12.103).

P12.15 Calculate the mean lifetime of the hydrogen-like $3s$ state.

P12.16 Show that when a quantum system has two or more operators that commute with the Hamiltonian but not with each other, in general the states of the system are degenerate. What general property of the system does this fact reflect?

P12.17 Prove that in the presence of a magnetic field, $m\hat{v}_i = \hat{p}_i - \frac{e}{c}A_i$, and the velocities $\hat{v}_i = \dot{x}_i$ satisfy the commutation rules

$$[\hat{v}_i, \hat{v}_j] = \frac{ie\hbar}{m^2 c}\varepsilon_{ijk}B_k. \tag{12.160}$$

P12.18 Determine the wavelength of the three Zeeman lines produced at the $3d \to 2p$ transition of an H atom placed in a 10^4 Gauss magnetic field.

Bibliographical Notes

For a detailed discussion of the Aharonov–Bohm effect (up to 1987), see Peshkin and Tonomura (1989). More recent discussions of the A-B question show that the physical explanation is still controversial; see, for example, Boyer (2000, 2023) and the review by Cohen et al. (2019).

13 Stationary Perturbation Theory and the Variational Method

In most cases of interest, there is no exact analytical solution to the Schrödinger equation for the given problem. In other cases, the exact solution can be found, but its structure is so complicated that it lacks transparency or becomes impractical. Nowadays, it is possible to resort to computational methods, which can be very powerful but have the disadvantage of hiding the qualitative behavior of the solution. In such cases, it is useful or even at times necessary to use a method that allows an approximate solution to be found. There is no universal recipe: in each instance the method to be used must be chosen carefully, because although there may be more than one method that can serve similar purposes, one of them may have advantages over the others for some specific reason.

In this chapter we will review two of the most useful approximation methods for time-independent problems: perturbation theory and the variational method, with applications of specific interest.

13.1 Introduction to Time-Independent Perturbation Theory

It is often the case that a problem can be successfully analyzed using the methods of perturbation theory if it is a minor modification of a problem whose analytical solution is known. For example, in the previous chapter we saw that for an atom subject to an external magnetic field, the Hamiltonian is $\widehat{H} = \widehat{H}_0 + \widehat{H}_{\mathrm{mag}}$, where $\widehat{H}_0$ is the Hamiltonian of the atom in the absence of the field and $\widehat{H}_{\mathrm{mag}} = (\mu_0/\hbar) \boldsymbol{B} \cdot \widehat{\boldsymbol{L}}$ represents the interaction between the orbiting electron and the field. In this case, as long as the magnetic field is not excessively high, it will probably only slightly alter the behavior due to $\widehat{H}_0$; therefore we can consider the initial problem as one due to $\widehat{H}_0$, but perturbed by the magnetic interaction. The energy levels and eigenfunctions of $\widehat{H}$ are then expected to be very close to the original ones, with small corrections due to the perturbation. This is precisely the idea behind the perturbation theory, which will be developed below.[1]

We start by writing the total Hamiltonian of the problem as $\widehat{H} = \widehat{H}_0 + \widehat{V}$, where $\widehat{V}$ is the perturbative potential, or the Hamiltonian of the perturbation. As the magnetic example illustrates, $\widehat{V}$ can depend on position as well as on other operators. In any

---

[1] Quantum perturbation theory has its origins in classical perturbation theory. The prototypical example of the latter is the solar system, in which the orbit of each planet can be studied separately, as if the other planets and other minor elements of the system did not exist. The combined effects of all these neglected bodies can normally be considered as small disturbances, which are to be calculated separately, one by one. It is on this idea that the astronomical perturbation theory was proposed and successfully developed over the last three centuries.

case, in this chapter we will focus on the study of stationary problems, so the equation to be solved is

$$\widehat{H}\Psi_n = (\widehat{H}_0 + \widehat{V})\Psi_n = E_n\Psi_n, \tag{13.1}$$

where both H_0 and V are time-independent. We write the unperturbed equation, whose solution is assumed to be completely known, as

$$\widehat{H}_0\psi_n = E_n^{(0)}\psi_n, \tag{13.2}$$

to distinguish the eigenfunctions ψ_n and the eigenvalues $E_n^{(0)}$ of the original Hamiltonian $\widehat{H}_0$ from the corresponding ones of the total Hamiltonian $\widehat{H}$.

In this section we restrict ourselves to the case of nondegenerate states, for which it is assumed that the wave function of the perturbed state n differs little from that of the corresponding unperturbed state n. Later we will see that this is not necessarily true for degenerate states (i.e., when more than one state corresponds to the same energy level), if the perturbation breaks the degeneracy. So for now, we propose that $\Psi_n = \psi_n + \delta\psi_n$, where $\delta\psi_n$ is the modification of ψ_n due to the perturbation, which we will consider to be "small."

The basic idea is to use the unperturbed eigenfunctions ψ_n as a basis for expanding the unknown function $\delta\psi_n$,

$$\Psi_n = \psi_n + \sum_k C_{nk}\psi_k. \tag{13.3}$$

We substitute this expression in Eq. (13.1), subtract Eq. (13.2) from the result, and write the total (perturbed) energy E_n in the form

$$E_n = E_n^{(0)} + \delta E_n, \tag{13.4}$$

to get

$$\sum_k C_{nk}E_k^{(0)}\psi_k + \widehat{V}\psi_n + \sum_k C_{nk}\widehat{V}\psi_k$$
$$= \sum_k C_{nk}E_n^{(0)}\psi_k + \delta E_n\psi_n + \delta E_n\sum_k C_{nk}\psi_k. \tag{13.5}$$

To transform this differential problem into an algebraic one we multiply the equation by ψ_l^* and integrate. Writing the matrix elements of V in the usual form $V_{nl} = \langle n|\widehat{V}|l\rangle$, we get

$$\sum_k V_{lk}C_{nk} + V_{ln} - (E_n^{(0)} - E_l^{(0)} + \delta E_n)C_{nl} = \delta E_n\delta_{nl}. \tag{13.6}$$

The index n denotes the state we are studying while l can take any possible value; thus, Eq. (13.6) represents a system of algebraic equations for the coefficients C_{nl} of the expansion (13.3) of the wave function and for the corrections δE_n to the energy. Although in general it is not possible to solve this system exactly (the equations are nonlinear and there may be a very large or even infinite number of them), we can draw some general conclusions as follows.

Taking $l = n$ in the preceding equation, we get

$$\sum_k V_{nk}C_{nk} + V_{nn} - \delta E_n C_{nn} = \delta E_n. \tag{13.7}$$

We now extract from the sum the term with $k = n$ and solve for δE_n, to obtain

$$\delta E_n = V_{nn} + \frac{\sum' V_{nk} C_{nk}}{1 + C_{nn}}, \tag{13.8}$$

where the prime means that the term with $k = n$ is to be excluded from the sum, $\sum' \equiv \sum_{k \neq n}$.

This expression shows that the effects of the perturbation on the energy are manifested both directly through the average potential term V_{nn}, and indirectly through the modification of the wave function represented by the coefficients C_{nk}. As we will see later (e.g., in Eqs. (13.18) and (13.19)), the direct effect on the energy is of first order, while the indirect effect is of at least second order in the perturbation, precisely because of its indirect nature (the coefficients C_{nk} generally depend on the perturbation at all orders).

We now take $l \neq n$ in (13.6) and eliminate δE_n using (13.8), to obtain, after minor rearrangements,

$$\sum_{k \neq n,l} V_{lk} C_{nk} + (1 + C_{nn}) V_{ln}$$

$$- \left(E_n^{(0)} - E_l^{(0)} + V_{nn} - V_{ll} + \frac{\sum' V_{nk} C_{nk}}{1 + C_{nn}} \right) C_{nl} = 0 \quad (n \neq l). \tag{13.9}$$

This (obviously nonlinear) system determines, in principle, the coefficients C_{nk} for $n \neq k$. To determine C_{nn} we impose the normalization condition on Ψ_n (assuming, of course, that the ψ_n are normalized),

$$\langle \Psi_n | \Psi_n \rangle = \langle \psi_n + \sum_k C_{nk}^* \psi_k | \psi_n + \sum_{k'} C_{nk'} \psi_{k'} \rangle$$

$$= 1 + C_{nn} + C_{nn}^* + \sum_k |C_{nk}|^2 = 1, \tag{13.10}$$

which leads to

$$|C_{nn}|^2 + 2\operatorname{Re} C_{nn} + \sum' |C_{nk}|^2 = 0. \tag{13.11}$$

Knowing C_{nk} for $n \neq k$, this equation determines the function $(\operatorname{Re} C_{nn})^2 + 2\operatorname{Re} C_{nn} + (\operatorname{Im} C_{nn})^2$. Since there are no more conditions to satisfy, we can conveniently take $\operatorname{Im} C_{nn} = 0$, that is, C_{nn} real. Thus, the preceding equation gives

$$1 + C_{nn} = \sqrt{1 - \sum' |C_{nk}|^2} \tag{13.12}$$

(the negative solution is unacceptable, since by hypothesis $|C_{nn}|$ is small compared to unity). Taken together, Eqs. (13.9) and (13.12) determine all the coefficients C_{kl} of the expansion, whereby the wave functions and (with the help of (13.8)) the energy eigenvalues are determined, in principle. However, the complexity of the previous system of equations, which is exact, makes it necessary in practically all cases to look for an approximate solution. In the following sections we present a systematic way of proceeding in perturbation theory.

A note of caution is in order. There are situations in which perturbation theory is not appropriate. The problem at hand is normally nonlinear, and nonlinear systems can exhibit unpredictable and unexpected behavior. In such cases perturbation theory may lead to unreliable results. At the very least, it is necessary to check the convergence of your result. Problems arise when even small perturbations of the system lead to large changes in the response, which is typical behavior for nonlinear systems.

13.2 Perturbation Theory for Nondegenerate States

As just mentioned, our aim is to develop a method that allows us to find approximate solutions to Eqs. (13.9) and (13.12). For the sake of clarity, it is convenient to rewrite the perturbing potential in the form $\lambda \widehat{V}$, where the auxiliary parameter λ (which will be made equal to one in all final results) helps us to determine the order of smallness of a given term by simple inspection of the factor λ^r, which indicates that the respective term is of order r. We express both the coefficient C_{nl} and the correction to the energy δE_n as the sum of the contributions of first, second, third, ... order, as follows,[2]

$$C_{nl} = \lambda C_{nl}^{(1)} + \lambda^2 C_{nl}^{(2)} + \lambda^3 C_{nl}^{(3)} + \dots \tag{13.13}$$

$$\delta E_n = \lambda \delta E_n^{(1)} + \lambda^2 \delta E_n^{(2)} + \lambda^3 \delta E_n^{(3)} + \dots \tag{13.14}$$

where the factor λ^r ($r = 1, 2, 3, \dots$) indicates the order of smallness of the term. Substituting in (13.12) we get (summing over k is understood)

$$1 + \lambda C_{nn}^{(1)} + \lambda^2 C_{nn}^{(2)} + \dots = \left[1 - \sum\nolimits' |\lambda C_{nk}^{(1)} + \lambda^2 C_{nk}^{(2)} + \dots |^2 \right]^{1/2}$$
$$= 1 - \tfrac{1}{2}\lambda^2 \sum\nolimits' |C_{nk}^{(1)}|^2 - \tfrac{1}{2}\lambda^3 \sum\nolimits' (C_{nk}^{(1)*} C_{nk}^{(2)} + C_{nk}^{(1)} C_{nk}^{(2)*}) + \dots \tag{13.15}$$

Equating the coefficients of each power of λ that appear on either side of this expression, we get the system

$$C_{nn}^{(1)} = 0; \quad C_{nn}^{(2)} = -\tfrac{1}{2} \sum\nolimits' |C_{nk}^{(1)}|^2; \quad \dots \tag{13.16}$$

We see that the first-order correction to C_{nn} is zero, the second-order correction is $-\tfrac{1}{2} \sum_{n \neq k} |C_{nk}^{(1)}|^2$, and so on.

To obtain the correction to the energy for different orders of approximation, we substitute Eqs. (13.13, 13.14, 13.16) into (13.8),

$$\delta E_n = \lambda \delta E_n^{(1)} + \lambda^2 \delta E_n^{(2)} + \dots$$
$$= \lambda V_{nn} + \lambda \sum\nolimits_k' V_{nk}(\lambda C_{nk}^{(1)} + \lambda^2 C_{nk}^{(2)} + \dots)(1 - \lambda^2 C_{nn}^{(2)} + \dots)$$
$$= \lambda V_{nn} + \lambda^2 \sum\nolimits_k' C_{nk}^{(1)} V_{nk} + \lambda^3 \sum\nolimits' C_{nk}^{(2)} V_{nk}$$
$$+ \lambda^4 \sum\nolimits_k' (C_{nk}^{(3)} - C_{nk}^{(1)} C_{nk}^{(2)}) V_{nk} + \dots \tag{13.17}$$

---

[2] This is equivalent to assuming that the solutions are analytic functions of the parameter λ, which is not necessarily true. In the latter case it usually means that the perturbative expansion is not convergent.

Equating the coefficients of each power of λ we obtain

$$\delta E_n^{(1)} = V_{nn} = \langle n|\widehat{V}|n\rangle, \qquad (13.18)$$

$$\delta E_n^{(2)} = {\sum}' C_{nk}^{(1)} \langle n|\widehat{V}|k\rangle, \qquad (13.19)$$

and so on for the higher orders. Formula (13.18) is without doubt the most important result of perturbation theory; it shows that, up to the first order, the correction to the energy of the state n is always given by the expectation value of the perturbing Hamiltonian, calculated with the *unperturbed* wave function of the corresponding state n.

We will analyze formula (13.19) after determining the corrections to the wave functions. To do so, we proceed as before, substituting Eqs. (13.13, 13.14, 13.16) into (13.9),

$$\lambda\left[V_{ln} - (E_n^{(0)} - E_l^{(0)})C_{nl}^{(1)}\right] + \lambda^2\left[\sum_{k\neq l,n} C_{nk}^{(1)}V_{lk} - (E_n^{(0)} - E_l^{(0)})C_{nl}^{(2)}\right.$$

$$\left. -(V_{nn} - V_{ll})C_{nl}^{(1)}\right] + \lambda^3\left[V_{ln}C_{nn}^{(2)} + \sum_{k\neq l,n} C_{nk}^{(2)}V_{lk} - (E_n^{(0)} - E_l)C_{nl}^{(3)}\right.$$

$$\left. -(V_{nn} - V_{ll})C_{nl}^{(?)} - {\sum}' V_{nk}^{(0)} C_{nk}^{(1)} C_{nl}\right] + \cdots = 0. \qquad (13.20)$$

For this system to be satisfied up to all orders of perturbation theory, the coefficient of each power of λ must again vanish separately. Up to the second order, we get

$$V_{ln} - (E_n^{(0)} - E_l^{(0)})C_{nl}^{(1)} = 0,$$

$$\sum_{k\neq n,l} V_{lk}C_{nk}^{(1)} - (E_n^{(0)} - E_l^{(0)})C_{nl}^{(2)} - (V_{nn} - V_{ll})C_{nl}^{(1)} = 0. \qquad (13.21)$$

If $E_n^{(0)} - E_l^{(0)} \neq 0$ for all l and n, that is, if the level n is not degenerate (or more generally, if there are no pairs of very close energy levels), the previous system can be solved satisfactorily, giving

$$C_{nl}^{(1)} = \frac{V_{ln}}{E_n^{(0)} - E_l^{(0)}}, \qquad (13.22)$$

$$C_{nl}^{(2)} = \frac{\sum_{k\neq n,l} V_{lk}C_{nk}^{(1)} - (V_{nn} - V_{ll})C_{nl}^{(1)}}{E_n^{(0)} - E_l^{(0)}}$$

$$= \frac{1}{E_n^{(0)} - E_l^{(0)}}\left[\sum_{k\neq n,l} \frac{V_{lk}V_{kn}}{E_n^{(0)} - E_k^{(0)}} + \frac{(V_{ll} - V_{nn})V_{ln}}{E_n^{(0)} - E_l^{(0)}}\right], \qquad (13.23)$$

and so on for higher-order corrections. Perturbation theory is usually only carried up to first or second order, so we will not write final results beyond what is required.

It is important to underline that Eq. (13.22) is inconsistent if there is degeneracy, since in such a case it would be necessary that $\langle l|\widehat{V}|n\rangle = 0$ for arbitrary $\widehat{V}$, which is obviously not true in general. We will develop the theory for the degenerate case later, but this observation already anticipates that it will be necessary to move to a basis that diagonalizes the perturbing Hamiltonian.

The formulas (13.22) and (13.23) allow us to calculate correctly normalized wave functions up to first and second order, respectively. We can now introduce (13.22) in (13.19) to obtain an explicit formula for the second-order energy correction

$$\delta E_n^{(2)} = \sum_{k \neq n} V_{nk} C_{nk}^{(1)} = \sum_{k \neq n} V_{nk} \frac{V_{kn}}{E_n^{(0)} - E_k^{(0)}}, \qquad (13.24)$$

which can be rewritten as follows, assuming that $\widehat{V}$ is Hermitian,

$$\delta E_n^{(2)} = \sum_{k \neq n} \frac{|V_{nk}|^2}{E_n^{(0)} - E_k^{(0)}}. \qquad (13.25)$$

This is also an important formula because it is often the case that the first-order correction is zero and the main effect on the energy levels is then of second order. When n refers to the ground state, each term of (13.25) will be negative, so this correction will be negative; that is, the ground-state energy up to the first-order is always in excess.

To show that this conclusion can be generalized to any ψ real, we write the perturbed wave function in the form $\Psi = \psi G$ and substitute into Eq. (13.1), subtracting the unperturbed equation (13.2) from the result. After dividing by G, we get

$$\psi \delta E = \widehat{V} \psi - \frac{\hbar^2}{2m} \left[\left(\frac{\nabla^2 G}{G} \right) \psi + 2 \left(\frac{\nabla G}{G} \right) \cdot \nabla \psi \right]. \qquad (13.26)$$

An integration by parts shows that

$$\int \psi^* \frac{\nabla G}{G} \cdot \nabla \psi \, dx = -\int \left[\psi \frac{\nabla G}{G} \cdot \nabla \psi^* + \psi^* \frac{\nabla^2 G}{G} \psi - \psi^* \left(\frac{\nabla G}{G} \right)^2 \psi \right] dx. \qquad (13.27)$$

If ψ is real (e.g., for s states), the first term on the right-hand side cancels the one on the left-hand side, whence

$$2 \int \psi^* \frac{\nabla G}{G} \cdot \nabla \psi \, dx = -\int \psi^* \left[\frac{\nabla^2 G}{G} - \left(\frac{\nabla G}{G} \right)^2 \right] \psi \, dx. \qquad (13.28)$$

With this result, the expectation value of (13.26) reduces to

$$\delta E = \left\langle V - \frac{\hbar^2}{2m} \left(\frac{\nabla G}{G} \right)^2 \right\rangle. \qquad (13.29)$$

From Eq. (13.26) it is easy to see that G can be taken real if V is real. Therefore, for ψ real, the first-order correction $\langle V \rangle$ to the energy is always in excess, since the correction up to all orders of perturbation theory,

$$(-\hbar^2/2m) \langle (\nabla G/G)^2 \rangle, \qquad (13.30)$$

is negative or zero.

Returning to the formula (13.25) for the second-order energy correction, we see that for $E_k^{(0)} > E_n^{(0)}$, the energy of state n decreases, while for $E_k^{(0)} < E_n^{(0)}$, E_n increases with respect to $E_n^{(0)}$; in both cases the levels E_n and E_k tend to move further apart. This phenomenon is known as *level repulsion*.

13.2.1 The harmonic oscillator in a uniform electric field

As an example of application of the previous results, we study the problem of a 1D HO immersed in a uniform and constant electric field applied in the direction of oscillation. The problem is interesting because its perturbative solution is simple and illustrative, and because it can be solved exactly, which allows the results to be compared. In addition, it makes us aware of the inherent advantages and limitations of the perturbative method, which can be summarized schematically as follows.

Suppose that perturbation theory has led us, up to a certain approximation, to a solution of the form $f(x) = 1 - x + \frac{1}{2}x^2$; this is clearly useful information, but from that alone it is impossible to know whether the exact solution is, say, $f_{\text{exac}}(x) = e^{-x}$, or some other function of x with the same first three terms. In other words, the behavior of the system for values of x approaching 1 remains hidden. It is clear that the closed (analytical) form gives much more information (both qualitatively and quantitatively) than its perturbative approximation, although the two may coincide numerically within the range of validity of the approximation, and the latter may be correct and sufficient for many purposes. Such limitations (and possibilities) are inherent in the perturbative method and must be taken into account in its applications.

13.2.1.1 Approximate perturbative solution

The Hamiltonian of the problem is

$$\widehat{H} = \widehat{H}_0 + \widehat{V} = \widehat{H}_0 - e\mathcal{E}x, \tag{13.31}$$

where $\mathcal{E}$ represents the intensity of the applied electric field and $\widehat{H}_0$ is the Hamiltonian of the oscillator of frequency ω. The matrix elements of the perturbing potential are $\langle n|\widehat{V}|m\rangle = -e\mathcal{E}\langle n|x|m\rangle$, where $|n\rangle$ is an eigenvector of $\widehat{H}_0$. We know from Chapter 9 that the only nonzero matrix elements are

$$\langle n|\widehat{V}|n-1\rangle = -\frac{1}{\sqrt{2}}e\mathcal{E}\alpha_0\sqrt{n}, \tag{13.32}$$

$$\langle n|\widehat{V}|n+1\rangle = -\frac{1}{\sqrt{2}}e\mathcal{E}\alpha_0\sqrt{n+1}, \tag{13.33}$$

where $\alpha_0 = \sqrt{\hbar/m\omega}$. Therefore, up to first order in the perturbation there is no change in the energy,

$$\delta E_n^{(1)} = \langle n|\widehat{V}|n\rangle = 0. \tag{13.34}$$

For the second-order energy correction, we obtain from Eq. (13.25), using (9.33) for $E_n^{(0)}$,

$$\delta E_n^{(2)} = {\sum}' \frac{|V_{nm}|^2}{E_n^{(0)} - E_m^{(0)}} = \frac{1}{\hbar\omega} {\sum}' \frac{|V_{nm}|^2}{n - m}$$

$$= \frac{1}{\hbar\omega}\left[|V_{n,n-1}|^2 - |V_{n,n+1}|^2\right] = \frac{e^2\mathcal{E}^2\alpha_0^2}{2\hbar\omega}(n - n - 1), \tag{13.35}$$

therefore,

$$\delta E^{(2)} \equiv \delta E_n^{(2)} = -\frac{e^2\mathcal{E}^2}{2m\omega^2}. \tag{13.36}$$

This result shows that the energy of all states decreases by the same amount $\frac{1}{2}m\omega^2 x_0^2$, where $x_0 = e\mathcal{E}/m\omega^2$; hence up to second order the spectral lines are not modified by the perturbation.

To obtain the first-order corrections to the wave functions we use (13.22),

$$C_{nl}^{(1)} = \frac{V_{ln}}{E_n^{(0)} - E_l^{(0)}} = \frac{V_{ln}}{\hbar\omega(n-l)}. \tag{13.37}$$

Thus, $C_{n,n-1}^{(1)} = -q\sqrt{n}$, $C_{n,n+1}^{(1)} = q\sqrt{n+1}$, where $q = e\mathcal{E}\alpha_0/\sqrt{2}\hbar\omega = x_0/\sqrt{2}\alpha_0$. Using (13.3), we obtain for the first-order perturbed wave function

$$\Psi_n = \psi_n - q\sqrt{n}\psi_{n-1} + q\sqrt{n+1}\psi_{n+1}, \quad q = \frac{x_0}{\sqrt{2}\alpha_0}. \tag{13.38}$$

Exercise E13.1. Comparison with the exact solution for the harmonic oscillator in a constant electric field

Find the exact solution for the oscillator perturbed by a constant electric field and compare with the approximate solution obtained earlier using perturbation theory.

Solution. To solve the problem exactly, we write the Hamiltonian in the form

$$\begin{aligned}\widehat{H} &= \frac{\widehat{p}^2}{2m} + \frac{1}{2}m\omega^2 x^2 - e\mathcal{E}x \\ &= \frac{\widehat{p}^2}{2m} + \frac{1}{2}m\omega^2\left(x - \frac{e\mathcal{E}}{m\omega^2}\right)^2 - \frac{e^2\mathcal{E}^2}{2m\omega^2},\end{aligned} \tag{13.39}$$

which shows that the effect of the field is to shift the equilibrium position of the oscillator by a distance $x_0 = e\mathcal{E}/m\omega^2$ and to add to all levels a constant negative term to the energy. The exact Schrödinger equation is (we take into account that $e^2\mathcal{E}^2/2m\omega^2 = m\omega^2 x_0^2/2$)

$$E_n\Psi_n = (E_n^{(0)} + \delta E_n)\Psi_n = \widehat{H}_0\Psi_n - \frac{1}{2}m\omega^2 x_0^2 \Psi_n, \tag{13.40}$$

where the relation between the (exact) perturbed wave function and the unperturbed one is given by

$$\Psi_n(x) = \psi_n(x - x_0). \tag{13.41}$$

Since the change to the variable $y = x - x_0$ does not modify the energy levels, we have $E_n^{(0)}\Psi_n = \widehat{H}_0\Psi_n$; subtracting this equation from (13.40) we get Eq. (13.36), but here it is obtained as an exact result. Expanding $\psi_n(x - x_0)$ in Taylor series around x and approximating to first order, we get

$$\Psi_n(x) = \psi_n(x - x_0) = \psi_n(x) - x_0\frac{d\psi_n}{dx}. \tag{13.42}$$

Equations (9.61, 9.62), based on the property of the Hermite polynomials $H_n' = 2nH_{n-1}$, give for the derivative:

$$\frac{d\psi_n}{dx} = \frac{1}{\sqrt{2}\alpha_0}\left(\sqrt{n}\psi_{n-1} - \sqrt{n+1}\psi_{n+1}\right). \tag{13.43}$$

Substituting this expression into the previous one, we obtain precisely (13.38). The comparison between Eqs. (13.38) and (13.41) clearly illustrates the power and simplicity of the perturbative method, as well as one of its intrinsic limitations.

A general lesson to be drawn from this exercise is that it is useful to pay attention to certain features of the problem that may suggest a simple way of solving it – such as the symmetries, or its relationship to a problem whose solution is known – before embarking on a complicated mathematical procedure that may not give the best result.

13.3 Perturbation Theory for Degenerate States

The exact equations of the theory developed in Section 13.1 and, in particular, Eqs. (13.8), (13.9), and (13.12) are obviously also applicable to the study of degenerate states. It is only in Section 13.2 when making the approximations in the first of the equations (13.14) that we introduced the hypothesis that limits the perturbative analysis to the nondegenerate case. To see this, we rewrite Eq. (13.6) for $n \neq l$, introducing the parameter λ and assuming that the states n and l are degenerate, that is, that $E_l^{(0)} = E_n^{(0)}$ for some l; to first order in the energy and the C_{nk}, we have

$$\lambda(V_{ln} + \sum_k \lambda V_{lk} C_{nk}^{(1)} - \lambda \delta E_n^{(1)} C_{nl}^{(1)}) = 0. \qquad (13.44)$$

This condition mixes first- and second-order terms and can only be satisfied if $V_{ln} = 0$, which is not generally true. However, if we allow the coefficients C_{nk} to have a zero-order component, the conflict disappears, as the resulting equation

$$V_{ln} + \sum_k V_{lk} C_{nk}^{(0)} - \delta E_n^{(1)} C_{nl}^{(0)} + \cdots = 0 \qquad (13.45)$$

has a solution (all terms explicitly written here are of first order). The presence of zero-order components in the C_{nk} coefficients means that in the case of degenerate states, the different (unperturbed) wave functions belonging to the same energy *become mixed.*

We can look at this behavior from a slightly different angle. If we construct new zero-order wave functions that diagonalize V, then the $V_{n'l'}$ (the prime denoting the new basis) are zero for $n' \neq l'$ and the conflict disappears. Physically, this means that the wave functions modified by the perturbation according to the formulas derived in Section 13.2 are not the original ψ_n (which were determined completely independently of $\widehat{V}$), but certain linear combinations of them, $\varphi_n = \sum A_{nl} \psi_l$, which are eigenfunctions of the *complete* Hamiltonian, that is, they are determined taking into account the perturbing potential. These are the *correct wave functions* for the perturbed system.

In order to avoid confusion, we introduce a slight change in notation. Suppose that the level n is degenerate and that there are g different states with the same energy $E_n^{(0)}$; we will denote these states by n_i, $i = 1, 2, \ldots g$, and focus our attention on them, since the nondegenerate levels will be treated as before, with the method presented in Section 13.2. The exact equation (13.6) is now rewritten for $n = n_i$ and $l = n_j$,

$$\sum_k V_{n_j k} C_{n_i k} + V_{n_j n_i} - \delta E_{n_i} C_{n_i n_j} = \delta E_{n_i} \delta_{n_i n_j}. \tag{13.46}$$

Further, we rewrite the coefficient $C_{n_i k}$ as the sum of a zero-order term (a mixture of the degenerate states) denoted with $a_{n_i k}$, and higher-order terms (containing all powers of λ) denoted now collectively with $B_{n_i k} = B_{n_i k}(\lambda)$,

$$C_{n_i k} = a_{n_i k} + B_{n_i k}. \tag{13.47}$$

Since we wish to mix only degenerate states, we take

$$a_{n_i k} = \begin{cases} a_{n_i k} & \text{if } k \in \{n_i\} \\ 0 & \text{otherwise.} \end{cases} \tag{13.48}$$

So, for example,

$$\sum_k V_{n_j k} C_{n_i k} = \sum_{n_s} V_{n_j n_s} a_{n_i n_s} + \sum_k V_{n_j k} B_{n_i k}. \tag{13.49}$$

We use these expressions to rewrite Eq. (13.46); at the same time, we add and subtract some terms to explicitly write all first-order contributions,

$$\lambda \sum_{n_s} V_{n_j n_s} a_{n_i n_s} + \lambda \sum_k V_{n_j k} B_{n_i k} + \lambda V_{n_j n_i} - \lambda \delta E^{(1)}_{n_i} a_{n_i n_j}$$
$$-(\delta E_{n_i} - \lambda \delta E^{(1)}_{n_i}) a_{n_i n_j} - \delta E_{n_i} B_{n_i n_j} = \lambda \delta E^{(1)}_{n_i} \delta_{n_i n_j} + (\delta E_{n_i} - \lambda \delta E^{(1)}_{n_i}) \delta_{n_i n_j}. \tag{13.50}$$

We can now split this equation into two parts, one of first order and another containing all the remaining terms, and cancel each of them separately. This gives

$$\sum_{n_s} V_{n_j n_s} a_{n_i n_s} + V_{n_j n_i} - \delta E^{(1)}_{n_i} a_{n_i n_j} = \delta E^{(1)}_{n_i} \delta_{n_i n_j}, \tag{13.51}$$

$$\lambda \sum_k V_{n_j k} B_{n_i k} - (\delta E_{n_i} - \lambda \delta E^{(1)}_{n_i}) a_{n_i n_j} - \delta E_{n_i} B_{n_i n_j}$$
$$= (\delta E_{n_i} - \lambda \delta E^{(1)}_{n_i}) \delta_{n_i n_j}. \tag{13.52}$$

Equation (13.52) determines the coefficients B_{nk} once the a_{nk} are known, that is, it serves to calculate both the corrections to the correct wave functions (i.e., to the appropriate mixtures of degenerate states), as well as the second- or higher-order corrections to the energy. This system is used *after* solving Eq. (13.51) for the correct (zero-order) wave functions and the first-order energy corrections. Since the solution of (13.52) is equivalent to the nondegenerate problem already discussed, we will not analyze it further here.

To solve Eq. (13.51), we rewrite it in terms of new coefficients A defined as

$$A_{n_i n_j} = \delta_{n_i n_j} + a_{n_i n_j}, \tag{13.53}$$

which gives

$$\delta E^{(1)}_{n_i} A_{n_i n_j} - \sum_{n_s} V_{n_j n_s} A_{n_i n_s} = 0. \tag{13.54}$$

This system of equations is a generalization to the degenerate case of Eq. (13.18), which gives the first-order correction to the energy in the nondegenerate case. In fact, if in (13.53) we take $A_{ij} = \delta_{ij}$, that is, if we cancel all the a's (which can only be done if the state is nondegenerate), Eq. (13.54) reduces for $i = j$ to $\delta E_{n_i}^{(1)} - V_{n_i n_i} = 0$, which is precisely (13.18). An expression for the energy correction in the degenerate case can be obtained by setting $i = j$ in (13.54),

$$\delta E_{n_i}^{(1)} = V_{n_i n_i} + \frac{\sum_{n_s \neq n_i} V_{n_i n_s} A_{n_i n_s}}{A_{n_i n_i}}. \tag{13.55}$$

However, this expression is not useful in practice, since in general the energy corrections are determined first, and these in turn are used to obtain the A coefficients. To do this in a systematic way, we rewrite (13.54) in the equivalent form

$$\sum_{n_s} \left[V_{n_j n_s} - \delta E_{n_i}^{(1)} \delta_{n_j n_s} \right] A_{n_i n_s} = 0. \tag{13.56}$$

For this homogeneous system of coupled linear equations to have a nontrivial solution, its determinant must be zero. Applying this condition, we obtain a *secular equation* of degree g (the notation has been simplified by writing $V_{n_i n_j} = V_{ij}$),

$$\begin{vmatrix} V_{11} - \delta E^{(1)} & V_{12} & \cdots & V_{1g} \\ V_{21} & V_{22} - \delta E^{(1)} & \cdots & V_{2g} \\ \vdots & \vdots & \ddots & \vdots \\ V_{g1} & V_{g2} & \cdots & V_{gg} - \delta E^{(1)} \end{vmatrix} = 0. \tag{13.57}$$

The solutions of this equation are the g values (equal or different) for $\delta E_{n_i}^{(1)}$, which are the first-order corrections to each of the g degenerate states corresponding to the energy $E_n^{(0)}$ (the order of the numbering is entirely arbitrary). By successively substituting each of these values of $\delta E_{n_i}^{(1)}$ into the system of equations (13.56), we obtain the corresponding coefficients of the A's. We then choose, say, $A_{n_i n_j}$, so that the zero-order functions are correctly normalized (the explicit condition is written later). In this way, the zero-order normalized wave functions and the corresponding first-order energy are fully determined. To continue the calculation to higher order in the perturbation, we use the system of equations (13.52), as already mentioned.

The solution of the secular equation is simplified in the frequent case where the perturbation preserves certain symmetries of the original Hamiltonian. In such a case there are operators, which we will call $\widehat{Q}$, that commute with both $\widehat{H}$ and $\widehat{H}_0$, and therefore with $\widehat{V} = \widehat{H} - \widehat{H}_0$. Consequently, the eigenfunctions of the perturbed Hamiltonian $\widehat{H}$ can be chosen as simultaneous eigenfunctions of $\widehat{Q}$, and therefore only wave functions belonging to the *same* eigenvalue q of $\widehat{Q}$ will be mixed. This means that in this basis the secular equation is factorized into as many independent factors as there are different values of q.

We conclude that an effective use of symmetry properties can help to reduce considerably the algebraic work involved in the perturbative solution of a degenerate problem. For example, suppose that a system with spherical symmetry is immersed in a uniform weak electric field oriented along the z-axis; since the symmetry around the z-axis is preserved, $\widehat{L}_z$ commutes with both Hamiltonians $\widehat{H}_0$ and $\widehat{H}$. In this case the perturbation will only mix wave functions corresponding to the same quantum number m, as will be illustrated in Section 13.4.

13.3.1 Meaning of the transformed basis; diagonalization of V

We will now study some properties of the correct wave functions φ_{n_i}, which are, by definition, the new zero-order wave functions. To do this we write

$$\Psi_{n_i} = \psi_{n_i} + \sum_k (a_{n_i k} + B_{n_i k})\psi_k = \psi_{n_i} + \sum_{n_s} a_{n_i n_s} \psi_{n_s} + \sum_k B_{n_i k}\psi_k$$

$$= \sum_{n_s} A_{n_i n_s} \psi_{n_s} + \sum_k B_{n_i k}\psi_k = \varphi_{n_i} + \chi_{n_i}, \tag{13.58}$$

where

$$\varphi_{n_i} = \sum_{n_s} A_{n_i n_s} \psi_{n_s} \tag{13.59}$$

are the correct (zero-order) wave functions and

$$\chi_{n_i} = \sum_k B_{n_i k}\psi_k \tag{13.60}$$

are the corrections to them. Considering ψ_{n_s} and φ_{n_i} as column vectors of g elements, we rewrite Eq. (13.59) in matrix form

$$\varphi = \widehat{A}\psi, \tag{13.61}$$

where $\widehat{A}$ is a square matrix of dimension $g \times g$ with elements $A_{n_i n_j}$. The condition that the φ are properly normalized gives $1 = \langle \varphi|\varphi \rangle = \langle \widehat{A}\psi|\widehat{A}\psi \rangle = \langle \psi|\widehat{A}^\dagger A|\psi \rangle$, therefore

$$\widehat{A}^\dagger \widehat{A} = 1, \tag{13.62}$$

that is, the matrix $\widehat{A}$ that transforms the initial basis ψ into the correct basis φ is unitary (see Problem P13.5). Explicitly, this means that

$$\sum_{n_s} A_{n_i n_s} A^*_{n_j n_s} = \delta_{ij} \quad (\widehat{A}\widehat{A}^\dagger = 1), \tag{13.63}$$

$$\sum_{n_s} A^*_{n_s n_i} A_{n_s n_j} = \delta_{ij} \quad (\widehat{A}^\dagger \widehat{A} = 1). \tag{13.64}$$

The matrix elements of $\widehat{V}$ in the correct basis are

$$\int dx\, \varphi^*_{n_j} \widehat{V} \varphi_{n_i} = \sum_{n_s, n_{s'}} \int dx\, A^*_{n_j n_{s'}} \psi^*_{n_{s'}} \widehat{V} A_{n_i n_s} \psi_{n_s}$$

$$= \sum_{n_s, n'_s} A^*_{n_j n_{s'}} A_{n_i n_s} V_{n_{s'} n_s}, \tag{13.65}$$

or, using Eqs. (13.56) and (13.63),

$$\int dx\, \varphi^*_{n_j} \widehat{V} \varphi_{n_i} = \delta E^{(1)}_{n_i} \sum_{n_{s'}} A^*_{n_j n_{s'}} A_{n_i n_{s'}} = \delta E^{(1)}_{n_i} \delta_{ij}. \tag{13.66}$$

We see that the transformed vector basis (13.61) is exactly the one that diagonalizes the perturbation Hamiltonian, as anticipated in the introductory discussion. This means that the g wave functions φ_{n_i}, which are eigenfunctions of the original Hamitonian H_0, are also eigenfunctions of $\widehat{H} = \widehat{H}_0 + \widehat{V}$, with eigenvalues $E_n^{(0)} + \delta E_{n_i}^{(1)}$,

$$\widehat{H}\varphi_{n_i} = (E_n^{(0)} + \delta E_{n_i}^{(1)})\varphi_{n_i}. \tag{13.67}$$

Mathematically, the problem we have solved is the diagonalization of the matrix $\widehat{V}$, that is, the construction of the eigenvectors φ_n of $\widehat{V}$. This is a very common problem in QM, and the procedure presented here is useful for dealing with it in general.

13.3.2 Two coupled harmonic oscillators

As an example of application of the theory just presented, we consider the problem of two harmonic oscillators oscillating on perpendicular axes and coupled by a bilinear potential. The problem has the double interest of providing a simple illustration of the perturbative method in the presence of degeneracy, and, at the same time, of accepting an exact solution, which allows us to compare results and get a better feeling of the merits and limitations of the perturbative method for the present more elaborate case.

13.3.2.1 Perturbative solution

We consider two oscillators of equal mass m and natural frequency ω, oscillating along the x and y axes, respectively, and coupled by a potential of the form

$$V = m\omega^2 \beta xy, \tag{13.68}$$

where β is a dimensionless coupling parameter that can be positive or negative. The complete Hamiltonian is

$$\widehat{H} = \frac{1}{2m}\left(\widehat{p}_x^2 + \widehat{p}_y^2\right) + \frac{1}{2}m\omega^2(x^2 + y^2) + m\omega^2 \beta xy. \tag{13.69}$$

For the unperturbed problem of the two independent oscillators we write the wave function as the product $\psi_{n_1}(x)\psi_{n_2}(y)$; the sum of their energies,

$$E_n^{(0)} = \hbar\omega(n_1 + n_2 + 1), \tag{13.70}$$

depends only on the total quantum number $n = n_1 + n_2$.

Since n_1 can take the values $0, 1, 2, \ldots, n$, while n_2 takes the values $n, n-1, n-2, \ldots, 0$, the degeneracy of the energy level E_n is of order $g = n+1$. The $n+1$ unperturbed degenerate states, denoted by $|n_i\rangle$, $i = 1, 2, 3, \ldots, n+1$, are

$$|n_i\rangle = |i-1\rangle|n-i+1\rangle. \tag{13.71}$$

The matrix element $V_{ij} = \langle n_i|V|n_j\rangle$ is therefore proportional to the product $\langle i-1|x|j-1\rangle\langle n-i+1|y|n-j+1\rangle$. Using Eqs. (9.58) and (9.59) we get

$$V_{i,i+1} = \tfrac{1}{2}\beta\hbar\omega\sqrt{i(n-i+1)}, \tag{13.72}$$

$$V_{i,i-1} = \tfrac{1}{2}\beta\hbar\omega\sqrt{(i-1)(n-i+2)}, \tag{13.73}$$

and all other elements V_{ij} are zero. With this preliminary material, we can apply perturbation theory.

To illustrate, we take the simplest case, $n = 1$. The degenerate states are $(n_1, n_2) = (0, 1)$ and $(1, 0)$, so i can take the values $i = 1, 2$. The corresponding matrix elements (13.72) and (13.73) are

$$V_{12} = V_{21} = \tfrac{1}{2}\beta\hbar\omega \equiv V_0, \tag{13.74}$$

and the secular equation (13.57) reduces to

$$\begin{vmatrix} -\delta E^{(1)} & V_0 \\ V_0 & -\delta E^{(1)} \end{vmatrix} = 0, \tag{13.75}$$

which gives $(\delta E^{(1)})^2 - V_0^2 = 0$ for the first-order corrections to the energy: thus,

$$\delta E^{(1)}_{1,2} = \pm V_0 = \pm\tfrac{1}{2}\beta\hbar\omega. \tag{13.76}$$

To obtain the correct wave functions we first take the case $i = 1$, with $\delta E_1^{(1)} = V_0$. Substituting into the system of equations (13.56) we get

$$-\delta E_1^{(1)} A_{11} + V_0 A_{12} = 0,$$
$$V_0 A_{11} - \delta E_1^{(1)} A_{12} = 0. \tag{13.77}$$

With the value of $\delta E_1^{(1)}$ selected and the wave functions normalized to 1, it follows that $A_{11} = A_{12} = \tfrac{1}{\sqrt{2}}$, whence, using (13.71), we get

$$|1\rangle = A_{11}|1_1\rangle + A_{12}|1_2\rangle = \tfrac{1}{\sqrt{2}}[|0\rangle|1\rangle + |1\rangle|0\rangle], \quad \delta E_1^{(1)} = V_0. \tag{13.78}$$

Analogously for the case $i = 2$, with $\delta E_2^{(1)} = -V_0$, we obtain $A_{21} = -A_{22} = \tfrac{1}{\sqrt{2}}$, so that

$$|2\rangle = A_{21}|1_1\rangle + A_{22}|1_2\rangle = \tfrac{1}{\sqrt{2}}[|0\rangle|1\rangle - |1\rangle|0\rangle], \quad \delta E_2^{(1)} = -V_0. \tag{13.79}$$

The interaction between the oscillators has broken the energy degeneracy expressed in (13.70), resulting in two energy levels separated by $2V_0$.

The next more complex case is $n = 2$, which is triply degenerate, with $i = 1, 2, 3$ corresponding to the unperturbed states $(n_1, n_2) = (0, 2), (1, 1), (2, 0)$, respectively. With the nonzero matrix elements given by

$$V_{12} = V_{21} = V_{23} = V_{32} = \tfrac{1}{\sqrt{2}}\beta\hbar\omega = \sqrt{2}V_0, \tag{13.80}$$

the secular equation is

$$\begin{vmatrix} -\delta E^{(1)} & \sqrt{2}V_0 & 0 \\ \sqrt{2}V_0 & -\delta E^{(1)} & \sqrt{2}V_0 \\ 0 & \sqrt{2}V_0 & -\delta E^{(1)} \end{vmatrix} = 0, \tag{13.81}$$

or explicitly, $\delta E^{(1)}[(\delta E^{(1)})^2 - 4V_0^2] = 0$. From the solutions of this equation it follows that the first-order energy corrections are

$$i = 1, \quad \delta E_1^{(1)} = -2V_0 = -\beta\hbar\omega, \tag{13.82}$$

$$i = 2, \quad \delta E_2^{(1)} = 0, \tag{13.83}$$

$$i = 3, \quad \delta E_3^{(1)} = 2V_0 = \beta\hbar\omega. \tag{13.84}$$

Substituting successively each of these values of $\delta E^{(1)}$ into (13.56) and solving for the coefficients A_{ij} we obtain, after normalizing each state vector to unity, that

$$|1\rangle = \tfrac{1}{2}\Big(|0\rangle|2\rangle - \sqrt{2}|1\rangle|1\rangle + |2\rangle|0\rangle\Big), \quad \delta E_1^{(1)} = -2V_0, \tag{13.85}$$

$$|2\rangle = \tfrac{1}{\sqrt{2}}\Big(|0\rangle|2\rangle - |2\rangle|0\rangle\Big), \quad \delta E_2^{(1)} = 0, \tag{13.86}$$

$$|3\rangle = \tfrac{1}{2}\Big(|0\rangle|2\rangle + \sqrt{2}|1\rangle|1\rangle + |2\rangle|0\rangle\Big), \quad \delta E_3^{(1)} = 2V_0. \tag{13.87}$$

In this case the original energy level has been split into three.

Note that in both cases, the correct wave functions – in addition to being orthonormal – turn out to be either symmetric or antisymmetric with respect to the exchange of oscillators 1 and 2, i.e., they have well-defined parity. This is an important feature, which we might have expected, since the two oscillators are identical and therefore their exchange should have no physical consequences.

Exercise E13.2. Exact solution of the two coupled harmonic oscillators

Solve the problem of two harmonic oscillators of equal mass and frequency, linearly coupled by the potential $V = m\omega^2 \beta xy$.

Solution. As shown in Section 9.5, the transformation from x, y to the coordinates of the (independent or decoupled) normal modes of oscillation is the usual way to proceed in the case of two similar classical, coupled harmonic oscillators. It is easy to see that if we take

$$x = \tfrac{1}{\sqrt{2}}(x_1 + x_2), \quad y = \tfrac{1}{\sqrt{2}}(x_1 - x_2), \tag{13.88}$$

the Hamiltonian (13.69) reduces to

$$\widehat{H} = \frac{\widehat{p}_1^2}{2m} + \frac{\widehat{p}_2^2}{2m} + \tfrac{1}{2}m\omega^2[(1+\beta)x_1^2 + (1-\beta)x_2^2]. \tag{13.89}$$

This is the Hamiltonian of two quasi-particles of mass m, oscillating independently with frequencies

$$\omega_1 = \omega\sqrt{1+\beta}, \quad \omega_2 = \omega\sqrt{1-\beta}. \tag{13.90}$$

Since we are now dealing with two independent oscillators, the solutions of the Schrödinger equation are of the form

$$\Psi_{ni}(x_1, x_2) = \psi_{i-1}(x_1; \omega_1)\psi_{n-i+1}(x_2; \omega_2), \tag{13.91}$$

and the energy is the sum of the respective energies,

$$\begin{aligned}
E_{ni} &= \hbar\left[\omega_1(i - 1 + \tfrac{1}{2}) + \omega_2(n - i + 1 + \tfrac{1}{2})\right] \\
&= \hbar\omega\left[(i - \tfrac{1}{2})\sqrt{1+\beta} + \left(n - i + \tfrac{3}{2}\right)\sqrt{1-\beta}\right] \\
&= E_n^{(0)}\sqrt{1-\beta} + \left(i - \tfrac{1}{2}\right)\left(\sqrt{1+\beta} - \sqrt{1-\beta}\right)\hbar\omega.
\end{aligned} \tag{13.92}$$

We have thus obtained the exact solution of the problem. To compare with the perturbative solution, we expand the above expression in a Taylor series around $\beta = 0$; for small values of β we obtain

$$E_n = \hbar\omega \left[n + 1 - \tfrac{1}{2}\beta(n + 2 - 2i) - \tfrac{1}{8}\beta^2(n + 1) + \ldots \right] \quad (13.93)$$

or rearranging,

$$E_{ni} = E_n^{(0)} \left(1 - \tfrac{1}{8}\beta^2 + \ldots \right) - \tfrac{1}{2}\beta\hbar\omega(n + 2 - 2i) + \ldots \quad (i = 1, 2, \ldots, n+1). \quad (13.94)$$

We see that the first-order corrections (linear in β) go from $-\tfrac{1}{2}\beta\hbar\omega n$ (for $i = 1$) to $\tfrac{1}{2}\beta\hbar\omega n$ (for $i = n + 1$), in equal steps of $\beta\hbar\omega$; they completely break the initial degeneracy. On the other hand, the second-order corrections are independent of i, that is, they are the same for all degenerate states. For $n = 1$ ($i = 1, 2$), (13.94) gives

$$\delta E_{1,2}^{(1)} = \pm \tfrac{1}{2}\beta\hbar\omega = \pm V_0, \quad (13.95)$$

while for $n = 2$ ($i = 1, 2, 3$),

$$\delta E_{1,2,3}^{(1)} = -2V_0, 0, 2V_0. \quad (13.96)$$

These results coincide with those provided by perturbation theory, Eqs. (13.76) and (13.82–13.84).

From this analysis it is clear that perturbation theory can only give a satisfactory answer for values of β small enough that the Taylor expansion of the radicals $\sqrt{1-\beta}$ and $\sqrt{1+\beta}$ to first or second order can be considered a good approximation. This limitation is due to the fact that only for $\beta \ll 1$ can the perturbation energy be considered small compared to the other energy terms of the Hamiltonian (in terms of their expectation values). However, it often happens that even when this is not entirely true, the information provided by perturbation theory is qualitatively acceptable, even up to higher orders. We will have a specific example of this situation later when we study the He atom, for which perturbation theory gives good qualitative information, despite the fact that the interaction energy between the electrons is comparable to that of each electron with the atomic nucleus. The basic criterion is that as long as the perturbation does not cause the energy levels to cross or come too close together (i.e., as long as they remain sufficiently differentiated), perturbation theory can give reliable results, at least qualitatively.

13.4 The Stark Effect

The preceding perturbation theory was developed by Schrödinger (and published in the third of his series of four foundational papers on QM) based on the classical perturbative techniques of Lord Rayleigh. For this reason it is known as Rayleigh–Schrödinger perturbation theory. Schrödinger applied it to the study of the Stark effect, making it the first quantum problem to be successfully solved using perturbative methods with his theory.[3]

In 1913, after a long search, J. Stark[4] observed that the lines of the H atom's Balmer series split into several components in the presence of a uniform electric field. This phenomenon, characteristic of all atoms and known since then as the *Stark effect*, could not be properly explained by Bohr's theory, although it was possible to understand some aspects of it. The explanation given by QM is simple and direct, and the results agree in detail with the observations, so that this phenomenon provided important support for the further development of Schrödinger's theory.

The perturbing potential is written as

$$V = -e\boldsymbol{E} \cdot \boldsymbol{r}, \tag{13.97}$$

where $\boldsymbol{E}$ is the electric field, which for simplicity is taken to be uniform and constant. The energy levels of the atom up to second order in the perturbation are, according to Eqs. (13.18), (13.19), and (13.22),

$$E_n = E_n^{(0)} - e\boldsymbol{E} \cdot \boldsymbol{r}_{nn} + e^2 \sum_{k \neq n} \frac{(\boldsymbol{E} \cdot \boldsymbol{r}_{nk})(\boldsymbol{E} \cdot \boldsymbol{r}_{kn})}{E_n^{(0)} - E_k^{(0)}}. \tag{13.98}$$

If n refers to a *nondegenerate* level, the unperturbed wave function ψ_n has well-defined parity because the Hamiltonian of a central problem is invariant to an inversion of the coordinates with respect to the origin, that is, $\widehat{H}_{0P} \equiv \widehat{P}\widehat{H}_0\widehat{P}^{-1} = \widehat{H}_0$ where $\widehat{P}$ is the parity operator (see Section 8.7.3). Hence it follows that

$$\boldsymbol{r}_{nn} = \langle n|\boldsymbol{r}|n\rangle = \langle n|(-\widehat{P}^\dagger \boldsymbol{r}\widehat{P})|n\rangle = -\langle n\widehat{P}|\boldsymbol{r}|\widehat{P}n\rangle = -\langle n|\boldsymbol{r}|n\rangle, \tag{13.99}$$

which implies that $\boldsymbol{r}_{nn} = 0$. Therefore, nondegenerate atomic (or molecular) states, including the ground state, do not show a linear Stark effect (proportional to the electric field) but only a quadratic Stark effect.

The H atom is a special case. As we have seen, for the Coulomb potential (and only for it) the excited energy levels are degenerate with respect to the angular-momentum quantum number l ($l = 0, 1, \ldots, n-1$). Since the n degenerate states have different parity given by $(-1)^l$, the matrix element of $\boldsymbol{r}$ need not vanish. The conclusion is that the excited H atom has a permanent electric dipole and therefore exhibits a first-order (linear) Stark effect.

To calculate the effect of the perturbation on the dipole moment of the atom, we write the perturbed dipole as

$$\boldsymbol{d} = e\int \Psi_n^* \boldsymbol{r} \Psi_n d^3r \tag{13.100}$$

and introduce the expression (13.3) for the perturbed wave function

$$\Psi_n = (1 + C_{nn})\psi_n + \sum_k{}' C_{nk}\psi_k, \tag{13.101}$$

---

[3] A little earlier, Heisenberg had solved the problem of a nonlinear HO as a first step in matrix mechanics.
[4] The German physicist Johannes Stark (1874–1957) was awarded the Nobel Prize in 1919 for his work on the splitting of atomic spectral lines in the presence of an electric field. The Stark effect is used today in applications as diverse as imaging the firing activity of neurons and measuring the pressure broadening of spectral lines in plasmas.

thus obtaining

$$d = e(1+C_{nn})^2 r_{nn} + e(1+C_{nn})\sum_{k\neq n}[C_{nk}r_{nk} + C^*_{nk}r_{kn}] + e\sum_{k\neq n}\sum_{k'\neq n} C^*_{nk}C_{nk'}r_{kk'}. \tag{13.102}$$

Using the perturbative expansions (13.14) and (13.16) and formula (13.22) for the first-order coefficients, we obtain

$$d = er_{nn} - e^2 \sum_{k\neq n} \frac{(r_{kn}\cdot E)r_{nk} + (r_{nk}\cdot E)r_{kn}}{E_n^{(0)} - E_k^{(0)}} + \ldots \tag{13.103}$$

The first term, which does not depend on the field, represents the permanent (or intrinsic) dipole moment,

$$d_0 = \langle n|er|n\rangle. \tag{13.104}$$

The remaining terms in (13.103) represent an induced dipole moment that is created by the presence of the field and disappears with it. This suggests introducing the polarizability tensor $\overleftrightarrow{\alpha}$ with components

$$\alpha_{ij} = \frac{\partial d_i}{\partial E_j}, \tag{13.105}$$

which measures the capacity of the system to acquire a dipole moment (i.e., to polarize) in different directions as a result of the action of the external field oriented in a given direction. From Eq. (13.103) we see that to lowest order, the polarizability is

$$\alpha_{ij}^{(0)} \equiv \alpha_{ij}(E=0) = e^2 \sum_{k\neq n} \frac{r_{i\,kn}r_{j\,nk} + r_{i\,nk}r_{j\,kn}}{E_k^{(0)} - E_n^{(0)}}. \tag{13.106}$$

The total dipole moment of the system is then

$$d = d_0 + \int_0^{\mathcal{E}} \overleftrightarrow{\alpha} \cdot dE = d_0 + \overleftrightarrow{\alpha}^{(0)} \cdot E + \ldots \tag{13.107}$$

and the interaction energy is

$$V = -\int_0^{\mathcal{E}} d \cdot dE = -d_0 \cdot E - \tfrac{1}{2}E \cdot \overleftrightarrow{\alpha}^{(0)} \cdot E + \ldots \tag{13.108}$$

These last two expressions agree with the first-order perturbative expansions (13.103) and (13.98), respectively, and help to better understand the physical meaning of the different terms.

13.5 The Dielectric Constant

In a dielectric medium the electrons do not flow through the material because they are bound to their atoms. Therefore, in response to an applied electric field, they can only produce a local polarization, quantified as the density of (induced) electric dipoles. An average macroscopic measure of the polarizability of a medium is the dielectric constant ϵ, defined by the equation

$$P = \frac{\epsilon - 1}{4\pi} E, \tag{13.109}$$

where P is the polarization density induced by the applied electric field E. Therefore, if we know the polarization produced by the field, using Eq. (13.109), we can determine its dielectric constant (or permittivity).

The problem is interesting for a number of reasons. From a physical point of view, it is a question of determining from the microscopic structure of the medium how the polarization, and hence the dielectric constant, varies with frequency, that is, of solving one of the most important problems in the theory of light scattering by a dielectric medium. From a methodological point of view, it is a problem in which it is more important to determine the modification of the wave function due to the perturbation than the modification of the atomic levels. Finally, from the point of view of perturbation theory, this is a case where the perturbation potential is time-dependent, so it is necessary to first eliminate the time dependence in order to apply the theory developed so far. (Chapter 17 develops the theory of time-dependent perturbations more generally.)

Exercise E13.3. Dielectric constant of a transparent material

Use first-order perturbation theory to determine the dielectric constant of a material that has no permanent electric dipole and is transparent to incident radiation.

Solution. We restrict the problem to one dimension, and consider incident radiation of sufficiently long wavelength so that we can assume that its electric vector does not vary appreciably in the dimensions of an atom; in this case we can write the incident electric field in the form $E = \mathcal{E} \cos \omega t$, where $\omega = 2\pi c/\lambda$, and the potential perturbing the atomic electrons becomes

$$V = e_0 \mathcal{E} x \cos \omega t \tag{13.110}$$

(we neglect the effects of the magnetic field). To remove the time dependence, we consider the atomic electrons initially in an unperturbed state, $\psi_n = \varphi_n e^{-i\omega_n t}$ and write the perturbed state as $\psi = \psi_n + \delta\psi$. Substituting these expressions into the time-dependent Schrödinger equation and expanding, we obtain to first order

$$\left(i\hbar \frac{\partial}{\partial t} - \widehat{H}_0\right) \delta\psi = \tfrac{1}{2} e_0 \mathcal{E} x \varphi_n (e^{-i(\omega_n - \omega)t} + e^{-i(\omega_n + \omega)t}). \tag{13.111}$$

This result suggests that we write the induced change in the wave function in the form

$$\delta\psi = u_- e^{-i(\omega_n - \omega)t} + u_+ e^{-i(\omega_n + \omega)t}. \tag{13.112}$$

By substituting, we can separate the preceding equation into two time-independent equations,

$$[\hbar(\omega_n \pm \omega) - \widehat{H}_0] u_\pm = \tfrac{1}{2} e_0 \mathcal{E} x \varphi_n. \tag{13.113}$$

First we write $u_- = \sum_k a_k \varphi_k$ and substitute, taking into account that $\widehat{H}_0 \varphi_k = E_k \varphi_k$; after multiplying the result by φ_l^* and integrating over all space we obtain for the coefficient a_k

$$a_k = -\frac{e_0 \mathcal{E}}{2\hbar} \frac{\langle k|x|n\rangle}{\omega_{kn} + \omega}, \quad \omega_{kn} \equiv \omega_k - \omega_n = \frac{E_k - E_n}{\hbar}. \tag{13.114}$$

The second solution is obtained by replacing ω by $-\omega$. Collecting the results and simplifying, we get

$$\psi = e^{-i\omega_n t}\left[\varphi_n - \frac{e_0 \mathcal{E}}{\hbar}\sum_k \frac{\langle k|x|n\rangle}{\omega_{kn}^2 - \omega^2}\varphi_k(\omega_{kn}\cos\omega t - i\omega\sin\omega t)\right]. \quad (13.115)$$

Note that the perturbation contains the coefficients x_{kn} for $k \neq n$, which represent the dipolar response of the atomic electrons to the field modes of frequency ω_{kn}; these are the modes that induce a polarization of the atoms and determine the refraction index of the medium. The average induced electric dipole moment of the perturbed atoms is $\langle d \rangle = -e_0 \int \psi^* x \psi d^3 x$. If N is the number of atoms of the dielectric per unit volume, the polarization vector is given by

$$P = N\langle d \rangle = \frac{2Ne_0^2 \mathcal{E}}{\hbar}\sum_k \frac{\omega_{kn}|\langle n|x|k\rangle|^2}{\omega_{kn}^2 - \omega^2}\cos\omega t. \quad (13.116)$$

Using Eq. (13.109), we obtain for the dielectric constant as a function of the frequency, in the dipole approximation,

$$\epsilon = 1 + \frac{8\pi Ne_0^2}{\hbar}\sum_k \frac{\omega_{kn}}{\omega_{kn}^2 - \omega^2}|\langle n|x|k\rangle|^2. \quad (13.117)$$

Note that according to this result, at resonance ($\omega = \omega_{kn}$) the value of the dielectric constant is infinite, which is obviously incorrect. By taking into account the dissipative effect due to the radiation reaction of the electrons (proportional to $\dddot{x}$) – as done, for example, in Section 10.2 – the denominator of Eq. (13.117) acquires a complex contribution $i\gamma(\omega)$, which leads to the expression

$$\sum_k \frac{f_{kn}}{\omega_{kn}^2 - \omega^2 + i\gamma} = \sum_k f_{kn}\frac{\omega_{kn}^2 - \omega^2 - i\gamma(\omega)}{\left(\omega_{kn}^2 - \omega^2\right)^2 + \gamma^2(\omega)}. \quad (13.118)$$

The real part of this expression now becomes finite for $\omega^2 = \omega_{kn}^2$ and small $\gamma(\omega)$ (instead of being infinite); the imaginary term, proportional to $\gamma(\omega)$, describes the losses in the dielectric.

Second, we note that it is customary to write (13.117) in the form

$$\epsilon = 1 + \frac{4\pi Ne_0^2}{m}\sum_k \frac{f_{kn}}{\omega_{kn}^2 - \omega^2}, \quad (13.119)$$

where f_{kn} are the so-called oscillator intensities, given by

$$f_{kn} = \frac{2m}{\hbar}\omega_{kn}|\langle n|x|k\rangle|^2. \quad (13.120)$$

In classical theory the oscillator intensities are integers, corresponding to the number of electrons per atom in the eigenfrequency state $\omega_\alpha = \omega_{kn}$ but this is not the case in quantum theory or experiment. In fact, the observation that the oscillator intensities were not integer led to the first doubts about the validity of the classical description.

13.6 *The Canonical Transformation Method

In the previous exposition we developed a specific procedure to carry out the perturbative analysis, which corresponds to the Rayleigh–Schrödinger method. An alternative way of doing a perturbative expansion is to go from the unperturbed to the perturbed problem by means of a unitary transformation. Although here we will illustrate this method as applied to perturbation theory, its use in solving complex problems is much broader, including cases where it is not possible to perform perturbative expansions. A classic example of the latter situation is Bogolyubov's[5] successful use of the method of canonical transformations to solve the superconductivity problem (for an illustration, see Problem P13.13).

We assume that there is a unitary operator $\widehat{S}$ that transforms the unperturbed basis $|n\rangle \equiv |\psi_n\rangle$ into the perturbed one $|\Psi_n\rangle$,

$$|\Psi\rangle = \widehat{S}|\psi\rangle. \tag{13.121}$$

From the Schrödinger equation it follows immediately that $\langle m|\widehat{H}_0 + \lambda\widehat{V}|\Psi_n\rangle = E_n\langle m|\Psi_n\rangle$, that is,

$$\lambda\langle m|\widehat{V}|\Psi_n\rangle = \left(E_n - E_m^{(0)}\right)\langle m|\Psi_n\rangle. \tag{13.122}$$

For $|m\rangle = |n\rangle$ it follows that

$$E_n - E_n^{(0)} = \lambda\frac{\langle n|\widehat{V}|\Psi_n\rangle}{\langle n|\Psi_n\rangle} = \lambda\frac{\langle n|\widehat{V}\widehat{S}|n\rangle}{\langle n|\widehat{S}|n\rangle} = \lambda\sum_k \frac{\langle n|\widehat{V}|k\rangle\langle k|\widehat{S}|n\rangle}{\langle n|\widehat{S}|n\rangle}, \tag{13.123}$$

from where

$$E_n = E_n^{(0)} + \lambda\langle n|\widehat{V}|n\rangle + \lambda\sum_k{}' \langle n|\widehat{V}|k\rangle\frac{\langle k|\widehat{S}|n\rangle}{\langle n|\widehat{S}|n\rangle}. \tag{13.124}$$

For $|m\rangle \neq |n\rangle$, we see from (13.122) that

$$\lambda\sum_{k\neq m}\langle m|\widehat{V}|k\rangle\langle k|\widehat{S}|n\rangle = \left(E_n - E_m^{(0)} - \lambda\langle m|\widehat{V}|m\rangle\right)\langle m|\widehat{S}|n\rangle. \tag{13.125}$$

By iteratively using (13.124) and (13.125) it is possible to determine the corrections to the energy and the matrix elements of $\widehat{S}$. For example, to obtain E_n to first order in λ it is sufficient to write $\widehat{S} = 1$ in (13.124), which gives

$$E_n = E_n^{(0)} + \lambda\langle n|\widehat{V}|n\rangle. \tag{13.126}$$

To go further we write $\widehat{S} = 1 + \lambda\widehat{S}_1 + \ldots$ in (13.125) and get

$$\lambda\langle m|\widehat{V}|n\rangle\langle n|1 + \lambda\widehat{S}_1|n\rangle = (E_n - E_m)\langle m|\lambda\widehat{S}_1|n\rangle, \tag{13.127}$$

that is,

$$\langle m|\widehat{S}_1|n\rangle = \lambda\frac{\langle m|\widehat{V}|n\rangle}{E_n - E_m}, \tag{13.128}$$

---

[5] Nicolai Nikolayevich Bogolyubov (1909–1992) was an outstanding Russian mathematician and theoretical physicist who made many first-rate contributions to statistical mechanics, quantum field theory, and the theory of dynamical systems.

which gives a second-order correction to the energy, and so on. Note that in the denominator we write $E_n - E_m^{(0)} - \lambda\langle m|\widehat{V}|m\rangle$ as $E_n - E_m$, using (13.126). To first order in λ it would have been sufficient to write $E_n^{(0)} - E_m^{(0)}$, as in the case of the Rayleigh–Schrödinger method, to obtain an expansion in powers of λ.

13.7 *The Feynman and Hellman Method

An alternative method can be obtained from a generalization of the *Feynman–Hellman formula*, which applies to a perturbative Hamiltonian of the form $\widehat{H} = \widehat{H}_0 + \lambda\widehat{V}$. This formula states that if $\widehat{F}(\lambda)$ is a Hermitian operator depending on a parameter λ, and $|\psi(\lambda)\rangle$, $f(\lambda)$ are its eigenvectors and eigenvalues, then

$$\langle\psi(\lambda)|\frac{\partial \widehat{F}}{\partial \lambda}|\psi(\lambda)\rangle = \frac{\partial f}{\partial \lambda}. \tag{13.129}$$

We first generalize this expression, to later apply it to the perturbative problem. Let $|\psi_n(\lambda)\rangle, |\psi_m(\lambda)\rangle$ be two eigenvectors of $\widehat{F}(\lambda)$, with eigenvalues f_n, f_m, respectively; the $|\psi\rangle$ are normalized to unity, so the product

$$\langle\psi_n(\lambda)|\psi_m(\lambda)\rangle = \delta_{nm} \tag{13.130}$$

does not depend on λ. Multiplying the eigenvalue equation

$$\widehat{F}(\lambda)|\psi_n(\lambda)\rangle = f_n(\lambda)|\psi_n(\lambda)\rangle \tag{13.131}$$

by $\langle\psi_m(\lambda)|$ and deriving with respect to λ, one gets

$$\frac{\partial\langle\psi_m|}{\partial\lambda}\widehat{F}|\psi_n\rangle + \langle\psi_m|\widehat{F}\frac{\partial|\psi_n\rangle}{\partial\lambda} + \langle\psi_m|\frac{\partial \widehat{F}}{\partial\lambda}|\psi_n\rangle = \frac{\partial f_n}{\partial\lambda}\delta_{nm}. \tag{13.132}$$

The sum of the first two terms can be written as follows,

$$\frac{\partial\langle\psi_m|}{\partial\lambda}\widehat{F}|\psi_n\rangle + \langle\psi_m|\widehat{F}\frac{\partial|\psi_n\rangle}{\partial\lambda} = f_n\frac{\partial\langle\psi_m|}{\partial\lambda}|\psi_n\rangle + f_m\langle\psi_m|\frac{\partial|\psi_n\rangle}{\partial\lambda}$$
$$= \tfrac{1}{2}(f_n + f_m)\frac{\partial}{\partial\lambda}\langle\psi_m|\psi_n\rangle + \tfrac{1}{2}(f_n - f_m)\left[\frac{\partial\langle\psi_m|}{\partial\lambda}|\psi_n\rangle - \langle\psi_m|\frac{\partial|\psi_n\rangle}{\partial\lambda}\right]. \tag{13.133}$$

Since the first term vanishes due to (13.130), we obtain the desired generalization after simplifying,

$$\langle\psi_m|\frac{\partial \widehat{F}}{\partial\lambda}|\psi_n\rangle + \tfrac{1}{2}(f_n - f_m)\left(\frac{\partial\langle\psi_m|}{\partial\lambda}|\psi_n\rangle - \langle\psi_m|\frac{\partial|\psi_n\rangle}{\partial\lambda}\right) = \frac{\partial f_n}{\partial\lambda}\delta_{nm}. \tag{13.134}$$

With $m = n$ we recover the Feynman–Hellman formula.

To apply this result to perturbation theory we take $\widehat{F}(\lambda) = \widehat{H}_0 + \lambda\widehat{V}$, so that $\partial\widehat{F}/\partial\lambda = \widehat{V}$; (13.134) gives then

$$\langle\psi_m|\widehat{V}|\psi_n\rangle + \tfrac{1}{2}(E_n - E_m)\left(\frac{\partial\langle\psi_m|}{\partial\lambda}|\psi_n\rangle - \langle\psi_m|\frac{\partial|\psi_n\rangle}{\partial\lambda}\right) = \frac{\partial E_n}{\partial\lambda}\delta_{nm}. \tag{13.135}$$

In particular, for $n = m$ we get

$$\frac{\partial E_n}{\partial\lambda} = \langle\psi_n|\widehat{V}|\psi_n\rangle, \tag{13.136}$$

while for $m \neq n$,

$$2\langle\psi_m|\widehat{V}|\psi_n\rangle + (E_n - E_m)\left(\frac{\partial\langle\psi_m|}{\partial\lambda}|\psi_n\rangle - \langle\psi_m|\frac{\partial|\psi_n\rangle}{\partial\lambda}\right) = 0. \qquad (13.137)$$

These exact expressions are widely used in quantum chemistry, in particular for the calculation of intermolecular forces. The perturbative expansions can be obtained from them in a systematic way, following a procedure similar to that used earlier (compare Eqs. (13.124) and (13.126)).

13.8 The Variational Method: Minimum Energy of the Quantum States

In 1870, while trying to solve the vibration problem of organ pipes, Lord Rayleigh observed that it was possible to determine the frequencies of the normal modes of oscillation of a system by means of a variational principle. In particular, he showed that the fundamental frequency of oscillation of a classical system is a minimum, in the sense that small variations in the amplitudes of the oscillations give rise to second-order changes in frequency. This Rayleigh principle, which is widely used in technical applications, was generalized by W. Ritz[6] in 1900, when he proposed a variational method for the approximate solution of an eigenvalue equation. To recognize the contributions of both scientists, the theory was later named the *Rayleigh–Ritz method*.

The Rayleigh–Ritz (or Ritz) method has been successfully applied in QM to a wide variety of systems, including atoms, molecules, and the scattering of particles and radiation. It is particularly useful in cases where perturbation theory fails because the full Hamiltonian is not close to one with known eigenvalues and eigenstates, such as in the case of polyatomic molecules.

To illustrate the possibilities of this method, we will first use it to determine an upper bound for the energy of the ground state of a quantum system, and then apply it to more complicated cases.

Let $\widehat{H}$ be the Hamiltonian of the system of interest and ψ_k the eigenfunction corresponding to the eigenvalue E_k (with ψ_k and E_k unknown),

$$\widehat{H}\psi_k = E_k\psi_k. \qquad (13.138)$$

The index k is chosen such that the energy increases with it, $E_k > E_{k'}$ for $k > k'$; the ground state is denoted by $k = 0$ and its energy is E_0. Let Φ be an arbitrary function normalized to unity, expressed in terms of the basis ψ_k,

$$\Phi = \sum_k a_k \psi_k, \qquad \sum_k |a_k|^2 = 1. \qquad (13.139)$$

Using Eq. (13.139), we obtain for the expectation value of H

$$\langle\Phi|\widehat{H}|\Phi\rangle = \sum_{m,n} a_m^* a_n \int \psi_m^* \widehat{H} \psi_n dx = \sum_{m,n} a_m^* a_n E_n \delta_{mn}$$

---

[6] Walther Ritz (1878–1909) was a Swiss painter, engineer, and theoretical physicist, known for several methods that bear his name. He died of tuberculosis at the age of 31.

$$= \sum_n |a_n|^2 E_n \geq E_0 \sum_n |a_n|^2, \qquad (13.140)$$

or, since $E_n \geq E_0$,

$$\langle \Phi | \widehat{H} | \Phi \rangle \geq E_0. \qquad (13.141)$$

This shows that the expectation value of $\widehat{H}$ calculated with an arbitrary function Φ (normalized to unity) is greater than, or at best equal to, the ground state energy. In other words, *the eigenfunction corresponding to the ground state is the (normalized) function that minimizes the expectation value.* We can use this result to obtain an upper bound for E_0; it suffices to make Φ depend on one or more parameters and minimize $\langle \Phi | \widehat{H} | \Phi \rangle$ with respect to each one of them. This procedure will give us the optimal value of the parameters and the best possible estimate of the eigenvalue sought, for a wave function of the proposed type.

The quality of the result depends to a large extent on the quality of the test function Φ, so its choice must be made judiciously. In general, one starts with functions that are mathematically easy to handle (analytically or numerically, as the case may be), and that have suitable properties for the given problem. In particular, for the ground state the function should have no nodes (with possible exceptions such as the one noted in Section 15.3, possess the correct parity, decrease exponentially at infinity in the case of bound states, and reflect the symmetries of the problem. If chosen properly, one or two parameters are usually sufficient to obtain reasonable approximations in simple cases.

The method can be extended to the excited states, taking in each case the test function orthogonal to those of the lower states. Thus, if the solutions up to state $n - 1$ are known, we write for state n

$$\Phi_n = \sum_{k \geq n} a_k \psi_k. \qquad (13.142)$$

With this selection, $\langle \psi_l | \Phi_n \rangle = 0$ for $l = 0, 1, \ldots n - 1$, and we readily obtain

$$\langle \Phi_n | \widehat{H} | \Phi_n \rangle = \sum_{k \geq n} |a_k|^2 E_k \geq E_n. \qquad (13.143)$$

If only approximate wave functions are known for $k \leq n - 1$, the accumulated errors mean that this method can only be used successfully for the first excited states. The treatment of highly excited states can then be done by resorting to the semiclassical (WKB) approximation.

The Ritz variational method can be extended to find approximate solutions to any eigenvalue equation, allowing an infinite number of degrees of freedom of a system to be reduced to a finite number, thus making analysis easier and more practical (see Exercise 13.44).

13.8.1 The electron distribution minimizes the mean energy

Equation $\widehat{H} \psi_k = E_k \psi_k$ implies that the spatial distribution of particles is just that which minimizes the mean energy of the system. This is an extremely interesting property of Schrödinger's theory, which holds in general because it is an eigenvalue problem. To see this, we write the expectation value of the energy

$$E = \langle \psi | \widehat{H} | \psi \rangle, \quad \text{with} \quad \langle \psi | \psi \rangle = 1, \qquad (13.144)$$

and introduce the modification $|\psi\rangle \to |\psi\rangle + |\delta\psi\rangle$, $\langle\psi| \to \langle\psi| + \langle\delta\psi|$, where $|\delta\psi\rangle$ and $\langle\delta\psi|$ are arbitrary, but such that the modified function is also normalized; to first order, the change in E is

$$\delta E = \langle\psi|\widehat{H}|\delta\psi\rangle + \langle\delta\psi|\widehat{H}|\psi\rangle, \quad \langle\delta\psi|\psi\rangle + \langle\psi|\delta\psi\rangle = 0. \tag{13.145}$$

The condition of stationarity of E means that $\delta E = 0$. Introducing the normalization condition with the method of undetermined Lagrange multipliers, we obtain

$$\langle\psi|(\widehat{H} - \lambda)|\delta\psi\rangle + \langle\delta\psi|(\widehat{H} - \lambda)|\psi\rangle = 0. \tag{13.146}$$

This condition must be satisfied for any pair $|\delta\psi\rangle$, $\langle\delta\psi|$ and it is met whenever both the Schrödinger equation and its adjoint are satisfied,

$$\widehat{H}|\psi\rangle = \lambda|\psi\rangle, \quad \langle\psi|\widehat{H} = \lambda\langle\psi|, \quad \lambda = \langle\widehat{H}\rangle = E. \tag{13.147}$$

This can be easily seen by choosing, in particular, $|\delta\psi\rangle = \varepsilon(\widehat{H} - \lambda)|\psi\rangle$ with ε infinitesimal and real, which gives $2\varepsilon\langle\psi|(\widehat{H} - \lambda)(\widehat{H} - \lambda)|\psi\rangle = 0$. So the fact that $|\psi\rangle$ is a solution of the Schrödinger eigenvalue equation guarantees that the mean energy at each discrete level has an extreme value (we have already seen that it is actually a minimum) against small arbitrary changes of the corresponding stationary distribution. Alternatively, we can say that the preceding is equivalent to a variational derivation of the Schrödinger equation.

There are two interesting observations here. The first is that nothing in the preceding derivation attaches a specific interpretation to the operator $\widehat{H}$. This means that *all quantum integrals of motion* (which are solutions of an eigenvalue equation) *are extremal*. The second is that the previous result explains why in first-order perturbation theory, the energy correction is simply $\delta E^{(1)} = \langle V \rangle$, where V is the perturbation Hamiltonian. The reason is that, being $E^{(0)}$ extremal with respect to first-order changes in $|\psi\rangle$, the modification of the wave function generated by the perturbation cannot affect $\langle\widehat{H}\rangle$ to first order.

13.8.2 Variational method applied to the H atom

As an illustrative exercise, we will apply the variational method to the H atom. The ground state of the H atom is an s-state ($l = 0$), so its wave function does not depend on either θ or ϕ, that is, $\psi_0 = \psi_0(r)$. Since it must have no nodes and should decay exponentially at infinity, we propose as a test function (with $\alpha \geq 0$)

$$\psi_0 = Ae^{-\alpha r}, \quad A = \left(\frac{\alpha^3}{\pi}\right)^{1/2}, \tag{13.148}$$

where the value of A was determined by normalizing ψ_0. This gives

$$\langle\widehat{H}\rangle = 4\pi\frac{\alpha^3}{\pi}\int_0^\infty \left[-\frac{\hbar^2}{2m}e^{-\alpha r}\nabla^2 e^{-\alpha r} - \frac{e^2}{r}e^{-2\alpha r}\right]r^2 dr = \frac{\hbar^2}{2m}\alpha^2 - e^2\alpha. \tag{13.149}$$

As a function of α this expression has its minimum nonzero value for

$$\alpha = \frac{me^2}{\hbar^2} = \frac{1}{a_0}, \tag{13.150}$$

where a_0 is the Bohr radius, which gives the correct value of the ground-state energy,

$$E_0 \equiv E_{1s} \leq \min\langle \widehat{H} \rangle = \langle \widehat{H}(1/a_0) \rangle = -\frac{e^2}{2a_0}. \tag{13.151}$$

The best approximation for the corresponding wave function is therefore

$$\psi_0 \equiv \psi_{1s} = \frac{1}{\sqrt{\pi a_0^3}} e^{-r/a_0}. \tag{13.152}$$

Due to the simplicity of the problem we were able to find the exact solution in this case; things are usually not that simple.

To gain further experience, we will determine the solution for the first excited s-state (for states with $l \neq 0$ the test function must be of the form $R(r)Y_l^m(\theta,\phi)$, which makes the problem more difficult, but still manageable). We choose the test function such that: (a) it is normalized to unity; (b) it is orthogonal to ψ_{1s}; (c) it has a node on the r-axis for $r > 0$; and (d) it decays exponentially. With these restrictions the simplest possibility is

$$\psi_{2s} = A\left(1 - \beta\frac{r}{a_0}\right) e^{-\alpha r/a_0}. \tag{13.153}$$

Condition (b) leads to $\beta = (1+\alpha)/3$, and from the normalization condition it follows that $A^2 = 3\alpha^5/\pi a_0^3(\alpha^2 - \alpha + 1)$; thus the test function (13.153) depends on a single variational parameter, α. The expectation value of $\widehat{H}$ is

$$\langle \widehat{H} \rangle = -\frac{e^2}{a_0}\left[\frac{\alpha^2}{2(\alpha^2 - \alpha + 1)} + \tfrac{7}{6}\alpha^2 + \tfrac{1}{2}\alpha\right], \tag{13.154}$$

and the condition that it takes a stationary value, $\partial\langle \widehat{H} \rangle/\partial\alpha = 0$, is fulfilled for $\alpha = 1/2$. With this choice we obtain

$$E_{2s} \leq -\frac{e^2}{8a_0}, \tag{13.155}$$

and the corresponding eigenfunction,

$$\psi_{2s} = \frac{1}{\sqrt{8\pi a_0^3}}\left(1 - \frac{r}{2a_0}\right) e^{-r/2a_0}. \tag{13.156}$$

Both results are again exact thanks to the relative simplicity of the problem, which leads to an excellent test function.

So far, the method has allowed us to obtain an upper bound on the eigenvalue, but it is also possible to obtain a lower bound, as follows. Let ψ_n and E_n be the eigenfunction and eigenvalue of the original problem, $\widehat{H}\psi_n = E_n\psi_n$, and ψ' and $E' = \langle \psi' | \widehat{H} | \psi' \rangle$ the respective approximate expressions obtained by applying the variational procedure. We introduce an error vector $|\chi\rangle$ and an estimated bound Δ, defined as

$$|\chi\rangle = (\widehat{H} - E')|\psi'\rangle, \quad \Delta = \langle \chi | \chi \rangle^{1/2}. \tag{13.157}$$

We have, on the one hand,

$$\Delta^2 = \langle \chi | \chi \rangle = \langle \psi' | (\widehat{H} - E')^2 | \psi' \rangle,$$
$$\Delta^2 = \langle \psi' | \widehat{H}^2 | \psi' \rangle - E'^2, \tag{13.158}$$

and on the other hand, inserting a spectral expansion,

$$\Delta^2 = \sum_{k,k'} \langle \psi' | (E_k - E')(E_{k'} - E') | \psi_k \rangle \langle \psi_k | \psi_{k'} \rangle \langle \psi_{k'} | \psi' \rangle$$

$$= \sum_{k} (E_k - E')^2 \langle \psi' | \psi_k \rangle \langle \psi_k | \psi' \rangle$$

$$= \sum_{k} (E_k - E')^2 |\langle \psi' | \psi_k \rangle|^2. \tag{13.159}$$

Since E' is a best estimate of E_n, the quantity $(E_k - E')^2$ should take a minimum value for $k = n$, so we can write

$$\Delta^2 \geq (E_n - E')^2 \sum_{k} |\langle \psi' | \psi_k \rangle|^2$$

$$= (E_n - E')^2 \sum_{k} \langle \psi' | \psi_k \rangle \langle \psi_k | \psi' \rangle$$

$$= (E_n - E')^2 \langle \psi' | \psi' \rangle = (E_n - E')^2. \tag{13.160}$$

Since $E' \geq E_n$, it follows that

$$\Delta \geq E' - E_n \quad \text{and} \quad \Delta \geq -(E' - E_n). \tag{13.161}$$

The second inequality is equivalent to $-\Delta \leq E' - E_n$, hence

$$E' - \Delta \leq E_n \leq E' + \Delta, \tag{13.162}$$

to which it must be added that $E_n \leq E'$. In deriving this result, the hypothesis that E_n is the eigenvalue closest to E' was essential; if this is not true, it is clear that the previous estimate is no longer reliable. More advanced textbooks on QM contain other methods that give more precise results for the eigenvalue bounds, although they are, of course, more complicated.

13.8.3 Other applications: Partial shielding of the Coulomb potential

Before concluding this section, we apply the variational method to a problem whose exact solution we do not know, so that it is no longer a trivial example. The aim is to determine the ground-state energy for a particle subject to a partially shielded Coulomb potential produced by a neutral H atom. The effective potential, obtained by calculating the average (over the atomic ground state) of the potential produced by the nucleus and that produced by the orbiting electron, has the form

$$V(r) = \frac{k}{r} e^{-\alpha r}, \tag{13.163}$$

where α depends on the structure of the atom, A reasonable choice of trial function is somewhat similar to that for the ground state of the H atom, since this guarantees a good approximation for small values of α. Once normalized, this function can be written as

$$\psi(r) = \left(\frac{\beta^3}{\pi}\right)^{1/2} e^{-\beta r}, \tag{13.164}$$

and the expectation value of the Hamiltonian becomes

$$E(\beta) = \langle \psi | \widehat{H} | \psi \rangle = \frac{\hbar^2 \beta^2}{2m} - \frac{4k\beta^3}{(\alpha + 2\beta)^2}. \tag{13.165}$$

The complicated dependence of this expression on β forces us to carry out a graphical or numerical analysis to determine its minimum. However, some conclusions can easily be drawn from Eq. (13.165). For example, $E(\beta) \leq 0$ for all values of β between 0 and a maximum (which can be easily determined), when $0 \leq \alpha \leq 1$ (in atomic units, with $k = 1$). For all these values of α and $\beta > 0$ there is a minimum of $E(\beta)$, which is the upper limit of the energy of the corresponding ground state. For example, for $\alpha = 0.5$ we get $\beta \simeq 0.9$ and $E_{\min} = E' \simeq -0.15$ a.u. (remember that for the H atom, $E_0 = -0.5$ a.u.). According to these calculations, the potential (13.163) ceases to have bound states at $\alpha = 1$, but a more precise analysis shows that this actually occurs around the value $\alpha \simeq 1.2$.

Once the value of β for which the minimum occurs has been obtained, that is, an upper bound for the ground-state energy has been found, and a lower bound can be obtained by combining Eqs. (13.162) and (13.158). To apply these, it is only necessary to calculate $\langle \psi' | \widehat{H}^2 | \psi' \rangle$; the calculation is a bit tedious, but straightforward. For example, for $\alpha = 0.5$ we get $\langle \psi' | \widehat{H}^2 | \psi' \rangle = 0.026$ a.u. and $\Delta = 0.059$ a.u., which means that $E_0 > -0.15 - 0.059$ a.u. in this case.

As mentioned earlier, the variational method is successfully applied to solve many important problems. When the number of parameters used is small and the problem is simple, analytical methods of solution are often possible. For systems with a larger number of components, computational algorithms have been developed that successfully use variational methods, with increasingly smaller margins of error. In Section 16.1 we will have the opportunity to introduce the variational method for the theoretical study of systems of general interest that do not have an exact solution, and in Section 16.3.3 we apply it in particular to the He atom.

Exercise E13.4. General use of the Rayleigh–Ritz method

The Ritz method for the approximate solution of an eigenvalue equation consists of expressing the eigenfunctions sought in terms of a conveniently truncated basis and requiring that the approximate eigenvalues computed with these functions are stationary with respect to small arbitrary changes in the development coefficients. Show that this procedure leads to a secular equation that solves the problem and that the smallest eigenvalue computed is an upper bound on the corresponding exact eigenvalue.

Solution. We write the equation to be solved in the form

$$\widehat{L} \Phi_k = \Lambda_k \Phi_k. \tag{13.166}$$

Instead of the exact problem (13.166) we propose the approximate problem $\widehat{L} \varphi_k = \lambda_k \varphi_k$, where the approximate eigenfunctions φ_k are constructed in terms of a proper basis $\{u_n\}$, but using only a finite number of terms,

$$\varphi_k = \sum_{n=0}^{N} c_{kn} u_n. \tag{13.167}$$

If L_{mn} denotes the matrix elements of the operator $\widehat{L}$ in the u representation, it follows from the previous two expressions that

$$\int \varphi_k^* \widehat{L} \varphi_k dx = \sum_{m,n}^N c_{km}^* c_{kn} L_{mn} = \lambda_k \sum_{m,n}^N c_{km}^* c_{kn} \delta_{mn}. \qquad (13.168)$$

λ_k is now a function of the coefficients c_{kl}, c_{kl}^*; requiring it to be stationary with respect to them (i.e., imposing the conditions $\partial \lambda_k / \partial c_{kl} = 0$, $\partial \lambda_k / \partial c_{kl}^* = 0$), we obtain by derivation of the previous equation that

$$\sum_n^N c_{kn} L_{mn} = \lambda_k \sum_n^N c_{kn} \delta_{mn}, \qquad (13.169)$$

which can be expressed as a secular equation

$$\sum_n^N (L_{mn} - \lambda_k \delta_{mn}) c_{kn} = 0, \qquad (13.170)$$

as requested. Solving this equation by the usual algebraic methods gives the approximate eigenvalues λ_k and the coefficients c_{kl} of the expansion (13.167). In the limit $N \to \infty$ the solution becomes exact.

To show that this method gives an upper bound on the lowest eigenvalue, we proceed as follows. First, we order the eigenvalues, both exact and approximate, from smallest to largest:

$$\Lambda_0 < \Lambda_1 < \Lambda_2 < \ldots, \quad \lambda_0 < \lambda_1 < \lambda_2 < \ldots \qquad (13.171)$$

Then we express the approximate functions in terms of the exact ones,

$$\varphi_k = \sum_m a_{km} \Phi_m. \qquad (13.172)$$

From the preceding equations, it follows that

$$\lambda_k = \frac{\int \varphi_k^* \widehat{L} \varphi_k dx}{\int \varphi_k^* \varphi_k dx} = \frac{\sum_{m,n} a_{km}^* a_{kn} \Lambda_n \delta_{mn}}{\sum_n |a_{kn}|^2} = \frac{\sum_n \Lambda_n |a_{kn}|^2}{\sum_n |a_{kn}|^2}. \qquad (13.173)$$

Since $\Lambda_n \geq \Lambda_0$, substituting in the previous expression each Λ_k by Λ_0 we see that $\lambda_k \geq \Lambda_0$. In particular, taking $k = 0$ gives $\lambda_0 \geq \Lambda_0$, which is precisely the requested result. In a similar way it is shown that

$$\lambda_1 \geq \Lambda_1 - (\Lambda_1 - \Lambda_0) \frac{|a_{10}|^2}{\sum_n |a_{1n}|^2}. \qquad (13.174)$$

Problems

P13.1 Show that if $\widehat{f}$ is a dynamical variable of a perturbed system, its first-order perturbed matrix elements are

$$f_{nm} = f_{nm}^{(0)} + \sum_{k \neq n} \frac{V_{nk} f_{km}^{(0)}}{E_n^{(0)} - E_k^{(0)}} + \sum_{k \neq m} \frac{V_{km} f_{nk}^{(0)}}{E_m^{(0)} - E_k^{(0)}}. \qquad (13.175)$$

P13.2 The potential of a one-dimensional anharmonic oscillator can be approximated by the expression

$$V = \tfrac{1}{2}m\omega^2 x^2 \left[1 + \alpha \frac{x}{x_0} + \beta \left(\frac{x}{x_0}\right)^2 \right], \tag{13.176}$$

where $x_0 = \sqrt{\hbar/m\omega}$. Use perturbation theory to determine the corrections to the energy up to second-order and the first-order wave functions. Use your results to calculate $\langle x \rangle, \langle \widehat{p} \rangle, \langle (\Delta x)^2 \rangle, \langle (\Delta \widehat{p})^2 \rangle$.

P13.3 A particle moves on a vertical circle of radius R; neglecting the effects of friction, but taking into account the force of gravity, the Hamiltonian of the system is

$$\widehat{H} = \frac{\widehat{L}_z^2}{2mR^2} + mgR\sin\varphi. \tag{13.177}$$

Determine the unperturbed solutions, as well as the first- and second-order corrections to the energy, treating the gravitational term as a perturbation. Under what conditions is this solution valid?

P13.4 Study the normal Zeeman effect of spinless particles using the methods of perturbation theory.

P13.5 Show that the matrix $\widehat{A}$ defined by Eq. (13.61) also satisfies the condition $\widehat{A}\widehat{A}^\dagger = 1$, which, taken together with $\widehat{A}^\dagger \widehat{A} = 1$, guarantees that $\widehat{A}$ is unitary.

P13.6 Explain why the linear Stark effect increases with the principal quantum number n.

P13.7 Calculate the Stark effect on the H atom for the levels with $n = 3$. Make maximum use of the symmetry properties of the system, using spherical coordinates.

P13.8 Calculate the intensities of the components of the H_α line of hydrogen, split by the linear Stark effect.

P13.9 Two coupled oscillators have the Hamiltonian

$$\widehat{H} = \frac{1}{2m}\left(\widehat{p}_1^2 + \widehat{p}_2^2\right) + \tfrac{1}{2}m\omega_1^2 x_1^2 + \tfrac{1}{2}m\omega_2^2 x_2^2 + m\omega^2 \beta x_1 x_2, \tag{13.178}$$

where $\omega^2 = \omega_1 \omega_2$ and β is arbitrary. Study this problem with perturbative methods when:

(a) ω_1/ω_2 is an irrational number;
(b) ω_1/ω_2 is a rational number.

Solve the problem exactly and compare results.

P13.10 Determine the emission spectrum of the coupled oscillators of the previous problem, and compare it with the corresponding spectrum of two independent oscillators.

P13.11 Consider a particle inside a 3D square well of side L. Find the corrections to the correct energy levels and wave functions of the problem perturbed by gravity.

P13.12 Two identical flat rotors are coupled so that their Hamiltonian is

$$\widehat{H} = A\left(p_{\theta_1}^2 + p_{\theta_2}^2\right) - B\cos(\theta_1 - \theta_2), \quad A, B > 0.$$

Determine the energy spectrum of the system and the corresponding zero-order eigenfunctions:

a) up to the linear terms in B, assuming that $B \ll A\hbar^2$;
b) reducing the problem to that of a harmonic oscillator if $B \gg A\hbar^2$.

P13.13 Consider the Hamiltonian

$$\widehat{H} = \hat{a}^\dagger \hat{a} + \tfrac{1}{2} + \eta(\hat{a}^\dagger \hat{a}^\dagger + \hat{a}\hat{a}), \quad |\eta| < \tfrac{1}{2}, \quad (13.179)$$

representing an oscillator perturbed by an interaction with coupling constant η. Consider the linear transformation

$$\hat{a} = \alpha \hat{\lambda} + \beta \hat{\lambda}^\dagger, \quad \hat{a}^\dagger = \alpha \hat{\lambda}^\dagger + \beta \hat{\lambda}, \quad (13.180)$$

with real α, β, and $\alpha^2 - \beta^2 = 1$, where $\hat{\lambda}, \hat{\lambda}^\dagger$ are two new operators that comply with the usual commutation rule (which allows us to equate them with creation and annihilation operators), $\left[\hat{\lambda}, \hat{\lambda}^\dagger\right] = \mathbb{I}$. This leads to a Hamiltonian without interaction

$$\widehat{H} = \sqrt{1 - 4\eta^2}\left(\hat{\lambda}^\dagger \hat{\lambda} + \tfrac{1}{2}\right), \quad (13.181)$$

if

$$\alpha^2 + \beta^2 + 4\eta\alpha\beta = \sqrt{1 - 4\eta^2},$$

$$\tfrac{1}{2} + 2\eta\alpha\beta + \beta^2 = \tfrac{1}{2}\sqrt{1 - 4\eta^2}. \quad (13.182)$$

This is the so-called *canonical Bogolyubov transformation*. Determine the operator that performs this transformation.

P13.14 To solve the harmonic oscillator problem with the variational method, the following test functions are proposed:

(i) $A_0 e^{-\alpha x^2}$ for the ground state;
(ii) $A_1 x e^{-\alpha x^2}$ for the first excited state;
(iii) $A_2(1 + bx^2)e^{-\alpha x^2}$ for the second excited state.

Justify this selection and determine the wave functions and the energy eigenvalues for the first three states.

P13.15 Study the problem of a particle bound by the Yukawa[7] potential,

$$V = -g^2 \frac{e^{-\alpha r}}{r}, \quad (13.183)$$

using the variational method. Find the maximum value of α for which there is at least one bound state. Go to the limit of the Coulomb potential and show that for this potential there is always a discrete spectrum. Find the optimal ground-state energy and wave function and in the Coulomb limit. *Hint:* For Coulomb solutions to be exact, it is convenient to use $\psi = Ae^{-\beta r}$ as a test function.

---

[7] Hideki Yukawa (1907–1981) was a Japanese physicist who won the Nobel Prize in Physics in 1949 for successfully predicting the existence of the meson. He is also known for introducing a potential of the form (13.183) as a model for the nuclear force.

P13.16 In nuclear physics, a truncated harmonic-oscillator potential $V = \frac{1}{2}m\omega^2 (x^2 - a^2)$ for $|x| \leq a$ and $V = 0$ for $|x| > a$ is often used as a model for an attractive well.

a) Use the variational method to estimate the energy of the ground state and the first excited state.

b) Estimate the preceding eigenvalues using the WKB method. Under what conditions can the well contain only one bound state?

Compare the results and determine which are more reliable.

P13.17 Use the variational method to show that every purely attractive 1D potential has at least one bound state, regardless of its depth. *Hint*: A Gaussian test function works well for this purpose.

Bibliographical Notes

An introduction to the perturbation theory of classical systems and its application to the H atom can be found in José and Saletan (1998) and in Corben and Stehle (1960), chapter 13. The latter treats the classical Zeeman effect.

The excellent book on the H and He atoms by Bethe and Salpeter (1957) contains a detailed discussion of the Stark effect in H, including a comparison with the experiment. For very strong fields, the problem is treated using the WKB method.

A detailed discussion of perturbation theory can be found in Merzbacher (1998), chapter 18.

Fernández (2020) takes a practical approach to perturbation theory that addresses its actual implementation.

For an advanced textbook on variational principles, see Basdevant (2023).

14 The Electron Spin. Entangled States

The introduction of spin in QM has far-reaching consequences. The most obvious of them have to do with the Pauli exclusion principle, which is responsible for the orbital structure of atoms and molecules, the subject of Chapter 16. More generally, as will be seen in Chapter 15, the statistics of quantum particles is determined by whether their spin is integer or half-integer.

This chapter is essentially devoted to the spin of the electron. The first part deals with its groundbreaking discovery, its introduction into quantum formalism, and some of its most important effects on atomic spectra, notably the anomalous Zeeman effect.

Historically, in QM spin has been considered as an angular momentum that particles can have by the mere fact of their existence, which is called "intrinsic" and does not allow any explanation. To address this shortcoming, in Section 4 we present a possible explanation for the origin of electron spin as a result of its interaction with the ZPF.

Section 5 introduces the entangled system of two particles with spin, which provides an opportunity for a discussion – necessarily schematic – of the Schrödinger cat and the Einstein–Podolsky–Rosen (EPR) thought experiment, as well as the Bell inequalities.

14.1 Discovery of the Electron Spin

In the years prior to the establishment of QM, there had been intensive efforts to explain the structure of atomic spectra. In particular, some progress had been made in describing the Zeeman effect, but a full explanation of the phenomenon had to wait until electron spin could be incorporated into Schrödinger's theory. As a reminder of the confusion that prevailed in those years, a certain form of the phenomenon, precisely the part that could not be explained in terms of classical models, is still called the *anomalous Zeeman effect* to distinguish it from what is known as the *normal Zeeman effect*, which corresponds to the part due to orbital angular momentum and was studied in Chapter 13.

A detailed analysis of the anomalous Zeeman effect led A. Landé[1] in 1921 to propose a modification of the classical formula for the gyromagnetic ratio, that is, the

---

[1] Alfred Landé (1888–1976) was a German-US theoretical physicist who carried out a deep and successful study of the anomalous Zeeman effect, in parallel with a search during forty years for an objectively realistic description of the quantum systems. He became a constant and fierce opponent of the Copenhagen interpretation and of the dualistic (particle and wave) view, and wrote several books on the foundations of the theory.

relationship between the magnetic moment of a circulating charge (e.g., an orbiting electron) and its corresponding angular momentum. Specifically, instead of the classical formula (12.139)

$$\frac{\mu_z}{L_z} = \frac{e}{2mc} = -\frac{e_0}{2mc}, \tag{14.1}$$

Landé proposed to write

$$\frac{\mu_z}{L_z} = -g\frac{e_0}{2mc}, \tag{14.2}$$

where g is a numerical factor. Landé himself was able to determine the value of g as a function of the quantum numbers of the atomic system on empirical grounds, following an exhaustive study of the known experimental data on the Zeeman effect, but without theoretical support. Furthermore, the formula for g had the surprising peculiarity of requiring the use of half-integer quantum numbers (see Eq. (14.41)).

Prior to Landé's work, in 1915, the formula (14.2) had been used experimentally in the case of electrons by Einstein in collaboration with the Dutch physicist Wander Johannes de Haas (1878–1960). The idea behind the experiment was to test Ampère's hypothesis that the magnetization of a ferromagnetic core is due to the elementary magnets generated by the currents in each of its molecules. To test the theory, Einstein and de Haas suspended a magnetizable cylinder from a quartz filament and applied an alternating current to a coil around the cylinder, ensuring that the frequency of the induced current matched the frequency of the torsional oscillation to take advantage of resonance. With this device they verified that the change in magnetization produces a similar change in angular momentum, so that μ_z/L_z remains constant and negative. However, as they focused on the order of magnitude of the phenomenon rather than its details, their reported numerical value was not very precise. Subsequent careful measurements of the Einstein–de Haas effect conducted by different authors [Barnett (1917); Stewart (1918)] gave $g = 2$. Therefore, the anomalous value 2 of the gyromagnetic ratio for the electrons of a ferromagnetic body was indeed known before Landé's proposal and could be taken as independent evidence for the necessity to adopt Eq. (14.2) instead of (14.1) in the general atomic case.

At about the same time (1921), O. Stern[2] proposed an experiment to verify the recently discovered angular-momentum quantization rules in the presence of a magnetic field. As we saw in Chapter 11, the component of the orbital momentum in the direction of the magnetic field can only have the value $\hbar m$, where m is an integer. Therefore, the projection μ_z of the magnetic moment can only have one of the $(2l + 1)$ eigenvalues given by

$$\mu_z = -\mu_0 m, \tag{14.3}$$

where μ_0 is the Bohr magneton. This surprising result, obtained by Sommerfeld in 1916 within the framework of Bohr's theory, became known as space quantization,[3] and it was this result that Stern proposed to verify. Following Stern's proposal, the

---

[2] Otto Stern (1888–1969) was a German–US American physicist who, with Walther Gerlach (1889–1979), contributed to the discovery of spin. He was awarded the Nobel Prize in Physics in 1943 for his contribution to the development of the molecular beam method and his discovery of the magnetic moment of the proton.
[3] This quantization displeased Stern so much that he proposed to verify it experimentally.

German physicist Walter Gerlach (1889–1979) carried out an experiment in which a beam of monovalent atoms was passed through a highly inhomogeneous magnetic field. According to electrodynamics, the force acting on a magnetic moment (a dipole or a circulating current) passing through a magnetic field $\boldsymbol{B}$ is given by

$$\boldsymbol{F} = \nabla(\boldsymbol{\mu} \cdot \boldsymbol{B}). \tag{14.4}$$

If the z axis is oriented along the field and y along the direction of the beam, the rapid precession of the magnetic moment around the axis causes the component μ_x to average to zero, so that the force acting on the magnetic moment in the z-direction can be approximated as

$$F_z = \mu_x \frac{\partial B_x}{\partial z} + \mu_z \frac{\partial B_z}{\partial z} \simeq \mu_z \frac{\partial B_z}{\partial z} = -\mu_0 \frac{\partial B_z}{\partial z} m = -F_0 m, \tag{14.5}$$

with $F_0 = \mu_0 \partial B_z / \partial z$. The angular momentum is randomly oriented, so classical theory would predict a uniform distribution for μ_z from $-\mu$ to $+\mu$, producing a single extended patch on the screen. However, since according to Eq. (14.5) the force on the dipole depends on the quantum number m, the beam must split into as many branches as there are different values of this number, that is, $2l + 1$; furthermore, since l is an integer according to Bohr's theory, $2l + 1$ should be odd.

The experiment was first carried out with atoms from group I of the periodic table (alkali metals H, Li, Na, K, Rb, Cs, Fr) in the s-state, for which $l = 0$ and $m = 0$. It turned out that *two* separate beams formed, indicating that the atoms had a nonzero magnetic moment in the s-state! With an angular momentum $\hbar s$ associated with the magnetic moment, $2s + 1 = 2$ gives $s = 1/2$. Although the result was consistent with Landé's empirical theory, the origin of the half-integer angular momentum remained unexplained.

In parallel with this series of works, various researchers were working on the origin and structure of the periodic table of elements. In particular, generalizing Bohr's views and developing a proposal by Edmund Stoner (1899–1968) concerning the possible number of stationary states of an atom in an external magnetic field, Pauli concluded in 1925 that it was possible to understand the structure of the periodic table by considering that each electronic orbital state was determined by four quantum numbers instead of the three known n, l, m, and by simultaneously assuming that each state could be occupied by only one electron. According to Pauli, the proposed fourth number should be no more than a dichotomous variable in order to avoid an excessive increase in the size of the atomic Hilbert space.

This momentous proposal, known as the *Pauli exclusion principle*, has far-reaching consequences, some of which we will have the opportunity to analyze in due course. In Chapter 6 we saw that the free electrons in a metal cannot all occupy the lowest energy level as a result of the exclusion principle. At this point, it is interesting to highlight the discovery of the need for a new quantum number to characterize the atomic state of an electron, which, according to Pauli, should have only two possible values. In fact, the exclusion principle not only solved the problem of the number of electrons that could be accommodated in each atomic shell[4] with a well-defined angular momentum,

---

[4] The organization of the atomic electrons into shells was first proposed by Bohr and later extended to all atoms.

but also gave a qualitative answer to a much more fundamental old question, namely, what prevents the electrons of a heavy atom (which can contain dozens of them) from all falling into the lowest-energy state, which would seem to be a more stable solution?[5]

In 1925, the young Dutch physicists G. E. Uhlenbeck[6] and S. Goudsmit[7] jointly proposed a reinterpretation of the previous results that resolved several of the difficulties in a very simple way. Instead of considering the new dichotomous quantum number as a characteristic of the state of the atom, they proposed to interpret it as an intrinsic angular momentum of the electron, with a value of 1/2 in units of $\hbar$. This is precisely the spin of the electron, whose formal quantum treatment was formulated shortly afterwards (1927) by Pauli himself, and whose description we will develop in what follows. The value 1/2 is uniquely fixed by the condition that the spin projection in a certain direction, identified with the new quantum number, can take only the two values $+\frac{1}{2}\hbar$ and $-\frac{1}{2}\hbar$, as required by the Pauli exclusion principle and by the Stern–Gerlach experiments.

14.2 The Quantum Theory of Spin

Since the electron spin is $\hbar/2$, we can apply to it the angular-momentum theory developed in Section 11.3 and express it in terms of the 2×2 Pauli matrices

$$\widehat{\boldsymbol{S}} = \tfrac{1}{2}\hbar\widehat{\boldsymbol{\sigma}}, \tag{14.6}$$

where $\widehat{\boldsymbol{\sigma}}$ is the 3D vector with components $\widehat{\sigma}_i$, the Pauli matrices,

$$\hat{\sigma}_1 = \begin{pmatrix} 0 & 1 \\ 1 & 0 \end{pmatrix}, \quad \hat{\sigma}_2 = \begin{pmatrix} 0 & -i \\ i & 0 \end{pmatrix}, \quad \hat{\sigma}_3 = \begin{pmatrix} 1 & 0 \\ 0 & -1 \end{pmatrix} \tag{14.7}$$

and

$$[\sigma_i, \sigma_j] = 2i\epsilon_{ijk}\sigma_k, \tag{14.8}$$

Furthermore, the Pauli matrices satisfy the relation

$$\sigma_i \sigma_j = \delta_{ij} + i\epsilon_{ijk}\sigma_k, \tag{14.9}$$

which shows in particular that the square of each one of them is the unit 2×2 matrix. Multiplying (14.9) by $\widehat{A}_i \widehat{B}_j$, where $\widehat{A}_i$ and $\widehat{B}_j$ are the Cartesian components of two arbitrary vector operators that commute with $\widehat{\boldsymbol{\sigma}}$, and adding over repeated indices, we obtain the useful formula

$$(\widehat{\boldsymbol{\sigma}} \cdot \widehat{\boldsymbol{A}})(\widehat{\boldsymbol{\sigma}} \cdot \widehat{\boldsymbol{B}}) = \widehat{\boldsymbol{A}} \cdot \widehat{\boldsymbol{B}} + i\widehat{\boldsymbol{\sigma}} \cdot (\widehat{\boldsymbol{A}} \times \widehat{\boldsymbol{B}}), \quad \text{for } \left[\widehat{\boldsymbol{A}},\widehat{\boldsymbol{\sigma}}\right] = \left[\widehat{\boldsymbol{B}},\widehat{\boldsymbol{\sigma}}\right] = 0. \tag{14.10}$$

---

[5] The present argument is a posteriori. In fact, Pauli's arguments (and Stoner's at the same time) concerned the anomalous Zeeman effect (discussed in Section 3) and, incidentally, the atomic configuration of rare gases. We recall that Pauli, guided by his positivist views, referred to the dichotomous variable he had discovered as an ambivalence, avoiding any use of models that he considered superfluous – a point of view that has permeated the literature until today. This issue is revisited in the next section.

[6] George Eugene Uhlenbeck (1900–1988) was a Dutch–US theoretical physicist, best known for having introduced together with Goudsmit, the electron spin as an intrinsic angular momentum.

[7] Samuel Abraham Goudsmit (1902–1978) was a Dutch–US theoretical physicist, famous for jointly proposing the concept of electron spin as an intrinsic angular momentum.

Since the electron spin is represented by a 2 × 2 matrix operator, the wave function of the electron with spin requires two components; this structure is called a spinor, and has the form $\begin{pmatrix} \psi_1 \\ \psi_2 \end{pmatrix}$, see Eq. (11.144). Spinors are mathematical objects that must be clearly distinguished from other objects with several components, such as a vector on the plane. In particular, the transformation properties under rotation of the coordinate system are so peculiar to spinors that the former are used to define the latter. As there is not enough space to develop this point fully, we will limit ourselves to explaining its basic aspects.

We recall that Eq. (8.167) is the general formula for the operator of rotation by the angle θ around the $\hat{\mathbf{n}}$ axis through the origin. With the angular-momentum operator $\hat{\mathbf{S}}$ given by Eq. (14.6), we obtain for the rotation of a spin 1/2

$$\widehat{T}_{\hat{\mathbf{n}}}(\theta) = e^{-\frac{i}{\hbar}\theta\hat{\mathbf{n}}\cdot\hat{\mathbf{S}}} = e^{-\frac{1}{2}i\theta\hat{\mathbf{n}}\cdot\hat{\boldsymbol{\sigma}}}. \tag{14.11}$$

Expanding the exponential into a power series and taking into account that according to (14.10) (see Eq. (14.107))

$$(\hat{\mathbf{n}}\cdot\hat{\boldsymbol{\sigma}})^{2k} = \mathbb{I}, \quad (\hat{\mathbf{n}}\cdot\hat{\boldsymbol{\sigma}})^{2k+1} = \hat{\mathbf{n}}\cdot\hat{\boldsymbol{\sigma}}, \tag{14.12}$$

we get

$$\widehat{T}_{\hat{\mathbf{n}}}(\theta) = \mathbb{I}\cos\frac{\theta}{2} - i(\hat{\mathbf{n}}\cdot\hat{\boldsymbol{\sigma}})\sin\frac{\theta}{2}. \tag{14.13}$$

Note the half angle, characteristic of spinors. In particular, for the operator representing an infinitesimal rotation around the direction $\hat{\mathbf{n}}$ by the angle $\delta\theta$, this gives

$$\widehat{T}_{\hat{\mathbf{n}}}(\delta\theta) = \mathbb{I} - \tfrac{1}{2}i\delta\theta\hat{\mathbf{n}}\cdot\hat{\boldsymbol{\sigma}}. \tag{14.14}$$

Now let Ψ be the spinor that describes the electron state of interest

$$\Psi = \begin{pmatrix} \psi_1 \\ \psi_2 \end{pmatrix}, \tag{14.15}$$

where ψ_1 and ψ_2 are one-component wave functions (solutions of the Schrödinger equation). According to the general theory of transformations discussed in Section 8.6, an infinitesimal rotation of the coordinate system described by $\widehat{T}_{\hat{\mathbf{n}}}(\delta\theta)$ transforms the Ψ spinor into Ψ', where

$$\Psi' = \widehat{T}_{\hat{\mathbf{n}}}(\delta\theta)\Psi = \left(1 - i\tfrac{1}{2}\delta\theta\hat{\mathbf{n}}\cdot\hat{\boldsymbol{\sigma}}\right)\Psi. \tag{14.16}$$

We see that the transformation is carried out with half the angle of rotation. This property defines spinors and essentially distinguishes them from tensors (scalars A, vectors A_i, second-rank tensors A_{ij}, etc.), since tensors are transformed by integral multiples of the angle. For example, a scalar remains unchanged after a rotation around the z axis, $A = A'$, while the x-component of a 3D vector $\mathbf{A}$ becomes $A'_x = A_x \cos\theta + A_y \sin\theta$, and so on. In contrast, the spinor Ψ becomes

$$\Psi' = \left(1 - i\tfrac{1}{2}\delta\theta\hat{\sigma}_3\right)\Psi = \begin{pmatrix} \left(1 - i\tfrac{1}{2}\delta\theta\right)\psi_1 \\ \left(1 + i\tfrac{1}{2}\delta\theta\right)\psi_2 \end{pmatrix} = \Psi - i\tfrac{1}{2}\delta\theta\begin{pmatrix} \psi_1 \\ -\psi_2 \end{pmatrix}. \tag{14.17}$$

An immediate consequence of this property is that a spinor is not invariant under a rotation by the angle 2π. Indeed, from (14.13) it follows that

$$\Psi' = \widehat{T}_z(2\pi)\Psi = (\cos\pi)\Psi = -\Psi. \tag{14.18}$$

Therefore a spinor cannot directly represent a physical quantity (an observable); the physical quantities must be bilinear combinations of spinors, which do remain invariant against rotations by an angle of 2π due to the double change of sign. This is not a limitation, since we know from the general rules of QM that the quantities of interest, such as the expectation values

$$\langle \widehat{A} \rangle = \int \Psi^\dagger \widehat{A} \Psi dx, \tag{14.19}$$

involve bilinear (quadratic) combinations of spinors.

It is clear that an expression like (14.19) is to be understood as a matrix formula, with $\widehat{A}$ a 2×2 matrix whose components may themselves be operators acting on the Hilbert space spanned by the ψ_i. For example, the particle density is

$$\Psi^\dagger \Psi = \begin{pmatrix} \psi_1^* & \psi_2^* \end{pmatrix} \begin{pmatrix} \psi_1 \\ \psi_2 \end{pmatrix} = \psi_1^* \psi_1 + \psi_2^* \psi_2; \tag{14.20}$$

the expectation value of $\widehat{S}_z$ is

$$\langle \widehat{S}_z \rangle = \tfrac{1}{2}\hbar \int dx \Psi^\dagger \widehat{\sigma}_z \Psi = \tfrac{1}{2}\hbar \int dx \begin{pmatrix} \psi_1^* & \psi_2^* \end{pmatrix} \begin{pmatrix} 1 & 0 \\ 0 & -1 \end{pmatrix} \begin{pmatrix} \psi_1 \\ \psi_2 \end{pmatrix}$$

$$= \tfrac{1}{2}\hbar \int dx(\psi_1^*\psi_1 - \psi_2^*\psi_2); \tag{14.21}$$

the expectation value of $\widehat{S}_x$ is

$$\langle \widehat{S}_x \rangle = \tfrac{1}{2}\hbar \int dx \begin{pmatrix} \psi_1^* & \psi_2^* \end{pmatrix} \begin{pmatrix} 0 & 1 \\ 1 & 0 \end{pmatrix} \begin{pmatrix} \psi_1 \\ \psi_2 \end{pmatrix} = \tfrac{1}{2}\hbar \int dx(\psi_1^*\psi_2 + \psi_2^*\psi_1), \tag{14.22}$$

and so on.

14.2.1 The Pauli equation

We now proceed to construct an evolution equation for Ψ which replaces (or rather, generalizes) the Schrödinger equation to the case of the electron with spin. The usual approach, introduced by Pauli, is to take into account that when an electron is placed in a magnetic field, its spin magnetic moment is coupled to the field. The magnetic moment in the z direction is $\widehat{\mu}_z = -g\frac{e_0}{2mc}\widehat{S}_z$, as follows from Eq. (14.2) by replacing $\widehat{L}_z$ with $\widehat{S}_z$. In vector notation we can then write, with $g = 2$ and $\mu_0 = e_0\hbar/2mc$ (the Bohr magneton),

$$\widehat{\boldsymbol{\mu}} = -2\frac{e_0}{2mc}\widehat{\mathbf{S}} = -\frac{e_0}{2mc}\hbar\widehat{\boldsymbol{\sigma}} = -\mu_0\widehat{\boldsymbol{\sigma}}. \tag{14.23}$$

This result shows that the magnitude of μ is (by definition) one Bohr magneton. The coupling to a field $\boldsymbol{B}$ then gives rise to the mutual energy

$$\widehat{V} = -\widehat{\boldsymbol{\mu}} \cdot \boldsymbol{B} = \mu_0 \boldsymbol{B} \cdot \widehat{\boldsymbol{\sigma}}. \tag{14.24}$$

This interaction term is to be added to the Hamiltonian of the Schrödinger equation, now written as a matrix equation for the spinor Ψ (see Eq. (14.28)). The resulting equation is called the Pauli equation: it correctly describes the behavior of the nonrelativistic electron with spin, as we will have the opportunity to see when we study, for example, the anomalous Zeeman effect in the following section.

The Pauli equation is a postulate that introduces the spin and the correct value of the intrinsic magnetic moment of the electron ad hoc, with $g = 2$. However, this equation can be derived from Dirac's fundamental theory of the relativistic electron, which naturally includes spin (see Chapter 19). In order not to remain on a purely phenomenological level, we will derive the Pauli equation from a more fundamental postulate than the mere introduction of the correct intrinsic magnetic moment, which then follows from the formalism (see the bibliographical note at the end of the chapter).

When an electron (more generally, a charged particle) is immersed in an electromagnetic field, the mechanical momentum $\boldsymbol{p}$ appearing in the Hamiltonian must be replaced by the canonical momentum $\boldsymbol{P} = \boldsymbol{p} - (e/c)\boldsymbol{A}$, as we have seen in Section 12.6. On the other hand, it follows from Eq. (14.10) that $(\hat{\boldsymbol{\sigma}} \cdot \hat{\boldsymbol{p}})^2 = \hat{p}^2$. This means that in the Schrödinger equation we can replace $\hat{p}^2$ by $(\hat{\boldsymbol{\sigma}} \cdot \hat{\boldsymbol{p}})^2$; each of the two components of the new wave function Ψ then satisfies the Schrödinger equation. We therefore postulate that to describe an electron with spin 1/2 in the presence of an external magnetic field, the substitution $\hat{p}^2 \to \hat{P}^2 \to (\hat{\boldsymbol{\sigma}} \cdot \hat{\boldsymbol{P}})^2$ must be made in the Hamiltonian.

The operator $\hat{\boldsymbol{p}}$ acts on the Hilbert space of the Schrödinger eigenfunctions, $\mathcal{H}_H$, while $\hat{\boldsymbol{\sigma}}$ acts on a newly introduced Hilbert space for the spin eigenfunctions, $\mathcal{H}_s$. The substitution thus expands the initial Hilbert space $\mathcal{H}_H$ to the product space $\mathcal{H}_H \otimes \mathcal{H}_s$ and transforms the Schrödinger scalar wave functions into two-component spinors. This is a simple way of extending the Hilbert space to include spinor functions, which is further justified by confirming that the resulting equations maintain their invariance against the Galileo group of transformations (3D translations and rotations), as is consistent with a nonrelativistic theory. Moreover, it is the nonrelativistic version of a procedure used to construct the Dirac equation, as discussed in Section 19.2.

According to the proposed procedure, the Pauli (or Pauli–Schrödinger) equation for an electron with spin 1/2 immersed in an electromagnetic field is

$$i\hbar \frac{\partial \Psi}{\partial t} = \frac{1}{2m} \left[\hat{\boldsymbol{\sigma}} \cdot \left(\hat{\boldsymbol{p}} - \frac{e}{c}\boldsymbol{A} \right) \right]^2 \Psi + V\Psi. \tag{14.25}$$

We will now rewrite this equation in a more suggestive way. To do this, we use Eqs. (14.9) and (14.10) to write

$$\left[\hat{\boldsymbol{\sigma}} \cdot \left(\hat{\boldsymbol{p}} - \frac{e}{c}\boldsymbol{A} \right) \right]^2 = \sum_{ij} \left(\hat{p}_i - \frac{e}{c}A_i \right) \left(\hat{p}_j - \frac{e}{c}A_j \right) \hat{\sigma}_i \hat{\sigma}_j$$

$$= \left(\hat{\boldsymbol{p}} - \frac{e}{c}\boldsymbol{A} \right)^2 + i\hat{\boldsymbol{\sigma}} \cdot \left(\hat{\boldsymbol{p}} - \frac{e}{c}\boldsymbol{A} \right) \times \left(\hat{\boldsymbol{p}} - \frac{e}{c}\boldsymbol{A} \right). \tag{14.26}$$

Since

$$\left(\hat{\boldsymbol{p}} - \frac{e}{c}\boldsymbol{A} \right) \times \left(\hat{\boldsymbol{p}} - \frac{e}{c}\boldsymbol{A} \right) = -\frac{e}{c}(\hat{\boldsymbol{p}} \times \boldsymbol{A} + \boldsymbol{A} \times \hat{\boldsymbol{p}})$$

$$= -\frac{e}{c}(-i\hbar \nabla \times \boldsymbol{A}) = \frac{i\hbar e}{c}\boldsymbol{B}$$

(in the Coulomb gauge $\nabla \cdot \mathbf{A} = 0$), substituting in the previous expression, we obtain

$$\left[\hat{\boldsymbol{\sigma}} \cdot \left(\hat{\boldsymbol{p}} - \frac{e}{c}\mathbf{A}\right)\right]^2 = \left(\hat{\boldsymbol{p}} - \frac{e}{c}\mathbf{A}\right)^2 - \frac{\hbar e}{c}\mathbf{B} \cdot \hat{\boldsymbol{\sigma}}. \tag{14.27}$$

This result allows us to write Eq. (14.25) as

$$i\hbar \frac{\partial \Psi}{\partial t} = \frac{1}{2m}\left(\hat{\boldsymbol{p}} - \frac{e}{c}\mathbf{A}\right)^2 \Psi + \left(V - \frac{\hbar e}{2mc}\mathbf{B} \cdot \hat{\boldsymbol{\sigma}}\right)\Psi, \tag{14.28}$$

which is a more common version of the Pauli equation. We see that the postulate $\hat{P}^2 \to (\hat{\boldsymbol{\sigma}} \cdot \hat{\boldsymbol{P}})^2$ leads naturally to the appearance of the interaction term $-\hat{\boldsymbol{\mu}} \cdot \mathbf{B}$ and assigns to the electron the intrinsic magnetic moment

$$\hat{\boldsymbol{\mu}} = -\mu_0 \hat{\boldsymbol{\sigma}}, \tag{14.29}$$

where $\mu_0 = \hbar e_0 / 2mc$ is the Bohr magneton. This result agrees with Eqs. (14.23) and (14.24), that is, it actually implies that the gyromagnetic ratio predicted for the electron spin contains the required factor of 2.

Expanding the Hamiltonian of (14.28) as is done, for example, in Section 12.7 and using $\hat{\boldsymbol{J}} = \hat{\boldsymbol{L}} + \hat{\boldsymbol{S}}$, we can give to the Pauli equation the alternative form,

$$i\hbar \frac{\partial \Psi}{\partial t} = \hat{H}_0 \Psi + \frac{\mu_0}{\hbar} \mathbf{B} \cdot (\hat{\boldsymbol{L}} + 2\hat{\boldsymbol{S}})\Psi. \tag{14.30}$$

It is important to note that the sum $\hat{\boldsymbol{L}} + 2\hat{\boldsymbol{S}} = \hat{\boldsymbol{J}} + \hat{\boldsymbol{S}}$ appears in the Hamiltonian, that is, that $\hat{\boldsymbol{S}}$ contributes twice as much as $\hat{\boldsymbol{L}}$ to the magnetic energy (and to the magnetic moment) because for $\hat{\boldsymbol{L}}$, $g = 1$, but for the spin, $g = 2$. This means, in particular, that the total magnetic moment $-\mu_0(\hat{\boldsymbol{L}} + 2\hat{\boldsymbol{S}})$ and the total angular momentum vector $\hat{\boldsymbol{L}} + \hat{\boldsymbol{S}}$ are not parallel.

14.3 Effects of Spin on Atomic Spectra

As mentioned in Section 14.1, it was not until the electron spin was introduced into Schrödinger's theory that the structure of atomic spectra in the presence of a magnetic field could be adequately explained. As we shall see, several other important properties of atomic spectra can only be accounted for when the spin of the electron and the spin of the proton are taken into account. Our first example is the anomalous Zeeman effect, that is, the splitting of spectral lines in paramagnetic materials with unpaired electron spins.

14.3.1 The anomalous Zeeman effect

Atoms whose total electronic spin is zero have integer angular momentum $\hat{\boldsymbol{J}} = \hat{\boldsymbol{L}}$ and the theory of the Zeeman effect discussed in Section 12.7 can be applied to them. The simplest example is He because it has two orbital electrons, whose spins can couple to 0 (or to 1; this topic is discussed in Section 16.3). However, if an atom has an electronic spin other than zero, the associated magnetic moment will also couple to the external magnetic field and contribute to the Zeeman effect, causing the spectral lines to split not into triplets, but into multiplets that can be of a more complicated structure, giving rise to the anomalous Zeeman effect.

Take an atom with Hamiltonian $\widehat{H}_0$, immersed in a weak, uniform magnetic field along the z-axis; the stationary Pauli–Schrödinger equation is

$$E\Psi = \widehat{H}_0 \Psi + \frac{\mu_0 B}{\hbar}(\widehat{J}_z + \widehat{S}_z)\Psi. \tag{14.31}$$

This equation is very complicated because $\widehat{S}_z$ represents the total electron spin operator, which can be $0, \frac{1}{2}, 1, \frac{3}{2}, \ldots$ and the Ψ wave function should have the corresponding number $(2S+1)$ of components. However, we can solve this problem with the help of a theorem from the theory of tensor operators, called the Wigner–Eckart theorem, which will not be proved here (see the references), but a simple version of it will be constructed now.

Equation (14.33) is a particular case of the Wigner–Eckart theorem. We can understand the meaning of this important theorem as follows. In a system with spherical symmetry, $\widehat{\boldsymbol{J}}$ determines the only direction of reference; it follows that the expectation value $\langle\widehat{\boldsymbol{A}}\rangle$ of any vector $\boldsymbol{A}$, if different from zero, is proportional to $\langle\widehat{\boldsymbol{J}}\rangle$, hence $\langle\widehat{\boldsymbol{A}}\rangle = C\langle\widehat{\boldsymbol{J}}\rangle$, and $\langle\widehat{\boldsymbol{A}}\cdot\widehat{\boldsymbol{J}}\rangle = C\langle\widehat{\boldsymbol{J}}^2\rangle$. Therefore, $C = \langle\widehat{\boldsymbol{A}}\cdot\widehat{\boldsymbol{J}}\rangle/\langle\widehat{\boldsymbol{J}}^2\rangle$, which, when substituted into the first expression, gives

$$\langle\widehat{A}_\alpha\rangle = \langle\widehat{J}_\alpha\rangle\left[\langle\widehat{\boldsymbol{A}}\cdot\widehat{\boldsymbol{J}}\rangle/\langle\widehat{\boldsymbol{J}}^2\rangle\right]. \tag{14.32}$$

The angular dependence of $\langle\widehat{A}_\alpha\rangle$ is therefore determined by $\langle\widehat{J}_\alpha\rangle$ (the other factors in (14.32) are scalars). Applying this result to our case, the matrix elements of $\widehat{\boldsymbol{A}}$ are given by

$$\langle jm'_j|\widehat{A}_q|jm_j\rangle = \frac{\langle jm_j|\widehat{\boldsymbol{A}}\cdot\widehat{\boldsymbol{J}}|jm_j\rangle}{\hbar^2 j(j+1)}\langle jm'_j|\widehat{J}_q|jm_j\rangle, \tag{14.33}$$

where $\widehat{A}_q$ and $\widehat{J}_q$ are the spherical components of the respective vectors, defined for $q = -1, 0, 1$ as

$$\widehat{A}_{\pm 1} = \tfrac{1}{\sqrt{2}}(\widehat{A}_x \pm i\widehat{A}_y), \ \widehat{A}_0 = \widehat{A}_z. \tag{14.34}$$

A comparison with (11.102) shows in particular that $\widehat{J}_1 = \widehat{J}_+$, $\widehat{J}_0 = \widehat{J}_z$, $\widehat{J}_{-1} = \widehat{J}_-$. Furthermore, the states $|jm_j\rangle$ and $|jm'_j\rangle$ may still correspond to different values n and n', respectively, of additional quantum numbers that may be required by the specific problem, but are independent of the angular description.

We now return to the problem of solving Eq. (14.31), and consider the case where B is not too large, so that we can treat the magnetic term as a perturbation. Since $\widehat{H}_0$ has spherical symmetry for an atom, we can take as unperturbed eigenfunctions the eigenstates of $\widehat{\boldsymbol{J}}^2$ and $\widehat{J}_z$, so that the first-order correction to the energy is given by

$$\delta E = \frac{\mu_0 B}{\hbar}\langle jm_j|\widehat{J}_z + \widehat{S}_z|jm_j\rangle. \tag{14.35}$$

The matrix element of $\widehat{J}_z$ is immediate,

$$\langle jm_j|\widehat{J}_z|jm_j\rangle = \hbar m_j. \tag{14.36}$$

To compute the matrix element of $\widehat{S}_z = \widehat{S}_0$ we use (14.33), which gives

$$\langle jm_j|\widehat{S}_z|jm_j\rangle = \frac{\langle jm_j|\hat{\boldsymbol{S}}\cdot\hat{\boldsymbol{J}}|jm_j\rangle}{\hbar^2 j(j+1)}\langle jm_j|\widehat{J}_z|jm_j\rangle$$
$$= \frac{m_j}{\hbar j(j+1)}\langle jm_j|\hat{\boldsymbol{S}}\cdot\hat{\boldsymbol{J}}|jm_j\rangle. \quad (14.37)$$

For the calculation of the remaining matrix element we proceed as follows. So far, the states are only required to be eigenstates of $\hat{\boldsymbol{J}}^2$ and $\widehat{J}_z$. But the required j can be obtained by combining different values of l and s, and since, by hypothesis, the atom possesses well-defined total spin, we must characterize the eigenstates with the quantum numbers j,l,s,m_j, that is, $|jm_j\rangle \to |jlsm_j\rangle$. Using this basis, the calculation is considerably simplified, since from $\hat{\boldsymbol{J}} = \hat{\boldsymbol{L}}+\hat{\boldsymbol{S}}$ it follows that $\hat{\boldsymbol{S}}\cdot\hat{\boldsymbol{J}} = \frac{1}{2}\left(\hat{\boldsymbol{J}}^2 + \hat{\boldsymbol{S}}^2 - \hat{\boldsymbol{L}}^2\right)$, so that

$$\langle jlsm_j|\hat{\boldsymbol{S}}\cdot\hat{\boldsymbol{J}}|jlsm_j\rangle = \frac{\hbar^2}{2}[j(j+1) + s(s+1) - l(l+1)]. \quad (14.38)$$

Substituting this value into (14.37), we obtain

$$\langle jm_j|\widehat{S}_z|jm_j\rangle = \hbar\frac{j(j+1) + s(s+1) - l(l+1)}{2j(j+1)}m_j. \quad (14.39)$$

Finally, adding (14.36) and (14.39) and substituting into (14.35), we obtain the correction to the energy due to the Zeeman effect,

$$\delta E = g\mu_0 B m_j, \quad (14.40)$$

where the Landé g factor is

$$g = 1 + \frac{j(j+1) + s(s+1) - l(l+1)}{2j(j+1)}, \quad (14.41)$$

in exact agreement with the empirical expression obtained by Landé before the introduction of QM.

Since g takes different values for different atomic states (j,l,s,m_j), the spectral lines can be decomposed into very complicated multiplets. Consider the simplest case, $s = 1/2$, when there are two different eigenstates of $\hat{\boldsymbol{L}}^2$ for a given j, namely $l = j + \frac{1}{2}$ and $l = j - \frac{1}{2}$. In both cases, Eq. (14.41) gives

$$g = \frac{j + \frac{1}{2}}{l + \frac{1}{2}}. \quad (14.42)$$

This is the case for the hydrogen atom, which we will use as an example. For $s_{1/2}$-states (the notation is L_J), it follows from (14.42) that $g = (1/2 + 1/2)/(0 + 1/2) = 2$; for $p_{1/2}$-states, $g = 2/3$; for $p_{3/2}$-states, $g = 4/3$; and so on. If g_0 is the Landé factor of the initial state and g that of the final state, the frequency of the radiation emitted in the atomic transition is

$$\omega = \omega_0 + (g_0 m_j^0 - g m_j)\omega_L, \quad (14.43)$$

where ω_0 is the emission frequency in the absence of the magnetic field, $\omega_L = \mu_0 B/\hbar$ is the Larmor frequency, and the quantum numbers m_j and m_j^0 are related by the selection rules $m_j - m_j^0 = 0, \pm 1$. Some transitions corresponding to the Lyman series are

14.3 Effects of Spin on Atomic Spectra

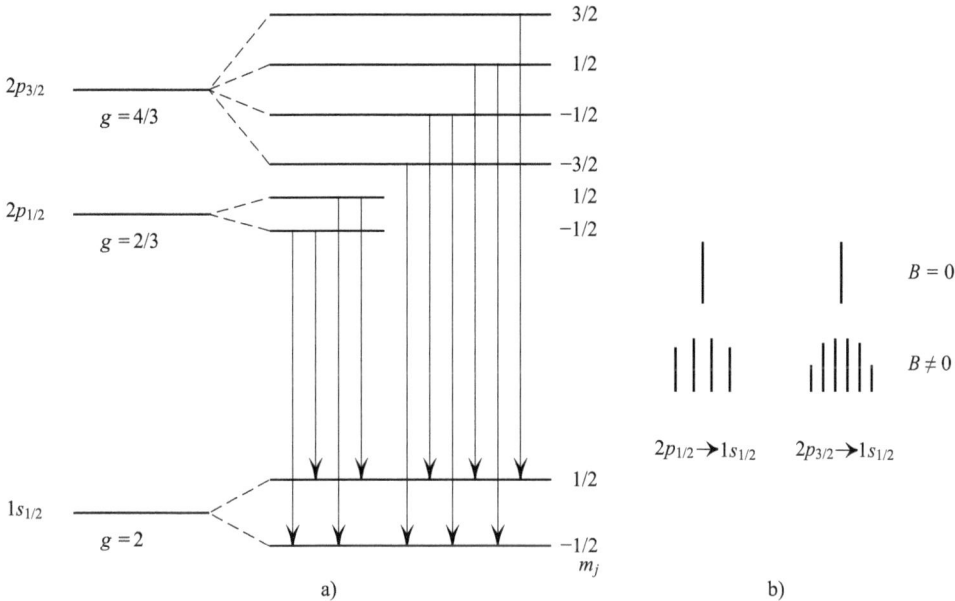

Figure 14.1 Diagram of the Zeeman spectrum showing the transitions between the lowest energy levels of the H atom. (a) The states $2p_{1/2}$ and $2p_{3/2}$ are separated by the spin–orbit coupling. (b) Line splitting for the allowed transitions $2p_{1/2} \to 1s_{1/2}$ and $2p_{3/2} \to 1s_{1/2}$.

shown in Fig. 14.1. It is clear that the complex dependence of the Landé factor on the atomic quantum numbers produces a variety of multiplets, the study of which provides information about j and l for the states involved in each transition.

When the total electron spin is zero (as in the case of He), the Landé factor reduces to $g = 1$; Eq. (14.40) reduces to (12.135) for the normal Zeeman effect, and the spectral lines split into the triplets predicted by the (classical) Lorentz theory. The electron spin is thus identified as the cause of the failure of the classical description of the Zeeman effect for atoms with an odd number of valence electrons.

Exercise E14.1. Atomic diamagnetism and paramagnetism
Investigate the diamagnetism and the paramagnetism of atoms immersed in a constant, homogeneous magnetic field.

Solution. According to Eq. (14.40) and up to linear terms in $\mathbf{A}$, the magnetic interaction energy is

$$\delta E_1 = \mu_0 B g m_j. \tag{14.44}$$

On the other hand, if in the Hamiltonian of Eq. (14.28) the quadratic terms in $\mathbf{A}$ are preserved, a second correction is obtained, which for the case of a uniform external magnetic field in the z-direction is

$$\delta E_2 = \frac{e^2}{2mc^2}\langle \mathbf{A}^2 \rangle = \frac{e^2}{8mc^2}\langle x^2 + y^2 \rangle B^2 = \frac{e^2 B^2}{12mc^2}\langle r^2 \rangle \tag{14.45}$$

(it was assumed that in the absence of A, the system possesses spherical symmetry, so $\langle x^2+y^2\rangle = \frac{2}{3}\langle r^2\rangle$). Therefore, the magnetic energy of the interaction between the atom and the field is

$$\Delta E_{\text{mag}} = \mu_0 g B m_j + \frac{e^2 B^2}{12mc^2}\langle r^2\rangle. \tag{14.46}$$

From this result it follows that the magnetic moment of the atom in the magnetic field is given by

$$\mu \equiv -\frac{\partial \Delta E_{\text{mag}}}{\partial B} = -\mu_0 g m_j - \frac{e^2 B}{6mc^2}\langle r^2\rangle. \tag{14.47}$$

The first term represents the intrinsic magnetic moment of the atom and characterizes its paramagnetic properties, while the second term, proportional to B, represents the induced magnetic moment and characterizes the diamagnetic properties of the atom ($\mu_{\text{mag}} < 0$). If N is Avogadro's number, the diamagnetic susceptibility per mole is

$$\chi_{\text{diam}} \equiv N \frac{\partial \mu_{\text{diam}}}{\partial B} = -\frac{e^2 N}{6mc^2}\langle r^2\rangle. \tag{14.48}$$

This term is always nonzero and negative, which means that all atoms have diamagnetic properties (i.e., they react against the external field). Equation (14.48) is called the Langevin formula. This formula allows us to determine $\langle r^2\rangle$ from the experimental value of χ_{diam}; the values obtained coincide with those provided by the kinetic theory of gases for atomic radii.

With regard to the paramagnetic moment, it should be noted that, since it is proportional to m_j, it can have either sign; however, in a state of thermodynamic equilibrium (characterized by the Boltzmann distribution) negative values of m_j prevail, since they correspond to lower energy states and the average over m_j is not zero. This leads to an average paramagnetic susceptibility (which tends to strengthen the external field) given by the Curie–Langevin formula:

$$\chi_{\text{param}} = \frac{\mu_0 g^2}{3kT} j(j+1). \tag{14.49}$$

If $j \neq 0$, χ_{param} is much larger than $|\chi_{\text{diam}}|$ and the atom has paramagnetic properties (e.g., hydrogen is always paramagnetic, since $j_{\min} = 1/2$). However, atoms with paired electron spins show diamagnetic properties (e.g., He in its ground state, or, more generally, the noble gases).

In a paramagnetic atom, the dipole moments tend to align in the direction of the field, inducing a field that strengthens the applied one. However, the alignment tends to be destroyed by thermal atomic motion, so the susceptibility should decrease with temperature; the Curie–Langevin formula (which is only valid at relatively low temperatures) shows that this is the case. A more accurate calculation predicts that the magnetization reaches a temperature-independent saturation value, corresponding to the case where all unpaired atomic magnetic dipoles are aligned with the external field; it is the intensity of the saturation field that increases with temperature. Exceptionally paramagnetic materials that retain their acquired magnetization when the external field is removed (Fe, Co, etc.) are called ferromagnetic materials.

14.3.2 *Fine and hyperfine structure of the H spectrum

The spin of the orbital electrons has a small but nonnegligible effect on the atomic spectrum, even in the absence of an external magnetic field. This is due to the fact that the electron in its orbital motion generates a magnetic field, to which the intrinsic magnetic moment of the electron is coupled. The resulting interaction term is commonly called the *spin–orbit interaction*. To this must be added a relativistic correction to the kinetic energy, which is of the same order of magnitude. The corresponding formulas are more conveniently derived in detail from the Dirac equation (see Chapter 19).

In general, the predictions of these formulas are in excellent agreement with the experimental data on the so-called fine spectrum. However, with the refinement of experimental techniques, discrepancies appeared on a smaller scale. In particular, in 1947 W. Lamb[8] together with Robert Retherford (1912–1981) observed a small separation between the $2s_{1/2}$ and $2p_{1/2}$ levels of the H atom, which is about one-tenth of the separation between the $2p_{1/2}$ and $2p_{3/2}$ levels. This Lamb shift can be explained as essentially due to the radiative effects of the electron, which were not taken into account in the previous theory or in Dirac's, but are fully accounted for in QED and in SED.[9]

There are even finer corrections to the atomic spectrum, called hyperfine corrections, which have various causes. The most important are related to the nucleus. We can think first of the effect on the atomic energy levels due to the finite size of the nucleus, which modifies the Coulomb potential in the nuclear region; but this correction, even for s-states, is at most of the order of 0.1% of the Lamb shift; see Eq. (14.111). A more important nuclear effect (in addition to the mass correction already made) comes from the magnetic moment associated with the spin of the nuclear proton, which generates a magnetic field to which the magnetic moment of the electron couples. This correction is of the order of $m/M \sim 10^{-3}$ times that due to the fine effects (relativistic and spinorial) and is even smaller than the Lamb effect (ca. 20%); it is therefore called a hyperfine correction.

Transitions in H atoms between the slightly excited state $1s_{1/2}$ with total spin $S = 1$ and the ground state $1s_{1/2}$ with $S = 0$ produce radiation of frequency $\omega \sim 1420$ MHz, of the order of the Lamb shift. This radiation, with a wavelength of 21.4 cm, has played an important role in the study of the distribution of H in intergalactic space. The reason for this is that the state with $S = 1$ is not produced by electromagnetic transitions (the radiative transition $S = 0$ to $S = 1$ is forbidden by the selection rule $\Delta l = 1$), but by collisions; therefore, the astronomical study of the 21 cm radiation produced when this state of H is de-excited allows us to know the density and the average temperature of the H clouds in interstellar space.

---

[8] Willis Eugene Lamb (1913–2008) was a US physicist who won the Nobel Prize in Physics in 1955 for his discoveries concerning the fine structure of the hydrogen spectrum. In the latter part of his career he devoted increasing attention to the field of quantum measurement. Lamb was openly critical of many of the trends in QM interpretation, and in particular of the use of the term *photon*.

[9] In Chapter 17 we will look at radiative corrections, including the Lamb shift.

Exercise E14.2. Effect of spin–spin interaction between nucleons
Consider the interaction between two equal nucleons in the CM system and determine the simplest form in which the central spin–spin interaction potential can be expressed.

Solution. The system is conveniently described in terms of the relative variables $\boldsymbol{r} = \boldsymbol{r}_1 - \boldsymbol{r}_2$, $\boldsymbol{p} = \frac{1}{2}(\boldsymbol{p}_1 - \boldsymbol{p}_2)$. The interaction potential between the nucleons can be modeled as the sum of a scalar contribution $V_0(r)$, a spin–spin coupling term $V_{ss}(\hat{\sigma}_1 \cdot \hat{\sigma}_2)$, plus additional spin-interaction terms. We shall calculate the contribution of the first two terms to the energy.

From $4\hat{\boldsymbol{S}}^2 = \hat{\sigma}_1^2 + \hat{\sigma}_2^2 + 2\hat{\sigma}_1 \cdot \hat{\sigma}_2$ it follows that $\hat{\sigma}_1 \cdot \hat{\sigma}_2 = 2\widehat{\boldsymbol{S}}^2 - 3$, so that for a potential of the form

$$V = V_0(r) + V_{ss}(\hat{\sigma}_1 \cdot \hat{\sigma}_2) \tag{14.50}$$

the Hamiltonian commutes with $\hat{\boldsymbol{L}}$ and $\hat{\boldsymbol{S}}$ and we will have eigenfunctions of $\hat{L}^2, \hat{L}_3, \widehat{\boldsymbol{S}}^2, \widehat{S}_3$. The total spin can take any of the values $S = 0, 1$, giving respectively

for $S = 0$, $\quad \hat{\sigma}_1 \cdot \hat{\sigma}_2 = -3, \quad V = V_0 - 3V_{ss};$
for $S = 1$, $\quad \hat{\sigma}_1 \cdot \hat{\sigma}_2 = +1, \quad V = V_0 + V_{ss}.$

The problem has been reduced to a Schrödinger problem for a spinless particle in a central potential whose value depends on the total spin.

14.3.3 Magnetic resonance

Atomic electrons or nuclei with nonzero spin can absorb or emit electromagnetic radiation in response to magnetic fields. This is because the spin magnetic moment precesses at a frequency proportional to the field that produces the Zeeman splitting (the Larmor frequency $\omega_L = \mu_0 B/\hbar$, in the case of the electron), and it is therefore possible to induce transitions between spin states via resonance coupling to an oscillating electromagnetic field related to this frequency.

The first induced transitions between nuclear Zeeman levels were obtained by I. Rabi[10] in 1933 using radiofrequency waves in an atomic beam. In 1945 the Swiss physicist Felix Bloch (1905–1983) and the US physicist Edward Purcell (1912–1997) independently performed the first nuclear magnetic resonance (NMR) experiments in condensed matter, directly detecting the response of the H proton to the electromagnetic field. Today, NMR has a wide range of applications in physics, chemistry, medicine, geology, and other fields.

In electron paramagnetic resonance (EPR) – first observed by the Soviet physicist Yuri K. Zavoysky in 1944 – a material composed of atoms or molecules carrying an unpaired electron is placed in a magnetic field that polarizes the material's bulk magnetic moment. When the material is exposed to the radiofrequency of the field, the

---

[10] Isidor Isaac Rabi (1898–1988) was a US physicist who won the Nobel Prize in Physics in 1944 for his discovery of nuclear magnetic resonance. He was also one of the first scientists in the United States to work on the cavity magnetron, which is used in microwave radar and microwave ovens (more generally, to generate microwaves).

spin is "flipped" by the transfer of energy from the field to the sample. EPR has made it possible to study phenomena such as the defects that give certain crystals their color, the formation of free radicals in bulk materials, and the behavior of free or conduction electrons in metals.

To simplify the analysis we consider the Hamiltonian for an electron spin in the presence of a magnetic field $\boldsymbol{B}$ given by Eq. (14.24), where $\boldsymbol{B}$ consists of two terms: the static field B_0 along the z-axis, and the radiofrequency field $B_1(t)$ along the x-axis, that is,

$$\widehat{H} = -\widehat{\boldsymbol{\mu}} \cdot (B_0 \hat{\mathbf{k}} + B_1 \hat{\mathbf{i}} \cos \omega t). \tag{14.51}$$

From Eq. (14.23) we have $\widehat{\boldsymbol{\mu}} = -\mu_0 \widehat{\boldsymbol{\sigma}}$, where the components of $\widehat{\boldsymbol{\sigma}}$ are the 2×2 Pauli matrices σ_i. Therefore, in matrix form $\widehat{H}$ becomes

$$\widehat{H} = \hbar \begin{pmatrix} \omega_L & \Omega \cos \omega t \\ \Omega \cos \omega t & -\omega_L \end{pmatrix}, \tag{14.52}$$

where

$$\omega_L = \mu_0 B_0 / \hbar, \quad \Omega = \mu_0 B_1 / \hbar. \tag{14.53}$$

The response to the oscillatory term is most conveniently described by splitting the Hamiltonian into zero- and first-order terms in Ω,

$$\widehat{H} = \widehat{H}_0 + \widehat{H}_1 = \hbar \omega_L \begin{pmatrix} 1 & 0 \\ 0 & -1 \end{pmatrix} + \hbar \Omega \cos \omega t \begin{pmatrix} 0 & 1 \\ 1 & 0 \end{pmatrix}, \tag{14.54}$$

and writing the first-order term in the interaction representation studied in Section 8.6.6,

$$\widehat{H}_{1I} = e^{i \widehat{H}_0 t / \hbar} \widehat{H}_1 e^{-i \widehat{H}_0 t / \hbar} = \hbar \Omega \cos \omega t \left(e^{i \omega_L t \widehat{\sigma}_z} \widehat{\sigma}_x e^{-i \omega_L t \widehat{\sigma}_z} \right). \tag{14.55}$$

After some matrix algebra, we get

$$e^{i \omega_L t \widehat{\sigma}_z} \widehat{\sigma}_x e^{-i \omega_L t \widehat{\sigma}_z} = \begin{pmatrix} 0 & e^{2i\omega_L t} \\ e^{-2i\omega_L t} & 0 \end{pmatrix}, \tag{14.56}$$

which introduced into (14.55) gives, using $\cos \omega t = \frac{1}{2} \left(e^{i\omega t} + e^{-i\omega t} \right)$ and neglecting the rapidly oscillating terms $e^{\pm i(\omega + 2\omega_L)t}$,

$$\widehat{H}_{1I} \simeq \frac{\hbar \Omega}{2} \begin{pmatrix} 0 & e^{-i(\omega - 2\omega_L)t} \\ e^{i(\omega - 2\omega_L)t} & 0 \end{pmatrix}. \tag{14.57}$$

This approximation is called the rotating-wave approximation, RWA. It is a very good approximation when the frequency ω of the applied field is close to $2\omega_L$, because then the system responds resonantly to the field, while the rapidly oscillating terms average out. As an effect of the resonant response, the field periodically transfers the population between the two Zeeman levels.

The name of the approximation derives precisely from the form of the Hamiltonian in the interaction picture, where only the part of the electromagnetic wave that approximately co-rotates is retained and the counter-rotating component is discarded. The RWA is widely used in magnetic resonance studies and in quantum dynamics of two-level systems more generally.[11]

---

[11] See Section 17.3.2 for a more detailed treatment of the RWA.

14.4 Modeling the Electron Spin

In the quantum literature it is common to refrain from assigning a specific model to the electron spin. This is because the electron is considered to be a point particle and it is not clear how a magnetic moment or an intrinsic rotation can be associated with a (mathematical) point. Nevertheless, there are various models that try to give it a specific physical meaning. The following is an example that is in line with the principles of SED, based on the experimental observation that electrons interact with circularly polarized modes of the radiation field.

The circularly polarized modes of the ZPF have, in addition to an energy $\frac{1}{2}\hbar\omega$ and a linear momentum $\frac{1}{2}\hbar\mathbf{k}$, an "intrinsic" angular momentum given by

$$J = \pm \frac{\hbar}{2}\hat{\mathbf{k}} \qquad (14.58)$$

along the direction of propagation $\hat{\mathbf{k}}$. The permanent interaction of the electron with the ZPF impresses on it an "internal" rotation, which can be regarded as a nonrelativistic version of the *zitterbewegung* discussed in Section 19.3 in connection with Dirac's equation.

To calculate the angular momentum of this rotation, we use the evolution equation for the average value of a dynamical function $\mathcal{G}(\mathbf{x},\mathbf{p})$ derived in Section 10.3, Eq. (10.86), and apply it to the Cartesian components $l_k = \epsilon_{ijk} x_i p_j$. This leads to the evolution equation for the average angular momentum (with the external torque given by $\mathbf{M} = \mathbf{x} \times \mathbf{f}$),

$$\frac{d}{dt}\langle \mathbf{L}\rangle = \langle \mathbf{M}\rangle + m\tau \langle \mathbf{x} \times \dddot{\mathbf{x}}\rangle - e^2 \langle \mathbf{x} \times \mathcal{D}\rangle, \qquad (14.59)$$

which in the quantum regime (i.e., in the Markovian limit) must be interpreted in terms of expectation values. For every Cartesian component we get

$$\frac{d}{dt}\langle \hat{L}_{ij}\rangle = \langle \hat{M}_{ij}\rangle + m\tau \langle \hat{x}_i \dddot{\hat{x}}_j - \hat{x}_j \dddot{\hat{x}}_i\rangle - \langle \hat{D}_{ij}^{px} - \hat{D}_{ji}^{px}\rangle. \qquad (14.60)$$

This equation reveals the mechanism that leads to a mean effective angular momentum through the antisymmetric part of the diffusion coefficient $\hat{D}^{px}$. In the absence of an external torque, the angular momentum is conserved when the following balance equation between the two radiative terms is satisfied,

$$m\tau \langle \hat{x}_i \dddot{\hat{x}}_j - \hat{x}_j \dddot{\hat{x}}_i\rangle = \langle D_{ij}^{px} - D_{ji}^{px}\rangle. \qquad (14.61)$$

For simplicity, we consider an isotropic HO with frequency ω in its ground state. Then $m\dddot{x}_j = -\omega^2 p_j$, and from (14.61) we have

$$\langle L_{ij}\rangle = \langle x_i p_j - x_j p_i\rangle = \frac{-1}{\tau\omega^2}\langle D_{ij}^{px} - D_{ji}^{px}\rangle. \qquad (14.62)$$

The usual numerical value $\langle L_{ij}\rangle = 0$ for the ground state represents the average net angular momentum induced by the full ZPF. However, since the electron responds separately to modes of left or right polarization, we should consider the separate effect of these modes on the angular momentum.

14.4 Modeling the Electron Spin

The circular polarizations are characterized by the vectors

$$\epsilon_{k\pm} = \frac{1}{\sqrt{2}} \left(\epsilon_{ki} \pm i\epsilon_{kj} \right), \tag{14.63}$$

with $\epsilon_{ki}, \epsilon_{kj}$ unit Cartesian vectors orthogonal to **k**. The appropriate variables to describe the response of the particle to the (circularly) polarized field modes are the spherical variables (x^+, x^-, x_k), which are given by

$$x^{\pm} = \tfrac{1}{\sqrt{2}} \left(x_i \mp i x_j \right), \tag{14.64}$$

$$x_i = \tfrac{1}{\sqrt{2}} \left(x^+ + x^- \right), \quad x_j = i\tfrac{1}{\sqrt{2}} \left(x^+ - x^- \right). \tag{14.65}$$

The nonzero matrix elements of the oscillator are $x_{i01} = \left(x_{01}^+ + x_{01}^- \right)/\sqrt{2}$; and so on. Further, since $x_{10}^{\pm} = \left(x_{01}^{\mp} \right)^*$, Eq. ((14.62)) gives

$$\langle L_{ij} \rangle_0 = m\omega_0 \left(x_{01}^+ x_{10}^- - x_{01}^- x_{10}^+ \right) = m\omega_0 \left(|x_{01}^+|^2 - |x_{01}^-|^2 \right). \tag{14.66}$$

In the ground state, $\langle L_{ij} \rangle_0 = 0$; hence the two terms, $|x_{01}^+|^2$ and $|x_{01}^-|^2$, contribute with equal magnitude and opposite sign to the k-component of the oscillator's angular momentum, as should be the case for a nonpolarized vacuum. These separate contributions are

$$\langle L_{ij} \rangle_0^+ = m\omega_0 |x_{01}^+|^2, \quad \langle L_{ij} \rangle_0^- = -m\omega_0 |x_{01}^-|^2. \tag{14.67}$$

Using $x_{i01} = x_{j01} = \sqrt{\hbar/(2m\omega_0)}$ for the harmonic oscillator, we get

$$m\omega_0 |x_{01}^{\pm}|^2 = \frac{\hbar}{2}, \tag{14.68}$$

whence the size of each separate contribution to the angular momentum in (14.66) is just $\hbar/2$. In order to distinguish this contribution from the orbital component of the angular momentum we write $\langle S_{ij} \rangle^{\pm}$ instead of $\langle L_{ij} \rangle_0^{\pm}$, so that

$$\langle S_{ij} \rangle^{\pm} = \pm \frac{\hbar}{2}. \tag{14.69}$$

Furthermore, by applying Eq. (10.86) to the square of the angular momentum, one obtains the correct quantum value for $\langle L^2 \rangle_0$, independent of the position coordinates and consisting of two terms due to each of the polarized ZPF modes (see the bibliographic notes at the end of the chapter),

$$\langle S^2 \rangle^+ = \langle S^2 \rangle^- = \frac{3}{4}\hbar^2. \tag{14.70}$$

The fact that the results for $\langle S_{ij} \rangle$ and $\langle S^2 \rangle$ do not depend on the oscillator's frequency ω_0, suggests that they hold in a more general case, and in particular for the free particle. We therefore conclude that this angular momentum induced by the ZPF can be identified with the spin of the electron. The term "intrinsic," which is usually attached to it, indicates its permanence, since the ZPF is always and everywhere present.

14.5 Quantum Correlations and Apparent Nonlocality

Nonlocality is in conflict with general physical principles. However, it is a fact that the quantum-mechanical *description* has nonlocal features. We have discussed one aspect of this issue in Chapter 5 in connection with quantum tunneling, and in Chapter 10 in connection with the quantum potential. Here we deal with another manifestation of nonlocality, specifically related to the (apparent) nonlocal (and noncausal) influence between objects – what Einstein called "spooky action at a distance." Such nonlocality manifests itself characteristically in the form of (quantum) correlations between dynamical variables of composite systems. Taken at face value, the Pauli principle, which prohibits two noninteracting electrons from occupying the same quantum state, is a reflection of such quantum nonlocality. Where do these nonlocalities come from, what is the underlying physics?

Long before the advent of QM, the concept of field was introduced by Faraday[12] to explain the apparent "action at a distance" of noncontact forces affecting the dynamics of particles. In classical physics, the neglect of the field (whether gravitational, electric, magnetic or other) as a mediator between the particles leads to an apparent nonlocality (taken by an "effective" one) resulting from the incomplete consideration of the system. It is therefore reasonable to assume that also in quantum physics, the nonlocality of the description may be the result of an incompleteness. In other words, that an explanation for the (apparently) nonlocal behavior of quantum systems can be found by taking into account the presence of a field that acts as a mediator between the particles.

This is precisely the idea behind the explanation of quantum nonlocality provided by SED, with the introduction of the radiation field, including the ZPF, as the mediator. The phenomenon of quantum correlation, which is conventionally regarded as a signature of nonlocality, is a particularly useful example to illustrate this role of the field. By being connected to the separate (not directly interacting) particles that make up a system, the field can act as a mediator and lead to causal correlations between them – which can persist as long as the system is not perturbed by external factors that can cause decoherence (see Section 15.5.7).

For the sake of clarity, the discussion will focus on the specific case of a system composed of two identical particles, that is, particles that, by virtue of having the same physical properties, exhibit the same behavior under the same circumstances. The results obtained for the bipartite case can then be extended to the analysis of a multipartite system of identical particles.

---

[12] Michael Faraday (1791–1867) was a most eminent English scientist dedicated to the study and development of electromagnetism and electrochemistry. He is one of the most brilliant and prolific scientists in history, yet he never went to school. In his laboratory he constructed (invented) the first electric motor, the first electric generator, and the first induction transformer. He introduced the notion of (electromagnetic) field in order to eliminate the appeal to actions at a distance from science. Maxwell turned this proposal into the most transcendental principle of science: nature is made up of matter and fields.

14.5.1 Bipartite system correlated by the background field

As a background to the study of the correlations between two particles, we recall how a single quantum particle responds to the background field, according to SED. In Section 10.4 we have seen that in the quantum regime, the kinematics of the particle is determined by the field modes with which the particle is in interaction, which by default are those of the ZPF; specifically, according to (10.119),

$$x_n(t) = x_{nn} + \sum_{k \neq n} x_{nk} a_{nk} e^{-i\omega_{kn} t} + \text{c.c.}, \quad p_n(t) = p_{nn} + \sum_{k \neq n} p_{nk} a_{nk} e^{-i\omega_{kn} t} + \text{c.c.}, \tag{14.71}$$

for any state n, with x_{nk} and p_{nk} the elements of matrices $\hat{x}$ and $\hat{p}$, respectively, satisfying the canonical commutator

$$[\hat{x}, \hat{p}] = i\hbar. \tag{14.72}$$

Let us now consider a system consisting of two equal, noninteracting particles, subject to a common binding potential. The fact that they are part of one and the same system means that they are also subject to the same background field. Furthermore, since the particles are identical, they satisfy the same equation of motion and therefore *they respond to common modes of the field.*

Both particles together have an energy $\mathcal{E}$ which is the sum of the individual energies $\mathcal{E}_n$ and $\mathcal{E}_m$ corresponding to states n and m. The total energy level is degenerate, since $\mathcal{E}_n$ may be the energy of particle 1 and $\mathcal{E}_m$ that of particle 2, or vice versa. To distinguish between the two possibilities we name them A and B, and write, for $m \neq n$ and $\mathcal{E}_n \neq \mathcal{E}_m$,

$$\mathcal{E}_A = \mathcal{E}_n + \mathcal{E}_m, \; \mathcal{E}_B = \mathcal{E}_m + \mathcal{E}_n, \; \mathcal{E}_A = \mathcal{E}_B = \mathcal{E}, \tag{14.73}$$

with

$$\mathcal{E}_n - \mathcal{E}_m = \hbar\omega_{nm}. \tag{14.74}$$

In Eq. (14.73) and the following expressions, the first term refers to particle 1 and the second term to particle 2; for the sake of simplicity in writing, the labels are omitted whenever this does not lead to confusion.

Our aim is to calculate the correlation between the variables $x(t)$ of the two particles. The product corresponding to state A is, according to Eq. (14.71),

$$(x_1 x_2)_A (t, \phi) = \sum_{k,l} x_{nk} a_{nk} e^{-i\omega_{kn} t} x_{ml} e^{i\gamma_{ml}\phi} a_{ml} e^{-i\omega_{lm} t}, \tag{14.75}$$

where $\gamma_{ml}\phi$ is a phase difference between the coupling of particle 2 and that of particle 1 to the respective field mode, the value of which will be given in due course. Similarly, the product corresponding to state B is

$$(x_1 x_2)_B (t, \phi) = \sum_{k,l} x_{mk} a_{mk} e^{-i\omega_{km} t} x_{nl} e^{i\gamma_{nl}\phi} a_{nl} e^{-i\omega_{ln} t}. \tag{14.76}$$

In Eq. (14.75), all products $a_{nk}a_{ml}$ average to zero except two: (i) when $k = n$ and $l = m$, and (ii) when $l = n$ and $k = m$, because $\langle a_{nk}a_{kn}\rangle = \langle a_{nk}a_{nk}^*\rangle = 1$ and hence also $a_{nn} = 1$. Therefore, the average contains only two terms,

$$\langle (x_1 x_2)_A \rangle = x_{nn} x_{mm} + x_{nm} x_{mn} e^{i\gamma_{mn}\phi}. \tag{14.77}$$

Similarly, the only products $a_{mk}a_{nl}$ in Eq. (14.76) that do not average to zero are: (i) when $k = m$ and $l = n$, and (ii) when $l = m$ and $k = n$, whence

$$\langle (x_1 x_2)_B \rangle = x_{mm} x_{nn} + x_{mn} x_{nm} e^{i\gamma_{nm}\phi}. \tag{14.78}$$

In the terms of the state vectors in the product Hilbert space,

$$|A\rangle = |n\rangle |m\rangle, \quad |B\rangle = |m\rangle |n\rangle, \tag{14.79}$$

where the left ket refers to particle 1 and the right ket to particle 2, Eqs. (14.77) and (14.78) have the form

$$\langle (x_1 x_2)_A \rangle = \langle A| \hat{x}_1 \hat{x}_2 |A\rangle + e^{i\gamma_{mn}\phi} \langle A| \hat{x}_1 \hat{x}_2 |B\rangle, \tag{14.80}$$

$$\langle (x_1 x_2)_B \rangle = \langle B| \hat{x}_1 \hat{x}_2 |B\rangle + e^{-i\gamma_{mn}\phi} \langle B| \hat{x}_1 \hat{x}_2 |A\rangle, \tag{14.81}$$

or regrouping,

$$\langle (x_1 x_2)_A \rangle = \langle A| \hat{x}_1 \hat{x}_2 \left| \left(A + e^{i\gamma_{mn}\phi} B \right) \right\rangle, \tag{14.82}$$

$$\langle (x_1 x_2)_B \rangle = e^{-i\gamma_{mn}\phi} \langle B| \hat{x}_1 \hat{x}_2 \left| \left(A + e^{i\gamma_{mn}\phi} B \right) \right\rangle. \tag{14.83}$$

Combining the two expressions and taking into account the invariance of the correlation with respect to the exchange of A and B, we obtain

$$\langle (x_1 x_2)_{AB} \rangle = \frac{1}{2} \left[\langle (x_1 x_2)_A \rangle + \langle (x_1 x_2)_B \rangle \right]$$
$$= \langle \Psi_{AB} | \hat{x}_1 \hat{x}_2 | \Psi_{AB} \rangle, \tag{14.84}$$

where

$$|\Psi_{AB}\rangle = \frac{1}{\sqrt{2}} \left| A + e^{i\gamma_{mn}\phi} B \right\rangle \tag{14.85}$$

is the state vector of the bipartite system with energy $\mathcal{E}_A = \mathcal{E}_B = \mathcal{E}$. Furthermore, since the exchange of A and B can at most change the sign of $|\Psi_{AB}\rangle$, the phase factor must be such that $e^{i\gamma_{mn}\phi} = \pm 1$, that is,

$$\gamma_{mn}\phi = n\pi, \quad n = 0, 1, 2, \ldots \tag{14.86}$$

which means that $|\Psi_{AB}\rangle$ is *either symmetric or antisymmetric* with respect to the exchange of A and B, depending on whether n is even or odd,

$$|\Psi_{AB}^{\pm}\rangle = \frac{1}{\sqrt{2}} |A \pm B\rangle. \tag{14.87}$$

In Chapter 15 it will be shown that the even or odd parity of $|\Psi_{AB}\rangle$ is crucial in determining the statistical properties of systems of identical particles.

According to Eq. (14.84), by taking the average of the product of functions $x_1(t)$ and $x_2(t)$, we have obtained the quantum expectation value of the product $\hat{x}_1 \hat{x}_2$ calculated in the state described by (14.87). With respect to the individual state vectors, this is a *nonfactorizable* vector. The cause of the nonfactorizability can be traced back to the second term on the right-hand side of Eqs. (14.80) and (14.81), which is due to the common field mode (nm), the only mode to which both particles respond coherently. In other words, it is a consequence of their being coupled to the same field mode (nm), *either in phase or in antiphase*.

Note that the Hilbert space to which $|\Psi_{AB}\rangle$ belongs is a new Hilbert space formed by the tensor product of the individual Hilbert spaces spanned by the sets of state vectors belonging to particles 1 and 2. This new Hilbert space is spanned by the set $\{|\Psi_{AB}\rangle\}$ of the two (orthonormal) eigenvectors corresponding to the total energy $\mathcal{E}_{AB}$, given by Eq. (14.87).

Thus, in addition to elucidating the physical mechanism that correlates the particles, SED explains the emergence of entangled state vectors. The procedure used here to calculate the correlation of x_1 and x_2 can be logically extended to other products of dynamical variables, as required.

14.5.2 Entanglement

A quantum system consisting of two or more particles can be in an entangled state, which has the property that it cannot be written down as a single product of the states of the individual particles. Entanglement – which Schrödinger originally called *Verschränkung* (interlacing) – involves a relationship between the components of the system, expressed by the appearance of characteristic correlations between their variables, as in the example of the previous section. This property has significant potential applications in quantum cryptography, quantum computing, particle interferometry, and teleportation, and has given rise to a whole new multidisciplinary branch called *quantum information*, which deals with information subject to the laws of QM. The current literature on this subject is growing rapidly, not least because entanglement has opened up the possibility of dense coding for calculation and remote transport of quantum information.

As these topics constitute a vast and promising field that is very active today, it is convenient to extend our introductory discussion of the bipartite entangled state a little.

The elementary quantum information unit is the qubit, which is an arbitrary (single-particle) unit vector in a 2D Hilbert space, $|\chi\rangle = a\,|+\rangle + b\,|-\rangle = a\binom{1}{0} + b\binom{0}{1} = \binom{a}{b}$, with $|a|^2 + |b|^2 = 1$. If we identify the one-particle eigenstates with $|n\rangle = \binom{1}{0}$ and $|m\rangle = \binom{0}{1}$, as in the previous case, the unit vector is $|\chi\rangle = a\,|n\rangle + b\,|m\rangle$, and the possible bipartite entangled eigenstates are, according to Eq. (14.87),

$$|\Psi_{\pm}\rangle = \frac{1}{\sqrt{2}}\left(|n\rangle\,|m\rangle \pm |m\rangle\,|n\rangle\right). \quad (14.88)$$

It is important to note that when the particles move independently, their individual states $|n\rangle, |m\rangle$ are well defined and can be combined to form the product states $|n\rangle\,|m\rangle$ and $|m\rangle\,|n\rangle$. It is only when both particles form a *single* system in a state characterized by a well-defined eigenvalue – $\mathcal{E} = \mathcal{E}_n + \mathcal{E}_m$ with $\mathcal{E}_n \neq \mathcal{E}_m$, in the preceding example – that they are entangled.

The two-state system of a spin-1/2 particle is a particularly simple case of a qubit. For this reason, the terminology of spin-1/2 particles is often used in quantum information; but note that there are other physical realizations of two-state systems, such as *quantum dot systems, superconducting quantum interference devices* (SQUID), and so on.

In Section 11.4 we introduced the (entangled, with total spin $S = 0$) singlet state formed by two spin-1/2 particles, Eq. (11.182),

$$|00\rangle = \tfrac{1}{\sqrt{2}}\Big(|+\rangle\,|-\rangle - |-\rangle\,|+\rangle\Big). \tag{14.89}$$

Such states have been widely used to exhibit peculiar quantum properties ever since they were discussed by Einstein, Podolsky, and Rosen in their famous 1935 EPR paper and by Schrödinger in the series of papers in which he introduced the equally famous cat that bears his name. The EPR paper challenged the notion – popular at the time and still held to by many today – of QM as a complete theory, in the (particular) sense that every element of physical reality has a counterpart in the theory; see Section 14.5.4. In the same year, at Einstein's suggestion, Schrödinger elaborated on entanglement and proposed the cat paradox, which is discussed in the following section.

As mentioned earlier, entangled states can involve more than two particles forming a composite system, resulting in different degrees of entanglement. A general n-particle entangled state occurs when each product state $|a\rangle\,|b\rangle\cdots|z\rangle$ satisfies the conditions specified by the problem as a whole, for example, by representing an eigenstate of a given total spin or total energy,

$$|\Psi\rangle = \sum_{(a,b,\ldots z)} C_{ab\ldots z}\,|a\rangle_1\,|b\rangle_2\cdots|z\rangle_n\,. \tag{14.90}$$

We see that unless all $C_{ab\ldots z}$ except one are zero, it is not possible to assign a (unique) state to the individual components; one then speaks of nonfactorizable states. Nonfactorizable states invariably imply nonclassical correlations.

14.5.3 Schrödinger's cat paradox

To expose the peculiar nature of nonfactorizable wave functions, Schrödinger took the quantum description to its extreme and applied it to a macroscopic system, for which it becomes manifestly paradoxical. Schrödinger concluded that it is in the entangled states that the definitive separation between the behavior of classical and quantum systems becomes apparent and the deepest conceptual problems arise, which is why these states must be studied exhaustively to unravel their mysteries.

The system analyzed by Schrödinger consists of a radioactive nucleus and a cat, all enclosed in a box. The box contains a detector that triggers a mechanism capable of poisoning the cat if the nucleus decays. Both the nucleus and the cat are then considered to be in one of two possible states: (a) for the nucleus, $|0\rangle_N$ if it has not decayed and $|1\rangle_N$ if it has already decayed; (b) for the cat, $|1\rangle_C$ if it is alive and $|0\rangle_C$ if it has died. Assuming the joint system behaves quantum mechanically, its wave function as a function of time is

$$|\Psi\rangle = C_{01}(t)\,|0\rangle_N\,|1\rangle_C + C_{10}(t)\,|1\rangle_N\,|0\rangle_C\,. \tag{14.91}$$

This is an "entangled state of cat+nucleus."

If, instead of a cat, another nucleus or atomic system were considered, the description would be perfectly acceptable. But describing the state of the cat in a superposition of dead and alive cats is an unwarranted extrapolation: multimolecular macroscopic entities like cats do not get entangled (with nuclei or whatever). Moreover, one must consider that not only the cat but also the detector and the poison dispenser are macroscopic entities.

The cat paradox is further accentuated by "measuring," that is, "looking inside the box," since we know from experience that we will always find a dead or alive cat, but never the superposition (14.91). In an attempt to escape the paradox, and on the assumption that $|\Psi\rangle$ describes the state of each individual system, the postulate of the collapse or reduction of the wave function is introduced, according to which it is instantly reduced (collapsed) by the act of observation to $|0\rangle_N |1\rangle_C$ if the nucleus has not yet decayed (and the cat is still alive), or to $|1\rangle_N |0\rangle_C$ if the nucleus has decayed and the cat is found dead.

From the ensemble perspective, however, there is no physical collapse, nor is the situation paradoxical. Even before measuring, the cat is either alive or dead. The observation in each individual instance (i.e., each element of the ensemble) gives information that can be used to construct a new description; for example, we can construct states representing separate ensembles of dead and alive cats. This happens with every statistical description (classical or quantum) as a result of using the new information. In other words, a statistical description is never complete nor does it exhaust the possibilities, so that the increase in information allows the description to be improved.

We conclude that what is paradoxical is not QM itself, but its application to a system that is obviously not quantum. It is inappropriate to conceive of a system consisting of a macroscopic object composed by a large number of particles, such as a real cat (plus the decaying atomic nucleus) as being in a superposition state as described by Eq. (14.91), called a Schrödinger cat state, because of the unavoidable lack of quantum coherence; see Section 15.5.7.

Schrödinger cats have become a paradigm of nonclassical, purely quantum mechanical states and a subject of intense investigation. Cat states with crystals weighing a few micrograms have already been produced; however, realizing a coherent superposition of larger objects in macroscopically distinct states, remains a challenge due to the inevitable decoherence induced by the environment.

14.5.4 The EPR *Gedankenexperiment*

For the sake of clarity in the following discussion, we will use a variation of the original EPR example given by Bohm and Aharonov, called EPRB. Consider a composite particle with spin zero at rest, which spontaneously decays into two particles, each with spin 1/2, moving in opposite directions along the x-axis. The wave function describing the spin state of the bipartite system is given by Eq. (14.89), which represents a scalar with spin 0, that is, with no privileged direction; the state is completely isotropic.

Let us consider this system from the EPR point of view. When the particles are so far apart that there is no interaction between them, we measure (with a Stern–Gerlach apparatus) the spin of particle 2 along an arbitrary direction z, perpendicular to x;

suppose the result is $+1/2$. This tells us that if the z-projection of the spin of particle 1 (which we have not measured) were measured, it would be $-1/2$. For EPR this result, predicted with probability 1, means that particle 1 has, in some appropriate sense, a spin $-1/2$ on the z-axis.

We now decide to measure spin 2 along the y-axis instead of the z-axis. Suppose we get $+1/2$ again. This means that the y-projection of spin 1 would be $-1/2$ with probability 1, a prediction that is just as legitimate as the previous one. This imaginary procedure leads us to conclude that the spin of particle 1 has (in some sense) a well-defined value simultaneously in both the y and the z directions. In QM, however, these values cannot coexist, since there are no simultaneous eigenstates of the noncommuting operators σ_y and σ_z. The conclusion of EPR is that not all objective properties of a quantum system are contained in the description provided by the wave function, so the theory is incomplete – like any statistical theory. The assertion of the incompleteness of QM is known as the EPR theorem.

At present, the same thought experiment and a similar analysis are commonly used to conclude, not that QM is incomplete, but that the dynamical variables do not possess well-defined values (or have no values at all) prior to their measurement, or, in plain and more frequent words, that it is the observation that precisely defines or assigns the value. The observer and his measuring instruments enter actively into the description, but none of this has any place in the Hamiltonian of the system.

14.5.5 The Bell inequalities

The EPR debate leads naturally to the question: Can the quantum description be completed? One of the root causes of the belief about the impossibility of completing the description is the no-hidden-variable theorem first formulated by J. von Neumann[13] in 1932. Shortly after its publication, in 1935, the proof was shown to be flawed by the young German mathematician Grete Hermann (1901–1984), but her work went largely unrecognized. After Bohm (1952) gave a constructive proof of the possibility of a hidden-variable (HV) theory, Bell rediscovered the error in von Neumann's proof in 1964 and began to formulate his famous inequalities.[14]

Thirty years after EPR, Bell (1964) used the thought-experiment method to establish his theorem, which has become a major attraction for theoretical and experimental work on the foundations of quantum theory. His aim was to obtain a result that would allow testing the hypothesis that the quantum description can be completed by introducing HV. The introduction of such variables would serve to restore determinism in the completed description. For example, if measuring the momentum of a particle in two equivalent experiments yields p_1 in one case and p_2 in the other, it is assumed that

---

[13] János (John) Lajos Neuman (1903–1957) was an outstanding Hungarian–US polymath who made fundamental contributions to mathematics, theoretical physics, computation, economics, cybernetics and other fields; he is considered one of the most important mathematicians of the twentieth century.

[14] Over the years, a number of theorems have appeared claiming the impossibility of constructing HV theories consistent with QM. They all ended up specifying the kind of HV that were forbidden or allowed. However, this is not a distinguishing feature of quantum theory; no-go theorems appear in many branches of knowledge, from social issues such as national censuses to mathematical questions (as is well known thanks to Gödel's theorems), with a relevant position in current information theory (see, e.g., no-go theorems in Wikipedia).

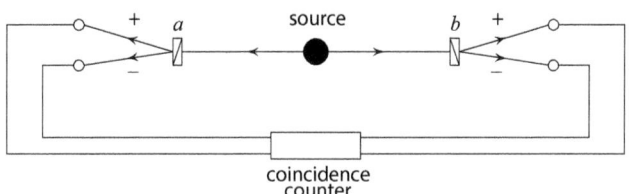

Figure 14.2 Diagram of Bell's test with coupled spin-1/2 particles in a singlet (total spin 0) state. The orientations a and b of the Stern–Gerlach apparatuses can be adjusted by the experimenter. The individual coincidences (++, --, +- and -+) registered are added to obtain the experimental correlation $C_E(a,b)$.

this is because the variables that determine the precise value of p in the experiment, which are hidden to the current quantum description, had a value λ_1 in the first case and a different value λ_2 in the second. If we had access to the λ variables (which would then no longer be hidden and belong to a more detailed theory than current QM) we would be able to predict or know exactly the value of p in each case.

Following Bell's line of argument, let us resort to the EPRB version and analyze the correlation of the projections of spins 1 and 2 along directions a, b, respectively, for a two-spin entangled system described by Eq. (14.89), see Fig. 14.2. The condition $S = 0$ clearly indicates that the two 1/2-spins are correlated; the purpose is to determine whether the quantum correlation can be explained in terms of HV and, if so, to draw some conclusion about the nature of these variables.

Let us first outline the experimental procedure. Since the singlet state is isotropic, we can make the z-axis coincide with the direction of a. Suppose we measure the projection of spin 1 along a; the result, A, will be either $+1$ or -1. We do the same with spin 2, this time measuring it along b; the result, B, will be either $+1$ or -1. The large separation between the two particles means that the measurements are independent and there is no a priori reason to consider the possibility of a mutual influence. From the experimental data collected after a large number N of measurements, we can calculate the quantity

$$C_E(a,b) = \lim_{N \to \infty} \frac{1}{N} \sum_{n=1}^{N} A_n B_n, \qquad (14.92)$$

where A_n, B_n are the results ± 1 of measurement n. This expression gives the experimental value of the correlation of the projections of spins 1 and 2 measured along a and b, respectively.

To establish an inequality, we consider four series of experiments that consist in measuring first in the directions a, b, then a, b', then a', b, and finally a', b'. Assuming that the values of the four outcomes A, A', B, B' are *predetermined* – that is, that the HV λ assigned to each and every particle is independent of the setup and of the measurement[15] – we write, for every λ,

$$S_E(a,a',b,b') = A(a)B(b) - A(a)B'(b') + A'(a')B(b) + A'(a')B'(b'), \qquad (14.93)$$

or in shorthand notation,

---

[15] More explicitly, this assumption means that the values of each outcome do not depend on the specific series (or measurement context), be it $(a,b), (a,b'), (a',b)$ or (a',b'). This assumption – which has important implications, as will be seen below – is often referred to in the literature as measurement independence or context independence.

$$S_E = A(B - B') + A'(B + B'). \tag{14.94}$$

Since $B' = \pm B$, one of the expressions in the parentheses always vanishes, which gives

$$S_E = \pm 2. \tag{14.95}$$

Therefore, the average of $|S_E|$ taken over many measurements must satisfy the inequality

$$\overline{|S_E|} = \overline{|A(B - B') + A'(B + B')|} \leqslant 2. \tag{14.96}$$

The theoretical equivalent of (14.92), assuming the existence of a set of variables λ that determine the result of each measurement, with $\rho(\lambda)$ being the corresponding probability, is

$$C_B(\boldsymbol{a}, \boldsymbol{b}) = \int_\Lambda A(\boldsymbol{a}, \lambda) B(\boldsymbol{b}, \lambda) \rho(\lambda) d\lambda, \tag{14.97}$$

where the subindex B refers to the correlation under the Bell conditions, that is, the assumption of deterministic HV and the requirement of factorization,

$$(AB)(\boldsymbol{a}, \boldsymbol{b}; \lambda) = A(\boldsymbol{a}, \lambda) B(\boldsymbol{b}, \lambda), \tag{14.98}$$

which implies that the probability distributions for the two particles are independent of each other: $A(\boldsymbol{a}, \lambda)$ does not depend on the direction $\boldsymbol{b}$ in which spin 2 is measured, and likewise $B(\boldsymbol{b}, \lambda)$ does not depend on $\boldsymbol{a}$, in accordance with the general principle of locality.[16] The procedure used to arrive at the inequality for $\overline{|S_E|}$, Eq. (14.96), can now be applied to (14.97) to obtain

$$\overline{|S_B|} = \left| C_B(\boldsymbol{a}, \boldsymbol{b}) - C_B(\boldsymbol{a}, \boldsymbol{b}') + C_B(\boldsymbol{a}', \boldsymbol{b}) + C_B(\boldsymbol{a}', \boldsymbol{b}') \right| \leq 2. \tag{14.99}$$

This result, called the CHSH inequality after its authors, J. Clauser,[17] M. Horne, A. Shimony and R. Holt, is a variant of the original Bell inequality (BI).[18]

On the other hand, the quantum formula for the correlation of the spin projections is obtained as the average value of the product $(\hat{\boldsymbol{\sigma}}_1 \cdot \boldsymbol{a})(\hat{\boldsymbol{\sigma}}_2 \cdot \boldsymbol{b})$, taken over the singlet state of spin zero given by Eq. (14.89),

$$C_Q(\boldsymbol{a}, \boldsymbol{b}) = \langle 00 | \hat{\boldsymbol{\sigma}}_1 \cdot \boldsymbol{a} \hat{\boldsymbol{\sigma}}_2 \cdot \boldsymbol{b} | 00 \rangle. \tag{14.100}$$

For the calculation of this quantity we may take, without loss of generality (because of the isotropy of the singlet state), the direction $\boldsymbol{a}$ along the z-axis. This diagonalizes the operator $\hat{\boldsymbol{\sigma}}_1 \cdot \boldsymbol{a}$, with $\langle \pm | \hat{\boldsymbol{\sigma}}_1 \cdot \boldsymbol{a} | \pm \rangle = \pm 1$, and therefore

$$C_Q(\boldsymbol{a}, \boldsymbol{b}) = \tfrac{1}{2} \Big(\langle - | \hat{\boldsymbol{\sigma}}_2 \cdot \boldsymbol{b} | - \rangle - \langle + | \hat{\boldsymbol{\sigma}}_2 \cdot \boldsymbol{b} | + \rangle \Big) = -b_z = -\cos\theta,$$

---

[16] As mentioned earlier, according to the general principle of locality there is no reason to consider a mutual influence between the results of two spins measured in spatially separated regions. The idea that two particles can interact instantaneously over a distance – accepted until now by not a few quantum physicists – was criticized by Einstein in a letter to M. Born, with the qualifier *spukhafte Fernwirkung* (spooky action at a distance).

[17] John Francis Clauser (1942) is a US-born theoretical and experimental physicist known for his contributions to the foundations of QM, for which he received the 2022 Nobel Prize.

[18] Several variants of the original Bell inequality have been proposed; here they are generally referred to as BI. The topic has undergone much development and has been the subject of considerable controversy and objection, including the derivation of concrete local and realistic counterexamples that violate Bell's inequalities.

or, in more general terms,
$$C_Q(\boldsymbol{a}, \boldsymbol{b}) = -\boldsymbol{a} \cdot \boldsymbol{b} = -\cos\theta_{ba}, \qquad (14.101)$$

where θ_{ba} is the angle formed by the unit vectors $\boldsymbol{a}$ and $\boldsymbol{b}$.

Bell's theorem is established by showing that the result (14.101) predicted by QM is incompatible with the CHSH inequality (14.99). This incompatibility can be checked by selecting certain directions $\boldsymbol{a}, \boldsymbol{a}', \boldsymbol{b}, \boldsymbol{b}'$. Specifically, for $\boldsymbol{a} = 90°, \boldsymbol{a}' = 0, \boldsymbol{b} = 225°$, $\boldsymbol{b}' = 135°$ (all lying on the yz plane), you get

$$\overline{|S_Q|} = \frac{4}{\sqrt{2}} = 2\sqrt{2}, \qquad (14.102)$$

which evidently is in contradiction with (14.99).

Numerous experiments over the past 50 years have confirmed beyond reasonable doubt the quantum prediction (14.101) and thus the violation of the BI. A prevailing conclusion is that QM cannot be completed with local HV, which in turn is commonly interpreted to mean that QM is incompatible with a local realistic interpretation. But this is not a foregone conclusion.

As we have seen earlier, a *central assumption* in the derivation of the BI is that the outcome of every measurement on particle 1 and on particle 2 is predetermined by a specific value of the HV λ; thus, for example, the value $A(a, \lambda)$ is independent of whether it is measured along with B or with B'. This assumption allowed grouping the terms in the inequality (14.96) or its equivalent in terms of correlations, the BI (14.99), which conflicts with QM (and with experiment). Therefore, (14.99) tells us that QM cannot be completed with (local) *deterministic* HV.

If the introduction of a distribution of deterministic HV cannot complete the quantum mechanical description, the natural conclusion is that a random element is still there. Throughout this text we have referred several times to the fluctuations inherent in the quantum mechanical description and have identified the ZPF as a source. Only a finer and deeper description, aimed at explaining the ultimate cause of this (then only apparent) randomness, could remove the indeterminism of the current quantum formalism.[19] Locality and realism are still alive; there is no "spooky action at a distance," but we have to be content with a statistical description for now.

Problems

P14.1 Show that
$$[\widehat{\boldsymbol{L}} \cdot \widehat{\boldsymbol{S}}, \widehat{J}_i] = 0, \qquad (14.103)$$

$$[\widehat{\boldsymbol{L}} \cdot \widehat{\boldsymbol{S}}, \widehat{\boldsymbol{J}}^2] = [\widehat{\boldsymbol{L}} \cdot \widehat{\boldsymbol{S}}, \widehat{\boldsymbol{L}}^2] = [\widehat{\boldsymbol{L}} \cdot \widehat{\boldsymbol{S}}, \widehat{\boldsymbol{S}}^2] = 0. \qquad (14.104)$$

---

[19] This is the idea behind the superdeterministic theories advocated by some authors, notably by G. 't Hooft (2016) in the framework of his cellular automaton theory.

P14.2 Any analytic function of the Pauli matrices can be written as a linear expression of these matrices plus the unit 2×2 matrix. Give the following expressions this form,

$$(1+\sigma_x)^n, \quad (1+\sigma_x)^{1/2}, \quad (a\sigma_x + b\sigma_y)^2; \tag{14.105}$$

$$\sigma_x^{-k}, \quad (a+\sigma_x)^{-1} \quad (a \neq 1). \tag{14.106}$$

P14.3 Show in detail that

$$e^{i\hat{\sigma} \cdot \hat{n}\theta} = \cos\theta + i\hat{\sigma} \cdot \hat{n}\sin\theta. \tag{14.107}$$

P14.4 Construct the spinors that are eigenfunctions of the operator

$$\widehat{S} = \widehat{S}_x \cos\theta + \widehat{S}_y \sin\theta \tag{14.108}$$

and show that their eigenvalues are $\lambda = \pm 1$. Apply your result to the particular cases $\theta = 0, \frac{\pi}{2}, \pi, 2\pi$.

P14.5 Solve the Pauli equation for a free particle. Which are the conserved quantities?

P14.6 Show that if the magnetic field varies only with time, the Pauli wave function can be factored into a spatial part that satisfies the Schrödinger equation with no magnetic field and a spinorial part $|\chi\rangle$, the solution of the equation

$$i\hbar \frac{\partial |\chi(t)\rangle}{\partial t} = \mu_0(\hat{\sigma} \cdot \boldsymbol{B})|\chi(t)\rangle. \tag{14.109}$$

P14.7 Show that

$$\langle JJ|\widehat{S}_z|JJ\rangle = \frac{J(J+1) + S(S+1) - L(L+1)}{2(J+1)}. \tag{14.110}$$

Hint: First prove that $\widehat{\boldsymbol{J}} \cdot \widehat{\boldsymbol{S}} = (\widehat{J}_z + 1)\widehat{S}_z + \widehat{S}_- \widehat{J}_+ + \widehat{J}_- \widehat{S}_+$, in units $\hbar = 1$.

P14.8 Show that the mass appearing in the relativistic correction to the kinetic energy of the electron of a hydrogen-like atom, $-p^4/8m^3c^2$, is the mass of the electron and not the reduced mass.

P14.9 Show that the correction to the energy of a hydrogen-like atom whose nucleus is modeled as a sphere of radius R_N with uniform charge distribution is given by

$$\delta E = \frac{2}{5} \frac{Z^4 e^2 R_N^2}{a_0^3 n^3} \delta_{l0}. \tag{14.111}$$

Note: In the nuclear region we can make the approximation $|\psi(r)|^2 \simeq |\psi(0)|^2$.

P14.10 Calculate the spin–orbit coupling effect on the energy of an isotropic 3D harmonic oscillator of mass m and frequency ω.

P14.11 Determine the energy eigenvalues and eigenfunctions for a charged particle (without spin) moving in a space occupied by an electric field and a magnetic field, both uniform and constant, whose directions are perpendicular to each other.

P14.12 Determine exactly the energy spectrum of a charged (spinless) isotropic oscillator immersed in a constant, uniform magnetic field.

P14.13 An electron is in a magnetic field that varies with time according to the law

$$B_x = B\sin\theta \cos\omega t, \quad B_y = B\cos\theta \sin\omega t, \quad B_z = B\cos\theta. \quad (14.112)$$

At time $t = 0$ the projection of the spin in the direction of the field is $+1/2$. Find the probability as a function of time that the particle is in the $-1/2$ spin state in the direction of the field, for $t > 0$.

P14.14 Three particles of spin 1/2, placed at the corners of an equilateral triangle, are described by the interaction Hamiltonian

$$\widehat{H} = \tfrac{1}{3}\lambda \left(\hat{\sigma}_1 \cdot \hat{\sigma}_2 + \hat{\sigma}_1 \cdot \hat{\sigma}_3 + \hat{\sigma}_2 \cdot \hat{\sigma}_3 \right). \quad (14.113)$$

List the energy levels and their degeneracy.

P14.15 Consider a neutron interference experiment in which the two beams have opposite polarizations on the z-axis, so that the superposition of the beams at the output of the instrument has the spin in the xy-plane. Under these conditions, if the spin of one of the two branches is inverted by means of a magnetic field, the final spin will be completely polarized on the z-axis. Show that this is so.

Bibliographical Notes

For a detailed discussion of entanglement and some of its applications, see Auletta, Fortunato, and Parisi (2009).

For Bohm's variant of the EPR thought experiment with an entangled spin pair (EPRB), see Bohm (1951), Bell (1964).

Von Neumann's theorem of 1932 was followed by a series of increasingly refined variants. For a survey, see Avis, Moriyama, and Owari (2009).

For a concrete derivation of the entangled spin pair with local HV that reproduces the quantum correlation and thus violates the BI, see Cetto (2022).

15 Identical-Particle Systems: Quantum Statistics. The Density Matrix

In previous chapters we studied some effects of the electron spin that could be treated as perturbations of an otherwise spinless system. This could lead one to believe that spin effects are mere corrections that can be taken into account or not, depending on the required precision of the calculation or the observation one wishes to explain. This conclusion is certainly wrong, since the spin (whether integer or half-integer) can have crucial effects that are not only impossible to obtain by approximate methods, but in many cases determine fundamental features of the system's behavior. In this chapter we will focus on such effects, collectively called exchange effects. Their existence is characteristic of states of two or more identical particles that are invariant with respect to their exchange. This provides an opportunity to briefly discuss some quantum collective phenomena that are a defining feature of condensed matter.

Most real quantum systems are composed of subsystems in different quantum states. For a systematic study of such composite systems – which are mixtures of pure states – it is necessary to adopt a statistical approach and to introduce the relevant concepts for their description. For this reason, in Section 15.5 we present an introduction to the density matrix as a convenient tool for dealing with statistical quantum mixtures. The chapter then concludes with a brief discussion of decoherence.

15.1 Multipartite Systems

We will begin our analysis by studying the structure of the Hamiltonian for a general multipartite system. This can contain from a few particles, such as the electrons of an atom, to an arbitrarily large number, such as the molecules in a macroscopic volume of a homogeneous gas.

The Hamiltonian of a system of N particles (identical or different, for the moment) subject to both internal interactions and external fields, is

$$\widehat{H} = -\hbar^2 \sum_{i=1}^{N} \frac{1}{2m_i} \nabla_i^2 + U(\{r_i\}). \tag{15.1}$$

In the absence of magnetic fields, which we will assume to be the case, U does not depend on the spin operators. We can write U as the sum of two terms

$$U(\{r_i\}) = \sum_i V_i^{\text{ext}}(r_i) + V(\{r_i\}), \tag{15.2}$$

where the potential due to the external fields, V_i^{ext}, is a function of the position of particle i, whereas the potential due to the internal interactions, V, depends normally on the relative coordinates,

$$V(\{r_i\}) = V(\{r_i - r_j\}). \tag{15.3}$$

This can be seen by turning off the external fields. The total momentum must then be conserved, which means that the momentum operator $\widehat{P}$ must commute with the Hamiltonian. Indeed, with

$$\widehat{P} = \sum_i \widehat{p}_i = -i\hbar \sum_i \nabla_i, \tag{15.4}$$

and V written in the form (15.3), we have

$$[\widehat{P}, H] = [\widehat{P}, V] = -i\hbar \sum_i \nabla_i V(\{r_i - r_j\}) \tag{15.5}$$

$$= -i\hbar(V' - V' + V' - V' + \cdots + V' - V') = 0.$$

The result is, in fact, obvious, since it reduces to the demand that the Hamiltonian of a system not acted upon by external fields must be invariant with respect to arbitrary translations of the origin of the coordinates. Hence, V can only depend on internal (relative) coordinates.

15.1.1 Equal-particle systems: Exchange degeneracy

Let us now consider a system where all particles are equal;[1] in this case, (15.1) becomes

$$\widehat{H} = \sum_i \widehat{H}_i(r_i) + V(\{r_i - r_j\}), \tag{15.6}$$

where $\widehat{H}_i(r_i)$ is the Hamiltonian of particle i moving in the common potential. Since V depends only on the relative coordinates, (15.6) is not affected by the exchange of particles (only the order in which the terms appear in the sum changes). We conclude that the Hamiltonian of a system of identical particles is invariant with respect to the permutation of any pair of them, simply because they are all identical.[2]

The invariance of the Hamiltonian with respect to permutations leads to a new type of degeneracy: the exchange degeneracy, which consists in the fact that all the eigenfunctions of $\widehat{H}$ obtained by pairwise exchange (interchange) of equal particles are degenerate (they all have the same energy). However, not all of them are invariant upon permutation of particles. Our main objective here is to precisely construct the simultaneous eigenfunctions of the Hamiltonian and permutation operators. Before we proceed, it is useful to obtain some preliminary results.

---

[1] Often the system contains one or more particles that are different from the rest, such as the nucleus of an atom and its electrons; those different particles are not considered in the discussion that follows.
[2] We are assuming that we can label the particles in order to identify and number them. This hypothesis is not as unobjectionable in the quantum case as it might seem at first sight, since the quantum (statistical) description is oblivious to any notion of individuality; a consistent description of identical particles is only one that does not identify them, but rather treats them with full equivalence, as will be seen later. This is the reason why quantum theory has only quantum statistics, that is, those based on the identity (actually statistical indifference) of particles. It is common in the quantum literature to find the rather subjective term indistinguishability to refer to this property.

Let $\widehat{P}_{ij}$ be the permutation operator that interchanges the particles i and j; the Hamiltonian $\widehat{H}_P$ with the permuted particles is $\widehat{H}_P = \widehat{P}_{ij}\widehat{H}\widehat{P}_{ij}^{-1}$. From the invariance condition $\widehat{H} = \widehat{H}_P$ we see that $\widehat{P}_{ij}$ and $\widehat{H}$ commute, $\widehat{P}_{ij}\widehat{H} = \widehat{H}\widehat{P}_{ij}$, which means that simultaneous eigenstates of $\widehat{H}$ and $\widehat{P}_{ij}$ can be constructed.

When $\widehat{P}_{ij}$ acts on a function, it interchanges all the variables of the particles i and j, including the spin variables. If $\xi_i = (r_i, \sigma_i)$ is the set of space and spin variables of particle i, we write

$$\widehat{P}_{12}f(\xi_1, \xi_2, \xi_3 \ldots) = f(\xi_2, \xi_1, \xi_3 \ldots), \tag{15.7}$$

and so on. The eigenvalues of $\widehat{P}_{ij}$ are easily determined as follows. Let $\Psi = \Psi(\xi_1, \xi_2, \xi_3 \ldots)$ be an eigenfunction of $\widehat{P}_{ij}$; from the eigenvalue equation

$$\widehat{P}_{ij}\Psi = \lambda_{ij}\Psi; \tag{15.8}$$

we can see that

$$\widehat{P}_{ij}^2\Psi = \lambda_{ij}P_{ij}\Psi = \lambda_{ij}^2\Psi = \Psi, \tag{15.9}$$

that is, after two successive exchanges of i and j, the original function is recovered. Therefore,

$$\lambda_{ij} = \pm 1. \tag{15.10}$$

Consequently, the eigenfunctions of $\widehat{P}_{ij}$ are either invariant or change sign upon permutation of particles i and j. In the first case the wave function is symmetric (S) and in the second, it is antisymmetric (A) with respect to particles i and j. If the wave function is S or A at $t = 0$, it will be so at all t, since λ is an integral of motion.[3]

To illustrate, consider the case of three particles numbered 1, 2 and 3, two of which are in state m with energy E_m and the other in state n with energy E_n. The energy of the system is $E = 2E_m + E_n$. However, the state vector is not well defined because we do not know which particle has which energy. By renumbering the particles in all possible ways (without distinguishing between those with the same energy E_m) we get three different vectors $|E_m E_m E_n\rangle, |E_m E_n E_m\rangle, |E_n E_m E_m\rangle$. The most general state vector is an arbitrary (normalized) linear combination of these,

$$|E\rangle = a|E_m E_m E_n\rangle + b|E_m E_n E_m\rangle + c|E_n E_m E_m\rangle. \tag{15.11}$$

This state vector does not have a well-defined symmetry with respect to $\widehat{P}_{12}$, for example, and does not in general satisfy condition (15.8).[4] This means that $|E\rangle$ changes upon the permutation even though physically nothing has been done; therefore it does

---

[3] In the quantum literature it is common to blame the indistinguishability of identical particles for the need to introduce the total (anti)symmetrization characteristic of quantum statistics. However, apart from the fact that this notion introduces an unwanted subjective element into the physical description, as already mentioned, there are classical situations where the distinguishability of equal components is irrelevant for statistical purposes. Consider, for example, statistics on the occupancy of hospital beds in a city, a case in which it matters only whether a patient occupies a bed at a given time, and not who occupies it. If, in one of the hospitals (or in all of them, for that matter) the use of floors by specialty is changed from one day to the next, the statistics do not change at all. This indistinguishability does not imply quantum statistics. What is important in the quantum case is that identical particles have the same relevant physical properties.

[4] A side note. This example illustrates a very common phenomenon, namely that the solution to a differential equation may have fewer symmetries than the equation it solves. In such cases it is usually said that there has been a spontaneous breaking of the symmetry (note the metaphorical use of the term

not represent a physical solution in general. We can construct physical solutions by demanding them to be either symmetric ($\lambda_{12} = 1$) or antisymmetric ($\lambda_{12} = -1$) eigenfunctions of $\widehat{P}_{12}$; it suffices to take $b = c$, with a arbitrary in the first instance,

$$|E\rangle_S = a|E_1 E_1 E_2\rangle + b(|E_1 E_2 E_1\rangle + b|E_2 E_1 E_1\rangle) = \widehat{P}_{12}|E\rangle_S, \tag{15.12}$$

and $c = -b$, with $a = 0$ in the second case,

$$|E\rangle_A = b(|E_1 E_2 E_1\rangle - |E_2 E_1 E_1\rangle) = -\widehat{P}_{12}|E\rangle_A. \tag{15.13}$$

However, the physical solutions must also be symmetric or antisymmetric with respect to $\widehat{P}_{13}$ and $\widehat{P}_{23}$. To complicate matters, the exchange operators do not necessarily commute with each other. For example,

$$\widehat{P}_{13}\widehat{P}_{12}\Psi(123) = \widehat{P}_{13}\Psi(213) = \Psi(231), \tag{15.14}$$

whereas

$$\widehat{P}_{12}\widehat{P}_{13}\Psi(123) = \widehat{P}_{12}\Psi(321) = \Psi(312). \tag{15.15}$$

In a sense, identical entities are being treated differently. This raises a key question: how do we construct the eigenfunction that describes the physical state of the system?

Let us consider first the simple case of only two identical particles, so that

$$\widehat{H}(\xi_1, \xi_2) = \widehat{H}(\xi_2, \xi_1). \tag{15.16}$$

Due to the symmetry of the problem, we can write the Schrödinger equation

$$\widehat{H}(\xi_1, \xi_2)\Psi(\xi_1, \xi_2) = E\Psi(\xi_1, \xi_2) \tag{15.17}$$

in the equivalent form, obtained by interchanging the two particles,

$$\widehat{H}(\xi_2, \xi_1)\Psi(\xi_2, \xi_1) = E\Psi(\xi_2, \xi_1) \tag{15.18}$$

and use Eq. (15.16) to get

$$\widehat{H}(\xi_1, \xi_2)\Psi(\xi_2, \xi_1) = E\Psi(\xi_2, \xi_1). \tag{15.19}$$

For Eqs. (15.17) and (15.19) to be consistent, it is required that

$$\Psi(\xi_2, \xi_1) = \pm \Psi(\xi_1, \xi_2), \tag{15.20}$$

which is Eq. (15.8) for this particular case. Therefore, the eigenfunctions of $\widehat{H}$ must be symmetrized. This is easily done by considering the symmetrized combinations $\Psi(\xi_1, \xi_2) + \Psi(\xi_2, \xi_1)$ and $\Psi(\xi_1, \xi_2) - \Psi(\xi_2, \xi_1)$ after normalization.

"spontaneous"). For example, the Schrödinger equation of the H atom has spherical symmetry, but only its s-state eigenfunctions have this symmetry.
A more interesting example, which helps to clarify the concept of spontaneous symmetry breaking, occurs in a high-temperature ferromagnetic system with spherical symmetry. If the temperature is slowly lowered beyond the Curie point, the initially disordered spins spontaneously order themselves (i.e., through their interactions) and the material acquires a certain magnetization. The system has undergone a phase transition, moving from its initial spherical state to a state with axial symmetry, that is, the phase transition is accompanied by a spontaneous reduction (breaking) of the symmetry. The density matrix (see Section 4) describing the final state has less symmetry than the Hamiltonian.

15.1.2 Symmetric and antisymmetric states

Similarly, in the general case of more than two particles, the total wave function must be either S or A. These fully symmetrized wave functions are simultaneous eigenfunctions of all exchange operators $\widehat{P}_{ij}$ with the same eigenvalue, so they all commute. It is clear that the states with all λ_{ij} equal (either $+1$ or -1) are those with the maximum symmetry that can be constructed. They are the only ones that treat equal particles in an equivalent way. Not surprisingly, these functions are the most physically interesting, as we will see in the next section.

The symmetric N-particle wave function can easily be constructed from the $N!$ degenerate wave functions $\Psi(\xi_i)$ as follows,

$$\Psi^S = C \sum_{P_k} \widehat{P}_k \Psi(\xi_1, \xi_2, \ldots, \xi_N), \tag{15.21}$$

where the sum runs over all possible pairwise permutations P_k; C is the normalization constant. The antisymmetric wave function is constructed in a similar way, but alternating the signs of each term according to the parity of the permutation. More specifically, if r_k is the number of pairs of particles that must be interchanged to achieve the permutation P_k, we have

$$\Psi^A = C \sum_k (-1)^{r_k} \widehat{P}_k \Psi(\xi_1, \xi_2, \ldots, \xi_N). \tag{15.22}$$

The coefficient $(-1)^{r_k}$ is 1 if P_k is even and -1 if P_k is odd; therefore, if we perform a single permutation of any pair of particles in (15.22), Ψ^A changes sign.

The wave functions (15.21) and (15.22) are generally very complicated, and it is difficult to say more about them without going into detail. However, there is one case of great practical importance where the wave functions can be expressed more explicitly, namely when the interaction between the particles is weak enough to be considered a perturbation. In this case, the unperturbed wave functions correspond to $V = 0$, and as Eq. (15.6) shows, the Hamiltonian reduces to the sum of one-particle Hamiltonians, so that Ψ can be written as the product of N one-particle eigenfunctions,

$$\Psi(\xi_1, \xi_2 \ldots, \xi_N) = \psi_{n_1}(\xi_1)\psi_{n_2}(\xi_2)\ldots\psi_{n_N}(\xi_N). \tag{15.23}$$

The N-particle wave function must then be symmetrized or antisymmetrized according to (15.21) or (15.22).

Exercise E15.1. Projectors for bipartite states

Let $|n_1 n_2\rangle = \frac{1}{\sqrt{2}}(|n_1\rangle|n_2\rangle + |n_2\rangle|n_1\rangle)$ if $|n_1\rangle \neq |n_2\rangle$, $|n_1 n_2\rangle = |n_1\rangle|n_2\rangle$ if $|n_1\rangle = |n_2\rangle$, describe a symmetric state of two identical particles. Prove that $\widehat{\Omega} = \sum_{n_1 \geq n_2} |n_1 n_2\rangle \langle n_1 n_2|$ is the projector on symmetric states.

Solution. For convenience we write the state $|n_1 n_2\rangle$ in the form

$$|n_1 n_2\rangle = \frac{C_n}{\sqrt{2}}(|n_1\rangle|n_2\rangle + |n_2\rangle|n_1\rangle), \tag{15.24}$$

$$C_n = \begin{cases} 1 & \text{when } n_1 \neq n_2, \\ 1/\sqrt{2} & \text{when } n_1 = n_2. \end{cases}$$

To study the effect of $\widehat{\Omega}$ on an arbitrary state $|\psi\rangle = |m_1 m_2\rangle$, we go to the x-representation,

$$\langle x_1 x_2 | \widehat{\Omega} | m_1 m_2 \rangle = \int dx'_1 dx'_2 \langle x_1 x_2 | \widehat{\Omega} | x'_1 x'_2 \rangle \langle x'_1 x'_2 | m_1 m_2 \rangle. \quad (15.25)$$

We first determine the matrix element

$$\langle x_1 x_2 | \widehat{\Omega} | x'_1 x'_2 \rangle \quad (15.26)$$

$$= \sum_{n_1 \geq n_2} \tfrac{1}{2} C_n^2 \langle x_1 x_2 | (|n_1\rangle |n_2\rangle + |n_2\rangle |n_1\rangle) (\langle n_1 | \langle n_2 | + \langle n_2 | \langle n_1 |) |x'_1 x'_2\rangle$$

$$= \sum_{n_1 \geq n_2} \tfrac{1}{2} C_n^2 \left(\varphi_{n_1}(x_1) \varphi_{n_2}(x_2) + \varphi_{n_2}(x_1) \varphi_{n_1}(x_2) \right)$$

$$\times \left(\varphi^*_{n_1}(x'_1) \varphi^*_{n_2}(x'_2) + \varphi^*_{n_2}(x'_1) \varphi^*_{n_1}(x'_2) \right)$$

$$= \tfrac{1}{2} \cdot \tfrac{1}{2} \cdot 4 \varphi_n(x_1) \varphi_n(x_2) \varphi^*_n(x'_1) \varphi^*_n(x'_2)$$

$$+ \sum_{n_1 > n_2} \tfrac{1}{2} \varphi_{n_1}(x_1) \varphi_{n_2}(x_2) \left(\varphi^*_{n_1}(x'_1) \varphi^*_{n_2}(x'_2) + \varphi^*_{n_1}(x'_2) \varphi^*_{n_2}(x'_1) \right)$$

$$+ \sum_{n_1 > n_2} \tfrac{1}{2} \varphi_{n_2}(x_1) \varphi_{n_1}(x_2) \left(\varphi^*_{n_2}(x'_1) \varphi^*_{n_1}(x'_2) + \varphi^*_{n_1}(x'_1) \varphi^*_{n_2}(x'_2) \right).$$

In the last term we swap n_1 and n_2, $\sum_{n_1 > n_2} \to \sum_{n_1 < n_2}$, and combine with the previous term to write

$$\langle x_1 x_2 | \widehat{\Omega} | x'_1 x'_2 \rangle \quad (15.27)$$

$$= \tfrac{1}{2} \sum_{n_1} \sum_{n_2} \varphi_{n_1}(x_1) \varphi_{n_2}(x_2) \left(\varphi^*_{n_1}(x'_1) \varphi^*_{n_2}(x'_2) + \varphi^*_{n_1}(x'_2) \varphi^*_{n_2}(x'_1) \right)$$

$$= \tfrac{1}{2} \sum_{n_1} \varphi_{n_1}(x_1) \varphi^*_{n_1}(x'_1) \sum_{n_2} \varphi_{n_2}(x_2) \varphi^*_{n_2}(x'_2)$$

$$+ \tfrac{1}{2} \sum_{n_1} \varphi_{n_1}(x_1) \varphi^*_{n_1}(x'_2) \sum_{n_2} \varphi_{n_2}(x_2) \varphi^*_{n_2}(x'_1)$$

$$= \tfrac{1}{2} \delta(x_1 - x'_1) \delta(x_2 - x'_2) + \tfrac{1}{2} \delta(x_1 - x'_2) \delta(x_2 - x'_1).$$

Introducing this result in (15.25), we get

$$\langle x_1 x_2 | \widehat{\Omega} | m_1 m_2 \rangle \quad (15.28)$$

$$= \tfrac{1}{2} \int dx'_1 dx'_2 \left(\delta(x_1 - x'_1) \delta(x_2 - x'_2) \right.$$

$$\left. + \delta(x_1 - x'_2) \delta(x_2 - x'_1) \right) \langle x'_1 x'_2 | m_1 m_2 \rangle$$

$$= \tfrac{1}{2} \langle x_1 x_2 | m_1 m_2 \rangle + \tfrac{1}{2} \langle x_2 x_1 | m_1 m_2 \rangle = \tfrac{1}{2} \left(\langle x_1 x_2 | + \langle x_2 x_1 | \right) | m_1 m_2 \rangle,$$

which can be written as

$$\langle x_1 x_2 | \widehat{\Omega} | \psi \rangle = \langle x | \psi^S \rangle, \quad (15.29)$$

where $|\psi^S\rangle$ is the symmetric component of $|\psi\rangle$.

15.2 Bosons and Fermions. The Pauli Principle

We have just seen that physical stationary states of identical particles are those in which all particles play an equal role. This observation can be elevated to the category of principle by demanding that identical particles receive identical treatment in the theory. The only situation consistent with the principle of equal treatment is that the eigenvalues of all exchange operators $\hat{P}_{ij}$ are well defined, that is, that the wave functions are S or A.[5] It is important to note that these N-particle wave functions are nonfactorizable, therefore they describe entangled states, as discussed in Section 14.5 in connection with the two-particle states.

Which of the two possibilities is realized depends on the type of particles in question. It turns out that:

(i) Particles with integer spin are described by S wave functions in the stationary state ($\lambda_{ij} = 1$). These particles are generically known as bosons, and include pions (spin 0), photons (spin 1), etc.

(ii) Particles with half-integer spin are described by A wave functions in the stationary state ($\lambda_{ij} = -1$). These particles are called fermions. They include electrons, protons, and neutrons (all of which have spin 1/2), etc.[6]

This fundamental and very important correspondence, known as the *spin-statistics theorem*, was postulated by Pauli in 1939 and there are no known exceptions to it. For example, atoms and molecules will have S or A wave functions depending on whether their total electronic spin is integer or half-integer, respectively. This property is considered a postulate because it cannot be derived from other principles within the framework of nonrelativistic QM.[7]

According to the spin-statistics theorem, the stationary states of a system of identical particles of spin s are those whose wave function Ψ is transformed according to

$$\hat{P}_{ij}\Psi(1, 2, \ldots, i, \ldots, j \ldots, N) = (-1)^{2s}\Psi(1, 2, \ldots, j, \ldots, i \ldots, N). \quad (15.30)$$

Let us take first the case of bosons. The wave function must be of the form of (15.21), with the normalization coefficient determined as follows: The states n_j can be the same or different. Assume there are N_1 particles in state n_1, N_2 in state n_2 and so on, with $N_1 + N_2 + \cdots = N$; the occupancy numbers N_i fully specify the state. The S wave function contains $N!/N_1!N_2!\ldots$ terms, so, considering the orthonormality of the functions $\psi_{n_j}(\xi_i)$, we write

---

[5] Heisenberg and Dirac independently considered the possibility of using fully symmetrized or antisymmetrized wave functions as early as 1926.

[6] The neutron has a magnetic moment, despite being electrically neutral. This was an early indication that the neutron is not an elementary particle; in fact, it is made up of quarks, which are charged particles. Neutral elementary particles with spin, such as neutrinos, photons and the Z boson, do not have a magnetic moment; we know they have a spin because of the conservation of angular momentum.

[7] We recall from Section 14.5.1 that SED provides an explanation for the antisymmetry of the electron wave function as a consequence of the antiphase pair coupling to the background field modes. More generally, the spin-statistics theorem can be derived within relativistic quantum field theory, which involves postulates that are not part of QM; see Tomonaga (1997).

$$\Psi^S = \sqrt{\frac{N_1! N_2! \cdots}{N!}} \sum_{P_k} \widehat{P}_k \psi_{n_1}(\xi_1) \psi_{n_2}(\xi_2) \cdots, \qquad (15.31)$$

where the N_i can take any value between 0 and N, with the restriction that the total sum be N.

In the case of fermions, the wave function must be of the form of (15.22), which can be written as

$$\Psi^A = C \begin{vmatrix} \psi_{n_1}(\xi_1) & \psi_{n_1}(\xi_2) & \cdots & \psi_{n_1}(\xi_N) \\ \psi_{n_2}(\xi_1) & \psi_{n_2}(\xi_2) & \cdots & \psi_{n_2}(\xi_N) \\ \vdots & \vdots & \ddots & \vdots \\ \psi_{n_N}(\xi_1) & \psi_{n_N}(\xi_2) & \cdots & \psi_{n_N}(\xi_N) \end{vmatrix}. \qquad (15.32)$$

This is the *Slater determinant*.[8] Permuting any two particles here means swapping the corresponding columns, which changes the sign of the determinant, that is, of Ψ, as it should. However, we observe an extremely important new phenomenon: if in (15.32) we make any two states equal (e.g., if we put $n_1 = n_2$), the determinant will have two equal rows and give zero. This is always the case: there is no (stationary) state of a physical system in which two (or more) equal fermions share the *same* quantum numbers. This most important result is the Pauli principle or exclusion principle referred to in Section 14.1, which allows us to understand the periodic table of the elements as a consequence of the atomic structure of matter, as will be shown in Section 16.2. Pauli was able to infer this principle from an analysis of atomic structure and other data; we will apply it later to precisely solve this problem, although we have used it previously to explain metallic conductivity (Section 6.3). Although the Pauli principle is a particular result of a more general principle, the one that assigns fully antisymmetrized functions to any stationary system of identical fermions, it continues to be called a principle.

From the total antisymmetry of the wave function it follows that the fermion occupancy number of each state can only be 0 (empty state) or 1 (fully occupied state). Therefore $N_i! = 1$ (since $0! = 1! = 1$) and the normalized wave function is obtained by putting into (15.32)

$$C = \frac{1}{\sqrt{N!}}, \qquad (15.33)$$

as corresponds to the case in which we have $N!$ different states of equal weights. Note that the S wave function can be obtained with the help of the Slater determinant for the A wave function, but using only $+$ signs in the expansion.

15.2.1 The two quantum statistics

Bose–Einstein (B–E) statistics. Consider an ideal gas of bosons in equilibrium; this could even be a "gas" of photons. In a given volume at arbitrary temperature $T > 0$, one finds a certain number of particles in each of the possible quantum states, which

---

[8] John Clarke Slater (1900–1976) was a US-born theoretical physicist and chemist, who introduced the important Slater determinants for physical and chemical calculations. He wrote several highly recommended textbooks (if you are lucky enough to find them).

gives the quantum gas a characteristic statistical behavior. In the electromagnetic case, for example, the equilibrium distribution is given by Planck's law. An extreme case of this distribution occurs at very low temperatures (of the order of millionths of a Kelvin), at which the particles tend to fill the ground state and move very slowly, so that the de Broglie wavelength associated with their motion, $\lambda = h/p$, can reach dimensions similar to those of the container. As a result, virtually all the waves associated with the particles in the volume superimpose, leading to a high correlation between most of the bosons in the system. When this almost complete correlation is reached, a *Bose–Einstein condensate* spontaneously emerges and the system can exhibit vivid macroscopic effects of coherence and interference, as discussed in Section 15.4.

Fermi–Dirac (F–D) statistics. An ideal gas of fermions in equilibrium (e.g., electrons) has a very different statistical behavior. Each fermion in the system occupies a different quantum state, as can be seen from the vanishing of the wave function (15.32) when $\xi_1 = \xi_2$. In other words, the probability of two identical fermions being in the same state in the same system is zero. This makes the statistical behavior of the fermion gas sui generis. We had the opportunity to study some basic features of the F–D statistics in Section 6.3 (see Figure 6.6) and to apply it to the calculation of the specific heat of metals.

A simple example will allow us to clarify the idea behind the different statistics. Consider a system of two identical particles that can be in either state n or m, and assume that their interaction can be neglected. There are thus four possible product states of independent particles: $|nn\rangle = |n\rangle |n\rangle$, $|nm\rangle = |n\rangle |m\rangle$, $|mn\rangle = |m\rangle |n\rangle$, and $|mm\rangle = |m\rangle |m\rangle$. In a macroscopic problem, these four states would be counted as four possible and different cases, according to the methods of classical statistical mechanics, which lead to the Maxwell–Boltzmann distribution (M–B statistics). Suppose instead that the particles are bosons; we must construct fully symmetric, mutually orthogonal state functions, which gives three different possible states,

$$|nn\rangle^S = |n\rangle |n\rangle, \tag{15.34}$$

$$|nm\rangle^S = \tfrac{1}{\sqrt{2}}(|n\rangle |m\rangle + |m\rangle |n\rangle), \tag{15.35}$$

$$|mm\rangle^S = |m\rangle |m\rangle. \tag{15.36}$$

For fermions, instead, we must construct totally antisymmetric combinations, of which there is only one,

$$|nm\rangle^A = \tfrac{1}{\sqrt{2}}(|n\rangle |m\rangle - |m\rangle |n\rangle). \tag{15.37}$$

Thus for the given conditions of the problem, the M–B (classical) statistics admits four different states, the B–E statistics admits three, and the F–D statistics admits only one.

If the two particles are in the same (m or n) state, this happens in 2 out of 4 cases (relative probability 1/2 for each one) in the M–B case; in 2 out of 3 cases (relative probability 2/3) in the B–E case, and not at all in the F–D case (probability 0). It is therefore more likely to find clusters of bosons in the same quantum state than classical statistics predicts, and impossible to find in the F–D case. This is expressed by saying that bosons in the same state "attract each other," while fermions "repel each other." These effective interactions are not explicitly contained in the Hamiltonian, but have been introduced into the preceding description in a phenomenological way, by symmetrization or antisymmetrization of the state functions.

The fact that the state function is a symmetric or antisymmetric sum of terms introduces quantum correlations between the particles, as we learned from the discussion in Section 14.5 and in particular from the case of bipartite singlet spin studied in 14.5.5. The following examples illustrate the influence of the different statistics on the correlation between the position variables of the particles.

Exercise E15.2. Correlation between position variables

Analyze the correlation between the position variables of two equal particles whose entangled wave function is given by

$$|nm\rangle = \frac{1}{\sqrt{2}}(|n\rangle|m\rangle + \lambda|m\rangle|n\rangle), \quad \lambda = \pm 1, \tag{15.38}$$

and discuss the effect of the parameter λ.

Solution. As a measure of the correlation between r_1 and r_2, we shall calculate the covariance, defined as

$$\text{cov}(r_1, r_2) = \langle (r_1 - \langle r_1 \rangle) \cdot (r_2 - \langle r_2 \rangle) \rangle. \tag{15.39}$$

After some simple algebra, we obtain

$$\text{cov}(r_1, r_2) = \lambda |\langle n|r|m\rangle|^2, \tag{15.40}$$

where we took into account that the expectation values are the same for particle 1 and particle 2.

Note that the particle positions are only correlated if the state functions $|n\rangle$ and $|m\rangle$ overlap, i.e., if there are values of r for which both functions are simultaneously different from zero. In other words, the symmetrization effects are only important when the overlap of the wave functions is important. The covariance is negative for the antisymmetric wave function and positive for the symmetric one, as would be expected: in the first case the particles tend to avoid each other, while in the second case they tend to approach each other.

Exercise E15.3. Mean square distance between particles

Consider two entangled particles whose orbital wave function is given by (15.38). Calculate the expectation value of the square of the distance between the particles and discuss the result as a function of λ.

Solution. Using Eq. (15.38), the calculation is straightforward and gives

$$\langle nm|(r_1 - r_2)^2|nm\rangle_\lambda \tag{15.41}$$
$$= \langle n|r^2|n\rangle + \langle m|r^2|m\rangle - 2\langle n|r|n\rangle \cdot \langle m|r|m\rangle - 2\lambda|\langle n|r|m\rangle|^2.$$

The first three terms on the right do not depend on the symmetrization and represent the expectation value of $r_1^2 + r_2^2 - 2r_1 \cdot r_2$ as if the particles were independent; therefore we can write

$$\langle nm|(r_1-r_2)^2|nm\rangle_\lambda = \langle nm|(r_1-r_2)^2|nm\rangle_{\lambda=0} - 2\lambda|\langle n|r|m\rangle|^2. \tag{15.42}$$

The term with $\lambda = 0$ would be the mean square distance if the system followed the M–B statistics. To this classical result we must add the positive quantity $2|\langle n|r|m\rangle|^2$ if $\lambda = -1$ (antisymmetric wave function) or subtract the same quantity if $\lambda = +1$ (symmetric wave function). This shows that bosons tend to stay closer together than classical particles, while equal-spin fermions tend to stay further apart.

15.3 Effects of Statistics on the Energy Spectrum

The correlations discussed in the previous section have important consequences on the atomic energy levels. Let us consider the case of two electrons whose spin-independent interaction can be taken as a perturbation. (This calculation is used in Section 16.3 in connection with the study of the He atom.) The individual orbital wave functions in the absence of interaction are $\psi_n(r)$ and $\psi_m(r)$; the fully symmetrized orbital wave functions are given by Eq. (15.38), which we rewrite as

$$\psi_\pm = \frac{1}{\sqrt{2}}\left[\psi_n(r_1)\psi_m(r_2) \pm \psi_m(r_1)\psi_n(r_2)\right]. \tag{15.43}$$

The complete wave function is the product of an orbital function and a spin factor, since by hypothesis, the Hamiltonian is spin-independent. This wave function must be antisymmetric because we are dealing with electrons, so one of its factors must be symmetric and the other antisymmetric. From Eqs. (11.175), (11.177), (11.181), and (11.182) we see that the spin factors are symmetric for $S = 1$ and antisymmetric for $S = 0$. Therefore, the sign $\pm$ in Eq. (15.43) can be written as $(-1)^S$, and $\psi_\pm = \psi_S$.

Now let $V(|r_1 - r_2|)$ be the perturbation potential. Since this potential commutes with $\widehat{P}_{12}$, it does not mix states of different permutational symmetry and we can treat the problem as if it were nondegenerate. Therefore, to first order in perturbation theory the energy of the symmetric orbital state ($S = 0$) is

$$E_+ = E^{(0)} + \langle\psi_+|V|\psi_+\rangle, \tag{15.44}$$

and for the state with $S = 1$ it is

$$E_- = E^{(0)} + \langle\psi_-|V|\psi_-\rangle. \tag{15.45}$$

As we learned from the previous exercises, in the region $|r_1-r_2| \approx 0$, where $V(|r_1-r_2|)$ is normally important, these matrix elements have very different values. It is therefore convenient to write the correction as the sum of two terms,

$$\langle\psi_\pm|V|\psi_\pm\rangle = J \pm K. \tag{15.46}$$

The J term, given by

$$J = \int |\psi_n(r_1)|^2|\psi_m(r_2)|^2 V d^3r_1 d^3r_2, \tag{15.47}$$

is usually interpreted as the expectation value of the interaction potential when particle 1 (not one of the particles) is in state n and particle 2 in state m. Since particle 1 is not in state n all the time, etc., this interaction energy has to be corrected. The correction term, the exchange energy, is given by

$$K = \int \psi_n^*(r_1)\psi_m^*(r_2) V \psi_n(r_2)\psi_m(r_1) d^3r_1 d^3r_2. \tag{15.48}$$

The total energy,[9]

$$E_\pm = E^{(0)} + J \pm K, \tag{15.49}$$

shows that there is an energy difference of $2K$ between the states with $S = 0$ and $S = 1$, an effect known as exchange splitting (see Problem 15.5.2).

The previous result can be rewritten as an effective-coupling effect between the spins of the electrons, introducing Dirac's exchange operator

$$\widehat{V}_{\text{exchange}} = -\tfrac{1}{2} K \left(1 + \frac{\widehat{s}_1 \cdot \widehat{s}_2}{\hbar^2} \right). \tag{15.50}$$

Indeed, from $\widehat{S}^2 = \widehat{s}_1^2 + \widehat{s}_2^2 + 2\widehat{s}_1 \cdot \widehat{s}_2$ we get

$$\frac{\widehat{s}_1 \cdot \widehat{s}_2}{\hbar^2} = 2S(S+1) - 3 = \begin{cases} -3 & \text{for } S = 0 \\ 1 & \text{for } S = 1 \end{cases} \tag{15.51}$$

and therefore the eigenvalues of the exchange operator are K for $S = 0$ and $-K$ for $S = 1$, in accordance with Eq. (15.49).

The following are three additional examples of gross effects on the energy spectrum of a composite system, formally produced by the exchange interaction.

1. Two-particle systems. Take first a two-boson system; its wave function is totally symmetric. On the other hand, we know that the wave function is symmetric only for even orbital momentum. Therefore, the solutions for odd l, although mathematically legitimate, are not realized.

Now take a two-spin 1/2 system; the total spin can be 0 or 1. In the first case (antisymmetric spinor wave function) the orbital wave function is symmetric and corresponds to even orbital momenta. In the spin 1 case, in turn, only wave functions corresponding to odd orbital momentum can be accepted. In both cases only some of the energy levels predicted by the Schrödinger equation are realized, depending on the total spin, even when the interaction does not depend on spin. This example shows that, in general, the spectrum of an atomic system depends on its total spin.

2. Three-particle systems. Although still simple to analyze, the tripartite case already shows the complications to be expected when dealing with a larger number of particles. We assume that the interaction between the particles is negligible and consider the 1D case for simplicity. The unsymmetrized orbital wave functions can then be expressed as the product of three single-particle wave functions

$$\psi_{lmn}(x_1, x_2, x_3) = \psi_l(x_1)\psi_m(x_2)\psi_n(x_3), \tag{15.52}$$

---

[9] Note that this division of the interaction energy into two parts J and K, although mathematically convenient, is physically meaningless. In fact, the calculation must be done with the joint probability that either particle is in state n and the other in state m. Therefore, it makes no sense to physically interpret any of the terms of the expression (15.46) separately, although this is very often done.

with total energy given by

$$E \equiv E_{lmn} = E_l + E_m + E_n. \tag{15.53}$$

If the particles are spin-zero bosons, the following cases may occur:
(i) $l = n = m$, $E = 3E_n$. The wave function (15.52) is already symmetric. This is the case in particular for the ground state of the system.
(ii) $l = n \neq m$. The only physically realizable state is

$$\psi^S_{nnm} = \frac{1}{\sqrt{3}}(\psi_{nnm} + \psi_{nmn} + \psi_{mnn}), \quad E = 2E_n + E_m. \tag{15.54}$$

This is the case for the system's first excited state.
(iii) l, n, m different. The only fully symmetric state that can be constructed from the six states with exchange degeneracy is

$$\psi^S_{nlm} = \frac{1}{\sqrt{6}}(\psi_{lnm} + \psi_{nml} + \psi_{mln} + \psi_{nlm} + \psi_{lmn} + \psi_{mnl}), \quad E = E_l + E_n + E_m. \tag{15.55}$$

Now consider three fermions with spin 1/2. In addition to the space coordinates $x_i, i = 1, 2, 3$, we use the spin coordinates S_{3i}; we denote the set of variables $(x_i, S_{3i}) = (\xi_i)$. With the individual wave function (for spin-independent Hamiltonians) written as the product of the orbital function $\psi_n(x_i)$ times the spin wave function $\chi_\pm(S_{3i})$,

$$\psi^\pm_n(\xi_i) = \psi_n(x_i)\chi_\pm(S_{3i}), \tag{15.56}$$

and the spin orientation denoted by a, b, c, the unsymmetrized wave function is

$$\psi^{abc}_{lnm} = \psi^a_l(\xi_1)\psi^b_n(\xi_2)\psi^c_m(\xi_3), \tag{15.57}$$

and the fully antisymmetric wave function is given by the Slater determinant,

$$(\psi^{abc}_{nlm})^A = \frac{1}{\sqrt{6}} \begin{vmatrix} \psi^a_l(\xi_1) & \psi^a_l(\xi_2) & \psi^a_l(\xi_3) \\ \psi^b_n(\xi_1) & \psi^b_n(\xi_2) & \psi^b_n(\xi_3) \\ \psi^c_m(\xi_1) & \psi^c_m(\xi_2) & \psi^c_m(\xi_3) \end{vmatrix}. \tag{15.58}$$

While the l, n, m orbital quantum numbers can be the same or different, the spin indices a, b, c cannot all be different (since they can only take the values ± 1); this fact has a profound effect on the spectrum, as shown in the following.
(i) $l = n = m$. Since necessarily at least two spin indices are equal, the determinant has at least two equal rows, so this state is not realized.
(ii) $l = n \neq m$. The equality of l and n requires that a and b are different. There are four states that satisfy this condition: $(+-+, +--, -++, -+-)$, corresponding to the energy $E = 2E_n + E_m$; one of these states would be, for example

$$(\psi^{-++}_{nnm})^A = \frac{1}{\sqrt{6}}(\psi^{-++}_{nnm} + \psi^{+-+}_{mnn} + \psi^{++-}_{nmn} - \psi^{++-}_{mnn} - \psi^{+-+}_{nnm} - \psi^{-++}_{nmn}). \tag{15.59}$$

All these states have spin projection 1/2 and are doubly degenerate. In particular, the ground state is one of them. We conclude that if the spin is nonzero, the orbital wave function of the ground state can have nodes.
(iii) l, n, m different. All possible combinations of spin indices are acceptable, so there are 2^3 degenerate states, corresponding to one spin-3/2 quadruplet and two spin-1/2 doublets. Among these states there are those with well-defined symmetry with respect

to the exchange of space or spin coordinates separately, such as the following state, which is symmetric in the spin indices and antisymmetric in the orbital indices,

$$\left(\psi_{lnm}^{+++}\right)^A = \frac{1}{\sqrt{6}}\left(\psi_{lnm}^{+++} + \psi_{nml}^{+++} + \psi_{mln}^{+++} - \psi_{nlm}^{+++} - \psi_{lmn}^{+++} - \psi_{mnl}^{+++}\right). \tag{15.60}$$

There are states that are antisymmetric only for the simultaneous exchange of both sets of indices, as in case (ii) above. The energy of these states is $E = E_l + E_n + E_m$.

3. The ground state of N bosons or fermions in a 1D infinite well. The individual wave functions and energies are given by

$$\psi_n(x) = \sqrt{\frac{2}{a}} \sin\frac{\pi n}{a}x, \quad E_n = \frac{\pi^2\hbar^2}{2ma^2}n^2 \equiv E_1 n^2. \tag{15.61}$$

For N bosons that do not interact with each other, the ground state corresponds to all particles with $n = 1$ and total energy $E = NE_1$, so that the energy per particle is

$$\frac{E}{N} = E_1. \tag{15.62}$$

Now suppose we have N fermions with spin 1/2. In this case there are only two electrons for each value of n (with two possible spin-1/2 states), so that in the ground state of the system all energy levels with n from 1 to $N/2$ are filled. The ground-state energy is therefore given by

$$E = 2E_1 \sum_{n=1}^{N/2} n^2 = \tfrac{1}{12}E_1 N(N+1)(N+2), \tag{15.63}$$

as follows from the formula $\sum_{k=1}^{n} k^2 = \tfrac{1}{6}n(n+1)(2n+1)$. If N is very large, we can write with good approximation

$$E = \tfrac{1}{12}E_1 N^3. \tag{15.64}$$

The average energy per particle can now be considerably higher than E_1, since it is proportional to the square of the total number of particles,

$$\frac{E}{N} = \tfrac{1}{12}E_1 N^2. \tag{15.65}$$

The energy of the top occupied level is the Fermi energy, as explained in Section 6.3,

$$E_F \equiv E_{\frac{N}{2}} = \tfrac{1}{4}E_1 N^2 = \frac{\pi^2\hbar^2}{8m}\rho^2. \tag{15.66}$$

The last equality was written for the 1D case, with the fermion density ρ defined as $\rho = N/a$ (see Eq. (6.91), which applies to the 3D case). As a physical example of this situation, we recall the problem discussed in Section 6.3, where we saw that the conduction electrons of a metal can be considered to a first approximation as forming a gas of particles moving in a common potential produced by the entire structure. An energy diagram of this metal at low temperatures is essentially that shown in Figure 6.6 for the 1D metal model at $T = 0$. More generally, the particle distribution as a function of temperature is discussed in the following section.

15.3.1 The boson and fermion distributions as a function of temperature

The formulas for both the fermion and the boson distributions are derived using the techniques of statistical mechanics; see Section 15.5.6 for an introduction. The distribution of fermions as a function of energy and temperature, called the Fermi–Dirac distribution,

$$\bar{n}_{FD}(E) = \frac{2}{e^{(E-E_F)/kT} + 1}, \tag{15.67}$$

is obtained taking into account the exclusion principle. At $T = 0$, all energy levels up to E_F are occupied and those above E_F are empty. With increasing temperature the curve rounds off as the excited electrons near E_F occupy higher energy states and leave lower energy states. This shows that at very low temperatures (close to absolute zero) the energy of a metal is essentially independent of its temperature and is close to the minimum energy.

The Bose–Einstein distribution, in turn, is of the general form

$$\bar{n}_{BE}(E) = \frac{1}{e^{(E-\mu)/kT} - 1} \tag{15.68}$$

with $E > \mu$, where μ is the chemical potential.[10] In the case of photons, $\mu = 0$ and Eq. (15.68) coincides with Planck's formula. A comparison with the F–D formula shows that the key difference is the sign of the 1 in the denominator.

As can be seen from Fig. 15.1, at very low energies the contrast between the distributions is high: while the bosons tend to concentrate, fermions tend to stay apart, as discussed earlier. For very large energies compared to kT, both distributions approach the classical M–B distribution, $\bar{n}(E) \sim e^{-E/kT}$.

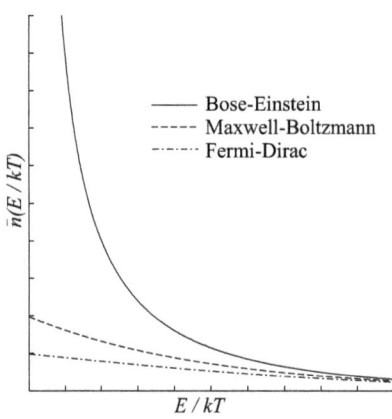

Figure 15.1 A comparison of the B–E, M–B, and F–D distribution functions at different temperatures, for $\mu/kT = -0.1$.

---

[10] The chemical potential of a given substance is the chemical energy per mole of the substance. The chemical potential marks the energy at which Bose–Einstein condensation occurs (see Section 15.4).

Exercise E15.4. The Bose distribution

Consider a mode of frequency ω of the radiation field in equilibrium with matter at temperature T. Using the quantum description in terms of photons, determine: (a) the probability $P_n(T)$ with which the (field) eigenstate n contributes to the state of equilibrium; (b) the dispersion σ_n^2 of the number of photons of frequency ω.

Solution. The probability of state n is given by

$$P_n = \frac{1}{Z} e^{-\beta E_n} = e^{-\beta \hbar \omega n}(1 - e^{-\beta \hbar \omega}),$$

where $Z = e^{-\beta E_0}/(1 - e^{-\beta \hbar \omega})$. Since the average number of photons is

$$\bar{n} = 1/\left(e^{\beta \hbar \omega} - 1\right),$$

it follows that $e^{\beta \hbar \omega} = 1 + (1/\bar{n})$, so $1 - e^{-\beta \hbar \omega} = 1/(\bar{n} + 1)$, which gives

$$P_n = \frac{1}{\bar{n}+1}\left(\frac{\bar{n}}{\bar{n}+1}\right)^n. \tag{15.69}$$

This result is known as the Bose distribution and applies to all systems consisting of bosons. A distribution of the form $P_n = N g^n$ (here $g = \bar{n}/(\bar{n}+1) = e^{-\beta \hbar \omega}$) is a geometric distribution.[11]

The mean square of n is obtained as follows.

$$\begin{aligned}\langle n^2 \rangle &= \sum n^2 P_n = P_0 \sum [n(n-1) + n] g^n \\ &= P_0 g^2 \sum n(n-1) g^{n-2} + \bar{n} \\ &= P_0 g^2 \frac{d^2}{dg^2} \frac{1}{1-g} + \bar{n} = \frac{2g^2}{(1-g)^2} + \bar{n},\end{aligned}$$

i.e.,

$$\langle n^2 \rangle = \sum n^2 P_n = 2\bar{n}^2 + \bar{n} = \bar{n}(2\bar{n} + 1). \tag{15.70}$$

The dispersion of the number n of photons of frequency ω is thus

$$\sigma_n^2 = \langle n^2 \rangle - \bar{n}^2 = \bar{n}^2 + \bar{n}. \tag{15.71}$$

Note that with $\overline{E} = \hbar \omega \bar{n}$, the result just obtained gives for the fluctuations of the energy

$$\sigma_E^2 = \hbar^2 \omega^2 \sigma_n^2 = \overline{E}^2 + \hbar \omega \overline{E}, \tag{15.72}$$

---

[11] Since $P_1 = P_0 g$ we can also write $P_n = P_0 (P_1/P_0)^n$, $P_{n+s} = P_n (P_1/P_0)^s$. This property can be taken as the definition of a geometric distribution. In particular, the probability that the ω modes contain n or more photons is

$$f(n) = \sum_{r=n}^{\infty} P_r = \sum_{s=0}^{\infty} P_{n+s} = P_n \sum_{s=0}^{\infty} (1/\xi)^s = P_n \frac{\xi}{\xi - 1},$$

where $\xi = P_0/P_1 > 1$.

therefore a linear and a quadratic term contribute to the fluctuations. At very high temperatures the quadratic term dominates over the linear one, so that in the classical limit we get the Rayleigh–Jeans distribution

$$\sigma_E^2 = \pi^4 c^6 \frac{\rho^2(\omega)}{\omega^4} = \overline{E}^2 = \hbar^2 \omega^2 \bar{n}^2 = k_B^2 T^2, \quad kT \gg \hbar\omega. \tag{15.73}$$

This allows us to identify the quadratic term as the classical prediction for the energy fluctuations of the Maxwell field. In the low-temperature limit the linear term is dominant,

$$\sigma_E^2 = \hbar\omega\overline{E} = \hbar^2\omega^2\bar{n}, \quad kT \ll \hbar\omega, \tag{15.74}$$

or

$$\sigma_n^2 = \bar{n}. \tag{15.75}$$

Einstein interpreted this result in 1905 as follows. Consider a gas composed of molecules moving independently with fixed energy $\mathcal{E} = \hbar\omega$; suppose that the number of these molecules in a unit volume follows a Poisson distribution (which is consistent with the independence hypothesis). The fluctuations of this number are then given by $\sigma_n^2 = \bar{n}$, and their contribution to the energy fluctuations is $\sigma_E^2 = \mathcal{E}\bar{n} = \hbar\omega\bar{n}$.[12] This led Einstein to propose that the linear term reflects an independent "granular" structure of the radiation field, that is, that at very low intensities the field must be considered as consisting of independent radiation quanta – the current photons.[13]

15.4 Collective Phenomena. Phonons and Bose-Einstein Condensates

When large numbers of atoms or electrons in a solid interact, their local quantum mechanics can be very complex. But sometimes strikingly simple collective properties appear, which cannot exist for single atoms, or even small collections of them: they emerge at a higher level. *Quantum collective phenomena* include the quantum Hall effect, superconductivity and magnetism in strongly correlated metals, B–E condensation of dilute gases, phonons and excitons, topologically ordered quantum states and quantum critical phenomena.

Here we briefly present some of the most important examples of quantum collective phenomena; for more in-depth discussions, the student is referred to modern textbooks on condensed matter.

15.4.1 Lattice dynamics: Phonons

Atoms vibrate in every material: in crystals, atomic clusters, or more complicated structures. In crystals, the vibrational modes are related to a number of properties, such as

---

[12] A Poisson distribution has the form $P(n) = e^{-a} a^n / n!$, and $\bar{n} = a, \sigma_n^2 = a = \bar{n}$. The relative fluctuations of the number n are given by $(\sigma_n/\bar{n})^2 = 1/\bar{n}$ and quickly become negligible as $\bar{n}$ grows.

[13] As noted in Section 2.7, these "simple" results provided the young Einstein with the basis for proposing the corpuscular (photonic) structure of a low-density radiation field.

the temperature dependence of the lattice parameters and elastic constants; they determine heat and electric conductivity, critical temperature, and other thermodynamic functions. Vibrational modes enhance the diffusion processes and play a crucial role in superconductivity, ferroelectricity, shape memory alloys, and so on.

The basic theory of the lattice dynamics was given by M. Born and K. Huang (1919–2005), who developed the theory for crystal lattices with translational symmetry. The interaction of electrons with vibrating atoms in the lattice can be conveniently treated as an interaction with *phonons*,[14] which are collective excitations, that is, excited states of the correlated vibrational modes of the structure.

For simplicity, consider a 1D crystalline lattice or linear chain consisting of a large number of particles N, which can be atoms or molecules. If the lattice is rigid enough, the atoms must exert forces on each other to maintain their equilibrium position. (These forces can be van der Waals forces, covalent bonds, electrostatic attractions, and others, all of which are ultimately due to the electric force.) The potential energy of the whole lattice is the sum of all pairwise potential energies $V(x_i - x_j)$.

In order to solve this many-body problem, two important approximations are usually made: the sum is performed only over neighboring atoms (which is legitimate because the fields produced by distant atoms are effectively screened out), and the potentials V are considered as harmonic-oscillator potentials (which is legitimate as long as the atoms remain close to their equilibrium positions). Assuming that all atoms have the same mass m and the same natural frequency of oscillation ω, the Hamiltonian for this system is

$$H = \frac{1}{2m}\sum_{i=1}^{N} p_i^2 + \frac{1}{2}m\omega^2 \sum_{i=j+1}(x_i - x_j)^2. \tag{15.76}$$

The normal coordinates and momenta in Fourier space, defined as

$$Q_k = \frac{1}{\sqrt{N}}\sum_l e^{ikal}x_l, \quad P_k = \frac{1}{\sqrt{N}}\sum_l e^{ikal}p_l, \tag{15.77}$$

where a is the lattice spacing, satisfy the commutation relations

$$[Q_k, P_{k'}] = \frac{1}{N}\sum_{l,m} e^{ikal}e^{-ik'am}[x_l, p_m], \tag{15.78}$$

$$[Q_k, Q_{k'}] = 0, \quad [P_k, P_{k'}] = 0. \tag{15.79}$$

The Hamiltonian can thus be written in wave vector space as (see Problem P15.10)

$$H = \frac{1}{2m}\sum_k P_k P_{-k} + \frac{m}{2}\sum_k \omega_k^2 Q_k Q_{-k}, \tag{15.80}$$

where

$$\omega_k = 2\omega \left|\sin\frac{ka}{2}\right|. \tag{15.81}$$

---

[14] The concept of the phonon was introduced in 1932 by I. Tamm. The name phonon is derived from the Greek word φωνή (sound), because long-wavelength phonons produce sound.

Equation (15.80) represents a set of uncoupled harmonic oscillators in the reciprocal space; these oscillators are the phonons, whose oscillation frequency depends on k. According to Eq. (15.81), the allowed values for k depend on the boundary conditions. If, for simplicity, periodic boundary conditions are imposed, the resulting quantization is

$$k_s = \frac{2\pi s}{Na}, \quad s = 0, \pm 1, \pm 2, \ldots \pm \frac{N}{2},$$

with the upper limit coming from the minimum wavelength, which is twice the lattice spacing a. The corresponding energy levels are given by

$$E_n = \left(\frac{1}{2} + n\right) \hbar \omega_k, \quad n = 0, 1, 2, \ldots.$$

The relation between frequency and wave vector, $\omega = \omega(k)$, is known from classical optics as a *dispersion relation*.

15.4.1.1 Dispersion relations

In general, dispersion can be caused either by geometric boundary conditions or by the interaction of the waves with the medium. Given the dispersion relation, the phase velocity and group velocity of each wave component can be calculated as a function of frequency.

The speed of propagation (group velocity) of an acoustic phonon, which is also the speed of sound in the lattice, is given by the slope of the dispersion relation, $d\omega_k/dk$. At low values of k (i.e., long wavelengths), the dispersion relation is nearly linear according to Eq. (15.81), and the speed of sound is approximately ωa. As a result, packets of phonons with different (but long) wavelengths can travel long distances through the lattice without breaking apart. This is the reason why sound propagates through solids without significant distortion. However, for large values of k, i.e., short wavelengths, this behavior fails due to the microscopic details of the lattice.

For a one-dimensional alternating array of two types of ions or atoms of mass m_1, m_2 connected by springs of spring constant K, two modes of vibration (or branches) result, namely

$$\omega_\pm^2 = K\left(\frac{1}{m_1} + \frac{1}{m_2}\right) \pm K\sqrt{\left(\frac{1}{m_1} + \frac{1}{m_2}\right)^2 - \frac{4\sin^2\frac{ka}{2}}{m_1 m_2}}, \quad (15.82)$$

where k is the wave vector of the vibration related to its wavelength by $k = 2\pi/\lambda$. In this case the dispersion relations exhibit two types of phonons, namely optical and acoustic, corresponding to the upper and lower curves in Fig. 15.2. More optical and acoustic dispersion branches appear in a unit cell with a larger number of atoms. Typical maximal phonon frequencies range from 10 to 30 THz (1 THz = 10^{12} s^{-1}).[15]

Since the phonons are described as harmonic oscillators, we can use the second quantization technique that is introduced in Section 17.6 and write the Hamiltonian of the system as

---

[15] Efficient heat transport by phonons in this frequency range is important for temperature control in nanoscale devices, where it replaces classical heat diffusion.

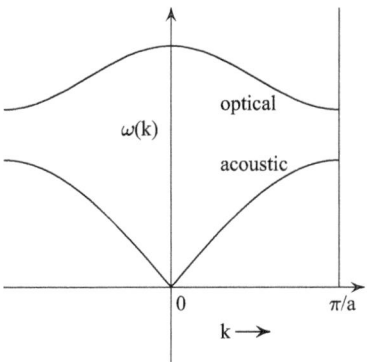

Figure 15.2 Phonon dispersion curves of a linear diatomic chain with two different types of atoms, showing an optical branch (upper curve) in addition to the acoustic branch (lower curve). The vertical axis is the energy or frequency of the phonon and the horizontal axis is the wave vector k.

$$\mathcal{H} = \sum_k \hbar\omega_k \left(b_k^\dagger b_k + \tfrac{1}{2} \right),$$

where $b_k^\dagger b_k = n_k$ is the occupation number operator and the $b_k^\dagger$ and b_k, respectively, create and destroy a single field excitation – a phonon – with an energy of $\hbar\omega_k$. Note that the ground state of each phonon has an associated energy of $\tfrac{1}{2}\hbar\omega_k$, indicating that the motion of the atoms is not frozen; the atoms move randomly about their equilibrium position. At $T > 0$, the thermally excited states of the lattice vibrations can be considered as a *phonon gas*. This means that the chemical potential for adding a phonon is zero, just as in the case of a photon gas inside a cavity. Therefore, the phonon gas obeys the B–E statistics, that is, the probability of having phonons of a given frequency in a given state is

$$n\left(\omega_{k,s}\right) = \frac{1}{e^{\hbar\omega_{k,s}/kT} - 1}, \tag{15.83}$$

where ω_{ks} is the frequency of the phonons in the state.

15.4.2 Superconductivity

Superconductivity occurs when electrical resistance disappears and magnetic fields are expelled from the material. Soon after K. Onnes[16] succeeded in liquefying helium, he discovered that the electrical resistance of mercury samples suddenly dropped to zero below the critical temperature of 4.15 K; this is the defining property of superconductivity. It was soon realized that superconductivity was quite common in metals at low temperatures – with a few exceptions. Twenty-two years later W. Meissner (1882–1974) found that when a magnetic field is applied to a superconducting sample, the magnetic flux is expelled from its bulk; this is the *Meissner effect*.

---

[16] H. Kamerlingh Onnes (1853–1926) was a Dutch physicist famous for his discovery of superconductivity. In 1908 he was the first to liquefy helium and cool it to nearly 1.5 Kelvin, for which he was awarded the Nobel Prize in Physics in 1913.

A breakthrough in understanding the structure of the superconducting ground state and the nature of its excitations was the BCS theory proposed in 1957,[17] which explained the transport properties in terms of a condensate of Cooper pairs.[18] In a superconductor, electrons are bound together into Cooper pairs by a weak attractive force. In a conventional superconductor, this attraction is caused by an exchange of phonons between the electrons. The evidence that phonons – the vibrations of the ionic lattice described above – are relevant for superconductivity is provided by the *isotope effect*, that is, the dependence of the superconducting critical temperature on the mass of the ions.

A few years later, B. D. Josephson predicted that a supercurrent could flow between two pieces of superconductor separated by a thin layer of insulator. The Josephson effect is exploited by superconducting quantum interference devices (SQUIDs) and is used to accurately measure the magnetic flux quantum $\Phi_0 = h/2e$ (see Section 15.4.4).

In superconductors there is a characteristic length scale ξ called the coherence length, which is a measure of the size of a Cooper pair (i.e., the distance between the two electrons that make up the pair). The electron near or at the Fermi surface moving through the lattice of a metal produces an attractive potential behind it over a distance of the order of 3×10^{-6} cm, which is large compared with the lattice spacing ($\sim 10^{-8}$ cm). In a low-temperature superconductor the coherence length is of the order of 10^{-4} cm, covering a few thousand atoms, while in the high-temperature superconductors it is only of the order of $10^{-8} - 10^{-6}$ cm.

Another important parameter is the London penetration depth (usually denoted as λ or λ_L), which characterizes the distance to which a magnetic field penetrates into the superconductor and becomes equal to e^{-1} times the magnetic field at the surface of the superconductor. Typical values of λ range from 50 to 500 nm. When $\lambda < \xi/\sqrt{2}$ the superconductor is of type I, when $\lambda > \xi/\sqrt{2}$ the superconductor is of type II.[19] In the first case the superconductivity is abruptly destroyed when a strong magnetic field is applied (as is usually the case with pure metals); type II superconductors have two critical magnetic fields and a more complex magnetic behavior.

Superconductivity is a vibrant field of research, especially since the advent of so-called high-temperature superconductors (with critical temperatures of the order of -140 C or lower at ambient pressure). Promising large-scale applications include electric power transmission, transformers, and other devices that would have significant economic and environmental benefits.

---

[17] Named after John Bardeen (1908–1991), Leon Cooper (1930), and Robert Schrieffer (1931–2019), who were awarded the Nobel Prize in 1972 "for their jointly developed theory of superconductivity." Bardeen had already been awarded the Nobel Prize in 1956 for his work with W. Shockley and W. H. Brattain on semiconductors, which led to the history-making invention of the transistor.

[18] *Superfluidity* in liquid ^{4}He, discovered by P. Kapitza and (independently) by J. Allen and D. Misener in 1937, had already been explained in 1938 by F. and H. London and L. Titza (with further elaborations by L. Landau) as a consequence of Bose–Einstein condensation.

[19] The coherence length ξ was introduced by Vitaly Ginzburg (1916–2009; Nobel Prize 2003) and L. Landau as part of their theory of superconductivity, according to which the electrons contributing to superconductivity form a superfluid. The ratio $\kappa = \lambda/\xi$ is known as the *Ginzburg–Landau parameter*.

15.4.3 Bose–Einstein condensate

A Bose–Einstein condensate (BEC) is a state of matter in which extremely cold atoms exhibit collective behavior and act like a single "super atom." To reach such a state, the system, composed of a few thousand bosons (atoms), must be very close to zero Kelvin ($\sim 10^{-9}$ K). Under such conditions, a large fraction of the bosons occupy the lowest quantum states in which microscopic quantum mechanical phenomena, in particular wave-function interference, become macroscopically visible. The low speeds of motion of the bosons imply a large de Broglie wavelength, as large as the dimensions of the container, which leads to extended correlations between the variables of the particles. Indeed, the BEC is in a highly entangled state, due to the fact that the particles are coherently distributed in space.

As mentioned in Chapter 2, this new phase of matter was discovered by Einstein in 1925, following a related suggestion by Bose, hence the name Bose–Einstein condensate. For almost 70 years, the technical difficulties involved in producing BEC kept them in the laboratory, but they are now being produced conventionally.

The basic BEC concepts, such as the critical temperature or the critical particle density, can be obtained by considering an ideal gas of noninteracting bosons. The transition to BEC occurs below the critical temperature T_c, which for a uniform 3D gas composed of particles with no apparent internal degrees of freedom is

$$T_c \simeq 3.3125 \frac{\hbar^2 n_c^{2/3}}{m k_B}, \tag{15.84}$$

where m is the boson mass and n_c is the particle density, given in terms of the thermal de Broglie length λ_c by

$$n_c \simeq 2.612 \frac{1}{\lambda_c^3}. \tag{15.85}$$

Since $n_c^{-1/3}$ is the average distance between particles, Eq. (15.85) means that condensation occurs when this distance becomes comparable to λ_c at the given temperature. While in air at atmospheric pressure and room temperature the number of molecules (N_2 and O_2) per cm^3 is 10^{19}, in a condensate the density is much lower, only 10^{12}–10^{13} atoms /cm^3.

A particularly spectacular quantum phenomenon occurs when the condensate is split into two separate parts, which are then brought closer until they overlap, producing a macroscopic quantum wave interference pattern. It is also possible to prepare a BEC with fermions by first pairing up fermions of opposite spin, similarly as the electrons do to form Cooper pairs.[20]

15.4.4 The quantum Hall effect

The Hall effect, discovered in 1879 by Edwin Hall (1855–1938), is a simple consequence of the motion of charged particles in a magnetic field. The underlying physics of the

---

[20] The superfluidity of two different phases of ^{3}He is due to the similar Bose–Einstein condensation of pairs of its fermion atoms.

Hall effect is the following: a current-carrying plate is subject to a constant, perpendicular magnetic field. If the field points in the z-direction and the electrons are restricted to move in the xy-plane, we get the coupled equations of motion

$$m\ddot{x} = -eB\dot{y}, \quad m\ddot{y} = eB\dot{x}. \tag{15.86}$$

The general solution is

$$x(t) = X - R\sin(\omega_B t + \phi), \quad y(t) = Y + R\cos(\omega_B t + \phi), \tag{15.87}$$

where $\omega_B = eB/m$ is the cyclotron frequency. When a constant current I is made to flow by applying an electric field in the x-direction, this current is deflected by the magnetic field and bent in the y-direction. In a finite material, this results in an accumulation of charge along the edge and an associated electric field in the y-direction. This continues until the electric field cancels out the bending due to the magnetic field and the electrons then only move in the x-direction. The induced electric field E_y is responsible for the Hall voltage V_H along the y-direction. The Hall coefficient is defined by

$$R_H = \frac{\rho_{xy}}{B}, \tag{15.88}$$

where ρ_{xy} is the off-diagonal component of the resistivity tensor. In the classical case, R_H depends only on the charge and density of the conducting particles, $R_H = 1/Ne$.

However, at low temperatures and in strong magnetic fields, the resistivity exhibits a peculiar behavior, as was first demonstrated by K. von Klitzing in 1980.[21] The electrons moving in the metal plate can be regarded as a confined two-dimensional electron gas experiencing quantum correlations. At low temperatures the Hall resistance becomes quantized,

$$R_H = \frac{\hbar}{ne^2}, \tag{15.89}$$

where n is an integer.[22] The Coulomb interactions lead to additional values of R_H with odd fractional numbers in the denominator, $n = 1/3, 2/3, 3/5, \ldots$; this is the *fractional quantum Hall effect*. The value of R_H remains strictly constant in a finite range of B.

These discoveries revealed the very rich physics of 2D systems at the nanoscale, which has since become an active field of theoretical and experimental research with important technological applications.

15.5 The Density Matrix. Pure States and Mixtures

As mentioned in Chapter 8, in nature we most often find quantum systems that require combinations of noninterfering states to describe the complete ensemble. Consider, for example, a variable A whose value A_i is well defined in each sub-ensemble i with

---

[21] Klaus von Klitzing (1943) is a German physicist best known for the discovery of the integer quantum Hall effect, for which he was awarded the Nobel Prize in Physics in 1985.

[22] It is a remarkable fact that the quantized resistance is determined only by fundamental constants. In fact, the original proposal by Klitzing et al. (1980) was to use the quantum Hall effect as a method for a highly accurate determination of the fine-structure constant $\alpha = e^2/\hbar c$.

weight w_i; its average value $\langle A \rangle = \sum_{i=1}^{N} w_i A_i$ depends on the relative weight of the independent subensembles. For instance, in a macroscopic volume of H_2 in equilibrium the molecules can have total electron spin 0 or 1, for the rest they are entirely equivalent. So there are two different but essentially equivalent subsystems coexisting, forming what we take to be a homogeneous gas. Their corresponding wave functions do not interfere (because they are independent subsystems) and the system as a whole must be described by something that treats them equally. In the more general case, we have a set of states with different weights,

$$\begin{array}{cccccc} \text{State} & |1\rangle & |2\rangle & |3\rangle & \cdots & |N\rangle \\ \text{Weight} & w_1 & w_2 & w_3 & \cdots & w_N \end{array}. \quad (15.90)$$

In such cases the system as a whole is not characterized by one wave function, but by several wave functions $|i\rangle$ and their corresponding statistical weights w_i. The system is then said to be in a mixture of pure states $|i\rangle$. In the H_2 example, we have a mixture of w_0 molecules with spin 0 and w_1 molecules with spin 1.

A simple and elegant tool to describe such situations is the density operator or density matrix, which was proposed simultaneously but independently in 1927 by Landau and von Neumann,

$$\widehat{\rho} = \sum_i w_i |i\rangle\langle i|. \quad (15.91)$$

In writing this expression it is not assumed that the states $|i\rangle$ are orthogonal, although normally they are.[23] It is often convenient to use a different basis (e.g., the eigenvectors of the operator $\widehat{A}$, or the energy eigenfunctions, and so on.). We therefore expand the state $|i\rangle$ in terms of a convenient basis $\{|n\rangle\}$ as

$$|i\rangle = \sum_n a_{in} |n\rangle, \qquad a_{in} = \langle n|i\rangle. \quad (15.92)$$

Now the average value over the complete ensemble is

$$\langle \widehat{A} \rangle = \sum_i w_i A_i = \sum_i w_i \langle i|\widehat{A}|i\rangle = \sum_i \sum_{m,n} w_i \langle n|a_{in}^* \widehat{A} a_{im}|m\rangle$$
$$= \sum_i \sum_{m,n} w_i a_{in}^* a_{im} \langle n|\widehat{A}|m\rangle. \quad (15.93)$$

Rewriting the coefficient of $\langle n|\widehat{A}|m\rangle$ in this expression as follows:

$$\sum_i w_i a_{in}^* a_{im} = \sum_i w_i \langle m|i\rangle\langle i|n\rangle = \langle m| \left(\sum_i w_i |i\rangle\langle i| \right) |n\rangle$$
$$= \langle m|\widehat{\rho}|n\rangle = \rho_{mn}, \quad (15.94)$$

and substituting in the preceding expression for $\langle \widehat{A} \rangle$, we get

$$\langle \widehat{A} \rangle = \sum_{n,m} \rho_{mn} A_{nm} = \sum_m (\widehat{\rho}\widehat{A})_{mm} = \operatorname{tr} \widehat{\rho}\widehat{A}. \quad (15.95)$$

---

[23] A notable exception is well known in quantum optics, where it is useful to express the density matrix in the Glauber representation in terms of coherent states, which are not orthogonal; see Section 9.4.3.

This is a very important result: it tells us that the ensemble average of $\widehat{A}$ is given by the trace of the product $\widehat{\rho}\widehat{A}$. Note that swapping A and ρ gives the same result for $\langle \widehat{A} \rangle$, even if they do not commute,

$$\langle \widehat{A} \rangle = \operatorname{tr} \widehat{\rho}\widehat{A} = \sum_{n,m} A_{nm} \rho_{mn} = \sum_{n} (\widehat{A}\widehat{\rho})_{nn} = \operatorname{tr} \widehat{A}\widehat{\rho}. \tag{15.96}$$

This is an instance of the general theorem that establishes that in the calculation of a trace it is possible to perform any cyclic exchange of the factors, even if they do not commute. For example, for noncommuting $\widehat{A}, \widehat{B}, \widehat{C}$, $\operatorname{tr} \widehat{A}\widehat{B}\widehat{C} = \operatorname{tr} \widehat{B}\widehat{C}\widehat{A} = \operatorname{tr} \widehat{C}\widehat{A}\widehat{B}$.

An alternative way of introducing the density matrix goes as follows. When the system is in the pure state $|i\rangle$, all w_k are zero except w_i. The average of $\widehat{A}$ is then simply given by

$$A_i = \langle i|\widehat{A}|i\rangle = \sum_n \langle i|\widehat{A}|n\rangle\langle n|i\rangle = \sum_n \langle n|i\rangle\langle i|\widehat{A}|n\rangle = \operatorname{tr} \widehat{P}_i \widehat{A}, \tag{15.97}$$

where $\widehat{P}_i = |i\rangle\langle i|$ is the projector onto the state $|i\rangle$, and $\{|n\rangle\}$ is any basis. Now we generalize this result to a mixture of states with different weights, introducing $\sum_n |n\rangle\langle n| = 1$,

$$\langle \widehat{A} \rangle = \sum_i w_i \langle i|\widehat{A}|i\rangle = \sum_i \sum_n w_i \langle i|\widehat{A}|n\rangle\langle n|i\rangle$$

$$= \sum_n \langle n| \sum_i w_i |i\rangle\langle i|\widehat{A}|n\rangle = \sum_n \langle n| \left(\sum_i w_i \widehat{P}_i\right) \widehat{A}|n\rangle$$

$$= \operatorname{tr} \sum_i w_i \widehat{P}_i \widehat{A}. \tag{15.98}$$

The density matrix $\widehat{\rho}$ can therefore be obtained from first principles by comparing Eqs. (15.95) and (15.98),

$$\langle A \rangle = \sum_i w_i A_i = \sum_i w_i \operatorname{tr} \widehat{P}_i \widehat{A} = \operatorname{tr}\left(\sum_i w_i \widehat{P}_i\right)\widehat{A} = \operatorname{tr}\widehat{\rho}\widehat{A}, \tag{15.99}$$

which gives

$$\widehat{\rho} = \sum_i w_i \widehat{P}_i = \sum_i w_i |i\rangle\langle i|, \tag{15.100}$$

This result is the direct generalization of (15.97) to mixed states.

15.5.1 The density matrix in quantum statistical mechanics

For a more comprehensive conceptual understanding, it is useful to see the density matrix from the point of view of statistical mechanics, a subject in which it plays an important role.

Any real physical system of interest is inevitably open, that is, it is necessarily in contact with its environment, with which it can exchange energy, momentum, and even particles.[24] Therefore, none of its dynamical variables can have a perfectly defined and

---

[24] In particular, there is no known way of getting rid of ubiquitous fields such as the ZPF and the gravitational field.

constant value. Even if the system is in equilibrium, only the mean values of the dynamical variables will be constant, but their instantaneous values will fluctuate constantly around their respective averages. The random, unpredictable nature of these fluctuations is due to the enormous complexity of the environment, which requires a large number of variables, often themselves subject to fluctuations, to fully characterize it.

Under these conditions, if we want to study a complex quantum system, which can be in different states depending on external conditions, we have to pay attention not only to the system, but also to those elements of the physical world to which it is related and which most strongly influence its behavior, at least from the point of view of our actual interest (for other studies, other external elements may be relevant). By including these elements as part of our "complete system," we can consider it to be disconnected from the rest of the physical world (i.e., we consider the influence of the rest of the universe to be negligible for our purposes). In this approximation, and by characterizing the state of the relevant external part with a set of vectors $|\alpha\rangle$, the state vector of the "complete system" can be written in the form

$$|\Psi\rangle = \sum_{i,\alpha} c_{i\alpha}|\alpha\rangle|i\rangle = \sum_i c_i(t)|i\rangle, \qquad (15.101)$$

where the random coefficients $c_i(t) = \sum_\alpha c_{i\alpha}(t)|\alpha\rangle$ are complicated combinations of the exterior states. This description corresponds to the case where the state $|\alpha\rangle$ contributes with fluctuating probability $|c_i|^2$ to the state $|\Psi\rangle$; while $|i\rangle$ characterizes the state of the system of interest, the c_i are determined by the outside with which it is in contact.

It is not difficult to establish the relationship between the coefficients c_i and the weights w_i. This relationship must be statistical, since the c_i fluctuate, whereas the weights w_i are fixed for a given situation. In fact, we should expect the number of particles that are instantly in a given state (e.g., with energy in the interval $(E_i, E_i + \Delta E_i)$), to vary erratically. The quantum expectation value of a variable $\widehat{A}$, here denoted $\langle \widehat{A} \rangle_{\text{QM}}$, is given by

$$\langle \widehat{A} \rangle_{\text{QM}} = \langle \Psi | \widehat{A} | \Psi \rangle = \sum_{i,j} c_i^* c_j \langle i | \widehat{A} | j \rangle. \qquad (15.102)$$

To determine $\langle \widehat{A} \rangle$ we still have to average this expression over the set of possible values of the c_i's, an operation that we will denote with an upper bar; note carefully that this is generally not a quantum operation. We thus obtain

$$\langle \widehat{A} \rangle = \overline{\langle \widehat{A} \rangle_{\text{QM}}} = \sum_{i,j} \overline{c_i^* c_j} \langle i | \widehat{A} | j \rangle. \qquad (15.103)$$

To go further, we need to make some assumptions about the statistical behavior of the coefficients c_i. Given the enormous number of degrees of freedom involved in determining the instantaneous value of each coefficient, the most natural hypotheses are the following:

a) Postulate of a priori weights; this means

$$\overline{c_i} = 0, \quad \overline{|c_i|^2} = w_i. \qquad (15.104)$$

b) Postulate of statistical independence; if $i \neq j$, then c_i and c_j are statistically independent, that is,

$$\overline{c_i^* c_j} = 0, \quad \text{if } i \neq j. \tag{15.105}$$

Combining the two expressions, we get

$$\overline{c_i^* c_j} = w_i \delta_{ij}. \tag{15.106}$$

The resulting weights

$$w_i = \overline{c_i^* c_i} \tag{15.107}$$

are determined by physical factors belonging to the external system and the interaction mechanism; they are, therefore, outside the theoretical apparatus related to the quantum system under study here, and so we take them as given. Their determination usually corresponds to other branches of physics; for example, statistical mechanics provides rules for determining these functions under specified conditions. Substituting (15.106) into (15.103) we get

$$\langle \widehat{A} \rangle = \sum_{i,j} w_i \delta_{ij} \langle i | \widehat{A} | j \rangle = \sum_i w_i A_i. \tag{15.108}$$

We have thus recovered Eq. (15.91) as a *statistical description* of the system of interest.

Formula (15.106) gives the desired relationship between the a priori weights and the coefficients c_i of the expansion of the state vector of the complete system. On the other hand, it follows from (15.91) (considering the basis as orthonormal) that

$$\langle i | \widehat{\rho} | j \rangle = \sum_k w_k \langle i | k \rangle \langle k | j \rangle \tag{15.109}$$

$$= \sum_k w_k \langle k | j \rangle \delta_{ik} = w_i \langle i | j \rangle = w_i \delta_{ij},$$

which introduced in (15.106) gives

$$\overline{c_i^* c_j} = \langle j | \widehat{\rho} | i \rangle. \tag{15.110}$$

This expression establishes the connection between the coefficients of the state function Ψ and the matrix elements of $\widehat{\rho}$, giving the latter a direct statistical meaning. Substituting this into Eq. (15.103), we obtain the fundamental formula (15.95),

$$\langle \widehat{A} \rangle = \sum_{i,j} \langle j | \widehat{\rho} | i \rangle \langle i | \widehat{A} | j \rangle = \text{tr} \, \widehat{\rho} \widehat{A}. \tag{15.111}$$

Equation (15.108) shows that the different states $|i\rangle$ contribute independently to the value of $\langle \widehat{A} \rangle$, as a result of the hypothesis of statistical independence (or random phases) of their coefficients, expressed in (15.105). In fact, with $\psi(x) = \sum c_i \varphi_i(x)$, the spatial density averaged over the external parameters is

$$\rho(x) = \overline{|\psi(x)|^2} = \sum_{i,j} \overline{c_i^* c_j} \varphi_i^* \varphi_j = \sum_i w_i |\varphi_i|^2 = \sum_i w_i \rho_i(x). \tag{15.112}$$

The ensemble is manifestly an incoherent superposition, that is, probabilities (not amplitudes) are added. If the c_i were fixed and no averaging over their realizations were required, the result would be

$$\rho(x) = \sum_{ij} c_i^* c_j \varphi_i^* \varphi_j = \sum_i |c_i|^2 \rho_i(x) + \sum_{i \neq j} c_i^* c_j \varphi_i^* \varphi_j, \quad (15.113)$$

which has all the interference terms with $i \neq j$ characteristic of the coherent superpositions (of amplitudes) representing the states described by a wave function. This confirms the need to introduce the density matrix if one wishes to be able to study incoherent systems, which are more complex than those described by a single state vector.

Before concluding this section, an additional observation should be made. The weights (a priori probabilities) w_i are determined, as we have seen, by considerations external to the system under study. Their origin can be either quantum or classical, and their determination is a completely different physical problem, although not unrelated to our immediate interest. The w_i are usually taken to be constant, so that the time evolution of the density matrix is determined solely by the states it contains, that is, by the quantum rules. Conceptually, this need not be the case; we could imagine, for example, that the outside of the system is changing because the temperature of the thermal bath is changing, or for some other reason. In such cases, the evolution of the system depends on the outside, and the problem that makes sense to study is that of the whole system. For this reason, we will normally restrict ourselves to the case where the external conditions remain unchanged and the weights maintain their constant value. From now on, we will assume that this is the case.

15.5.2 Basic properties of the density matrix

All the basic properties of the density matrix follow from its definition (15.91) and from the basic rule for calculating expectation values, Eq. (15.95). Taking here $\widehat{A} = 1$, we get

$$\operatorname{tr} \widehat{\rho} = 1. \quad (15.114)$$

This result is a direct consequence of the condition

$$\sum_i w_i = 1, \quad (15.115)$$

as follows from (15.91) in combination with (15.114),[25]

$$\sum_i w_i = 1 = \sum_i w_i \langle i | i \rangle = \operatorname{tr} \sum_i w_i |i\rangle\langle i| = \operatorname{tr} \hat{\rho}.$$

Since, by definition, the weights w_i are positive and real,

$$w_i \geq 0, \quad w_i^* = w_i, \quad (15.116)$$

---

[25] If the states $|i\rangle$ are not orthogonal, Eq. (15.115) is not necessarily valid.

$\widehat{\rho}$ is automatically self-adjoint,

$$\widehat{\rho}^\dagger = \left(\sum_i w_i |i\rangle\langle i|\right)^\dagger = \sum_i w_i |i\rangle\langle i| = \widehat{\rho}. \tag{15.117}$$

This result also follows from Eq. (15.94), which shows that $\rho_{mn} = \rho_{nm}^*$. From this expression we get

$$\rho_{nn} = \sum_i w_i |a_{in}|^2 \geq 0.$$

Since (15.114) establishes that $\sum_n \rho_{nn} = 1$, it follows that the diagonal elements of the density matrix are not only real but also satisfy the following constraint,

$$0 \leq \rho_{nn} \leq 1. \tag{15.118}$$

It follows that if the mixture contains N independent pure states, it is characterized by a Hermitian matrix of dimension $N \times N$, positive definite[26] and with trace one. Since this matrix contains $N^2 - 1$ real parameters, these are the ones needed to specify the state of the mixture characterized by $\widehat{\rho}$. The square matrix $N \times N$ contains N^2 complex elements; the hermiticity condition reduces the number of independent real parameters to N^2, and the condition on the trace reduces this number to $N^2 - 1$. For example, an arbitrary mixture of spin-1/2 states contains two independent pure states ($N = 2$), corresponding to spin down and spin up. Therefore, the polarization state of these particles (electrons, nucleons, etc.) is specified by three parameters that can be chosen as the components of the polarization vector.

Any positive definite Hermitian matrix of trace 1 can be taken as a density matrix. The proof is as follows. Let $\widehat{\rho}'$ be an operator that satisfies these requirements; in its own representation we can do the spectral decomposition, so if ρ_n' are the eigenvalues of $\widehat{\rho}'$, we write $\widehat{\rho}' = \sum_n \rho_n' |n\rangle\langle n|$; since the trace is 1, we have that $\sum_n \rho_n' = 1$, and since the eigenvalues of a Hermitian operator are real (and in this case positive definite by hypothesis), it is also true that $0 \leq \rho_n' \leq 1$. Consequently, the eigenvalues ρ_n' have the properties required to be used as statistical weights and $\widehat{\rho}'$ can be interpreted as a statistical operator.

The matrix $\widehat{\rho}$ also has the following property: in the representation in which it is diagonal, $\rho_{nm} = \rho_n \delta_{mn}$, we may write

$$\text{tr}\,\widehat{\rho}^2 = \sum_{n,m} \rho_{nm}\rho_{mn} = \sum_n \rho_n^2 \leq \left(\sum_n \rho_n\right)^2 = (\text{tr}\,\rho)^2 = 1. \tag{15.119}$$

Since the trace is representation-independent, we get that, in general,

$$\text{tr}\,\widehat{\rho}^2 \leq 1. \tag{15.120}$$

The equal sign applies only to pure states.

---

[26] A matrix ρ is positive definite if for any vector with components A_n, $\sum \rho_{nm} A_n^* A_m > 0$ (i.e., $A^\dagger \rho A > 0$). That this is true here is immediately clear from Eq. (15.94), from which it follows that $\sum_{nm} \rho_{nm} A_n^* A_m = \sum_i w_i \left|\sum_n a_{in} A_n\right|^2 \geq 0$. The eigenvalues of a positive definite matrix are nonnegative.

On the other hand, it can be shown that when changing the representation, the matrix $\widehat{\rho}$ is transformed according to the general rules of transformation of an operator. In fact, when going from the $|i\rangle$ representation to a $|p\rangle$ representation by means of the unitary operator $\widehat{U}$, $|p\rangle = \widehat{U}|i\rangle$, we get

$$\widehat{\rho} = \sum_i w_i |i\rangle\langle i| = \sum_i w_i \widehat{U}^{-1}|p\rangle\langle p|\widehat{U} = \widehat{U}^{-1}\left(\sum_i w_i |p\rangle\langle p|\right)\widehat{U}. \tag{15.121}$$

In the last equality we took into account that the eigenvalues of an operator do not change with the representation. The quantity in parentheses is the density matrix in the new representation, which we call $\widehat{\rho}'$, $\widehat{\rho} = \widehat{U}^{-1}\widehat{\rho}'\widehat{U}$, and inversion confirms the usual rule,

$$\widehat{\rho}' = \widehat{U}\widehat{\rho}\widehat{U}^{-1}. \tag{15.122}$$

From here it is easy to show that not only the eigenvalues, but, in general, the expectation values of an operator are independent of the representation. Indeed, by denoting with A' the variables in the new representation, we get

$$\langle \widehat{A}' \rangle = \operatorname{tr} \widehat{\rho}'\widehat{A}' = \operatorname{tr} \widehat{U}\widehat{\rho}\widehat{U}^{-1}\widehat{U}\widehat{A}\widehat{U}^{-1} = \operatorname{tr} \widehat{U}\widehat{\rho}\widehat{A}\widehat{U}^{-1} = \operatorname{tr} \widehat{\rho}\widehat{A} = \langle \widehat{A} \rangle. \tag{15.123}$$

In the derivation we used $\operatorname{tr} \widehat{A}\widehat{B} = \operatorname{tr} \widehat{B}\widehat{A}$.

Before going further, and by way of illustration, it is useful to apply the general theory to the particular case of the coordinate representation (assuming this makes sense in the case we are interested in). We write $\widehat{\rho}$ in the form

$$\widehat{\rho} = \sum_{i,m,n} w_i a_{im} a_{in}^* |m\rangle\langle n| = \sum_{m,n} \rho_{mn} |m\rangle\langle n|, \tag{15.124}$$

where

$$\rho_{mn} = \sum_i w_i a_{im} a_{in}^*. \tag{15.125}$$

From here we obtain

$$\rho(x', x) \equiv \langle x'|\widehat{\rho}|x\rangle = \sum_{m,n} \rho_{mn} \langle x'|m\rangle\langle n|x\rangle, \tag{15.126}$$

and denoting with $\varphi_n(x) = \langle x|n\rangle$ the wave function of state $|n\rangle$, we obtain the important result

$$\rho(x', x) = \sum_{m,n} \rho_{mn} \varphi_m(x') \varphi_n^*(x), \tag{15.127}$$

which gives for the expectation value of $\widehat{A}$

$$\langle \widehat{A} \rangle = \operatorname{tr} \widehat{\rho}\widehat{A} = \int dx \langle x|\widehat{\rho}\widehat{A}|x\rangle. \tag{15.128}$$

Writing the integrand in the form

$$\langle x|\widehat{\rho}\widehat{A}|x\rangle = \int dx' \langle x|\rho|x'\rangle\langle x'|\widehat{A}|x\rangle = \int dx' \rho(x, x')\langle x'|\widehat{A}|x\rangle, \tag{15.129}$$

we get

$$\langle \widehat{A} \rangle = \int dx\, dx' \rho(x, x')\langle x'|\widehat{A}|x\rangle. \tag{15.130}$$

This is just the specific form of the general rule $\langle \widehat{A} \rangle = \mathrm{tr}\left(\widehat{\rho}\widehat{A}\right)$ for $\widehat{\rho}$ and $\widehat{A}$ continuous matrices. In general, we can write the matrix element of $\widehat{A}$ – which acts on x – as

$$\langle x'|\widehat{A}|x\rangle = \widehat{A}_x \langle x'|x\rangle = \widehat{A}_x \delta(x'-x), \tag{15.131}$$

where the subscript x means that $\widehat{A}$ acts only on the variable x. For example, setting $\widehat{A} = 1$ yields $\langle x'|1|x\rangle = \delta(x'-x)$, which when introduced in (15.130) confirms that the trace of ρ is 1,

$$1 = \int dx dx' \rho(x,x') \delta(x'-x) = \int \rho(x,x) dx. \tag{15.132}$$

Similarly, with $\widehat{A} = f(x)$, an arbitrary function of x, we have $\langle x'|f(x)|x\rangle = f(x)\delta(x'-x)$ and

$$\langle f(x)\rangle = \int f(x)\rho(x,x) dx. \tag{15.133}$$

From these results it follows that the spatial density of particles is given by the diagonal element of $\rho(x,x')$,

$$\rho(x) = \rho(x,x). \tag{15.134}$$

On the other hand, for $\widehat{A} = \widehat{p}$, from (15.131) we obtain

$$\langle x'|\widehat{p}|x\rangle = -i\hbar \frac{\partial}{\partial x}\delta(x'-x), \tag{15.135}$$

from which, after integration by parts, we get

$$\langle \widehat{p}\rangle = -i\hbar \int dx dx' \rho(x,x') \frac{\partial}{\partial x}\delta(x'-x) = i\hbar \int dx \left[\frac{\partial \rho(x,x')}{\partial x}\right]_{x'=x}. \tag{15.136}$$

15.5.3 Time evolution of the density matrix

In order to construct the equation for the time evolution of $\widehat{\rho}$ we write the state $|i(t)\rangle$ at time t in terms of the state $|i(t_0)\rangle$ at an initial time t_0 and apply the evolution operator $\widehat{S}(t,t_0)$,

$$|i(t)\rangle = \widehat{S}(t,t_0)|i(t_0)\rangle, \quad \widehat{S}(t) = e^{-i\widehat{H}t/\hbar}. \tag{15.137}$$

In turn, the evolution of $\widehat{S}$ satisfies the rule

$$i\hbar \frac{\partial \widehat{S}}{\partial t} = \widehat{H}\widehat{S}. \tag{15.138}$$

Equation (15.137) allows us to relate $\widehat{\rho}(t)$ to $\widehat{\rho}(t_0)$, assuming constant weights (the result is immediate using Eq. (15.122) with $\widehat{U} = \widehat{S}$),

$$\widehat{\rho}(t) = \sum_i w_i |i(t)\rangle\langle i(t)| = \sum_i w_i \widehat{S}|i(t_0)\rangle\langle i(t_0)|\widehat{S}^\dagger$$

$$= \widehat{S}(t,t_0)\left[\sum_i w_i |i(t_0)\rangle\langle i(t_0)|\right]\widehat{S}^\dagger(t,t_0) = \widehat{S}(t,t_0)\widehat{\rho}(t_0)\widehat{S}^\dagger(t,t_0). \tag{15.139}$$

By taking the partial derivative with respect to time, we get, using Eq. (15.138) and its adjoint,

$$\frac{\partial \widehat{\rho}}{\partial t} = \frac{\partial \widehat{S}}{\partial t} \widehat{\rho}(t_0)\widehat{S}^\dagger + \widehat{S}\widehat{\rho}(t_0)\frac{\partial \widehat{S}^\dagger}{\partial t}$$
$$= \frac{1}{i\hbar}\left[\widehat{H}\,\widehat{S}\widehat{\rho}(t_0)\widehat{S}^\dagger - \widehat{S}\widehat{\rho}(t_0)\widehat{S}^\dagger\widehat{H}\right] = \frac{1}{i\hbar}\left[\widehat{H}\widehat{\rho}(t) - \widehat{\rho}(t)\widehat{H}\right], \qquad (15.140)$$

that is,

$$i\hbar\frac{\partial \widehat{\rho}}{\partial t} = [\widehat{H},\widehat{\rho}]. \qquad (15.141)$$

This is known as the von Neumann equation. Note that the sign in this expression is the opposite of that corresponding to the equation of motion for a dynamical variable in Heisenberg's description, Eq. (8.102). The reason is that (15.140) actually means that $i\hbar\,(d\widehat{\rho}/dt) = i\hbar\,(\partial\widehat{\rho}/\partial t) + [\widehat{\rho},\widehat{H}] = 0$, that is,

$$\frac{d\widehat{\rho}}{dt} = 0, \qquad (15.142)$$

which is the quantum version of Liouville's theorem.[27] If we define the Liouville operator as $\hat{L}\cdot = [\widehat{H},\cdot]$, we get $\hat{L}\widehat{\rho} = [\widehat{H},\widehat{\rho}]$, and the von Neumann equation takes the form

$$i\hbar\frac{\partial \widehat{\rho}}{\partial t} = \hat{L}\widehat{\rho}, \qquad (15.143)$$

which is known as the quantum Liouville equation. In the particular, but very important case that the Hamiltonian is time-independent, we have $\widehat{S}(t,t_0) = e^{-\frac{i}{\hbar}\widehat{H}(t-t_0)}$, whence

$$\widehat{\rho}(t) = e^{-\frac{i}{\hbar}\widehat{H}(t-t_0)}\widehat{\rho}(t_0)e^{\frac{i}{\hbar}\widehat{H}(t-t_0)}. \qquad (15.144)$$

In the energy representation $\widehat{H}|n\rangle = E_n|n\rangle$, one gets

$$\langle n|\rho(t)|n'\rangle = \langle n|e^{-\frac{i}{\hbar}\widehat{H}(t-t_0)}\rho(t_0)e^{\frac{i}{\hbar}\widehat{H}(t-t_0)}|n'\rangle$$
$$= e^{-i\omega_{nn'}(t-t_0)}\langle n|\widehat{\rho}(t_0)|n'\rangle, \qquad (15.145)$$

indicating that the time evolution of $\widehat{\rho}(t)$ is reduced in this representation to an oscillation of the nondiagonal matrix elements, with constant diagonal matrix elements.

From Eq. (15.140) it follows that if $\widehat{\rho}$ depends only on integrals of motion, the system is in equilibrium, $\partial\widehat{\rho}/\partial t = 0$. This result has its classical counterpart, which states that any distribution that depends only on integrals of motion corresponds to a state of equilibrium. This parallelism results from the fact that the evolution equation for the density matrix (15.140) can be regarded as the quantum analog of the Liouville equation, as we have seen. However, it must still be emphasized that the latter equation corresponds to a description in phase space, whereas in the quantum problem the evolution takes place in Hilbert space.

---

[27] In its classical Hamiltonian version, Liouville's theorem says that the number of states contained in a given region of phase space is conserved.

Exercise E15.5. Definition of a stationarity time
Assume a "stationarity time" T_s defined by the relation

$$\frac{1}{T_s^2} = \langle (\Delta \widehat{\rho})^2 \rangle, \tag{15.146}$$

where $\widehat{\rho}$ is the density matrix. Discuss the validity of this definition and show that for any quantum system,

$$T_s^2 \langle (\Delta \widehat{H})^2 \rangle \geq \hbar^2. \tag{15.147}$$

When does this equality hold?

Solution. Although for a nonstationary system, $\dot{\widehat{\rho}} \neq 0$, the mean value $\langle \dot{\widehat{\rho}} \rangle$ is always zero (see Problem P15.19). This suggests using the dispersion of $\dot{\widehat{\rho}}$ as a measure of the degree of stationarity of the system, as in Eq. (15.146). Since T_s is given directly by the density matrix and does not refer to a particular dynamical variable, this stationarity time characterizes the (state of the) system globally. Note carefully the difference with times that refer to a specific variable, which may also satisfy Heisenberg's inequalities, but are related to the evolution of that variable.

Taking into account that $\langle \dot{\widehat{\rho}} \rangle = 0$, and using $\widehat{\rho} = \sum_n w_n |n\rangle \langle n|$, we get

$$\langle (\Delta \dot{\widehat{\rho}})^2 \rangle = \langle \dot{\widehat{\rho}}^2 \rangle = \operatorname{tr} \widehat{\rho} \dot{\widehat{\rho}}^2 = \sum_{m,n} \langle m|w_n|n\rangle \langle n|\dot{\widehat{\rho}}^2|m\rangle$$

$$= \sum_n w_n \langle n|\dot{\widehat{\rho}}^2|n\rangle = \sum_{n,k} w_n \langle n|\dot{\widehat{\rho}}|k\rangle \langle k|\dot{\widehat{\rho}}|n\rangle. \tag{15.148}$$

Using now the evolution equation for the density matrix (15.140), we note that

$$i\hbar \langle n|\dot{\widehat{\rho}}|k\rangle = \langle n|\widehat{H}\widehat{\rho} - \widehat{\rho}\widehat{H}|k\rangle \tag{15.149}$$

$$= \sum_m \langle n|\widehat{H}w_m|m\rangle \langle m|k\rangle - \sum_m \langle n|w_m|m\rangle \langle m|\widehat{H}|k\rangle = (w_k - w_n)H_{nk}.$$

Substituting, we can write

$$\langle (\Delta \dot{\widehat{\rho}})^2 \rangle = \frac{1}{\hbar^2} \sum_{n,k} w_n (w_n - w_k)^2 H_{nk} H_{kn} \leq \frac{1}{\hbar^2} \sum_{n,k \neq n} w_n H_{nk} H_{kn}. \tag{15.150}$$

The inequality follows from the fact that $(w_n - w_k)^2 \leq 1$, since $0 \leq w_k \leq 1$. On the other hand, the following relations hold:

$$\langle (\Delta \widehat{H})^2 \rangle = \sum_n w_n H_{nn}^2 - \left(\sum_n w_n H_{nn} \right)^2$$

$$= \sum_n w_n \langle (\Delta \widehat{H})^2 \rangle_n + \left[\sum_n w_n (H_{nn})^2 - \left(\sum_n w_n H_{nn} \right)^2 \right]$$

$$\geq \sum_n w_n \langle (\Delta \widehat{H})^2 \rangle_n, \tag{15.151}$$

where $\langle (\Delta \widehat{H})^2 \rangle_n$ is the dispersion of the energy in the pure state n. Since the dispersion can be written as the sum over the nondiagonal elements,

$$\langle(\Delta\widehat{H})^2\rangle_n = \langle n|\widehat{H}^2|n\rangle - \left(\langle n|\widehat{H}|n\rangle\right)^2$$
$$= \sum_k \langle n|\widehat{H}|k\rangle\langle k|\widehat{H}|n\rangle - \langle n|\widehat{H}|n\rangle\langle n|\widehat{H}|n\rangle = \sum_{k\neq n} H_{nk}H_{kn}, \quad (15.152)$$

we get

$$\langle(\Delta\widehat{H})^2\rangle \geq \sum_{n,k\neq n} w_n \widehat{H}_{nk}\widehat{H}_{kn}. \quad (15.153)$$

Substituting into the expression for the dispersion of $\widehat{\rho}$ and using the definition of T_s, Eq. (15.146), we obtain Eq. (15.147). From this it follows that the equals sign occurs if and only if the system is in a pure state.

Exercise E15.6. Evolution equation for $\widehat{\rho}$ in the interaction representation

Write the equation of evolution of a quantum ensemble described by a density matrix in the interaction description.

Solution. If V is the interaction potential, the Hamiltonian can be written in the form $\widehat{H} = \widehat{H}_0 + \widehat{V}$, so the density matrix satisfies the Heisenberg equation

$$i\hbar\frac{\partial\widehat{\rho}}{\partial t} = [\widehat{H},\widehat{\rho}] = \widehat{H}_0\widehat{\rho} - \widehat{\rho}\widehat{H}_0 + \widehat{V}\widehat{\rho} - \widehat{\rho}\widehat{V}. \quad (15.154)$$

In the interaction description we have

$$\widehat{\rho}_{\text{int}} = e^{i\widehat{H}_0 t/\hbar}\widehat{\rho}e^{-i\widehat{H}_0 t/\hbar}, \quad (15.155)$$

$$\widehat{H}_{0\text{int}} = \widehat{H}_0, \quad \widehat{V}_{\text{int}} = e^{i\widehat{H}_0 t/\hbar}\widehat{V}e^{-i\widehat{H}_0 t/\hbar}, \quad (15.156)$$

and the time derivative gives

$$i\hbar\frac{\partial\widehat{\rho}_{\text{int}}}{\partial t} = i\hbar\left(\frac{i}{\hbar}\widehat{H}_0\widehat{\rho}_{\text{int}} - \widehat{\rho}_{\text{int}}\frac{i}{\hbar}\widehat{H}_0\right) + i\hbar e^{i\widehat{H}_0 t/\hbar}\frac{\partial\widehat{\rho}}{\partial t}e^{-i\widehat{H}_0 t/\hbar}. \quad (15.157)$$

Using Eq. (15.154) and simplifying, the desired result is obtained,

$$i\hbar\frac{\partial\widehat{\rho}_{\text{int}}}{\partial t} = [\widehat{V}_{\text{int}},\widehat{\rho}_{\text{int}}]. \quad (15.158)$$

This is the quantum Liouville equation in the interaction representation; if we replace the commutator multiplied by $1/i\hbar$ by a Poisson bracket, we obtain the classical Liouville equation in the interaction description.

15.5.4 Pure states

From the preceding discussion it is clear that the description of a physical system in terms of a density matrix is the most general quantum description available and includes as a special case the description of pure states.

When the system is in the pure state $|p\rangle$, the mixture is reduced to only this state. The pure state is therefore defined by the condition

$$w_i = \delta_{pi} \tag{15.159}$$

and the corresponding density matrix $\hat{\rho} = \sum_i w_i |i\rangle\langle i| = \sum_i \delta_{pi} |i\rangle\langle i|$ coincides with the projector for the state $|p\rangle$,

$$\hat{\rho} = |p\rangle\langle p| = \hat{P}_p. \tag{15.160}$$

For example, in the coordinate representation, the density matrix of the pure state $\psi_p(x)$ is

$$\rho_p(x',x) = \langle x'|p\rangle\langle p|x\rangle = \psi_p(x')\psi_p^*(x). \tag{15.161}$$

Note that in particular

$$\rho_p(x,x) = \psi_p(x)\psi_p^*(x) = \rho(x). \tag{15.162}$$

From Eq. (15.160) it follows that

$$\hat{\rho}^2 = |p\rangle\langle p|p\rangle\langle p| = |p\rangle\langle p| = \hat{\rho}, \tag{15.163}$$

so the density matrix of a pure state is idempotent. This property is used to define (construct) a pure state.

Since an algebraic relation between matrices also holds for their eigenvalues, the eigenvalues ρ' of the density matrix of a pure state satisfy $\rho'^2 = \rho'$, so they can only be 0 and 1.

For a mixture, $\operatorname{tr}\hat{\rho}^2 \leq 1$; the equal sign only applies if it is a pure state ($\operatorname{tr}\hat{\rho} = 1$ always). This allows us to define a degree of impurity of a mixture as

$$\mathcal{I} \equiv 1 - \operatorname{tr}\hat{\rho}^2 = \operatorname{tr}\left(\hat{\rho} - \hat{\rho}^2\right) = \operatorname{tr}\hat{\rho}(1-\hat{\rho}), \tag{15.164}$$

representing the expectation value of $\langle 1 - \hat{\rho}\rangle$, which is equal to 0 for a pure state and tends to 1 as the impurity increases (when $\operatorname{tr}\hat{\rho}^2 \ll 1$). Equation (15.163) can be written as $\sum w_n^2 |n\rangle\langle n| = \sum w_n |n\rangle\langle n|$, whence $w_n^2 = w_n$, that is, an idempotent density matrix can contain weights of value 0 or 1 only.

The extreme opposite of the pure case is when the complete lack of information about the possible state of any of the N subsystems leads to an equal (a priori) weighting of all of them,

$$\operatorname{tr}\hat{\rho}^2 = \sum_n^N w_n^2 = \sum^N \frac{1}{N^2} = \frac{1}{N}. \tag{15.165}$$

For $N \to \infty$ this quantity can be arbitrarily small. This is the minimum value that $\operatorname{tr}\hat{\rho}^2$ can take. In fact, if two of the weights take the values $\frac{1}{N} + \delta$ and $\frac{1}{N} - \delta$ (the trace is one), there is a change in $\operatorname{tr}\hat{\rho}^2$ given by $\left(\frac{1}{N}+\delta\right)^2 + \left(\frac{1}{N}-\delta\right)^2 - \frac{2}{N^2} = 2\delta^2 \geq 0$, and the value of $\operatorname{tr}\hat{\rho}^2$ increases for any $\delta \neq 0$.

15.5.5 Spin polarization

When the density matrix involves a small number of parameters, it is possible, and even convenient, to express it in terms of results directly accessible from the experiment. This possibility often feeds an operational view for the construction of the density matrix that pervades (and even distorts) the literature on the subject. The following topic is an instructive example of considerable physical interest, since it deals with the construction of a density matrix to describe the polarization states of particles with spin 1/2. The procedure is applicable to any problem for which a two-dimensional Hilbert space is appropriate.

We write the spin operator in terms of the Pauli matrices, namely

$$\hat{S} = \tfrac{1}{2}\hbar\hat{\sigma}. \tag{15.166}$$

A general ensemble of spin-1/2 particles is in a superposition of the different possible polarization states, so the expectation value of the spin is

$$\langle\hat{S}\rangle = \tfrac{1}{2}\hbar\langle\hat{\sigma}\rangle = \tfrac{1}{2}\hbar \boldsymbol{P}. \tag{15.167}$$

This defines the polarization vector,

$$\boldsymbol{P} \equiv \langle\hat{\sigma}\rangle. \tag{15.168}$$

We now express the polarization vector as the average over the density matrix of the corresponding polarization state,

$$\boldsymbol{P} = \langle\hat{\sigma}\rangle = \operatorname{tr}\hat{\rho}\hat{\sigma}. \tag{15.169}$$

To construct the density matrix that goes into this formula, we take into account that the three Pauli matrices plus the identity matrix 2×2 form a basis for the representation of the operators acting in a 2D space; then $\hat{\rho}$, which has the dimension 2×2, can be written in the form

$$\hat{\rho} = a_0\mathbb{I} + a_1\hat{\sigma}_1 + a_2\hat{\sigma}_2 + a_3\hat{\sigma}_3. \tag{15.170}$$

Taking the trace of this expression and using $\operatorname{tr}\hat{\sigma}_i = 0$, we get $\operatorname{tr}\hat{\rho} = 1 = a_0\operatorname{tr}\mathbb{I} = 2a_0$, so that $a_0 = 1/2$. Now taking the i component of (15.169) and introducing (15.170), we get

$$P_i = \operatorname{tr}\hat{\rho}\,\hat{\sigma}_i = \operatorname{tr}\left[a_0\hat{\sigma}_i + \sum_k a_k\hat{\sigma}_k\hat{\sigma}_i\right] = \sum_k a_k \operatorname{tr}\hat{\sigma}_k\hat{\sigma}_i = 2a_i. \tag{15.171}$$

The latter result follows from the fact that if i and k are different, the product $\hat{\sigma}_i\hat{\sigma}_k$ is another Pauli matrix, therefore the only contributing term is $i = k$, with $\hat{\sigma}_i^2 = \mathbb{I}$. This gives $a_i = P_i/2$. Collecting results, we get the revealing expression

$$\hat{\rho} = \tfrac{1}{2}(\mathbb{I} + \boldsymbol{P}\cdot\hat{\sigma}), \tag{15.172}$$

or, in explicit matrix form,

$$\hat{\rho} = \tfrac{1}{2}\begin{pmatrix} 1 + P_3 & P_1 - iP_2 \\ P_1 + iP_2 & 1 - P_3 \end{pmatrix}. \tag{15.173}$$

We see that knowing the three components P_i of the polarization vector is sufficient to completely determine the density matrix. This means that the results of three independent measurements (three experiments) are sufficient to determine the spin density matrix. The pragmatic application of this point of view provides the basis for the operational interpretation of the density matrix mentioned earlier.

Exercise E15.7. Density operator and spin polarization

A spin polarized beam is in a pure state described by the spinor $|p\rangle = \begin{pmatrix} \sqrt{1/3} \\ \sqrt{2/3} \end{pmatrix}$. Find the corresponding density matrix and spin polarization vector.

Solution. In terms of the basis

$$|+\rangle = \begin{pmatrix} 1 \\ 0 \end{pmatrix}, \quad |-\rangle = \begin{pmatrix} 0 \\ 1 \end{pmatrix}, \tag{15.174}$$

the pure state represented by $|p\rangle$ is

$$|p\rangle \equiv \begin{pmatrix} \sqrt{1/3} \\ \sqrt{2/3} \end{pmatrix} = \sqrt{\tfrac{1}{3}}\,|+\rangle + \sqrt{\tfrac{2}{3}}\,|-\rangle. \tag{15.175}$$

Its density matrix is therefore given by

$$\hat{\rho} = |p\rangle\langle p| = \begin{pmatrix} \sqrt{1/3} \\ \sqrt{2/3} \end{pmatrix} \begin{pmatrix} \sqrt{1/3} & \sqrt{2/3} \end{pmatrix} = \tfrac{1}{3}\begin{pmatrix} 1 & \sqrt{2} \\ \sqrt{2} & 2 \end{pmatrix}. \tag{15.176}$$

Since it is a pure state, this matrix must be idempotent, and indeed it is,

$$\hat{\rho}^2 = \tfrac{1}{9}\begin{pmatrix} 3 & 3\sqrt{2} \\ 3\sqrt{2} & 6 \end{pmatrix} = \tfrac{1}{3}\begin{pmatrix} 1 & \sqrt{2} \\ \sqrt{2} & 2 \end{pmatrix} = \hat{\rho}. \tag{15.177}$$

On the other hand, this density matrix can be written in terms of the polarization vector $\mathbf{P} = \langle \hat{\boldsymbol{\sigma}} \rangle$ in the form (15.173). Comparing (15.176), we get

$$\tfrac{1}{2}(1 + P_3) = \tfrac{1}{3}, \quad \tfrac{1}{2}(1 - P_3) = \tfrac{2}{3}; \tag{15.178}$$

$$\tfrac{1}{2}(P_1 - iP_2) = \tfrac{\sqrt{2}}{3}, \quad \tfrac{1}{2}(P_1 + iP_2) = \tfrac{\sqrt{2}}{3}. \tag{15.179}$$

The components of the polarization vector are therefore

$$P_1 = \tfrac{2\sqrt{2}}{3}, \quad P_2 = 0, \quad P_3 = -\tfrac{1}{3}, \tag{15.180}$$

that is,

$$\mathbf{P} = \left(\tfrac{2\sqrt{2}}{3},\, 0,\, -\tfrac{1}{3}\right). \tag{15.181}$$

This vector lies in the xz-plane and its magnitude is $P^2 = 1$. The latter is to be expected, since every pure state with spin 1/2 is fully polarized in one direction.

15.5.6 The density matrix and quantum statistics

We have seen that the density matrix allows us to make the most general description possible of an open quantum system and that it can be considered as the quantum counterpart of the classical distribution function, since its evolution equation is the quantum version of the Liouville equation. We will now use this perspective to build a theory of systems that require both statistical and quantum treatment, such as those that obey quantum statistics. We will therefore deal simultaneously with classical statistics and the statistical properties of quantum systems.

15.5.6.1 The canonical and grand canonical ensembles

Let us consider a system of macroscopic dimensions – which we will call a macrosystem – composed of a large number of (micro) systems, such as a volume of gas, and so on. If the macrosystem is in equilibrium at a fixed temperature T and E_n represents the eigenvalue of the Hamiltonian $\widehat{H}$ of one of its component subsystems, then the probability that it is in the energy state E_n is given by the Maxwell–Boltzmann distribution

$$w_n = \frac{1}{Z} e^{-\beta E_n}, \tag{15.182}$$

with $\beta = 1/kT$, where $Z = \sum_n e^{-\beta E_n}$ is the partition function, such that $\sum w_n = 1$. Note that each state contributes to the value of Z with a number of terms equal to the order of the degeneracy of the state. The density matrix now takes the form (for simplicity, we consider nondegenerate states)

$$\widehat{\rho} = \frac{1}{Z} \sum_n e^{-\beta E_n} |n\rangle\langle n|, \tag{15.183}$$

where $|n\rangle$ represents one of the eigenvectors of $\widehat{H}$, $\widehat{H}|n\rangle = E_n|n\rangle$. Note that we have used classical statistics to describe the states of the macrosystem, and now we apply quantum statistics to the result. Note also that for the partial eigenstates $|k\rangle$ of the Hamiltonian we get

$$\widehat{\rho}|k\rangle = \frac{1}{Z} \sum_n e^{-\beta \widehat{H}} |n\rangle \langle n|k\rangle = \frac{1}{Z} e^{-\beta E_k} |k\rangle = w_k |k\rangle, \tag{15.184}$$

that is, they are simultaneously eigenstates of the density matrix, with eigenvalue equal to the statistical weight of the state. It is possible to rewrite Eq. (15.183) independently of any representation, since

$$\widehat{\rho} = \frac{1}{Z} \sum_n e^{-\beta \widehat{H}} |n\rangle\langle n| = \frac{1}{Z} e^{-\beta \widehat{H}} \sum_n |n\rangle\langle n|, \tag{15.185}$$

and taking into account that $|n\rangle$ is a member of a complete set,

$$\widehat{\rho} = \frac{1}{Z} e^{-\beta \widehat{H}}. \tag{15.186}$$

The ensemble of systems described by this density matrix is by definition, a canonical ensemble. A canonical ensemble is the statistical ensemble that describes the possible states of a mechanical system in thermal equilibrium at a fixed temperature. Since

$\hat{\rho}$ depends only on the Hamiltonian, which is an integral of motion, it is an ensemble in equilibrium, whose thermodynamic properties are determined by the partition function Z. Since

$$\operatorname{tr}\hat{\rho} = 1 = \frac{1}{Z}\operatorname{tr} e^{-\beta\hat{H}}, \qquad (15.187)$$

the partition function is

$$Z = \operatorname{tr} e^{-\beta\hat{H}}. \qquad (15.188)$$

Substituting this result into Eq. (15.186) we get

$$\hat{\rho} = \frac{e^{-\beta\hat{H}}}{\operatorname{tr} e^{-\beta\hat{H}}}. \qquad (15.189)$$

The weights w_n are given by Eq. (15.183),

$$w_n = \frac{e^{-\beta E_n}}{\sum_n e^{-\beta E_n}}. \qquad (15.190)$$

It is important to note that the sum in the denominator of this expression is over the states, not over the eigenvalues, which is not the same when there is degeneracy.

If the ensemble has the additional property of allowing the number of particles in the system to vary, it is called a grand canonical ensemble. The density matrix then has the form

$$\hat{\rho} = \frac{1}{Z} e^{\beta(\mu\hat{n}-\hat{H})}, \quad Z = \sum_n e^{\beta(\mu\hat{n}-\hat{H})}, \qquad (15.191)$$

where μ is the chemical potential and $\hat{n}$ is the number operator.

Without going further into statistical quantum mechanics, which is the subject of a more advanced course, it is important to emphasize here that, with the proposed methods, it is possible to construct a statistical description of quantum systems without introducing the notion of phase space.

Exercise E15.8. The Wigner distribution function

As shown in Section 3.3.1, the Weyl transform of a function of two variables $f(x, y)$ is $\frac{1}{\pi\hbar}\int_{-\infty}^{\infty} f(x-z, y+z)e^{-2ipz/\hbar}dz$. Show that the Wigner distribution function, defined as the Weyl transform of the density matrix, has properties that allow it to be interpreted as the quantum equivalent of a probability density in phase space.

Solution. The Wigner distribution $P_W(x, p)$ associated with the density matrix $\hat{\rho}$, which in the coordinate representation is $\rho(x', x'')$, is obtained by writing $x'' = x - z$, $x' = x + z$ and performing a Fourier transformation on z (the Fourier variable is $k = p/\hbar$). Therefore,

$$P_W(x, p) = \frac{1}{\pi\hbar} \int_{-\infty}^{\infty} \langle x+z|\hat{\rho}|x-z\rangle e^{-2ipz/\hbar} dz. \qquad (15.192)$$

For a pure state, $\langle x'|\hat{\rho}|x''\rangle = \psi^*(x'')\psi(x')$ and (15.192) reduce to

$$P_W(x, p) = \frac{1}{\pi\hbar} \int_{-\infty}^{\infty} \psi^*(x-z)\psi(x+z) e^{-2ipz/\hbar} dz. \qquad (15.193)$$

To prove that $P_W(x, p)$ plays the role of a probability density in phase space, we first calculate the marginal probabilities. Integrating over p, we get

$$\int P_W(x, p) dp = \frac{1}{\pi \hbar} \int \rho(x + z, x - z) e^{-2ipz/\hbar} dp$$

$$= \frac{1}{2\pi} 2\pi \int \rho(x + z, x - z) \delta(z) = \rho(x, x) = \rho(x). \tag{15.194}$$

To calculate $\int_{-\infty}^{\infty} P_W(x, p) dx$ it is preferable to first obtain the density matrix in momentum space. Taking $\phi_i(p) = \tilde{\psi}_i(p)$, we obtain

$$\rho(p, p') = \sum_i w_i \varphi_i(p) \varphi_i^*(p') = \frac{1}{2\pi \hbar} \int \left[\sum_i w_i \psi_i(x) \psi_i^*(x') \right] e^{-\frac{i}{\hbar}(px - p'x')} dx dx', \tag{15.195}$$

that is,

$$\rho(p, p') = \frac{1}{2\pi \hbar} \int \rho(x, x') e^{-\frac{i}{\hbar}(px - p'x')} dx dx'. \tag{15.196}$$

The probability in momentum space is therefore given by

$$P(p) = \rho(p, p) = \frac{1}{2\pi \hbar} \int \rho(x', x'') e^{-\frac{i}{\hbar}p(x' - x'')} dx' dx''$$

$$= \frac{1}{2\pi \hbar} \int \rho(x + z, x - z) e^{-2ipz/\hbar} 2 dx dz, \tag{15.197}$$

so, comparing with (15.192), we get

$$\int_{-\infty}^{\infty} P_W(x, p) dx = P(p). \tag{15.198}$$

Since $\rho(x)$ and $P(p)$ are both normalized to unity, it follows from (15.194) or (15.198) that

$$\int_{-\infty}^{\infty} dx \int_{-\infty}^{\infty} dp \, P_W(x, p) = 1. \tag{15.199}$$

The properties (15.194), (15.198), and (15.199) suggest that $P_W(x, p)$ can be considered the quantum equivalent of the probability density in phase space. This is a frequent and useful application of the Wigner distribution; however, care must be taken with this interpretation, since the function $P_W(x, p)$ can take negative values.

A variety of distributions are found in the literature, although the Wigner distribution is the most widely used. In particular, it is used extensively in the development of stochastic optics pioneered by T. W. Marshall and E. Santos in the 1980s in order to understand the role of the ZPF in quantum optics experiments.

15.5.7 Decoherence

For many years, quantum measurement theory has attempted to provide a convincing answer to the problem posed by the intertwining of the states of the instrument used for measurement and those of the microsystem being measured.

Suppose a microsystem is described by an initial wave function which is a coherent superposition of all possible eigenstates of a certain variable we want to measure. Since a single result is obtained, after the measurement the microsystem is assigned a single state, namely an eigenstate of the measured variable. After many similar measurements, the ensemble will no longer be in a pure state, but will be described by a statistical mixture (a density matrix) of the different outcomes, with their respective probabilities.[28] This transition is not unitary and therefore does not conform to all the laws of QM. Because of its importance for the present discussion, we will briefly present the central idea here.

Consider a quantum system described in terms of the set of eigenstates $\{|\varphi_k\rangle\}$ of a particular operator whose value is to be measured with an instrument A, characterized by eigenstates $|A_k\rangle$ and eigenvalues a_k.[29] If the meter reads a_1 after the measurement, the microsystem is said to be in state $|\varphi_1\rangle$, and so on. Assume that the initial state of the whole system (microsystem + instrument) is described by the vector

$$|\Psi\rangle = \sum_k c_k |A_k\rangle |\varphi_k\rangle. \tag{15.200}$$

This pure state corresponds to the density matrix

$$\hat{\rho}_{\text{in}} = |\Psi\rangle\langle\Psi| = \sum_{k,k'} c_k c_{k'}^* (|A_k\rangle\langle A_{k'}|)(|\varphi_k\rangle\langle\varphi_{k'}|). \tag{15.201}$$

The different variants of measurement theory propose that carrying out the measurement is equivalent in one way or another to averaging over the states of the instrument, which is taken to correspond to the partial trace over the orthonormal vectors $|A_l\rangle$, thus obtaining a reduced density matrix,

$$\hat{\rho}_{\text{red}} = \sum_l \langle A_l |\Psi\rangle\langle\Psi| A_l\rangle = \sum_l \sum_{k,k'} c_k c_{k'}^* (\langle A_l |A_k\rangle\langle A_{k'}| A_l\rangle)(|\varphi_k\rangle\langle\varphi_{k'}|)$$

$$= \sum_k |c_k|^2 |\varphi_k\rangle\langle\varphi_k|. \tag{15.202}$$

This result tells us that the measurement will give the value k with probability $|c_k|^2$, and so on. The point is that the reduced $\hat{\rho}_{\text{red}}$ no longer describes a pure state, but a mixture. This is clear from the absence of interference terms in $\hat{\rho}_{\text{red}}$. The transition $\hat{\rho}_{\text{in}} \to \hat{\rho}_{\text{red}}$ corresponds to what is called reduction or collapse of the state vector (or the wave function, if you prefer). But this transition is not unitary, since a unitary evolution transforms a pure state into another pure state and never into a mixture. We see that the operation of taking the partial trace over a portion of the whole system, thereby reducing a density matrix, is not part of Schrödinger's theory.

Entangled states such as the one described by Eq. (15.200) are, as we have seen, very common in the description of composite quantum systems. The entangled components can often be more than two, and all of them can correspond to microsystems governed by quantum laws. Because they are pure, entangled states imply coherence

---

[28] The theory of the density matrix and the notion of pure states are discussed in Section 15.5. A previous reading of this section is useful for a thorough understanding of the following discussion.

[29] This may viewed as an extrapolation of the cat example, with the consequences it implies.

between the different terms that make up the wave function, as is clear from the example of Eq. (15.201) with the presence of interference terms. These states are ubiquitous in quantum theory, but they are not observed in macroscopic systems because of the huge number of components, which means that coherence must be lost in the transition from micro to macro (in particular to a cat; see Section 14.5.3). We have already seen that a system governed by quantum laws evolves in a unitary manner, so such a transition process cannot occur within the framework of this theory. In short, QM is not the place for reduced matrices; they belong to other fields, such as information theory.

In an attempt to resolve this contradiction – and to avoid having to introduce the unconvincing notion of (nonlocal) collapse – it has been argued that an isolated microsystem is an idealization and that, in reality, every system coexists with an extremely complex and chaotic environment. Therefore, a realistic description should be done using a state vector or a density matrix given in its simplest form by Eqs. (15.200) and (15.201), where now the states $|A_k\rangle$ refer to the environment in which the microsystem is inserted. Averaging over the states of the environment is justified not only by their expected relevance, but more fundamentally by the fact that their chaotic nature makes them unknown, inviting us to average over their possible realizations. By averaging, we get a reduced result as given by Eq. (15.202), where all initial coherence has been erased. The density matrix thus obtained refers to description-independent probabilities, typical of macroscopic systems.

The enormous complexity of the environment means that the eigenvalues of interest are effectively a continuum, so that any coupling of the microsystem to the environment produces important changes. For example, estimates of the effects of the cosmic microwave background radiation at a temperature as low as 3 Kelvin show that it is capable of producing decoherence in times as short as 10^{-23} s, thus apparently, but only apparently, settling the problem of achieving consistency between quantum predictions and empirical data.

We conclude that appealing to our ignorance of the specific and detailed conditions of a part of the total system does not solve the problem in principle. Nor is it solved by considering that the real state is the entangled one, but that for all practical purposes this can be interpreted as if the reduction had taken place. A bona fide description of the system must include all accessible information about the whole system (such as that given by Eqs. (15.200) and (15.201)), and cannot be arbitrarily replaced by the reduced description unless such a reduction has a direct physical meaning and corresponds to an irreversible effect.

Although there are several variants of the decoherence program, none of them has gained sufficient acceptance to be considered a solution to the problem. An attempt of a slightly different nature is to modify the Schrödinger equation itself by adding nonlinear terms in such a way that the behavior of microsystems is practically unaffected, but a loss of coherence is obtained when the body attains macroscopic dimensions. An ingenious and remarkable version of such tentative theories, known as the spontaneous localization model, is due to Ghirardi, Rimini, and Weber (1986). However, it should be clear that these theories are ad hoc, since the form of the added stochastic nonlinear terms, as well as the value of their parameters, are chosen to obtain the desired results, so that they can only be considered as provisional.

Problems

P15.1 Show that the particle exchange operator $\widehat{P}_{ij}$ is Hermitian and that it commutes with $\widehat{P}_{nm}$ only when (i, j) and (n, m) refer to different pairs of particles (there is no common particle).

P15.2 Show that the operators
$$\widehat{P}_{ij}^{\pm} \equiv \tfrac{1}{2}\left(1 \pm \widehat{P}_{ij}\right) \tag{15.203}$$
are projectors. What is the effect of these operators on a fully symmetric state function $\Psi(\xi_1, \ldots, \xi_i, \xi_j, \ldots, \xi_N)$?

P15.3 Obtain the solution (15.49) using perturbation theory for degenerate systems. The degenerate wave functions are considered to be $\psi_{\pm}$.

P15.4 Show that if $\psi_{Nn}(x_1, x_2)$ is the wave function of the two linearly coupled oscillators (see Eq. (9.125)), then
$$\widehat{P}_{12}\psi_{Nn} = (-1)^n \psi_{Nn}, \tag{15.204}$$
where N, n represent the excitation quantum number of the center of mass and relative motion, respectively.

P15.5 The Hamiltonian of a system of three linearly coupled identical spin-zero bosons is
$$\widehat{H} = \sum_{i=1}^{3} \frac{\widehat{p}_i^2}{2m} + \sum_{i=1}^{3} \tfrac{1}{2}m\omega^2 x_i^2 + \sum_{j>i}\sum_{i=1}^{3} \tfrac{1}{2}m\omega^2\beta\left(x_i - x_j\right)^2. \tag{15.205}$$

In terms of the normal coordinates
$$Z = \tfrac{1}{3}(x_1 + x_2 + x_3); \quad z_1 = x_1 - x_2; \quad z_2 = x_3 - \tfrac{1}{2}(x_1 + x_2), \tag{15.206}$$
the Hamiltonian takes the form
$$\widehat{H} = \frac{\widehat{P}^2}{2M} + \tfrac{1}{2}M\omega^2 Z^2 + \frac{\widehat{p}_{z_1}^2}{2m_1} + \tfrac{1}{2}m_1\omega_1^2 z_1^2 + \frac{\widehat{p}_{z_2}^2}{2m_2} + \tfrac{1}{2}m_2\omega_2^2 z_2^2, \tag{15.207}$$
where
$$\widehat{P} = \widehat{p}_1 + \widehat{p}_2 + \widehat{p}_3; \quad \widehat{p}_{z_1} = \tfrac{1}{2}(\widehat{p}_1 - \widehat{p}_2); \quad \widehat{p}_{z_2} = \tfrac{2}{3}(\widehat{p}_3 - \frac{\widehat{p}_1 + \widehat{p}_2}{2});$$
$$\omega_1^2 = \omega_2^2 = \omega^2(1 + 3\beta); \quad M = 3m; \quad m_1 = \tfrac{1}{2}m; \quad m_2 = \tfrac{2}{3}m. \tag{15.208}$$

(a) Find the solutions of the Schrödinger equation.
(b) Determine the eigenvalues of the Hamiltonian.
(c) Indicate which states are physically realizable.
(d) Show that solutions with exchange degeneracy are not necessarily orthogonal.

P15.6 Consider a system of four uncoupled, equal and collinear harmonic oscillators. Construct the wave functions and specify the energy eigenvalues of the physically realizable steady states when:

(a) All four particles are bosons with spin zero.
(b) All four particles are fermions with spin 1/2.

P15.7 The deuteron (consisting of a proton and a neutron) has spin 1. List the possible states of spin and total angular momentum of a system of two deuterons in a state of angular momentum L.

P15.8 Three particles of spin zero are rigidly joined to form an equilateral triangle that rotates on a circle of radius r. Determine the rotational levels of this system.

P15.9 Let $\widehat{F}$ be an operator of the form

$$\widehat{F}(1, 2, \ldots, N) = \widehat{f}(1) + \widehat{f}(2) + \cdots + \widehat{f}(N) = \sum_{i=1}^{N} \widehat{f}(i), \quad (15.209)$$

where $\widehat{f}(i)$ operates only on the coordinates of particle i in a system of N identical particles. Show that

$$\left\langle \Psi^A \mid \widehat{F} \mid \Psi^A \right\rangle = \sum_{i=1}^{N} \langle \psi_i \mid \widehat{f}(i) \mid \psi_i \rangle, \quad (15.210)$$

where Ψ^A is the antisymmetric wave function (15.32) and the wave functions ψ_i of a particle are orthonormal.

P15.10 Prove that the lattice Hamiltonian (15.76) transforms into (15.80) in the reciprocal space.

P15.11 Prove that in the limit $m_2 \to m_1$, Eq. (15.82) is reduced to (15.81).

P15.12 Prove that $\mathrm{tr}\,\widehat{A}\widehat{B}\widehat{C} = \mathrm{tr}\,\widehat{C}\widehat{A}\widehat{B}$ and that $\mathrm{tr}\left[\widehat{A}, \widehat{B}\right]\widehat{C} = \mathrm{tr}\,\widehat{A}\left[\widehat{B}, \widehat{C}\right]$.

P15.13 State whether the following operators can be taken as density matrices and whether they describe or can describe pure states and under what conditions:

$$\begin{pmatrix} \frac{1}{2} & ia \\ -ia & \frac{1}{2} \end{pmatrix}, \quad \begin{pmatrix} \frac{1}{1+b^2} & e^{i\alpha} \\ e^{-i\alpha} & \frac{b^2}{1+b^2} \end{pmatrix}, \quad \begin{pmatrix} 0.3 & iZ\beta \\ iZ\beta & 0.7 \end{pmatrix}. \quad (15.211)$$

The parameters a, b, Z, α are real, but β can be complex.

P15.14 A physical system can be in two independent states. Show that the most general density matrix describing this situation has the form

$$\widehat{\rho} = \begin{pmatrix} a & be^{i\varphi} \\ be^{-i\varphi} & 1-a \end{pmatrix}. \quad (15.212)$$

Determine the requirements that a and b must meet in general and the conditions under which the described state is pure.

P15.15 Prove that the operators $\Lambda_n^{\pm} = (1 \pm \widehat{\sigma} \cdot \widehat{\mathbf{n}})/2$ are idempotent and mutually orthogonal, that is, that they are projection operators. Study their action on a density matrix, both for pure states and for mixtures.

P15.16 Construct the density operator and the polarization vector that correspond to the pure state described by the spinor $\begin{pmatrix} \sqrt{1/3} \\ \sqrt{2/3} \end{pmatrix}$.

P15.17 A physical system can be in three states $|1\rangle, |2\rangle, |3\rangle$ with probabilities $\frac{1}{2}, \frac{3}{8}, \frac{1}{8}$, respectively. Build the corresponding density matrix. How many additional conditions can be imposed?

P15.18 Show that the free-particle density matrix in the momentum representation is
$$\rho(p, p') = \delta(p - p')e^{-\beta p^2/2m}. \tag{15.213}$$

P15.19 Show that $\langle \widehat{\hat{\rho}} \rangle = 0$ is always true.

P15.20 Consider a mixture of the form $\hat{\rho} = \lambda \hat{\rho}_1 + (1 - \lambda) \hat{\rho}_2$, $0 < \lambda < 1$. Prove that the dispersion of a generic dynamic variable $\hat{A}$ satisfies the condition
$$\left\langle (\Delta_\rho \hat{A})^2 \right\rangle \geq \lambda \left\langle (\Delta_{\rho_1} \hat{A})^2 \right\rangle + (1 - \lambda) \left\langle (\Delta_{\rho_2} \hat{A})^2 \right\rangle.$$
When does the equality hold?

Bibliographical Notes

Tomonaga's Story of Spin (1997) is especially enjoyable to read. In particular, it contains a detailed discussion of the relationship between spin and statistics.

For a comprehensive theoretical discussion of phonons, see Srivastava (2022).

For an early authoritative explanation of superconductivity and Cooper pairs based on physical intuition, see Weisskopf (1979).

A classic reference on superconductivity is Tinkham (1996).

There are a number of textbooks on condensed matter which contain extensive descriptions of the quantum Hall effect. Several very good lecture notes on the quantum Hall effect can be found on www.damtp.cam.ac.uk/user/tong/qhe.html.

For a review of the Wigner distribution theory, see Hillery et al. (1984).

For the application of the Wigner distribution in stochastic optics, see Marshall and Santos (1988) and Casado et al. (2019).

For an extensive review of the many dynamical models proposed over the years to elucidate quantum measurements, see Allaverdyan et al. (2013). This review paper also contains a flexible and realistic model to account for the emergence of classicality in the measurement process.

16 Atoms and Molecules

This chapter begins by showing how QM explains the structure and basic properties of multielectron atoms and their position in the periodic table of elements. It then looks at the different types of forces that bind atoms together to form molecules, and at the long-range intermolecular forces that are responsible for many of the chemical properties of matter, and more generally play a crucial role in physics, chemistry and biology. In the final section, the properties of atomic nuclei are shown to reveal an internal structure with periodicities reminiscent of those of atoms.

When we turn our attention to the structure of the atom, we encounter a complex mathematical problem (all too familiar from classical physics), namely the many-body problem. Dealing with molecules, which are usually made up of many atoms, introduces further complications. But little by little, and year by year, our ability to deal with such problems has improved thanks to the development of various specially designed approximation methods, the most important of which are discussed in this chapter, with a focus on the simplest, diatomic molecules. Of course, things have changed radically with the advent of the computer; relatively simple systems, such as atoms with not too many electrons, can now be studied with precision. In quantum chemistry, however, in particular when dealing with systems of larger size, computational approaches face significant challenges due to the complexity arising from the exponential growth of the system's wave function with each added particle. More modern computational techniques using efficient quantum algorithms are expected to have runtimes and resource requirements that scale polynomially with system size and desired accuracy, making it possible to carry out complex calculations involving large molecules.

16.1 Multielectron Atoms. The Hartree–Fock Method

Solving the problem of a system of many electrons in the presence of nuclei is central to much of physics and chemistry. Since the many-body problem does not in general admit of an exact solution, approximate methods must be used to study such systems. The most important such method for solving the Schrödinger equation for systems with more than one electron, and in general for complicated atomic, molecular, and nuclear problems, is the self-consistent field method, which was proposed in its intuitive form by Hartree[1] in 1928 and further developed by Fock in 1930 taking into account the antisymmetry of the electron wave functions.

---

[1] Douglas Rayner Hartree (1897–1958) was an English mathematician and physicist, best known for developing numerical analysis and its application to atomic structure calculations.

Taking the atomic case as an example, the basic idea of the Hartree–Fock method is that it considers that each electron moves in a mean field produced by the other electrons and the nucleus and, in turn, contributes to the effective field felt by the other electrons. As they are simpler (though less accurate), we will first derive Hartree's atomic equations and then correct them.

Consider an atom with N electrons. The Hamiltonian of the kth electron coupled to the nucleus is $\widehat{H}_0(\mathbf{r}_k)$, but the Coulomb repulsion couples it to the rest of the electrons, and so the total Hamiltonian of the atom (neglecting spin–orbit coupling) becomes

$$H = \sum_k \widehat{H}_0(\mathbf{r}_k) + \sum_{k<l} V(\mathbf{r}_k, \mathbf{r}_l) = \sum_k \widehat{H}_0(\mathbf{r}_k) + \tfrac{1}{2} \sum_{l \neq k} V(\mathbf{r}_k, \mathbf{r}_l). \tag{16.1}$$

To apply the variational method (see Section 13.8), we propose as the initial test function the product of N independent-particle wave functions,

$$\psi(\mathbf{r}_1, \ldots, \mathbf{r}_N) = \varphi_1(\mathbf{r}_1)\varphi_2(\mathbf{r}_2)\ldots\varphi_N(\mathbf{r}_N). \tag{16.2}$$

The expectation value of the Hamiltonian calculated with this wave function is

$$\langle \widehat{H} \rangle = \sum_k \int \varphi_k^* \widehat{H}_0 \varphi_k \, d^3 r_k + \tfrac{1}{2} \sum_{l \neq k} \int \varphi_k^* \varphi_l^* V(\mathbf{r}_k, \mathbf{r}_l) \varphi_k \varphi_l \, d^3 r_k \, d^3 r_l. \tag{16.3}$$

We want this quantity to be stationary (in fact, to be a minimum) against arbitrary small changes of the test function. If the φ_k^* are varied, leaving the φ_k fixed, this implies

$$\delta \langle \widehat{H} \rangle = \sum_k \int \delta\varphi_k^* \left[\widehat{H}_0 + \sum_{l \neq k} \langle l | V(k, l) | l \rangle \right] \varphi_k \, d^3 r_k = 0. \tag{16.4}$$

However, the $\delta\varphi_k^*$ are not independent, since from the condition that the norm of each wave function does not change it follows that

$$\int \delta\varphi_k^* \varphi_k \, d^3 r_k = 0. \tag{16.5}$$

To take these restrictions into account when calculating $\delta \langle \widehat{H} \rangle$ we apply the method of indeterminate Lagrange multipliers by multiplying (16.5) by the undetermined constant E_k and subtracting the sum of all these expressions from Eq. (16.4). This gives

$$\sum_k \int \delta\varphi_k^* \left[\widehat{H}_0 + V_k - E_k \right] \varphi_k \, d^3 r_k = 0, \tag{16.6}$$

with

$$V_k = \sum_{l \neq k} \langle l | V(k, l) | l \rangle = \sum_{l \neq k} \int \varphi_l^* V(\mathbf{r}_k, \mathbf{r}_l) \varphi_l \, d^3 r_l. \tag{16.7}$$

$V_k(\mathbf{r}_k)$ represents the average potential at the position of electron k, produced by the other electrons. Since in (16.6) the arbitrary variations $\delta\varphi_k^*$ can already be considered independent, the following system of Hartree equations must be satisfied

$$(\widehat{H}_0 + V_k)\varphi_k = E_k \varphi_k. \tag{16.8}$$

Each Hartree equation is the Schrödinger equation for an atomic electron moving in the field produced by the nucleus (interaction contained in $\widehat{H}_0$) and the mean-field potential V_k produced by the other electrons. We see that the procedure has allowed us to formally transform the problem of N coupled particles into N one-particle problems.

The main drawback of this method is that the exclusion principle has not been taken into account. This was corrected by Fock[2], replacing the test function (16.2) by the corresponding fully antisymmetrized function, including the spinor dependence. In the notation used in Chapter 15,

$$\Psi(\xi_1,\ldots,\xi_N) = \sum_{P_k} \widehat{P}_k a_{P_k} \varphi_1(\xi_1)\varphi_2(\xi_2)\ldots\varphi_N(\xi_N). \tag{16.9}$$

The calculation is complicated, but it can be done explicitly. In the absence of spin–orbit coupling, one obtains a system of orbital equations analogous to the previous one, the so-called Fock–Dirac equations or Hartree–Fock equations,

$$(\widehat{H}_0 + V_k)\varphi_k - \sum_{l \neq k} V_{kl}\varphi_l = E_k\varphi_k, \tag{16.10}$$

where

$$V_{kl} = \int \varphi_k^* V(k,l) \varphi_l \, d^3 r_l = \langle k|V(k,l)|l\rangle \tag{16.11}$$

is the exchange potential, the meaning of which was discussed in Section 15.3. A comparison with the Hartree equations shows that the latter are obtained from the Fock–Dirac equations by neglecting the exchange effects.

Both the Hartree and the Fock–Dirac equations are integrodifferential, since the effective potentials are only known once the solutions are known. The numerical procedure proposed by Hartree to solve them consists of finding a self-consistent solution by iteration, as follows. An initial selection of test functions $\varphi_k^{(0)}$ is used for a first approximate calculation of the potentials V_k and V_{kl}. These are introduced into the system of equations, which is solved to obtain a new approximation to the wave functions, $\varphi_k^{(1)}$. With these new values, the effective potentials are recalculated and the process is repeated, to obtain $\varphi_k^{(2)}$. The procedure is continued until the resulting wave functions match (within acceptable errors) those entered in the previous step. Once these solutions are determined, the energy is calculated using the formula

$$E = \sum_k E_k - \tfrac{1}{2} \sum_k \sum_{l \neq k} \left[\langle kl|\widehat{V}(k,l)|kl\rangle - \langle kl|\widehat{V}(k,l)|lk\rangle \right], \tag{16.12}$$

which is obtained by taking the dot product of Eq. (16.10) with φ_k and summing over k. The E_k represent an approximation to the ionization energy associated with each electron.

This method was successfully applied by Fock as early as 1934 to the Li and Na atoms, and is widely used today to solve problems of atomic structure. Assuming that each electron moves in the mean central field produced by the nucleus and the other orbiting electrons, the electronic states can be described with the quantum numbers n, l, m and s_3 used for the hydrogen-like states. This means that the atom is considered to have a well-defined total angular momentum and, in the approximation in which spin–orbit coupling can be neglected, both the orbital and the spinorial atomic

---

[2] Vladimir Alexandrovich Fock (or Fok) (1898–1974) was an outstanding Russian-Soviet theoretical physicist and mathematician; about a dozen subjects are named in his honor.

momentum are well defined (this is particularly the case for light atoms). However, the fact that the effective potential differs significantly from a simple Coulomb potential implies that the energy of the electron depends both on the principal quantum number n (defined as $l + 1+$ number of nodes of the radial function) and on the orbital momentum l. Thus, the energy of the atom depends in a rather complicated way on the electronic quantum numbers.

On the other hand, the requirement that the atom must not contain two electrons in the same state (n, l, m, s_3), according to the exclusion principle, implies that as we add electrons to successively build the different atoms, we must increase the value of n from the lowest possible value $n = 1$ to a maximum that depends on the total number of electrons to be accommodated. From experience with hydrogen-like wave functions, we know that the energy of the electron increases as n increases. This fact, which is universal (for fixed l), leads to the observation that the physical and chemical properties of atoms do not change continuously as the atomic number Z (the number of protons in the nucleus) increases. When the addition of a new electron makes it necessary to assign a higher value to n, an obvious discontinuity appears. This observation is verified in practice and forms the basis of the periodic table of the elements proposed by Mendeleev[3] in 1869, as we shall see in the next section. With the advent of QM the periodic table took on the value of a structural arrangement.

The conclusion is that the classification of atomic electrons according to the value of the principal quantum number n is fundamental and expresses in synthetic terms a set of atomic properties.

16.1.1 Atomic shell structure

The preceding discussion suggests the introduction of the concept of (atomic) shells, a notion proposed by Bohr: all the electrons with the same n are found in the same shell. A shell is full (or closed) if it contains as many electrons as the exclusion principle allows. The shells are denoted by the corresponding value of n, or alternatively by the notation introduced by X-ray spectroscopy, namely,

n	1	2	3	4	5	6	7
Shell	K	L	M	N	O	P	Q

Within each shell, electrons can have integer values of l ranging from 0 to $n-1$, forming the subshells s, p, d, f, with $l = 0, 1, 2, 3$, respectively. To determine the number of electrons that fit in a subshell, we note that in each state with fixed n, l, m there can be two electrons, associated with the two possible values of s_3. Therefore, the total number of electrons in the subshells s, p, d, f, is 2, 6, 10, 14, respectively. If all atomic wave functions were hydrogen-like, this would give $2n^2$ electrons per shell. However, this is not the case in real atoms, where only the subshells s, p, d, and f exist, so a shell never contains more than 32 (2+6+10+14) electrons when it is full, as shown in Table 16.1.

---

[3] Dmitri Ivanovich Mendeleyev (better known as Dmitri Mendeleev) (1834–1907) was a Russian chemist, best known for his observation and development of the periodic classification of the (chemical) elements. He knew and classified about 70 elements and predicted the existence of another 10.

Table 16.1 Electronic configuration and spectroscopic notation of the atomic elements. Transition metal groups: (1) Fe group, (2) Pd group, (3) lanthanoid or rare-earth group, (4) Pt group, (5) actinoid group.

Z	Element	Configuration	SN	$n=1$ $l=s$	2 $s\ p$	3 $s\ p\ d$	4 $s\ p\ d\ f$	5 $s\ p\ d\ f$	6 $s\ p\ d\ f$	7 $s\ p$		Shell
1	H	$1s$	$^2S_{1/2}$	1								K
2	He	$1s^2$	$^1S_0$	2								
3	Li	He$2s$	$^2S_{1/2}$	2	1							
4	Be	He$2s^2$	$^1S_0$	2	2							
5	B	He$2s^2 2p$	$^2P_{1/2}$	2	2 1							
⋮												
10	Ne	He$2s^2 2p^6$	$^1S_0$	2	2 6							
11	Na	Ne$3s$	$^2S_{1/2}$	2	2 6	1						M
12	Mg	Ne$3s^2$	$^1S_0$	2	2 6	2						
13	Al	Ne$3s^2 3p$	$^2P_{1/2}$	2	2 6	2 1						
⋮												
18	Ar	Ne$3s^2 3p^6$	$^1S_0$	2	2 6	2 6						
19	K	Ar$4s$	$^2S_{1/2}$	2	2 6	2 6	1					N
20	Ca	Ar$4s^2$	$^2S_0$	2	2 6	2 6	2					
21	Sc	Ar$fs^2 3d$	$^2D_{3/2}$	2	2 6	2 6 1	2				1)	
⋮												
30	Zn	Ar$4s^2 3d^{10}$	$^1S_0$	2	2 6	2 6 10	2					
31	Ga	Ar$4s^2 3d^{10} 4p$		2	2 6	2 6 10	2 1					
⋮												
36	Kr	Ar$4s^2 3d^{10} 4p^6$	$^1S_0$	2	2 6	2 6 10	2 6					
37	Ru	Kr$5s$	$^2S_{1/2}$	2	2 6	2 6 10	2 6	1				O
38	Sr	Kr$5s^2$	$^1S_0$	2	2 6	2 6 10 26	2					
39	Y	Kr$5s^2 4d$	$^2D_{3/2}$	2	2 6	2 6 10	2 6 1	2			2)	
⋮												
48	Cd	Kr$5s^2 4d^{10}$	$^1S_0$	2	2 6	2 6 10	2 6 10	2				
49	In	Kr$5s^2 4d^{10} 5p$	$^2P_{1/2}$	2	2 6	2 6 10	2 6 10	2 1				
⋮												
54	Xe	Kr$5s^2 4d^{10} 5p^6$	$^1S_0$	2	2	2 6 10		2 6 10	2 6			
55	Cs	Xe$6s$	$^2S_{1/2}$	2	2 6	2 6 10	2 6 10	2 6	1			P
56	Ba	X3$6s^2$	$^1S_0$	2	2 6	2 6 10	2 6 10	2 6	2			
57	La	Xe$6s^2 5d$	$^2D_{3/2}$	2	2 6	2 6 10	2 6 10	2 6	2			
58	Ce	Xe$6s^2 4f 5d$	$^3H_5$	2	2 6	2 6 10	2 6 10 1	2 6	2		3)	
⋮												
71	Lu	Xe$6s^2 4f^{14} 5d$	$^2D_{3/2}$	2	2 6	2 6 10	2 6 10 14	2 6 1	2			
72	Hf	Xe$6s^2 4f^{14} 5d^2$	$^3F_2$	2	2 6	2 6 10	2 6 10 14	2 6 2	2		4)	
⋮												
80	Hg	Xe$6s^2 4f^{14} 5d^{10}$	$^1S_0$	2	2 6	2 6 10	2 6 10 14	2 6 10	2			
81	Tl	Xe$6s^2 4f^{14} 5d^{10} 6p$	$^2P_{1/2}$	2	2 6	2 6 10	2 6 10	14	2 6 10	2 1		
⋮												
86	Rn	Xe$6s^2 4f^{14} 5d^{10} 6p^6$	$^1S_0$	2	2 6	2 6 10	2 6 10 14	2 6 10	2 6			
87	Fr	Rn$7s$	$^2S_{1/2}$	2	2 6	2 6 10	2 6 10 14	2 6 10	2 6	1		Q
88	Ra	Rn$7s^2$	$^1S_0$	2	2 6	2 6 10	2 6 10 14	2 6 10	2 6	2		
89	Ac	Rn$7s^2 6d$	$^2D_{3/2}$	2	2 6	2 6 10	2 6 10 14	2 6 10	2 6 1	2		
90	Th	Rn$7s^2 6d^2$	$^3F_2$	2	2 6	2 6 10	2 6 10 14	2 6 10	2 6 2	2		
91	Pa	Rn$7s^2 5f^2 6d$	$^4K_{11/2}$	2	2 6	2 6 10	2 6 10 14	2 6 10	2 6 2	2	5)	
⋮												
103	Lw	Rn$7s^2 5f^{14} 6d$		2	2 6	2 6 10	2 6 10 14	2 6 10 14	2 6 1	2		

The order in which the layers and subshells are filled as Z increases depends on the interactions between the electrons, with the tendency to fill the lower energy levels first. Thus, the first shells to be filled are $n = 1, 2, 3, \ldots$, but as n increases, additional values of l appear and since the energy of the electrons depends on both n and l, there are states that appear in the reverse order to that expected, as shown in Table 16.1. These inversions are important, as we will see later, because filling the $3d, 4d, 5d$ subshells produces the transition metals (38 in total), while the filling of the $4f, 5f$ subshells (which opens space for 14 elements each) produces the lanthanides and actinides, respectively. The inversions are due to the fact that the outer shell of an atom in its ground state can only contain s and p electrons, so the d and f subshells are filled as inner subshells, increasing their energy.

The notation used in Table 16.1 to specify the electronic configuration of the orbitals is $(nl)^k$ or simply nl^k to refer to k equivalent electrons in the nl state. In the spectroscopic notation a so-called term $^{2S+1}L_j$ is introduced, which specifies the quantum numbers of the angular state of the atom. For example, the ground state of H is $^2S_{1/2}$, while that of He is $^1S_0$. The correspondence between the electronic configuration and the spectroscopic name (the term) is not straightforward and is determined by means of Hund's empirical rules proposed by Friedrich Hund (1896–1997) during 1925, according to which the state with the largest s and the largest L (for this s) has the lowest energy.

16.2 The Periodic Table of the Elements

Table 16.2 contains all the presently known chemical elements up to number 118,[4] the heaviest element. The table shows how the electronic subshells fill up as the atomic number increases, forming the nine traditional groups (the columns of the table, numbered from 0 to VIII; the present-day, upgraded table contains 18 groups), and seven periods (the rows of the table, numbered from 1 to 7, corresponding to the shells K, L,... Q). The elements are arranged in the order of increasing atomic number Z, and each group contains chemically similar elements.

As we move from left to right in the table, that is, as the layers are filled, there is a gradual evolution of the chemical and physical properties of the elements, showing the periodicities discovered by Mendeleev. We will briefly discuss these properties in what follows and refer the interested reader to specialized texts for more detailed description.

Most of the elements in the table (about 80%) are metals; another seven are considered metalloids (or semi-metals, such as Si, Ge, etc., which have properties between metals and nonmetals); the rest are nonmetals. Metals are good conductors of heat and electricity (have high conductivity), while nonmetals are insulators (poor conductors of heat and electricity). A semiconductor is a material with a conductivity between conductors and insulators, as discussed in Section 6.3. In chemistry a metal is defined

---

[4] For decades, the table contained the "natural" elements, the last one being ^{92}U (uranium); now the table contains several additional elements up to (currently) ^{118}Og (oganesson), an extremely short-lived noble gas produced at the Joint Institute for Nuclear Research (JINR) in Dubna, Russia. In fact, all the "new" elements are produced in large accelerators.

Table 16.2 Periodic table of the elements. For each shell, the orbitals beyond the inert gas shell, that is, beyond the upper closed shell, are given. For example, the electronic configuration of $_{11}$Na is (Ne)(3s)=$(1s)^2(2s)^2(2p)^6(3s)$.

Periodic Table of Elements

1 H 1.0080																	2 He 4.0026
3 Li 6.94	4 Be 9.0122											5 B 10.81	6 C 12.011	7 N 14.007	8 O 15.999	9 F 18.998	10 Ne 20.180
11 Na 22.990	12 Mg 24.305											13 Al 26.982	14 Si 28.085	15 P 30.974	16 S 32.06	17 Cl 35.45	18 Ar 39.95
19 K 39.098	20 Ca 40.078	21 Sc 44.956	22 Ti 47.867	23 V 50.942	24 Cr 51.996	25 Mn 54.938	26 Fe 55.845	27 Co 58.933	28 Ni 58.693	29 Cu 63.546	30 Zn 65.38	31 Ga 69.723	32 Ge 72.63	33 As 74.922	34 Se 78.971	35 Br 79.904	36 Kr 83.798
37 Rb 85.468	38 Sr 87.62	39 Y 88.906	40 Zr 91.224	41 Nb 92.906	42 Mo 95.95	43 Tc [98]	44 Ru 101.07	45 Rh 102.91	46 Pd 106.42	47 Ag 107.87	48 Cd 112.41	49 In 114.82	50 Sn 118.71	51 Sb 121.76	52 Te 127.60	53 I 126.90	54 Xe 131.29
55 Cs 132.91	56 Ba 137.33	57-71	72 Hf 178.49	73 Ta 180.95	74 W 183.84	75 Re 186.21	76 Os 190.23	77 Ir 192.22	78 Pt 195.08	79 Au 196.97	80 Hg 200.59	81 Tl 204.38	82 Pb 207.2	83 Bi 208.98	84 Po [209]	85 At [210]	86 Rn [222]
87 Fr [223]	88 Ra [226]	89-103	104 Rf [267]	105 Db [268]	106 Sg [269]	107 Bh [270]	108 Hs [269]	109 Mt [277]	110 Ds [281]	111 Rg [282]	112 Cn [285]	113 Nh [286]	114 Fl [290]	115 Mc [290]	116 Lv [293]	117 Ts [294]	118 Og [294]

Symbol — atomic number / atomic standard weight

lanthanoids

57 La 138.91	58 Ce 140.12	59 Pr 140.91	60 Nd 144.24	61 Pm [145]	62 Sm 150.36	63 Eu 151.96	64 Gd 157.25	65 Tb 158.93	66 Dy 162.50	67 Ho 164.93	68 Er 167.26	69 Tm 168.93	70 Yb 173.05	71 Lu 174.97

actinoids

| 89
Ac
[227] | 90
Th
232.04 | 91
Pa
231.04 | 92
U
238.03 | 93
Np
[237] | 94
Pu
[244] | 95
Am
[243] | 96
Cm
[247] | 97
Bk<
[247] | 98
Cf
[251] | 99
Es
[252] | 100
Fm
[257] | 101
Md
[258] | 102
No
[259] | 103
Lr
[262] |
|---|---|---|---|---|---|---|---|---|---|---|---|---|---|---|

as an element that can easily form positive ions by losing electrons (cations) and tends to form metallic bonds. A metallic bond is formed by the valence electrons moving freely through the metal lattice. The valence electrons are those in the outermost shell of an atom, so they can be transferred to or shared with another atom to form bonds. Chemically, metals react with oxygen to form a metal oxide; a metal oxide reacts with water to form a base.

To the left of the table, essentially in the columns Ia–IIIa, are the most electropositive metals, containing one to three electrons in their outer layer, which they can easily transfer in an electrovalent bond, thus becoming ions (cations) with a noble-gas configuration. Those in column Ia are the so-called alkali metals (Li, Na, K, ..., used in the Stern–Gerlach experiments precisely because they have only one outer electron and a ground state $^2S_{1/2}$). The alkaline earth metals, contained in group IIa, are all divalent. At the other end of the table, especially at the top, are the nonmetals (acid-forming elements); they need one or more electrons to close their outer shell, so they participate with negative ionic valence. These properties are even more pronounced in the halogens (F, Cl, Br, I, At), contained in column VIIb, which, lacking a single electron to complete their outer shell, easily form ionic compounds with alkali metals, known as salts (LiF, NaCl, KCl, etc.). Among the halogens, F is the most electronegative element (the concept of electronegativity, introduced by Pauling,[5] refers to the ability of an atom in a molecule to attract electrons). The complete transfer of valence electrons between atoms gives rise to ionic bonding between two oppositely charged ions. For example, in NaCl, the Na loses its single outer electron (becoming Na^+), which is captured by the Cl to complete its outer shell (becoming Cl^-).

From columns Ia to VIIb of the table there is a gradual transition from the more metallic to the less metallic elements. The elements in the middle region of the table have intermediate properties. In particular, in the first two periods, C and Si (which belong to group IVb) have four electrons in their outer shell (valence 4) and do not donate or accept electrons, they share pairs of electrons, forming covalent rather than ionic compounds (see Section 16.4).

The number of electrons an atom gives up or gains (number of valence bonds) to achieve a stable molecular configuration is called valence. Covalency is the maximum number of covalent bonds an atom can form within a molecule using its empty orbitals; for example, the covalency of carbon is 4.

The longer periods (4, 5, 6) in the intermediate region contain the transition metals, which include the series of Fe (from Sc to Zn), Pd (from Y to Cd), and Pt (from Hf to Tl) (the classification varies somewhat from author to author). The transition metals differ from the electropositive metals in several properties, such as the multiplicity of valences, the formation of covalent bonds, and so on. The metals of groups Ib–IIIb have one to three electrons more than a pseudo-rare gas (i.e., a configuration corresponding to a rare gas plus a complete d-subshell). However, the metals of group Ib (Cu, Ag, and Au, which belong to the so-called noble metals) can use electrons from the d-subshell to form compounds, making them di- and trivalent, as well as

---

[5] Linus Pauling (1901–1994) was a prominent US-born chemical engineer, chemist, biochemist and notorious peace activist for nuclear disarmament; he was awarded the Nobel Prize for Chemistry in 1954 and the Nobel Peace Prize in 1962.

monovalent. Group IVb and Vb elements include semimetals (or semiconductors), which have transitional properties between one class and another.

The rare gases, members of group 0, form a separate class. These elements, also known as noble gases, are chemically inert because all their layers are closed. Finally, we have the internal transition metals, the lanthanides and actinides, which are chemically equivalent to La and Ac, respectively, because they are formed by filling an inner layer close to the nucleus, while the outer layer (also incomplete) remains essentially the same. The 14 lanthanides extend from lanthanum ($Z=57$) to lutetium ($Z=71$); the 14 actinides start with actinium ($Z=89$) and end with lawrencium ($Z=103$).

There are alternative forms of the periodic table. Among the best known is the grouping based on electronic structure proposed by Thomsen (1895) and Bohr (1922); it is a geometric rearrangement of Table 16.2, relatively easy to understand from the point of view of the previous discussion.

16.3 The Helium Atom

The theoretical study of the He atom – the simplest element after H – is of great importance, since it represents a crucial test for Schrödinger's theory, as Bohr's theory could not deal with it. The He atom has been studied extensively and, although it can only be solved by approximate methods, highly accurate solutions are now available. The remarkable agreement between the theoretical and experimental results confirms beyond doubt the validity of the theoretical description. More specifically, it confirms that the electromagnetic interaction is the only one that plays a relevant role in atomic structure.

To solve the general n-electron problem, one proceeds as follows. The motion of the center of mass (CM) is irrelevant to the analysis, so it can be omitted from the description. If $\mathbf{r}_{ei}$ is the position of electron i ($i=1, 2,\ldots, n$) with mass m_e, and $\mathbf{r}_0$ is the position of the nucleus with mass M, then the coordinate of the center of mass is given by

$$\mathbf{R} = \frac{M\mathbf{r}_0 + m_e \sum_{i}^{n} \mathbf{r}_{ei}}{M + nm_e}. \tag{16.13}$$

The Schrödinger equation for the system of $n + 1$ particles is rewritten in terms of the relative coordinates $\mathbf{r}_i = \mathbf{r}_{ei} - \mathbf{r}_0$, under the hypothesis that the potential depends only on the latter. The resulting equation contains the CM motion in separable form and can be eliminated as was done, for example, in Section 12.1 in connection with the H problem. The result is a Schrödinger equation for the relative motion of n bodies of reduced mass $m = \left(M^{-1} + \sum m_i^{-1}\right)^{-1}$,

$$-\frac{\hbar^2}{2m} \sum_{i=1}^{n} \nabla_i^2 \psi - \frac{\hbar^2}{M} \sum_{i>j}^{n} \left(\frac{\partial^2}{\partial x_i \partial x_j} + \frac{\partial^2}{\partial y_i \partial y_j} + \frac{\partial^2}{\partial z_i \partial z_j}\right) \psi + V\psi = E\psi. \tag{16.14}$$

This equation shows that the movement of the nucleus produces two different effects. The first consists in replacing the mass of the electron by its reduced mass and, although it depends on the isotope, it is the same for all energy levels. The second effect exists only if $n > 1$ (so it occurs in He, but not in H) and generates a correction to the energy that is given to first order in perturbation theory by (see Eq. (13.18))

$$\delta E = \frac{1}{M}\sum_{i>j}\langle p_{x_i}p_{x_j} + p_{y_i}p_{y_j} + p_{z_i}p_{z_j}\rangle, \qquad (16.15)$$

and therefore depends on the state under consideration. In the case of the He atom, for example, this correction essentially only affects the p-states and also depends on the symmetry of the orbital wave function, that is on the total spin. In other words, it is a small additional correction due to the interaction between the electrons, which is essentially an exchange effect. However, since it is a small disturbance (of the order of m/M), we will disregard it from now on and consider only the mass correction.

In the approximation just discussed, the Hamiltonian of the helium-like atom with two electrons and arbitrary Z (neglecting relativistic, spin-orbit, and other small corrections) is

$$\widehat{H} = \frac{\hat{p}_1^2}{2m} + \frac{\hat{p}_2^2}{2m} - \frac{Ze^2}{r_1} - \frac{Ze^2}{r_2} + \frac{e^2}{|\mathbf{r}_1 - \mathbf{r}_2|}. \qquad (16.16)$$

The eigenvalues and eigenfunctions of this Hamiltonian are determined by approximation methods. The most accurate results are obtained using the variational method but, as we will see later, even a crude perturbative approximation gives acceptable results. Before carrying out detailed calculations, however, we will determine the qualitative structure of the He spectrum. To do this, we neglect in Eq. (16.16) the repulsive interaction potential between the two electrons, thereby transforming the Hamiltonian into the sum of two hydrogen-like Hamiltonians. In this (zero-order) approximation the energy of the system is the sum of the energies of two hydrogen-like electrons, which we will consider in the states (n_1, l_1, m_1) and (n_2, l_2, m_2) (for the moment we ignore the electron spin),

$$E^{(0)} = E_{n1} + E_{n2} = -\tfrac{1}{2}\alpha^2 mc^2 \left(\frac{Z^2}{n_1^2} + \frac{Z^2}{n_2^2}\right) \quad (\alpha^2 mc^2 = \frac{e^2}{a_0}). \qquad (16.17)$$

In particular, for the ground-state energy this approximation gives

$$E = 2E_1 = -\tfrac{1}{2}\alpha^2 mc^2 \cdot 2Z^2 = -108.8 \text{ eV} \qquad (Z = 2),$$

which is eight times the energy of the H ground state, therefore the He ground state is very deep (this will not change qualitatively with subsequent corrections). The first excited state corresponds to $n_1 = 1, n_2 = 2$ and gives

$$E' = E_1 + E_2 = -\tfrac{1}{2}\alpha^2 mc^2 \cdot \tfrac{5}{4}Z^2 = -68.0 \text{ eV},$$

while the energy of the excited state with $n_1 = n_2 = 2$ is

$$E'' = 2E_2 = -\tfrac{1}{2}\alpha^2 mc^2 \cdot \tfrac{1}{2}Z^2 = -27.2 \text{ eV}.$$

The ionization energy of He is obtained as the difference between the energy of the He^+ ion (the free electron is considered to be at rest) and the energy of the ground state, that is,

$$I = E_1 - 2E_1 = -E_1 = \tfrac{1}{2}\alpha^2 mc^2 Z^2 = 54.4 \text{ eV}. \qquad (16.18)$$

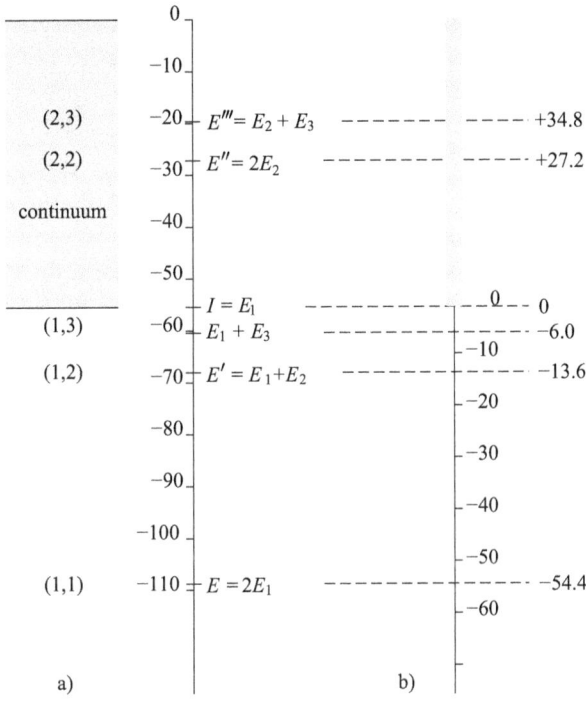

Figure 16.1 Spectrum of a helium-like atom, neglecting the Coulomb repulsion between the electrons. In (a) the energy scale corresponds to the case where the energy of the infinitely excited state ($n_1, n_2 \to \infty$) is zero. In (b) the zero of the energy is set at the continuum threshold; the ground-state energy is then equal to $-I$, and for positive energies the atom ionizes.

These results are shown in Fig. 16.1(a). When an energy $\geq I$ is supplied to an atom in its ground state, one of its electrons is released; the released electron can have any energy, so the new system (ion + emitted electron) is in the continuum. As usual, we fix the origin of the energy at the threshold of the continuum, that is, at 54.4 eV above the ground-state energy (as in Fig. 16.1(b)). Note that an interesting phenomenon occurs: atoms with two outer electrons (and, in particular, the helium-like atom) have excited states such as $(2s)^2$ which are in the continuum (positive energies). The excited states of interest are therefore reduced to those with one $1s$ electron and the other in the nl orbital ($n \neq 1$), which have the continuum as their asymptotic limit.

The fact that all the other excited states are in the continuum gives rise to characteristic effects, such as the Meitner–Auger or self-ionization effect (discovered by Lise Meitner[6] in 1922 and by Auger[7] shortly afterwards), which consists in the fact that, once one such state is produced (e.g., by absorption of radiation), the system can decay without emitting radiation into a He^+ ion and an emitted electron. From now on, we will limit the study of the He spectrum to the discrete states, so we will assume that one

---

[6] Lise Meitner (1878–1968) was an Austrian-born physicist whose influential work forced her to flee Germany for Sweden in 1938, where she discovered the power of the nuclear fission reaction.

[7] Pierre Auger (1899–1993) was a French physicist who worked in the fields of atomic, nuclear, and cosmic-ray physics.

of the electrons is in the 1s state. Since the other electron is in one of the $|nlm\rangle$ states, we can use its quantum numbers to characterize the state of the atom and write the energy in the form E_{nl} (this energy must also depend on the total electronic spin, 0 or 1, as we will see later). With this notation, Eq. (16.17) becomes

$$E_{nl} \simeq -\tfrac{1}{2}\alpha^2 mc^2 \left(Z^2 + \frac{Z^2}{n^2} \right). \tag{16.19}$$

It is possible to improve this expression (at least for large values of n and l) without having to go deeper into the calculation, by the following considerations. If n and l are large enough, the nl orbital will be concentrated in regions much further away from the nucleus than the inner 1s orbital. Therefore, while the entire nuclear charge Z acts on the inner electron, only the charge $Z - 1$ acts on the outer electron. Consequently, instead of (16.19) we write

$$E_{nl} \simeq -\tfrac{1}{2}\alpha^2 mc^2 \left(Z^2 + \frac{(Z-1)^2}{n^2} \right), \tag{16.20}$$

expressing the shielding effect of the nucleus by the electron. The result of a variational calculation confirms the soundness of this consideration, as we will see later.

16.3.1 Perturbative solution

To improve the calculation of the He spectrum, we will first use a perturbative method. The choice of the unperturbed Hamiltonian is not trivial and depends largely on the purpose of the calculation. If we are interested in simplicity rather than precision, the obvious choice is to consider the potential between the electrons as the perturbation, as was done before; the minimum-energy state then corresponds to $(1s)^2$. For the total wave function to be antisymmetric, the spinor factor must be antisymmetric; the system is therefore in the state $S = 0$ (antiparallel spins). The first-order correction to the ground-state energy is then given by

$$\delta E_{10}^{(1)} = \int d^3 r_1 d^3 r_2 \Psi_{10}^*(\mathbf{r}_1, \mathbf{r}_2) \frac{e^2}{|\mathbf{r}_1 - \mathbf{r}_2|} \Psi_{10}(\mathbf{r}_1, \mathbf{r}_2), \tag{16.21}$$

where

$$\Psi_{10}(\mathbf{r}_1, \mathbf{r}_2) = \psi_{100}(\mathbf{r}_1)\psi_{100}(\mathbf{r}_2)\chi_0, \tag{16.22}$$

with χ_0 the spinor wave function, which in obvious notation takes the form

$$\chi_0 = \frac{1}{\sqrt{2}} \left[\chi_+(1)\chi_-(2) - \chi_-(1)\chi_+(2) \right]. \tag{16.23}$$

Since the interaction does not contain spin operators, and $|\chi_0|^2 = 1$, the energy correction reduces to

$$\delta E_{10}^{(1)} = e^2 \int |\psi_{100}(\mathbf{r}_1)|^2 \frac{1}{|\mathbf{r}_1 - \mathbf{r}_2|} |\psi_{100}(\mathbf{r}_2)|^2 d^3 r_1 d^3 r_2. \tag{16.24}$$

Inserting here the hydrogen-like ground-state functions and rearranging, we get

$$\delta E_{10}^{(1)} = \frac{e^2 Z^6}{\pi^2 a_0^6} \int_0^\infty r_1^2 e^{-2Zr_1/a_0} dr_1 \int_0^\infty r_2^2 e^{-2Zr_2/a_0} dr_2$$
$$\times \int_{\Omega_1} d\Omega_1 \int_{\Omega_2} d\Omega_2 \frac{1}{(r_1^2 + r_2^2 - 2r_1 r_2 \cos\theta)^{1/2}}, \quad (16.25)$$

where θ is the angle formed by $\mathbf{r}_1$ and $\mathbf{r}_2$. Aligning the z axis in the direction of $\mathbf{r}_1$, we have (with $x = \cos\theta$)

$$\int d\Omega_1 \int \frac{d\Omega_2}{|\mathbf{r}_1 - \mathbf{r}_2|} = 4\pi \int_0^{2\pi} d\varphi \int_{-1}^1 \frac{dx}{(r_1^2 + r_2^2 - 2r_1 r_2 x)^{1/2}}$$
$$= \frac{4\pi^2}{r_1 r_2}(r_1 + r_2 - |\mathbf{r}_1 - \mathbf{r}_2|), \quad (16.26)$$

whence

$$\delta E_{10}^{(1)} = \frac{8e^2 Z^6}{a_0^6} \int_0^\infty r_1^2 e^{-2Zr_1/a_0}$$
$$\times \left[\frac{1}{r_1} \int_0^{r_1} r_2^2 e^{-2Zr_1/a_0} dr_2 + \int_{r_1}^\infty r_2 e^{-2Zr_2/a_0} dr_2 \right] dr_1. \quad (16.27)$$

This result can be obtained directly if we remember that the electrostatic potential outside an isotropic charge distribution is equal to that produced by a point charge at the center of the distribution, while inside the distribution it is constant and equal to that of the surface. Carrying out the radial integrals, we obtain

$$\delta E_{10}^{(1)} = \tfrac{5}{4} Z \cdot \tfrac{1}{2} \alpha^2 mc^2, \quad (16.28)$$

so that the ground-state energy is to first order

$$E_{10}^{(1)} = -\tfrac{1}{2}\alpha^2 mc^2 (2Z^2 - \tfrac{5}{4} Z). \quad (16.29)$$

This gives for the He atom

$$E_{10}^{(1)} = -74.8 \text{ eV}, \quad (16.30)$$

which differs from the experimental value $E_{10}^{\text{exp}} = -78.97$ eV by 5–6%. The result is stimulating, considering the crudeness of the calculation. A higher-order perturbative calculation is extremely tedious, so a variational method is used instead for a better estimate.

To estimate the variational corrections to the energy of the excited states, we must first construct the zero-order wave functions. Since one electron is in the $1s$ state and the other in the nl state with $n \neq 1$, the orbital wave function must be the symmetric combination of ψ_{100} and ψ_{nlm} if $S = 0$, and the antisymmetric one if $S = 1$, where S represents the total spin of the two orbital electrons. Therefore we write

$$\Psi_{nl}^\pm = \psi_\pm(\mathbf{r}_1, \mathbf{r}_2) \chi_S, \quad (16.31)$$

where

$$\psi_\pm = \tfrac{1}{\sqrt{2}} [\psi_{100}(\mathbf{r}_1)\psi_{nlm}(\mathbf{r}_2) \pm \psi_{100}(\mathbf{r}_2)\psi_{nlm}(\mathbf{r}_1)] \quad (16.32)$$

(the + sign corresponds to $S=0$ and the − sign, to $S=1$). From here (and even from (16.22)) we see that the He wave function describes in each case an entangled state, with all its consequences.[8] Moreover, these states are described as waves in a six-dimensional configuration space. It is clear that if ψ is taken as a probability amplitude, the dimensionality of space does not in principle pose a problem; but the peculiarities associated with entanglement remain.

The first-order energy correction can be written in the form

$$\delta E_{nl}^{(1)} = J_{nl} \pm K_{nl}, \qquad (16.33)$$

where J_{nl} and K_{nl} are given by

$$J_{nl} = e^2 \int |\psi_{100}(\mathbf{r}_1)|^2 \frac{1}{|\mathbf{r}_1 - \mathbf{r}_2|} |\psi_{nlm}(\mathbf{r}_2)|^2 d^3 r_1 d^3 r_2, \qquad (16.34)$$

$$K_{nl} = e^2 \int \psi_{100}^*(\mathbf{r}_1) \psi_{nlm}^*(\mathbf{r}_2) \frac{1}{|\mathbf{r}_1 - \mathbf{r}_2|} \psi_{100}(\mathbf{r}_2) \psi_{nlm}(\mathbf{r}_1) d^3 r_1 d^3 r_2. \qquad (16.35)$$

Both integrals can be calculated in closed form (the calculation is laborious and will not be done here) and are positive (this is obvious for J).

16.3.2 Orthohelium and parahelium

According to Eq. (16.33), the energy spectrum will depend on the total spin state. The states with $S = 0, 1$ correspond to parahelium and orthohelium respectively. As seen in Fig. 16.2, the orthohelium states lie below the corresponding parahelium states; this is a particular example of the empirical Hund rule mentioned earlier. We can understand this by noting that in the orthohelium the parallel electronic spins tend to repel each other, so the average distance between the electrons is greater than in the parahelium; this reduces the electrostatic repulsion between them and lowers their energy relative to the parahelium states. It can be shown that the transition $l = 0 \to l = 0$ is strictly forbidden for all orders of multipole moments, so both the $1s2s$ state of the parahelium and the $2s$ state of the orthohelium are metastable. Here we have a concrete example of a system whose ground state (the 1s2s of orthohelium) has nodes, since the surface $r_1 = r_2$ is a nodal surface (this possibility was mentioned in Section 15.3).

The existence of orthohelium and parahelium is a manifestation of exchange effects, that is, of how purely electrostatic interactions give rise to effects that can be described in terms of spin (which results in a somewhat artificial description), due to the fact that the spatial distribution of the particles is conditioned by the spin coupling.

To give the exchange effects a more concrete physical meaning, we will describe how the electrons periodically exchange roles when their energies are comparable. To do this, we take as the general orbital solution to first order in perturbation theory the function

$$\Psi(t) = a_+ \psi_+(t) + a_- \psi_-(t), \qquad (16.36)$$

where

$$\psi_\pm(t) = \psi_\pm(\mathbf{r}_1, \mathbf{r}_2) e^{-i\omega_\pm t}. \qquad (16.37)$$

[8] See the discussion in Section 14.5.

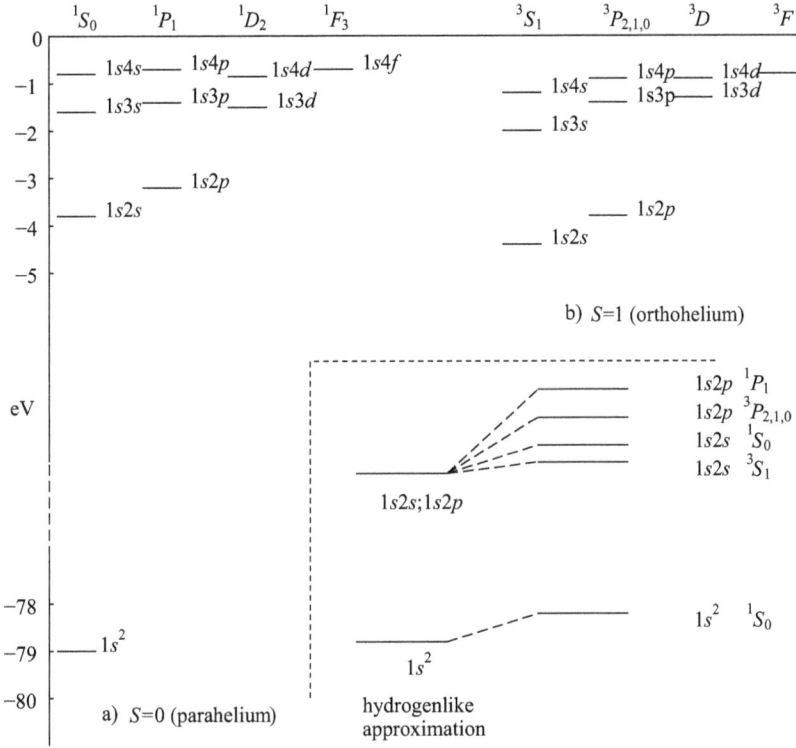

Figure 16.2 He energy levels. Parahelium (a) and orthohelium (b) energy spectra are shown schematically, indicating both the electronic configuration and the atomic state. The insert shows how the hydrogen-like levels separate due to exchange. Because of spin–orbit coupling and other corrections, levels like $^3P_{2,1,0}$ are not really degenerate.

The perturbed energies $\hbar\omega_\pm$ are given to first order by

$$\omega_\pm \equiv \omega \pm \delta\omega \equiv \frac{E_{nl}^{(0)} + J_{nl} \pm K_{nl}}{\hbar}. \tag{16.38}$$

To examine the time behavior of Ψ, let us take $a_+ = a_- = 1/\sqrt{2}$. At $t = 0$, $\Psi(0) = \psi_{10}(1)\psi_{nl}(2)$ describes the situation in which (statistically speaking) the electrons 1 are in the $1s$ state and the electrons 2 are in the nl state. At a later time t, we have

$$\Psi(t) = e^{-i\omega t}\left[\psi_{10}(1)\psi_{nl}(2)\cos\delta\omega\, t - i\psi_{10}(2)\psi_{nl}(1)\sin\delta\omega\, t\right]. \tag{16.39}$$

Therefore, at the time t_1 given by

$$t_1 = \frac{\pi}{2\delta\omega} = \frac{\pi\hbar}{2K_{nl}}, \tag{16.40}$$

we have $\Psi(t) = -ie^{-i\omega t_1}\psi_{10}(2)\psi_{nl}(1)$, which describes the situation where the electrons 2 are in the $1s$ state and the electrons 1 are in the nl state. that is, the electrons have exchanged their roles. The exchange time t_1 is determined by K_{nl}; the larger this integral (i.e., the greater the overlap between the functions ψ_{10} and ψ_{nl}), the more frequent the exchange. If $K_{nl} = 0$, each electron remains in its state without exchanging energy

($t_1 = \infty$); in this case we are dealing with two essentially independent systems. Intuitively, if one electron is in a highly excited state and therefore very far from the nucleus, and the other is in the inner $1s$ state, the interaction between the electrons is so small that no significant exchange can be expected.[9]

To get a more quantitative idea of this exchange phenomenon, we take $n = 2$ and $l = 0$, and proceed as we did to evaluate the integral in (16.25), which yields

$$K_{20} \simeq \left(\frac{4}{27}\right)^2 \frac{Ze^2}{a_0}, \tag{16.41}$$

whence from Eq. (16.40), ($\alpha = e^2/\hbar c \sim 1/137$ is the fine-structure constant)

$$t_1 \simeq \left(\frac{3}{2}\right)^6 \frac{\pi a_0}{\alpha c} \sim 10^{-15}\text{s}. \tag{16.42}$$

Since αc is of the order of the electron's orbital velocity and πa_0 is of the order of the length of the orbit, t_1 is of the order of the electron's rotation period; this indicates that the relatively large overlap between ψ_{10} and ψ_{20} causes the electrons to switch roles at almost every rotation. However, with $n = 10$, $l = 0$, t_1 spans years; in this case each electron remains in its state without interfering with the other, and the orbital solution is essentially reduced to the product $\psi_{10}(1)\psi_{nl}(2)$ ($n \gg 1$).

16.3.3 Variational solution

To demonstrate the possibilities of the variational method for the study of atomic systems, we will apply it to the case of the helium-like atom in its ground state. As a simple test function, it is customary to use the expressions (16.22) and (13.148), but treating the coefficient in the exponential as a variational parameter,

$$\psi_{10} = \frac{1}{\pi}\left(\frac{A}{a_0}\right)^3 \exp\left[-\frac{A}{a_0}(r_1 + r_2)\right] \tag{16.43}$$

(we have omitted the spin factor). This choice is equivalent to not fixing the value of Z beforehand, or replacing $Z = 2$ by another value Z^* corresponding to the partially shielded nuclear charge, which is determined variationally. This idea was used in writing (16.20), with $Z^* = Z$ for the inner electron and $Z^* = Z - 1$ for the outer one. If, as we propose with the choice (16.43) for the test function, the value of Z^* must be common to both electrons, we can anticipate that it will be close to $Z - \frac{1}{2}$.

The expectation value of the Hamiltonian (16.16) calculated with the test function (16.43) is

$$\langle \widehat{H} \rangle = \frac{e^2}{a_0}\left[A^2 - (2Z - 5/8)A\right] = -\alpha^2 mc^2 A(2Z - 5/8 - A). \tag{16.44}$$

---

[9] Atoms with one or more highly excited electrons (where the level n can be even higher than 100) are called *Rydberg atoms*, and they have several peculiar properties, including long decay times, large electric dipole moments, and wave functions that approximate classical orbits; the outermost (valence) electrons are subject to a potential similar to that of an H atom. Rydberg atoms are currently the subject of intense research because of their many applications, including radio-frequency sensing, quantum optics, quantum computing, quantum simulation, and astrophysics. For example, Rydberg spectral lines have been detected corresponding to H atoms and C atoms several microns in diameter.

16.3 The Helium Atom

Table 16.3 Ionization energy of some atoms.

Ion	Experiment	Variational method One parameter	Many parameters
He	0.90372	0.85	0.90369
Li$^+$	2.7799	2.72	2.77991
Be^{+++}	5.656	5.50	5.65556
C^{++++}	14.41	14.35	14.4063

Minimizing this expression with respect to A gives $A = Z - 5/16$, whence

$$E_{10} \leq \min\langle \widehat{H} \rangle = -\alpha^2 mc^2 \left(Z - \tfrac{5}{16}\right)^2, \tag{16.45}$$

that is, the effective nuclear charge is $Z^* = Z - 5/16$, confirming our prediction. For $Z = 2$ this gives $E_{10} < -77.4$ eV, which is within 2% of the experimental result -78.97 eV.

What is measured in the laboratory is not the energy E_{10}, but the ionization potential, which, as we have seen, is given by $I = E_0 - E_{10}$, where E_0 represents the energy of the hydrogen-like ion. It is customary to express the results in atomic units (au), in which $e^2/a_0 = \alpha^2 mc^2 = 1$ (i.e., $e = m = \hbar = 1$ and $c = 1/\alpha \approx 137$). With this convention, we obtain for the energy

$$E_{10} = -\left(Z - \tfrac{5}{16}\right)^2 = -2.85 \text{ au} \quad \text{for } Z = 2, \tag{16.46}$$

and for the ionization potential

$$I = \left(Z - \tfrac{5}{16}\right)^2 - \tfrac{1}{2}Z^2 = \tfrac{1}{2}\left(Z^2 - \tfrac{5}{4}Z + \tfrac{25}{128}\right) = 0.85 \text{ au} \quad \text{for } Z = 2; \tag{16.47}$$

the current experimental value is 0.9036 au. Variational calculations with several tens of parameters give results that differ from experiment by less than 10^{-6} au. Table 16.3 shows the results obtained with the variational method for other ions ("many parameters" refers to calculations made with up to 50 parameters).

It is useful to compare the result (16.46) with those obtained using the Hartree and Hartree–Fock methods, which were originally used precisely to study the ground state of He. With the Hartree method, a product wave function of two orbitals gives $E_{10} = -2.862$ au, while the introduction of the Pauli principle reduces this value to $E_{10} = -2.876$ au, which differs from the experimental value (-2.903 au) by less than 1%. The main advantage of Hartree's method applied to helium-like atoms is that it gives the most accurate wave functions in the form of a product of two orbitals, whereas the variational method gives a better result for the ionization potential.

Exercise E16.1. Magnetic properties of parahelium
Determine the magnetic properties of parahelium in its ground state.

Solution. The ground state of parahelium has zero total angular momentum, so it is a diamagnetic gas. The diamagnetic properties are conveniently described in terms of the magnetic susceptibility, given by the Langevin formula Eq. (14.48),

$$\chi_{\text{diam}} = -\frac{e^2 N}{6mc^2}\langle r_1^2 + r_2^2\rangle. \tag{16.48}$$

Since the atom is in its ground state, to calculate the required expectation values we can use as the wave function the one provided by the variational method, Eq.(16.43), using the previously determined value for A, $A = Z^* = 27/16$. This gives

$$\langle r_1^2 + r_2^2\rangle = \frac{2a_0^2}{Z^{*2}}, \tag{16.49}$$

so that the diamagnetic susceptibility per gram atom is

$$\chi_{\text{diam}} = -\frac{e^2 N a_0^2}{3mc^2 Z^{*2}} \simeq -1.67 \times 10^{-6}. \tag{16.50}$$

This result compares favorably with the experimental value $\chi_{\text{exp}} = -1.9 \times 10^{-6}$.

16.4 Molecular Structure

16.4.1 Nature of the chemical bond

In Section 16.1 we saw that in simple compounds (e.g., inorganic salts) the valence with which each of the elements intervenes is determined by the number of electrons that each one of the atoms gives up (positive ionic valence) or gains (negative ionic valence). However, this picture of the chemical bond as the result of electrostatic attraction between two ions, or between an ion and a neutral atom polarized by the presence of the ion, is not sufficient to explain the vast majority of chemical bonds. In particular, it cannot explain the existence of structures as simple as the homonuclear diatomic molecules H_2, O_2 or N_2. In such cases we speak of covalent or homopolar bonds, and the valence responsible for them is called spin valence.

We seem to be faced with the need to construct two different theories of chemical bonding, involving two types of atomic valence. In reality, however, a single notion of valence suffices, and we can speak of a generic type of chemical bond, the extreme cases of which, namely purely ionic or purely covalent bonds, are never realized in the strict sense. In order to prepare the ground for a more general discussion of this important topic, which is at the basis of theoretical chemistry, we will first study the simplest of all molecules, the H_2 molecule and its ion, H_2^+.

16.4.2 The molecular hydrogen ion H_2^+

The existence of the H_2^+ on indicates that the electrostatic attraction exerted simultaneously by the electron on the two nuclei can be sufficient to overcome the electrostatic repulsion between them: the orbital electron is the molecular cement.

To understand qualitatively how a H_2^+ molecule is formed, suppose that a neutral H atom (in its ground state) and a proton (which can be thought of as an H^+ atom)

16.4 Molecular Structure

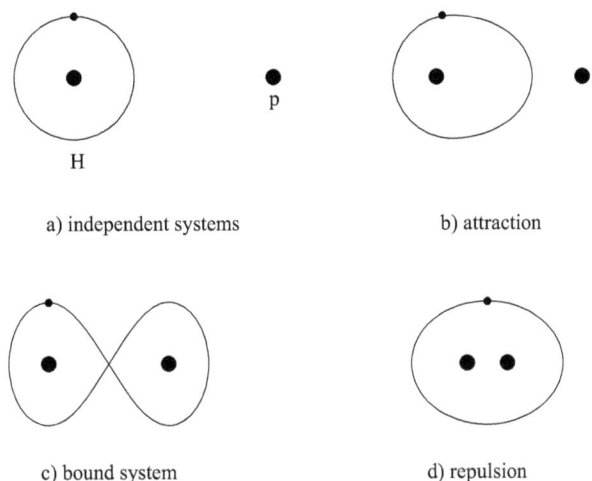

Figure 16.3 The effects of a proton approaching a neutral H atom are shown schematically. The polarization induced by the proton leads to an initial attraction between the atoms until a stable molecule is formed. At very small distances beyond which a common orbit is established, the repulsion between the nuclei becomes the dominant force.

approach each other from a great distance. As a result, the electron orbit is deformed by the attraction between the proton and the electron, as shown schematically in Fig. 16.3. If the two subsystems are close enough, the strong attraction of the proton to the electron can cause a situation similar to that shown in Fig. 16.3(c), where the electron is trapped by both nuclei, following a complicated orbit.[10]

As we will see later, under these conditions the system remains bound. As the nuclei come closer together, as in Fig. 16.3(d), the electron no longer travels in between the nuclei, and the internuclear repulsion exceeds the weak attraction of the electron. As a result, the protons tend to separate until they return to a situation like the previous one, where the average attractive and repulsive forces are in balance.

Suppose that the H_2^+ molecule has been formed and is in a stationary state, as shown in Fig. 16.3(c). Roughly speaking, we can say that the orbital electron spends half its time in the state $\psi_A(r)$ around nucleus A and the rest of its time in the same state, but around nucleus B, which we denote by $\psi_B(r)$. This crude *statistical* description of the molecule is represented by the wave function

$$\Psi(r) = N[\psi_A(r) + \psi_B(r)]\chi. \tag{16.51}$$

Wave functions such as this, constructed using a linear combination of atomic orbitals, are called molecular orbitals, and this method of describing the molecular state is known as the LCAO method (where the abbreviation stands for linear combination of atomic orbitals).

---

[10] The illustrations in Fig. 16.3 are conventional and do not imply that each electron follows a smooth orbit, but only that there is an orbital motion, probably very complicated and random, which can be represented on average by a smooth and periodic closed curve, like the one shown in the illustration.

16.4.3 The H$_2$ molecule

The previous observations serve as a guide to describing the H$_2$ molecule. To simplify the task, we will assume that in the CM system the two nuclei are at rest; their relative motion can be introduced later as a perturbation (see Section 16.6). So we simply write the product of the molecular orbitals (MO) of the two electrons, which is equivalent to neglecting the interaction between them. We also need to introduce the spinor wave function χ and a normalization factor N,

$$\Psi(r_1, r_2) = N[\psi_A(r_1) + \psi_B(r_1)][\psi_A(r_2) + \psi_B(r_2)]\chi_0. \tag{16.52}$$

This wave function describes a situation where each electron is close enough to one of the nuclei for a part of the time and the motion of the two electrons is essentially independent.

We have explicitly written the spin factor χ_0 corresponding to the total electron spin $S = 0$ because the orbital factor in (16.52) is invariant under the exchange of electrons, so the spin factor must be antisymmetric for the total wave function to be antisymmetric. This is a special case of a general empirical rule which states that the ground state of the molecules of virtually all chemically stable compounds of elements of the main groups has zero spin; notable exceptions are O$_2$ ($S = 1$), and NO, NO$_2$, ClO$_2$ ($S = 1/2$). However, since we are not considering spin–orbit coupling, the choice of a particular spin factor is practically irrelevant. Later we will have the opportunity to verify that the proposed choice (antisymmetric spinor wave function) is the correct one.

For the treatment of H$_2^+$ the wave function (16.51) is the only one allowed to zero order. However, for molecules with more than one valence electron, including H$_2$, an alternative method of constructing the wave function to zero order (i.e., neglecting the interactions between the valence electrons and between the nuclei) can be used. This method, called mesomerism or valence bonding, introduced by Heitler[11] and London[12] in 1927, considers the molecule as a structure made up of atoms. We can easily explain the idea using the example of H$_2$. Half the time electron 1 is bound to nucleus A, in the atomic orbital $\psi_A(1)$, while the other electron is bound to nucleus B, in the atomic orbital $\psi_B(2)$; this composite state is described by $\psi_I = \psi_A(1)\psi_B(2)$. The rest of the time, the electrons will have swapped roles (not their energies) and the state will be described by $\psi_{II} = \psi_B(1)\psi_A(2)$. Therefore, the wave function that describes the molecule statistically for all times can be approximated by[13]

$$\Psi(1,2) = N(\psi_I + \psi_{II})\chi_0 = N[\psi_A(1)\psi_B(2) + \psi_B(1)\psi_A(2)]\chi_0, \tag{16.53}$$

where N is the normalization constant (which varies from case to case). Each of the two terms in this expression is a valence orbital.

---

[11] Walter Heinrich Heitler (1904–1981) was a German physicist, best known for his contributions to quantum electrodynamics.
[12] Fritz Wolfgang London (1900–1954) was a German-born physicist who made fundamental contributions to the theories of chemical bonding and of intermolecular forces. He was unsuccessfully nominated several times for the Nobel Prize in Chemistry.
[13] We have once again restricted ourselves to symmetric orbitals and antisymmetric spinors. In this case we have an additional argument in favor of this choice, namely that the antisymmetric orbital combination has a nodal surface at $r_1 = r_2$, so that it does not belong to the ground state, which is the one we are examining.

The two preceding descriptions are approximate, in that they both neglect all interactions except those that bind the electrons. Both take into account, in different ways, the basic effect of the proximity of the atoms, that is, the fact that each valence electron "belongs" half the time to one nucleus. Because of the different way in which the exchange is introduced, these methods are not equivalent in detail, but they give numerical results that are as accurate as if a sufficiently large number (several tens) of variational parameters were used. To illustrate the difference between the two methods, we expand the MO wave function Eq. (16.52),

$$\Psi(1,2) = N_1[\psi_A(1)\psi_A(2) + \psi_B(1)\psi_B(2)]\chi_0$$
$$+ N_2[\psi_A(1)\psi_B(2) + \psi_B(1)\psi_A(2)]\chi_0. \quad (16.54)$$

The first term describes two complementary situations: (a) the two electrons are bound to nucleus A and nucleus B has none; (b) the two electrons are bound to B and nucleus A has none. In both cases there is a negative ion (a proton with two orbital electrons) and a positive one (a bare proton); these represent (purely) ionic bonds. The second term on the right is precisely the Heitler–London wave function and represents the (purely) covalent coupling, that is, the sharing of the electron pair between the two atoms. In the case of the H_2 molecule, ionic bonding (the first term) is almost nonexistent, so the wave function (16.53) gives a better description than (16.52) when N_1 and N_2 are taken as equal; but the opposite happens if both parameters are determined variationally. In most cases, it is precisely a combination of ionic and covalent binding, such as Eq. (16.54), that correctly describes the physical situation. Since the LCAO method (with free coefficients) allows both types of bonds to be generated simultaneously, it is widely used in the theoretical study of molecules, usually in combination with the self-consistent field method.

16.4.4 Energy calculations for H_2^+ and H_2

We now carry out the explicit calculations, based on the preceding discussion. To simplify the procedure, we will first consider the H_2^+ ion and take the adiabatic approximation, in which the nuclear motion is neglected. The Hamiltonian of the system is

$$\widehat{H} = -\frac{\hbar^2}{2m}\nabla^2 - \frac{e^2}{|r - r_A|} - \frac{e^2}{|r - r_B|} + \frac{e^2}{|r_A - r_B|}, \quad (16.55)$$

where r_A, r_B, r represent the position of the two nuclei and of the electron respectively. With nucleus A at the origin of coordinates and nucleus B at the position R, the Hamiltonian takes the form

$$\widehat{H} = -\frac{\hbar^2}{2m}\nabla^2 - \frac{e^2}{r} - \frac{e^2}{|r + R|} + \frac{e^2}{R}. \quad (16.56)$$

Although the Schrödinger equation with this Hamiltonian is solvable in elliptic coordinates, we will take a more elementary approach by applying the variational method with a single free parameter (Wang, 1928). As a test function we take the one given by Eq. (16.51) with

$$\psi_A(r) = \left(\frac{1}{\pi a_0^3}\right)^{1/2} e^{-r/a_0}, \tag{16.57}$$

and similarly for B. To be more general, we will examine the predictions obtained not only with the function (16.51), but also with the one with the antisymmetric orbital factor. So instead of (16.51) we write

$$\Psi_\pm(\mathbf{r}, \mathbf{R}) = N_\pm[\psi_A(\mathbf{r}, \mathbf{R}) \pm \psi_B(\mathbf{r}, \mathbf{R})]\chi, \tag{16.58}$$

with the normalization factor given by

$$N_\pm = \langle A \pm B | A \pm B \rangle^{-1/2} \equiv [2 \pm 2S(R)]^{-1/2}, \tag{16.59}$$

where S depends on the internuclear distance, which plays the role of the variational parameter and must be selected so that the energy of the system is a minimum. Specifically, S measures the overlap between A and B,

$$S(R) = \langle A | B \rangle \tag{16.60}$$

and its value is

$$S = \frac{1}{\pi a_0^3} \int \exp\left[-\frac{|\mathbf{R} - \mathbf{r}|}{a_0} - \frac{r}{a_0}\right] d^3r. \tag{16.61}$$

The calculation of the energy of H_2^+ and H_2 requires the solution of several integrals of this type or more complicated ones. Starting from the formula

$$\frac{e^{-kr}}{r} = \frac{1}{2\pi^2} \int \frac{e^{i\mathbf{q}\cdot\mathbf{r}}}{q^2 + k^2} d^3q, \tag{16.62}$$

the validity of which should be checked by direct evaluation, and differentiating (16.54) with respect to k, one gets

$$e^{-kr} = \frac{k}{\pi^2} \int \frac{e^{i\mathbf{q}\cdot\mathbf{r}}}{(q^2 + k^2)^2} d^3q, \tag{16.63}$$

which can be used to write the hydrogen-like wave function as an integral over q,

$$\psi_A(\mathbf{r}) = \left(\frac{1}{\pi a_0}\right)^{5/2} \int \frac{e^{i\mathbf{q}\cdot\mathbf{r}}}{(q^2 + 1/a_0^2)^2} d^3q. \tag{16.64}$$

Equation (16.61) becomes thus

$$\begin{aligned}
S &= \left(\frac{1}{\pi a_0}\right)^5 \int \int \frac{e^{i\mathbf{q}\cdot\mathbf{r}}}{(q^2 + 1/a_0^2)^2} \cdot \frac{e^{i\mathbf{q}'\cdot(\mathbf{R}-\mathbf{r})}}{(q'^2 + 1/a_0^2)^2} d^3q\, d^3q'\, d^3r \\
&= \frac{8}{\pi^2 a_0^5} \int \frac{e^{i\mathbf{q}'\cdot\mathbf{R}}\delta(\mathbf{q} - \mathbf{q}')}{(q^2 + 1/a_0^2)^2 (q'^2 + 1/a_0^2)^2} d^3q\, d^3q' \\
&= \frac{8}{\pi^2 a_0^5} \int \frac{e^{i\mathbf{q}\cdot\mathbf{R}}}{(q^2 + 1/a_0^2)^4} d^3q.
\end{aligned} \tag{16.65}$$

To obtain this result, we first integrated over r; the integral over q is carried out as usual, taking the z axis in the direction of R,

$$S = \left[1 + \frac{R}{a_0} + \frac{1}{3}\left(\frac{R}{a_0}\right)^2\right] e^{-R/a_0}. \tag{16.66}$$

For integrals involving the Coulomb potential we can use Eq. (16.62); thus, for example,

$$\frac{1}{|R - r|} = \frac{1}{2\pi^2} \int \frac{e^{iq\cdot(R-r)}}{q^2} d^3q. \tag{16.67}$$

With this we conclude our mathematical digression and return to the central theme. The expectation value of the Hamiltonian (16.56) is (with $\langle A|\widehat{H}|A\rangle = \langle B|\widehat{H}|B\rangle$ and $\langle A|\widehat{H}|B\rangle = \langle B|\widehat{H}|A\rangle$),

$$\langle \widehat{H} \rangle = N_\pm^2 \langle A \pm B|\widehat{H}|A \pm B\rangle$$
$$= \frac{1}{1 \pm S}[\langle A|\widehat{H}|A\rangle \pm \langle A|\widehat{H}|B\rangle]. \tag{16.68}$$

Using E_1 to denote the ground-state energy of the H atom, we write

$$\langle A|\widehat{H}|A\rangle = \langle A|\left(-\frac{\hbar^2}{2m}\nabla^2 - \frac{e^2}{r}\right) + \frac{e^2}{R} - \frac{e^2}{|R+r|}|A\rangle$$
$$= E_1 + \frac{e^2}{R} - e^2\langle A|\frac{1}{|R+r|}|A\rangle$$
$$= E_1 + e^2\left(1 + \frac{R}{a_0}\right)\frac{\exp(-2R/a_0)}{R}. \tag{16.69}$$

Likewise,

$$\langle A|\widehat{H}|B\rangle = \left(E_1 + \frac{e^2}{R}\right)S - e^2\langle A|\frac{1}{|R+r|}|B\rangle \tag{16.70}$$
$$= \left(E_1 + \frac{e^2}{R}\right)S - \frac{e^2}{a_0}\left(1 + \frac{R}{a_0}\right)\exp(-R/a_0).$$

Substituting these values and the value of S from (16.66), we obtain with $x \equiv R/a_0$,

$$\langle \widehat{H} \rangle = E_1 + \frac{e^2}{R}\frac{(1+x)e^{-2x} \pm (1 - 2x^2/3)e^{-x}}{1 \pm S}. \tag{16.71}$$

This result is characteristic: the chemical forces decrease exponentially with the internuclear distance. As expected, the energy is reduced to that of an H atom as the distance between the nuclei increases indefinitely. The energy of the system has as its upper bound the minimum of (16.71),

$$E \leq \min\langle \widehat{H} \rangle. \tag{16.72}$$

The complexity of the expression for $\langle \widehat{H} \rangle$ makes it preferable to use numerical methods to determine the value of x that minimizes it. The results for the symmetric (+) and antisymmetric (−) selections of the wave function (16.58) are shown in Fig. 16.4. We note that for the antisymmetric solution there is no minimum for the energy at finite internuclear distances and that this exceeds E_1 for all R, so that the minimum energy

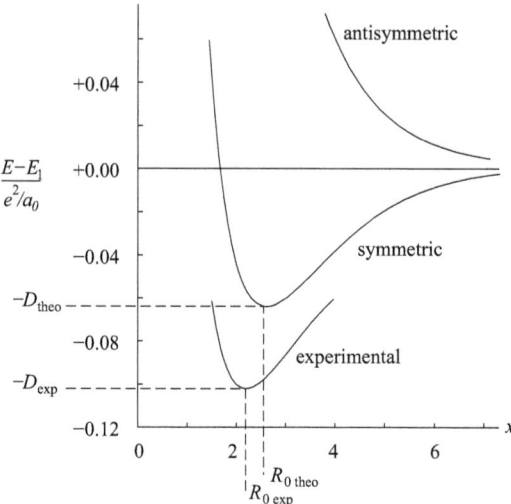

Figure 16.4 Curves showing the energy of the H_2^+ molecule (in units of e^2/a^2) as function of the internuclear distance.

state corresponds to infinitely separated nuclei, that is, it is a repulsive state. The symmetric solution, on the other hand, has a minimum with an energy lower than that of the H atom, so the formation of the molecule is energetically advantageous and the system binds. The equilibrium distance read from the curve is $R_0 = 5a_0/2$, which is about 1.2 Å. The binding energy of the molecule is then (D is the dissociation energy, i.e., the work required to separate the molecule into its constituent atoms)

$$E_{\text{bind}} \equiv -D = E_{\min} - E_1 = -0.065 \frac{e^2}{a_0} = 0.13 E_1 = -1.76 \,\text{eV}. \tag{16.73}$$

The corresponding experimental values are 1.06 Å and $D = 2.79$ eV, as can be seen in Fig. 16.4. Qualitatively, the results are very satisfactory, since the theory explains the formation of the molecule; quantitatively, they are poor, since the discrepancy between theoretical and experimental values is considerable. However, when multiple parameters are introduced, the variational method accurately predicts the experimental results.

The H_2 molecule can be treated in the same way. In this case the Hamiltonian is

$$\widehat{H} = \frac{\widehat{p}_1^2}{2m} + \frac{\widehat{p}_2^2}{2m} - \frac{e^2}{r_{1A}} - \frac{e^2}{r_{1B}} - \frac{e^2}{r_{2A}} - \frac{e^2}{r_{2B}} + \frac{e^2}{r_{12}} + \frac{e^2}{R}, \tag{16.74}$$

where the notation is that given in Fig. 16.5. The wave function should be taken with its orbital part symmetric according to the results with the H_2^+ ion, and can be chosen in the form of (16.52) (a product of molecular orbitals) or (16.53) (a sum of valence orbitals). In both cases the electrons are in a singlet state. With the MO method, the expectation value of the Hamiltonian becomes

$$\langle \widehat{H} \rangle = 2 E_{H^+}(R) - \frac{e^2}{R} + \langle \frac{e^2}{r_{12}} \rangle. \tag{16.75}$$

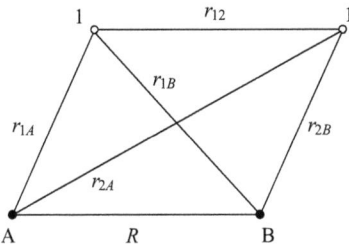

Figure 16.5 Coordinates of the particles that make up an H_2 molecule.

The explicit form of the electron repulsion term is relatively complicated and will not be given here. What is important is that this expression is qualitatively analogous to that obtained for the H_2^+ ion; the numerical analysis fixes $R = 0.85$ Å as the equilibrium distance, which corresponds to the binding energy $E = -2.68$ eV. The experimental values are 0.74 Å and -4.74 eV, so the agreement between this approximation and the experiment is only qualitatively satisfactory. Using the Heitler–London method, that is, the wave function (16.53), the final results obtained are $R = 0.80$ Å and $E = -3.14$ eV, which do not represent a significant improvement. In both cases many variational parameters are required to obtain a reasonably accurate description. We conclude that the introduction of the second electron only changes the details, as long as the two electrons forming the covalent bond have antiparallel (paired) spins.

16.4.5 Valence bonds

If a third H atom approaches an H_2 molecule, they will repel each other regardless of the direction of their spins, that is, the H_3 molecule will not form. In such cases, the valence of the molecule is said to be saturated, so that it cannot form any further bonds. This phenomenon does not occur at the classical level because classically no force is saturated. For example, the Sun can gravitationally attract any body that approaches it, regardless of how many are already in its gravitational field.

Valence saturation, a common quantum phenomenon, explains why there are no molecules like H_3 or why noble gases are chemically inert. It can be understood by considering that the valence electrons of an atom are the unpaired outer electrons. If an atom has n of these electrons, its maximum spin is $S_{max} = n/2$; therefore the valence of a configuration is twice the maximum possible spin of that configuration (this does not apply to transition elements, which may have specific properties). For example, nitrogen in its ground state has the configuration $1s^2 2s^2 2p^3$ and spin 3/2; the 1s and 2s electrons are paired, and there are three p electrons with parallel spin, so the valence is 3 (see Fig. 16.6). However, the weakly excited state with the same electron configuration but two paired $2p$ electrons, shown in Fig. 16.6, has spin 1/2; in this configuration N is monovalent. We see that valence is a property of the state of the system and can be multiple for a given element. Therefore, the chemical properties of an atom are fundamentally determined by the maximum spin valence of its lowest states. We will analyze the elements of the periodic table on this basis.

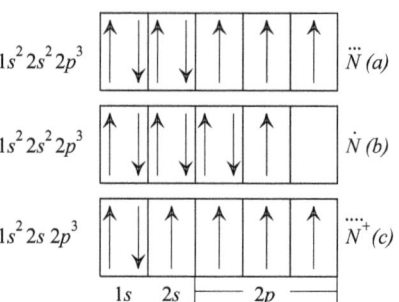

Figure 16.6 In the external configuration $2s^2 2p^3$, N has valence 3 in its ground state (quadruplet) and 1 in the excited state (doublet). In configuration $2s 2p^3$ its valence is 5 (4 spin and 1 ionic).

As can be seen from Tables 16.1 and 16.2, the elements of Group I (which includes the six alkali metals Li, Na, K, Rb, Cs and Fr) have only one outer electron, which is necessarily unpaired ($S = 1/2$), so their valence is 1. The spin can only increase when an electron leaves a closed shell, producing highly excited states that do not form stable molecules; hence all these elements are exclusively monovalent.

Group II elements (Be, Mg, Ca, Sr, Ba, Ra) have two outer paired s-electrons ($S = 0$) in their ground state, so their valence is 0 and they do not form compounds in this state. However, they have a slightly excited state in which one s-electron goes to the p state, changing the s^2 configuration to sp; in this case the electrons are unpaired ($S = 1$) and the valence is 2.

A similar situation occurs with the elements of group III (containing the first group of transition metals Bo, Al, etc.), which in their ground state have the $s^2 p$ outer configuration with $S = 1/2$ and a valence of 1; however, the slightly excited sp^2 configuration corresponds to a valence of 3, so that these elements (with the exception of B and Al) are mono- and trivalent.

The elements of group IV have the outer configuration $s^2 p^2$ ($S = 1$) in the ground state and the slightly excited configuration sp^3 ($S = 2$), so they have valences of 2 and 4; however, C normally has only the valence 4 (except in the compound CO), a fact on which organic chemistry is based. Similarly, the elements of groups V and VI have the valences 1, 3, 5 and 0, 2, 4, 6 respectively.

The atoms of group VII (which includes the halogens) have the external configuration $s^2 p^5$ in their ground state with $S = 1/2$, so they are monovalent, although through their excited configurations they can have the additional valences 3, 5, 7 (with the exception of F, which is exclusively monovalent). For example, chlorine acts with valence 1, 3, 5, 7 in the compounds HCl, $HClO_2$, $HClO_3$ and $HClO_4$, respectively.

The elements of group 0 (noble gases; it now contains the element 118 oganesson, the heaviest element of the table so far) have the outer configuration s^2 ($S = 0$); again, since spin can only be obtained by extracting a closed-shell electron, a process that requires too much energy, the only valence they have is 0, which explains the chemical inactivity of these gases, and the name of the group.

The transition elements can act with valences that differ by 1 and not only by 2, as is the case with the main group elements; this can occur, for example, by the transition of f electrons to s or p states. The noble metals (Cu, Ag, Au, Pl) have similar properties to those of the transition metals. For example, Cu has a $4s$ electron that is easily released, giving it alkali-metal properties. However, it has slightly excited states in which a $3d$ electron goes to the $4s$ or $4p$ subshell, where it tends to lose two electrons to form the cupric ion; in these compounds the d subshell is incomplete and the metal tends to behave as befits a transition element.

So far it has not been necessary to distinguish between the homopolar and the heteropolar nature of the bond; the difference between them is purely quantitative, and in practice any intermediate case can occur. There are, of course, situations where the nature of the bond is clearly defined. For example, in the third case shown in Fig. 16.6 the N atom will have lost an s electron, producing an ionic valence which, when added to the four spin valences of the $2s2p^3$ outer configuration, gives this configuration the valence 5 characteristic of group V elements.

16.4.6 Valence bonding and chemical forces

The above discussion allows us to understand some other properties of chemical forces in a simple way. For example, once two valence electrons from different atoms form a bond (and pair up in a singlet), they become chemically inactive, leading to saturation. Another consequence of the scheme is that a molecule will normally have zero spin. However, there are also unsaturated covalent bonds, formed between two atoms by the overlap of half-filled valence atomic orbitals of each atom containing an unpaired electron. Their appearance can be explained using the example of the simplest metal, Li: its electronic configuration is $1s^2 2s$, so its outer shell is analogous to that of H and it forms the molecule Li_2. However, it happens that the $1s^2 2p$ configuration has almost the same energy, which makes this molecule radically different from H_2, whose first excited state is very high. If we now bring a third Li atom close to the Li_2 molecule, a bond is formed with the electron in the $2p$ atomic orbital, allowing the exclusion principle to be reversed in this case. In fact, this process can be repeated many times, as there are a large number of atomic orbitals of virtually the same energy. It is these unsaturated bonds that give rise to metallic crystals in which each atom is bonded to its neighbors, sharing each valence electron with several of them. In particular, this is what allows extra charges (extra electrons) to move very easily from atom to atom, giving metals their characteristic electrical properties.

It is an experimental fact that molecules have a well-defined spatial structure. For example, the H_2O molecule is triangular, the CO_2 molecule is linear, the NH_3 molecule is pyramidal, etc. The quantum theory of valence allows us to explain these structures using the concept of directed valence associated with p orbitals. Let us take N as an example to illustrate the idea. The three valence p electrons can be described by three linearly independent orbitals, denoted p_x, p_y, and p_z, whose angular wave functions are given by

$$\langle \theta, \varphi | p_x \rangle = \tfrac{1}{\sqrt{2}} \left(-Y_1^1 + Y_1^{-1} \right) = \sqrt{\tfrac{3}{4\pi}} \sin\theta \cos\varphi, \qquad (16.76)$$

$$\langle \theta, \varphi | p_y \rangle = i\tfrac{1}{\sqrt{2}} \left(Y_1^1 + Y_1^{-1} \right) = \sqrt{\tfrac{3}{4\pi}} \sin\theta \sin\varphi, \qquad (16.77)$$

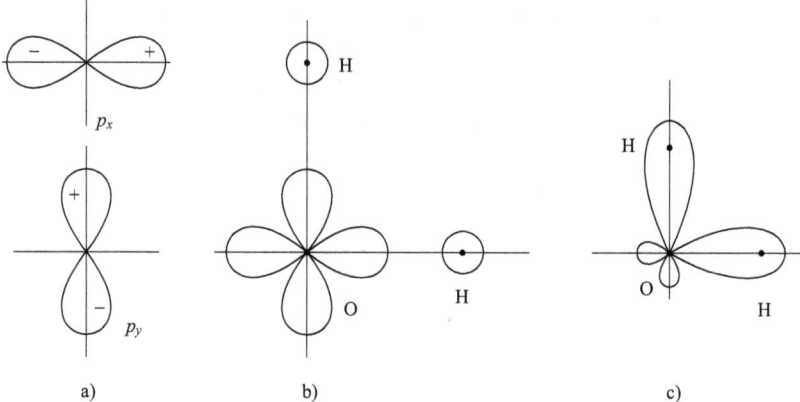

Figure 16.7 The water molecule. In (a) the p_x and p_y atomic orbitals of O are shown schematically; in (b) the O and H atomic orbitals are shown when they are far apart (the overlap is zero). At distances where the overlap is significant, the atomic orbitals combine as shown in (c) to form the molecular orbitals. In the resulting water molecule, the two H atoms are not at 90° to each other; the repulsion between them increases this angle to 104.45°.

$$\langle \theta, \varphi | p_z \rangle = Y_1^0 = \sqrt{\frac{3}{4\pi}} \cos \theta. \tag{16.78}$$

The directions in which the electron density has a maximum when its spatial distribution is given by these functions are mutually orthogonal. It follows that the directions of the three bonds formed by the valence electrons will be orthogonal, since the overlap with the electrons of atoms approaching in these directions will be maximum. In particular, the NH_3 molecule should then have a pyramidal structure with the three H atoms at the base and the N atom at the apex, and an angle of 90° between the three NH bonds. Experiment shows that this angle is 107.3°; the difference is explained by the electrostatic repulsion between the H atoms, which allows the N atom to move closer to the base of the pyramid.

The principle of maximum overlap used in the previous explanation is at the basis of all stereochemistry. Take the water molecule as a second example. The two outer 2p electrons of O are in orthogonal orbitals, such as p_x and p_y, so the H bonds will be maximal along these directions, as shown in Fig. 16.7. If the effect of the 2s electrons on the molecule is taken into account by using as orbitals not the p states, but linear orthogonal combinations of s and p states, so as to minimize the energy of the bonded system (a procedure called hybridization of atomic orbitals, proposed by Pauling), it turns out that the hybridized orbitals form an angle of 109.6° with each other. The remaining discrepancy with the experiment (104.45°) is due to the fact that the 2s electrons are not fully equivalent to the 2p electrons because of their lower energy, which implies partial hybridization. The hybridization procedure is essential to explain, for example, why the four valences of carbon are entirely equivalent, even though the configuration that produces them is $2s2p^3$: the s electron and the 3 p electrons form four sp^3 hybrid orbitals in a highly symmetric tetragonal structure, with a bond angle of 109.6°. This is precisely the structure of methane, CH_4.

Due to the great difficulty of the subject, there is still no satisfactory theory of the covalent bonding of complex molecules, and it is sometimes necessary to limit oneself to qualitative considerations. On the other hand, in order to describe these systems, the quantum laws must be supplemented with additional elements that take into account the influence of the specific properties of the compounds under consideration.

Exercise E16.2. Angular momentum selection rules for diatomic molecules
Based on qualitative considerations, discuss the selection rules for rotational angular momentum of homonuclear diatomic molecules.

Solution. The eigenfunctions of the Hamiltonian are symmetric or antisymmetric with respect to the geometric center of the molecule, so the electron density is symmetric; consequently a homonuclear molecule does not have an electric dipole moment. It follows from this that there are no electric dipole transitions in these molecules, so the basic radiation spectrum is produced by magnetic dipole or electric quadrupole transitions, whose selection rules for L are $\Delta L = 0, \pm 1, \pm 2$. However, it is possible to eliminate the rule $\Delta L = \pm 1$ from the following considerations. The symmetry of the total molecular wave function, including the nuclear spins, is defined by the rules of maximum symmetrization; since the spins are not affected during molecular transitions (spin–orbit and spin–spin couplings are negligible), the symmetry of the orbital function cannot change, which requires that the parity of L be preserved, that is, radiative transitions can only involve even changes in L. Then $\Delta L = 0, \pm 2$ are the requested selection rules.

16.5 Effects of Nuclear Motion on Diatomic Molecules

To analyze the effects of the nuclear motion on molecular spectra, we write the Schrödinger equation for a diatomic molecule in the form

$$[\widehat{T}_r + \widehat{T}_R + V(r, R)]\Psi(r, R) = E\Psi(r, R), \tag{16.79}$$

where $\widehat{T}_r$ is the total kinetic energy of the electrons, $\widehat{T}_R$ that of the nuclei, and $V(r, R)$ the total potential energy, which includes the repulsions between nuclei, the attractions between them and the electrons, and the repulsions between the latter; all other effects are neglected. A perturbative treatment of the problem can be made by considering the nuclei fixed (all Rs constant), so that the Schrödinger equation reduces to

$$[\widehat{T}_r + V(r, R)]\psi_n = \mathcal{E}_n(R)\psi_n, \tag{16.80}$$

where $\mathcal{E}_n(R)$ is the total electronic energy. The approximation method introduced by Born and Oppenheimer[14] in 1927 starts by using the solutions of Eq. (16.80) to expand the solutions of (16.79) in the form

---

[14] Julius Robert Oppenheimer (1904–1967) was a US-born theoretical physicist who made significant contributions to quantum mechanics and nuclear physics. Recruited to work on the Manhattan Project during World War II, he later campaigned against the nuclear arms race.

$$\Psi(r, R) = \sum_m \varphi_m(R)\psi_m(r, R). \tag{16.81}$$

After some simplifications one obtains

$$\sum_m [\widehat{T}_R \varphi_m \psi_m + \mathcal{E}_m(R)\varphi_m \psi_m] = E \sum_m \varphi_m \psi_m. \tag{16.82}$$

Since

$$\widehat{T}_R \varphi_m(R)\psi_m(r, R) = \psi_m \widehat{T}_R \varphi_m + \varphi_m \widehat{T}_R \psi_m - \frac{\hbar^2}{M}(\nabla_R \varphi_m) \cdot (\nabla_R \psi_m), \tag{16.83}$$

where M is the reduced nuclear mass (half the mass of each nucleus for a homonuclear molecule), and in the region close to equilibrium the electronic functions ψ vary slowly with R (see Figure 16.4), Eq. (16.82) reduces to

$$\sum_m \psi_m [\widehat{T}_R \varphi_m + \mathcal{E}_m(R)\varphi_m] = E \sum_m \psi_m \varphi_m. \tag{16.84}$$

Multiplying by $\psi_n^\dagger$ and integrating over all electronic variables, one gets a Schrödinger equation for the nuclear motion,

$$\widehat{T}_R \varphi_n + \mathcal{E}_n(R)\varphi_n = E\varphi_n, \tag{16.85}$$

which shows that the electronic energy $\mathcal{E}_n(R)$ of the state under consideration is the potential that determines the nuclear motion. This potential has a shape similar to that shown in Figure 16.4, and its dependence on the parameter R is very complicated. For this reason, the study of nuclear motion with Eq. (16.85) is done by approximating $\mathcal{E}_n(R)$ with a reasonably constructed function that reproduces its basic properties and is mathematically simple at the same time. A well-known approximation is the Morse[15] potential (1929),

$$\mathcal{E}(R) = -D + D\left[e^{-(R-R_0)/a} - 1\right]^2, \tag{16.86}$$

where D represents the dissociation energy and R_0 is the equilibrium distance at which the potential is a minimum, $\mathcal{E}'(R_0) = 0$. The Morse potential gives good results for reasonable deviations from R_0 but cannot be used to study the behavior of the system for very large R, where the interaction potential decreases as R^{-6} instead of exponentially, as explained in Section 16.6.1.

In a diatomic molecule, the relative motion of the nuclei occurs around R_0 as long as it is not in a highly excited vibrational state. The potential can then be expanded in a Taylor series around R_0. From (16.86) to second order it follows that

$$\mathcal{E}_n(R) = -D + \tfrac{1}{2} M \omega_0^2 (R - R_0)^2, \tag{16.87}$$

with

$$D = -\mathcal{E}_n(R_0), \tag{16.88}$$

$$M\omega_0^2 = \left(\frac{\partial^2 \mathcal{E}_n}{\partial R^2}\right)_{R=R_0}, \tag{16.89}$$

---

[15] Philip McCord Morse (1903–1985) was a US-born theoretical physicist who pioneered operations research. He wrote several outstanding physic texts, among which Morse and Feshbach (1953) stands out, having been the bible of theoretical physics for many decades.

so that from (16.85) one gets

$$\frac{\hbar^2}{2M}\nabla^2\varphi + [E + D - \tfrac{1}{2}M\omega_0^2(R - R_0)^2]\varphi = 0. \tag{16.90}$$

The nuclear motion has thus been separated into two independent components: a rigid-body rotation (studied in Section 12.3), and oscillations about the equilibrium distance with the characteristic frequency ω_0 given by Eq. (16.89). Since the potential is central, we can immediately write the radial equation for the function $u(R) = R\varphi(R)$,

$$\frac{d^2u}{dR^2} + \frac{2M}{\hbar^2}\left[E + D - \frac{1}{2}M\omega_0^2(R - R_0)^2 - \frac{\hbar^2 L(L+1)}{2MR^2}\right]u = 0, \tag{16.91}$$

where L is the angular momentum of the rigid rotation of the molecule. It is convenient to pass from R to $r = R - R_0$; in the centrifugal term we can take $R^{-2} = (R_0 + r)^{-2} \simeq R_0^{-2}$, so that (16.91) reduces to the equation for a harmonic oscillator,

$$\frac{d^2u}{dr^2} + \frac{2M}{\hbar^2}\left[E + D - \frac{\hbar^2 L(L+1)}{2MR_0^2} - \frac{1}{2}M\omega_0^2 r^2\right]u = 0. \tag{16.92}$$

The total energy is then

$$E = -D + \hbar\omega_0\left(K + \tfrac{1}{2}\right) + B\hbar L(L+1), \quad (K = 0, 1, 2, \ldots) \tag{16.93}$$

with

$$B = \frac{\hbar}{2MR_0^2} = \frac{\hbar}{2I}. \tag{16.94}$$

The molecule dissociates when $E > 0$; this means that it has a finite number of discrete energy levels, those that satisfy the inequality

$$\hbar\omega_0\left(K + \tfrac{1}{2}\right) + B\hbar L(L+1) \geq D. \tag{16.95}$$

Physically, this limitation results from the fact that if the molecule rotates with very high angular momentum, the centrifugal force overcomes the binding forces, or if the vibrational energy is too high ($K \gg 1$), the atoms move too far apart.

From Eq. (16.93) we see that the energy of the molecule can be expressed as the sum of an electronic, a vibrational and a rotational contribution,

$$E = E_{\text{elect}} + E_{\text{vib}} + E_{\text{rot}}, \tag{16.96}$$

with

$$E_{\text{rot}} \ll E_{\text{vib}} \ll |E_{\text{elect}}| = D. \tag{16.97}$$

To confirm these inequalities, we proceed as follows. Let a be a characteristic dimension of the molecule; the kinetic energy of the electrons is at least of the order of that required by Heisenberg's inequality, $p^2/2m \sim \hbar^2/2ma^2$. Hence

$$|E_{\text{elect}}| \sim \frac{\hbar^2}{ma^2}. \tag{16.98}$$

With this estimate we can write $\partial^2 \mathcal{E}/\partial R^2 \sim \hbar^2/ma^4$, whence from (16.89),

$$E_{\text{vib}} \sim \hbar\omega_0 \sim \sqrt{\frac{m}{M}}|E_{\text{elect}}|. \tag{16.99}$$

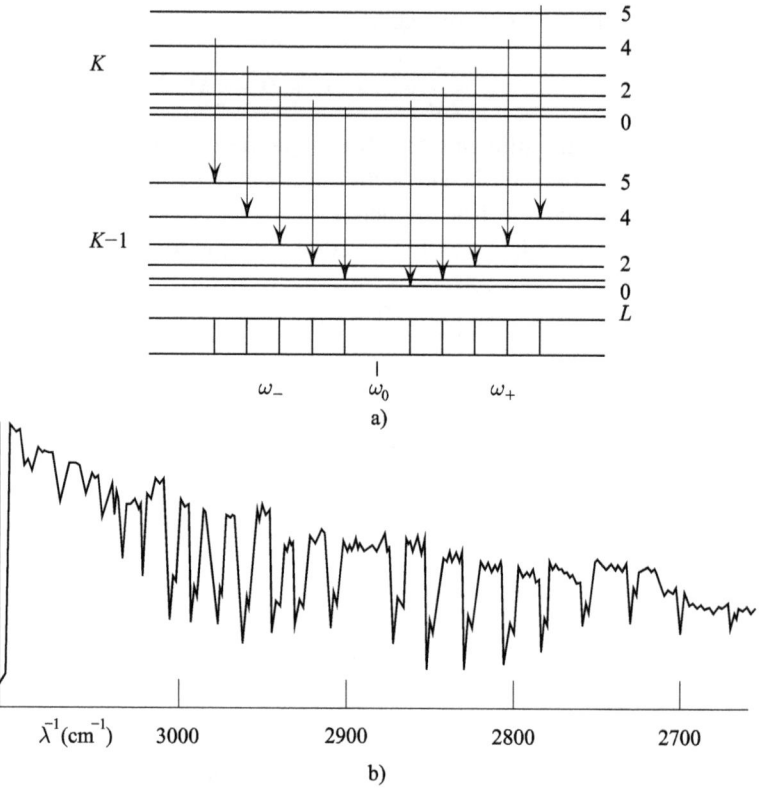

Figure 16.8 Vibrational–rotational spectrum of a diatomic molecule. In (a) the basic structure of the spectrum is shown as a succession of equidistant lines, in which the line corresponding to the frequency ω_0 is missing, since the transition $\Delta l = 0$ is forbidden. In (b) an absorption curve of electromagnetic radiation by HCl molecules is shown (the abscissa corresponds to $\lambda^{-1} = \omega/2\pi c$), with no line of frequency ω_0. The double peaks in the minima are due to the presence of two Cl isotopes.

Finally, we have

$$E_{\text{rot}} \sim \frac{\hbar^2}{Ma^2} = \frac{m}{M} \frac{\hbar^2}{ma^2} = \frac{m}{M} |E_{\text{elect}}|. \tag{16.100}$$

The vibrational-rotational emission spectrum results from the fact that during transitions the vibrational quantum number K must decrease (by 1), but L can increase ($L \to L+1$) or decrease ($L \to L-1$), since the system is deenergized anyway due to the lost vibrational energy; hence the radiated frequencies are

$$\omega^{\pm} = \frac{E(K,L) - E(K-1, L\pm 1)}{\hbar} = \omega_0 + \begin{cases} -2B(L+1) \\ +2BL \end{cases} \tag{16.101}$$

as shown in Fig. 16.8(a). Experimental analysis of this spectrum (in molecules such as CO, HCl, etc.) allows the determination of important molecular properties, such as moments of inertia, the isotopic composition of the nuclei, and so on.

A diatomic molecule can have a spectrum in the visible range that is much more complicated than those we have derived, consisting of more or less broad bands (band

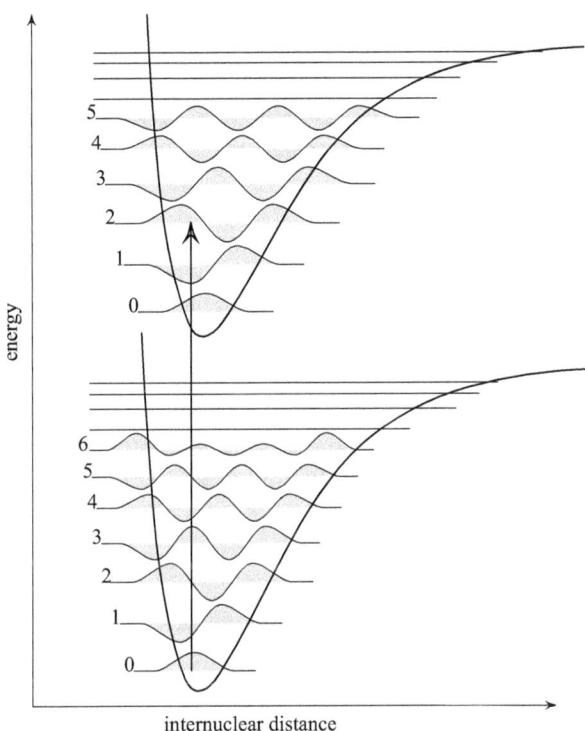

Figure 16.9 In a diatomic molecule the potential energy curves for the lower and upper electronic states are different. According to the Franck–Condon principle, the transitions are "vertical," that is, the nuclei do not move while the electrons are excited; note that the electronic transition is accompanied by a change in the vibrational quantum number.

spectrum), often formed by a large number of very close lines that start abruptly at a certain point. The treatment of this problem is greatly simplified by the introduction of the Franck–Condon principle. This principle states that, due to the large difference between the nuclear and electronic masses, the nuclei practically do not move while an electronic transition takes place (Fig. 16.9). Once a transition has taken place, the binding forces change and so does the internuclear distance, which modifies the molecule's moment of inertia; this change can excite vibrational and rotational states by energy exchange. As a result, the frequencies of the spectral lines become

$$\omega = \omega_1 + \omega_{LL'}, \tag{16.102}$$

with

$$\omega_1 = \frac{E_n - E_{n'}}{\hbar} \pm \omega_0, \tag{16.103}$$

$$\omega_{LL'} = BL(L+1) - B'L'(L'+1). \tag{16.104}$$

This gives rise to three independent branches, with frequencies

$$\omega^+ = \omega_1 + \omega_{LL-1} = \omega_1 + (B - B')L^2 + (B + B')L, \tag{16.105}$$

$$\omega^0 = \omega_1 + \omega_{LL} = \omega_1 + (B - B')L(L+1), \tag{16.106}$$

$$\omega^- = \omega_1 + \omega_{LL+1} = \omega_1 + (B - B')(L+1)^2 - (B + B')(L+1). \tag{16.107}$$

Exercise E16.3. The specific heat of a diatomic gas
Based on the results just presented, discuss the behavior of the specific heat of diatomic gases as a function of temperature.

Solution. Each molecule of a diatomic gas has 6 degrees of freedom: 3 translational, 2 rotational, and 1 vibrational. In the classical limit (at high temperatures) each degree of freedom contributes to the specific heat per molecule with the amount $\frac{1}{2}k$ (k is the Boltzmann constant), so the specific heat per molecule is given by $c_V = 3k$. However, when the average energy $\frac{3}{2}kT$ associated with the translational motion of the molecule is less than the energy difference between two adjacent vibrational levels, the collisions between molecules due to thermal agitation cannot excite the vibrational modes. The molecule then behaves like a rigid body with only 5 degrees of freedom; that is, the vibrations are frozen and the specific heat is only $\frac{5}{2}k$. The temperature at which the vibrational modes thaw is

$$\tfrac{3}{2}kT_{\text{vib}} = \hbar\omega. \tag{16.108}$$

Substituting into this expression the value of the vibrational frequency of H_2, we find that $T_{\text{vib}} \sim 10^3 K$, whereas for Cl_2 the vibrations contribute already at room temperature. These results ($c_V = \frac{5}{2}k$ and $3k$, respectively) are in agreement with the experiment (Fig. 16.10).

Similarly, we can define a threshold (defrost) temperature for the rotations using the equation

$$\tfrac{3}{2}kT_{\text{rot}} = \frac{\hbar^2}{2I}. \tag{16.109}$$

For any temperature below T_{rot} the diatomic molecule will behave with only 3 (translational) degrees of freedom, that is, as monatomic with specific heat $\frac{3}{2}k$. For hydrogen, for example, $T_{\text{rot}} \sim 50K$. Furthermore, taking into account that at low temperatures the statistics of the system are quantum and not classical, we have to use the quantum theory of specific heats, which predicts that c_V tends to zero with temperature, as seen in Section 2.3.2. The general character of the specific heat per molecule for a diatomic

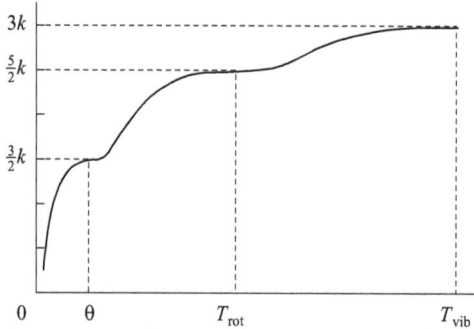

Figure 16.10 Specific heat of a diatomic gas as a function of temperature. The Debye temperature θ is a property of the material (more rigid materials have a higher θ).

gas is summarized in Figure 16.10, where θ (the Debye temperature) is a characteristic parameter of the gas.

16.6 Intermolecular Forces. Casimir and van der Waals Forces, Measurable Effects of the ZPF

Intermolecular forces are much weaker than the intramolecular forces of attraction, but they play a central role in determining the physical properties of aggregates of molecules, such as their boiling point, melting point, density, and enthalpies of fusion and evaporation,[16] and are also responsible for most of the chemical properties of matter.

In order from strongest to weakest, intermolecular forces can exist between an ion and a dipole (this is why, for example, some ionic compounds dissolve in water); H bonding (between the hydrogen attached to an electronegative atom in a molecule, such as O, N, or F, and another electronegative atom); between permanent dipoles, or between induced dipoles.

When two molecules come very close together, so that their electron clouds overlap, electrostatic forces arise between them, such as those associated with the chemical bonds. These chemical forces, which occur when the centers of the molecules are separated by 3 A or less, tend to be attractive and decrease rapidly (exponentially) with increase in distance; they are therefore considered to be short-range.

At long distances, on the other hand, there are attractive forces between molecules, generated by the coupling of their dipoles or multipoles, whose potential energy decreases with intermolecular distance, usually as R^{-6} or faster. Because of their relatively long range, these forces are critical in determining the properties of real gases and of some liquids and solids, including surface tension, friction, viscosity, vapor pressure, evaporation, and solubility. More generally, intermolecular forces play a crucial role in physics, chemistry, and biology; without intermolecular forces there would be virtually no solids or liquids; no nanostructures, biological molecules, or cellular structures could form.

Dipole–dipole interactions occur between polar molecules.[17] These forces arise from the alignment of the permanent or induced dipoles, enhancing the intensity of the attractive interactions. Polar molecules, such as water and ammonia, have an uneven distribution of electrons, resulting in positive and negative ends. The positive end of one molecule is attracted to the negative end of another, forming a weak electrostatic interaction.

Hydrogen bonding, a subset of dipole–dipole interactions, involves a particularly strong type of molecular attraction. It occurs when H atoms from one molecule form strong bonds with highly electronegative atoms, such as N, O, or F, in neighboring

---

[16] An enthalpy is a thermodynamic quantity equal to the total heat content of the system, i.e., the sum of its internal energy and the product pV.

[17] A polar molecule has the effective structure of an electric dipole due to the accumulation of charge on one side, and of the opposite charge in the other side.

molecules. The presence of H bonding is responsible for various extraordinary properties of water, including its high boiling point, density anomaly, and surface tension. It also plays a vital role in the structures and functions of biomolecules such as DNA and proteins.

16.6.1 Van der Waals forces

Van der Waals interactions are ubiquitous, existing between atoms, molecules, colloidal particles, and macroscopic objects. Van der Waals[18] forces include various types of attractive and repulsive long-range interatomic or intermolecular forces, resulting from *transient* shifts in electron density. Unlike ionic or covalent bonds, these attractions do not arise from a chemical electronic bond; they arise from the interaction between uncharged atoms or molecules, are comparatively weak, and vanish at larger distances between molecules. They lead not only to phenomena such as the cohesion of condensed phases, the physical absorption of gases and the self-assembly of nanomaterials, but also to a universal force of attraction between macroscopic bodies.

A molecule with a permanent dipole can *induce* a dipole in a similar neighboring molecule and cause mutual (induced dipole–dipole) attraction. This interaction is called the *Debye force*, named after P. Debye. A simple example is the interaction between the HCl molecule and the noble gas Ar. In this system, the Ar atom becomes a dipole as its electrons are attracted to the H side of HCl or repelled from the Cl side. The angle-averaged HCl–Ar interaction is given by the equation

$$V = \frac{-d_1^2 \alpha_2}{16\pi^2 \varepsilon_0^2 \varepsilon_r^2 R^6}, \tag{16.110}$$

where α_2 is the polarizability of the atom. This force is much weaker than a dipole–dipole interaction, but stronger than the London dispersion force.

The most important van der Waals forces are the *dispersion forces*,[19] the basic theory of which was formulated by London et al. (1926–1930). The London dispersion forces arise from attractive interactions between the *instantaneously induced* dipoles of molecules. Polarization can be induced either by a polar molecule or by the repulsion of negatively charged electron clouds in non-polar molecules. They are the weakest intermolecular forces, and yet they produce important phenomena, such as the tendency of colloidal particles to aggregate, which in turn may be the basis for other effects, such as coagulation, and so on. They are the most important component of the van der Waals forces because all materials are polarizable, whereas the Debye forces require permanent dipoles. The term dispersion forces derives from their association with optical dispersion.

The dispersion forces are traditionally considered to be purely quantum, that is, not explainable by classical physics. From the usual quantum point of view, two neutral polarizable bodies attract each other because the quantum fluctuations of one of them induce a dipole moment in the other, which in turn induces a dipole moment in the first

---

[18] Johannes van der Waals (1837–1923) was a Dutch theoretical physicist, famous for his pioneering work on the equation of state for real gases, for which he was awarded the Nobel Physics Prize in 1910.
[19] The term dispersion refers to the relation between energy and momentum.

one. The distance dependence of these dispersion forces can be estimated by noting that the dipole field of the first particle falls off as R^{-3} so that the strength of the induced dipole has this dependence, and the interaction energy of two dipoles also varies as R^{-3}, giving an overall R^{-6} dependence.

Exercise E16.4. Attractive van der Waals force between a neutral atom and a conducting plate

Calculate the interaction energy between a neutral atom and a perfectly conducting flat plate as a function of the distance, considering the interaction potential between the atom and its image.

Solution. Let d be the distance between the atom and the plate. If the plate is on the xy-plane, the interaction potential between the atom and its image is obtained from Eq. (16.115) by setting $e_2 = -e_1 = -e$, $x_2 = x_1 = x$, $y_2 = y_1 = y$, $z_2 = -z_1 = z$, $R = 2d$. Therefore,

$$\mathcal{H} = -\frac{e^2}{8d^3}(x^2 + y^2 + 2z^2). \tag{16.111}$$

For an H atom in its ground state, the first-order interaction energy from perturbation theory is

$$\mathcal{E} = \langle 100|\mathcal{H}|100\rangle = -\frac{e^2}{8d^3}4a_0^2 = E_1\left(\frac{a_0}{d}\right)^3, \tag{16.112}$$

where E_1 is the energy of the ground state. We see that the force with which the wall attracts the neutral atom is proportional to $1/d^4$ and is due to the dipole interaction with the image itself. Note that for the interaction between two independent dipoles it is required to use second-order perturbation theory, since $\langle 0|x_i|0\rangle = 0$, which produces a potential $\sim 1/d^6$; however, for dipole–plate the first-order theory suffices, which generates the much stronger interaction $\sim 1/d^3$.

16.6.1.1 Van der Waals force between two H atoms

The theoretical study of intermolecular forces is best carried out using the variational method introduced in Section 13.8. To keep things simple, we will consider the long-distance interaction between two H atoms; the generalization to the case of molecules is straightforward in principle, but can become very complex in practice.

Because of the large distance between the two atoms, we do not need to consider exchange effects (the Pauli principle applies in this case to each atom separately, whereas, if the overlap were important, it would apply to the system as a whole).[20] If r_1 and r_2 represent the position coordinates of the electrons with

---

[20] There is no rigorous formulation of the conditions under which the Pauli principle ceases to apply (or, more generally, when exchange effects cease to be important). Intuitively, if the wave functions of the constituents of the physical system have a significant overlap, the system forms a unit and the Pauli principle should apply (for coupled systems this happens at small distances, or on the order of the dimensions of the smaller of the constituents). On the other hand, if the overlap is negligible (e.g., at large distances), we can consider the parts to be (essentially) independent, and the Pauli principle applies to each of them separately.

respect to their own nucleus and $\boldsymbol{R}$ is the vector connecting them, we can write the Hamiltonian as

$$\widehat{H} = \widehat{H}_1(\boldsymbol{r}_1) + \widehat{H}_2(\boldsymbol{r}_2) + \widehat{\mathcal{H}}(\boldsymbol{r}_1, \boldsymbol{r}_2, \boldsymbol{R}), \tag{16.113}$$

where the term $\widehat{\mathcal{H}}$ is due to the Coulomb interaction between the charge distributions of the two atoms. If we describe these charge distributions by means of their multipole moments, the interaction between the neutral atoms is given in a first approximation by the coupling of their instantaneous electric dipoles,

$$\widehat{\mathcal{H}} = \frac{e^2}{R^3}\left[\boldsymbol{r}_1 \cdot \boldsymbol{r}_2 - \frac{3(\boldsymbol{r}_1 \cdot \boldsymbol{R})(\boldsymbol{r}_2 \cdot \boldsymbol{R})}{R^2}\right]. \tag{16.114}$$

The next expansion terms, corresponding to dipole–quadrupole coupling (proportional to R^{-4}), quadrupole–quadrupole coupling (proportional to R^{-5}), and so on, are negligible at large distances compared to the dipole–dipole contribution.

To simplify the calculation, it is useful to take the z-axis in the direction of $\boldsymbol{R}$; $\widehat{\mathcal{H}}$ is then written as

$$\widehat{\mathcal{H}} = \frac{e^2}{R^3}\left[x_1 x_2 + y_1 y_2 - 2 z_1 z_2\right]. \tag{16.115}$$

Assume that the atoms are in their ground state; if $\widehat{\mathcal{H}}$ is taken as a perturbation, the unperturbed wave function is $\psi_0 = \psi_0(\mathbf{r}_1)\psi_0(\mathbf{r}_2)$. Since the expectation value of the electric dipole moment of each atom in its ground state is zero, perturbation theory gives a zero correction to the first order, so the calculation must be carried over to the second order; this can be done, though not without some difficulty. Here we solve the problem using the variational method proposed by H. R. Hassé in 1930. We write the wave function in the form

$$\psi = \psi_0(\mathbf{r}_1, \mathbf{r}_2)(1 + v), \tag{16.116}$$

where the unknown function $v = v(\boldsymbol{r}_1, \boldsymbol{r}_2, \boldsymbol{R})$ is determined by requiring that $\langle \widehat{H} \rangle$ be a minimum. Since this wave function is not normalized, we use the expression $E \leq \langle \psi | \widehat{H} | \psi \rangle / \langle \psi | \psi \rangle$ to get (for v real),

$$\left(1 + 2\langle v \rangle + \langle v^2 \rangle\right) E \leq E_0 + 2 E_0 \langle v \rangle + 2\langle v \mathcal{H} \rangle + \langle v \widehat{H} v \rangle, \tag{16.117}$$

where all the expectation values are calculated using the unperturbed function ψ_0; in particular, E_0 is the sum of the energies of the two independent atoms. This expression shows that the problem is simplified by taking

$$v = \lambda \mathcal{H} \tag{16.118}$$

where λ is the variational parameter; substituting, we get

For intermediate distances, the situation is confusing and, for the moment, each case must be examined specifically.

$$E \leq \frac{E_0 + 2\lambda\langle\mathcal{H}^2\rangle + \lambda^2\langle\mathcal{H}\widehat{H}_0\mathcal{H}\rangle}{1+\lambda^2\langle\mathcal{H}^2\rangle}. \tag{16.119}$$

A direct calculation shows that $\langle\mathcal{H}\widehat{H}_0\mathcal{H}\rangle$ vanishes, so we can write

$$E \leq \frac{E_0 + 2\lambda\langle\mathcal{H}^2\rangle}{1+\lambda^2\langle\mathcal{H}^2\rangle} = E_0 + \frac{\lambda(2-\lambda E_0)\langle\mathcal{H}^2\rangle}{1+\lambda^2\langle\mathcal{H}^2\rangle}. \tag{16.120}$$

The energy correction given by this expression has a minimum if $E_0 < 0$, which is the case for the real root of

$$(1+\lambda^2\langle\mathcal{H}^2\rangle)(1-\lambda E_0) = \lambda^2\langle\mathcal{H}^2\rangle(2-\lambda E_0). \tag{16.121}$$

Due to the smallness of the van der Waals forces, it is expected that $\langle\mathcal{H}^2\rangle \ll E_0^2$. Furthermore, it follows from Eq. (16.118) that for the perturbation to be small, it must be true that $|\lambda\mathcal{H}| \ll 1$, and therefore that $\lambda^2\langle\mathcal{H}^2\rangle \ll 1$. Taking this into account, the previous equation can be written to a sufficient approximation as $(1-\lambda E_0) = 0$, that is,

$$\lambda = \frac{1}{E_0}. \tag{16.122}$$

This value of λ gives

$$E \leq E_0 + \frac{\langle\mathcal{H}^2\rangle}{E_0}, \tag{16.123}$$

and an approximation to the ground-state wave function close to the optimum is

$$\psi = \psi_0(\mathbf{r}_1)\psi_0(\mathbf{r}_2)\left(1+\frac{\mathcal{H}}{E_0}\right). \tag{16.124}$$

The expectation value of $\mathcal{H}^2$ is easily determined by noting that in the ground state, $\langle x\rangle = 0$ and $\langle x^2\rangle = \frac{1}{3}\langle r^2\rangle = a_0^2$,

$$\langle\mathcal{H}^2\rangle = \frac{e^4}{R^6}\langle x_1^2 x_2^2 + y_1^2 y_2^2 + 4z_1^2 z_2^2\rangle = \frac{6e^4 a_0^4}{R^6}. \tag{16.125}$$

Since the ground-state energy of two H atoms is $E_0 = -e^2/a_0$, we finally have

$$E \leq E_0 - \frac{6e^2 a_0^5}{R^6} = E_0\left(1 + 6\left(\frac{a_0}{R}\right)^6\right). \tag{16.126}$$

A more careful variational calculation shows that the correct value of the energy shift is obtained by writing 6.47 instead of the coefficient 6.

16.6.1.2 Retarded dispersion forces

At long distances, the R^{-6} dependence just obtained does not fully explain the observed behavior; this failure is usually attributed to the neglect of retardation effects. In 1948, Casimir and Polder showed that in the case of retardation effects due to the finite speed of propagation of electromagnetic interactions, Eq. (16.126) must be multiplied by a function of R which is unity at short distances, but becomes asymptotically C/R, so that the potential energy of the interaction varies as R^{-7} instead of R^{-6} at long distances. Such forces are called *retarded van der Waals–London forces* or *retarded dispersion* forces.

16.6.2 Van der Waals and Casimir forces explained by SED

The Debye interactions can be understood in terms of the concepts of classical electrostatics. On the other hand, the van der Waals interaction, the Casimir forces and related effects are generally considered to owe their existence to the quantum fluctuations and thus to be quantum in origin. However, a more physically transparent picture can be obtained by considering these forces as a direct consequence of the existence of the real fluctuating vacuum, that is, the ZPF.

A simple way of expressing the idea behind the calculations is as follows. The ZPF in free space is an isotropic and homogeneous field with an average energy $\frac{1}{2}\hbar\omega$ per normal mode. But the presence of microscopic or macroscopic bodies – be they molecules, conducting plates, or the walls of a cavity – affects the field, as required by the laws of electrodynamics. Thus, the modes in a given region of space depend on the geometry and electric and magnetic properties of the bodies present, resulting in effective interactions between them, mediated by the ZPF. One such example was discussed in Section 10.2.1, namely the Casimir force between two neutral plates. In the following we briefly discuss the *retarded* Casimir effect from the same perspective.

16.6.2.1 Calculation of the Casimir–Polder force

As an illustration, we refer here to Boyer's (1969) calculation of the Casimir–Polder effect between two neutral polarizable bodies. The bodies attract each other because the dipole induced by the fluctuating field in one induces a dipole moment in the other, which in turn induces a dipole moment in the first. The R^{-6} energy term results from the interaction of these dipoles.

Let us focus on one of the modes of the field with frequency ω and energy $\frac{1}{2}\hbar\omega$: the presence of the two polarizable bodies produces a position-dependent change in the natural frequency and a concomitant change in the total energy of the ZPF, leading to an effective potential $U(R)$ between the particles. In short, the Casimir–Polder effect is the result of the modification of the dielectric constant K_0 of the vacuum to a higher value K, so the energy of the modes varies from $E_0 \sim K_0^2$ to $E \sim K^2$. Since in its turn the energy of a field mode is proportional to its frequency $E \sim \omega$, the frequency of each mode varies from ω_0 to $\omega = (K_0/K)^2 \omega_0$.

To calculate $U(R)$, consider a conducting cavity containing particle A into which particle B is introduced. The second particle shifts the frequencies of the normal modes of oscillation, changing the total energy $\sum \frac{1}{2}\hbar\omega$ to $\sum \frac{1}{2}(\hbar\omega + \delta\omega)$. The shifts $\delta\omega$ are calculated taking into account the retardation. After a lengthy but transparent calculation, the result for the potential function is

$$U(R) = [-23(\alpha_A\alpha_B + \beta_A\beta_B) + 7((\alpha_A\beta_B + \beta_A\alpha_B))]\hbar c/4\pi R^7,$$

where α and β are the electric and magnetic polarizabilities, respectively. The R^{-7} potential arises thus as a result of the retarded effect of the second induced dipole on the first. This formula is identical to the one obtained by Feinberg and Sucher (1968) using quantum dispersion-theory techniques, but the SED calculation is considerably simpler and physically clear.

16.7 Nuclear Structure. Extending the Periodic Table of the Elements

The systematic study of the properties of atomic nuclei shows that they have periodicities reminiscent of those of atoms. For example, the binding energy per nucleon has very pronounced maxima when the number of neutrons or protons is 2, 8, 20, 28, 50, 82, and so on. Nuclei with a number of protons or neutrons equal to one of the preceding "magic numbers" are particularly stable and consequently their relative abundance is large. Using the notation $_ZX^A$, where A is the mass number (number of protons plus neutrons), some magic nuclei are $_2$He, $_8$O^{16}, $_{20}$Ca^{40}, $_{82}$Pb. An example of particular stability is Pb, because while its isotope Pb^{204} accounts for only 1.4% of all existing lead, the "doubly magic" isotope Pb^{208} accounts for more than 52% (doubly magic means that the nucleus contains a magic number of protons and of neutrons at the same time).

The set of periodically recurring properties that distinguish magic (and doubly magic) nuclei suggests the possibility of describing the nucleus with a shell model similar to that of the atom; this is the basis of the theory developed by Göppert-Mayer and Jensen around 1950.[21] In a nuclear version of the Hartree–Fock method, each nucleon is considered to move in the common central potential produced by the rest of the nucleons; if this is the case, the choice of a reasonable short-range potential to model the effective potential of the self-consistent field should allow us to derive the magic numbers, taking into account the exclusion principle. This would be the nuclear transcription of the process that led to the atomic "magic numbers" 2 (He), 10 (Ne), 18(Ar),... using the Coulomb interaction, the independent-particle approximation and the Pauli principle.

As a first approximation we model the nuclear potential as a 3D rectangular well. This is adequate to describe the basic properties of the deuteron, $_1$H^2, precisely because it is a short-range potential, which is one of the essential properties of the nuclear potential. The energy spectrum of particles bound by an infinite spherical potential well is shown in Table 16.4, where N_s represents the number of particles in the subshell $N_s = 2(2l + 1)$, and N the total (cumulative) number of particles. With the layer divisions suggested in Table 16.4, this model predicts the magic numbers 2, 8, 20, 40, 58,... the first three of which do indeed coincide with the empirical magic numbers, but for the rest the two sequences differ considerably. An analogous calculation using an isotropic harmonic oscillator potential yields the magic number sequence 2, 8, 20, 40, 70, 112,..., which is no better than the preceding one. The conclusion is that a short-range central potential cannot correctly predict all nuclear magic numbers, although it qualitatively agrees with the experimental situation. To improve the model it is necessary to introduce a noncentral potential, in particular the spin–orbit interaction.

---

[21] Maria Göppert-Mayer (1906–1972) was an outstanding German–US nuclear physicist. She developed a mathematical model for the structure of nuclear shells, which explained why certain numbers of nucleons result in particularly stable configurations. She was the second woman, after Marie Curie, to receive the Nobel Prize in Physics, which she shared in 1963 with Wigner and J. Hans D. Jensen (1907–1973) for their development of the nuclear shell model.

Table 16.4 Parameters associated with the subshells for a particle bound by an infinite spherical potential well. The energy is expressed in units of $h^x/8\pi^y ma^2$.

Subshell	E	N_s	N	Magic number
1s	9.9	2	2	
				2
1p	20.1	6	8	
				8
1d	33.2	10	18	
2s	39.5	2	20	
				20
2p	59.7	6	40	
				40
1g	67.0	18	58	
...	...	...	...	...

16.7.1 Introducing spin–orbit interaction

The results discussed earlier indicate that in order to correctly predict the magic numbers, one must partially break the degeneracy in l. The additional potential must depend on the orbital momentum, which suggests the introduction of a spin–orbit potential. The spin–orbit coupling can be considered to result from the interaction of the intrinsic (normal + anomalous) magnetic moment of the particle with the magnetic field produced by its own orbital motion. The formula

$$V_{SL} = -\frac{\hbar}{4m^2c^2}\left\langle\frac{1}{r}\frac{dV}{dr}\right\rangle \hat{\boldsymbol{\sigma}} \cdot \hat{\boldsymbol{L}}, \quad (16.127)$$

is derived directly from the Dirac equation; in Section 19.2 it is obtained for the specific case of the relativistic harmonic oscillator. The expectation values obtained from Eq. (16.127) are

$$\langle \hat{V}_{SL}\rangle = \begin{cases} V_0 l & \text{if } j = l + \frac{1}{2}, \\ -V_0(l+1) & \text{if } j = l - \frac{1}{2} \end{cases} \quad (16.128)$$

where

$$V_0 = -\frac{\hbar^2}{4m^2c^2}\left\langle\frac{1}{r}\frac{dV}{dr}\right\rangle. \quad (16.129)$$

The experimental data show that V_0 must be taken as positive, so that the levels with $j = l + 1/2$ fall below the levels with $j = l - 1/2$ (this is the opposite sign to that of atomic electrons). Fig. 16.11 shows the results obtained by assigning an appropriate value to this constant using the following phenomenological central potential, which lies between the square and the harmonic potentials,

$$V = \frac{V_1}{1 + e^{(r-R)/a}}. \quad (16.130)$$

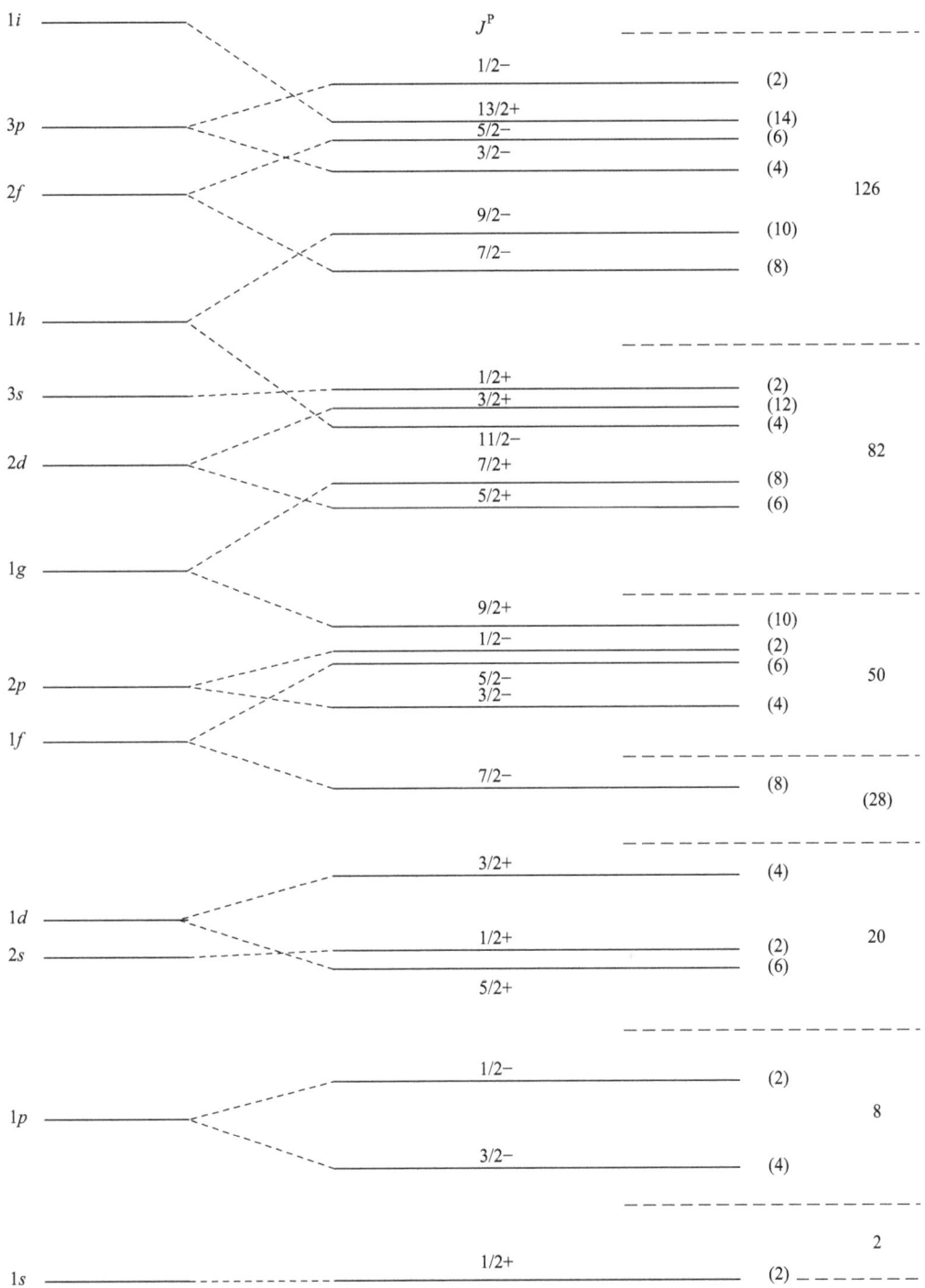

Figure 16.11 Energy levels of nucleons in an atomic nucleus according to the shell model. On the left are the values of (nl) and in the center those of J^P (P = parity). The dotted lines separate the layers. The penultimate column shows the number of neutrons or protons that fit into each subshell, and the column to the right shows the magic numbers predicted by this theory, which coincide with those found empirically (28 is sometimes considered "semimagic").

In this expression the parameters V_1 and a are constant, while the radius of the nucleus R depends on the mass number A. From Fig. 16.11 we can see that the theory correctly predicts not only the magic numbers, but also the order in which the subshells with well-defined values of the total angular momentum J and of the parity P should appear: the first shell is filled with the nucleons in the state $1s_{1/2}$ ($J^P = 1/2$), the second shell is filled with equal particles in the states $1p_{3/2}$ and $1p_{1/2}$ ($J^P = 3/2$ and $1/2$, respectively), and so on.

The magic numbers of the nuclear shell model suggested the possibility of enhanced nuclear stability for some superheavy isotopes, particularly around atomic number 114 and neutron number 184, which would be doubly magic. This led to the hypothesis of the existence of an *island of stability* that could contain nuclides with half-lives of thousands, perhaps even millions of years. To this date (2025), the synthesis of nuclides too unstable to occur naturally on Earth has expanded the periodic table to 118 elements. While the lighter transuranic elements have found uses, the isotopes of those beyond ^{103}Lr (lawrencium), the so-called superheavy elements, are too unstable to exist outside the laboratory.

In its finished form, the shell model allows many properties of atomic nuclei to be correctly determined, despite its obvious approximate character and its phenomenological nature. Since the details of this subject are beyond the scope of the present text, we will not go into them here, but refer the interested reader to modern standard texts on nuclear physics.

Problems

P16.1 The neutral atoms F, Ca, and Rb have 9, 20, and 37 electrons, respectively. What is their electronic configuration in the ground state? What fundamental physical and chemical properties can be predicted for them?

P16.2 Prove directly that $\langle \text{ortho}|\mathbf{r}_1 + \mathbf{r}_2|\text{para}\rangle$ is zero for the He atom.

P16.3 In Section 16.3.3 we used the variational method to find the ground-state energy of helium using the trial function (16.43),

$$\psi = \frac{1}{\pi a_0^3} \exp\left(-\frac{r_1 + r_2}{a_0}\right), \tag{16.131}$$

where a_0 is the variational parameter. Try to modify this test function so that the error in the ground state energy does not exceed 0.5% of the experimental value. -78.97 eV.

P16.4 Use the wave function (16.131) to determine the electrostatic field generated by the helium atom in the space around it.

P16.5 Consider an isotropic 3D harmonic oscillator. Prove that in the N shell (N is the principal quantum number) there are $(N+1)(N+2)$ equal particles with spin 1/2. Use this information to show that the nuclear magic numbers predicted by the harmonic oscillator model are 2, 8, 20, 40, 70, and 112.

16.7 Nuclear Structure. Extending the Periodic Table of the Elements

P16.6 At large distances the binding potential of the H_2 molecule decreases exponentially, while the van der Waals force decreases as R^{-7}. What is the reason for this apparent discrepancy?

P16.7 Find the coefficients of the expansion of the electronic energy $\mathcal{E}(R)$ when it is modeled with the Morse potential

$$\mathcal{E}(R) = D[e^{-2(R-R_0)/a} - 2e^{-(R-R_0)/a}], \qquad (16.132)$$

up to and including the fourth order.

P16.8 Find the order of magnitude of the wavelength of the radiation emitted during a $K = 1 \to K = 0$ vibrational transition in a LiH molecule. Assume that the equilibrium distance between the atoms in a diatomic molecule is similar to the distance between the atoms in a crystal of the same substance. *Note*: The LiH lattice is cubic and the crystal has an approximate density of 0.83 g/cm^{-3}.

(Solution: $\lambda \sim 1.8 \times 10^{-3}$ cm)

P16.9 Based on the absorption spectrum of HCl vapor in the near infrared shown in Figure 16.8(b), prove that the equilibrium distance between the H and Cl atoms of this molecule is of the order of 1.3×10^{-8} cm.

P16.10 Find the dissociation energy of the D_2 molecule, using the fact that the dissociation energy and the minimum vibrational energy of the H_2 molecule are equal to 4.46 eV and 0.26 eV, respectively.

P16.11 Prove that the molecule He_2^{3+} does not exist.

P16.12 An empirical formula for the potential of the NaCl molecule is

$$V = -\frac{e^2}{4\pi\varepsilon_0 R} + Ae^{-R/a}, \qquad (16.133)$$

where R is the internuclear distance. The equilibrium internuclear distance is 2.5 Å and the dissociation energy is 3.6 eV. Find the value of A and a/R_0, and explain the physical meaning of the parameters A and a.

P16.13 Compare the rotational and vibrational levels of an HCl molecule calculated with the potential

$$V(r) = 4\alpha\left[\left(\frac{d}{r}\right)^{12} - \left(\frac{d}{r}\right)^{6}\right], \quad \alpha = 3.1 \times 10^{-12} \text{ eV}, \quad d = 3.3 \text{ Å}. \quad (16.134)$$

Hint: Since the equilibrium position r_0 depends weakly on angular momentum, the effective potential can be expanded around r_0. It is convenient to introduce $b = \hbar^2/md^2$ and express the potential in terms of b and $x_0 = r_0/d$.

P16.14 The vibrational frequency of the CO molecule in its lowest state is $\nu_0 = 2 \times 10^{13}$ Hz. What is the wavelength of the radiation emitted by the lowest vibrational excitation? What is the probability that the first vibrational state is excited? Compare your result with the probability that the CO molecule is in its vibrational ground state when the temperature is 300 K.

Bibliographical Notes

For a general treatment of the chemical bond, we recommend the classic book by Linus Pauling (1960).

For the latest release of the periodic table (dated 4 May 2022), see https://iupac.org/what-we-do/periodic-table-of-elements/#a10.

For an interactive periodic table of the isotopes, see https://ciaaw.org/periodic-table-isotopes.html

For a more detailed discussion of intermolecular forces, we refer the reader to specialist texts in physical chemistry, such as Israelachvili (2011).

For a comprehensive account of Casimir effect experiments, see Lamoreux (2005).

17 Time-Dependent Perturbations. Field Quantization and Second Quantization

In the usual quantum framework, the electromagnetic field acting on matter is considered a perturbation. Since the radiation field oscillates, we need tools for the treatment of time-dependent perturbations. The theory developed in Section 17.1 will allow us to deal more generally with the problem of a quantum system subject to a time-dependent perturbation.

We then apply the theory to the interaction of atomic matter with the radiation field and derive what are known in quantum theory as radiative corrections. Section 17.2 is devoted to a detailed discussion of atomic transitions – both induced and spontaneous – with some important applications. The formal introduction of field quantization in Section 17.3 allows us to push the theory beyond the usual limits of QM and to make contact with (nonrelativistic) quantum electrodynamics (QED) and quantum optics.

Section 17.4 presents an alternative, direct derivation of the radiative corrections using SED, where the matter–field interaction is present ab initio. In the same spirit, Section 17.5 provides a physical justification for the quantization of the field (done formally in QED), by deriving the conventionally postulated field operators and their basic commutators.

The chapter ends with a brief introduction to second quantization, an elegant and powerful formalism for describing and analyzing quantum many-body systems.

17.1 Time-Dependent Perturbation Theory

We saw in Chapter 13 that a quantum system changes its state when it is perturbed. Assuming that the state ψ_n is one of the eigenstates of $\widehat{H}_0$, the state Ψ, an eigenstate of the perturbed Hamiltonian $\widehat{H} = \widehat{H}_0 + V$, is written in the general form

$$\Psi_n = \psi_n + \sum_k C_{nk} \psi_k,$$

where the coefficient $C_{nk} = \langle \psi_k | \Psi_n \rangle$ gives the amplitude of the state k of $\widehat{H}_0$ contained in Ψ_n. Therefore, $P_{n \to k} \equiv |C_{nk}|^2$ is the probability that the state of the system changes from ψ_n to Ψ_k as a result of the perturbation. This example shows that one of the most important effects of a perturbation is to produce transitions between states.

So far we have dealt with time-independent perturbations; we will now extend the theory to the time-dependent case by writing the Schrödinger equation

$$i\hbar \frac{\partial \Psi}{\partial t} = \widehat{H}\Psi = (\widehat{H}_0 + \widehat{\mathcal{H}}(t))\Psi, \qquad (17.1)$$

where $\widehat{\mathcal{H}}(t)$ represents the perturbative Hamiltonian. We know that the general solution of the unperturbed problem

$$i\hbar \frac{\partial \psi}{\partial t} = \widehat{H}_0 \psi \tag{17.2}$$

can be written in the form

$$\psi(x,t) = \sum_k C_k e^{-iE_k t/\hbar} \varphi_k(x), \tag{17.3}$$

where φ_k are the eigenfunctions of $\widehat{H}_0$ and E_k are the corresponding eigenvalues and the coefficients C_k of the expansion are given by

$$C_k = \int \varphi_k^*(x) \psi(x,0) \, dx. \tag{17.4}$$

To solve Eq. (17.1) we apply the method of variation of parameters, which consists in writing the solutions in the form (17.3), but with the constants C_k replaced by respective functions of time to be determined,

$$\Psi = \sum_k C_k(t) e^{-iE_k t/\hbar} \varphi_k(x). \tag{17.5}$$

Substituting into Eq. (17.1) and simplifying gives

$$i\hbar \sum_k \dot{C}_k(t) e^{-iE_k t/\hbar} \varphi_k(x) = \sum_k C_k(t) e^{-iE_k t/\hbar} \widehat{\mathcal{H}}(t) \varphi_k(x),$$

and multiplying by φ_l^* and integrating over x, we get

$$\dot{C}_l(t) = -\frac{i}{\hbar} \sum_k C_k(t) e^{i(E_l - E_k)t/\hbar} \mathcal{H}_{lk}, \tag{17.6}$$

where

$$\mathcal{H}_{lk} = \int \varphi_l^* \widehat{\mathcal{H}}(t) \varphi_k \, dx. \tag{17.7}$$

Integrating (17.6) over time, assuming that the perturbation is applied at time $t = 0$, we obtain

$$C_l(t) = C_l(0) - \frac{i}{\hbar} \sum_k \int_0^t C_k(t') e^{i\omega_{lk} t'} \mathcal{H}_{lk} \, dt', \tag{17.8}$$

where $\omega_{kn} = (E_k - E_n)/\hbar$. From (17.4) it follows that the initial value of $C_l(t)$ is

$$C_l(0) = \int \varphi_l^*(x) \Psi(x,0) \, dx. \tag{17.9}$$

We will limit ourselves to the special case in which the system, before being perturbed, is in an eigenstate of $\widehat{H}_0$, $\Psi(x,0) \equiv \Psi_n(x,0) = \varphi_n(x)$; it follows from (17.9) that $C_k(0) = \delta_{nk}$. Since $\Psi_n(t)$ is the wave function we are looking for, we add the index n to the coefficients of the expansion (17.5) and write

$$\Psi_n(t) = \sum_k C_{nk}(t) e^{-iE_k t/\hbar} \varphi_k(x), \quad C_{nk}(0) = \delta_{nk}. \tag{17.10}$$

To solve the system of integral equations (17.8), which is exact, we perform a perturbative expansion of the coefficients,

$$C_{nk}(t) = C_{nk}(0) + C_{nk}^{(1)}(t) + C_{nk}^{(2)}(t) + \cdots . \tag{17.11}$$

Substituting into Eq. (17.8) gives

$$
\begin{aligned}
C_{nl}(t) &= C_{nl}(0) - \frac{i}{\hbar} \sum_k \int_0^t \left(C_{nk}(0) + C_{nk}^{(1)}(t') + \cdots \right) e^{i\omega_{lk}t'} \mathcal{H}_{lk} dt' \\
&= C_{nl}(0) - \frac{i}{\hbar} \int_0^t e^{i\omega_{ln}t'} \mathcal{H}_{ln} dt' - \frac{i}{\hbar} \sum_k \int_0^t \left(C_{nk}^{(1)}(t') + \cdots \right) e^{i\omega_{lk}t'} \mathcal{H}_{lk} dt',
\end{aligned}
\tag{17.12}
$$

from which it follows that

$$C_{nk}^{(1)}(t) = -\frac{i}{\hbar} \int_0^t e^{i\omega_{kn}t'} \mathcal{H}_{kn}(t') dt', \tag{17.13}$$

$$
\begin{aligned}
C_{nl}^{(2)}(t) &= -\frac{i}{\hbar} \sum_k \int_0^t \left(C_{nk}^{(1)}(t') + \cdots \right) e^{i\omega_{lk}t'} \mathcal{H}_{lk} dt' \\
&= \left(-\frac{i}{\hbar}\right)^2 \sum_k \int_0^t dt' \int_0^{t'} dt'' e^{i\omega_{lk}t'} \mathcal{H}_{lk}(t') e^{i\omega_{kn}t''} \mathcal{H}_{kn}(t''),
\end{aligned}
\tag{17.14}
$$

and so on. $C_{nl}^{(2)}(t)$ is often rewritten in a more convenient form, taking into account that the unperturbed states, eigenstates of $\widehat{H}_0$, form a complete basis,

$$
\begin{aligned}
C_{nl}^{(2)}(t) &= \left(-\frac{i}{\hbar}\right)^2 \sum_k \int_0^t dt' \int_0^{t'} dt'' \\
&\quad \times \langle l| e^{iE_l t'/\hbar} \widehat{\mathcal{H}}(t') e^{-iE_k t'/\hbar} |k\rangle \langle k| e^{iE_k t''/\hbar} \widehat{\mathcal{H}}(t'') e^{-iE_n t''/\hbar} |n\rangle \\
&= \left(-\frac{i}{\hbar}\right)^2 \sum_k \int_0^t dt' \int_0^{t'} dt'' \\
&\quad \times \langle l| e^{i\widehat{H}_0 t'/\hbar} \widehat{\mathcal{H}}(t') e^{-i\widehat{H}_0 t'/\hbar} |k\rangle \langle k| e^{i\widehat{H}_0 t''/\hbar} \widehat{\mathcal{H}}(t'') e^{-i\widehat{H}_0 t''/\hbar} |n\rangle ,
\end{aligned}
\tag{17.15}
$$

that is,

$$C_{nk}^{(2)}(t) = -\frac{1}{\hbar^2} \int_0^t dt' \int_0^{t'} dt'' \langle k| e^{i\widehat{H}_0 t'/\hbar} \widehat{\mathcal{H}}(t') e^{i\widehat{H}_0(t''-t')/\hbar} \widehat{\mathcal{H}}(t'') e^{-i\widehat{H}_0 t''/\hbar} |n\rangle . \tag{17.16}$$

In Eq. (17.15), the matrix elements of the interaction Hamiltonian appear in the Heisenberg picture relative to the unperturbed Hamiltonian, that is, in the interaction representation, as briefly explained in Section 8.6.6. The first-order approximation, which is sufficient in many cases of interest, is valid as long as the coefficients $C_{nk}(t)$ do not differ much from their initial value.

Equations (17.13)–(17.16) show that the perturbation produces transitions to the states k connected to the initial state n by the matrix of the perturbing potential or its products. After an important example of induced vibrations, we will separately consider two general cases of great interest.

Exercise E17.1. Rabi oscillations in EPR

Use Eq. (17.6) to study the time evolution of the population of the two energy levels in electron (paramagnetic) spin resonance.

Solution. As shown in Section 14.3.3, resonant transitions occur when electrons are subject to a constant magnetic field $B_z = B_0$ and a perpendicular field $\boldsymbol{B}_1$ that oscillates with frequency very close to the transition frequency between levels.

Here, in contrast to Section 14.3.3, we consider the uniform field as part of the unperturbed Hamiltonian and the oscillating field as a perturbation. The unperturbed Hamiltonian is then

$$\widehat{H}_0 = 2\frac{e_0 B_0}{2mc}\widehat{S}_z = 2\omega_L \widehat{S}_z = \tfrac{1}{2}\hbar\omega_0 \hat{\sigma}_z, \tag{17.17}$$

where $\omega_L = e_0 B_0/2mc$ is the Larmor frequency and $\omega_0 = 2\omega_L$ twice this frequency (or cyclotron frequency); the unperturbed eigenenergies are $E_1 = \tfrac{1}{2}\hbar\omega_0$, $E_2 = -\tfrac{1}{2}\hbar\omega_0$, and the unperturbed states are the eigenvectors of $\widehat{S}_z$.

Using the rotating-wave approximation (RWA), we assume that the rotating field of interest (the other component is neglected) rotates against the clock in the xy-plane and has the form

$$\boldsymbol{B}_1(t) = B_1 \left(\mathbf{i}\cos\omega t + \mathbf{j}\sin\omega t\right), \tag{17.18}$$

so that the perturbation Hamiltonian is

$$\widehat{\mathcal{H}}(t) = \omega_B \left(\cos\omega t\, \widehat{S}_x + \sin\omega t\, \widehat{S}_y\right), \tag{17.19}$$

or, in explicit matrix form (compare with Eq. (14.52)),

$$\widehat{\mathcal{H}}(t) = \tfrac{1}{2}\hbar\omega_B \begin{pmatrix} 0 & e^{-i\omega t} \\ e^{i\omega t} & 0 \end{pmatrix}, \tag{17.20}$$

with $\omega_B = 2\mu_0 B_1/\hbar$.

As only two levels are active, the system of Eq. (17.6) is reduced to

$$i\hbar\frac{dC_1}{dt} = C_1 \mathcal{H}_{11} e^{i\omega_{11}t} + C_2 \mathcal{H}_{12} e^{i\omega_{12}t},$$
$$i\hbar\frac{dC_2}{dt} = C_1 \mathcal{H}_{21} e^{i\omega_{21}t} + C_2 \mathcal{H}_{22} e^{i\omega_{22}t}, \tag{17.21}$$

with $\omega_{11} = \omega_{22} = 0$ and $\omega_{12} = -\omega_{21} = \omega_0$. From Eq. (17.20), the nondiagonal matrix elements of $\widehat{\mathcal{H}}(t)$ are $\mathcal{H}_{12} = \mathcal{H}_{21}^* = \tfrac{1}{2}\hbar\omega_B e^{-i\omega t}$, so that

$$\frac{dC_1}{dt} = -\tfrac{1}{2}i\omega_B e^{-i(\omega-\omega_0)t} C_2, \tag{17.22}$$

$$\frac{dC_2}{dt} = -\tfrac{1}{2}i\omega_B e^{i(\omega-\omega_0)t} C_1. \tag{17.23}$$

To solve this system it is convenient to decouple the equations; this is easily done by deriving the first and in the second substituting the result to eliminate C_2. This gives

$$\frac{d^2 C_1}{dt^2} + i(\omega-\omega_0)\frac{dC_1}{dt} + \left(\tfrac{1}{2}\omega_B\right)^2 C_1 = 0, \tag{17.24}$$

the general solution of which is

$$C_1 = Ce^{-i(\omega-\omega_0)t/2} \sin\left(\tfrac{1}{2}\Omega t + \alpha\right), \tag{17.25}$$

$$\Omega^2 = (\omega - \omega_0)^2 + \omega_B^2. \tag{17.26}$$

The coefficient C_2 follows immediately from Eq. (17.22),

$$C_2 = \frac{C}{\omega_B} e^{i(\omega-\omega_0)t/2} \left[i\Omega \cos\left(\tfrac{1}{2}\Omega t + \alpha\right) + (\omega - \omega_0)\sin\left(\tfrac{1}{2}\Omega t + \alpha\right)\right]. \tag{17.27}$$

The quantities $|C_1(t)|^2$ and $|C_2(t)|^2$ determine the instantaneous populations of the respective levels. If initially only the lower level is populated ($C_1 = 0, C_2 = 1$), we must put $\alpha = 0$, $C = -i\omega_B/\Omega$, and we are left with

$$|C_1(t)|^2 = 1 - |C_2(t)|^2 = \left(\frac{\omega_b}{\Omega}\right)^2 \sin^2 \tfrac{1}{2}\Omega t.$$

These results show that transitions between the two states occur periodically with a frequency Ω that depends on the coupling, as follows from (17.26). These transitions are known as Rabi oscillations. At resonance, $\omega = \omega_0$, and Ω reaches its minimum value, $\Omega_{\text{res}} = \omega_B$; in this case both levels are populated on average with equal probability,

$$|C_1(t)|^2 = \sin^2 \tfrac{1}{2}\Omega t, \quad |C_2(t)|^2 = \cos^2 \tfrac{1}{2}\Omega t.$$

17.1.1 Perturbations acting for finite times

We now consider the case of a potential V applied to the system at $t = 0$ and disappearing after some time. For times t sufficiently larger than the duration of V (e.g., for $t = \infty$), the coefficients C will have reached their final value and the wave function will be a superposition of eigenstates of the original Hamiltonian $\hat{H}_0$, $\Psi_n = \sum_k C_{nk}(\infty)e^{-iE_kt/\hbar}\varphi_k(x)$. Since $C_{nk}(\infty)$ is the amplitude of the state k contained in Ψ_n, the transition probability from the initial state n to the state k during the perturbation is to first order given by Eq. (17.13),

$$w_{nk} = |C_{nk}^{(1)}(\infty)|^2 = \frac{1}{\hbar^2} \left| \int_0^\infty V_{kn} e^{i\omega_{kn}t} dt \right|^2, \quad k \neq n. \tag{17.28}$$

Suppose that the potential V changes very little over time intervals of the order of ω_{kn}^{-1}; in this case, due to the oscillations of the integrand, the integral is small. We conclude that when a perturbation varies adiabatically, that is, very slowly, a system in a nondegenerate stationary state remains in it. If there is degeneracy between states k and n, the previous argument does not hold, because $\omega_{kn} = 0$ and transitions between these states will occur since they do not imply an energy exchange. In this case, the preceding theory is not satisfactory, but the problem can be solved explicitly starting from the exact equation (17.8) (see problem P17.1).

The opposite extreme case, corresponding to a pulse of height V and width Δt in the limit where Δt tends to zero but the area $V\Delta t$ remains constant, is more difficult to analyze, but allows us to observe an interesting phenomenon. Assuming that the pulse is applied at time t_1, we obtain directly from Eq. (17.13) with $\int_0^\infty dt \Rightarrow \int_{t_1}^{t_1+\Delta t} dt$, that

$$C_{nk}^{(1)} = \frac{V_{kn}}{\hbar \omega_{kn}} e^{i\omega_{kn} t_1} \left(1 - e^{i\omega_{kn} \Delta t}\right), \quad \omega_{kn} \neq 0. \tag{17.29}$$

The second term of this expression gives the (first-order) coefficients of the expansion of Ψ_n at time $t + \Delta t$ of the complete Hamiltonian $\widehat{H}$ in terms of the eigenfunctions φ_n of $\widehat{H}_0$ and, therefore, does not contribute to the transitions between the states n and k. The transition probability from the initial state n to the state k is then given by the first term in Eq. (17.29) and is

$$w_{nk} = \frac{|V_{kn}|^2}{\hbar^2 \omega_{kn}^2} = \frac{|V_{kn}|^2}{(E_n - E_k)^2}. \tag{17.30}$$

We can verify that Eq. (17.30) is correct by analyzing the meaning of each of the two terms in Eq. (17.29). In general, whatever the size of the pulse, if we know the eigenfunctions Ψ_k of $\widehat{H}$ and φ_n of $\widehat{H}_0$, we can calculate the transition probability between these states by using the formula

$$w_{nk} = \left| \int \Psi_k^* \varphi_n dx \right|^2. \tag{17.31}$$

In the particular case of a small pulse, the eigenvalues of $\widehat{H}$ essentially coincide with those of $\widehat{H}_0$, so we can write $\widehat{H}_0 \varphi_n = E_n \varphi_n$ and $\widehat{H} \Psi_k \approx E_k \Psi_k$; from the first of these equations and the conjugate of the second, it follows that

$$\int (\varphi_n \widehat{H}^* \Psi_k^* - \Psi_k^* \widehat{H}_0 \varphi_n) dx = \int (\Psi_k^* \widehat{H} \varphi_n - \Psi_k^* \widehat{H}_0 \varphi_n) dx = \int \Psi_k^* V \varphi_n dx$$
$$= \int (\varphi_n E_k \Psi_k^* - \Psi_k^* E_n \varphi_n) dx = (E_k - E_n) \int \Psi_k^* \varphi_n dx, \tag{17.32}$$

and from here we get to first order

$$\int \Psi_k^* \varphi_n dx = \frac{\int \Psi_k^* V \varphi_n dx}{E_k - E_n} \simeq \frac{\int \varphi_k^* V \varphi_n dx}{E_k - E_n} = \frac{V_{kn}}{\hbar \omega_{kn}}, \tag{17.33}$$

since for a small perturbation the eigenfunctions of $\widehat{H}$ and $\widehat{H}_0$ are equal to zero order; from here Eq. (17.30) follows immediately. The term we neglected when we took $\Psi_k^* \to \varphi_k^*$ corresponds to the additional term in (17.29), which we must also neglect there when calculating w_{nk}.

Let us now consider the case of a potential V which is switched on at $t = 0$ and switched off after a long time t. From Eqs. (17.11) and (17.13), it follows that

$$C_{nn} = 1 - \frac{i}{\hbar} \int_0^t V_{nn} dt = 1 - \frac{i V_{nn}}{\hbar} t \simeq e^{-i V_{nn} t / \hbar}, \tag{17.34}$$

which shows that the first-order energy correction is V_{nn}, in agreement with the time-independent perturbation theory. Analogously,

$$C_{nk} = -\frac{i V_{kn}}{\hbar} \int_0^t e^{i\omega_{kn} t'} dt' = \frac{V_{kn}}{\hbar \omega_{kn}} (1 - e^{i\omega_{kn} t}). \tag{17.35}$$

As before, the second term of this expression is the one that transforms the stationary states of $\hat{H}_0$ into the corresponding stationary states of $\hat{H}$. Taking this into account, the term $C_{nk}^{(1)} = V_{kn}/\hbar\omega_{kn} = V_{kn}/(E_k - E_n)$ gives the first-order correction to the wave function, in correspondence with the time-independent perturbation theory. The transition probability between states n and k at time t is given by

$$w_{nk} = |C_{nk}|^2 = \frac{|V_{kn}|^2}{\hbar^2 \omega_{kn}^2} \left|1 - e^{i\omega_{kn}t}\right|^2 = \frac{4|V_{kn}|^2}{\hbar^2} \frac{\sin^2 \frac{1}{2}\omega_{kn}t}{\omega_{kn}^2}. \tag{17.36}$$

For nondegenerate states, w_{nk} remains small for all time if $|V_{kn}|$ is small; conversely, if $|V_{kn}|$ is large, (17.36) holds only for times small enough for $|V_{kn}|t \ll \hbar$ to hold. If the state is degenerate, then the amplitude of the transition between degenerate states ($\omega_{kn} = 0$) increases with time, and the previous expression holds only for short times, as in the previous case of large perturbations. From this we conclude that the smallest perturbation is sufficient to produce transitions between the different degenerate states.

The transition probability between states n and k per unit time is defined as

$$W_{nk} \equiv \frac{|C_{nk}|^2}{t} = \frac{4|V_{kn}|^2}{\hbar^2} \frac{\sin^2 \frac{1}{2}\omega_{kn}t}{\omega_{kn}^2 t}. \tag{17.37}$$

This function is only significantly different from zero for $|\omega_{kn}|t \leq \pi$, that is, $|E_n - E_k|t \leq \pi\hbar$, and approaches zero with time as shown in Fig. 17.1. Thus, while for very short times the perturbation can cause transitions between states whose energies differ by an arbitrary amount, as time progresses, the most favored transitions are those between states of increasingly closer energies, until at $t \to \infty$ the net transition is between states of equal energy. In this limit, the function behaves like a Dirac delta function,

$$\lim_{t \to \infty} \frac{\sin^2 \omega t}{\pi t \omega^2} = \delta(\omega). \tag{17.38}$$

To convince ourselves of this, we first note that if $\omega \neq 0$, the limit is zero, but for $\omega = 0$ it has an infinite value; in addition, the function is normalized to unity,

$$\frac{1}{\pi} \int_{-\infty}^{\infty} \frac{\sin^2 \omega t}{\omega^2 t} d\omega = 1.$$

It is clear from Fig. 17.1 that in the limit $t \to \infty$, the behavior of the function is precisely that of $\delta(\omega)$. Therefore, we can write

$$\lim_{t \to \infty} \frac{\sin^2 \frac{1}{2}\omega_{kn}t}{\omega_{kn}^2 t} = \frac{\pi}{4}\delta(\tfrac{1}{2}\omega_{kn}) = \tfrac{1}{2}\pi\hbar\delta(E_n - E_k).$$

Substituting this result into Eq. (17.37), we get for the probability per unit time at large times

$$W_{nk} = \frac{2\pi}{\hbar}|V_{kn}|^2\delta(E_n - E_k). \tag{17.39}$$

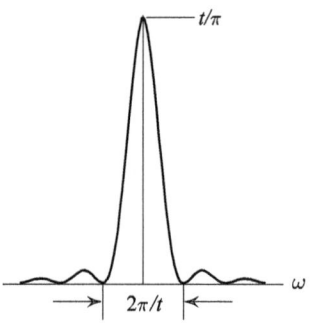

Figure 17.1 A constant perturbation V is applied during a time t. The graph shows how the transition probability per unit time between two states varies as a function of ω and becomes very sharp around $\omega = 0$ for large times.

17.1.2 Fermi's golden rule

The most important use of Eq. (17.39) is when the state k belongs to a set of final states in the continuum; for example, in problems of scattering by a short-range potential V (where the final state corresponds to a range of free particles), or when an essentially free particle is produced as the result of the interaction. In such cases, within a small interval ΔE_k of final energies there is a number $\Delta N(E_k)$ of states with different momentum, polarization, and so on, which must be taken into account. Our interest must then not be in W_{nk}, but in the quantity $\sum_{\Delta E_k} W_{nk}$ to which all final states with energy within ΔE_k contribute. If the number of states with energy in the interval $(E_k, E_k + dE_k)$ is $dN(E_k)$, we write

$$W_{nk} \to \sum_{\Delta E_k} W_{nk} \to \int_{\Delta N(E_k)} W_{nk} dN(E_k). \tag{17.40}$$

On the other hand, if $\rho(E_k)$ is the density of final states with energy E_k, that is, if the number of states $dN(E_k)$ is given by

$$dN(E_k) = \rho(E_k) dE_k, \tag{17.41}$$

(17.40) is written

$$W_{nk} = \frac{2\pi}{\hbar} \int_{E_k - \frac{1}{2}\Delta E_k}^{E_k + \frac{1}{2}\Delta E_k} |V_{nk}|^2 \delta(E_n - E_k) \rho(E_k) dE_k, \tag{17.42}$$

that is,

$$W_{nk} = \frac{2\pi}{\hbar} |V_{nk}|^2 \rho(E_k), \quad E_k = E_n. \tag{17.43}$$

This formula was called the *golden rule* by Fermi because of its numerous and important applications in the theory of transitions. Comparing Eq. (17.43) with (17.39), we see that the infinitely narrow density $\delta(E_n - E_k)$ has been replaced by the realistic density $\rho(E_k)$ of finite width. Equation (17.43) is also valid for describing first-order transitions in the reverse sense, that is, from the continuum to a stationary state. An important application of the golden rule will be discussed in Section 18.2.

17.1.3 Periodic perturbations: Resonant transitions

An important use of time-dependent perturbation theory is for periodic potentials of the form $V \cos \omega t$, which can be expressed as linear combinations of the complex Hamiltonians

$$\widehat{\mathcal{H}}^{\pm} = \widehat{V}^{\pm} e^{\pm i\omega t}. \tag{17.44}$$

As we will see in what follows, each of these two terms produces a different effect, so there is no interference between them and they can be considered separately (as was done in Exercise E17.1). Substituting into the general formula (17.13), we obtain for the coefficient of the perturbative expansion to first order

$$c_{nk}^{(1)}(t) = -\frac{i}{\hbar} V_{kn}^{\pm} \int_0^t e^{i(\omega_{kn} \pm \omega)t'} dt'. \tag{17.45}$$

From this it follows that all the results of the previous section apply to the present case with the substitution $\omega_{kn} \to \omega_{kn} \pm \omega$. In particular, the transition probability per unit time is

$$W_{nk} = \frac{2\pi}{\hbar} |V_{nk}^{\pm}|^2 \rho(E_k), \tag{17.46}$$

where the energies of the system in the initial and final states are related by $\omega_{kn} \pm \omega = 0$, that is,

$$E_k = E_n \mp \hbar \omega. \tag{17.47}$$

Equation (17.37) is written in this case (for finite time) as

$$W_{nk}^{\pm} = \frac{4}{\hbar^2} |V_{kn}^{\pm}|^2 \frac{\sin^2 \frac{1}{2}(\omega_{kn} \pm \omega)t}{(\omega_{kn} \pm \omega)^2 t}. \tag{17.48}$$

If one of the conditions $\omega_{kn} \pm \omega \simeq 0$ is fulfilled, the corresponding probability $W_{nk}^{\pm}$ increases with time and the transition becomes highly probable. In these cases we speak of resonant transitions, which are the ones most favored by the system.

Exercise E17.2. Power absorbed by a canonical ensemble of molecules
Consider a system of molecules that is perturbed by a field of frequency ω. The unperturbed state i with energy E_i contributes to the system with weight w_i. Find an expression in the Heisenberg description for the average power that the system absorbs from the field.

Solution. With the perturbation potential written in the form $\widehat{\mathcal{H}}(t) = V \cos \omega t$, it follows from the preceding discussion that the probability per unit time of a transition from state n to state k is given by

$$W_{nk} = \frac{1}{4} \frac{2\pi}{\hbar^2} |V_{nk}|^2 \left[\delta(\omega_{kn} - \omega) + \delta(\omega_{kn} + \omega) \right]. \tag{17.49}$$

Since $\hbar \omega_{kn} W_{nk}$ is the corresponding average energy absorbed per second, summing over k, we obtain the average power absorbed by the system due to all transitions induced by the perturbation. We can calculate the average power change in two equivalent ways:

(a) by summing over the absorptions from state n and subtracting the emissions leading to state n, or

(b) by summing over the absorptions from state n formally described by the two contributions V^+, V^-. Equation (17.50) shows the formal equivalence of the two methods.

Since state n occurs with probability w_n, the total average power absorbed by the system is

$$\overline{W}_{abs} = \sum_n w_n \sum_k \hbar \omega_{kn} W_{nk}$$
$$= \frac{\pi}{2\hbar} \sum_{k,n} w_n \omega_{kn} |V_{nk}|^2 [\delta(\omega_{kn} - \omega) + \delta(\omega_{kn} + \omega)]$$
$$= \frac{\pi \omega}{2\hbar} \sum_{k,n} (w_n - w_k) |V_{nk}|^2 \delta(\omega_{kn} - \omega). \tag{17.50}$$

This result is expressed in Schrödinger's description. To move to the Heisenberg description we write the delta function in terms of an integral over time and combine with the matrix elements of $\widehat{V}$ in the following form,

$$|\widehat{V}_{nk}|^2 \delta(\omega_{kn} - \omega) = \frac{1}{2\pi} \int_{-\infty}^{\infty} \langle n|\widehat{V}|k\rangle \langle k|\widehat{V} \exp [i(\omega_{kn} - \omega)t]|n\rangle dt$$
$$= \frac{1}{2\pi} \int_{-\infty}^{\infty} \langle n|\widehat{V}|k\rangle \langle k| \exp(i\frac{E_k}{\hbar}t)\widehat{V} \exp(-i\frac{E_n}{\hbar}t)|n\rangle \exp(-i\omega t) dt$$
$$= \frac{1}{2\pi} \int_{-\infty}^{\infty} \langle n|\widehat{V}|k\rangle \langle k| \exp(i\frac{\widehat{H}_0}{\hbar}t)\widehat{V} \exp(-i\frac{\widehat{H}_0}{\hbar}t)|n\rangle \exp(-i\omega t) dt, \tag{17.51}$$

where $\widehat{H}_0$ is the Hamiltonian of the unperturbed system. This result can be rewritten directly in terms of the matrix elements of $\widehat{V}$ in the Heisenberg description, $\widehat{V}_H \equiv V(t) = \exp(i\widehat{H}_0 t/\hbar)\widehat{V} \exp(-i\widehat{H}_0 t/\hbar)$. Recalling Heisenberg's general equation of evolution

$$i\hbar \frac{d\widehat{F}_H}{dt} = \left[\widehat{F}_H, \widehat{H}_H\right], \tag{17.52}$$

we get

$$\overline{W}_{abs} = \frac{\omega}{4\hbar} \sum_{n,k} \int_{-\infty}^{\infty} (w_n - w_k)\langle n|\widehat{V}(0)|k\rangle \langle k|\widehat{V}(t)|n\rangle \exp(-i\omega t) dt. \tag{17.53}$$

To give a more condensed form to this expression we note that

$$\sum_{n,k} w_n \langle n|\widehat{V}(0)|k\rangle \langle k|\widehat{V}(t)|n\rangle = \sum_n w_n \langle n|\widehat{V}(0)\widehat{V}(t)|n\rangle \equiv \langle \widehat{V}(0)\widehat{V}(t)\rangle, \tag{17.54}$$

where $\langle \widehat{A}\rangle$ represents the average $\sum_n w_n \langle n|\widehat{A}|n\rangle$, as explained in Chapter 15. Analogously,

$$\sum_{n,k} w_k \langle n|\widehat{V}(0)|k\rangle \langle k|\widehat{V}(t)|n\rangle = \sum_k w_k \langle k|\widehat{V}(t)\widehat{V}(0)|k\rangle = \langle \widehat{V}(t)\widehat{V}(0)\rangle. \tag{17.55}$$

Therefore,

$$\overline{\dot{W}}_{\text{abs}} = \frac{\omega}{4\hbar} \int_{-\infty}^{\infty} \langle \widehat{V}(0)\widehat{V}(t) - \widehat{V}(t)\widehat{V}(0) \rangle \exp(-i\omega t) dt. \tag{17.56}$$

One particular case deserves special consideration. As we have seen in Chapter 15, in a canonical ensemble at temperature T the weights w_k are given by the Maxwell–Boltzmann distribution, that is, $w_k = C \exp(-E_k/kT)$, so that

$$w_n - w_k = w_n[1 - \exp(-\hbar\omega_{kn}/kT)]. \tag{17.57}$$

Substituting in Eq. (17.53) and proceeding as before, we get

$$\overline{\dot{W}}_{\text{abs}} = \frac{\omega}{4\hbar}[1 - \exp(-\hbar\omega/kT)] \int_{-\infty}^{\infty} \langle \widehat{V}(0)\widehat{V}(t) \rangle \exp(-i\omega t) dt. \tag{17.58}$$

17.2 Radiative Transitions

Any periodic perturbation, whatever its physical nature, tends to produce resonant transitions between the various atomic states. The case of greatest interest to us here – though not the only physically interesting one – is when the perturbation is due to an external electromagnetic field. In such a case $\hbar\omega$ is the energy associated with the mode of frequency ω, which is added to or subtracted from the initial atomic energy to obtain the energy of the final state of the atom, as follows from Eq. (17.47). In the first case, that is, when $E_k = E_n + \hbar\omega$, which corresponds to the perturbation $\widehat{\mathcal{H}}^-$, the atom absorbs the energy $\hbar\omega$ from the field (this is the induced absorption). On the contrary, in the case where the perturbation is described by $\widehat{\mathcal{H}}^+$, we have $E_k = E_n - \hbar\omega$, which means that the atom gives off the energy $\hbar\omega$ to the field, giving rise to an induced emission. With the methods developed in Section 17.1, we can study such resonant atomic transitions induced by an external radiation field.

17.2.1 Induced emission and absorption of radiation

We saw in Section 12.7 that by neglecting the quadratic effects of the electromagnetic field (assuming that the field is not too strong relative to the energies involved), the Hamiltonian of the minimal interaction between an atomic electron and the field, reduced to its linear term, is

$$\widehat{\mathcal{H}} = -\frac{e}{mc} \boldsymbol{A}(\boldsymbol{r},t) \cdot \widehat{\boldsymbol{p}}, \tag{17.59}$$

with the Coulomb gauge $\nabla \cdot \boldsymbol{A} = 0$, so $\boldsymbol{A}$ and $\widehat{\boldsymbol{p}}$ commute. In source-free regions, the vector potential of a field mode of frequency ω can be written in the form

$$\boldsymbol{A} = \boldsymbol{A}_-(\boldsymbol{r})e^{-i\omega t} + \boldsymbol{A}_+(\boldsymbol{r})e^{i\omega t}, \tag{17.60}$$

where $\boldsymbol{A}_+ = \boldsymbol{A}_-^*$, since $\boldsymbol{A}$ is real. According to the previous section, the first term corresponds to energy absorption and the second to energy emission. The $\boldsymbol{A}_+, \boldsymbol{A}_-$ are solutions of the Helmholtz equation,

$$\nabla^2 A_\pm + \frac{\omega^2}{c^2} A_\pm = 0.$$

Atomic transitions are known to involve polarized field modes. We therefore take a component along a given polarization ϵ and write

$$A = A_0 \epsilon [e^{i(k \cdot r - \omega t)} + e^{-i(k \cdot r - \omega t)}], \tag{17.61}$$

where A_0 is a real and constant vector and k the wave vector along the direction of propagation and perpendicular to ϵ (the free electromagnetic field, by being massless admits only two degrees of polarization). The corresponding electric and magnetic vectors are

$$E = -\frac{1}{c} \frac{\partial A}{\partial t} = -2 A_0 k \epsilon \sin(k \cdot r - \omega t), \tag{17.62}$$

$$B = \nabla \times A = 2 A_0 (\epsilon \times k) \sin(k \cdot r - \omega t), \tag{17.63}$$

which, after averaging over time, gives an energy density per unit volume,

$$u = \frac{1}{8\pi} \overline{(E^2 + B^2)}^t = \frac{A_0^2}{2\pi} k^2, \tag{17.64}$$

from which the energy content is

$$\mathcal{E} = V u = \frac{A_0^2 V}{2\pi} k^2, \tag{17.65}$$

if u is considered homogeneous. This energy is associated to the single frequency $\omega = kc$, so that we can use Planck's formula for the spectral energy density, Eq. (2.22), and write

$$\rho(\omega) = \frac{\omega^2}{2\pi^2 c^3} \mathcal{E}. \tag{17.66}$$

Note that we have divided by two, since in (2.22) the two possible polarizations are considered, while here we have set a direction for ϵ. With $\mathcal{E}$ given by (17.65), we get

$$\rho(\omega) = \frac{V \omega^4}{4\pi^3 c^5} A_0^2. \tag{17.67}$$

In the (quantum) literature it is common to find these results expressed in terms of photons, associating the energy $\mathcal{E}$ contained in V with an average number N of photons of frequency ω, that is, $\mathcal{E} = N \hbar \omega$. Although this language is foreign to the present treatment, which describes the field in classical (nonquantized) terms, we use it here for later convenience and express the energy density as

$$\rho(\omega) = \frac{\hbar \omega^3}{2\pi^2 c^3} N. \tag{17.68}$$

It is important to distinguish between the spectral energy density $\rho(\omega)$ and the energy density of final states $\rho(E_k)$, as is discussed in what follows.

Coming back to our subject we will first discuss induced emission processes. The corresponding interaction Hamiltonian is, from (17.59)–(17.61),

$$\widehat{\mathcal{H}}^+ = -\frac{e}{mc} A_+(r,t) \cdot \widehat{p} = \widehat{V}^+ e^{i\omega t}, \tag{17.69}$$

where

$$\widehat{V}^+ = -\frac{e}{mc}A_0 e^{-i\mathbf{k}\cdot\mathbf{r}}\boldsymbol{\epsilon}\cdot\widehat{\mathbf{p}}. \qquad (17.70)$$

Equation (17.46) gives, for the probability of emission per unit time,

$$W_{nk}^{\text{em}} = \frac{2\pi}{\hbar}|\langle n|\widehat{V}^+|k\rangle|^2 \rho(E_k) = \frac{2\pi e^2 A_0^2}{m^2 \hbar c^2}|\langle n|e^{-i\mathbf{k}\cdot\mathbf{r}}\boldsymbol{\epsilon}\cdot\widehat{\mathbf{p}}|k\rangle|^2 \rho(E_k). \qquad (17.71)$$

Since the system emits a photon, there will be more photons in the final state. Normalizing within a box of volume V, the number of photons contained in d^3p is

$$d^3n = \frac{V}{(2\pi\hbar)^3}d^3p = \frac{Vp^2}{(2\pi\hbar)^3}dp\,d\Omega_p. \qquad (17.72)$$

We now rewrite this expression in the form

$$d^3n = \frac{Vp^2}{(2\pi\hbar)^3}\frac{dp}{d\mathcal{E}}d\Omega_p\,d\mathcal{E}.$$

The density of final states within the solid angle $d\Omega_p$ is obtained from Eq. (17.41), which gives

$$d\rho(\mathcal{E}) = \frac{d^3n}{d\mathcal{E}} = \frac{Vp^2}{(2\pi\hbar)^3}\frac{dp}{d\mathcal{E}}d\Omega_p. \qquad (17.73)$$

This expression is used as a starting point to calculate the density of final states in the continuum in each particular case. Since for photons, as is our case, $dp/d\mathcal{E} = d(\mathcal{E}/c)/d\mathcal{E} = 1/c$ and $p = \hbar\omega/c$, we have

$$d\rho(\mathcal{E}) = \frac{Vp^2}{8\pi^3\hbar^3 c}d\Omega_p = \frac{V\omega^2}{8\pi^3\hbar c^3}d\Omega_p. \qquad (17.74)$$

Substituting this result into Eq. (17.71) and using (17.67), we get

$$dW_{nk}^{\text{em}} = \frac{\pi e^2 \rho(\omega)}{m^2 \hbar^2 \omega^2}\left|\langle n|e^{-i\mathbf{k}\cdot\mathbf{r}}\boldsymbol{\epsilon}\cdot\widehat{\mathbf{p}}|k\rangle\right|^2 d\Omega_p. \qquad (17.75)$$

Now we focus on the matrix element. The values of $\mathbf{r}$ that contribute significantly to the integral are of the order of the atomic radius a or less, due to the exponentially decreasing shape of the wave functions. This means that if k is not too large, we can approximate the exponential in (17.75) to unity. This is the long-wavelength approximation, and as already mentioned in Section 2.2, it is perfectly valid for light atoms and radiation in the visible range; characteristic values are $a \sim 10^{-8}$ cm; $\lambda = 1/k \sim 10^{-5}$ cm, that is, $ka \sim 10^{-3}$. We can therefore consider the field to be independent of the coordinates,

$$\mathbf{A}_+(\mathbf{r},t) = A_0 \widehat{\boldsymbol{\epsilon}} e^{i\omega t}.$$

This uniform field inside the atom corresponds to the (time-dependent) interaction potential $V = -e\mathbf{E}_+\cdot\mathbf{r} = -\mathbf{E}_+\cdot\mathbf{d}$ describing the coupling of the atomic electric dipole $\mathbf{d} = e\mathbf{r}$ to the electric field. We see therefore that the long-wavelength approximation leads to the electric dipole transitions discussed previously. Eq. (17.75) is then reduced to

$$dW_{nk}^{\text{em}} = \frac{\pi e^2 \rho(\omega)}{m^2 \hbar^2 \omega^2}|\langle n|\boldsymbol{\epsilon}\cdot\widehat{\mathbf{p}}|k\rangle|^2 d\Omega_p. \qquad (17.76)$$

We use the Heisenberg equation to calculate $\langle k|\hat{\boldsymbol{p}}|n\rangle = (m/i\hbar)\langle k|[\boldsymbol{r},H]|n\rangle$ and get

$$\langle k|\boldsymbol{\epsilon}\cdot\hat{\boldsymbol{p}}|n\rangle = im\omega_{nk}\langle k|\boldsymbol{\epsilon}\cdot\boldsymbol{r}|n\rangle. \tag{17.77}$$

Further, taking into account that the atom resonates precisely at the transition frequency, $\omega = \omega_{nk}$, we get

$$dW_{nk}^{\text{em}} = \frac{\pi e^2}{\hbar^2}\rho(\omega)|\langle n|\boldsymbol{\epsilon}\cdot\boldsymbol{r}|k\rangle|^2 d\Omega_p. \tag{17.78}$$

When considering photons emitted with any polarization, one must average over the two possible values of $\boldsymbol{\epsilon}$, giving them equal weight. The average of $|\langle n|\boldsymbol{\epsilon}\cdot\boldsymbol{r}|k\rangle|^2$ can be written as $\sum_{ij}\overline{\epsilon_i\epsilon_j}\langle n|x_i|k\rangle\langle k|x_j|n\rangle$, with

$$\overline{\epsilon_x} = \overline{\epsilon_y} = \overline{\epsilon_z} = 0, \quad \overline{\epsilon_x^2} = \overline{\epsilon_y^2} = \overline{\epsilon_z^2} = \tfrac{1}{3}\overline{\epsilon^2} = \tfrac{1}{3}. \tag{17.79}$$

Moreover, transitions with different polarizations are uncorrelated, so

$$\overline{\epsilon_i\epsilon_j} = \tfrac{1}{3}\delta_{ij}, \tag{17.80}$$

whence

$$\sum_{ij}\overline{\epsilon_i\epsilon_j}\langle n|x_i|k\rangle\langle k|x_j|n\rangle = \tfrac{1}{3}\sum_i \langle n|x_i|k\rangle\langle k|x_i|n\rangle = \tfrac{1}{3}|\langle n|\boldsymbol{r}|k\rangle|^2. \tag{17.81}$$

Substituting this result into Eq. (17.78) gives

$$dW_{nk}^{\text{em}} = \frac{\pi e^2}{3\hbar^2}\rho(\omega)|\langle n|\boldsymbol{r}|k\rangle|^2 d\Omega. \tag{17.82}$$

Since the radiation is assumed to be isotropic, we can immediately perform the angular integration and thus obtain the probability of a dipolar transition per unit time from state n to state k with emission of frequency $\omega = \omega_{nk}$ in any direction and with any polarization,

$$W_{nk}^{\text{em}} = \frac{4\pi^2 e^2}{3\hbar^2}\rho(\omega)|\langle n|\boldsymbol{r}|k\rangle|^2. \tag{17.83}$$

It is clear from the preceding that the derivation of the formulas for induced absorption follows a similar pattern. Again assuming that the field is isotropic, we obtain for the probability of absorption of a photon of frequency $\omega = \omega_{nk}$ in any direction and with any polarization, since $|\langle n|\boldsymbol{r}|k\rangle|^2 = |\langle k|\boldsymbol{r}|n\rangle|^2$,

$$W_{nk}^{\text{abs}} = \frac{4\pi^2 e^2}{3\hbar^2}\rho(\omega)|\langle n|\boldsymbol{r}|k\rangle|^2. \tag{17.84}$$

17.2.2 Selection rules for dipole transitions

Equations (17.83) and (17.84) are important results: they show that an external radiation field has the same capacity to induce atomic emissions and absorptions, and they allow us to calculate the transition probability per unit time – which determines the intensity of the corresponding spectral lines – as a function of the spectral density of the field and the atom's electric dipole moment.

Note that for electric dipole transitions to occur, $\langle n|r|k\rangle$ must be different from zero. These selection rules for allowed transitions play an important role in determining the shape of atomic and molecular spectra and thus provide useful information about the system, such as the symmetries and the associated conservation laws.

For example, in Chapter 12 we obtained $|\Delta l| = 1$, $|\Delta m| = 0, 1$, which tells us that for central problems (where the orbital angular momentum is a conserved quantity) the angular momentum of the atom or molecule must change by one (in units of $\hbar$) at each transition; this is due to the fact that the emitted or absorbed photon carries an angular momentum (spin) equal to 1. In Section 9.3 we obtained $|\Delta n| = 1$ for the harmonic oscillator, due to the fact that a linear system oscillating at frequency ω only interacts with field modes of that frequency. If the potential $V(x)$ is an even function of x (which is often the case), only dipole transitions between eigenstates of different parity are allowed, due to the odd parity of the dipole moment, as illustrated in Section 2.5 (see Exercise 17.44).

17.2.3 Multipole transitions

To study the effect of the higher multipole moments, we do a Taylor series expansion of the exponential in (17.75),

$$e^{-i\boldsymbol{k}\cdot\boldsymbol{r}} = 1 - i\boldsymbol{k}\cdot\boldsymbol{r} - \tfrac{1}{2}(\boldsymbol{k}\cdot\boldsymbol{r})^2 + \ldots \qquad (17.85)$$

The first term corresponds to the dipole; for the second term we need to calculate the matrix element of the operator

$$(\boldsymbol{\epsilon}\cdot\widehat{\boldsymbol{p}})(\boldsymbol{k}\cdot\boldsymbol{r}) = \tfrac{1}{2}\left[(\boldsymbol{\epsilon}\cdot\widehat{\boldsymbol{p}})(\boldsymbol{k}\cdot\boldsymbol{r}) + (\boldsymbol{\epsilon}\cdot\boldsymbol{r})(\boldsymbol{k}\cdot\boldsymbol{p})\right] + \tfrac{1}{2}(\boldsymbol{k}\times\boldsymbol{\epsilon})\cdot(\boldsymbol{r}\times\widehat{\boldsymbol{p}}). \qquad (17.86)$$

In the first term we replace the matrix element of $\widehat{\boldsymbol{p}}$ by that of $\boldsymbol{r}$ as before (with the auxiliary introduction of a sum over intermediate states) and in the last term we introduce the magnetic vector using (17.63) instead of the product $\boldsymbol{k}\times\boldsymbol{\epsilon}$, and the orbital angular momentum instead of $\boldsymbol{r}\times\widehat{\boldsymbol{p}}$. This gives an expression of the form

$$\langle(\boldsymbol{\epsilon}\cdot\widehat{\boldsymbol{p}})(\boldsymbol{k}\cdot\boldsymbol{r})\rangle = C_1\langle(\boldsymbol{\epsilon}\cdot\widehat{\boldsymbol{r}})(\boldsymbol{k}\cdot\boldsymbol{r})\rangle + C_2\langle\boldsymbol{B}\cdot\hat{\boldsymbol{L}}\rangle. \qquad (17.87)$$

The first term obviously refers to the effect of the electric quadrupole moment ($E2$ transitions), while the second generates transitions due to the magnetic dipole moment of the atom ($M1$ transitions). In general, if the transitions are due to a moment of order 2^l, the transitions are El or Ml, depending on the case.

Continuing with this procedure, we could, in principle, calculate the transition probabilities associated with any multipole and the corresponding selection rules. This is a subject for more advanced or specialized texts and we will not insist on it here. However, we can estimate the relative order of magnitude of the different contributions without the need for explicit calculations. We do this by expanding the exponential in the form

$$e^{-i\boldsymbol{k}\cdot\boldsymbol{r}} = \sum(-i)^l j_l(kr) P_l(\cos\theta), \qquad (17.88)$$

where θ is the angle formed by the vectors k and r. For small values of the argument we can approximate the spherical Bessel functions as

$$j_l(kr) = \frac{(kr)^l}{(2l+1)!!} + \cdots \qquad (17.89)$$

Since $ka \ll 1$, only the lowest multipoles contribute significantly to the transition probability. For example, in the frequent case cited above with $ka \simeq 10^{-3}$, if we take the transition probability $E1$ as 1, that of the quadrupole transition will be $(10^{-3})^2$, a million times smaller. Therefore, the spectral lines produced by multipole transitions are much fainter, and normally visible only when the corresponding dipole transitions are absent, i.e., when they are excluded by a selection rule.

17.2.4 The Einstein A and B coefficients

In 1916, Einstein published an elegant derivation of Planck's distribution law as an additional argument in favor of his photonic theory of the radiation field, based on a statistical study of equilibrium between matter and radiation. Schematically, this important part of the derivation goes as follows.

Consider, for simplicity, an ensemble of two-level atoms with $E_n > E_k$, subject to a radiation field with spectral density $\rho(\omega)$.[1] The resonant field-atom interaction produces transitions between states k and n, at the frequency $\omega = |\omega_{kn}|$; each atom is in one of the two possible stationary states. When equilibrium has been reached at temperature T, there are N_n and N_k atoms in states n and k, respectively, with the relative populations given by the Boltzmann factor,[2]

$$N_n/N_k = e^{-(E_n - E_k)/kT}. \qquad (17.90)$$

Transitions from k to n take place when the atom absorbs energy from the field and becomes excited. Since these processes can only occur in the presence of the field, Einstein proposed to write the mean number dn_{abs} of field-induced absorptions occurring during dt in the form

$$dn_{\text{abs}} = N_k B_{kn} \rho(\omega, T) dt. \qquad (17.91)$$

The absorption probability per unit time and unit spectral density of the field, B_{kn}, is an intrinsic property of the atom and can only be determined through the quantum study of the atom-field interaction.

We can treat the downward transitions in a similar way, by writing the mean number of field-induced emissions during dt as $N_n B_{nk} \rho(\omega, T) dt$. To this we must add the number of spontaneous emissions, those that occur in the absence of an external field, which Einstein wrote as $N_n A_{nk} dt$, where A_{nk} is the probability per unit time that an atom will decay freely from n to k. This is a "spontaneous" (i.e., not induced) process, independent of the external field. Consequently, the total number of emissions during dt is

$$dn_{\text{em}} = [N_n B_{nk} \rho(\omega, T) + N_n A_{nk}] dt. \qquad (17.92)$$

---

[1] There may of course be other energy levels, but our focus will be on the transitions between levels E_n and E_k. A modern quantum-mechanical approach to the two-level atom is discussed in Section 3.
[2] We take the degeneracy to be 1 for simplicity; the result is essentially independent of this consideration.

In equilibrium, the average number of absorptions during dt must be equal to the average number of emissions, so that[3]

$$N_k B_{kn} \rho = N_n B_{nk} \rho + N_n A_{nk}. \tag{17.93}$$

If we introduce Eq. (17.72), which holds in thermal equilibrium, we get

$$e^{-E_k/kT} B_{kn} \rho = e^{-E_n/kT} (B_{nk} \rho + A_{nk}). \tag{17.94}$$

Assuming that ρ grows indefinitely with increasing temperature T (which is not only intuitively clear, but is also confirmed by the final result), (17.94) gives

$$B_{kn} = B_{nk}, \tag{17.95}$$

which means that the probabilities of induced energy absorption or emission are equal. This is a first important prediction of the theory, which is confirmed by the quantum treatment of the problem, as shown at the end of the previous section. Taking this into account, we obtain from (17.94) that

$$\rho = \frac{A_{nk}/B_{nk}}{e^{(E_n - E_k)/kT} - 1}. \tag{17.96}$$

Using Wien's displacement law

$$\rho(\omega, T) = \omega^3 f(\omega/T), \tag{17.97}$$

where f is a universal function, we infer that

$$\frac{A_{nk}}{B_{nk}} = \lambda \omega^3 \tag{17.98}$$

and

$$E_n - E_k = \hbar \omega, \tag{17.99}$$

where λ and $\hbar$ are coefficients to be determined. Note that with this derivation, Bohr's hypothesis about atomic transitions between energy levels is no longer a hypothesis. Substituting these results in (17.96) we get

$$\rho = \frac{\lambda \omega^3}{e^{\hbar \omega/kT} - 1}. \tag{17.100}$$

By comparing with the Rayleigh–Jeans approximate formula $\rho = (\omega^2/\pi^2 c^3) kT$, which is known to work well at high temperatures, we get

$$\lambda = \frac{\hbar}{\pi^2 c^3}, \tag{17.101}$$

which introduced into (17.100) gives Planck's blackbody radiation law. Further, from (17.98) and (17.101) we get

$$A_{nk} = \frac{\hbar \omega^3}{\pi^2 c^3} B_{nk}. \tag{17.102}$$

This result is extremely important because it shows that when a quantum system responds to an external electromagnetic field ($B_{nk} \neq 0$), it will also decay without

---

[3] This property is called detailed balance, because it leads to a state of equilibrium at each frequency separately; historically, this notion was introduced by Einstein precisely in this context.

external intervention ($A_{nk} \neq 0$), emitting energy until it reaches its ground state. This proves that the excited states of an atomic system are unstable, i.e., they decay "spontaneously."

The above was the first derivation of the Planck distribution free from logical inconsistencies or arbitrary adjustments. The lack of quantum formalism in 1916 prevented Einstein from establishing the formulas for the A and B coefficients separately.

17.2.5 Spontaneous emission of radiation: a semiclassical approach

In Section 2.1 the quantum-mechanical formulas for the induced transition probabilities were derived by considering the external radiation field as a perturbation. According to Eqs. (17.83) and (17.84), the Einstein B coefficient, which gives the probability of induced transitions per unit time and per unit spectral density, is

$$B_{nk} = \frac{4\pi^2 e^2}{3\hbar^2} |\langle k|\mathbf{r}|n\rangle|^2. \tag{17.103}$$

However, the Schrödinger equation dictates that the excited states are stationary in the absence of an external field, so there is no spontaneous decay in QM. This contradiction cannot be resolved within the Schrödinger formalism, which does not take full account of radiative interactions. In Sections 17.3 and 17.4 it is shown that a proper treatment of the matter-field system, including the vacuum field, satisfactorily resolves the contradiction. As shown in the following exercise, a partial solution is obtained by using semiclassical arguments to derive a formula for the A coefficient of spontaneous decay, assuming that it is due to the radiation reaction of the electron.

Exercise E17.3. Semiclassical approach to spontaneous emission
Use the classical Larmor formula for the power radiated by a nonrelativistic point charge, to derive a quantum expression for the Einstein A coefficient.

Solution. The classical formula for the average energy emitted per unit time by an electric dipole $e\mathbf{r}$ is

$$\frac{d\mathcal{E}}{dt} = \frac{2e^2}{3c^3}\overline{(\ddot{\mathbf{r}})^2}^t, \tag{17.104}$$

where the symbol $^{-t}$ means time averaging. To give this expression a quantum form we proceed as follows.

We consider an ensemble of atoms that can make transitions between states k and n, and describe the state of such an ensemble as a superposition of both states, $\psi = C_n \psi_n + C_k \psi_k$, with $|C_n|^2 + |C_k|^2 = 1$. We then replace the acceleration $\ddot{\mathbf{r}}$ by its expectation value, which is given by

$$\langle \psi|\ddot{\mathbf{r}}|\psi\rangle = \int (C_n^* \psi_n^* + C_k^* \psi_k^*)\ddot{\mathbf{r}}(C_n \psi_n + C_k \psi_k)d^3x \tag{17.105}$$

$$= |C_n|^2 \langle n|\ddot{\mathbf{r}}|n\rangle + |C_k|^2 \langle k|\ddot{\mathbf{r}}|k\rangle + C_n^* C_k \langle n|\ddot{\mathbf{r}}|k\rangle + C_n C_k^* \langle k|\ddot{\mathbf{r}}|n\rangle.$$

Applying the Heisenberg equation twice, we get

$$\langle n|\ddot{\mathbf{r}}|k\rangle = -\left(\frac{E_n - E_k}{\hbar}\right)^2 \langle n|\mathbf{r}|k\rangle = -\omega^2 \langle n|\mathbf{r}|k\rangle, \qquad (17.106)$$

where $\omega = \omega_{nk}$. Atomic states usually have well-defined parity, even or odd, so the integrand in

$$\langle n|x_i|n\rangle = \int_{-\infty}^{\infty} \psi_n^* x_i \psi_n dx_i \qquad (17.107)$$

($i = 1, 2, 3$) is odd and the integral vanishes. This means that normally an atom has no dipole electric moment in its ground state. The preceding expectation value reduces therefore to (there are two equivalent contributions)

$$\langle \psi|\ddot{\mathbf{r}}|\psi\rangle = 2\text{Re}\left(C_n^* C_k\right) \langle n|\ddot{\mathbf{r}}|k\rangle. \qquad (17.108)$$

Since the coefficients are of order unity, we can write

$$|\langle \psi|\ddot{\mathbf{r}}|\psi\rangle|^2 = C_0^2 |\langle n|\ddot{\mathbf{r}}|k\rangle|^2, \qquad (17.109)$$

where C_0^2 is a real constant of the order of unity. Substituting in Eq. (17.104) we get

$$d\mathcal{E} = \frac{2e^2 \omega^4}{3c^3} C_0^2 |\langle n|\mathbf{r}|k\rangle|^2 dt. \qquad (17.110)$$

On the other hand, the mean energy radiated during dt is the energy $\hbar\omega$ radiated in each transition multiplied by the mean number of transitions $A_{nm}dt$, so that

$$A_{nk} = \frac{2e^2 \omega^3 C_0^2}{3\hbar c^3} |\langle k|\mathbf{r}|n\rangle|^2. \qquad (17.111)$$

With B_{nk} given by (17.103), we get

$$A_{nk} = \frac{\hbar \omega^3}{2\pi^2 c^3} \frac{C_0^2}{2} B_{nk}, \qquad (17.112)$$

which compares well with the exact result, Eq. (17.102), except for the indeterminate numerical factor of order 1. In Sections 17.3 and 17.4 we return to this point to show that a complete treatment of the matter–field system gives the correct formulas for the A and B coefficients, with $C_0^2 = 2$.

Before concluding this section, we note that since A_{nk} is the probability per unit time for a spontaneous transition to occur between states n and k, its inverse is the mean time it takes for such a transition to occur. This quantity is defined as the mean lifetime $T_{n \to k}$ of the excited state n,

$$T_{n \to k} = A_{nk}^{-1}. \qquad (17.113)$$

However, it is often the case that the system can decay from state n to any one of several states k, in which case each different decay mode contributes to shortening the mean lifetime. In such a case the mean lifetime of the excited state n takes the form

$$T_n = \left(\sum_{E_k < E_n} A_{nk}\right)^{-1}. \qquad (17.114)$$

Radiative lifetimes of excited atomic states fall typically in the range of $10^{-8}-10^{-10}$ s, with the exception of metastable states (states that do not decay via a dipole transition), which can have lifetimes on the order of seconds or longer.

In what follows we apply these results to a specific example for illustrative purposes.

Exercise E17.4. Mean lifetimes for an infinite well

Using the electric dipole selection rules for a 1D infinite well, find the mean lifetime for spontaneous transitions from the excited state $n = 4$.

Solution. We saw in Chapter 5 that the eigenfunctions of $\widehat{H}$ in this case are $\varphi_n = \sqrt{\frac{2}{a}} \sin \frac{\pi n}{a} x$. This gives for the matrix element of x

$$\langle n'|x|n\rangle = \int \varphi_{n'} x \varphi_n dx = \frac{4a}{\pi^2} \frac{nn'}{(n^2-n'^2)^2}[(-1)^{n-n'}-1].$$

Therefore, spontaneous dipole emissions exist for all $n > n'$ such that $\Delta n \equiv n - n' =$ odd.

When the initial state is $n = 4$, the emission spectrum will have as dominant lines those corresponding to the frequencies $3\omega_1, 5\omega_1, 7\omega_1, 15\omega_1$, where $\omega_1 = E_1/\hbar = \pi^2\hbar/2ma^2$, see Eq. (4.73). The lines produced by higher-order transitions are superimposed on this spectrum, but are much fainter.

The initial state $n = 4$ can decay either to the state $n = 3$, or directly to the ground state ($n = 1$), so that its total probability of decay is the sum of the probabilities of each of the possible transitions (because they are mutually exclusive events), $A_{4\text{tot}} = A_{43} + A_{41}$. Therefore, the mean lifetime of the state $n = 4$ is, according to Eq. (17.114),

$$T_4 = (A_{43} + A_{41})^{-1},$$

with $A_{4k} = \frac{4e^2\omega^3}{3\hbar c^3}|\langle k|x|4\rangle|^2$.

17.2.6 A basic introduction to the laser

An important application of the previous results is the laser. We have seen in Section 17.2.1 that an electromagnetic field of frequency ω stimulates atomic emissions of that frequency. Therefore, by bringing atoms into a suitable excited state and exposing them to a field of frequency equal to the emission frequency, it is possible to generate intense, coherent and (practically) monochromatic light. The device developed on this basis is the laser (acronym for Light Amplification by Stimulated Emission of Radiation).

As an example, consider a gas laser using a He–Ne mixture; these gases are used because the 2^1s_0 and 2^3s_1 levels of He[4] coincide with the excited levels $(2p)^5(5s)$ and $(2p)^5(4s)$ of Ne, respectively, and since He is easily excited and eventually decays into

---
[4] The notation for the electronic configuration of atoms is explained in Section 16.1. Essentially it is $^{2s+1}L_j$, plus the atomic shell (a number).

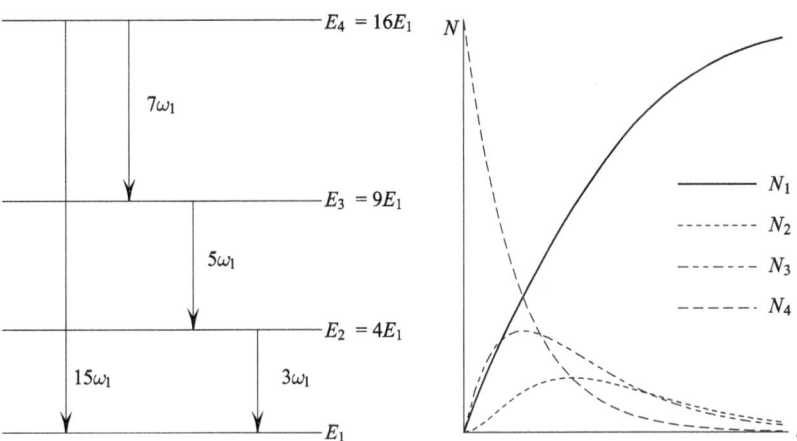

Figure 17.2 (a) Transitions between energy levels in a He–Ne laser. (b) Diagram of a He–Ne laser tube, 9–10 inches long. Before the invention of low-cost diode lasers, red He–Ne lasers were widely used in barcode scanners.

such states, collisions with Ne will transfer energy to it and leave many of its atoms in the said excited states (see Fig. 17.2). Thus, after an electric discharge, the population of excited Ne atoms will be abnormally high – an effect called population inversion. The Ne atoms decay to the $(2p)^5(4p)$ and $(2p)^5(3p)$ states ($5s \rightarrow 4p$ and $4s \rightarrow 3p$ transitions); the emitted light is reflected by mirrors at either end of the laser as shown schematically in Fig. 17.2, creating a strong field of the appropriate frequency to further stimulate emission. The process is self-stimulating, until finally the light becomes intense enough to escape through one of the mirrors, which is semi-transparent. The red emission line at 632.8 nm has a high spectral purity, long coherence length, and excellent spatial quality, making this laser a useful source for holography and as a wavelength reference for spectroscopy, albeit at the cost of very low efficiency (less than 0.025%).

Shortly after the successful operation of the first ruby and He–Ne lasers in 1960, it was demonstrated that a semiconductor diode could function as a laser. In light-emitting diodes (LEDs), which are driven either by voltage or by optical pumping, a photon of energy equal to the recombination energy between electrons and electron holes (see Section 6.3) can cause recombination by stimulated emission; the gain increases as the number of electrons and holes injected across the diode junction increases. The size of LEDs, their efficiency, and their ability to produce light of different wavelengths, make them extremely useful; but unlike a laser, the light emitted by an LED is neither spectrally coherent nor highly monochromatic.

As already explained, stimulated or induced emission was unknown in Einstein's time; in fact, it was only with his work of 1916 that the term appeared – along with the notion of spontaneous emission – as another product of Einstein's intuition, which, despite its importance, as we have just seen, is almost never cited. In a peculiar twist of meaning, recognition of his introduction of stimulated emission often leads to Einstein being credited as the inventor of the laser. (Just as the "invention" of the atomic bomb is often attributed to him as a result of his famous mass-energy equivalence formula $E = mc^2$.)

17.2.7 The photoelectric effect

Another example of important application of time-dependent perturbation theory is the photoelectric effect, a process in which an atom absorbs a photon of sufficient energy to be ionized. In order not to complicate matters, we will only consider the case of an H atom in the ground state, whose ionization energy is $I = Z^2 me^4/2\hbar^2 \simeq 13.6$ eV, and photons with $\hbar\omega \gg I$. From this relation and from the conservation of energy, which we can write in the approximate form $\hbar\omega \simeq p^2/2m$ (assuming W to be negligible in Einstein's equation $\hbar\omega - W \simeq p_{max}^2/2m$), it follows that the speed at which the electron is ejected from the atom must be high enough for it to ensure that

$$\frac{v}{c} \gg \frac{Ze^2}{\hbar c} = Z\alpha \quad \left(v = \frac{p}{m}\right). \tag{17.115}$$

We will assume that the radiation field can be described classically, while the electron obeys quantum laws. This approach differs substantially from the one discussed in Section 2.3.1, where it was shown that Einstein used the photoelectric effect precisely to support his fundamental proposal that the field must be quantized. The present calculation shows that the photoelectric effect can also be explained as being generated by the interaction of the quantized atom with the classical radiation field. The point of such a calculation is that the fundamental equation of the photoelectric effect – an example of Einstein's heuristic view – namely $\hbar\omega = T_{max} + W$, can be rewritten in the form

$$T_{max} + W = E_{out} - E_{in}, \tag{17.116}$$

where E_{out} and E_{in} are the final and initial atomic electron energies, respectively, so that $E_{out} - E_{in}$ is the energy available for the photoelectric effect. Writing it in terms of the Bohr transition frequency $\hbar\omega$, gives Einstein's formula. However, in writing the fundamental postulate in the form of (17.116), the quantization of the field is not used, but the Schrödinger equation enters instead. Several authors, including Mandel, Sudarshan, and Wolf (1964) and Lamb and Scully (1968), have argued that the correct form of this equation is precisely (17.116) and not Einstein's original, which happens to be derived, a consequence of the quantization of matter. Regardless of which principle one takes, the preceding discussion shows that it is legitimate to consider a classical radiation field and use perturbation theory to derive the probability of the photoelectric effect, which we will do here.

By considering the atom initially in its ground state, which we denote by $|0\rangle$, the wave function is

$$\langle r|0\rangle = \psi_{100}(r) = \sqrt{\frac{Z^3}{\pi a_0^3}} e^{-Zr/a_0}. \tag{17.117}$$

Since the emitted electron has a very large energy relative to the ionization energy, we can consider it to be free, so the wave function of the final state, which we denote by $|p\rangle$, is

$$\langle r|p\rangle = \frac{1}{\sqrt{V}} \exp\left(\frac{i\bm{p}\cdot\bm{r}}{\hbar}\right). \tag{17.118}$$

17.2 Radiative Transitions

A more precise theory would have to take into account that the electron is not really free but moves in the Coulomb field of the ionized atom. With the previously discussed meaning of the quantity N, the transition probability per unit time is (see Eq. (17.46))

$$W = \frac{2\pi}{\hbar}|\langle p|V^-|0\rangle|^2\rho(E_p) = \frac{4\pi^2 e^2 N}{m^2 V\omega}|\langle p|e^{i\mathbf{k}\cdot\mathbf{r}}\boldsymbol{\epsilon}\cdot\hat{\mathbf{p}}|0\rangle|^2\rho(E_p). \tag{17.119}$$

The density of final states $\rho(E_p)$ refers in this case to that of free electrons, for which $dE = v\,dp$ (both classically and relativistically), resulting in (see Section 7.1.5)

$$d\rho(E_p) = \frac{Vp^2}{8\pi^3\hbar^3}\frac{dp}{dE_p}d\Omega = \frac{Vm^2 v}{8\pi^3\hbar^3}d\Omega, \tag{17.120}$$

whence

$$dW = \frac{Ne^2 v}{2\pi\hbar^3\omega}|\langle p|e^{i\mathbf{k}\cdot\mathbf{r}}\boldsymbol{\epsilon}\cdot\hat{\mathbf{p}}|0\rangle|^2 d\Omega. \tag{17.121}$$

The integral in the matrix element

$$\langle p|e^{i\mathbf{k}\cdot\mathbf{r}}\hat{\mathbf{p}}|0\rangle = -i\hbar\sqrt{\frac{Z^3}{\pi V a_0^3}}\int \exp[-i(\mathbf{p}/\hbar - \mathbf{k})\cdot\mathbf{r}]\nabla\exp\left(-\frac{Zr}{a_0}\right)d^3r$$

$$= \sqrt{\frac{Z^3}{\pi V a_0^3}}(\mathbf{p} - \hbar\mathbf{k})\int \exp\left[-\frac{Zr}{a_0} - i(\frac{\mathbf{p}}{\hbar} - \mathbf{k})\cdot\mathbf{r}\right]d^3r \tag{17.122}$$

is solved using the formula

$$\int e^{-br+i\mathbf{q}\cdot\mathbf{r}}d^3r = \frac{4\pi}{q}\int_0^\infty re^{-br}\sin qr\,dr = \frac{8\pi b}{(b^2+q^2)^2}, \tag{17.123}$$

where $b = Z/a_0$ and

$$\mathbf{q} = (\mathbf{k} - \frac{\mathbf{p}}{\hbar}). \tag{17.124}$$

Since $\boldsymbol{\epsilon}\cdot\mathbf{k} = 0$, we get

$$\langle p|e^{i\mathbf{k}\cdot\mathbf{r}}\boldsymbol{\epsilon}\cdot\hat{\mathbf{p}}|0\rangle = 8\sqrt{\frac{\pi a_0^3}{Z^3 V}}\frac{\boldsymbol{\epsilon}\cdot\mathbf{p}}{\left(1+\frac{a_0^2 q^2}{Z^2}\right)^2}, \tag{17.125}$$

and substituting into Eq. (17.121), we finally have

$$\frac{dW}{d\Omega} = \frac{32 e^2 a_0^3 v}{Z^3 \hbar^3 \omega}\frac{(\boldsymbol{\epsilon}\cdot\mathbf{p})^2}{(1+a_0^2 q^2/Z^2)^4}\frac{N}{V}. \tag{17.126}$$

To give this result a more convenient form we observe that from $\hbar\omega \approx p^2/2m$ it follows that $p^2 \approx 2m\hbar\omega = 2mc\hbar k \gg p\hbar k$, thus $mc \gg p$; from this same relation it also follows that $2\hbar k/p = p/mc = v/c$. Since $p \gg \hbar k$, we can write from (17.124)

$$q^2 \approx (p^2/\hbar^2) - 2(pk\cos\theta/\hbar) = (p^2/\hbar^2)(1 - \frac{v}{c}\cos\theta), \tag{17.127}$$

where θ is the angle between $\mathbf{k}$ and $\mathbf{p}$. Further, from condition (17.115) it follows that $pa_0/Z\hbar = (1/Z\alpha)(v/c) \gg 1$, so that

$$1 + \frac{q^2 a_0^2}{Z^2} \simeq \frac{q^2 a_0^2}{Z^2} = \frac{p^2 a_0^2}{Z^2\hbar^2}\left(1 - \frac{v}{c}\cos\theta\right). \tag{17.128}$$

On the other hand, $\boldsymbol{\epsilon} \cdot \boldsymbol{p} = p \sin\theta \sin\phi$, where ϕ is the angle between $\boldsymbol{\epsilon}$ and the component of $\boldsymbol{p}$ perpendicular to $\boldsymbol{k}$. Substituting all these expressions into equation (17.126), we get

$$\frac{d\sigma}{d\Omega} \equiv \frac{(dW/d\Omega)}{j_{\text{inc}}} = \frac{V}{Nc}\frac{dW}{d\Omega} = \frac{32\alpha\hbar}{mck}\left(\frac{\hbar Z}{a_0 p}\right)^5 \frac{\sin^2\theta \sin^2\phi}{\left(1 - \frac{v}{c}\cos\theta\right)^4}. \tag{17.129}$$

We have divided $dW/d\Omega$ by the incident photon flux $j_{\text{inc}} = v_{\text{inc}}\rho_{\text{inc}} = c(N/V)$ to obtain the mean number of photoelectrons emitted in the element of solid angle $d\Omega$ per unit of incident flux. This quantity, which we have denoted as $d\sigma/d\Omega$, is called the differential cross section (of the photoelectric effect in this case). Due to the normalization introduced by dividing by j_{inc}, the differential cross section is independent of the illumination intensity and other incidental details of the experimental setup, so it is used as a properly normalized measure of the effective cross section.[5] The expressions obtained show that at low energies the photoelectrons are emitted preferentially in the polarization direction (in which $(\boldsymbol{\epsilon} \cdot \boldsymbol{p})^2$ takes its maximum value p^2), of the absorbed electric field. As the energy increases, the denominator $\left(1 - \frac{v}{c}\cos\theta\right)^4$ shifts this maximum in the forward direction (defined by the photoelectron). Furthermore, the cross section increases with the atomic number as Z^5 and decreases with the energy of the incident photon as $k^{-1}p^{-5} \sim \omega^{-7/2}$. All of these quantum predictions for the photoelectric effect are well verified in experiment.

17.3 Radiative Corrections: The Full Picture

We have seen in Section 17.2 that QM leads naturally to Einstein's (stimulated transition) B coefficients, but to obtain a reasonably approximate formula for the (spontaneous transition) A coefficients we resorted to arguments borrowed from classical physics. We will now proceed to derive the A_{nk} from a fundamental theory.

17.3.1 Dirac: The Formal Field Quantization

In 1927, Dirac was able to weave the photon concept into the fabric of the new QM and to describe the interaction of photons with matter. He found that the formula for the spontaneous transitions could be obtained by using a quantized field instead of a classical field to perturb the atomic electron. Dirac took the phases and energies of the field modes and interpreted them as operators related by commutators. Nowadays it is more common to quantize the (complex) Fourier amplitudes of the vector potential. The quantization of the field is thus formally achieved by replacing the classical oscillators describing the field by quantum oscillators, that is, by rewriting Eq. (17.61) in the form

$$\widehat{\boldsymbol{A}} = \widehat{a}\,\boldsymbol{\epsilon}\, A_-(\boldsymbol{r})e^{-i\omega t} + \widehat{a}^\dagger \boldsymbol{\epsilon}\, A_+(\boldsymbol{r})e^{i\omega t}, \tag{17.130}$$

---

[5] This important concept is discussed in detail in Section 18.1.

where $A_{\pm}(r) = A_0 e^{\mp i k \cdot r}$, the operators $\widehat{a}$ and $\widehat{a}^{\dagger}$ (corresponding to the mode of frequency ω) satisfy the commutators

$$[a, \widehat{a}^{\dagger}] = 1, \tag{17.131}$$

and operators pertaining to different field modes commute with each other. Note that these are the same as the lowering and raising operators used in Section 9.4 to describe the (mechanical) HO of frequency ω. In the present case, it is appropriate to speak of annihilation and creation because that is what $\widehat{a}$ and $\widehat{a}^{\dagger}$ do: applying $\widehat{a}$ decreases the number of photons by 1, and applying $\widehat{a}^{\dagger}$ increases it by 1. As in the case of the HO, n successive applications of $\widehat{a}$ to an initially excited state $|n\rangle$ of a given mode of the field eventually produces the state with zero photons, that is, the electromagnetic vacuum (or vacuum fluctuations, as it is called in quantum theory). Conversely, n successive applications of $\widehat{a}^{\dagger}$ to a field mode initially in the vacuum state $|0\rangle$ will bring the field to the excited state $|n\rangle$.

The issue of spontaneous transitions is solved as follows. As we can see from Eq. (17.130), A_- is accompanied by $\widehat{a}$ and A_+ is accompanied by $\widehat{a}^{\dagger}$, so using Eq. (17.68), we can replace N by the expectation value $\langle N|\widehat{a}^{\dagger}\widehat{a}|N\rangle = N$ in Eq. (17.84) for the probability of resonant absorption of photons in the form

$$W_{nk}^{\text{abs}} = \frac{4e^2\omega^3}{3\hbar c^3} N |\langle n|r|k\rangle|^2, \tag{17.132}$$

while in the formula for the probability of emission, Eq. (17.83), we must introduce $\langle N|\widehat{a}\widehat{a}^{\dagger}|N\rangle = \langle N|\widehat{a}^{\dagger}\widehat{a} + 1|N\rangle = N + 1$,

$$W_{nk}^{\text{em}} = \frac{4e^2\omega^3(N+1)}{3\hbar c^3} |\langle n|r|k\rangle|^2. \tag{17.133}$$

In the absence of an external field, that is, with zero photons ($N = 0$), this gives

$$W_{nk}^{\text{em}}(N = 0) \equiv A_{nk} = \frac{4e^2\omega^3}{3\hbar c^3} |\langle n|r|k\rangle|^2 = \frac{\hbar\omega^3}{\pi^2 c^3} B_{nk}, \tag{17.134}$$

in exact agreement with Eq. (17.102). This result confirms that the quantization of the radiation field – involved in the transition to quantum electrodynamics (QED) – is sufficient to obtain the coefficient for the spontaneous decay of atomic excited states, among other important phenomena that will be discussed in what follows.

We thus see that if the quantum properties of the field are properly taken into account, one obtains a theory that is free of difficulties (well, not absolutely free, as will be discussed later). Formally, we can say that we have a complete quantum theory.

Note that the transformation of the vector potential A into an operator transforms Maxwell's equations from relations between functions into (equivalent) relations between operators. The vectors E and B also become operators (linear in $\widehat{a}$ and $\widehat{a}^{\dagger}$) which do not commute with each other or with the field Hamiltonian $(1/8\pi)(\widehat{E}^2 + \widehat{B}^2)$, which now takes the form

$$\widehat{H}_{k\lambda} = \hbar\omega \left(\widehat{a}_{k\lambda}^{\dagger} \widehat{a}_{k\lambda} + \tfrac{1}{2} \right) \tag{17.135}$$

for every field mode of frequency ω, propagation direction $\mathbf{k}$ and polarization λ. The step from classical electrodynamics to QED is thus entirely analogous and just as fundamental (and therefore has similar implications) as the step from classical mechanics to QM.

An exposition of the physical rationale for this transformation is left for Section 17.4.

Exercise E17.5. Field commutators in the Coulomb gauge

Find the values of the field commutators, $\left[\hat{A}_i(\mathbf{r}), \hat{A}_j(\mathbf{r}')\right]$, $\left[\hat{A}_i(\mathbf{r}), \hat{E}_j(\mathbf{r}')\right]$, where i, j refer to the Cartesian field components in the Coulomb gauge, expressed in terms of the annihilation and creation operators $\hat{a}$ and $\hat{a}^\dagger$ of different field modes (m, n).

Solution. Let us consider two linear field operators $\hat{F}$ and $\hat{G}$, represented by

$$F_i = \sum_m \left[f_{im} \hat{a}_m + f_{im}^* \hat{a}_m^\dagger \right], \tag{17.136}$$

$$G_j = \sum_n \left[g_{jn} \hat{a}_n + g_{jn}^* \hat{a}_n^\dagger \right]. \tag{17.137}$$

Their commutator is $\left[\hat{F}_i, \hat{G}_j\right]$.

$= \sum \left\{ f_{im} g_{jn} [\hat{a}_m, \hat{a}_n] + f_{im} g_{jn}^* [\hat{a}_m, \hat{a}_n^\dagger] + f_{im}^* g_{jn} [\hat{a}_m^\dagger, \hat{a}_n] + f_{im}^* g_{jn}^* [\hat{a}_m^\dagger, \hat{a}_n^\dagger] \right\}$; with the usual commutation rules (see (17.131)) this reduces to

$$\left[\hat{F}_i, \hat{G}_j\right] = \sum_m \left[f_{im} g_{jm}^* - f_{im}^* g_{jm} \right]. \tag{17.138}$$

For $\hat{F}_i = \hat{A}_i(\mathbf{r})$ we have

$$f_{im} = A_{\omega_m} \epsilon_{im} e^{i\mathbf{k}_m \cdot \mathbf{r}}, \tag{17.139}$$

and similarly for the rest. Thus,

$$\left[\hat{A}_i(\mathbf{r}), \hat{A}_j(\mathbf{r}')\right] = -\sum_{\mathbf{k}\lambda} A_\omega^2 \epsilon_i \epsilon_j (e^{i\mathbf{k}\cdot\mathbf{r}} e^{-i\mathbf{k}\cdot\mathbf{r}'} - e^{-i\mathbf{k}\cdot\mathbf{r}} e^{i\mathbf{k}\cdot\mathbf{r}'}) = 0, \tag{17.140}$$

since the modes with $\mathbf{k}$ and $-\mathbf{k}$ contribute equally but with opposite sign.

Things change in the case of the commutator $\left[\hat{A}_i(\mathbf{r}), \hat{E}_j(\mathbf{r}')\right]$ due to the extra factor $i\omega$,

$$\left[\hat{A}_i(\mathbf{r}), \hat{E}_j(\mathbf{r}')\right] = -\sum_{\mathbf{k}\lambda} i\omega A_\omega^2 \epsilon_i \epsilon_j (e^{i\mathbf{k}\cdot\mathbf{r}} e^{-i\mathbf{k}\cdot\mathbf{r}'} + e^{-i\mathbf{k}\cdot\mathbf{r}} e^{i\mathbf{k}\cdot\mathbf{r}'})$$

$$= -2i \sum_{\mathbf{k}\lambda} \frac{\hbar}{2\epsilon_o L^3} \epsilon_i \epsilon_j e^{i\mathbf{k}\cdot(\mathbf{r}-\mathbf{r}')}.$$

The two polarization vectors and the normalized vector $\mathbf{k} = \mathbf{k}/|\mathbf{k}|$ form an orthonormal basis, so that $\sum_\lambda \epsilon_i \epsilon_j = \delta_{ij} - \frac{k_i k_j}{k^2} = \delta_{ij}^\perp$, and

$$\left[\hat{A}_i(\mathbf{r}), \hat{E}_j(\mathbf{r}')\right] = -2i \sum_{\mathbf{k}} \frac{\hbar}{2\epsilon_o L^3} \delta_{ij}^\perp e^{i\mathbf{k}\cdot(\mathbf{r}-\mathbf{r}')}.$$

When there is a continuum of field modes, the sum is replaced by the corresponding integral; thus,
$\left[\hat{A}_i(r), \hat{E}_j(r')\right] = \frac{\hbar}{i\epsilon_o} \frac{1}{(2\pi)^3} \int d^3k \delta_{ij}^\perp e^{ik\cdot(r-r')}$. The function $\delta_{ij}^\perp = (\delta_{ij} - \frac{k_i k_j}{k^2})$ is the transverse delta in reciprocal space; in real space it is given by

$$\delta_{ij}^\perp(r - r') = \frac{1}{(2\pi)^3} \int d^3k \delta_{ij}^\perp e^{ik\cdot(r-r')}$$

(see Mathematical Appendix A), and therefore

$$\left[\hat{A}_i(r), \hat{E}_j(r')\right] = \frac{\hbar}{i\epsilon_o} \delta_{ij}^\perp(r - r').$$

17.3.2 The two-level atom: Jaynes–Cummings model

We are now in a position to make a fully quantum treatment of atoms interacting with an electromagnetic field. To simplify the discussion, we will consider a two-level atom and restrict our attention to its dipole interaction with a single field mode $k\lambda$ of frequency ω, which we take to be close to the atomic transition frequency ω_0. This would be the case, for example, if the atoms were exposed to a laser whose radiation can be regarded as monochromatic to a good approximation.

The Hamiltonian of the system composed of an atomic electron interacting with the quantized field mode in the Coulomb gauge ($\nabla \cdot A = 0$) is then (see Section 12.7)

$$\hat{H} = \hat{H}_0 + \hat{H}_R + \hat{H}_I, \qquad (17.141)$$

with

$$\hat{H}_0 = \frac{\hat{p}^2}{2m} + \hat{V}(r), \qquad (17.142)$$

$$\hat{H}_R = \hbar\omega \left(\hat{a}_{k\lambda}^\dagger \hat{a}_{k\lambda} + \tfrac{1}{2}\right), \qquad (17.143)$$

and, neglecting the quadratic term of the field as usual,

$$\hat{H}_I = -\frac{e}{mc} \hat{A} \cdot \hat{p}, \qquad (17.144)$$

with

$$\hat{A}(r, t) = \sqrt{\frac{2\pi\hbar c^2}{V\omega}} \epsilon_{k\lambda} \left(\hat{a}_{k\lambda} e^{ik\cdot r - i\omega_k t} + \hat{a}_{k\lambda}^\dagger e^{-ik\cdot r + i\omega_k t}\right), \qquad (17.145)$$

where the coefficient has been selected such that the energy of the field mode is $\mathcal{E} = \hbar\omega$, hence,

$$\hat{H}_I = -\frac{e}{mc}\sqrt{\frac{2\pi\hbar c^2}{V\omega}} \left(\hat{a}_{k\lambda}(t) + \hat{a}_{k\lambda}^\dagger(t)\right) \hat{p} \cdot \epsilon_{k\lambda}. \qquad (17.146)$$

The $\hat{a}_{k\lambda}, \hat{a}_{k\lambda}^\dagger$ operate on the Hilbert space of the field oscillators, while particle operators act on their own Hilbert space, so that the Hilbert space of the whole system is the direct product of both spaces (now subspaces). Therefore, the components of the system's state vector will be written as products of two vectors, one in each of these subspaces. For the sake of simplicity, we will omit the indices $k\lambda$ that denote the field mode.

If we call $|i\rangle$ the eigenstates of the atomic Hamiltonian, we can express the latter in the form

$$\hat{H}_0 = \sum_i \sum_j |i\rangle \langle i| \hat{H}_{at} |j\rangle \langle j| = \sum_{i,j} \left(\langle i| \hat{H}_{at} |j\rangle\right) |i\rangle \langle j| = \sum_{i,j} E_j \delta_{ij} |i\rangle \langle j| \quad (17.147)$$

or, introducing the elementary operators $\hat{\sigma}_{ij} \equiv |i\rangle \langle j|$,

$$\hat{H}_0 = \sum_i E_i |i\rangle \langle i| = \sum_i E_i \hat{\sigma}_{ii}. \quad (17.148)$$

The momentum operator in its turn takes the form

$$\hat{p} = \sum_i \sum_j |i\rangle \langle i| \hat{p} |j\rangle \langle j| = \sum_{i,j} p_{ij} |i\rangle \langle j| = \sum_{i,j} p_{ij} \hat{\sigma}_{ij}. \quad (17.149)$$

With this, the total Hamiltonian becomes

$$\hat{H} = \sum_i E_i \hat{\sigma}_{ii} + \hbar\omega \left(\hat{a}^\dagger \hat{a} + \tfrac{1}{2}\right) - \hbar \sum_{i,j} C_{ij} \left(\hat{a} + \hat{a}^\dagger\right) \hat{\sigma}_{ij}, \quad (17.150)$$

where

$$C_{ij} = \frac{e}{m}\sqrt{\frac{2\pi}{V\hbar\omega}} p_{ij} \cdot \boldsymbol{\epsilon} = ie\sqrt{\frac{2\pi}{V\hbar\omega}} \omega_{ij} x_{ij} \cdot \boldsymbol{\epsilon}. \quad (17.151)$$

The last equality follows directly from the Heisenberg equations of motion, which give $p_{ij} = im\omega_{ij}x_{ij}$.

Our atom has only two active states, which we call $|1\rangle = |-\rangle$ and $|2\rangle = |+\rangle$, where $E_2 > E_1$ and $\omega_{21} = \omega_0 = (E_2 - E_1)/\hbar$. So the corresponding completeness relation reduces to $\hat{\sigma}_{11} + \hat{\sigma}_{22} = |+\rangle\langle+| + |-\rangle\langle-| = P_2 + P_1 = \mathbb{I}$, where P_i ($i = 1, 2$) are projection operators, and from Eq. (17.148) we get

$$\sum_i E_i \hat{\sigma}_{ii} = E_1 \hat{\sigma}_{11} + E_2 \hat{\sigma}_{22} = \tfrac{1}{2} E_1 \left(\mathbb{I} + \hat{\sigma}_{11} - \hat{\sigma}_{22}\right) + \tfrac{1}{2} E_2 \left(\mathbb{I} + \hat{\sigma}_{22} - \hat{\sigma}_{11}\right)$$

$$= \tfrac{1}{2}(E_1 + E_2) + \tfrac{1}{2}(E_2 - E_1)\left(\hat{\sigma}_{22} - \hat{\sigma}_{11}\right), \quad (17.152)$$

that is,

$$\sum_i E_i \hat{\sigma}_{ii} = E_0 \hat{\sigma}_0 + \tfrac{1}{2}\hbar\omega_0 \hat{\sigma}_3, \quad (17.153)$$

where E_0 is the mean energy of the two states; $\hat{\sigma}_0 \equiv \mathbb{I}$ is the unit matrix, and

$$\hat{\sigma}_3 = \hat{\sigma}_{22} - \hat{\sigma}_{11} = |+\rangle\langle+| - |-\rangle\langle-|. \quad (17.154)$$

Note that this operator describes the difference in population between states, since $\langle\hat{\sigma}_3\rangle = P_2 - P_1$. Assuming (as is usual) $x_{11} = x_{22} = 0$, $x_{12} = x_{21}$, and hence $p_{12} = -p_{21}$, (17.149) reduces to

$$\sum_{i,j} C_{ij} \hat{\sigma}_{ij} = C_{12}\hat{\sigma}_{12} + C_{21}\hat{\sigma}_{21} = -C_{21}\left(\hat{\sigma}_{12} - \hat{\sigma}_{21}\right) = -C\hat{\sigma}_2, \quad (17.155)$$

where

$$C = -iC_{21} = e\omega_0 \sqrt{\frac{2\pi}{V\hbar\omega}} x_{12} \cdot \epsilon \tag{17.156}$$

and

$$\hat{\sigma}_2 = i\left(\hat{\sigma}_{12} - \hat{\sigma}_{21}\right) = i\left(|-\rangle\langle+| - |+\rangle\langle-|\right). \tag{17.157}$$

Thus, (17.150) takes the form

$$\widehat{H} = E_0\hat{\sigma}_0 + \tfrac{1}{2}\hbar\omega_0\hat{\sigma}_3 + \hbar\omega\left(\hat{a}^\dagger\hat{a} + \tfrac{1}{2}\right) + \hbar C\left(\hat{a} + \hat{a}^\dagger\right)\hat{\sigma}_2. \tag{17.158}$$

In most applications, the additive constants E_0 and $\tfrac{1}{2}\hbar\omega$ do not play a dynamical role, so that the Hamiltonian can be redefined by subtracting these quantities,[6]

$$\widehat{H} = \tfrac{1}{2}\hbar\omega_0\hat{\sigma}_3 + \hbar\omega\hat{a}^\dagger\hat{a} + i\hbar C\left(\hat{a} + \hat{a}^\dagger\right)\left(\hat{\sigma} - \hat{\sigma}^\dagger\right), \tag{17.159}$$

where

$$\hat{\sigma} \equiv \hat{\sigma}_{12} = |-\rangle\langle+|, \quad \hat{\sigma}^\dagger \equiv \hat{\sigma}_{21} = |+\rangle\langle-| \tag{17.160}$$

are the atomic raising and lowering operators, which act on states $|+\rangle$ and $|-\rangle$ as usual,

$$\hat{\sigma}|-\rangle = |-\rangle\langle+|-\rangle = 0, \quad \hat{\sigma}|+\rangle = |-\rangle\langle+|+\rangle = |-\rangle, \tag{17.161}$$

$$\hat{\sigma}^\dagger|-\rangle = |+\rangle\langle-|-\rangle = |+\rangle, \quad \hat{\sigma}^\dagger|+\rangle = |+\rangle\langle-|+\rangle = 0. \tag{17.162}$$

We note that

$$[\hat{\sigma}, \hat{\sigma}^\dagger] = -\hat{\sigma}_3; \quad [\hat{\sigma}, \hat{\sigma}_3] = 2\hat{\sigma}, \quad [\hat{\sigma}^\dagger, \hat{\sigma}_3] = -2\hat{\sigma}^\dagger, \tag{17.163}$$

and more generally,

$$[\hat{\sigma}_{ij}, \hat{\sigma}_{kl}] = [\,|i\rangle\langle j|, |k\rangle\langle l|\,] = \delta_{jk}\hat{\sigma}_{il} - \delta_{il}\hat{\sigma}_{kj}. \tag{17.164}$$

Since the field and particle operators at equal times act on different spaces, their commutators are zero, for example, $[\hat{\sigma}_{ij}(t), \hat{a}(t)] = 0$. However, this is no longer true at different times, due to the interaction between the parts of the system, and the commutation relations must then be derived using the equations of motion, which give

$$\dot{\hat{a}} = \frac{1}{i\hbar}[\hat{a}, \widehat{H}] = -i\omega\hat{a} + C\left(\hat{\sigma} - \hat{\sigma}^\dagger\right). \tag{17.165}$$

$$\dot{\hat{\sigma}} = \frac{1}{i\hbar}[\hat{\sigma}, \widehat{H}] = -i\omega_0\hat{\sigma} + C\left(\hat{a} + \hat{a}^\dagger\right)\hat{\sigma}_3, \tag{17.166}$$

$$\dot{\hat{\sigma}}_3 = \frac{1}{i\hbar}[\hat{\sigma}_3, \widehat{H}] = -2C\left(\hat{a} + \hat{a}^\dagger\right)\left(\hat{\sigma} + \hat{\sigma}^\dagger\right), \tag{17.167}$$

and their adjoints.

The interaction Hamiltonian in Eq. (17.159) has four terms,

$$\left(\hat{a} + \hat{a}^\dagger\right)\left(\hat{\sigma} - \hat{\sigma}^\dagger\right) = \hat{a}\hat{\sigma} - \hat{a}\hat{\sigma}^\dagger + \hat{a}^\dagger\hat{\sigma} - \hat{a}^\dagger\hat{\sigma}^\dagger. \tag{17.168}$$

---

[6] This simplification is common in the literature. However, it should be borne in mind that the term associated with the zero-point energy, which in some cases may play a physical role, has been omitted.

From Eqs. (17.165) and (17.166) it follows that in the absence of coupling, $\hat{a}(t) = \hat{a}(0)e^{-i\omega t}$ and $\hat{\sigma}(t) = \hat{\sigma}(0)e^{-i\omega_0 t}$. Therefore, for sufficiently weak coupling, the terms $\hat{a}\hat{\sigma}$ and $\hat{a}^\dagger \hat{\sigma}^\dagger$ oscillate rapidly with frequency $\omega_0 + \omega$, while the terms $\hat{a}\hat{\sigma}^\dagger$ and $\hat{a}^\dagger \hat{\sigma}$ oscillate with significantly lower frequency $\omega_0 - \omega$, which vanishes at exact resonance. In the rotating-wave approximation introduced in Section 14.3.3, the nonresonant contributions are neglected and the Hamiltonian becomes reduced to

$$\widehat{H} = \tfrac{1}{2}\hbar\omega_0 \hat{\sigma}_3 + \hbar\omega \hat{a}^\dagger \hat{a} + i\hbar C \left(\hat{a}^\dagger \hat{\sigma} - \hat{a}\hat{\sigma}^\dagger \right). \tag{17.169}$$

This is known as the Jaynes–Cummings model. The term $\hat{a}^\dagger \hat{\sigma}$ contains a field-creation operator and an atom-lowering operator; the interaction $\hat{a}\hat{\sigma}^\dagger$ in turn describes the annihilation of a photon and the excitation of the atom. In other words, atomic excitation always involves the annihilation of a photon (term $\hat{a}\hat{\sigma}^\dagger$) and de-excitation involves the production of a photon (term $\hat{a}^\dagger \hat{\sigma}$). This would not be the case if the previously neglected terms $\hat{a}\hat{\sigma}$ and $\hat{a}^\dagger \hat{\sigma}^\dagger$ were retained. Because of this property, at resonance ($\omega = \omega_0$) the operator

$$\hat{N}(t) = \tfrac{1}{2}\hat{\sigma}_3 + \hat{a}^\dagger(t)\hat{a}(t) \tag{17.170}$$

is an integral of motion, with eigenvalues $N \pm \tfrac{1}{2}$ (N is the number of equivalent photons). For more details see Milonni (2013), chapter 6.

The equations of motion of the Jaynes–Cummings model are the Heisenberg equations for the operators $\hat{a}, \hat{a}^\dagger, \hat{\sigma}_3, \hat{\sigma}$ and $\hat{\sigma}^\dagger$. Using the commutation relations established earlier, one obtains

$$\widehat{\dot{a}} = -i\omega a + C\hat{\sigma}, \tag{17.171}$$

$$\widehat{\dot{\sigma}} = -i\omega \sigma + C\hat{a}\hat{\sigma}_3, \text{ and} \tag{17.172}$$

$$\widehat{\dot{\sigma}}_3 = -2C \left(\hat{a}^\dagger \hat{\sigma} - \hat{a}\sigma^\dagger \right) \tag{17.173}$$

and their adjoints. These equations, as well as the Jaynes–Cummings model, are used in a wide variety of problems in atomic physics, solid-state physics, quantum information circuits, quantum optics, and related fields.

17.4 The Radiative Corrections According to SED. Atomic Lifetimes and Lamb Shift

In QED, the vacuum fluctuations are the source of radiative corrections in an already quantized system, that is, these corrections are considered to be a primary effect of the vacuum. In contrast, according to SED, the vacuum fluctuations are responsible for quantizing the system in the first place, so that the radiative corrections appear as a secondary effect of the fluctuating vacuum.

In the following, we will take advantage of the physical transparency of the SED framework to derive the formulas for the radiative corrections.

17.4.1 Radiative transitions

In this section we will obtain a formula for the spontaneous transition probability, using the results of Section 10.3. To do this, we recall the SED evolution equation for a constant of motion $\xi(x, p)$, Eq. (10.92),

$$\frac{d}{dt}\langle \xi \rangle = m\tau \left\langle \dddot{x}_i \frac{\partial \xi}{\partial p_i} \right\rangle - e^2 \left\langle \frac{\partial \xi}{\partial p_i} \hat{D}_i \right\rangle, \qquad (17.174)$$

with

$$e^2 \hat{D}_i = D_{ij}^{px} \frac{\partial}{\partial x_j} + D_{ij}^{pp} \frac{\partial}{\partial p_j}. \qquad (17.175)$$

Applied to the particle Hamiltonian $H(p, x) = (1/2m)p^2 + V(x)$, Eq. (17.174) gives, after integrating by parts the second term on the r.h.s.,

$$\frac{d}{dt}\langle H \rangle = \tau \langle \dddot{x} \cdot p \rangle + \frac{1}{m}\text{tr}\langle D^{pp} \rangle, \qquad (17.176)$$

where $\tau = 2e^2/3mc^3$. We will work in 1D, for simplicity in the exposition. To calculate the last term we use Eq. (10.84),

$$D^{pp}(t) = e^2 \int_{-\infty}^{t} ds \left. \frac{\partial p(t)}{\partial p(s)} \right|_{x^{(0)}} \langle E(s)E(t) \rangle,$$

where the field correlation $\langle E(s)E(t) \rangle$ is given by

$$\langle E(s)E(t) \rangle = \varphi(t - s) = \frac{4\pi}{3}\int_0^{\infty} d\omega \rho(\omega) \cos \omega(t - s), \qquad (17.177)$$

according to Eq. (10.46), and $\varphi(t - s)$ is the spectral energy function of the ZPF. This gives

$$D^{pp}(t) = e^2 \int_{-\infty}^{t} ds \varphi(t - s) \left. \frac{\partial p(t)}{\partial p(s)} \right|_{x^{(0)}}, \qquad (17.178)$$

where the derivative inside the integral, $\partial p(t)/\partial p(s) = \{x(s), p(t)\}$, must be replaced by the corresponding commutator divided by $i\hbar$, according to (10.125); hence,

$$D^{pp}(t) = \frac{e^2}{i\hbar} \int_{-\infty}^{t} ds\, \varphi(t - s) [\hat{x}(s), \hat{p}(t)]. \qquad (17.179)$$

When the system in a stationary state n, Eq. (17.176), takes the form

$$\frac{d}{dt}\langle \hat{H} \rangle_n = \tau \left\langle n \left| \dddot{\hat{x}} \cdot \hat{p} \right| n \right\rangle + \frac{1}{m}\left\langle n \left| \hat{D}^{pp} \right| n \right\rangle, \qquad (17.180)$$

with

$$\left\langle n \left| \hat{D}^{pp} \right| n \right\rangle = \frac{e^2}{i\hbar} \int_{-\infty}^{t} ds\, \varphi(t - s) \langle n |[\hat{x}(s), \hat{p}(t)]| n \rangle. \qquad (17.181)$$

Writing the matrix elements of $\hat{x}(s), \hat{p}(t)$ as $x_{kn}(s) = x_{kn}\exp(i\omega_{kn}s)$, $p_{kn}(t) = im\omega_{kn}x_{kn}\exp(i\omega_{kn}t)$, we get

$$\langle n |[\hat{x}(s), \hat{p}(t)]| n \rangle = 2im \sum_k \omega_{kn} |x_{kn}|^2 \cos \omega_{kn}(t - s), \qquad (17.182)$$

which, when introduced into (17.181), gives

$$\langle n | D^{pp} | n \rangle = \frac{2me^2}{\hbar} \sum_k \omega_{kn} |x_{kn}|^2 \int_{-\infty}^t ds\, \varphi(t-s) \cos \omega_{kn}(t-s). \qquad (17.183)$$

Further, using (17.177) with $\rho_0(\omega) = \hbar\omega^3/2\pi^2 c^3$ for the ZPF and taking into account that

$$\int_{-\infty}^t ds\, \cos\omega(t-s)\cos\omega_{kn}(t-s) = \frac{\pi}{2}[\delta(\omega-\omega_{kn}) + \delta(\omega+\omega_{kn})], \qquad (17.184)$$

we finally obtain

$$\langle n | D^{pp} | n \rangle = -m^2\tau \sum_k \omega_{nk}^4 |x_{nk}|^2 \, \text{sign}\,\omega_{nk}. \qquad (17.185)$$

For the radiation reaction term in (17.180), again using $x_{kn}(t) = x_{kn}\exp(i\omega_{kn}t)$, we obtain

$$\tau(\widehat{\dot{x}}\,\hat{p})_{nn} = -m\tau \sum_k \omega_{nk}^4 |x_{nk}|^2. \qquad (17.186)$$

Note that Eq. (17.185) contains a mixture of positive and negative terms, whereas in (17.186) all contributions have the same sign. The two contributions taken together give

$$\frac{d\langle H\rangle_n}{dt} = -m\tau \sum_k \omega_{nk}^4 |x_{nk}|^2 (1 - \text{sign}\,\omega_{kn})$$

$$= -2m\tau \sum_{k<n} \omega_{nk}^4 |x_{nk}|^2. \qquad (17.187)$$

This result confirms that while the ground state ($n=0$) is strictly stationary – due to the equilibrium between the radiation reaction and the diffusive term – the excited states ($n>0$) are not stationary; transitions from state n to lower-energy states k take place by the combined action of the ZPF and the radiation reaction (both factors contributing in equal measure), without the intervention of an external radiation field. Writing the average energy loss per unit time as

$$\frac{d\langle H\rangle_n}{dt} = \sum_{k<n} \hbar\omega_{nk} W_{nk}, \qquad (17.188)$$

we get for the probability of spontaneous emissions from n to k per unit time

$$W_{nk} = \frac{4e^3}{3\hbar c^3} \omega_{nk}^3 |x_{nk}|^2, \qquad (17.189)$$

in exact agreement with the QED result, Eq. (17.134). It is clear that these expressions have a meaning only in a statistical sense: they refer to an ensemble of systems prepared initially in the same excited state n, which have a certain probability per unit time to decay, determined by the specific properties of the system. However, which transition will occur in each instance, or which member of the ensemble will decay at a given time, is a matter of random circumstances.

17.4.2 Breakdown of energy balance

One might now ask whether there is a background field $g(\omega)$ which, added to the ZPF, is capable of keeping the particles in an excited state n all the time. With the total field represented by a spectral energy density $\rho(\omega) = \rho_0(\omega)g(\omega)$, where $g(\omega) > 1$ is an even function of ω, Eq. (17.190) must be replaced by

$$\langle n | D^{pp} | n \rangle = -m^2\tau \sum_k g(\omega_{nk})\omega_{nk}^4 |x_{nk}|^2 \, \text{sign}\omega_{nk} \qquad (17.190)$$

so that instead of (17.187) we have

$$\frac{d\langle H\rangle_n}{dt} = -m\tau \sum_k \omega_{nk}^4 |x_{nk}|^2 (1 - g(\omega_{kn})\text{sign}\omega_{kn}). \qquad (17.191)$$

Note that the terms in parentheses have different signs depending on whether ω_{kn} refers to an upward or a downward transition (i.e., $k > n$ or $k < n$). Therefore, there is no way to satisfy detailed energy balance in general. However, as we will see later, there is one particular system that can remain in balance in an excited state, namely an ensemble of harmonic oscillators.

Exercise E17.6. Detailed energy balance condition for the harmonic oscillator
Determine the spectral density of the background field that can maintain an ensemble of oscillators in an excited state n.

Solution. The HO is exceptional in the sense that all the $|\omega_{nk}|$ contributing to the sum in (17.191) are equal in value and coincide with the oscillator frequency ω. Using the HO expressions

$$|x_{nk}|^2 = \frac{\hbar}{2m\omega}\left[(n+1)\delta_{k,n+1} + n\delta_{k,n-1}\right], \qquad (17.192)$$

we obtain for first term on the right side of Eq. (17.191)

$$-m\tau \sum_k \omega_{nk}^4 |x_{nk}|^2 = -\tfrac{1}{2}\hbar\tau\omega^3(2n+1), \qquad (17.193)$$

and for the second term

$$m\tau \sum_k \omega_{nk}^4 |x_{nk}|^2 g(\omega_{kn})\text{sign}\omega_{kn} = \tfrac{1}{2}\hbar\tau\omega^3 g_n(\omega). \qquad (17.194)$$

Therefore,

$$\frac{d\langle H\rangle_n}{dt} = -\tfrac{1}{2}\hbar\tau\omega^3 \left[2n+1 - g_n(\omega)\right], \qquad (17.195)$$

which indicates that detailed balance holds when $g_n(\omega) = 2n + 1$. In other words, if the HO in an excited state n is embedded in a field with spectral energy density

$$\rho(\omega) = \rho_0(\omega)(2n+1), \qquad (17.196)$$

there are as many absorptions as there are emissions per unit time, all with the same frequency ω, so that the average energy of the oscillators does not change. This result should not come as a surprise, since this field has precisely an energy per normal mode

$\hbar\omega(2n+1)/2$, equal to the energy of the mechanical oscillators with which it is in equilibrium.

It is important to note that even in the case where detailed balance holds, the excited states are *not* stationary (the energy is stationary, not the states): up and down transitions take place continuously, at a rate such that the average energy remains constant. In what follows, we derive the general formulas for the corresponding transition rates.

17.4.3 Atomic lifetimes

Consider an atom prepared in a state n, exposed to a radiation field with spectral energy density $\rho(\omega) = \rho_0(\omega)g(\omega)$ as earlier. It is convenient to separate the additional (external) field from the ZPF,

$$\rho(\omega) = \rho_0(\omega) + \rho_a(\omega), \quad \rho_a(\omega) = \rho_0(\omega)g_a(\omega), \tag{17.197}$$

so that Eq. (17.191) can be written as

$$\frac{d\langle H\rangle_n}{dt} = -m\tau \sum_k \omega_{nk}^4 |x_{nk}|^2 \left[1 - (1+g_a(\omega_{nk}))\mathrm{sign}\,\omega_{kn}\right]$$

$$= m\tau \sum_k \omega_{nk}^4 |x_{nk}|^2 \left[(g_a)_{\omega_{kn}>0} - (2+g_a)_{\omega_{kn}<0}\right]. \tag{17.198}$$

The term proportional to g_a represents the absorptions ($k > n$) and the second one, proportional to $2 + g_a$, the emissions ($k < n$). It is clear from this expression that there can only be absorption if there is an additional field, $\rho_a \neq 0$, which can therefore be identified with the *photonic* field. The emissions, on the other hand, can be either "spontaneous" (in the presence of only the ZPF, as in the previous section) or stimulated by the photonic field.

In terms of the A and B coefficients, Eq. (17.198) takes the form

$$\frac{d\langle H\rangle_n}{dt} = \frac{d}{dt}\langle H\rangle_n^{\text{sp em}} + \frac{d}{dt}\langle H\rangle_n^{\text{ind em}} + \frac{d}{dt}\langle H\rangle_n^{\text{ind abs}}$$

$$= \sum_{k>n} \hbar\omega_{kn} B_{kn}^{\text{abs}} \rho_a(\omega_{nk}) \tag{17.199}$$

$$- \sum_{k<n} \hbar\omega_{nk} \left[A_{nk} + B_{nk}^{\text{em}} \rho_a(\omega_{nk})\right],$$

or, using (17.197),

$$\frac{d\langle H\rangle_n}{dt} = \sum_k \hbar|\omega_{nk}| \left[\left(B_{kn}^{\text{abs}}\rho_0 g_a\right)_{\omega_{kn}>0} - \left(A_{nk} + B_{nk}^{\text{em}}\rho_0 g_a\right)_{\omega_{kn}<0}\right], \tag{17.200}$$

whence

$$A_{nk} = \frac{4e^2\omega_{nk}^3}{3\hbar c^3}|x_{nk}|^2, \quad (n>k) \tag{17.201}$$

and

$$B_{kn}^{\text{abs}} = B_{nk}^{\text{em}} = \frac{4\pi^2 e^2}{3\hbar^2}|x_{nk}|^2 \equiv B_{nk}, \tag{17.202}$$

in agreement with the respective qed formulas. Notice in particular the factor 2 in the relation

$$\frac{A_{nk}}{B_{nk}} = \frac{\hbar |\omega_{nk}|^3}{\pi^2 c^3} = 2\rho_0(\omega_{nk}). \qquad (17.203)$$

The structure of Eq. (17.198) indicates that one should actually write $2\rho_0 = \rho_0(1 - \text{sign}\,\omega_{kn})$, $\omega_{kn} < 0$, in order to keep track of the origin of each of the two terms: one is due to the vacuum fluctuations and the other to the Larmor radiation. They turn out to be of equal magnitude, as noted in relation to Eq. (17.187). In fact, it is precisely their equality that leads to the exact balance between them when the system is in its ground state.[7]

17.4.4 The Lamb shift

Another important radiative correction, which is considered a major success of QED despite its relative smallness, is the shift of some atomic energy levels due to a residual effect of the vacuum, known as the Lamb shift. The following is a rather straightforward and transparent calculation of the Lamb shift, using the SED tools.

To calculate the radiative energy shift we take $\mathcal{G} = xp$ in the equation of evolution for $\langle \mathcal{G} \rangle$, Eq. (10.86),

$$\frac{d}{dt} \langle xp \rangle = \left\langle \frac{dxp}{dt} \right\rangle_{\text{nr}} + m\tau \left\langle \dddot{x} \frac{\partial xp}{\partial p} \right\rangle - e^2 \left\langle \frac{\partial xp}{\partial p} \hat{D} \right\rangle, \qquad (17.204)$$

with $\hat{D}$ given by Eq. (17.175), which gives

$$\frac{d}{dt} \langle xp \rangle_n = \frac{1}{m} \left\langle n \left| p^2 \right| n \right\rangle + \langle n |xf| n \rangle + m\tau \langle n |x \dddot{x}| n \rangle + \left\langle n \left| D^{px} \right| n \right\rangle. \qquad (17.205)$$

This equation is a time-dependent version of the classical virial theorem $\langle n |p^2/m| n \rangle - \langle n |r \cdot \nabla V| n \rangle = 0$, with two additional terms that represent radiative corrections to the average energy, given by $\delta E_n = -\frac{m\tau}{2} \langle n |x \dddot{x}| n \rangle - \frac{1}{2} \langle n |D^{px}| n \rangle$. The first term does not contribute under stationarity, since

$$\langle n |x \dddot{x}| n \rangle = -\frac{1}{2} \langle n |\dot{x} \ddot{x} + \ddot{x} \dot{x}| n \rangle = \frac{1}{2} \frac{d}{dt} \langle \dot{x}^2 \rangle_n = 0. \qquad (17.206)$$

Thus, the Lamb shift is due entirely to the additional fluctuations impressed upon the electrons through their dipolar coupling to the ZPF,

$$\delta E_n = -\frac{1}{2} \langle n | D^{px} | n \rangle = -\frac{1}{2} e \langle n |xE| n \rangle. \qquad (17.207)$$

To calculate this term we write

$$D^{px}(t) = e \langle xE \rangle_0 = e^2 \int_{-\infty}^{t} ds \left. \frac{\partial x(t)}{\partial p(s)} \right|_{x^{(0)}} \langle E(s)E(t) \rangle_0, \qquad (17.208)$$

---

[7] It is often found in the literature that all spontaneous decay is attributed to one or the other of these two sources, more commonly to Larmor radiation. In QED the two contributions are equal, *provided* a symmetric operator order is used. Related discussions can be found in Davydov (2013), Dalibard et al. (1982), and Milonni (2013).

which, after we write the Poisson bracket in terms of the commutator, gives

$$D^{px}(t) = \frac{e^2}{i\hbar} \int_{-\infty}^{t} ds\, \varphi(t-s) [\widehat{x}(s), \widehat{x}(t)]. \tag{17.209}$$

With the matrix elements of $\widehat{x}$ given by $x_{kn}(t) = x_{kn} \exp(i\omega_{kn} t)$, we get

$$\langle n | [\widehat{x}(s), \widehat{x}(t)] | n \rangle = 2i \sum_k |x_{kn}|^2 \sin \omega_{kn}(t-s), \tag{17.210}$$

which when introduced into (17.209), gives

$$\langle n | D^{px} | n \rangle = \frac{2e^2}{\hbar} \sum_k |x_{kn}|^2 \int_{-\infty}^{t} ds\, \varphi(t-s) \sin \omega_{kn}(t-s).$$

Using (17.177) with $\rho_0(\omega) = \hbar \omega^3 / 2\pi^2 c^3$ for the ZPF and taking into account that

$$\int_{-\infty}^{t} ds\, \cos(t-s) \sin \omega_{kn}(t-s) = \frac{\omega_{kn}}{\omega_{kn}^2 - \omega^2}, \tag{17.211}$$

we obtain

$$\delta E_n = -\frac{1}{2} \langle n | D^{px} | n \rangle = -\frac{2e^2}{3\pi c^3} \sum_k |x_{nk}| \omega_{kn} \int_0^\infty d\omega \, \frac{\omega^3}{\omega_{kn}^2 - \omega^2}. \tag{17.212}$$

The *observable* Lamb shift $\delta E_{\text{L}n}$ is obtained by subtracting from δE_n the free-particle contribution, δE_{fp}, obtained in the limit of continuous electron energies (when ω_{kn} can be ignored compared to ω in the denominator),

$$\delta E_{\text{fp}} = \frac{2e^2}{3\pi c^3} \sum_m |x_{nm}|^2 \omega_{mn} \int_0^\infty d\omega\, \omega = \frac{e^2 \hbar}{\pi m c^3} \int_0^\infty d\omega\, \omega. \tag{17.213}$$

The last equality follows from the Thomas–Reiche–Kuhn sum rule $\sum_m |x_{nm}|^2 \omega_{mn} = 3\hbar/2m$. Hence,

$$\delta E_{\text{L}n} = \delta E_n - \delta E_{\text{fp}} = -\frac{2e^2}{3\pi c^3} \sum_k |x_{nk}|^2 \omega_{kn}^3 \int_0^\infty d\omega \, \frac{\omega}{\omega_{kn}^2 - \omega^2}, \tag{17.214}$$

in accordance with the (nonrelativistic) QED formula. Note that the integral is logarithmically divergent. The problem is solved as in QED by introducing a cutoff frequency equal to the Compton frequency of the electron $\omega_C = mc^2/\hbar$, thus giving[8]

$$\delta E_{\text{L}n} = \frac{2e^2}{3\pi c^3} \sum_k |x_{nk}|^2 \omega_{kn}^3 \ln \left| \frac{mc^2}{\hbar \omega_{kn}} \right|. \tag{17.215}$$

The numerical evaluation of this result gives 1057.833 MHz (in units of h) or 4.372×10^{-6} eV for the energy difference between the $2s_{1/2}$ and $2p_{1/2}$ H levels, which corresponds almost exactly to the experimental value observed by Lamb.

---

[8] Note that the logarithmic divergence here is a result of the previous approximations. If no approximation is made in the radiation reaction term in the denominator of Eq. (17.214), instead of ω^2, ω^4 will appear, and the integral will converge (and lead to a correct prediction).

in agreement with the respective qed formulas. Notice in particular the factor 2 in the relation

$$\frac{A_{nk}}{B_{nk}} = \frac{\hbar |\omega_{nk}|^3}{\pi^2 c^3} = 2\rho_0(\omega_{nk}). \tag{17.203}$$

The structure of Eq. (17.198) indicates that one should actually write $2\rho_0 = \rho_0(1 - \text{sign}\omega_{kn})$, $\omega_{kn} < 0$, in order to keep track of the origin of each of the two terms: one is due to the vacuum fluctuations and the other to the Larmor radiation. They turn out to be of equal magnitude, as noted in relation to Eq. (17.187). In fact, it is precisely their equality that leads to the exact balance between them when the system is in its ground state.[7]

17.4.4 The Lamb shift

Another important radiative correction, which is considered a major success of QED despite its relative smallness, is the shift of some atomic energy levels due to a residual effect of the vacuum, known as the Lamb shift. The following is a rather straightforward and transparent calculation of the Lamb shift, using the SED tools.

To calculate the radiative energy shift we take $\mathcal{G} = xp$ in the equation of evolution for $\langle \mathcal{G} \rangle$, Eq. (10.86),

$$\frac{d}{dt} \langle xp \rangle = \left\langle \frac{dxp}{dt} \right\rangle_{\mathrm{nr}} + m\tau \left\langle \dddot{x} \frac{\partial xp}{\partial p} \right\rangle - e^2 \left\langle \frac{\partial xp}{\partial p} \hat{\mathcal{D}} \right\rangle, \tag{17.204}$$

with $\hat{\mathcal{D}}$ given by Eq. (17.175), which gives

$$\frac{d}{dt} \langle xp \rangle_n = \frac{1}{m} \langle n | p^2 | n \rangle + \langle n | xf | n \rangle + m\tau \langle n | x \dddot{x} | n \rangle + \langle n | D^{px} | n \rangle. \tag{17.205}$$

This equation is a time-dependent version of the classical virial theorem $\langle n | p^2/m | n \rangle - \langle n | r \cdot \nabla V | n \rangle = 0$, with two additional terms that represent radiative corrections to the average energy, given by $\delta E_n = -\frac{m\tau}{2} \langle n | x \dddot{x} | n \rangle - \frac{1}{2} \langle n | D^{px} | n \rangle$. The first term does not contribute under stationarity, since

$$\langle n | x \dddot{x} | n \rangle = -\frac{1}{2} \langle n | \dot{x} \ddot{x} + \ddot{x} \dot{x} | n \rangle = \frac{1}{2} \frac{d}{dt} \langle \dot{x}^2 \rangle_n = 0. \tag{17.206}$$

Thus, the Lamb shift is due entirely to the additional fluctuations impressed upon the electrons through their dipolar coupling to the ZPF,

$$\delta E_n = -\frac{1}{2} \langle n | D^{px} | n \rangle = -\frac{1}{2} e \langle n | xE | n \rangle. \tag{17.207}$$

To calculate this term we write

$$D^{px}(t) = e \langle xE \rangle_0 = e^2 \int_{-\infty}^{t} ds \left. \frac{\partial x(t)}{\partial p(s)} \right|_{x(0)} \langle E(s)E(t) \rangle_0, \tag{17.208}$$

---

[7] It is often found in the literature that all spontaneous decay is attributed to one or the other of these two sources, more commonly to Larmor radiation. In QED the two contributions are equal, *provided* a symmetric operator order is used. Related discussions can be found in Davydov (2013), Dalibard et al. (1982), and Milonni (2013).

which, after we write the Poisson bracket in terms of the commutator, gives

$$D^{px}(t) = \frac{e^2}{i\hbar} \int_{-\infty}^{t} ds \, \varphi(t-s) \, [\widehat{x}(s), \widehat{x}(t)]. \tag{17.209}$$

With the matrix elements of $\widehat{x}$ given by $x_{kn}(t) = x_{kn} \exp(i\omega_{kn}t)$, we get

$$\langle n \, | [\widehat{x}(s), \widehat{x}(t)] | \, n \rangle = 2i \sum_k |x_{kn}|^2 \sin \omega_{kn}(t-s), \tag{17.210}$$

which when introduced into (17.209), gives

$$\langle n \, | D^{px} | \, n \rangle = \frac{2e^2}{\hbar} \sum_k |x_{kn}|^2 \int_{-\infty}^{t} ds \, \varphi(t-s) \sin \omega_{kn}(t-s).$$

Using (17.177) with $\rho_0(\omega) = \hbar\omega^3/2\pi^2 c^3$ for the ZPF and taking into account that

$$\int_{-\infty}^{t} ds \, \cos(t-s) \sin \omega_{kn}(t-s) = \frac{\omega_{kn}}{\omega_{kn}^2 - \omega^2}, \tag{17.211}$$

we obtain

$$\delta E_n = -\frac{1}{2} \langle n \, | D^{px} | \, n \rangle = -\frac{2e^2}{3\pi c^3} \sum_k |x_{nk}|^2 \omega_{kn} \int_0^\infty d\omega \, \frac{\omega^3}{\omega_{kn}^2 - \omega^2}. \tag{17.212}$$

The *observable* Lamb shift δE_{Ln} is obtained by subtracting from δE_n the free-particle contribution, δE_{fp}, obtained in the limit of continuous electron energies (when ω_{kn} can be ignored compared to ω in the denominator),

$$\delta E_{fp} = \frac{2e^2}{3\pi c^3} \sum_m |\mathbf{x}_{nm}|^2 \omega_{mn} \int_0^\infty d\omega \, \omega = \frac{e^2 \hbar}{\pi m c^3} \int_0^\infty d\omega \, \omega. \tag{17.213}$$

The last equality follows from the Thomas–Reiche–Kuhn sum rule $\sum_m |\mathbf{x}_{nm}|^2 \omega_{mn} = 3\hbar/2m$. Hence,

$$\delta E_{Ln} = \delta E_n - \delta E_{fp} = -\frac{2e^2}{3\pi c^3} \sum_k |\mathbf{x}_{nk}|^2 \omega_{kn}^3 \int_0^\infty d\omega \, \frac{\omega}{\omega_{kn}^2 - \omega^2}, \tag{17.214}$$

in accordance with the (nonrelativistic) QED formula. Note that the integral is logarithmically divergent. The problem is solved as in QED by introducing a cutoff frequency equal to the Compton frequency of the electron $\omega_C = mc^2/\hbar$, thus giving[8]

$$\delta E_{Ln} = \frac{2e^2}{3\pi c^3} \sum_k |\mathbf{x}_{nk}|^2 \omega_{kn}^3 \ln \left| \frac{mc^2}{\hbar \omega_{kn}} \right|. \tag{17.215}$$

The numerical evaluation of this result gives 1057.833 MHz (in units of h) or 4.372×10^{-6} eV for the energy difference between the $2s_{1/2}$ and $2p_{1/2}$ H levels, which corresponds almost exactly to the experimental value observed by Lamb.

[8] Note that the logarithmic divergence here is a result of the previous approximations. If no approximation is made in the radiation reaction term in the denominator of Eq. (17.214), instead of ω^2, ω^4 will appear, and the integral will converge (and lead to a correct prediction).

17.4.5 External effects on the radiative corrections

We note from Eq. (17.208) that the Lamb shift is due to the coupling of the (instantaneous) electric dipole moment of the atom to the electric component of the background field. But the background field can be modified externally, for example, by placing the atom in a reflective cavity that changes the field's boundary conditions and thus affects the distribution of the normal field modes, or by adding a photonic field to the vacuum. As a result, the Lamb shift is modified accordingly (Problem P17.11).

From the formulas obtained for the emission and absorption probabilities in Section 17.4.3, it is clear that the atomic radiation lifetimes are also modified by the addition of a photonic field or by placing the atom in a reflective cavity.

Such "environmental" effects have been studied and tested experimentally for more than 50 years, leading to the development of *cavity quantum electrodynamics* (CQED). The techniques developed to create and measure CQED states are now being applied to the development of quantum computers and other modern systems. The calculations are typically done within the framework of QED, although they can usually be done more easily within the framework of SED, as shown earlier, which reproduces the (nonrelativistic) QED formulas with the advantage of physical transparency.

17.5 Canonical Field Quantization Clarified

In Section 17.3.1, the field was quantized *à la* Dirac, that is, by replacing the classical field variables a, a^* by the annihilation and creation operators $a, \widehat{a}^\dagger$ satisfying the commutator (17.131). This formal quantization procedure, introduced as a postulate, was instrumental in the development of QED and is used more generally to quantize other fields. In the following we give a physical rationale for the replacement of variables by operators to describe the quantized radiation field.

17.5.1 Describing a field component in interaction with matter

We have seen in previous sections that in radiative transitions, atomic matter interacts resonantly with field modes of well-defined frequencies ω_{nk}. Any radiative transition that changes the state of the atom is accompanied by a concomitant change in the state of the field involved. To describe the effect *on the radiation field* resulting from this process, we will focus on a *single* frequency $\omega_{nk} \equiv \omega$ of the field.[9]

We call $q_n(t)$, $p_n(t)$ the canonical variables of the field in a stationary state n and express them as linear combinations of the normal field amplitudes (a, a^*) introduced in Section 10.4.1.[10] Using roman letters to distinguish the field variables from those of the atomic system (which were expressed in italics), we have

---

[9] As in the case of QM, our aim is to reveal the quantization – in this case of the field modes – as the primary effect of this interaction; the secondary effects can then be treated as perturbations.

[10] This applies as long as there are no nonlinear effects, produced by nonlinear optical materials or at very high light intensities ($> 10^8$ Vm^{-1}) such as those provided by lasers. The first nonlinear optical effect to be predicted was two-photon absorption, by Maria Göppert Mayer for her PhD in 1931.

$$q_n(t) = \sum_{n'} q_{nn'} a_{nn'} e^{-i\omega_{n'n}t} + \text{c.c.}, \quad p_n(t) = \sum_{n'} p_{nn'} a_{nn'} e^{-i\omega_{n'n}t} + \text{c.c.}, \quad (17.216)$$

where the amplitudes $a_{nn'}, a^*_{nn'}$ connect state n of the field with other possible states n' of the field, and $q_{nn'}, p_{nn'}$ are the respective *response coefficients*. Since the field oscillates with a single frequency ω, it follows that $\omega_{n'n} = \pm\omega$ and consequently, only two terms connecting state n with a state n' are different from zero. Since there are no intermediate states, we denote the upper state (corresponding to $\omega_{n'n} = +\omega$) by $n' = n+1$, and the lower state (corresponding to $\omega_{n'n} = -\omega$) by $n' = n-1$, so that Eqs. (17.216) become

$$q_n(t) = q_{nn+1} a_+ e^{-i\omega t} + q_{nn-1} a_- e^{i\omega t} + \text{c.c., and} \quad (17.217)$$

$$p_n(t) = -i\omega q_{nn+1} a_+ e^{-i\omega t} + i\omega q_{nn-1} a_- e^{i\omega t} + \text{c.c.} \quad (17.218)$$

From $p_n(t) = \dot{q}_n(t)$, we have

$$\omega q_{nn+1} - i p_{nn+1} = 0, \quad \omega q_{nn-1} + i p_{nn-1} = 0. \quad (17.219)$$

17.5.2 From field variables to field operators

In analogy with the procedure followed in Chapter 10, which led to the quantum mechanical canonical commutator $[\hat{x}, \hat{p}] = i\hbar$ we now take the Poisson bracket of the field variables given by Eqs. (17.217) and (17.218). The canonical Poisson bracket must satisfy

$$\{q_n(t), p_n(t)\}_{\{q,p\}} = 1. \quad (17.220)$$

Since the normal amplitudes (a, a^*) are related to the (canonical) quadratures (q, p) by the transformation

$$\omega q = \sqrt{\hbar\omega/2}(a + a^*), \quad p = -i\sqrt{\hbar\omega/2}(a - a^*), \quad (17.221)$$

the transformed Poisson bracket reads

$$\{q_n(t), p_n(t)\}_{\{a,a^*\}} \equiv \{q_n(t), p_n(t)\} = i\hbar. \quad (17.222)$$

From (17.217) and (17.218) we see that the only pairs of normal amplitudes involved in the Poisson bracket are (a_+, a^*_+) and (a_-, a^*_-); this gives explicitly

$$\sum_{n'=n\pm 1} \left(q_{nn'} p^*_{nn'} - p_{nn'} q^*_{nn'} \right) = i\hbar. \quad (17.223)$$

By using Eq. (17.217) we get

$$q^*_{nn'}(\omega_{n'n}) = q_{n'n}(\omega_{nn'}), \quad p^*_{nn'}(\omega_{n'n}) = p_{n'n}(\omega_{nn'}), \quad a^*_{nn'}(\omega_{n'n}) = a_{n'n}(\omega_{nn'}), \quad (17.224)$$

so that Eq. (17.223) can be written in the alternative form

$$\sum_{n'=n\pm 1} (q_{nn'} p_{n'n} - p_{nn'} q_{n'n}) = i\hbar. \quad (17.225)$$

By identifying $q_{nn'}$ and $p_{n'n}$ as the elements of matrices $\hat{q}$ and $\hat{p}$, respectively, Eq. (17.225) becomes

$$[\hat{q}, \hat{p}]_{nn} = [\hat{q}, \hat{p}] = i\hbar, \quad (17.226)$$

for any state n of the field. Therefore the canonical field commutator is the Poisson bracket of $q_n(t)$, $p_n(t)$, with respect to the normal amplitudes $(a_\pm, a^*_\pm)$ that take the field from any state n to states n±1.

From (17.219) we have $p_{nn'} + i\omega_{n'n}q_{nn'} = 0$; therefore, the matrix $\hat{a}$ and its adjoint, defined as

$$\hat{a} = \frac{1}{\sqrt{2\hbar\omega}}(\omega\hat{q} + i\hat{p}), \quad \hat{a}^\dagger = \frac{1}{\sqrt{2\hbar\omega}}(\omega\hat{q} - i\hat{p}), \tag{17.227}$$

have off-diagonal elements $a_{nn'}$, $a^*_{nn'}$ immediately above and below the main diagonal, that is, they play the role of *annihilation* and *creation* operators respectively. Combining Eqs. (17.226) and (17.227) gives for their commutator

$$\left[\hat{a}, \hat{a}^\dagger\right] = 1, \tag{17.228}$$

which is exactly Eq. (17.131), the cornerstone of the quantum theory of radiation. Note, however, that, according to this result, it is not the mode amplitudes a, a^* that are quantized; rather, the operators $\hat{a}$, $\hat{a}^\dagger$ contain the *response coefficients* of the modes involved in the matter–field interaction.

17.5.3 The interaction Hilbert space

In principle, any problem involving matter–field interaction forces us to work in an extended Hilbert space which is the product of the particle and field Hilbert spaces. In principle this means introducing an infinite-dimensional Hilbert space to account for the continuous nature of the field spectrum. Fortunately, in practice, for the study of matter–field exchange processes the field components are limited to those that are relevant in each case, depending on the problem at hand, taking into account that different field components do not interact among themselves and remain uncorrelated. Therefore, it makes sense to complete the quantum field formalism by considering only one field component, as earlier, and introducing the state vectors living in a single Hilbert space, on which the respective $\hat{a}$, $\hat{a}^\dagger$ operators act. To each field mode corresponds a pair of operators $\hat{a}, \hat{a}^\dagger$ acting on its own Hilbert space, different from the Hilbert space of other field modes and of the particle interacting with it.

From here on, we will revert to conventional notation and write the field operators as $\hat{a}, \hat{a}^\dagger$ instead of $\hat{a}, \hat{a}^\dagger$; however, the reader should bear in mind the physical meaning of these operators, which, as seen earlier, represent the response of the field in matter–field exchange processes.

In Dirac's notation, $|n\rangle$ stands for the state vector n, with $|0\rangle$ representing the ground state, and $\langle n|$ stands for the adjoint of $|n\rangle$. As we saw earlier the effect of $\hat{a}$ or $\hat{a}^\dagger$ on $|n\rangle$ is to lower or raise the number n by one,

$$\hat{a}|n\rangle = \sqrt{n}|n-1\rangle, \quad \hat{a}^\dagger|n\rangle = \sqrt{n+1}|n+1\rangle.$$

These are the basic elements used to describe how the field of frequency ω, interacting with matter, exchanges radiative energy in well-defined quantities $\pm\hbar\omega$.

By writing the Hamiltonian operator in *symmetrized* form,

$$\hat{H} = \frac{1}{2}\left(\omega^2\hat{q}^2 + \hat{p}^2\right) = \frac{\hbar\omega}{2}\left(\hat{a}\hat{a}^\dagger + \hat{a}^\dagger\hat{a}\right), \tag{17.229}$$

we get the well-known result for the expectation value of the energy,

$$\langle n|\hat{H}|n\rangle = \mathcal{E}_n = \left(n + \frac{1}{2}\right)\hbar\omega, \tag{17.230}$$

so that the ground-state energy is indeed $\mathcal{E}_0 = \hbar\omega/2$, and $|\mathcal{E}_n - \mathcal{E}_{n\pm 1}| = \hbar\omega$, as observed by Planck; see Section 2.9.

On its part, the *normal* operator order $\hat{a}^\dagger \hat{a}$ gives the total number n of photons available, that is, those that can be absorbed in *photodetection* processes, and conversely, the *antinormal* order $\hat{a}\hat{a}^\dagger$ gives the total probability of *photoemission* (both spontaneous and induced), as shown in Section 17.2.

Note that Eqs. (17.229)–(17.230), like all the previous ones, refer to a field component that exchanges energy *as a result of its interaction with quantized matter*; they tell us nothing about the (quantum or nonquantum) nature of the *free* radiation field. Consider, for example, the synchrotron radiation produced by highly accelerated orbiting electrons. When such radiation strikes a detector, a resonant interaction of the atoms of the detector at a well-defined frequency ω can lead to the absorption of a corresponding energy $\hbar\omega$; the energy absorbed by an atom is quantized, but we cannot ascertain that the energy emitted by the (free) accelerated electrons is.

By now the student may have realized that the image of an isolated atom or quantum particle in empty space is an entelechy, as previously discussed in Chapter 2. Even assuming that the system can be cooled to zero temperature, thereby eliminating the photonic radiation, the atom would still be surrounded by the field in its ground state, the ZPF. The inclusion of the ZPF transforms the apparently mechanical nature of quantum systems into a truly electrodynamic one, even in the absence of external fields. From the perspective gained by SED, we can say that quantum dynamics is a distinctive extension of classical physics into the stochastic domain. This counters the extended schizophrenic view of the world, in which its quantum and classical parts occupy mutually exclusive compartments (with the nanoworld somewhere in between).

17.6 *Second Quantization. An Introduction

Fully symmetrized or antisymmetrized wave functions correctly describe the behavior of the quantum system to which they refer, but, as we have seen in Chapter 15, they tend to be remarkably complex and unwieldy. It suffices to consider a case as simple as that of three equal particles, for the treatment of which it is necessary to deal with 6 (=3!) components of the wave function, to perceive the complexity that is soon reached when the number of particles increases. In addition to this practical difficulty, it turns out that for many purposes the wave function thus constructed contains much irrelevant or superfluous information, which, if eliminated, would simplify and clarify the description.

What we need in many cases is merely the description of a system containing identical particles that satisfy this or that statistics. The second-quantization method leads precisely to a synthetic description of the states of several identical particles, which makes it particularly suitable for carrying out the required calculations. What is being quantized are the fields that describe quantum particles, hence the name *second-quantization theory*. In fact, quantum field theory is a second-quantized theory.

A common way of second-quantizing Schrödinger's theory is to draw on the experience of using the creation and annihilation operators of the harmonic oscillator; however, in this brief introduction to the subject we will appeal to considerations that are more general in nature and therefore help to get a broader vision.

The starting point is the observation – which we will not prove here – that the Schrödinger equation can be derived from a Lagrangian density, that is, that there exists a Lagrangian density whose equations of motion (the corresponding Euler-Lagrange equations) are the Schrödinger equation and its adjoint. If we have a Lagrangian density[11] we can use it to construct the canonical momentum density conjugate to the generalized coordinate of the field. Using this procedure, it is found that by taking the wave function $\psi(x,t)$ as the generalized coordinate, the canonical conjugate momentum density becomes $\pi(x,t) = i\psi^*(x,t)$. As expected, the second-quantization method transforms *wave functions* into *operators*, that is,

$$\psi(x,t) \to \widehat{\psi}(x,t), \quad \pi(x,t) = i\psi^*(x,t) \to \widehat{\pi}(x,t) = i\widehat{\psi}^\dagger(x,t). \tag{17.231}$$

This is the standard procedure to quantize a field. However, since the commutation rules are different for bosonic and fermionic fields, one must study the two cases separately.

17.6.1 Second quantization of the Schrödinger field for bosons

The canonical variables $\widehat{\psi}(x,t)$ and $\widehat{\pi}(x',t)$ for a bosonic field must satisfy the commutation rule (taking $\hbar = 1$)

$$\left[\widehat{\psi}(x,t), \widehat{\pi}(x',t)\right] = i\delta(x - x'), \tag{17.232}$$

or equivalently with $\pi(x,t) = \psi^\dagger(x,t)$,

$$\left[\widehat{\psi}(x,t), \widehat{\psi}^\dagger(x',t)\right] = \delta(x - x'). \tag{17.233}$$

In addition,

$$\left[\widehat{\psi}(x,t), \widehat{\psi}(x',t)\right] = 0, \quad \left[\widehat{\psi}^\dagger(x,t), \widehat{\psi}^\dagger(x',t)\right] = 0. \tag{17.234}$$

As the fields evolve, it may well happen that some variables become dependent on the others; for example, $\widehat{\psi}(x',t')$ can depend on $\widehat{\psi}(x,t)$ for $t' > t$. The commutators for $t' \neq t$ will follow from the dynamics.

The quantized field $\widehat{\psi}(x,t)$ satisfies the Schrödinger equation

$$i\frac{\partial \widehat{\psi}(x,t)}{\partial t} = -\frac{1}{2m}\nabla^2 \widehat{\psi}(x,t) + V(x)\widehat{\psi}(x,t) \equiv \widehat{\mathcal{H}}\widehat{\psi}(x,t), \tag{17.235}$$

where the operator $\widehat{\mathcal{H}}$ is the Hamiltonian density. Let us consider the eigenstates $u_k(x)$ of $\widehat{\mathcal{H}}$,

$$\widehat{\mathcal{H}} u_k(x) = E_k u_k(x), \tag{17.236}$$

---

[11] A field is a continuous system, so it is necessary to use the Lagrangian density and not simply the Lagrangian.

as forming a complete orthonormal basis $\{u_k(\boldsymbol{x})\}$ that allows us to expand the $\widehat{\psi}$ state in the form

$$\widehat{\psi}(\boldsymbol{x},t) = \sum_k \widehat{a}_k u_k(\boldsymbol{x}) e^{-iE_k t}. \qquad (17.237)$$

The coefficients $\widehat{a}_k$ of the expansion are operators, in line with the operator structure of the field; they are given by

$$\widehat{a}_k = \int u_k^*(\boldsymbol{x}) \widehat{\psi}(\boldsymbol{x},t) e^{iE_k t} d^3 x. \qquad (17.238)$$

The commutation rules (17.233) and (17.234) now allow us to determine the commutation rules for the coefficients $\widehat{a}_k$. Taking into account the orthonormality of $\{u_k(\boldsymbol{x})\}$ we have

$$\left[\widehat{a}_k, \widehat{a}_{k'}^\dagger\right] = \int d^3 x \int d^3 x' u_k^*(\boldsymbol{x}) u_{k'}(\boldsymbol{x}') e^{i(E_k - E_{k'})t} \left[\widehat{\psi}(\boldsymbol{x},t), \widehat{\psi}^\dagger(\boldsymbol{x}',t)\right]$$
$$= \int d^3 x \, u_k^*(\boldsymbol{x}) u_{k'}(\boldsymbol{x}) e^{i(E_k - E_{k'})t} = \delta_{kk'}; \qquad (17.239)$$

thus,

$$\left[\widehat{a}_k, \widehat{a}_{k'}^\dagger\right] = \delta_{kk'}. \qquad (17.240)$$

Analogously, we get

$$[\widehat{a}_k, \widehat{a}_{k'}] = 0, \quad \left[\widehat{a}_k^\dagger, \widehat{a}_{k'}^\dagger\right] = 0. \qquad (17.241)$$

These commutation rules are what we expected for bosonic fields. In fact, one can go the other way and postulate Eqs. (17.240) and (17.241), to derive the canonical rules (17.233) and (17.234). The two procedures are equivalent and it is a matter of taste rather than of principle which one is chosen. Given what we learned in Chapter 9, we know that the newly derived rules assign to the operators $\widehat{a}_k^\dagger$ and $\widehat{a}_k$ the character of creation and annihilation operators for bosons, respectively.

To determine the commutation rule for arbitrary times we make the following calculation, which is now direct,

$$\left[\widehat{\psi}(\boldsymbol{x},t), \widehat{\psi}^\dagger(\boldsymbol{x}',t')\right] = \sum_{k,k'} \left[\widehat{a}_k, \widehat{a}_{k'}^\dagger\right] u_{k'}^*(\boldsymbol{x}') u_k(\boldsymbol{x}) e^{-iE_k t + iE_{k'} t'} =$$
$$= \sum_k u_k^*(\boldsymbol{x}') u_k(\boldsymbol{x}) e^{-iE_k(t-t')}. \qquad (17.242)$$

But this is precisely the expansion of the propagator (Green's function) of the Schrödinger equation for the problem at hand, as follows by comparison with Eq. (7.145). Taking $t = t'$, and remembering that $\sum_k u_k^*(\boldsymbol{x}') u_k(\boldsymbol{x}) = \delta(\boldsymbol{x} - \boldsymbol{x}')$, we recover Eqs. (17.233) and (17.234).

The Hamiltonian of the system follows from its density,

$$\widehat{H} = \int \widehat{\psi}^\dagger(\boldsymbol{x},t) \widehat{\mathcal{H}} \widehat{\psi}(\boldsymbol{x},t) d^3 x \qquad (17.243)$$
$$= \sum_{k,k'} \int d^3 x \, e^{i(E_k - E_{k'})t} \widehat{a}_k^\dagger \widehat{a}_{k'} u_k^*(\boldsymbol{x}) \widehat{\mathcal{H}} u_{k'}(\boldsymbol{x})$$
$$= \sum_{k,k'} e^{i(E_k - E_{k'})t} E_{k'} \widehat{a}_k^\dagger \widehat{a}_{k'} \int d^3 x \, u_k^*(\boldsymbol{x}') u_k(\boldsymbol{x}),$$

17.6 *Second Quantization. An Introduction

whence

$$\widehat{H} = \sum_k E_k \widehat{a}_k^\dagger \widehat{a}_k. \tag{17.244}$$

This result, which holds for any external potential, describes the Hamiltonian as the energy of independent (noninteracting) particles, each one with its own energy E_k. We have thus obtained a description in terms of the states of the system instead of the modes of the field, just as when we quantized the radiation field in Section 17.3. If we introduce the number operators

$$\widehat{N}_k = \widehat{a}_k^\dagger \widehat{a}_k, \tag{17.245}$$

we see that they all commute with the Hamiltonian and with each other,

$$[\widehat{N}_k, \widehat{H}] = 0, \tag{17.246}$$

$$[\widehat{N}_k, \widehat{N}_{k'}] = 0. \tag{17.247}$$

This means that $\widehat{H}, \{\widehat{N}_k\}$ form a maximal set of commuting operators (MSCO) and can be used with a complete basis for describing the system. This basis is denoted by kets of the form $|n_1, n_2, n_3, \cdots\rangle$, each of which describes a state of $n = n_1+n_2+n_3+\cdots$ particles, with n_1 particles in state 1, n_2 in state 2, etc., and $n_k = 0, 1, 2, 3, \cdots$. The set of numbers $\{n_1, n_2, n_3, \cdots | n = n_1 + n_2 + n_3 + \cdots\}$ is known as a partition of n, and the set of all possible partitions represents the set of possible states of the system. The space constructed on this basis is called *Fock space* or *space of occupation numbers*; the representation is called *number representation* and its vectors can be constructed using Eq. (7.145) to get

$$|n_1, n_2, n_3, \cdots\rangle = \frac{1}{(n_1! n_2! \cdots)^{1/2}} \left(\widehat{a}_1^\dagger\right)^{n_1} \left(\widehat{a}_2^\dagger\right)^{n_2} \cdots |0\rangle, \tag{17.248}$$

where the vacuum state $|0\rangle$ represents the state without particles. Since it is not possible to reduce any occupancy number from the empty state, it must be true that

$$\widehat{a}_k |0\rangle = 0. \tag{17.249}$$

Note the economy of the description, since $|n_1, n_2, n_3, \cdots\rangle$ specifies only the occupancy number in each energy eigenstate, independent of any spatial location.

Exercise E17.7. Heisenberg equation for the quantized boson field

Show that the Heisenberg equation for the Schrödinger field operator defined in Eq. (17.237) for bosons,

$$i\hbar \frac{\partial \widehat{\psi}}{\partial t} = [\widehat{\psi}, \widehat{H}], \tag{17.250}$$

where $\widehat{H}$ is the Hamiltonian operator given by (17.243), agrees with the Schrödinger equation for $\widehat{\psi}$. Derive the continuity equation for the particle density operator $\widehat{\rho}(x) = \widehat{\psi}^\dagger \widehat{\psi}$. What does the result mean? With the particle number operator given by

$$\widehat{N} = \int \widehat{\psi}^\dagger(x) \widehat{\psi}(x) d^3x, \tag{17.251}$$

show that $\widehat{N} = \sum_k \widehat{N}_k$, where $\widehat{N}_k = \hat{a}_k^\dagger \hat{a}_k$.

Solution. We have

$$\left[\widehat{\psi}(x), \widehat{H}\right] = \int \left[\widehat{\psi}(x), \widehat{\psi}^\dagger(x')\widehat{\mathcal{H}}(x')\widehat{\psi}(x')\mathrm{d}^3 x'\right], \tag{17.252}$$

where $\mathcal{H}$ represents the Hamiltonian density appearing in Eq. (17.235),

$$\widehat{\mathcal{H}}(x) = -\frac{\hbar^2}{2m}\nabla^2 + V(x). \tag{17.253}$$

The calculation is considerably simplified by applying the identity

$$[\widehat{A}, \widehat{B}\widehat{C}] = \widehat{B}[\widehat{A}, \widehat{C}] + [\widehat{A}, \widehat{B}]\widehat{C}, \tag{17.254}$$

which gives

$$\left[\widehat{\psi}(x), \widehat{H}\right] = \int \mathrm{d}^3 x' \widehat{\psi}^\dagger(x') \left[\widehat{\psi}(x), \widehat{\mathcal{H}}(x')\widehat{\psi}(x')\right]$$
$$+ \int \mathrm{d}^3 x' \left[\widehat{\psi}(x), \widehat{\psi}^\dagger(x')\right] \widehat{\mathcal{H}}(x')\widehat{\psi}(x'). \tag{17.255}$$

The first term on the right-hand side is zero, since

$$\left[\widehat{\psi}(x), \widehat{\mathcal{H}}(x')\widehat{\psi}(x')\right] = \widehat{\mathcal{H}}(x')\left[\widehat{\psi}(x), \widehat{\psi}(x')\right] = 0, \tag{17.256}$$

so that from (17.234) we get

$$\left[\widehat{\psi}(x), \widehat{H}\right] = \int \mathrm{d}^3 x' \delta(x - x') \widehat{\mathcal{H}}(x')\widehat{\psi}(x'), \tag{17.257}$$

that is, the Schrödinger equation

$$i\hbar \frac{\partial \widehat{\psi}}{\partial t} = \widehat{\mathcal{H}}(x)\widehat{\psi}(x). \tag{17.258}$$

Note that this evolution equation is precisely the corresponding Heisenberg equation for the quantized field.

We now take the time derivative of the particle density $\hat{\rho}(x) = \widehat{\psi}^\dagger \widehat{\psi}$ using the previous result,

$$\frac{d}{dt}\hat{\rho}(x) = \frac{d\widehat{\psi}^\dagger}{dt}\widehat{\psi} + \widehat{\psi}^\dagger \frac{d}{dt}\widehat{\psi} = -\frac{i\hbar}{2m}\left[\left(\nabla^2 \widehat{\psi}^\dagger\right)\widehat{\psi} - \widehat{\psi}^\dagger\left(\nabla^2 \widehat{\psi}\right)\right]$$
$$= -\frac{i\hbar}{2m}\nabla \cdot \left[\left(\nabla \widehat{\psi}^\dagger\right)\widehat{\psi} - \widehat{\psi}^\dagger\left(\nabla \widehat{\psi}\right)\right], \tag{17.259}$$

which leads to the continuity equation

$$\frac{d}{dt}\hat{\rho}(x) = -\nabla \cdot \hat{j} \tag{17.260}$$

for the current density operator

$$\hat{j} = -\frac{i\hbar}{2m}\left[\widehat{\psi}^\dagger\left(\nabla \widehat{\psi}\right) - \left(\nabla \widehat{\psi}^\dagger\right)\widehat{\psi}\right]. \tag{17.261}$$

The equal-time particle density operators commute at any pair of points (see problem P17.12),

$$[\hat{\rho}(x), \hat{\rho}(x')] = 0. \tag{17.262}$$

Because of its fundamental importance, this property could have been used as a requirement to quantize the field.

We now take the number operator

$$\widehat{N} = \int \widehat{\psi}^\dagger(x)\widehat{\psi}(x)\mathrm{d}^3 x = \int \rho(x)\mathrm{d}^3 x \qquad (17.263)$$

and reduce this expression by steps similar to those that led to (17.244). It is sufficient to suppress the Hamiltonian operator, which leads to

$$\widehat{N} = \sum_k \widehat{N}_k = \sum_k \widehat{a}_k^\dagger \widehat{a}_k. \qquad (17.264)$$

To determine $\widehat{N}$ we must first determine $\widehat{a}_k(t) \equiv \widehat{a}_k e^{-iE_k t}$. This is easily done by considering Schrödinger's equation (17.258) and using Eq. (17.237),

$$i\hbar \frac{\partial}{\partial t} \sum_k \widehat{a}_k u_k(x) e^{-iE_k t} = i\hbar \frac{\partial}{\partial t} \sum_k \widehat{a}_k(t) u_k(x) = i\hbar \sum_k \frac{d\widehat{a}_k(t)}{dt} u_k(x)$$

$$= \sum_k \widehat{a}_k(t)\widehat{\mathcal{H}}(x) u_k(x) = \sum_k \widehat{a}_k(t) E_k u_k(x), \qquad (17.265)$$

that is,

$$i\hbar \frac{d\widehat{a}_k}{dt} = E_k \widehat{a}_k(t). \qquad (17.266)$$

This equation describes the evolution of the field operator for a free field, that is, when the possible interactions between the particles that make up the system are not taken into account. By introducing such interactions, the equation is modified and the different modes (the k indices) interact and become coupled, so that the number of particles in each of the k states is no longer constant.

Note: Analogous results, mutatis mutandis, are obtained for fermions using anticommutators, as is demonstrated in the next section.

17.6.2 Second quantization of the Schrödinger–Pauli field for fermions

The quantization of the Schrödinger–Pauli field in the case of fermions is done in the same way as for bosons, with the only difference that the commutators are replaced by anticommutators.[12] If $\widehat{b}_k, \widehat{b}_k^\dagger$ are the operators for fermionic fields and their anticommutators are

$$\left\{\widehat{b}_k, \widehat{b}_{k'}^\dagger\right\} \equiv \widehat{b}_k \widehat{b}_{k'}^\dagger + \widehat{b}_{k'}^\dagger \widehat{b}_k, \qquad (17.267)$$

the appropriate (anti)commutation rules are

---

[12] The observation that it is sufficient to replace commutators by anticommutators to make the transition from B–E to F–D statistics is due to the German mathematician and theoretical physicist Pascual Jordan (1902–1980).

$$\{\widehat{b}_k, \widehat{b}_{k'}^\dagger\} = \delta_{kk'}, \tag{17.268}$$

$$\{\widehat{b}_k, \widehat{b}_{k'}\} = 0, \quad \{\widehat{b}_k^\dagger, \widehat{b}_{k'}^\dagger\} = 0. \tag{17.269}$$

Writing $\widehat{b}_k$ instead of $\widehat{a}_k$ everywhere leads to the theory for fermions, so the process will not be repeated here. From (17.269) it follows that

$$\{\widehat{b}_k, \widehat{b}_{k'}\} = \widehat{b}_k \widehat{b}_{k'} + \widehat{b}_{k'} \widehat{b}_k = 0, \tag{17.270}$$

and taking $k' = k$, we obtain $\widehat{b}_k^2 = 0$; similarly, $\left(\widehat{b}_k^\dagger\right)^2 = 0$. Now using (17.268) with $k' = k$ we can write the following equation for the number operator

$$\widehat{n}_k^2 = \widehat{b}_k^\dagger \widehat{b}_k \widehat{b}_k^\dagger \widehat{b}_k = \widehat{b}_k^\dagger \left(1 - \widehat{b}_k^\dagger \widehat{b}_k\right) \widehat{b}_k = n_k - \left(\widehat{b}_k^\dagger\right)^2 \left(\widehat{b}_k\right)^2, \tag{17.271}$$

which means that

$$\widehat{n}_k^2 = \widehat{n}_k. \tag{17.272}$$

The eigenvalues of the number operator therefore satisfy the equation $n_k^2 - n_k = 0$, that is, the occupancy numbers n_k can only take the values 0, 1, as corresponds to the F-D statistics. Since the state vector is (taking into account that $n_k! = 1$ for fermions)

$$|n_1, n_2, n_3, \cdots\rangle = \left(\widehat{b}_1^\dagger\right)^{n_1} \left(\widehat{b}_2^\dagger\right)^{n_2} \cdots |0\rangle, \tag{17.273}$$

from the anticommutation of the operators it follows that $|n_1, n_2, n_3, \cdots\rangle$ is antisymmetric with respect to the permutation of pairs of particles (or labels). We also have that

$$\widehat{b}_k |n_1, n_2, \cdots n_k, \cdots\rangle = (-1)^{n_k} \sqrt{n_k} |n_1, n_2, \cdots 1 - n_k, \cdots\rangle, \tag{17.274}$$

$$\widehat{b}_k^\dagger |n_1, n_2, \cdots n_k, \cdots\rangle = (-1)^{n_k} \sqrt{1 - n_k} |n_1, n_2, \cdots 1 - n_k, \cdots\rangle. \tag{17.275}$$

These results exhibit $\widehat{b}_k^\dagger$ as a creation operator, since if the initial state is empty ($n_k = 0$), the final state will be complete ($n_k = 1$) and vice versa. Similarly, b_k acts as an annihilation operator.

17.6.3 The two quantum statistics

The formalism introduced earlier can be applied to derive the B-E and the F-D statistics, with the help of the density matrix formalism developed in Chapter 15; as this is a subject for a more advanced course, we will skip the derivation but present the main results because of their importance. Using the commutation rule $[\widehat{a}, \widehat{a}^\dagger] = 1$ in the B-E case and the anticommutation rule $\{\widehat{a}, \widehat{a}^\dagger\}$ in the F-D case, one obtains

$$\mathrm{tr}\left(e^{-\beta \widehat{H}} \widehat{a}^\dagger \widehat{a}\right) = \mathrm{tr}\left(\widehat{a} e^{-\beta \widehat{H}} \widehat{a}^\dagger\right) = e^{-\beta \hbar \omega} \mathrm{tr}\left(e^{-\beta \widehat{H}} \widehat{a} \widehat{a}^\dagger\right) \tag{17.276}$$

$$= e^{-\beta \hbar \omega} \mathrm{tr}\left(e^{-\beta \widehat{H}} \left(1 \mp \widehat{a}^\dagger \widehat{a}\right)\right),$$

where the upper sign corresponds to fermions and the lower sign to bosons. From here it follows, multiplying by $e^{\beta \hbar \omega}$ and solving, that

$$(e^{\beta \hbar \omega} \pm 1) \mathrm{tr}\, e^{-\beta \widehat{H}} \widehat{a}^\dagger \widehat{a} = \mathrm{tr}\, e^{-\beta \widehat{H}} = Z_\pm, \tag{17.277}$$

where Z is the partition function, so that

$$\bar{n} = \langle \hat{a}^\dagger \hat{a} \rangle = \frac{1}{Z_\pm} \text{tr}\, e^{-\beta \hat{H}} \hat{a}^\dagger \hat{a} = \frac{1}{e^{\beta\hbar\omega} \pm 1} = \frac{e^{-\beta\hbar\omega}}{1 \pm e^{-\beta\hbar\omega}}. \quad (17.278)$$

With the plus sign in the denominator, this formula applies to a gas of fermions, and with the minus sign, it applies to massless bosons, as discussed in Section 15.3.

Problems

P17.1 Show that an adiabatic perturbation produces periodic transitions in a system that has two degenerate states.

P17.2 Show that if the potential $V(t) = \hat{A}\delta(t)$ is applied to a system in its ground state, the probability of a transition to any excited state is

$$w = \frac{1}{\hbar^2}\left(\langle 0|A^2|0 \rangle - \langle 0|A|0 \rangle^2 \right). \quad (17.279)$$

P17.3 A harmonic oscillator in its ground state is perturbed by the sudden application of a uniform electric field of intensity ϵ (not necessarily small), which remains constant from that moment on. Show that the probability of transition to the excited state n is given by the Poisson distribution

$$W_{0n} = \frac{\xi_0^{2n}}{2^n n!} e^{-\xi_0^2/2}, \quad (17.280)$$

where $\xi_0 = x_0/\sqrt{\hbar/m\omega}$ and $x_0 = e\epsilon/m\omega^2$. Find the mean value of n.

P17.4 A physical system having only the states $|1\rangle$ and $|2\rangle$, degenerate and orthogonal, is perturbed by the action of a potential V that is applied in the interval $(0, T)$. Assuming (for simplicity) that the matrix elements of V are $V_{11} = V_{22} = 0$, $V_{12} = V_{21} = V_0$, determine the exact probability of the transition from $|1\rangle$ to $|2\rangle$, if the system was initially in the state $|1\rangle$. Then use perturbation theory to determine the effect of the perturbation to first order and compare the results. When is the perturbative solution valid?

P17.5 Calculate in detail the Einstein coefficient B for resonant absorption processes and show that the Einstein relation $B_{nk} = B_{kn}$ is satisfied.

P17.6 Show that the probability per unit time that an atom makes a spontaneous quadrupole transition between states n and n' is

$$A^{(2e)}_{nn'} = \frac{\omega^5_{nn'}}{90\hbar c^5} \sum_{i,j} |\langle n|Q_{ij}|n' \rangle|^2, \quad (17.281)$$

where Q_{ij} are the components of the electric quadrupole moment,

$$Q_{ij} = e\left(3x_i x_j - r^2 \delta_{ij}\right), \quad i,j = 1,2,3. \quad (17.282)$$

P17.7 Find the selection rules for electrical quadrupole transitions of (a) particles confined in a 1D infinite rectangular well; (b) particles in a 1D HO well; (c) a flat rigid rotor.

P17.8 Determine the direction in which an electron is most likely to be emitted by the absorption of 50-keV γ radiation.

P17.9 Calculate the probability of an electron in an atom making a transition between stationary states under the influence of a charged particle passing close to the atom.

P17.10 A 1D HO originally at rest is subjected to a force $f(t)$. Find the equation of motion of the center of the wave packet using first-order perturbation theory and compare with the exact result.

P17.11 The free-particle Lamb effect can be identified with the contribution of the (free-space) ZPF to the particle's energy, given by Eq. (17.213)

$$\delta E_{\text{fp}} = \frac{e^2 \hbar}{\pi m c^3} \int_0^{\omega_C} d\omega\, \omega, \tag{17.283}$$

where the usual cutoff frequency $\omega_C = mc^2/\hbar$ (equal to Compton's frequency of the electron) has been introduced. Assume that the distribution of ZPF modes can be reduced by introducing the particle in a cavity that filters out all modes of frequency ω smaller than Ω without altering the isotropy of the field and write down the formula for the resulting energy shift.

(a) Estimate the minimum value of Ω needed to produce an energy shift of the order of the atomic Lamb effect.

(b) Estimate the size of the cavity needed to filter out the frequencies $\omega < \Omega$ and compare it with the size of the H atom. Considering 5 nm as the minimum size of current solid-state nanoelectronic devices, could such an energy shift be detectable? Discuss your answer.

P17.12 Use the fact that the normal field amplitudes used in Section 17.5 are independent from each other to prove that $[\hat{q}, \hat{p}]_{nn'} = i\hbar \delta_{nn'}$.

P17.13 By making repeated use of Eq. (17.254), determine the commutator of the densities $[\hat{\rho}(x), \hat{\rho}(x')]$, where $\hat{\rho}(x)$ is the boson density operator.

Bibliographical Notes

For physically transparent and accessible introductions to quantum electrodynamics, see Power (1964), Cohen-Tannoudji, Dupont-Roc, and Grynberg (1997), and Milonni (2013).

An alternative account of the photoelectric effect, showing that it can be explained equally well without invoking the photon, is Lamb and Scully (1969).

For calculations of radiative corrections in the SED framework with the advantage of added physical transparency, see, for example, Cetto and de la Peña (2012).

18 Elastic Scattering Theory

The quantum theory of scattering is a complex subject and is covered in more detail in a course on quantum field theory. In a first approximation, a collision or scattering is a process in which two particles, initially very far apart (and therefore not interacting), approach each other, interact (exchanging energy and momentum), and finally move apart. In this chapter we will introduce the main concepts and derive the most important formulas for elastic scattering, in a nonrelativistic framework. This will allow us to appreciate how the information obtained from scattering experiments allows us to explore the intricacies of quantum particles that are otherwise inaccessible.

We have been able to use the Schrödinger equation to describe atomic and molecular systems because we know the theory of electromagnetic interactions in great detail. However, if we want to use this equation to describe nuclear systems or, more generally, interactions between elementary particles – assuming that it is still valid for such levels of organization of matter – the first difficulty is that these interactions are not necessarily electromagnetic. There are four known fundamental interactions: the electromagnetic and the gravitational, already known in classical physics, plus the so-called strong (or nuclear) and weak interactions. At present, the Standard Model of particle interactions includes the strong, weak and electromagnetic interactions, but not the gravitational ones. Research aimed at integrating all four interactions into a general theory of fields and matter is currently very active. The fact that we are talking about a model rather than a theory is a recognition that there is still plenty of room (and need) for fundamental research.

18.1 Collision Phenomena

Due to the microscopic nature of the basic quantum corpuscles, the only method available to investigate their structure and interactions – to determine their dependence on the various parameters and variables such as the type of particles involved, energies, charges, orbital momenta, spins, relative distances, and so on – is to produce collisions between them and then analyze the outcomes under controlled conditions. The example of Rutherford's historic experiment, which led to the establishment of the atomic planetary model, shows that scattering experiments, even against unknown targets, can be used not only to find out the nature and characteristics of the interaction, but also to determine the structure of the particles involved in the scattering process.

These observations justify a careful study of Schrödinger's theory for the case of dispersion, that is, when the spectrum is continuous, which usually corresponds to positive total energy of the system. From a physical point of view, a simple general situation is as follows. A beam of particles with controlled and known momentum and energy is directed against a target of (possibly) known structure (a very thin metallic plate deposited on a suitable substrate, a gas, a crystal, etc.), which is at rest in the laboratory.[1] As a result of the collisions occurring in the target, the bombarding particles (which are independent of each other, assuming a low density of the beam) are scattered and leave the target in all possible directions and states.

Elastic processes – the only ones we will study – occur when the particles are scattered with the same energy as they hit; inelastic processes occur when the target is excited or ionized, or some particles are transformed into others, and so on. Whatever the process, the particles emerging from the interaction are received by suitable detectors placed very far from the interaction region (in what we will call the asymptotic region, where the interaction is null or negligible), which record each particle that hits them (in the form of traces on photographic film, or in a cloud or bubble chamber, or as light emission in a scintillation crystal, etc.). In addition to the obvious information about the average number of particles that hit the detector per second per unit area, it is often possible to deduce some of the properties of the recorded particle (nature, mass, charge, momentum, spin, energy, etc.). These are the data provided by the experimenter, and they constitute the material for further theoretical analysis of the problem.

18.1.1 Laboratory and center-of-mass frames

The first thing to note is that the description of the experiment is naturally carried out in a coordinate system linked to the laboratory, in which the target is (usually) at rest (this is the so-called laboratory L frame), while the theoretical description is more simply done in the reference frame in which the center of mass of the entire physical system is at rest (CM frame), since any uniform motion of the CM is irrelevant. When the target particle is very massive with respect to the projectile (e.g., atomic nuclei bombarded by electrons), both reference systems are almost coincident, since the CM remains very close to the target at all times, practically blending in with it. However, as is often the case, when the target and projectile masses are comparable, the two descriptions differ substantially. For example, from the kinematics of the collision between two particles it follows that the scattering angles θ_L in the L frame and θ_{CM} in the CM frame, as shown in Fig. 18.1, are related by (problem P18.1)

$$\tan \theta_{1_L} = \frac{\sin \theta_{CM}}{\cos \theta_{CM} + m_1/m_2}. \qquad (18.1)$$

This formula shows that the scattering angles measured in both systems are practically the same if $m_1/m_2 \ll 1$, but if this condition is not fulfilled, they can be very different. For example, the backscatter in the CM system ($\theta_{CM} = 180°$) corresponds to a deviation of $\theta_L = 90°$ of the incident particle in the laboratory frame when $m_1 = m_2$. On the other hand, part of the energy available in the L system is associated with the

---

[1] There are important exceptions to this rule. In collision rings, for example, two beams are accelerated in opposite directions and eventually brought to a head-on collision. This device greatly increases the energy available for the reaction by avoiding the energy required to accelerate the center of mass.

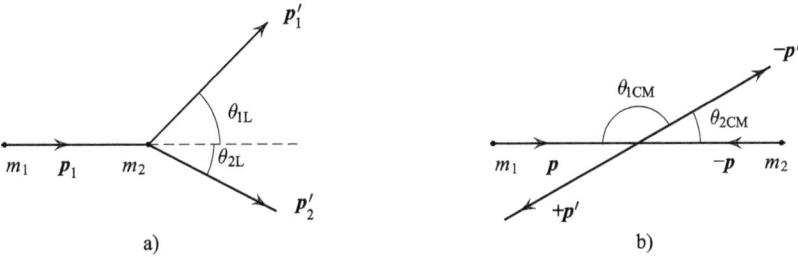

Figure 18.1 Diagram of an elastic collision between particles 1 and 2, seen from (a) the laboratory frame, (b) the CM frame.

CM motion and is not used for the scattering. Thus, the collision energy is less than that used to accelerate the particles according to the formula

$$E_{CM} = \left(1 - \frac{m_1}{M}\right) E_L, \quad M = m_1 + m_2. \tag{18.2}$$

We see that the loss can be significant for comparable masses. A very important result is that which relates the number of particles $n(\theta_{CM})$ that hit per unit area per unit time on a detector whose aperture $d\Omega_{CM}$ and angular position are defined in the CM system, to the corresponding quantity defined in the L system, $n(\theta_L)$ (see problem P18.2),

$$n(\theta_L) = \frac{\left[1 + 2\frac{m_1}{m_2}\cos\theta_{CM} + \left(\frac{m_1}{m_2}\right)^2\right]^{3/2}}{1 + \frac{m_1}{m_2}\cos\theta_{CM}} n(\theta_{CM}). \tag{18.3}$$

In the following we will work directly in the CM system; for simplicity we will omit the CM indices, which are to be understood.

18.1.2 Scattering amplitude and scattering cross section

In elastic collisions, the wave functions are eigenfunctions of the Hamiltonian corresponding to the energy E determined by the bombardment conditions. From the radial Schrödinger equation

$$\left(-\frac{\hbar^2}{2m}\frac{d^2}{dr^2} + V(r) + \frac{\hbar^2 l(l+1)}{2mr^2}\right) u(r) = E u(r) \tag{18.4}$$

we see that if the product $V(r)u(r)$ vanishes at infinity, then the asymptotic behavior of the radial function u at infinity is $u \sim e^{ikr}$, with $k = \sqrt{2mE}/\hbar$; this asymptotic behavior is obtained if we restrict ourselves to potentials such that $rV(r) \to 0$ for $r \to \infty$, which are called short-range potentials. In the following, V will represent a short-range potential. This excludes the Coulomb potential from the analysis, although we will have the opportunity to include it in our results later on.

When the interaction is due to a short-range potential, we can consider that it takes place entirely within a sphere of finite radius R, so that for distances greater than R, the motion is free. In particular, both the incident and the target particles can then be considered as free. In this case, the kinetic energy of the relative motion of the

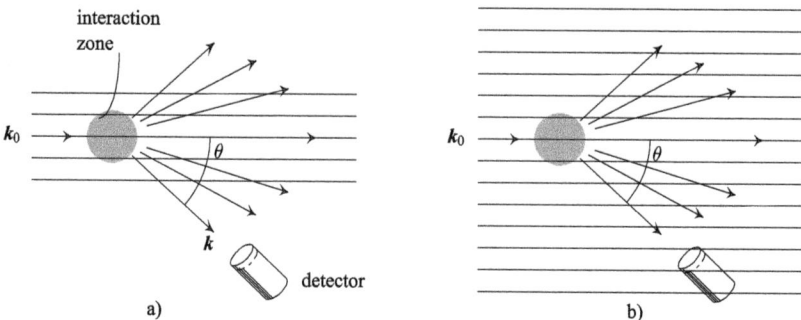

Figure 18.2 A typical scattering experiment seen from the laboratory frame. The target produces a short-range potential of radius R. (a) In the physical situation, the particle beam has a finite width. (b) The simplest theoretical description assumes that the incident beam corresponds to a plane monochromatic wave filling the whole space. The nonscattered particles should not be counted, as they are only part of the (idealized) description.

free particles is the total energy of the system. Since the incident particles move along parallel trajectories before entering the interaction region, as shown in Fig. 18.2(a), the orbital momentum about the origin, which is placed at the center of the target (where we assume the interaction to originate) varies from particle to particle, so the wave function will be a superposition of all possible orbital momentum states. Here we restrict ourselves to the case of spinless particles.

Based on the previous discussion, the wave function describing the ensemble of particles scattered in the asymptotic region ($r \to \infty$) can be written as the product of the asymptotic radial function $R(r) = u(r)/r$ with an angular wave function, which is not an angular momentum eigenstate and which we will denote by $f(\theta, \varphi)$. We therefore write

$$\psi_{\text{disp}} = f(\theta, \varphi) \frac{u(r)}{r} = f(\theta, \varphi) \frac{e^{ikr}}{r} \quad (r \to \infty), \tag{18.5}$$

where $f(\theta, \varphi)$ is the *scattering amplitude* and

$$k = \frac{\sqrt{2mE}}{\hbar}, \quad E > 0. \tag{18.6}$$

Since the detector is always placed very far away from the target to ensure that the scattering process is completed, we use this wave function to calculate the number of particles that are counted, to avoid having to describe the scattering process in more detail. The complete wave function has an additional component corresponding to the particles not scattered by the target, as can be seen in Fig. 18.2(b). Since these form a packet of free particles ψ_{inc}, we can write the total wave function away from the target as

$$\psi = \psi_{\text{inc}} + f(\theta, \varphi) \frac{e^{ikr}}{r}. \tag{18.7}$$

It is possible to considerably simplify the theoretical description by replacing ψ_{inc} with a plane wave whose energy is equal to the mean energy of the packet, as shown in Fig. 18.2(b). Using the conventional normalization of one incident particle per unit

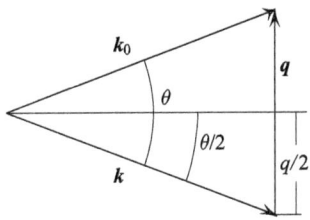

Figure 18.3 The vector q represents the momentum transferred from the particle to the target.

volume ($\rho_{\text{inc}} = 1$), we write ψ as follows, with the z-axis along the direction of incidence,[2]

$$\psi = e^{ikz} + f(\theta, \varphi) \frac{e^{ikr}}{r}. \tag{18.8}$$

The parameter k in the first term is the magnitude of the momentum of the incident particles $\mathbf{k}_0 = \hat{\mathbf{a}}_z k$; the k in the scattered component (spherical wave) is the magnitude of the momentum $\mathbf{k}$ of the scattered particles, which, for very large distances from the target, is essentially oriented along $\hat{\mathbf{a}}_r = \mathbf{r}/r$. Since the dispersion is elastic, $|\mathbf{k}_0| = |\mathbf{k}| = k$, but the two vectors have different directions, so that there is a momentum transfer $\mathbf{q}$ given by

$$\mathbf{k}_0 = \mathbf{k} + \mathbf{q}. \tag{18.9}$$

The quantity q is easily determined from Fig. 18.3, where θ is the scattering angle in the CM frame; it is clear that

$$q = 2k \sin \frac{\theta}{2}. \tag{18.10}$$

This is the momentum with which the target will move if it is initially at rest.

To count the number of particles hitting the detector we use (18.8) without the plane-wave contribution (counters are never placed exactly in front of the target). Take a small area dA; if the detector is placed at a large distance r from the target and its aperture is $d\Omega$ ester-radians, then $dA = r^2 d\Omega$; see Fig. 18.4. In the time dt all particles within the solid angle $d\Omega$ that are at a distance not greater than $dr = v\,dt$, where v is the average velocity of the particle flow, $v = \hbar k/m$, will enter the detector.

Since, on the other hand, the density of scattered particles is $|\psi_{\text{sc}}|^2$, the number of particles hitting the detector during dt is

$$dn = \rho dV = |\psi_{\text{sc}}|^2 dV = |\psi_{\text{sc}}|^2 r^2 v d\Omega dt. \tag{18.11}$$

The average number of particles registered is ηdn, where η is the counting efficiency of the detector; in the following we will use $\eta = 1$, which is often far from reality. Obviously, the count depends not only on the time during which the particles are recorded, but also on how many have been sent to the target during that time. In order to make the final result independent of these circumstantial elements, it is customary to report

---

[2] To study the scattering of photons by atoms, it is necessary to build correct asymptotic states, taking into account that the radiation and the atom are never disconnected; see Cohen–Tannoudji, Dupont–Roc, and Grynberg (1992).

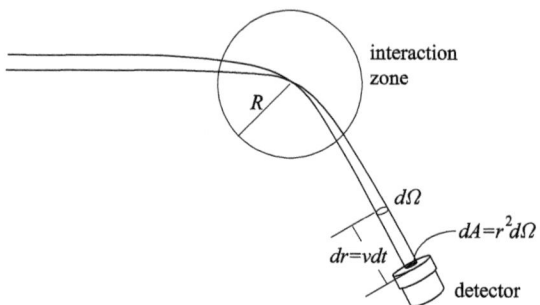

Figure 18.4 All the particles inside the cone with the base $r^2 d\omega$ and the height $dr = vdt$ will hit the detector within the time interval dt.

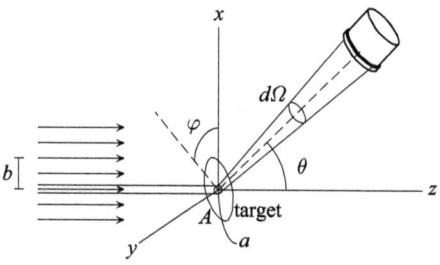

Figure 18.5 The fraction of particles incident on the target area A are scattered toward the detector. Alternatively, the particles incident on the effective target area $a = d\sigma$ will hit the detector. b is the impact parameter.

the value of dn properly normalized. Since the number of particles hitting the target per unit area and per unit time is $j_{\text{inc}} = \rho_{\text{inc}} v_{\text{inc}} = \hbar k/m$ (remember that the normalization $\rho_{\text{inc}} = 1$ was adopted), it is customary to report the quantity

$$d\sigma \equiv \frac{dn/dt}{j_{\text{inc}}}, \tag{18.12}$$

which is the number of particles detected per unit time per unit incident flow. The quantity $d\sigma$ is called differential *scattering cross section* (or *effective cross section*), because it has area dimensions,

$$\left[\frac{|\psi|^2 dV}{\rho_{\text{inc}} v dt}\right] = \left[\frac{1}{\rho_{\text{inc}} v dt}\right] = L^3 L^{-1} T T^{-1} = L^2. \tag{18.13}$$

Substituting dn and $j_{\text{inc}} = \hbar k/m$ in Eq. (18.12) and using Eq. (18.5), we obtain the important formula for the differential cross section,

$$\frac{d\sigma}{d\Omega} = |f(\theta,\varphi)|^2. \tag{18.14}$$

To give this effective scattering area a geometric sense, we turn to Fig. 18.5. In this figure, A represents the area of the target, while a represents the effective scattering area in the direction (θ,φ), that is, all particles falling on a have been deflected toward the detector with aperture $d\Omega$. On the other hand, if N is the number of particles hitting the target in one second, the incident flux is

$$j_{\text{inc}} = \frac{N}{A}. \tag{18.15}$$

Now let n be the number of particles that hit the detector in one second. From the definition of a, it follows that $n/M = a/A$, that is,

$$a = \frac{n}{N}A = \frac{n}{N/A} = \frac{n}{j_{\text{inc}}}. \tag{18.16}$$

But the ratio n/j_{inc} is exactly the scattering cross section given by (18.12), therefore a is the same as the effective cross section,

$$a = d\sigma. \tag{18.17}$$

18.2 The Born Approximation

For Eq. (18.14) to be useful it is necessary to determine the scattering amplitude, which is equivalent to solving the Schrödinger equation, at least for the asymptotic region. This can be done in different ways, depending on the specific conditions. Here we will study the Born approximation, which is applicable when the collision energy is high, or when the scattering potential is low, or when both conditions are met. To do this, we write the Schrödinger equation for the problem

$$\left(-\frac{\hbar^2}{2m}\nabla^2 + V\right)\psi = E\psi = \frac{\hbar^2 k^2}{2m}\psi \tag{18.18}$$

in the equivalent form

$$(\nabla^2 + k^2)\psi = \frac{2m}{\hbar^2}V\psi. \tag{18.19}$$

This can be transformed into an integral equation by the simple expedient of considering the right side as a source $\rho(r)$, so that

$$\frac{2m}{\hbar^2}V\psi = -4\pi\rho; \tag{18.20}$$

Equation (18.19) becomes thus a Helmholtz equation with a source,

$$(\nabla^2 + k^2)\psi = -4\pi\rho. \tag{18.21}$$

Comparing with Exercise E7.3, we see that the solution of this equation can be written in the form

$$\psi_{\text{sc}}(r) = \int G(r, r')\rho(r')\, d^3r' = -\frac{m}{2\pi\hbar^2}\int G(r, r')V(r')\psi(r')\, d^3r', \tag{18.22}$$

where $G(r, r')$ is the Green function of the operator $\nabla^2 + k^2$, that is, a solution of (18.21), which can be written as a linear combination of the functions G_+ and G_-, see Eqs. (7.87) and (7.88). In the present case, the asymptotic scattering solution is an *outgoing* spherical wave e^{ikr}/r (see (18.5)), of the form of G_+. So we take $G = G_+ = e^{ikR}/R$, with $R = |r - r'|$, and get

$$\psi_{\text{sc}}(r) = -\frac{m}{2\pi\hbar^2}\int \frac{e^{ik|r-r'|}}{|r-r'|}V(r')\psi(r')\, d^3r'. \tag{18.23}$$

This is the required integral equation, which replaces the Schrödinger differential equation plus the boundary condition (outgoing wave at infinity). To this expression we must add an arbitrary solution of the homogeneous equation $(\nabla^2 + k^2)\psi = 0$. To obtain (18.8) we choose the solution e^{ikz}, so that

$$\psi(\mathbf{r}) = e^{ikz} - \frac{m}{2\pi\hbar^2} \int \frac{e^{ik|\mathbf{r}-\mathbf{r}'|}}{|\mathbf{r}-\mathbf{r}'|} V(\mathbf{r}')\psi(\mathbf{r}')d^3r'. \tag{18.24}$$

It is important to note that the ψ appearing in the integrand of (18.24) is the complete one.

The potential $V(\mathbf{r}')$ in (18.23) and (18.24) is nonzero only in the region where $r' \leq R \ll r$. Therefore, the denominator can be approximated as $|\mathbf{r}-\mathbf{r}'| \simeq r$, while in the exponential we have to go one order further in the approximation and write

$$k|\mathbf{r}-\mathbf{r}'| = kr - \mathbf{k}\cdot\mathbf{r}', \quad |\mathbf{r}-\mathbf{r}'| \doteq r - \mathbf{r}'\cdot\hat{\mathbf{a}}_r, \tag{18.25}$$

where $\hat{\mathbf{a}}_r$ is a unit vector along the direction of $\mathbf{r}$. With these approximations, the wave function in the asymptotic region becomes

$$\psi(\mathbf{r}) = e^{ikz} - \frac{m}{2\pi\hbar^2}\frac{e^{ikr}}{r}\int V(\mathbf{r}')\psi(\mathbf{r}')e^{-i\mathbf{k}\cdot\mathbf{r}'}d^3r'. \tag{18.26}$$

Comparing with Eq. (18.8), we obtain the following exact, but formal, expression for the scattering amplitude:

$$f(\theta,\varphi) = -\frac{m}{2\pi\hbar^2}\int V(\mathbf{r})\psi(\mathbf{r})e^{-i\mathbf{k}\cdot\mathbf{r}}d^3r. \tag{18.27}$$

The original problem has thus been transformed into one of solving the integral equation (18.26) for $\psi(\mathbf{r})$. Since it is not normally possible to do this in closed form, it is customary to use an iteration method assuming that ψ_{sc} is small relative to ψ_{inc}.[3] This will be the case if (a) the particles are very energetic and cross the interaction region very quickly, so that they are slightly deflected from their initial trajectory by the scattering centers, or (b) the scattering potential is weak, or finally (c) both occur. Then the ψ in the integrand can be approximated by $\psi(\mathbf{r}) \simeq e^{ikz} = e^{i\mathbf{k}_0\cdot\mathbf{r}}$, giving the asymptotic wave function

$$\psi(\mathbf{r}) = e^{ikz} - \frac{m}{2\pi\hbar^2}\frac{e^{ikr}}{r}\int V(\mathbf{r}')e^{i\mathbf{q}\cdot\mathbf{r}'}d^3r', \tag{18.28}$$

where $\mathbf{q} = \mathbf{k}_0 - \mathbf{k}$ is the transferred momentum, see Eq. (18.9). This gives for the scattering amplitude in the first Born approximation

$$f(\theta,\varphi) = -\frac{m}{2\pi\hbar^2}\int V(\mathbf{r})e^{i\mathbf{q}\cdot\mathbf{r}}d^3r. \tag{18.29}$$

We see that, up to a numerical factor, the scattering amplitude is given in this approximation by the Fourier transform of the scattering potential. This is a very useful result and we will have the opportunity to apply it to some important problems. For now, let us use it to draw some general conclusions.

---

[3] Think of light passing through a window pane; most of it passes through the glass and only a small percentage (usually less than 5%) is reflected, that is, scattered.

18.2.1 Scattering by a periodic potential

We first consider the case of a periodic potential, which can be written in the form

$$V(r) = \sum A_n \exp 2\pi i \left[\frac{x}{a}n_1 + \frac{y}{b}n_2 + \frac{z}{c}n_3\right] = \sum A_n \exp i g_n \cdot r. \tag{18.30}$$

This can be, for example, the potential generated by a crystal lattice with parameters (a, b, c), similar to the Kronig and Penney model discussed in Section 6.3 or any other periodic structure of atomic dimensions; g_n is the corresponding reciprocal lattice, $g_n = 2\pi \left(\frac{n_1}{a}i + \frac{n_2}{b}j + \frac{n_3}{c}k\right)$. Applying Eq. (18.29), we get

$$f(\theta, \varphi) = -\frac{m}{2\pi \hbar^2} \sum A_n \int e^{i(q+g_n)\cdot r'} d^3 r' = -\frac{4\pi^2 m}{\hbar^2} \sum A_n \delta(q+g_n), \tag{18.31}$$

which indicates that the momentum transferred during the interaction is quantized and can only take the discrete values

$$-\hbar q_n = \hbar (k_n - k_0) = \hbar g_n = 2\pi \hbar \left(\frac{n_1}{a}i + \frac{n_2}{b}j + \frac{n_3}{c}k\right). \tag{18.32}$$

We conclude that when electrons (quantum corpuscles in general) are scattered by a periodic structure, the momentum transferred in the direction of the periodicity is quantized. This result was first formulated by W. Duane (1872–1935) in 1923, shortly before the advent of QM, as a principle derived from the old quantization rules. A. Landé and other authors even considered it to be the basic principle of quantization at that time, now it can be seen as a consequence of Schrödinger's theory.

It should be clear that what matters in the preceding derivation is the existence of a periodicity capable of influencing the behavior of matter, and not the nature of that periodicity. This means that different sources of periodicity can lead to a similar quantization of the exchanged momentum. An interesting example is the phenomenon known as the Kapitza[4]–Dirac effect, which consists of the diffraction of a molecular beam by a stationary electromagnetic wave, that is, the inverse phenomenon of the diffraction of light by a grating.

18.2.2 Scattering by a weak potential

Returning to the general case, let us now consider a problem in which the scattering potential is so weak that Eq. (18.29) holds even at very low energies, when $kR \ll 1$. It follows from Eq. (18.10) that in this case $qR \ll 1$ will also hold, and therefore the exponent of the integral in (18.29) will remain small throughout the integration region where $V(r')$ is different from zero; neglecting it, we can write

$$f(\theta, \varphi) \simeq f(0), \tag{18.33}$$

---

[4] Pyotr Leonidovich Kapitsa (1894–1984) was an outstanding Russian–Soviet physicist. His series of experiments to study liquid helium culminated in the discovery of superfluidity in 1937. In 1978 Kapitsa was awarded the Nobel Prize in Physics for his fundamental inventions and discoveries in the field of low-temperature physics.

where

$$f(0) = -\frac{m}{2\pi\hbar^2}\int V(r)d^3r = -\frac{m}{2\pi\hbar^2}\overline{V}R^3 \tag{18.34}$$

is the forward scattering amplitude, in all cases (even at high energies); $\overline{V}$ is an average potential such that $\int V(r)d^3r \equiv \overline{V}R^3$, and R is the range of the potential.

According to (18.33), at low energies the scattering produced by a short-range potential is isotropic. The *total cross section*, defined as the integral over all directions of the differential cross section – that is, as the total number of particles scattered in all directions for each particle incident on a unit area of the target – and given by

$$\sigma \equiv \int_\Omega \frac{d\sigma}{d\Omega}d\Omega = \int |f(\theta,\varphi)|^2 d\Omega, \tag{18.35}$$

is thus independent of the energy for low energies,

$$\sigma = \int |f(\theta,\varphi)|^2 d\Omega \simeq 4\pi |f(0)|^2 = \frac{1}{\pi}\left(\frac{m\overline{V}R^3}{\hbar^2}\right)^2. \tag{18.36}$$

The case of high energies corresponds to $kR \gg 1$; except within a narrow cone where the angle θ is small, which we exclude, also $qR \gg 1$ and the exponential of the integral in Eq. (18.29) oscillates violently, vanishing the integral. The effective contribution to the integral occurs within the cone for which $qR = 2kR\sin(\theta/2) \leq 1$ holds. Since $kR \gg 1$, this condition is only satisfied for angles θ for which $\sin\theta \leq 1/kR \ll 1$, so the differential section has a sharp peak around the direction of incidence of width $1/kR$, and vanishes for larger scattering angles. We conclude that the scattering produced by a short-range potential is highly anisotropic at high bombardment energies, as shown in Fig. 18.6. In this case we can approximate as follows:

$$\sigma \simeq 2\pi \int_0^{1/kR} |f(\theta)|^2 \sin\theta d\theta \simeq 2\pi |f(0)|^2 \int_0^{1/kR} \theta d\theta$$

$$= \frac{\pi |f(0)|^2}{k^2R^2} = \frac{1}{4\pi}\left(\frac{m\overline{V}R^2}{\hbar^2 k}\right)^2, \tag{18.37}$$

that is,

$$\sigma = \frac{1}{8\pi}\left(\frac{\overline{V}R^2}{\hbar}\right)^2 \frac{m}{E}. \tag{18.38}$$

We see that at high energies the scattering cross section decreases with the energy. This is because the interaction time decreases as the bombardment energy increases.

The region of validity of the Born approximation can be evaluated by noting that the approximation is satisfactory only in the case where the total section σ is small with respect to the geometric area presented by the target,

$$\sigma \ll \pi R^2, \tag{18.39}$$

since otherwise ψ_{sc} cannot be neglected in (18.26) and (18.27), This condition can alternatively be expressed as a condition on $\overline{V}$ as follows. For low energies we substitute σ given by Eq. (18.36) into (18.39) to obtain

$$|\overline{V}| \ll \frac{\pi\hbar^2}{mR^2}, \tag{18.40}$$

while at high energies Eq. (18.38) requires that

$$|\overline{V}| \ll \frac{2\pi\hbar v}{R} = \frac{\pi\hbar^2}{mR^2} 2kR. \tag{18.41}$$

The first condition (18.40) is generally stronger than the second, since if it is satisfied, the approximation is valid for all bombardment energies. On the other hand, the second condition is always satisfied for sufficiently high energies, so that kR is as large as necessary; hence the Born approximation is considered to be a high-energy approximation.

Exercise E18.1. Scattering by a 1D delta potential

A particle is scattered by a one-dimensional delta potential at the origin. Find the wave function, the scattering amplitude and the energy of the bound state.

Solution. Assuming that the beam hits the scattering potential $K\delta(z)$ from the left of the z-axis, the wave function is given by the 1D version of Eq. (18.22) with $k = p/\hbar$,

$$\psi(z) = e^{ikz} + \frac{2m}{\hbar^2} \int_{-\infty}^{+\infty} G(z,z')V(z')\psi(z')dz', \tag{18.42}$$

where $G(z,z')$ is the outgoing Green function of the operator $(d^2/dx^2) + k^2$,

$$G(z) = \int_{-\infty}^{+\infty} \frac{e^{ik'z}}{k'^2 - k^2 + i\varepsilon}dk' = -\frac{i}{2k}e^{ik|z|}. \tag{18.43}$$

Therefore,

$$\psi(z) = e^{ikz} - \frac{imK}{\hbar^2 k} \int_{-\infty}^{+\infty} e^{ik|z-z'|}\delta(z')\psi(z')dz'$$

$$= e^{ikz} - \frac{imK}{\hbar^2 k}\psi(0)e^{ik|z|} = e^{ikz} - \frac{imK}{\hbar^2 k + imK}e^{ik|z|}. \tag{18.44}$$

This result shows that for $k \to 0$ virtually all the particles are reflected by the potential and the transmission coefficient goes to zero. Instead, for $k \to \infty$ the scattering potential becomes transparent.

To construct the scattering amplitude for the 1D case we consider the limit $|z| \to \infty$ and write, with $k' = k(z/|z|)$,

$$\psi(z) = e^{ikz} + e^{ik|z|}f(k',k), \tag{18.45}$$

$$f(k',k) = -\frac{im}{\hbar^2 k} \int_{-\infty}^{+\infty} e^{-ikz'}V(z')\psi(z')dz'. \tag{18.46}$$

Comparing with (18.44), we obtain for the dispersion by the delta potential

$$f(k',k) = -\frac{imK}{\hbar^2 k + imK}. \tag{18.47}$$

This scattering amplitude has a pole at $p = \hbar k = -imK/\hbar$, corresponding to the energy

$$E = p^2/2m = -\frac{mK^2}{2\hbar^2}. \tag{18.48}$$

Therefore the delta potential has a single bound state with energy E given by Eq. (18.48).

The result just derived, which is an example of the general property that establishes that the poles of the scattering amplitude for negative energies correspond to bound states, can be obtained with the simple methods used in Chapter 5. To do this, it is enough to write the wave function in the form

$$\psi(z) = Ae^{ikz} + Be^{-ikz}, \quad z < 0,$$
$$\psi(z) = Ce^{ikz}, \quad z \geq 0.$$

From the continuity conditions at the origin of the wave function and its derivative (its discontinuity at $z = 0$ is obtained by integrating the Schrödinger equation in a small interval around the origin) we arrive at

$$\frac{B}{A} = \frac{mK}{i\hbar^2 k - mK}, \quad \frac{C}{A} = \frac{i\hbar^2 k}{i\hbar^2 k - mK}.$$

For a bound state the energy is negative; in this case $\hbar k = \sqrt{-2m|E|} \equiv i\alpha$, where $\alpha = \sqrt{2m|E|} \geq 0$. The scattering amplitude is then (compare with Eq. (18.47))

$$\frac{B}{A} = -\frac{mK}{\hbar\alpha + mK}$$

and has a single pole at $\alpha = -mK/\hbar$, in agreement with Eq. (18.48); see also problem P5.5.

18.2.3 Scattering by a central potential: Rutherford cross section

A general expression for the scattering amplitude for central potential problems is obtained by performing the angular integrals in Eq. (18.29). With the z-axis of the reference frame along the vector $\mathbf{q}$, $\mathbf{q} \cdot \mathbf{r} = qr\cos\theta$ and we get

$$f(\theta) = -\frac{m}{2\pi\hbar^2} \int_0^\infty r^2 dr \int_0^\pi \sin\theta' d\theta' \int_{-\pi}^\pi d\varphi' V(r) e^{iqr\cos\theta'}$$
$$= -\frac{m}{\hbar^2} \int_0^\infty r^2 V(r) dr \int_{-1}^1 e^{iqrx} dx, \tag{18.49}$$

or

$$f(\theta) = -\frac{2m}{\hbar^2 q} \int_0^\infty rV(r) \sin qr\, dr, \tag{18.50}$$

which is the desired result.

Let us consider the case, which is particularly important in atomic physics, where V represents a shielded Coulomb potential between the charges $Z_1 e$ and $Z_2 e$. The shielding effect in an atom is due to the fact that the orbital electrons partially shield or insulate the nucleus, as discussed in Section 16.3. A quantum evaluation shows that this effect can be taken into account by introducing the shielding factor $e^{-\alpha r}$, where α depends on the structure of the atom. Substituting $V_0 R = Z_1 Z_2 e^2$ and $R = \alpha^{-1}$ in (18.53), we obtain, for the Coulomb scattering cross section,

$$\frac{d\sigma}{d\Omega} = \left(\frac{2m Z_1 Z_2 e^2 R^2}{\hbar^2}\right)^2 \frac{1}{(1 + 4k^2 R^2 \sin^2 \theta/2)^2}. \tag{18.51}$$

For sufficiently high energies, $kR \gg 1$ and this expression becomes independent of R (for θ not too small),

$$\frac{d\sigma}{d\Omega} = \frac{Z_1^2 Z_2^2 m^2 e^4}{4 p^4 \sin^4 \theta/2} \quad (\theta \neq 0). \tag{18.52}$$

Since this result no longer depends on $\alpha = R^{-1}$, we can apply it to the limiting case $\alpha = 0$, that is, to the Coulomb potential. Therefore, outside the cone $\theta < 1/kR = \alpha/k \to_{\alpha \to 0} 0$, Eq. (18.52) gives the angular distribution produced by a Coulomb potential at sufficiently high energies $kR = k/\alpha \gg 1$. This is precisely the formula derived by Rutherford for the scattering cross section of α particles by heavy atoms using purely classical methods; for this reason it is called the *Rutherford cross section*. This happy coincidence was made possible by the fact that the quantum formula (18.52) is independent of $\hbar$.

The Coulomb potential is long-range, so the preceding methods are only valid for high energies. However, the Coulomb scattering problem has an exact solution in a parabolic coordinate system, which once again leads to the formula (18.52). We will not go into this problem here, but refer the student to the literature on the subject.

Exercise E18.2. Scattering by a Yukawa potential

The Yukawa potential

$$V = V_0 \frac{e^{-r/R}}{r/R} = V_0 R \frac{e^{-r/R}}{r} \tag{18.53}$$

is important for at least two reasons: first, it represents the static approximation to the nuclear interaction potential, and second, it describes – with a simple change of parameters – the partially shielded Coulomb potential (and in the limit $R \to \infty$ the Coulomb potential itself). Apply the previous results to the study of the dispersion produced by the Yukawa potential.

Solution. Applying the general formula (18.50) to the potential (18.53), we obtain

$$f(\theta) = -\frac{2m}{\hbar^2 q} V_0 R \int_0^\infty e^{-r/R} \sin qr \, dr = -\frac{2m V_0}{\hbar^2} \frac{R^3}{1 + q^2 R^2}. \tag{18.54}$$

For the differential cross section we thus get

$$\frac{d\sigma}{d\Omega} = |f(\theta)|^2 = \frac{a^2}{(1 + q^2 R^2)^2}, \tag{18.55}$$

where we introduced the abbreviation

$$a = \frac{2m V_0 R^3}{\hbar^2}. \tag{18.56}$$

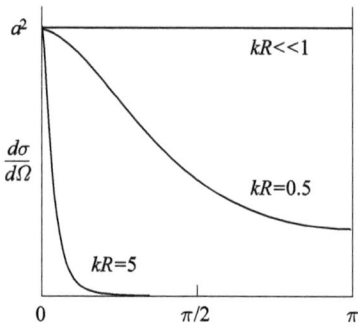

Figure 18.6 Differential cross section for a short-range potential as a function of the scattering angle and the parameter kR. At very low energies the cross section is isotropic, while at very high energies the particles are only slightly dispersed. The maximum value of the differential cross section is a^2.

Fig. 18.6 shows how the differential cross section varies as a function of the scattering angle, for different values of kR; these results exemplify the characteristic general behavior of short-range potentials, as discussed earlier. Using formula (18.10) for q we see that the total cross section is

$$\sigma = 2\pi \int_0^\pi d\theta \sin\theta \frac{a^2}{(1+k^2R^2 \cdot 4\sin^2\theta/2)^2}$$

$$= 2\pi a^2 \int_0^2 \frac{dz}{(1+2k^2R^2z)^2} = \frac{4\pi a^2}{1+4k^2R^2}. \tag{18.57}$$

At low energies σ has the approximate value $4\pi a^2$, which shows that $2a$ plays the role of an effective scattering radius; at high energies σ is inversely proportional to the bombardment energy, $E \sim k^2$, as expected.

18.3 Form Factors

In the previous analysis we considered the scattering potential to be point-like, but many important sources of interaction have a more or less extended structure, which means that the scattering potential has a significant spatial distribution. This is the case, for example, of the interaction of an electron (which we shall continue to consider as point-like) with an atom which has an essentially point-like positive nucleus, but whose orbital electrons produce an average charge density that may extend over a few Bohr radii and consequently affect the motion of the electron in a different way from more or less localized charges.

Recall that Rutherford used the scattering of α particles by heavy atoms to test Thomson's model of the atom as a distributed positive charge with electrons embedded in it, and concluded that the model was invalid. We will see in what follows how we can treat this type of situation, apply the results to the Rutherford case, and determine when the distributed charge of the electrons matters and when it does not.

18.3 Form Factors

We will look at the electrostatic case for the sake of simplicity. Considering a point charge e (the projectile) interacting with a distributed electric charge of density $e'\rho(r)$, where the density ρ is normalized to unity, $\int \rho d^3r = 1$, the interaction potential is

$$V(r) = ee' \int \frac{\rho(r')}{|r - r'|} d^3r', \tag{18.58}$$

where the integral extends over the volume of the extended charge. Substituting in Eq. (18.29) (with an obvious change in the names of the variables), we get

$$f(\theta) = -\frac{mee'}{2\pi\hbar^2} \int\int \frac{\rho(r')}{|r - r'|} e^{iq\cdot r} d^3r d^3r'. \tag{18.59}$$

This expression is conveniently rewritten with the change of variable $r - r' = R$,

$$f(\theta) = -\frac{mee'}{2\pi\hbar^2} \int \frac{e^{iq\cdot R}}{R} d^3R \int \rho(r') e^{iq\cdot r'} d^3r'. \tag{18.60}$$

The second integral,

$$F(q) \equiv \int \rho(r) e^{iq\cdot r} d^3r, \tag{18.61}$$

which is the Fourier transform of the charge distribution of the target, is generically called the *form factor* (in the present case it is the electric form factor). If $\rho(r)$ has spherical symmetry, the form factor depends only on q; we will assume that this is the case in what follows. The scattering amplitude is then

$$f(\theta) = -\frac{mee'}{2\pi\hbar^2} F(q) \int \frac{e^{iq\cdot r}}{r} d^3r. \tag{18.62}$$

To express this result in terms of the scattering amplitude produced by a point-like potential, we note that ee'/r is the interaction potential between two point charges e and e', so the previous result can be rewritten as (cf. Eq. (18.29))

$$f(\theta) = F(q) f_{\text{point}}(\theta). \tag{18.63}$$

It is clear that this result is general in the sense that it gives the relationship between the point scattering amplitude and the corresponding amplitude for the problem with structure. It follows that the differential scattering cross section of these two problems is generally related as follows,

$$\left.\frac{d\sigma}{d\Omega}\right|_{\text{ext}} = |F(q)|^2 \left.\frac{d\sigma}{d\Omega}\right|_{\text{point}}. \tag{18.64}$$

This very important result shows that by comparing the scattering produced by a target with structure (which determines $d\sigma/d\Omega|_{\text{ext}}^{\text{exp}}$) and the theoretical cross section calculated for a point target (which requires prior knowledge of the interaction potential), we can determine the form factor (or, where appropriate, the form factors) of the potential source distribution, that is, we can study the internal structure of the target–projectile system, given the interaction (see problem P18.8).

18.3.1 Atomic form factor

As an example we will study the dispersion of electrons produced by an atom. The atomic charge density can be written as the sum of a point-like contribution from the nucleus $Ze_0\delta(r)$ and the electronic contribution that corresponds to a charge $-Ze_0$ distributed over the entire volume of the atom, with density $-e_0 Z\rho_{\text{elect}}(r)$, so that

$$Ze_0\rho(r) = Ze_0[\delta(r) - \rho_{\text{elect}}(r)]. \tag{18.65}$$

The atomic (electric) form factor is then

$$F(q) = \int d^3r [\delta(r) - \rho_{\text{elect}}(r)] e^{i q \cdot r} = 1 - F_{\text{elect}}(q). \tag{18.66}$$

The form factor associated with the spatial density of electrons is given by

$$F_{\text{elect}} = \int d^3 r \rho_{\text{elect}}(r) e^{i q \cdot r}, \tag{18.67}$$

where $\rho_{\text{elect}} = |\psi|^2$ is the spatial density of orbital electrons normalized to unity,

$$\int \rho_{\text{elect}}(r) d^3 r = 1. \tag{18.68}$$

The scattering cross section of a charged particle by an atom is then given by

$$\frac{d\sigma}{d\Omega} = |1 - F_{\text{elect}}(q)|^2 \left(\frac{d\sigma}{d\Omega}\right)_{\text{Rutherford}}. \tag{18.69}$$

We will analyze two limiting cases of this expression. First, consider q small, $qa \ll 1$, where a is the atomic radius. This can be either because the bombardment energy is small $ka \ll 1$, or for very small angles, $\theta \ll \hbar/mav \equiv v_0/v$, where $v_0 = \hbar/ma = \alpha c$ can be interpreted as a characteristic speed of the orbital motion of the electrons; in particular for $\theta \ll 1$, $qa \approx ka\theta \ll 1$, $\theta \ll 1/ka$. In this case we can expand the exponential in Eq. (18.67) into a power series, keeping only the first few terms, to obtain

$$F_{\text{elect}} = \int \rho_{\text{elect}}(r)[1 + i q \cdot r - \tfrac{1}{2}(q \cdot r)^2 + \ldots] d^3 r =$$
$$= 1 + \frac{i q}{Ze} \cdot d - \tfrac{1}{6} q^2 \int r^2 \rho_{\text{elect}}(r) d^3 r, \tag{18.70}$$

where d is the electric dipole moment of the atom, which we should take to be zero in the general case. In the third integral we took into account that for an isotropic distribution,

$$\overline{(q \cdot r)^2} = \overline{q^2 z^2} = \tfrac{1}{3} q^2 \overline{(x^2 + y^2 + z^2)} = \tfrac{1}{3} q^2 \overline{r^2}. \tag{18.71}$$

Therefore, if r_0 is the mean square value of the distance of the electrons from the atomic nucleus, that is,

$$r_0^2 \equiv \int r^2 \rho_{\text{elect}}(r) d^3 r, \tag{18.72}$$

we get up to second order

$$F_{\text{elect}} = 1 - \tfrac{1}{6} q^2 r_0^2. \tag{18.73}$$

We see that at low energies the form factor of the electron distribution is very close to unity. This is because the orbital electrons have time to "average" the effects of the projectile, producing a result close to what they would produce if they were concentrated at the origin. Substituting into Eq. (18.69), we get

$$\frac{d\sigma}{d\Omega} = \left(\tfrac{1}{6}q^2 r_0^2\right)^2 \left(\frac{d\sigma}{d\Omega}\right)_{\text{Rutherford}} \qquad (qr_0 \ll 1). \tag{18.74}$$

We see that the scattering cross section is considerably reduced by the orbiting electrons, whose effects tend to cancel out that of the nucleus. Introducing here the value of the Rutherford cross section given by Eq. (18.52) and the value of q given by Eq. (18.10), the angular dependence of both factors cancels out and we obtain

$$\frac{d\sigma}{d\Omega} = \left(\frac{Zme_0^2 r_0^2}{3\hbar^2}\right)^2. \tag{18.75}$$

This result shows that studying the small-angle scattering of slow electrons gives information about the spatial distribution of electrons in a complex atomic or molecular system.

The case $qa \gg 1$ occurs at high energies ($ka \gg 1$) and scattering angles greater than v_0/v. In this case the exponential in Eq. (18.67) oscillates rapidly and the integral becomes small compared to unity; neglecting it, Eq. (18.66) gives $F(q) = 1$, which leads to

$$\frac{d\sigma}{d\Omega} = \frac{d\sigma}{d\Omega}\bigg|_{\text{Rutherford}}. \tag{18.76}$$

In this case the spatial structure of the atom is not manifested in the scattering, that is, only the nuclear point charge is effective in the scattering. This was precisely the case in Rutherford's experiments.

For the H atom we can proceed without any approximation. In the ground state

$$\rho_{\text{elect}}(r) = \frac{1}{\pi a_0^3} e^{-2r/a_0}, \tag{18.77}$$

and the electronic form factor is

$$F_{\text{elect}}(q^2) = \frac{1}{\pi a_0^3} \int e^{-2r/a_0 + i\mathbf{q}\cdot\mathbf{r}} d^3 r = \frac{1}{\left(1 + \tfrac{1}{4}q^2 a_0^2\right)^2}. \tag{18.78}$$

Substituting in (18.69) and simplifying, we get

$$\frac{d\sigma}{d\Omega} = \left(\frac{Zme^2 a_0^2}{\hbar^2}\right)^2 \left(\frac{1 + \tfrac{1}{8}a_0^2 q^2}{1 + \tfrac{1}{4}a_0^2 q^2}\right)^2. \tag{18.79}$$

This cross section is weakly dependent on q and thus on the scattering angle.

18.4 Partial-Wave Expansion. Introducing the S Matrix

Using Born's method, we have obtained an approximate expression for the scattering amplitude. We will now present a method, called *partial waves*, which allows us to obtain an exact expression for $f(\theta, \varphi)$. The basic idea is to decompose this function into an infinite sum, each term of which is associated with a given value of the orbital momentum. The method is quite complicated when it comes to applying it to noncentral potentials (e.g., the problem of particle scattering by a crystalline lattice), so in practice it is mainly used with central potentials, which is the only case we will consider here.

In Sections 5.4 and 6.2 we studied particle scattering by a 1D potential and showed that a significant effect of the potential is a phase shift with respect to the wave in the absence of the scattering potential. Here we will show that this is true for any potential and draw some general conclusions for central potentials. It is highly recommended to read Section 12.1 in preparation for this exposition.

Since, by hypothesis, the potential is short-range, the term $V(r)u$ in the Schrödinger equation is negligible in the asymptotic region $r \to \infty$ and the equation for a given l reduces to the free-particle equation,

$$\left(-\frac{d^2}{dr^2} + \frac{l(l+1)}{r^2} - k^2\right) u = 0, \quad r \to \infty, \tag{18.80}$$

the solution of which is

$$R(r) = \frac{u(r)}{r} = A_l j_l(kr) + B_l n_l(kr). \tag{18.81}$$

We rewrite this in the equivalent form

$$R = \alpha_l [j_l(kr) \cos \delta_l - n_l(kr) \sin \delta_l]. \tag{18.82}$$

To investigate the behavior of this solution at infinity we use the asymptotic expressions for the spherical Bessel and Neumann functions (see Section A.1 of the Appendix A). For $\rho \gg l$ we have

$$j_l(\rho) \approx \frac{1}{\rho} \sin\left(\rho - \frac{\pi}{2}l\right), \tag{18.83}$$

$$n_l(\rho) \approx -\frac{1}{\rho} \cos\left(\rho - \frac{\pi}{2}l\right). \tag{18.84}$$

Therefore, in the asymptotic region we get

$$u = rR \approx \frac{\alpha_l}{k} \left[\sin\left(kr - \frac{\pi}{2}l\right) \cos \delta_l + \cos\left(kr - \frac{\pi}{2}l\right) \sin \delta_l\right], \tag{18.85}$$

that is,

$$u(kr) \approx \frac{\alpha_l}{k} \sin\left(kr - \frac{\pi}{2}l + \delta_l\right). \tag{18.86}$$

In the absence of the scattering potential we have $\delta_l = 0$ and

$$u_0(kr) = \frac{A}{k} \sin\left(kr - \frac{\pi}{2}l\right). \tag{18.87}$$

Comparing the last two expressions we see that apart from a global factor α_l/A, the potential produces a *dephasing* of the radial wave function, equal to δ_l. For this reason, δ_l is called the *phase shift (or dephasing) of the wave l* (i.e., of the l component of the wave function).

18.4 Partial-Wave Expansion. Introducing the S Matrix

The integration constants A_l and B_l must be determined from the continuity conditions of the solution, that is, the proposed solution for $r \to \infty$ must be smoothly tied to the regular solution at the origin. In terms of an amplitude C_l and the phase δ_l, the solution is

$$R(r) = \frac{C_l}{kr} \sin\left(kr - \frac{\pi}{2}l + \delta_l\right), \quad kr \to \infty. \tag{18.88}$$

In the absence of the scattering potential, (18.88) holds for all r, including the origin. Since the solution must be regular and $n_l(0)$ is infinite, in this case we must take $B_l = 0$, thus $\delta_l = \arctan(-B_l/A_l) = 0$. In other words, δ_l represents the phase shift produced by the scattering potential on the l component.

Suppose that the incident beam has axial symmetry with respect to the z-axis. In this case the wave function does not depend on the azimuthal angle (which is equivalent to taking $m_l = 0$ for all l), the spherical harmonics reduce to Legendre polynomials, and we can write the asymptotic solution in the form

$$\psi = \sum_{l=0}^{\infty} \frac{C_l}{kr} \sin\left(kr - \frac{\pi}{2}l + \delta_l\right) P_l(\cos\theta). \tag{18.89}$$

This expression is an expansion in Legendre polynomials of the previously constructed solution, Eq. (18.8); equating them and isolating the scattering amplitude, we obtain

$$f(\theta) = -re^{ik(z-r)} + \sum_l \frac{C_l}{k} e^{-ikr} \sin\left(kr - \frac{\pi}{2}l + \delta_l\right) P_l(\cos\theta). \tag{18.90}$$

To determine the coefficients C_l we observe that they must be chosen so that the right-hand side of this equation does not depend on r, that is, it reduces to a function of θ. To introduce this condition, we have to express the plane wave e^{ikz} in terms of Legendre polynomials, which we can do with the help of the equation

$$e^{ikz} = \sum_{l=0}^{\infty} i^l (2l+1) j_l(kr) P_l(\cos\theta). \tag{18.91}$$

Since we are only interested in the value of this expression for $r \to \infty$, we can write instead of the spherical Bessel function its asymptotic value given by $j_l(kr) \simeq \frac{1}{kr} \sin(kr - \frac{\pi}{2}l)$, which gives

$$e^{ikz} = \sum_l i^l (2l+1) \frac{\sin\left(kr - \frac{\pi}{2}l\right)}{kr} P_l(\cos\theta), \quad kr \to \infty. \tag{18.92}$$

Introducing this expansion in (18.90) and writing the trigonometric functions in terms of exponentials, we obtain

$$f(\theta) = \frac{1}{2ik} \sum_l [C_l e^{i\delta_l} - i^l(2l+1)] e^{-i\frac{\pi}{2}l} P_l(\cos\theta)$$

$$- \frac{1}{2ik} e^{-2ikr} \sum_l \left[C_l e^{-i\delta_l} - i^l(2l+1)\right] e^{i\frac{\pi}{2}l} P_l(\cos\theta). \tag{18.93}$$

In order to remove the dependence on r the second sum must be canceled, which gives

$$C_l = i^l(2l+1) e^{i\delta_l}. \tag{18.94}$$

Substituting this value in the same Eq. (18.93) we obtain, taking into account that $i^l e^{-i\frac{\pi}{2}l} = (ie^{-i\frac{\pi}{2}})^l = 1$,

$$f(\theta) = \frac{1}{2ik} \sum_{l=0}^{\infty} (2l+1)(e^{2i\delta_l} - 1)P_l(\cos\theta) \equiv \sum_{l=0}^{\infty} f_l(\theta). \tag{18.95}$$

This important result, which is exact, forms the basis for all further analysis in terms of phase shifts. It can be expressed in various forms, which are obtained by using one of the following equivalent expressions

$$\frac{e^{2i\delta_l} - 1}{2i} = e^{i\delta_l} \sin\delta_l = a_l = k f_l = \frac{S_l(k) - 1}{2i}. \tag{18.96}$$

Both a_l and f_l are referred to interchangeably as the amplitude of the partial wave l; $S_l(k)$ is the element l of the scattering matrix S and is given by[5]

$$S_l(k) = e^{2i\delta_l}; \tag{18.97}$$

obviously $SS^\dagger = 1$, so S is unitary. A complete knowledge of the phases δ_l (or, equivalently, of the partial amplitudes a_l, or of the S matrix) as functions of energy, specifies the scattering amplitude $f(\theta)$ and this in turn allows us, in principle, to determine the scattering potential (e.g., by means of an inverse Fourier transformation if we use the Born approximation). In practice, however, this program requires overcoming major difficulties. This is a very simple example of the S matrix as a scattering matrix, the theory of which is highly relevant to more advanced studies.

18.4.1 The optical theorem

An expression for the total (elastic) scattering cross section can be obtained from Eq. (18.95). We have from (18.35) that

$$\sigma = \frac{2\pi}{k^2} \int_0^\pi d\theta \sin\theta \sum_{l,l'} (2l+1)(2l'+1) a_l^* a_{l'} P_l(\cos\theta) P_{l'}(\cos\theta); \tag{18.98}$$

taking into account the orthogonality of the Legendre functions

$$\int_{-1}^{1} P_l(x) P_{l'}(x) dx = \frac{2}{2l+1} \delta_{ll'}, \tag{18.99}$$

we obtain the formula

$$\sigma = \frac{4\pi}{k^2} \sum_{l=0}^{\infty} (2l+1)|a_l|^2 = \frac{4\pi}{k^2} \sum_{l=0}^{\infty} (2l+1) \sin^2\delta_l. \tag{18.100}$$

Since the maximum value of l satisfies $l_{max} \leq kR$, we see that the total cross section has the upper bound

$$\sigma_{max} \leq \frac{4\pi}{k^2} \sum_{l=0}^{l_{max}} (2l+1) = \frac{4\pi}{k^2} (l_{max} + 1)^2 \lesssim 4\pi R^2. \tag{18.101}$$

---

[5] An example of the use of the S matrix to describe elastic dispersion in a 1D problem was seen in Section 5.3.

From (18.95) it follows that

$$\mathrm{Im} f(0) = \frac{1}{k} \sum_l (2l+1) \sin^2 \delta_l P_l(1) \qquad (18.102)$$

and since $P_l(1) = 1$ for all l, comparing with (18.100), we obtain

$$\sigma = \frac{4\pi}{k} \mathrm{Im} f(0). \qquad (18.103)$$

This relation between the total cross section and the imaginary part of the forward scattering amplitude is called the *optical theorem* and is more general than what the previous derivation suggests (it is valid, e.g., for inelastic processes if σ represents the total section $\sigma = \sigma_{\mathrm{elast}} + \sigma_{\mathrm{inel}}$). This theorem is so called because it goes back to the optical work of Lord Rayleigh; in QM it was established in 1932 by E. Feenberg (1906–1977), but it is also known as the Bohr–Peierls–Placzek relation.

The meaning of the optical theorem can be easily understood by calculating the total flux of particles through a spherical surface in the asymptotic region centered on the origin of the coordinates (see problem P18.13),

$$J = -\frac{i\hbar}{2m} \int_\Omega (\psi^* \nabla \psi - \psi \nabla \psi^*) r^2 d\Omega = V\left(\sigma - \frac{4\pi}{k} \mathrm{Im} f(0)\right) \quad (r \to \infty). \qquad (18.104)$$

Since the process is elastic, the total number of particles is conserved and their net flux through the surface must vanish; this immediately leads to the optical theorem. On the other hand, if the phase shifts δ_l were not real, the optical theorem would not be satisfied as demonstrated here; we conclude that the unitary property of the scattering matrix S is a consequence of the flux conservation, as can be shown directly (problem P18.12).

From Eq. (18.95) it follows that the scattering amplitude can be written in the general form

$$f(\theta) = A_0 + A_1 \cos\theta + A_2 \cos^2\theta + \ldots = \sum_{n=0}^{\infty} A_n \cos^n \theta, \qquad (18.105)$$

so the differential cross section will have an analogous structure,

$$\frac{d\sigma}{d\Omega} = \sum_{n=0}^{\infty} B_n \cos^n \theta, \qquad (18.106)$$

which is a sum of terms produced by each value of l plus all possible interference terms. From the analysis of the experimental differential cross section, it is possible to determine the coefficients B_n of this expansion (which must be done for each energy value!) and, from them, the phase shifts as a function of energy. It is these latter data that have to be interpreted theoretically, which makes their theoretical determination a problem of prime importance. The problem is generally very complex and various methods have been developed to solve it, including numerical integration of the Schrödinger equation.

In particular, when formulas such as those given by the Born approximation are acceptable, it is possible to obtain theoretical expressions for the phase shifts for any model potential. For example, if we accept that the expression (18.29) is satisfactory, then integrating with respect to the angular variables for a central potential, we obtain

$$f(\theta) = -\frac{2m}{\hbar^2 q} \int_0^\infty r V(r) \sin qr \, dr \tag{18.107}$$

(in this formula the scattering angle enters through q). To give it the form of Eq. (18.95) we have to expand $f(\theta)$ into Legendre polynomials, which can be done using the formula taken from the theory of Bessel functions,

$$\frac{\sin qr}{qr} = \sum_{l=0} (2l+1) j_l^2(kr) P_l(\cos\theta) \quad \left(q \equiv 2k\sin\frac{\theta}{2}\right). \tag{18.108}$$

Inserting this expression into the previous one gives

$$f(\theta) = -\frac{2m}{\hbar^2} \int_0^\infty dr\, r^2 V(r) \sum_l (2l+1) j_l^2(kr) P_l(\cos\theta). \tag{18.109}$$

If we now compare the coefficients of P_l of this expression and (18.95), we obtain

$$a_l = e^{i\delta_l} \sin\delta_l = -\frac{2mk}{\hbar^2} \int_0^\infty V(r) j_l^2(kr) r^2 dr$$

$$= -\frac{\pi m}{\hbar^2} \int_0^\infty V(r) J_{l+1/2}^2(kr) r \, dr, \tag{18.110}$$

where we have used the relation between cylindrical and spherical Bessel functions, Eq. (A.62),

$$j_l(\rho) = \sqrt{\frac{\pi}{2\rho}} J_{l+1/2}(\rho). \tag{18.111}$$

For small phases it follows from (18.96) that $a_l \simeq \delta_l$, which gives an expression for the phase shifts when they are small,

$$\delta_l = -\frac{2mk}{\hbar^2} \int_0^\infty r^2 j_l^2(kr) V(r) dr. \tag{18.112}$$

This result allows us to verify that for an attractive potential ($V < 0$) the phase shifts are positive, while for repulsive potentials ($V > 0$), the δ_l take negative values.

Exercise E18.3. Scattering by an impenetrable sphere

Calculate the differential cross section for particles scattered by a perfectly rigid and impenetrable sphere of radius a.

Solution. We can model the scattering effect of the sphere with the central potential

$$V(r) = \begin{cases} \infty, & r \leq a, \\ 0, & r > a. \end{cases} \tag{18.113}$$

The condition of total impenetrability can alternatively be expressed as the boundary condition corresponding to total reflection, $\psi(r)|_{r=a} = 0$. The radial Schrödinger equation is

$$R'' + \frac{2}{r} R' + \left(k^2 - \frac{l(l+1)}{r^2}\right) R = 0, \quad k^2 = \frac{2mE}{\hbar^2}, \quad r \geq a, \tag{18.114}$$

and $R = 0$ for $r < a$. The solution can be written in terms of the spherical Bessel functions. In this case it is convenient to take $j_l(kr)$ and $h_l^{(1)}(kr)$ as independent solutions; the spherical Hankel function of the first kind (see Section A.1),

$$h_l^{(1)}(x) = \sqrt{\frac{\pi}{2x}} H_{l+1/2}^{(1)}(x), \tag{18.115}$$

behaves asymptotically as an outgoing spherical wave, as follows from (A.55),

$$h_l^{(1)}(kr) \xrightarrow{kr \to \infty} (-i)^{l+1} \frac{e^{ikr}}{kr}. \tag{18.116}$$

This is precisely the property that makes this function especially useful in scattering problems. Therefore, we write for $r > a$,

$$\psi = \sum_{l=0}^{\infty} \left((2l+1) i^l j_l(kr) + C_l h_l^{(1)}(kr) \right) P_l(\cos\theta). \tag{18.117}$$

The coefficient of the first term has been chosen appropriately to reproduce the incident plane wave, as will become clear later. The boundary condition on the surface of the sphere can only be satisfied if the coefficient of each Legendre polynomial vanishes separately at $r = a$. This means that each of the equations

$$(2l+1) i^l j_l(ka) + C_l h_l^{(1)}(ka) = 0 \tag{18.118}$$

must be satisfied separately; from them we obtain the value of the coefficients C_l. Substituting in the function ψ of Eq. (18.117), we obtain

$$\psi = \sum_{l=0}^{\infty} (2l+1) i^l \left[j_l(kr) - \frac{j_l(ka)}{h_l^{(1)}(ka)} h_l^{(1)}(kr) \right] P_l(\cos\theta)$$

$$= e^{ikz} - \sum_{l=0}^{\infty} (2l+1) i^l \frac{j_l(ka)}{h_l^{(1)}(ka)} h_l^{(1)}(kr) P_l(\cos\theta). \tag{18.119}$$

The second equality was obtained using the expansion of the plane wave into spherical waves and justifies the previous choice of the coefficients of the j_l functions. This is an exact solution of the Schrödinger equation whose asymptotic behavior is given by

$$\psi(r,\theta)|_{r \to \infty} = e^{ikz} + \frac{i}{k} \frac{e^{ikr}}{r} \sum_{l=0}^{\infty} (2l+1) \frac{j_l(ka)}{h_l^{(1)}(ka)} P_l(\cos\theta). \tag{18.120}$$

Comparing this expression with the general form of the asymptotic solution of the scattering problem, Eq. (18.8), we obtain the following exact expression for the scattering amplitude by an impenetrable sphere,

$$f(\theta) = \frac{i}{k} \sum_{l=0}^{\infty} (2l+1) \frac{j_l(ka)}{h_l^{(1)}(ka)} P_l(\cos\theta), \tag{18.121}$$

which gives for the differential cross section

$$\frac{d\sigma}{d\Omega} = \frac{1}{k^2} \left| \sum_{l=0}^{\infty} (2l+1) \frac{j_l(ka)}{h_l^{(1)}(ka)} P_l(\cos\theta) \right|^2 \tag{18.122}$$

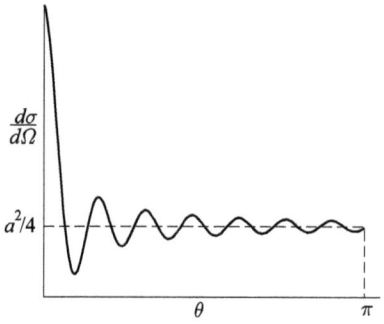

Figure 18.7 Differential cross section for a rigid sphere as a function of the scattering angle.

and for the total cross section (see Eq. (18.100))

$$\sigma = \frac{4\pi}{k^2} \sum_{l=0}^{\infty} (2l+1) \left| \frac{j_l(ka)}{h_l^{(1)}(ka)} \right|^2 . \qquad (18.123)$$

The terms that make up the differential cross section combine to give a series of maxima and minima as the angle θ varies. For $ka \ll 1$ there is a very sharp maximum in the forward direction and rapidly decreasing oscillations around the classical differential section $a^2/4$, as shown in Fig. 18.7. Since $j_l(ka)/h_l^{(1)}(ka) \sim (ka)^{2l+1}$, the dominant contribution comes from the s-wave. In this case, and as usual at low energies, the scattering by the short-range central potential is isotropic,

$$\frac{d\sigma}{d\Omega} = a^2 \qquad (18.124)$$

and the total cross section is four times the classical cross section of a perfectly rigid sphere or twice the total cross section of a perfectly absorbing sphere,

$$\sigma = 4\pi a^2. \qquad (18.125)$$

18.5 *Resonant Scattering

When the potential is attractive and allows the existence of bound states in the continuum, a resonance phenomenon can occur at low energies, which manifests itself as a significant increase in the scattering cross section around the resonance energy. This phenomenon is very important and is due to the formation of states that are bound by the potential but unstable. The topic is part of a more advanced course; however, because of its importance, the main results are presented here without derivation.

In the region of a resonance, one of the phases, let's say δ_l, changes rapidly with energy by the amount π (and passes through the angle $\pi/2$, where its tangent becomes

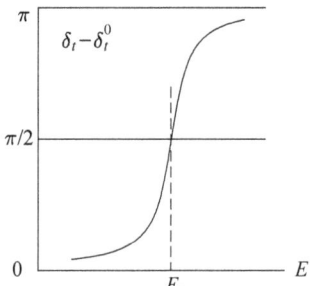
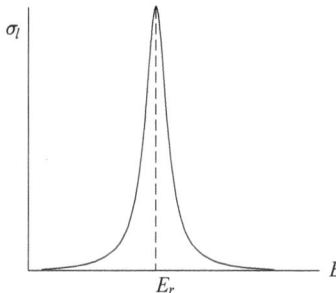

Figure 18.8 The phase shift δ_l and the cross section σ as functions of the energy E, close to a resonance.

infinite). If the value of this δ_l very far from resonance is denoted by δ_l^0, it can be written approximately as

$$\tan(\delta_l - \delta_l^0) = \frac{\Gamma_l/2}{E_r - E}. \tag{18.126}$$

The parameter Γ_l measures the width of the resonance and E_r is the resonance energy, that is, the energy at which the difference $\delta_l - \delta_l^0$ takes the value $\pi/2$. At this energy $\text{sen}^2(\delta_l - \delta_l^0)$ reaches its maximum value of 1, so that the cross section, which is approximately given by the dominant orbital momentum term l according to Eq. (18.100),

$$\sigma \simeq \frac{4\pi}{k^2}(2l+1)\text{sen}^2(\delta_l - \delta_l^0) = \frac{4\pi}{k^2}(2l+1) \quad \text{for} \quad E = E_r, \tag{18.127}$$

also reaches a maximum which is normally very sharp. The characteristic behavior of δ_l and σ around a resonance is shown in Fig. 18.8.

A formula for the scattering amplitude can be derived using the general equation (18.95) and assuming that the other phases are small, the result being

$$f(\theta) = f_l^0(\theta) - \frac{2l+1}{k}\frac{\Gamma_l/2}{E - E_r + i\Gamma_l/2}e^{2i\delta_l^0}P_l(\cos\theta) \tag{18.128}$$

where $f_l^0(\theta)$ is the scattering amplitude away from the resonance. This expression clearly separates the *external* or *potential scattering amplitude* (the nonresonant contribution) from the *internal* or *resonant scattering amplitude*. Since δ_l^0 is usually also small (see Fig. 18.9), the amplitude in the resonance region is often written in the form

$$f(\theta) = -\frac{2l+1}{k}\frac{\Gamma_l/2}{E - E_r + i\Gamma_l/2}P_l(\cos\theta). \tag{18.129}$$

The total cross section is then given by

$$\sigma_l = \frac{4\pi}{k^2}(2l+1)\sin^2\delta_l = \frac{4\pi}{k^2}(2l+1)\frac{g_l^2}{1+g_l^2},$$

that is,

$$\sigma_l = \frac{4\pi}{k^2}(2l+1)\frac{\Gamma_l^2/4}{(E-E_r)^2 + \Gamma_l^2/4}. \tag{18.130}$$

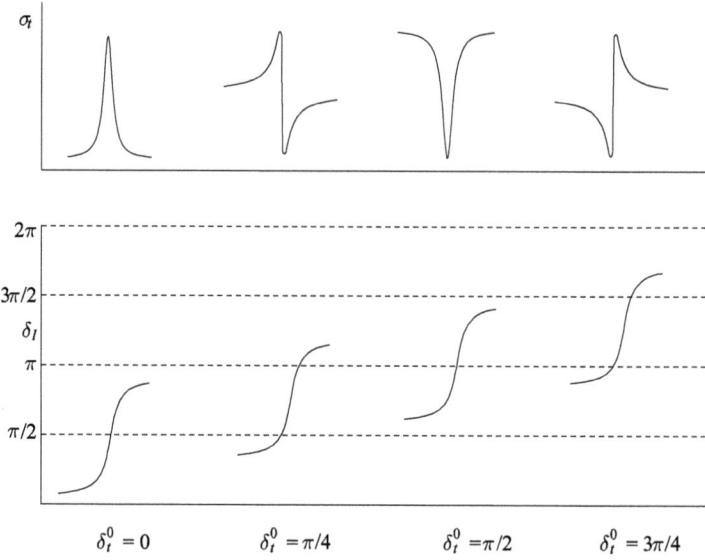

Figure 18.9 a) At energies close to resonance, the phase changes abruptly by π and the differential cross section reaches a peaked maximum when the background phase is small. (b) However, for large values of δ^0, there is no resonance; there may even be antiresonance.

This expression is a particular case of the Breit–Wigner formula for the resonant cross section. It shows that when $|E - E_r| = \Gamma_l/2$, the cross section is half its maximum; therefore, Γ_l measures the width of the resonance curve and is called the *width of the resonance*. Resonance curves of the form of Eq. (18.130) as a function of E are called *Lorentzian*.[6]

The preceding results admit of an interesting interpretation. We saw in Chapter 12 that the asymptotic solution of the Schrödinger equation for $r \to \infty$ has the general form

$$u = A(E)e^{-\alpha r} + B(E)e^{\alpha r}, \qquad (18.131)$$

where $\alpha \equiv -ik = \sqrt{-2mE}/\hbar$ is real for bound states ($E < 0$) and imaginary for dispersive states. Rewriting this solution in the usual form of scattering theory, as

$$u = C(E)\sin(kr + \delta_l), \qquad (18.132)$$

---

[6] A Lorentzian distribution has no moments after the second (they are all infinite). In mathematics, the Lorentzian distribution is written in the form

$$P(x) = \frac{1}{\pi}\frac{1}{1+x^2}$$

and is known as the Cauchy distribution. It is usual to characterize this distribution by its width Γ,

$$P_\Gamma(x) = \frac{1}{\pi}\frac{\Gamma}{x^2+\Gamma^2}.$$

However, its scale invariance

$$P_\Gamma(\lambda x)d(\lambda x) = P_{\Gamma/\lambda}(x)dx,$$

combined with the absence of moments after the second, imply the *nonexistence of a natural scale* associated with this distribution, so the fluctuations of x occur equally at all scales.

gives $e^{2i\delta_l} = -\frac{A}{B}$. Therefore, the amplitude f_l can be written in the form (see Eq. (18.96))

$$f_l = \frac{e^{2i\delta_l} - 1}{2ik} = -\frac{1}{2ik}\frac{A+B}{B}. \tag{18.133}$$

When the energy is negative (i.e., for bound states), α is positive and the second term of Eq. (18.131) diverges, so it is necessary to take $B(E) = 0$. But from (18.133) we see that the amplitude f_l has a single pole at $B = 0$, that is, the scattering amplitude, considered as a function of the energy, has single poles at the energies corresponding to the bound states.[7]

Returning now to resonant scattering, we see from Eq. (18.129) that $f(\theta, E)$ has a single pole for the *complex* energy

$$E = E_r - \frac{i}{2}\Gamma_l. \tag{18.134}$$

This suggests interpreting a resonance as a bound state, but with positive energy and metastable, meaning that it decays "spontaneously" with a relatively long lifetime. This can be seen by remembering that the stationary wave function is multiplied by $e^{-iEt/\hbar}$, which in this case is $e^{-iEt/\hbar} = e^{-iE_r t/\hbar} e^{-\Gamma t/2\hbar}$, so that the probability density *decreases* with time as $\rho \sim e^{-\Gamma t/\hbar}$; the state becomes depopulated, with a lifetime given by

$$T = \frac{\hbar}{\Gamma}. \tag{18.135}$$

In other words, $\Gamma/\hbar$ is the decay probability of the resonance per unit time.

The conclusion is that a resonance can be interpreted as an unstable composite particle, that is, as a quasi-stationary state of the target particle and the projectile, with well-defined quantum numbers (rest mass $E_r c^{-2}$, spin l, etc.) and whose stability is measured by Γ. Based on these results, a resonance is often formally defined as a single pole in the scattering amplitude (or in the S matrix) for a complex energy.

Exercise E18.4. Scattering by a spherical barrier at low energies

Compare the total scattering cross section produced by a uniform spherical barrier of radius a and height V_0 calculated by means of an expansion in partial waves, with that obtained in the Born approximation at low energies ($ka \ll 1$).

Solution. The elastic scattering cross section at low energies is essentially given by the s-wave contribution,

$$\sigma = 4\pi a^2 \left(1 - \frac{\tanh Ka}{Ka}\right)^2, \tag{18.136}$$

where $K = \sqrt{2mV_0}/\hbar$. On the other hand, the Born approximation gives the following amplitude for this interaction potential, using Eq. (18.50) with $\sin qr \simeq qr$,

$$f(\theta) = -\frac{2mV_0}{\hbar^2 q}\int_0^a qr^2 dr = -\frac{2mV_0 a^3}{3\hbar^2}, \tag{18.137}$$

---

[7] A similar argument was found in Chapter 5 from the properties of the transmission coefficient.

so the cross section predicted by this approximation is

$$\sigma = 2\pi \int_0^\pi |f(\theta)|^2 \operatorname{sen}\theta \, d\theta = \pi \left(\frac{4mV_0 a^3}{3\hbar^2}\right)^2 = \pi \left(\frac{2}{3}K^2 a^3\right)^2. \qquad (18.138)$$

For the Born approximation to be applicable, it must be true that $\sigma \ll \pi a^2$; from (18.138) we see that this is the case only if $4mV_0 a^2/3\hbar^2 = \frac{2}{3}K^2 a^2 \ll 1$, that is, if $Ka \ll 1$. When this condition is met, we can approximate $\tanh Ka \simeq Ka - \frac{1}{3}(Ka)^3$ and Eq. (18.136) reduces to

$$\sigma = 4\pi a^2 \cdot \left(\tfrac{1}{3}K^2 a^2\right)^2 = \pi \left(\tfrac{2}{3}K^2 a^3\right)^2, \qquad (18.139)$$

which agrees with the Born approximation, as expected.

Problems

P18.1 Derive the relations (18.1) and (18.2) that allow the transition between the laboratory and CM reference frames in a two-particle problem.

P18.2 Derive Eq. (18.3) which relates the number of particles per unit area in both reference frames.

P18.3 Prove that for a collision process of the type $P_1 + P_2 \to P_3 + P_4$, which can be elastic or inelastic so that the quantity $Q \equiv E_{\text{kyn.final}} - E_{\text{kyn.initial}}$ can be different from zero, in Eq. (18.3) the factor $\gamma = m_1/m_2$ must be replaced by

$$\gamma = \left(\frac{m_1}{m_2}\frac{m_3}{m_4}\frac{E_r}{E_r + Q}\right)^{1/2}, \quad E_r = \tfrac{1}{2}m_{\text{red}}v^2. \qquad (18.140)$$

P18.4 A particle of mass m collides elastically with a particle of mass M at rest (in the laboratory). Suppose that the scattering has spherical symmetry in the CM frame. What is the angular distribution of the target particles in the laboratory frame?

P18.5 A beam of particles with density n_0 part/cm^2 is fired at a target with total effective cross section σ. Show that if the thickness of the target is d, the beam that crosses it comes out with density $n = n_0 e^{-\mu d}$, where $\mu = N\sigma$ is the so-called linear attenuation coefficient and N is the number of collision centers that the target contains in the unit volume.

P18.6 Show that the differential scattering cross section in the Born approximation due to a spherical barrier of radius R and constant height V_0 is

$$\frac{d\sigma}{d\Omega} = \left(\frac{2mV_0}{\hbar^2}\right)^2 \frac{(\operatorname{sen}qR - qR\cos qR)^2}{q^6}. \qquad (18.141)$$

P18.7 For the Ramsauer–Townsend effect to occur (see Section 5.3.2), the cross section must vanish at very small bombardment energies (which can be assumed to be zero). Using a uniform spherical well as a model for the interaction potential, find the value that should be assigned to the product $a^2 V_0$. *Hint*: Consider at least the s and p waves.

P18.8 Find the differential cross section for the case where both the target and the projectile have structure.

P18.9 Protons with energy of 0.3 MeV are scattered by a thin sheet of aluminum. The number of backscattered protons is observed to be 0.96 times the predicted value for a Coulomb potential. Interpreting this discrepancy as due to shielding effects and assuming that the corresponding potential change appreciably affects only the s wave, determine the change in the value of δ_0. Indicate whether the interaction is attractive or repulsive.

P18.10 Consider scattering by a uniform spherical well of radius a and (large) depth V_0. Show that the condition for bound states to exist is

$$k_1 a = \left(n + \tfrac{1}{2}(l+1)\right)\pi, \qquad (18.142)$$

where $k_1^2 = 2m(E+V_0)/\hbar^2$.

P18.11 By solving the radial Schrödinger equation with the scattering potential as a perturbation, derive Eq. (18.110) for the phase shift when it is small,

$$\delta_l = -\frac{\pi m}{\hbar^2}\int_0^\infty rV(r)J^2_{l+1/2}(kr)\,dr. \qquad (18.143)$$

Note that with this procedure we avoid using (18.109).

P18.12 Show that the condition: incident wave flux = outgoing wave flux, implies that $|\widehat{S}|^2 = 1$.

P18.13 Show that the flux J through a spherical surface of radius $r \to \infty$ is given by

$$J = v\left(\sigma - \frac{4\pi}{k}\operatorname{Im} f(0)\right). \qquad (18.144)$$

P18.14 Determine the scattering amplitude and the differential cross section in Born's first approximation for the potential $V_0\delta(r-R)$.

P18.15 Find the differential and total cross section in Born's first approximation for the Gaussian potential $V_0\exp(-\alpha^2 r^2)$.

P18.16 Particles of mass m are incident on a central potential of the form

$$V(r) = \begin{cases} \infty, & r < a, \\ -V_0, & a \le r \le 2a, \\ 0, & r > 2a, \end{cases} \qquad (18.145)$$

$V_0 > 0$. Show that the cross section σ is such that

$$\lim_{v\to 0}\sigma = 4\pi\left[k_0^{-1}\tan(k_0 a) - 2a\right]^2, \qquad (18.146)$$

with $k_0 = \left[2mV_0/\hbar^2\right]^{1/2}$.

P18.17 The scattering amplitude for a certain interaction is given by

$$f(\theta) = \frac{1}{k}\left(e^{ika}\operatorname{sen} ka + 3ie^{2ika}\cos\theta\right), \qquad (18.147)$$

where a is the characteristic length of the interaction potential and k is the wave number of the incident particles. Find the differential cross section for s waves for this interaction.

P18.18 Determine the phase shifts δ_l produced by the potential $V(r) = g/r^2$. Find the effective differential section and verify that it is inversely proportional to the energy.

Bibliographical Notes

A very useful reference on scattering theory is Newton (1982).

The textbook by Afnan (2011) contains the exact solution of the Coulomb scattering problem in parabolic coordinates.

An extended modern account of scattering theory, including the field of ultracold matter that has been of so much interest in recent years, is Friedrich (2013).

A reference of historical value on the important Fermi experiments on meson-nucleon systems is Fermi (1955).

19 Relativistic Equations. An Introduction

This chapter is intended as an introduction to relativistic quantum mechanics, with the aim of providing the student with a basic knowledge of the subject and a link to more advanced courses in relativistic quantum theory.

So far we have been dealing with nonrelativistic QM, which works well in the context of Galilean relativity. When the Schrödinger equation is modified to make it consistent with special relativity, the result is the Klein–Gordon equation. Such a modification gives better qualitative, though not quantitative, agreement with experiment.

On the other hand, introducing the spin of the electron into the Schrödinger equation gives the Pauli equation, as we saw in Section 14.2. In this way, some important spin effects are incorporated into Schrödinger's theory.

When special relativity and spin are considered together, one obtains Dirac's equation. This equation applies only to particles with spin 1/2; the relativistic description of particles (or fields) with spin 1 is given by the Proca equation, that of particles of spin 3/2 by the Rarita–Schwinger equation, and so on. Of all these, the most important is undoubtedly Dirac's equation, since it allows us to obtain those corresponding to higher values of spin, as well as describing the behavior of the most interesting particles, namely the electron and the nucleons, the fundamental constituents of matter.

19.1 The Klein–Gordon Equation

We begin by constructing the Klein–Gordon equation for the relativistic description of spinless particles. To do this we start from the observation that Schrödinger's equation can be obtained simply by "quantizing" the relativistic energy expression $E = \mathbf{p}^2/2m + V$, by replacing E by $i\hbar \partial/\partial t$ and $\hat{\mathbf{p}}$ by $-i\hbar \nabla$, which leads to

$$i\hbar \frac{\partial \psi}{\partial t} = -\frac{\hbar^2}{2m} \nabla^2 \psi + V\psi. \tag{19.1}$$

Since the operators $\widehat{E}$ and $\hat{\mathbf{p}}$ are the components of the four-vector $-i\hbar(\partial_i, -ic\partial_4)$,[1] the substitution retains its validity when we move to the relativistic description.

---

[1] The notation used in this chapter is as follows. The Cartesian components of a tensor in space-time are denoted by Greek indices, $x_\mu, A_\mu, F_{\mu\nu}$, and so on, with $\mu = 1, 2, 3, 4$. The spatial components are indicated with a Latin index, $x_i, A_i, F_{ij}, F_{i\lambda}$, and so on. In Minkowski notation, the fourth component of the four-vectors is imaginary; for example,

$$\mathbf{x} = (x, y, z, ict), \quad \mathbf{A} = (A_x, A_y, A_z, iA_0), \tag{19.2}$$

$$\partial_\mu = \frac{\partial}{\partial x_\mu} = \left(\frac{\partial}{\partial x}, \frac{\partial}{\partial y}, \frac{\partial}{\partial z}, \frac{\partial}{i\partial ct} \right) = \left(\nabla, -\frac{i}{c}\frac{\partial}{\partial t} \right). \tag{19.3}$$

Therefore,

$$-i\hbar (\partial_i, -ic\partial_4) = \left(-i\hbar \nabla_i, i\hbar c \frac{\partial}{\partial ct} \right) = (\hat{p}_i, \widehat{E}). \tag{19.4}$$

To obtain an equation consistent with relativity, it is sufficient to use the corresponding expression for the energy, restricting ourselves first to the case of a free particle, $V = 0$. Using this procedure, we obtain the Klein–Gordon equation for a free particle from $E^2 = m^2c^4 + \boldsymbol{p}^2c^2$,

$$\nabla^2 \psi - \frac{1}{c^2}\frac{\partial^2 \psi}{\partial t^2} - \frac{m^2c^2}{\hbar^2}\psi = 0. \tag{19.5}$$

The first two terms, corresponding to the wave equation, can be grouped as the d'Alambertian of ψ, which gives

$$\Box \psi - \frac{m^2c^2}{\hbar^2}\psi = 0. \tag{19.6}$$

This equation was derived simultaneously and independently by many authors, including Schrödinger (the first), de Broglie, and of course Klein and Gordon.[2]

19.1.1 Basic properties of the Klein–Gordon equation

Let us look at some basic properties of the Klein–Gordon equation. To simplify the writing, we put $p_\mu = -i\hbar\partial_\mu$, where the index μ takes the four values $\mu = 1, 2, 3, 4$, with $x_4 = ict$. Then $\hbar^2 \Box = -p_\mu p_\mu$ and the free-particle equation can be written in the form

$$p_\mu p_\mu \psi + m^2c^2 \psi = 0. \tag{19.7}$$

With $\varkappa = mc/\hbar$, we get

$$\partial_\mu \partial_\mu \psi - \varkappa^2 \psi = 0 \tag{19.8}$$

and its conjugate

$$\partial_\mu \partial_\mu \psi^* - \varkappa^2 \psi^* = 0. \tag{19.9}$$

If we multiply these equations by ψ^* and ψ respectively and subtract the results, we get a conservation equation,

$$\partial_\mu j_\mu = 0, \tag{19.10}$$

where the current four-vector $j_\mu = (\boldsymbol{j}, ij_0)$ is defined as (A is a constant to be determined)

$$j_\mu = A(\psi^* \partial_\mu \psi - \psi \partial_\mu \psi^*). \tag{19.11}$$

The fourth component is $j_4 = ij_0$, with

$$j_0 \equiv c\rho = (A/c)(\psi^* \partial \psi/\partial t - \psi \partial \psi^*/\partial t); \tag{19.12}$$

with this identification and in three-dimensional notation, (19.10) takes the usual form of the continuity equation,

---

[2] Oskar Benjamin Klein (1894–1977) was a Swedish theoretical physicist; Walter Gordon (1893–1932) was a German physicist.

$$\frac{\partial \rho}{\partial t} + \nabla \cdot \boldsymbol{j} = 0. \tag{19.13}$$

In general, an equation of the form

$$\partial_\mu \theta_{\mu\alpha\beta...} = 0 \tag{19.14}$$

is a conservation equation, meaning that the fourth component of the tensor $\theta_{\mu\alpha\beta...}$ is a conserved quantity. In particular, Eq. (19.10) is equivalent to Eq. (19.13) for the conserved density ρ, given by the expression

$$\rho = \frac{A}{c^2}\left(\psi^* \frac{\partial \psi}{\partial t} - \psi \frac{\partial \psi^*}{\partial t}\right). \tag{19.15}$$

Since this expression has no definite sign, we cannot interpret it as the density of particles, although it is customary to give it the meaning of charge density (choosing the constant A accordingly); $\boldsymbol{j}$ is then the electric current density. Note, however, that the time evolution of the system could, in principle, change the sign of ρ.[3] For example, if we write $\psi = R\,e^{iS}$, using (19.15), we obtain

$$\rho = \frac{A}{c^2}R^2(2i\dot{S}) = -\frac{\hbar e}{mc^2}R^2 \dot{S}, \tag{19.16}$$

and it is clear that its sign depends on that of $\dot{S}$, which can change with the evolution (the appropriate value of A has already been inserted).

To set the coefficient A, we have taken into account that in the nonrelativistic limit Schrödinger's theory has to be recovered. Considering the special case where ψ refers to a stationary state $\psi = e^{-iEt/\hbar}\varphi(\mathbf{x})$ and substituting in (19.11), we get $\rho = -2i\frac{AE}{\hbar c^2}\psi^*\psi$, which in the nonrelativistic limit ($E \to mc^2$) gives $\rho = -2i\frac{Am}{\hbar}\psi^*\psi$. For this quantity to agree with the charge density of the nonrelativistic theory, $\rho = e\psi^*\psi$, we have to set $A = i\hbar e/2m$. From (19.11) we thus obtain the current density

$$j_\mu = \frac{i\hbar e}{2m}(\psi^* \partial_\mu \psi - \psi \partial_\mu \psi^*), \tag{19.17}$$

whose spatial components are the same as in the nonrelativistic case, although its fourth component, given by Eq. (19.15), differs from the corresponding nonrelativistic expression.

As a solution to the Klein–Gordon equation for the free particle, we propose

$$\psi = A e^{-iEt/\hbar + i\mathbf{k}\cdot\mathbf{x}}, \quad \mathbf{k} = \mathbf{p}/\hbar. \tag{19.18}$$

Introducing this into (19.6) we see that the equation is satisfied if $E = \pm E_0$, with $E_0 = +\sqrt{m^2c^4 + c^2\mathbf{p}^2} \equiv \hbar\omega$, The theory thus predicts the coexistence of states with positive and negative energy.[4] To see this in more detail, let us consider the two solutions

$$\psi_+ = A_+ e^{-i\omega t + i\mathbf{k}\cdot\mathbf{x}}, \quad \psi_- = A_- e^{i\omega t + i\mathbf{k}\cdot\mathbf{x}}, \tag{19.19}$$

---

[3] This problem was the main argument that led Schrödinger to abandon this equation for its nonrelativistic, first-order time derivative version.

[4] Models can be built that allow only positive energies, although they are somewhat artificial; see, for example, Dirac (1971).

where ψ_+ corresponds to $E = \hbar\omega > 0$ and ψ_- to $E = -\hbar\omega < 0$. Substituting these into the expression for the density, we get

$$\rho_+ = +\frac{eE_0}{mc^2}|A_+|^2, \quad \rho_- = -\frac{eE_0}{mc^2}|A_-|^2, \tag{19.20}$$

so that with the present convention the solutions of positive energy correspond to particles with charge of one sign (that of e), while those of negative energy correspond to charges of opposite sign. Incidentally, we note that something similar happens with any additive quantum number. This leads us to conclude that the Klein–Gordon equation contains solutions for both particles and their antiparticles, as we will see in more detail later.

19.1.2 Klein–Gordon equation with minimal coupling

To introduce the electromagnetic field, we use the principle of minimal coupling (see Section 12.7) in its relativistic version; this means that we have to replace the momentum-energy operator $\hat{p}_\mu$ by the corresponding canonical operator, $\hat{p}_\mu - (e/c)A_\mu$. This substitution in Eq. (19.7) gives the Klein–Gordon equation for particles subject to the action of an electromagnetic potential,

$$\left(\hat{p}_\mu - \frac{e}{c}A_\mu\right)\left(\hat{p}_\mu - \frac{e}{c}A_\mu\right)\psi + m^2c^2\psi = 0. \tag{19.21}$$

To apply this equation to the atomic case we introduce the following conditions:
(a) In the Lorentz gauge, $\mathbf{A} = 0$ and $A_4 = -V = Ze^2/r$.
(b) The atom is in a stationary state, so we can write $\psi(\mathbf{x}, t) = e^{-iE't/\hbar}\psi(\mathbf{x})$. The total energy is then $E' = E + mc^2$, where mc^2 is the rest energy and $E \ll mc^2$.
Under these conditions, we obtain from Eq. (19.21)

$$\nabla^2\psi + \frac{1}{\hbar^2c^2}\left[\left(E + mc^2 + \frac{Ze^2}{r}\right)^2 - m^2c^4\right]\psi = 0. \tag{19.22}$$

Let us first consider the nonrelativistic approximation that occurs for $|E + Ze^2/r|^2 \ll mc^2$. The term in square brackets reduces in this case to $2mc^2(E + Ze^2/r)$ to the first order, so that (19.22) takes the approximate form

$$\nabla^2\psi + \frac{2m}{\hbar^2}\left(E + \frac{Ze^2}{r}\right)\psi = 0, \tag{19.23}$$

which is exactly the Schrödinger equation. Therefore, we can consider Eq. (19.21) as the relativistic version of Schrödinger's equation.[5]

Returning to the problem of the relativistic atom, we need to solve Eq. (19.22). This can be done exactly by writing

$$\psi = R(r)Y_l^m(\theta, \varphi); \tag{19.24}$$

the resulting radial equation is

$$\left(\nabla_r^2 - \frac{l(l+1) - \alpha^2Z^2}{r^2} + \frac{2B}{r} - A\right)R = 0, \tag{19.25}$$

---

[5] This transition from the Klein–Gordon equation to the Schrödinger equation was precisely the path that Schrödinger followed to arrive for the first time at the nonrelativistic version of his equation.

where

$$\nabla_r^2 = \frac{1}{r}\frac{\partial^2}{\partial r^2}r, \quad A = \frac{m^2c^2}{\hbar^2}\left[1 - \left(1 + \frac{E}{mc^2}\right)^2\right], \quad B = \frac{Zme^2}{\hbar^2}\left(1 + \frac{E}{mc^2}\right), \quad (19.26)$$

and $\alpha = e^2/\hbar c \approx 1/137$. Equation (19.22) has just the form of the radial Schrödinger equation for Kepler's problem, although with different parameters. The comparison with the nonrelativistic results shows that for the boundary conditions

$$R \xrightarrow[r\to 0]{} \text{const} < \infty, \quad R \xrightarrow[r\to\infty]{} 0 \qquad (19.27)$$

to be satisfied, the relation between the coefficients A and B must be such that

$$\frac{B}{\sqrt{A}} - \sqrt{(l+\tfrac{1}{2})^2 - Z^2\alpha^2} - \tfrac{1}{2} = k, \quad k = 0, 1, 2, \ldots \qquad (19.28)$$

Since both coefficients must be real, the theory only applies to problems that meet the condition $Z\alpha < \tfrac{1}{2}$, that is, to light or medium atoms. From the relation between A and B and taking into account Eq. (19.22), we obtain for the energy (including the rest energy mc^2),

$$E' = \frac{mc^2}{\sqrt{1 + \frac{\alpha^2 Z^2}{\left(k+\tfrac{1}{2}+\sqrt{(l+\tfrac{1}{2})^2-Z^2\alpha^2}\right)^2}}}. \qquad (19.29)$$

Using the old definition $n = k + l + 1$, this expression shows that $E_{nl} = E' - mc^2$ depends on n and l, that is, the degeneracy on l has been broken. For the case $Z\alpha \ll 1$ we get the approximate result

$$E_{nl} = -\frac{R\hbar Z^2}{n^2}\left[1 + \frac{\alpha^2 Z^2}{n^2}\left(\frac{n}{l+\tfrac{1}{2}} - \tfrac{3}{4}\right) + \ldots\right], \qquad (19.30)$$

with R the Rydberg constant. The first term coincides with the Schrödinger solution and the second represents the relativistic correction; this term is the one that causes the breakdown of the degeneracy between different values of l. For example, between the 2s and 2p levels of the hydrogen atom there is an energy difference $E_{21} - E_{20}$, which corresponds to the transition frequency

$$\omega = \frac{E_{21} - E_{20}}{\hbar} = \tfrac{1}{6}R\alpha^2. \qquad (19.31)$$

This result differs from the experimental value by a factor of 1/3. In other words, the relativistic correction to Schrödinger's theory improves the qualitative properties of the predicted spectrum, but it is not sufficient. We will see later that this discrepancy is due to the fact that it does not take into account the effects of the electron spin on the energy levels, which turn out to be of the same order of magnitude (and opposite sign) as the relativistic effects. The equation that takes into account both the relativistic and the spinorial corrections is the one that can reproduce the empirical energy values; this is Dirac's equation.

19.2 The Dirac Equation

Dirac's original intention in proposing his equation for the electron was to simultaneously solve several problems that made the Klein–Gordon theory inadequate. One is the unsatisfactory shape of the particle density, another is the need to introduce the electron spin manually, and a third is to avoid the negative energy solutions characteristic of a relativistic theory.

Dirac's aim was to make the theory generate spin by itself, rather than introducing it phenomenologically from outside, in addition to providing a suitable solution to the other problems mentioned. This attitude is reminiscent of Einstein's approach to Newton's theory of gravity in classical mechanics, where the gravitational force is introduced by hand and its form chosen to reproduce the observed results. In both cases it was a matter of principle. However, precisely because it is characteristic of a relativistic theory to predict negative energies, the latter requirement has to be reformulated in an appropriate way, as we will see next.

The unsatisfactory expression (19.15) for the particle density arises from the fact that the Klein–Gordon equation contains a second-order time derivative. One way of avoiding the second derivative and preserving the Lorentz invariance of the theory is to linearize (to "extract the square root of") the formula $E^2 = c^2 p^2 + m^2 c^4$, that is, to propose that there exists a first-order linear relationship between the operators $\widehat{E}$ and $\widehat{p}$, so that we can write

$$\widehat{E} = c\boldsymbol{\alpha} \cdot \widehat{\boldsymbol{p}} + \beta m c^2, \tag{19.32}$$

while demanding that its square be consistent with the relativistic expression. Applied to the wave function Ψ, this relationship leads to the following equation, with $\boldsymbol{\alpha}$ and β constant,

$$\widehat{E}\Psi = c\boldsymbol{\alpha} \cdot \widehat{\boldsymbol{p}}\Psi + \beta m c^2 \Psi. \tag{19.33}$$

As we will see in more detail later, this equation also has negative energy solutions, so to solve the difficulty,[6] Dirac proposed to introduce the notion of positron or, more generally, of *antimatter*. The negative root of the equation $E^2 = c^2 p^2 + m^2 c^4$, $E = -E_0$ then does not refer to negative energies of the electron – since it always has positive energy $E \geq mc^2$ – but to particles below the energy level 0. It is therefore proposed that all levels with $E < 0$ are normally occupied by electrons, forming the *Dirac sea*, so that electrons cannot fall into such states due to the exclusion principle. When one of the electrons in the Dirac sea is excited and enters a positive energy state, this electron and the hole appear simultaneously in the sea; the hole is then identified as an antielectron, that is, a positron, since it corresponds to the *absence* of a negative charge, which is observed as a positive effective charge; similarly, the absence of negative energy appears as positive energy, and so on for the other properties of the electron. In this way, the negative energy states correspond to antimatter.

---

[6] The main difficulty with the existence of negative energy solutions is that particles with positive energy would spontaneously make transitions to states of increasingly negative energy, radiating the excess energy. The purpose of the Dirac sea is to put a limit on such processes.

In 1932, two years after Dirac's proposal, Anderson[7] experimentally found the positron in plates traced by cosmic rays. The prediction of antimatter was the first major prediction of the Dirac equation. Modern quantum field theory treats particles and antiparticles completely symmetrically, so the notion of a Dirac sea has become obsolete.

Introducing the expressions $\hat{p} = -i\hbar\nabla$, $\hat{E} = i\hbar\partial/\partial t$, we can write the Dirac equation explicitly as

$$i\hbar \frac{\partial \Psi}{\partial t} = -i\hbar c \boldsymbol{\alpha} \cdot \nabla\Psi + \beta mc^2 \Psi. \tag{19.34}$$

The four quantities α_i, β have to be chosen so that the relativistic laws are fulfilled. To determine them, we rewrite Eq. (19.33) using a convenient noncovariant provisional notation (exclusive for this calculation), in the form $E = c\alpha_\mu p_\mu$, with $p_0 = mc$, $\alpha_0 = \beta$. Note carefully that α_μ, p_μ defined in this way are *not* four-vectors, since, $p_0 = mc$ is a scalar, not the fourth component of a four-vector. Squaring and taking into account that the components p_μ commute among themselves, we obtain

$$E^2 = c^2 \alpha_\mu p_\mu \alpha_\nu p_\nu = c^2 p_\mu p_\nu \alpha_\mu \alpha_\nu = c^2 p_\mu p_\nu \tfrac{1}{2}\left(\alpha_\mu \alpha_\nu + \alpha_\nu \alpha_\mu\right), \tag{19.35}$$

where we have symmetrized by interchanging the indices ν, μ and taking the half sum. Since E^2 must be equal to $c^2 \boldsymbol{p}^2 + c^2 p_0^2 = c^2 p_\mu p_\mu$, it follows that we must write

$$\alpha_\mu \alpha_\nu + \alpha_\nu \alpha_\mu = 2\delta_{\mu\nu}. \tag{19.36}$$

This shows that that the α_μ should anticommute, so they cannot be simple numbers, but rather represent numerical matrices. Since the coefficients α_i, β in Eq. (19.34) are matrices, the functions Ψ must have several components to make sense. To propose a first-order equation for the electron, therefore, requires expanding the dimensionality of the space of solutions Ψ.

In order to construct the matrices that satisfy (19.36) and describe the spin, we tentatively consider that the Pauli matrices satisfy a similar equation

$$\hat{\sigma}_i \hat{\sigma}_j + \hat{\sigma}_j \hat{\sigma}_i = 2\delta_{ij}, \tag{19.37}$$

for $i, j = 1, 2, 3$: but when completing the basis with the unit matrix it turns out that the latter, combined with some $\hat{\sigma}_i$, does not satisfy, but commutes with, Eq. (19.36). Therefore, we cannot identify the α_i with the Pauli matrices and we have to look for a space of higher dimensionality. Since there are two degrees of freedom associated with the spin $\tfrac{1}{2}$ and two possible signs for the energy, there can be four different states for a given E^2, so a wave function with at least four linearly independent components is required; this means that the matrices α_μ must be of dimension at least 4×4. In the general case, for a particle with spin S, the matrices must be of dimension $N \times N$, with $N = 2(2S+1) = 4(S + \tfrac{1}{2})$. The simplest Dirac equation is therefore one of rank 4, corresponding to spin 1/2. Restricting ourselves to this case and choosing $\alpha_0 = \beta$, we rewrite Eq. (19.34) in the (classical) Dirac form,

$$i\hbar \frac{\partial \Psi}{\partial t} = -i\hbar c \boldsymbol{\alpha} \cdot \nabla\Psi + mc^2 \beta \Psi \equiv \hat{H}\Psi. \tag{19.38}$$

---

[7] Carl David Anderson (1905–1991) was a US-born physicist, who won the 1936 Nobel Prize for his discovery (observation and identification) of the positron. In 1936 he discovered the muon, a subatomic particle 207 times more massive than the electron (and ten times lighter than the proton).

Multiplying this equation from the left by β, taking into account that $\beta^2 = \alpha_\mu^2 = \mathbb{I}$, as follows from (19.36), introducing four new matrices γ_μ defined as

$$\gamma_0 = -i\gamma_4 = -i\beta = -i\alpha_0, \quad \gamma_i = -i\beta\alpha_i, \tag{19.39}$$

and dividing by $\hbar c$ gives (matrices γ_α and their relatives are not usually marked as operators, as there is no risk of confusion),

$$(\gamma_0 \partial_0 + \gamma_i \partial_i)\Psi + \frac{mc}{\hbar}\Psi = 0, \tag{19.40}$$

or since $\gamma_0 \partial_0 + \gamma_i \partial_i = \gamma_\mu \partial_\mu$,

$$\gamma_\mu \partial_\mu \Psi + \frac{mc}{\hbar}\Psi = 0. \tag{19.41}$$

This is the most common and convenient form of Dirac's equation for a free particle.[8] Since $mc/\hbar$ is a scalar, $\gamma_\mu \partial_\mu$ must also be a scalar, so γ_μ is a four-vector, and Eq. (19.41) is explicitly covariant under Lorentz transformations. From (19.38) we see that the Dirac Hamiltonian

$$\widehat{H} = -i\hbar c\boldsymbol{\alpha}\cdot\boldsymbol{\nabla} + \beta mc^2 \tag{19.42}$$

is a matrix of dimension 4×4, so the functions Ψ are *bispinors or spinors of four components*.

Exercise E19.1. The Dirac oscillator

The Dirac Hamiltonian is obtained by "taking the square root" of the Klein–Gordon–Hamiltonian. Use this idea to construct a linear Hamiltonian describing an isotropic 3D harmonic oscillator and show that it is indeed an oscillator with some peculiarities.[9]

Solution. The proposal is to "linearize" the operator (leaving aside the mass term)

$$\boldsymbol{p}^2 + m^2\omega^2\boldsymbol{r}^2 \Rightarrow [\boldsymbol{\alpha}\cdot(\boldsymbol{p} + im\omega\boldsymbol{r})][\boldsymbol{\alpha}\cdot(\boldsymbol{p} - im\omega\boldsymbol{r})], \tag{19.43}$$

using an idea similar to the preceding, but now including a linear coupling term in $\boldsymbol{r}$. It is clear that the terms $\pm im\omega\boldsymbol{r}$ can still be multiplied by any (scalar or pseudoscalar) matrix that anticommutes with the α_i matrices without affecting the factorization. The detailed analysis of this possibility leads to the conclusion that this factor must be taken as the β matrix, as this is the only way to guarantee that the resulting equation has the appropriate Lorentz covariance properties. This can be understood by recalling that the mass term in the free-particle Dirac equation itself, Eq. (19.34), has the form βm, so the m of the interaction terms must be accompanied by the factor β, $\pm im\omega\beta\boldsymbol{r}$. From these considerations we conclude that the proposed equation must be

$$i\hbar\frac{\partial\Psi}{\partial t} = \widehat{H}\Psi = c\boldsymbol{\alpha}\cdot(\boldsymbol{p} - im\omega\boldsymbol{r})\Psi + mc^2\beta\Psi. \tag{19.44}$$

It is easy to verify that this Hamiltonian is self-adjoint. One way to make physical sense of Eq.(19.44) is to consider the term $im\omega\beta\boldsymbol{r}$ as due to the coupling of the particle to an external (e.g., electric) field that grows linearly in the region of interest.

---

[8] There are variants in the literature which may differ from the present notation by signs or factors such as $\pm i$ in the definition of the fundamental matrices or the metric.
[9] This topic is discussed in detail in Moshinsky and Smirnov (1996).

To prove that Eq.(19.44) describes an oscillator and to study its properties, we calculate the square of the Hamiltonian, which, after some simplifications, gives (with $\hbar = c = 1$)

$$\widehat{H}^2 = \left[\alpha \cdot (p - im\omega r) + m\beta\right]^2 \qquad (19.45)$$
$$= \alpha_i \alpha_j p_i p_j - m^2 \omega^2 \alpha_i \beta \alpha_j \beta x_i x_j + m^2$$
$$- im\omega \left(\alpha_i \alpha_j \beta p_i x_j + \alpha_i \beta \alpha_j x_i p_j\right),$$

that is,

$$\widehat{H}^2 = p^2 + m^2 \omega^2 r^2 + m^2 - (4\boldsymbol{L}\cdot\boldsymbol{S} + 3)m\omega\beta, \qquad (19.46)$$

where $\boldsymbol{L}$ is the orbital angular momentum and $\boldsymbol{S} = \boldsymbol{\sigma}/2$ is the electron spin. This (squared) Hamiltonian effectively corresponds to that of a HO with spin–orbit coupling, plus irrelevant constants. An interesting exercise, left to the reader, is to show that the total angular momentum $\boldsymbol{J} = \boldsymbol{L} + \boldsymbol{S}$ commutes with $\widehat{H}^2$ and is therefore conserved; separately, neither $\boldsymbol{L}$ nor $\boldsymbol{S}$ is conserved.

19.2.1 The Dirac γ_μ matrices

We now move on to the study of the properties of the Dirac γ_μ matrices. The complete Dirac algebra consists of 16 linearly independent matrices of dimension 4×4; later the matrices γ_5 (one pseudoscalar) and $\sigma_{\mu\nu}$ (the six components of an antisymmetric tensor) are introduced in addition to the unit matrix. The remaining four matrices can be taken as $\gamma_5 \gamma_\mu$ (components of a pseudo-four-vector). This set is closed with respect to multiplication and constitutes a Clifford algebra, that is, its elements anticommute. (We recall that a pseudotensor is a tensor whose components change sign when rotated by 2π.)

From (19.39) we see that $\alpha_i = i\beta\gamma_i$. On the other hand, from (19.36) it follows that $\alpha_\mu \alpha_\nu = -\alpha_\nu \alpha_\mu$ for $\mu \neq \nu$; taking here $\mu = i$, $\nu = 0$, with $\alpha_0 = \beta$ we get $\gamma_i \beta = -i\beta\alpha_i\beta = i\beta^2 \alpha_i = i\alpha_i$, that is,

$$\gamma_i \beta = -\beta \gamma_i \qquad (19.47)$$

or, again with (19.39),

$$\gamma_i \gamma_4 + \gamma_4 \gamma_i = 0. \qquad (19.48)$$

Taking now $\mu = i, \nu = j$, we obtain

$$\alpha_i \alpha_j + \alpha_j \alpha_i = 2\delta_{ij} = -\beta \gamma_i \beta \gamma_j - \beta \gamma_j \beta \gamma_i = \beta^2 [\gamma_i \gamma_j + \gamma_j \gamma_i], \qquad (19.49)$$

that is,

$$\gamma_i \gamma_j + \gamma_j \gamma_i = 2\delta_{ij}. \qquad (19.50)$$

Finally, for $\mu = \nu = 0$ we get $\alpha_0 \alpha_0 + \alpha_0 \alpha_0 = 2 = 2i\gamma_0 i\gamma_0$, or

$$\gamma_4 \gamma_4 + \gamma_4 \gamma_4 = 2. \qquad (19.51)$$

Equations (19.48–19.51) show that the matrices γ_μ satisfy a relationship similar to (19.36),[10]

$$\gamma_\mu \gamma_\nu + \gamma_\nu \gamma_\mu = 2\delta_{\mu\nu}. \tag{19.52}$$

An important property of the γ_μ matrices is that they all have zero trace. This follows immediately from Eq. (19.52) with $\nu \neq \mu$, multiplying it on the left by γ_μ and calculating the trace, which gives

$$\operatorname{tr} \gamma_\mu^2 \gamma_\nu + \operatorname{tr} \gamma_\mu \gamma_\nu \gamma_\mu = 2\operatorname{tr} \gamma_\mu^2 \gamma_\nu = 2\delta_{\mu\nu}\operatorname{tr} \gamma_\mu = 0 \quad (\mu \neq \nu). \tag{19.53}$$

Since $\gamma_\mu^2 = 1$, it follows that for each μ,

$$\operatorname{tr} \gamma_\mu = 0. \tag{19.54}$$

19.2.2 The van der Waerden equation: Helicity operator

The preceding derivation of the Dirac equation, although very popular, is purely formal and it seems to be a matter of luck that it applies to the electron. For this reason, it is convenient to present a second derivation in which the physical content of the postulates is more transparent. To introduce the spin, we will make the substitution $p^2 \to (\boldsymbol{\sigma} \cdot \boldsymbol{p})^2$ successfully used in the nonrelativistic case to take account of the electron spin; see Section 14.2. Now we will use it in a relativistic context to arrive at a theory that contains both spin 1/2 and relativity by construction.

Our starting point is once again the equation $E^2/c^2 - \boldsymbol{p}^2 = m^2 c^2$, which with the substitution $\boldsymbol{p}^2 \to (\boldsymbol{\sigma} \cdot \boldsymbol{p})^2$ and the transition to operators becomes

$$\frac{\widehat{E}^2}{c^2} - (\hat{\boldsymbol{\sigma}} \cdot \widehat{\boldsymbol{p}})(\hat{\boldsymbol{\sigma}} \cdot \widehat{\boldsymbol{p}}) = m^2 c^2 \tag{19.55}$$

or, taking advantage of the fact that $\widehat{E}$ commutes with $\widehat{\boldsymbol{p}}$ and with the Pauli operators,

$$\left(\frac{\widehat{E}}{c} - \hat{\boldsymbol{\sigma}} \cdot \widehat{\boldsymbol{p}}\right)\left(\frac{\widehat{E}}{c} + \hat{\boldsymbol{\sigma}} \cdot \widehat{\boldsymbol{p}}\right) = m^2 c^2. \tag{19.56}$$

Using the quantum expressions for $\widehat{E}$ and $\widehat{\boldsymbol{p}}$ and applying the equation to a spinor ψ (of two components), we obtain a second-order equation in terms of 2×2 matrices,

$$\left(i\hbar\partial_0 + i\hbar\hat{\boldsymbol{\sigma}} \cdot \nabla\right)\left(i\hbar\partial_0 - i\hbar\hat{\boldsymbol{\sigma}} \cdot \nabla\right)\psi = m^2 c^2 \psi. \tag{19.57}$$

This is known as the *van der Waerden equation*. Because of the way this equation was derived, it clearly describes relativistic particles of spin 1/2. To show that it is just another form of the Dirac equation, we introduce the pair of spinors (of two components) $\phi^{(L)}$ and $\phi^{(R)}$, defined as

$$\phi^{(L)} = \psi, \tag{19.58}$$

$$\phi^{(R)} = \frac{1}{mc}\left(i\hbar\partial_0 - i\hbar\hat{\boldsymbol{\sigma}} \cdot \nabla\right)\phi^{(L)}. \tag{19.59}$$

---

[10] If two sets of matrices γ_μ and γ'_μ satisfy (19.52), there is always a nonsingular matrix $\hat{S}$ such that $\hat{S}\gamma_\mu \hat{S}^{-1} = \gamma'_\mu$. This so-called fundamental Pauli theorem leads to the invariance of Dirac's equation with respect to the change of representation.

From here and (19.57) it follows that these spinors satisfy the system of equations

$$\left(i\hbar\hat{\boldsymbol{\sigma}} \cdot \nabla - i\hbar\partial_0\right) \phi^{(L)} = -mc\phi^{(R)},$$
$$\left(i\hbar\hat{\boldsymbol{\sigma}} \cdot \nabla + i\hbar\partial_0\right) \phi^{(R)} = mc\phi^{(L)}. \tag{19.60}$$

Each of these equations has two components, so that in total we have four first-order equations for four functions, as is the case with Dirac's equation.

Before proceeding further, it is convenient to make a brief comment to establish the physical meaning of the spinors $\phi^{(R)}$ and $\phi^{(L)}$, which is interesting and useful. We consider the case $m = 0$, which decouples Eqs. (19.60),

$$\left(i\hbar\hat{\boldsymbol{\sigma}} \cdot \nabla - i\hbar\partial_0\right) \phi^{(L)} = 0, \quad \left(i\hbar\hat{\boldsymbol{\sigma}} \cdot \nabla + i\hbar\partial_0\right) \phi^{(R)} = 0. \tag{19.61}$$

Introducing $|\boldsymbol{p}| = E/c$ with $E > 0$ in the first equation, for eigenstates of E and $\boldsymbol{p}$ we get $(\hat{\boldsymbol{\sigma}} \cdot \hat{\boldsymbol{p}} + |\boldsymbol{p}|) \phi^{(L)} = 0$, that is,

$$\frac{\hat{\boldsymbol{\sigma}} \cdot \hat{\boldsymbol{p}}}{|\boldsymbol{p}|} \phi^{(L)} = -\phi^{(L)}, \tag{19.62}$$

and a similar equation for $\phi^{(R)}$, with the sign reversed. The operator

$$\hat{\mathfrak{h}} = \hat{\boldsymbol{\sigma}} \cdot \frac{\hat{\boldsymbol{p}}}{|\boldsymbol{p}|}, \tag{19.63}$$

called the *helicity operator*, describes the projection of the spin operator in the direction of the momentum; the negative eigenvalue in Eq. (19.62) indicates that the spin is antiparallel to $\boldsymbol{p}$ and that the helicity of the $\phi^{(L)}$ state is -1. Similarly, the solution for $\phi^{(R)}$ corresponds to the state of spin parallel to $\boldsymbol{p}$ and helicity $+1$. In summary, we see that for $m = 0$ and $S = 1/2$ there are two independent helicity states with values ± 1. For massive particles, the solutions are in general linear combinations of the two helicity states, or what is equivalent, they are not helicity eigenstates.

Returning to Eqs. (19.61) and (19.60) and taking first their difference and then their sum, we get the system of equations

$$-i\hbar\partial_0 \Psi_A - i\hbar\hat{\boldsymbol{\sigma}} \cdot \nabla \Psi_B = -mc\Psi_A, \tag{19.64}$$

$$i\hbar\partial_0 \Psi_B + i\hbar\hat{\boldsymbol{\sigma}} \cdot \nabla \Psi_A = -mc\Psi_B, \tag{19.65}$$

for the functions

$$\Psi_A = \phi^{(R)} + \phi^{(L)}, \quad \Psi_B = \phi^{(R)} - \phi^{(L)}. \tag{19.66}$$

These linear combinations are clearly arbitrary, and others could well have been adopted. This observation is important, since this freedom is at the basis for the possibility of constructing different representations of the Dirac matrices, as we will see later. If we rewrite the preceding system in matrix form, we get

$$\begin{pmatrix} -i\hbar\partial_0 & -i\hbar\hat{\boldsymbol{\sigma}} \cdot \nabla \\ i\hbar\hat{\boldsymbol{\sigma}} \cdot \nabla & i\hbar\partial_0 \end{pmatrix} \begin{pmatrix} \Psi_A \\ \Psi_B \end{pmatrix} = -mc \begin{pmatrix} \Psi_A \\ \Psi_B \end{pmatrix}. \tag{19.67}$$

We now construct a four-component bispinor (Dirac spinor) as

$$\Psi = \begin{pmatrix} \Psi_A \\ \Psi_B \end{pmatrix} = \begin{pmatrix} \phi^{(R)} + \phi^{(L)} \\ \phi^{(R)} - \phi^{(L)} \end{pmatrix}, \tag{19.68}$$

so that Eq. (19.67) becomes

$$\left[-i\hbar \begin{pmatrix} \mathbb{I} & \mathbb{O} \\ \mathbb{O} & -\mathbb{I} \end{pmatrix} \partial_0 + \hbar \sum_i \begin{pmatrix} \mathbb{O} & -i\hat{\sigma}_i \\ i\hat{\sigma}_i & \mathbb{O} \end{pmatrix} \partial_i \right] \Psi = -mc\Psi. \qquad (19.69)$$

Introducing the 4×4 matrices

$$\gamma_i = \begin{pmatrix} \mathbb{O} & -i\hat{\sigma}_i \\ i\hat{\sigma}_i & \mathbb{O} \end{pmatrix}, \quad \gamma_4 = \begin{pmatrix} \mathbb{I} & \mathbb{O} \\ \mathbb{O} & -\mathbb{I} \end{pmatrix}, \qquad (19.70)$$

and taking into account that $-i\partial_0 = \partial_4$, we obtain

$$\left(\gamma_\mu \partial_\mu + \frac{mc}{\hbar} \right) \Psi = 0, \qquad (19.71)$$

which is precisely the Dirac equation, (19.41).

The γ_μ matrices have been written here directly in terms of the Pauli matrices. If instead of $\Psi_A = \phi^{(R)} + \phi^{(L)}$ and $\Psi_B = \phi^{(R)} - \phi^{(L)}$ other linear combinations had been used, a different form for the γ_μ matrices would have been obtained, and thus a different representation of them. The specific representation used here, in addition to satisfying the condition (19.52), is Hermitian, as can be seen from its expression in terms of Pauli matrices. We will come back to this point later.

19.3 Properties of the Dirac Equation. The *Zitterbewegung*

After the Maxwell–Lorentz equations of classical electrodynamics, Dirac's equation was the next major step in the investigation of the properties of the electron. In the following we will introduce the elements needed to construct the conservation equation for the current density. We will then prove that Dirac's equation describes particles with spin 1/2 more generally, and discuss the *zitterbewegung*, the fast oscillation of the particle around its classical (relativistic) trajectory, which Schrödinger predicted in 1930 from his analysis of the solutions of the Dirac equation.

19.3.1 The Dirac adjoint and the continuity equation

Applying to Eq. (19.41) the operator $\gamma_\nu \partial_\nu - mc/\hbar$ from the left, we obtain

$$\left(\gamma_\nu \partial_\nu - \frac{mc}{\hbar} \right) \left(\gamma_\mu \partial_\mu + \frac{mc}{\hbar} \right) \Psi = \left(\gamma_\nu \gamma_\mu \partial_\nu \partial_\mu - \frac{m^2 c^2}{\hbar^2} \right) \Psi = 0, \qquad (19.72)$$

which can be written symmetrically as

$$\left[\tfrac{1}{2} \left(\gamma_\mu \gamma_\nu + \gamma_\nu \gamma_\mu \right) \partial_\mu \partial_\nu - \frac{m^2 c^2}{\hbar^2} \right] \Psi = 0. \qquad (19.73)$$

This expression is simplified with the help of (19.52), so that

$$\left(\Box - \frac{m^2 c^2}{\hbar^2} \right) \Psi = 0. \qquad (19.74)$$

The operator in parentheses acts on each of the four components of the spinor separately, which means that each component of the Dirac equation is a solution of the Klein–Gordon equation. This is not to say, however, that *any* solution of the Klein–Gordon equation is appropriate, since the bispinors are constrained by the Dirac equation.

Given a spinor Ψ, the *Dirac adjoint* is defined as

$$\overline{\Psi} = \Psi^\dagger \gamma_4 \qquad (19.75)$$

or explicitly, in the preceding representation of the Dirac matrices,

$$\overline{\Psi} = (\psi_1^* \quad \psi_2^* \quad -\psi_3^* \quad -\psi_4^*). \qquad (19.76)$$

Writing Dirac's equation in the form

$$\gamma_i \partial_i \Psi + \gamma_4 \partial_4 \Psi + \frac{mc}{\hbar}\Psi = 0 \qquad (19.77)$$

and taking its adjoint (with $\partial_4^* = -\partial_4$ since $x_4 = ict$), we get

$$\partial_i \Psi^\dagger \gamma_i - \partial_4 \Psi^\dagger \gamma_4 + \frac{mc}{\hbar}\Psi^\dagger = 0. \qquad (19.78)$$

To recover from the sign change that occurred in the fourth component, we multiply from the right by γ_4,

$$\partial_i \Psi^\dagger \gamma_i \gamma_4 - \partial_4 \Psi^\dagger \gamma_4 \gamma_4 + \frac{mc}{\hbar}\Psi^\dagger \gamma_4 = 0. \qquad (19.79)$$

We now take advantage of the fact that γ_4 commutes with itself, but anticommutes with the γ_i; rewriting the result in terms of the Dirac adjoint, we get

$$-\partial_i \Psi^\dagger \gamma_4 \gamma_i - \partial_4 \Psi^\dagger \gamma_4 \gamma_4 + \frac{mc}{\hbar}\Psi^\dagger \gamma_4 = -\partial_i \overline{\Psi}\gamma_i - \partial_4 \overline{\Psi}\gamma_4 + \frac{mc}{\hbar}\overline{\Psi} = 0, \qquad (19.80)$$

which leads to the *adjoint Dirac equation*,

$$-\partial_\mu \overline{\Psi}\gamma_\mu + \frac{mc}{\hbar}\overline{\Psi} = 0. \qquad (19.81)$$

This result can be used to derive the continuity equation. To do this, we multiply the Dirac equation on the left by $\overline{\Psi}$, its adjoint by Ψ from the right, and take the difference,

$$\overline{\Psi}\gamma_\mu \partial_\mu \Psi + (\partial_\mu \overline{\Psi})\gamma_\mu \Psi = 0; \qquad (19.82)$$

this can be written as

$$\partial_\mu(\overline{\Psi}\gamma_\mu \Psi) = 0, \qquad (19.83)$$

which has the form of a conservation equation $\partial_\mu j_\mu = 0$ if the current density is chosen to be proportional to $\overline{\Psi}\gamma_\mu \Psi$. For this expression to have the correct nonrelativistic limit, it is usual to write

$$j_\mu = ic\overline{\Psi}\gamma_\mu \Psi, \qquad (19.84)$$

that is, using $i\gamma_4 \gamma_i = \alpha_i$ (see Eq. (19.39)),

$$j_\mu = (c\Psi^\dagger \boldsymbol{\alpha}\Psi, ic\Psi^\dagger \Psi), \qquad (19.85)$$

where the first term in the parentheses represents the spatial component j and the last the density ρ multiplied by ic,

$$\rho = \Psi^\dagger \Psi = \psi_1^* \psi_1 + \psi_2^* \psi_2 + \psi_3^* \psi_3 + \psi_4^* \psi_4, \tag{19.86}$$

$$j = c \Psi^\dagger \boldsymbol{\alpha} \Psi. \tag{19.87}$$

The problem of the sign of ρ has thus been solved, since the expression obtained for ρ is analogous to the nonrelativistic one and legitimately corresponds to a density of particles independent of their charge, unlike what happens with the Klein–Gordon equation. The expression for j, Eq. (19.87), holds a surprise for us, since $\boldsymbol{\alpha} c$ plays the role of the velocity operator for the free particle, and the eigenvalues of the matrices α_i are ± 1 (note from (19.5) that $\alpha^2 = \mathbb{I}$). We will return to this point later to study its meaning; in the meantime, note that the expression for j is a scalar in spin space.

19.3.2 The spin of the electron

From the preceding derivations – in particular the van der Waerden derivation – we know that Dirac's equation describes particles with spin 1/2, but it is instructive to prove this directly. For simplicity, we will limit ourselves to the proof for a free particle, which can be generalized to the case of a central potential. Consider the Dirac Hamiltonian for a free particle,

$$\widehat{H} = c\boldsymbol{\alpha} \cdot \boldsymbol{p} + \beta m c^2, \tag{19.88}$$

and the orbital angular momentum operator,

$$\widehat{L}_3 = -i\hbar \epsilon_{3ij} x_i \partial_j. \tag{19.89}$$

The commutator

$$[\widehat{H}, \widehat{L}_3] = [c\widehat{\boldsymbol{\alpha}} \cdot \widehat{\boldsymbol{p}}, \widehat{L}_3] = -i\hbar c (\alpha_1 \widehat{p}_2 - \alpha_2 \widehat{p}_1) \tag{19.90}$$

is nonzero and therefore $\widehat{L}_3$ is not an integral of the motion. Since in this case there must be an angle operator that is conserved, we need an operator that completes $\widehat{L}_3$ and gives the conserved quantity. By applying σ_3 to the Hamiltonian, one obtains after some algebra (which we do not reproduce here)

$$[\widehat{H}, \sigma_3] = 2 i c (\alpha_1 \widehat{p}_2 - \alpha_2 \widehat{p}_1), \tag{19.91}$$

which combined with Eq. (19.90) gives

$$\left[\widehat{H}, \widehat{L}_3 + \tfrac{1}{2}\hbar\sigma_3\right] = 0. \tag{19.92}$$

Hence the quantity (be careful not to confuse j_i with J_i)

$$J_3 = \widehat{L}_3 + \widehat{S}_3 = \widehat{L}_3 + \tfrac{1}{2}\hbar\sigma_3, \tag{19.93}$$

with

$$\widehat{S}_3 = \tfrac{1}{2}\hbar\sigma_3, \tag{19.94}$$

is an integral of motion,

$$[\widehat{H}, \widehat{J}_3] = 0. \tag{19.95}$$

We conclude that the spin operator of the particle (the quantity added to L to complete the angular momentum) is given by $S = (\hbar/2)\sigma$, where σ is a Dirac matrix.

Sometimes it is convenient to use spin tensors, which are operators built from the matrices γ_μ that describe the spin of the relativistic electron. It is common to write these in the form

$$\sigma_{\mu\nu} = -\tfrac{1}{2}i(\gamma_\mu\gamma_\nu - \gamma_\nu\gamma_\mu), \tag{19.96}$$

or else

$$\sigma_{\mu\nu} = -i\gamma_\mu\gamma_\nu, \quad \mu \neq \nu, \tag{19.97}$$

$$= 0, \quad \mu = \nu. \tag{19.98}$$

The spatial components of these tensors are identified with the spin components; the meaning of the other components will be seen later.

19.3.3 *The *zitterbewegung*

We will now analyze some properties of the velocity vector $\dot{x}_k$ for a free particle in order to better understand the meaning of Eq. (19.87). We begin by noting that the commutator of the free-particle Dirac Hamiltonian $\widehat{H}$ with $\widehat{p}_k$ vanishes, hence

$$\dot{\widehat{p}}_k = \frac{i}{\hbar}\left[\widehat{H}, \widehat{p}_k\right] = 0. \tag{19.99}$$

This means that $\widehat{p}_k$ is an integral of the motion for the free particle. However, things are more complicated for the velocity, as follows by writing Heisenberg's equation for x_k,

$$\dot{\widehat{x}}_k = \frac{i}{\hbar}\left[\widehat{H}, \widehat{x}_k\right] = \frac{ic}{\hbar}\alpha_i\left[\widehat{p}_i, \widehat{x}_k\right] = c\alpha_i\delta_{ik} = c\alpha_k. \tag{19.100}$$

This result confirms that αc is the velocity operator,

$$\dot{\widehat{x}} = c\alpha, \tag{19.101}$$

for any free particle. Combining with Eq. (19.87) we see that we can write $j = \Psi^\dagger\dot{\widehat{x}}\Psi$. Since the eigenvalues of α_i are ± 1, the free Dirac particle always moves at speed c. In order to understand how this is possible, Schrödinger introduced the concept of *zitterbewegung*, which refers to a very fast (at speed c) vibrating or dancing motion of the electron around its trajectory, and which we translate as *tremor*. Let us see how this idea is justified and how the motion is described.

While the components of the momentum are constants of motion, the $\dot{x}_k$ are not (they are given by operators). This suggests that p_k can be regarded as the local average (over very short times) of $m\dot{x}_k$; in other words, that $\dot{x}_k$ contains more detailed information about the instantaneous motion than p_k, as we will confirm a little later (see Eq. (19.103)). To make the difference between the operators p_k and $m\dot{x}_k$ explicit, we again use Heisenberg's equation for α_k and write

$$\dot{\alpha}_k = \frac{1}{c}\ddot{x}_k = \frac{i}{\hbar}\left[\widehat{H}, \alpha_k\right] = \frac{2i\,c}{\hbar}\widehat{p}_k - \frac{2i}{\hbar}\alpha_k\widehat{H}. \tag{19.102}$$

From here and from $[\alpha_i, \alpha_j] \neq 0$, with $\alpha_i \sim \dot{x}_i$, we conclude that the velocity components of the free particle do not commute with each other and are not conserved, that is, they fluctuate. To study the changes in velocity we integrate Eq. (19.102) for an eigenstate of H and p_k, which allows us to treat these quantities as constant; the solution is

$$\dot{x}_k = c\alpha_k = \frac{c^2 p_k}{E} + \left(\dot{x}_k(0) - \frac{c^2 p_k}{E}\right) e^{-2iEt/\hbar} \tag{19.103}$$

and its integral over time gives

$$x_k = x_k(0) + \frac{c^2 p_k}{E} t + \frac{i\hbar c}{2E}\left(\frac{\dot{x}_k(0)}{c} - \frac{cp_k}{E}\right) e^{-2iEt/\hbar}. \tag{19.104}$$

The first two terms in this equation represent the classical solution for the relativistic free particle and the third is an additional term that oscillates in time. The amplitude of these oscillations around the classical trajectory is of the order of $\Delta x \sim \hbar c/E$ and therefore depends on the energy. These oscillations constitute the *zitterbewegung*.[11] For a particle that is not moving very fast (apart from *zitterbewegung*), $E \approx mc^2$ and so $\Delta x \sim \hbar/mc = \lambda_C$, the Compton wavelength (which for the electron is $\sim 10^{-12}$ m). At such energies the oscillation frequency of the *zitterbewegung* is $2E/\hbar \sim 2mc^2/\hbar \sim 10^{21}$ s^{-1}.

If in Eq. (19.103) we average over some cycles of the tremor (as happens in practice when measuring velocity or position), the term in parentheses vanishes, but the classical term is not altered, and we obtain

$$\overline{v_k} = c\overline{\alpha_k} = \frac{c^2}{E} p_k. \tag{19.105}$$

This result confirms that the relation between $\overline{v_k}$ and p_k is the relativistic one ($m\overline{v_k}$ is practically equal to p_k for $E \simeq m_0 c^2$). Note, however, that for positive energy states ($E \geq m_0 c^2 > 0$), $\langle v_k \rangle$ and p_k have the same sense, but for $E \leq m_0 c^2$ ($E < 0$), these quantities have opposite senses. From a classical point of view, with $p = mv$ (which corresponds to the limit $E = m_0 c^2$), the last case implies a negative mass. Another important observation is that Δx decreases with energy, while the frequency of the oscillation increases with it, which means that for ultrarelativistic particles the oscillation occurs at a very high frequency and negligible amplitude. In other words, the ultrarelativistic electron behaves as a point particle, with the effective structure produced by the tremor disappearing.

19.4 The Free Particle

We now turn to study the solution of the Dirac equation for a free particle. We propose to express the solution of the corresponding equation

$$(\gamma_\mu \partial_\mu + \frac{mc}{\hbar})\Psi = 0 \tag{19.106}$$

---

[11] The *zitterbewegung* is not an exclusive result of the Dirac equation, but rather a property of relativistic (not necessarily quantum) dynamics when restricted only by the constraints imposed by the Poincaré invariance group. It should therefore be seen as a relativistic prediction for spin particles, regardless of the details of the description.

in terms of plane waves in the form $\Psi = e^{-i\frac{E}{\hbar}t}\psi(r)$, with $\psi(r)$ a bispinor. We substitute Eq. (19.67) into the Dirac equation written in terms of the spinors Ψ_A, Ψ_B and cancel the common time factor; this gives the pair of coupled equations

$$c\sigma' \cdot p\Psi_B = (E - mc^2)\Psi_A, \tag{19.107}$$

$$c\sigma' \cdot p\Psi_A = (E + mc^2)\Psi_B. \tag{19.108}$$

In the reference system where the particle is at rest ($p = 0$) these equations reduce to

$$(E - mc^2)\Psi_A = 0, \tag{19.109}$$

$$(E + mc^2)\Psi_B = 0. \tag{19.110}$$

If $E > 0$, the first equation holds with $E = mc^2$ and any Ψ_A, but it follows from the second equation that Ψ_B must be zero; if $E < 0$, Ψ_A must be equal to zero and $E = -mc^2$. This allows us to identify Ψ_A as the spinor describing positive energy particles and Ψ_B as the corresponding spinor for negative energy states; the Dirac bispinors are therefore

$$\begin{pmatrix} \Psi_A \\ 0 \end{pmatrix}, \quad E > 0, \quad \begin{pmatrix} 0 \\ \Psi_B \end{pmatrix}, \quad E < 0. \tag{19.111}$$

To uniquely determine the bispinors $\begin{pmatrix} \Psi_A \\ \Psi_B \end{pmatrix}$ for $p \neq 0$, it is usual to introduce some additional condition. The simplest is to require that they be eigenfunctions of σ_3, which commutes with $\widehat{H}$ for the particle at rest. According to Eq. (19.91), the same consideration holds for any value of p_3 but with $p_1 = p_2 = 0$ and hence for the eigenstates of the helicity operator, which becomes $\mathfrak{h} = \sigma_3$ according to (19.63). Therefore $\Psi_A^{(+)} = \begin{pmatrix} 1 \\ 0 \end{pmatrix}$ corresponds to the eigenvalue $s_3 = 1$ and $\Psi_A^{(-)} = \begin{pmatrix} 0 \\ 1 \end{pmatrix}$ to $s_3 = -1$. The eigenfunctions Ψ_B for negative energy are constructed analogously. We thus obtain the following eigenfunctions of $\widehat{H}$ and σ_3 for $p = 0$,

$$\begin{pmatrix} 1 \\ 0 \\ 0 \\ 0 \end{pmatrix} e^{-imc^2t/\hbar}, (E > 0, s_3 = 1); \quad \begin{pmatrix} 0 \\ 1 \\ 0 \\ 0 \end{pmatrix} e^{-imc^2t/\hbar}, (E > 0, s_3 = -1);$$

$$\tag{19.112}$$

$$\begin{pmatrix} 0 \\ 0 \\ 1 \\ 0 \end{pmatrix} e^{imc^2t/\hbar}, (E < 0, s_3 = 1); \quad \begin{pmatrix} 0 \\ 0 \\ 0 \\ 1 \end{pmatrix} e^{imc^2t/\hbar}, (E < 0, s_3 = -1). \tag{19.113}$$

Since in experiments with electrons we have $E > 0$, the first pair of solutions is generally used.

To study the particle moving in the laboratory system we can either apply a Lorentz transformation to (19.112) or solve Eqs. (19.107) and (19.108) directly. For this we propose a solution of the type

$$\Psi = \begin{pmatrix} u_A(p) \\ u_B(p) \end{pmatrix} e^{i(-Et + p \cdot x)/\hbar}, \tag{19.114}$$

where u_A and u_B are spinors. From (19.107) and (19.108) we get

$$u_A(p) = \frac{c}{E - mc^2} \sigma' \cdot p\, u_B(p), \tag{19.115}$$

$$u_B(p) = \frac{c}{E + mc^2} \sigma' \cdot p\, u_A(p). \tag{19.116}$$

Combining these two equations we get a third one that is satisfied by either u_A or u_B, which indicates that there are different solutions whose selection depends on additional conditions of the problem. It is often proposed to take u_A simply as $u_A = \begin{pmatrix} 1 \\ 0 \end{pmatrix}$, which agrees with the solution for the free-particle eigenstate of σ_3'; then the second solution must be

$$u_B(p) = \frac{c}{E + mc^2} \sigma' \cdot p \begin{pmatrix} 1 \\ 0 \end{pmatrix}, \tag{19.117}$$

and the components of $u_B = \begin{pmatrix} u_B^{(1)}(p) \\ u_B^{(2)}(p) \end{pmatrix}$ become

$$u_B^{(1)} = \frac{c}{E + mc^2} p_3, \quad u_B^{(2)} = \frac{c}{E + mc^2} (p_1 + ip_2). \tag{19.118}$$

Since

$$\sigma' \cdot p = \sigma_1' p_1 + \sigma_2' p_2 + \sigma_3' p_3 = \begin{pmatrix} p_3 & p_1 - ip_2 \\ p_1 + ip_2 & -p_3 \end{pmatrix}, \tag{19.119}$$

this solution corresponds to $E > 0$ and positive spin projection along the z-axis; for the negative spin projection we take $u_A = \begin{pmatrix} 0 \\ 1 \end{pmatrix}$. Note that u_A is large compared to u_B for nonrelativistic particle velocities and positive energies (when $E + mc^2 \approx 2mc^2$). The two solutions with negative energy are obtained in an analogous way; in this case u_A becomes small compared to u_B. The final set of solutions is

$$u_1(p) = N \begin{pmatrix} 1 \\ 0 \\ +\frac{cp_3}{E+mc^2} \\ +\frac{c(p_1+ip_2)}{E+mc^2} \end{pmatrix}, \quad u_2(p) = N \begin{pmatrix} 0 \\ 1 \\ \frac{c(p_1-ip_2)}{E+mc^2} \\ -\frac{cp_3}{E+mc^2} \end{pmatrix}, \tag{19.120}$$

$$u_3(p) = N \begin{pmatrix} -\frac{cp_3}{|E|+mc^2} \\ -\frac{c(p_1+ip_2)}{|E|+mc^2} \\ 1 \\ 0 \end{pmatrix}, \quad u_4(p) = N \begin{pmatrix} -\frac{c(p_1-ip_2)}{|E|+mc^2} \\ \frac{cp_3}{|E|+mc^2} \\ 0 \\ 1 \end{pmatrix}. \tag{19.121}$$

These spinors are solutions of the equation

$$(i\gamma_\mu p_\mu + mc)u_r(p) = 0, \quad r = 1, 2, 3, 4, \tag{19.122}$$

which is the Fourier transform of the Dirac equation for the free particle. Each u_r is a column of four components, which in general is not an eigenfunction of σ_3 except when $p_1 = p_2 = 0$, as we have already noted.

From the previous discussion we conclude that there are only two cases where solutions of the free-particle Dirac equation exist which are eigenfunctions of $\boldsymbol{\sigma} \cdot \hat{\mathbf{n}}$, namely
(i) when $p = 0$,
(ii) when $\hat{\mathbf{n}} = \boldsymbol{p}/|\boldsymbol{p}|$.

These solutions are indeed eigenfunctions of the helicity operator. To verify this we start from the relation

$$[\widehat{H}, \boldsymbol{a} \cdot \boldsymbol{\sigma}] = -2ic\boldsymbol{\alpha} \cdot (\boldsymbol{p} \times \boldsymbol{a}) \tag{19.123}$$

and note that when $p = 0$ or when $\boldsymbol{a}$ is parallel to $\boldsymbol{p}$, the commutator vanishes (see problem P19.5).

To normalize the solution of Dirac's equation, we write

$$u^{(r)\dagger} u^{(r)} = C \tag{19.124}$$

and fix the constant C with an appropriate criterion. Two commonly used are

(a) Normalization to unity,

$$C = 1, \quad N = \sqrt{\frac{|E| + mc^2}{2|E|}}, \tag{19.125}$$

(b) Lorentz-invariant normalization,

$$C = \frac{|E|}{mc^2}, \quad N = \sqrt{\frac{|E| + mc^2}{2mc^2}}. \tag{19.126}$$

Exercise E19.2. The free particle with positive energy

Construct the solutions of the Dirac equation for a free particle with positive energy and helicity states ± 1.

Solution. We propose a solution of the form (in units of $\hbar = 1$)

$$\Psi = u(\boldsymbol{p}) e^{i(\boldsymbol{p} \cdot \boldsymbol{x} - Et)}, \tag{19.127}$$

with $E > 0$. The helicity operator is

$$\mathfrak{h} = \frac{\boldsymbol{\sigma} \cdot \boldsymbol{p}}{p} = \frac{1}{p} \sum_k \sigma_k p_k. \tag{19.128}$$

Using the standard representation of the Pauli matrices σ_k, the eigenvalue equation for the helicity applied to the spinor Ψ, $\mathfrak{h}\Psi = h\Psi$, $h = \pm 1$, is decomposed into equal systems of two equations, one for u_1, u_2 and a similar one for u_3, u_4, which means that the small spinor (with components u_3, u_4) is proportional to the large one (with components u_1, u_2). For the latter we obtain

$$p_3 u_1 + (p_1 - ip_2) u_2 = hp u_1, \tag{19.129}$$

$$(p_1 + ip_2) u_1 - p_3 u_2 = hp u_2. \tag{19.130}$$

It is convenient to simplify the expressions that follow by writing the momentum components in terms of the angles that determine their direction,

$$p_1 \pm ip_2 = p \sin\theta e^{\pm i\varphi}, \quad p_3 = p \cos\theta. \tag{19.131}$$

Equations (19.129) and (19.130) then reduce to

$$u_1 \cos\theta + u_2 \sin\theta e^{-i\varphi} = hu_1, \quad u_1 \sin\theta e^{i\varphi} - u_2 \cos\theta = hu_2, \tag{19.132}$$

which gives

(a) for $h = +1$, $\quad u_{2,4} = \tan\tfrac{1}{2}\theta e^{i\varphi} u_{1,3}$,
(b) for $h = -1$, $\quad u_{2,4} = -\cot\tfrac{1}{2}\theta e^{i\varphi} u_{1,3}$.

To find the components $u_{1,3}$ we substitute Ψ and Eqs. (19.131) into the Dirac equation, giving the system

$$u_4 \sin\theta e^{-i\varphi} + u_3 \cos\theta - \varkappa u_1 = 0,$$
$$u_3 \sin\theta e^{+i\varphi} - u_4 \cos\theta - \varkappa u_2 = 0,$$
$$-u_2 \sin\theta e^{-i\varphi} - u_1 \cos\theta + \frac{1}{\varkappa} u_3 = 0,$$
$$-u_1 \sin\theta e^{+i\varphi} + u_2 \cos\theta + \frac{1}{\varkappa} u_4 = 0, \tag{19.133}$$

where

$$\varkappa = \frac{cp}{E + mc^2}. \tag{19.134}$$

By inserting Eqs. (19.132) here, the system is reduced to $u_3 = h\varkappa u_1$. The still missing relation between the amplitudes is determined by the normalization, which is given by Eq. (19.124), $u^\dagger(p)u(p) = 1$. (There are a number of uses here.) This gives

(a) for $h = +1$,

$$u_+(p) = \frac{1}{\sqrt{1+\varkappa^2}} \begin{pmatrix} \cos\tfrac{1}{2}\theta e^{-i\varphi/2} \\ \sin\tfrac{1}{2}\theta e^{+i\varphi/2} \\ \varkappa\cos\tfrac{1}{2}\theta e^{-i\varphi/2} \\ \varkappa\sin\tfrac{1}{2}\theta e^{+i\varphi/2} \end{pmatrix},$$

(b) for $h = -1$,

$$u_-(p) = \frac{1}{\sqrt{1+\varkappa^2}} \begin{pmatrix} \sin\tfrac{1}{2}\theta e^{-i\varphi/2} \\ -\cos\tfrac{1}{2}\theta e^{+i\varphi/2} \\ -\varkappa\sin\tfrac{1}{2}\theta e^{-i\varphi/2} \\ \varkappa\cos\tfrac{1}{2}\theta e^{+i\varphi/2} \end{pmatrix}.$$

The structure of the solution confirms that it corresponds to positive energy, that is, to electrons.

19.5 *The Dirac Particle in an External Electromagnetic Field

We now turn to the study of Dirac's equation for the case where the particle is under the action of an external electromagnetic field. By introducing minimal coupling, we obtain Dirac's equation in the presence of an electromagnetic field,

$$\gamma_\nu \left(\partial_\mu - \frac{ie}{\hbar c} A_\mu\right)\Psi + \frac{mc}{\hbar}\Psi = 0. \tag{19.135}$$

To transform this equation we need a relativistic generalization of Eq. (14.10),

$$(\hat{\boldsymbol{\sigma}} \cdot \hat{\boldsymbol{A}})(\hat{\boldsymbol{\sigma}} \cdot \hat{\boldsymbol{B}}) = \hat{\boldsymbol{A}} \cdot \hat{\boldsymbol{B}} + i\hat{\boldsymbol{\sigma}} \cdot (\hat{\boldsymbol{A}} \times \hat{\boldsymbol{B}}), \qquad [\hat{\boldsymbol{A}}, \hat{\boldsymbol{\sigma}}] = 0, \qquad (19.136)$$

which we obtain as follows. We rewrite Eq. (19.96) in the form $\gamma_\mu \gamma_\nu = \gamma_\nu \gamma_\mu + 2i\sigma_{\mu\nu}$, and combine with $\gamma_\mu \gamma_\nu = -\gamma_\nu \gamma_\mu + 2\delta_{\mu\nu}$ to obtain

$$\gamma_\mu \gamma_\nu = \delta_{\mu\nu} + i\sigma_{\mu\nu}. \qquad (19.137)$$

Now let A_μ be an operator that commutes with γ_μ; by multiplying (19.137) with $A_\mu B_\nu$ on the right, we get

$$(\gamma_\mu A_\mu)(\gamma_\nu B_\nu) = A_\mu B_\mu + i\sigma_{\mu\nu} A_\mu B_\nu. \qquad (19.138)$$

On the other hand, if we exchange the indices, we see that

$$\sigma_{\mu\nu} A_\mu B_\nu = \sigma_{\nu\mu} A_\nu B_\mu = -\sigma_{\mu\nu} A_\nu B_\mu, \qquad (19.139)$$

due to the antisymmetry of $\sigma_{\mu\nu}$. Combining the two expressions, we get

$$\sigma_{\mu\nu} A_\mu B_\nu = \tfrac{1}{2}\sigma_{\mu\nu}(A_\mu B_\nu - A_\nu B_\mu), \qquad (19.140)$$

so that Eq. (19.138) can be written in the more convenient form,

$$(\gamma_\mu A_\mu)(\gamma_\nu B_\nu) = A_\mu B_\nu + \tfrac{1}{2}i\sigma_{\mu\nu}(A_\mu B_\nu - A_\nu B_\mu). \qquad (19.141)$$

This is the desired generalization. In particular, for $A = B$ we get

$$(\gamma_\mu A_\mu)^2 = A_\mu A_\mu + \tfrac{1}{2}i\sigma_{\mu\nu}[A_\mu, A_\nu]. \qquad (19.142)$$

Returning to our task, we apply from the left to Eq. (19.135) the operator $(\gamma_\nu B_\nu - mc/\hbar)$, with $B_\mu = \partial_\mu - (ie/\hbar c) A_\mu$,

$$\left(\gamma_\nu B_\nu - \frac{mc}{\hbar}\right)\left(\gamma_\mu B_\mu + \frac{mc}{\hbar}\right)\Psi = \left[(\gamma_\nu B_\nu)(\gamma_\mu B_\mu) - \frac{m^2 c^2}{\hbar^2}\right]\Psi = 0. \qquad (19.143)$$

We now apply (19.142), substituting B_μ for its original expression; the result is an equation that contains the Klein–Gordon equation plus additional terms describing the coupling between the particle's spin tensor and the external field,

$$\left(\partial_\mu - \frac{ie}{\hbar c}A_\mu\right)^2 \Psi - \frac{m^2 c^2}{\hbar^2}\Psi + \tfrac{1}{2}i\sigma_{\mu\nu}\left[\partial_\mu - \frac{ie}{\hbar c}A_\mu, \partial_\nu - \frac{ie}{\hbar c}A_\nu\right]\Psi = 0. \qquad (19.144)$$

The commutator that appears in this expression can be rewritten as follows,

$$[,] = -\frac{ie}{\hbar c}(\partial_\mu A_\nu + A_\mu \partial_\nu - \partial_\nu A_\mu - A_\nu \partial_\mu)$$

$$= -\frac{ie}{\hbar c}[(\partial_\mu A_\nu) - (\partial_\nu A_\mu)] = -\frac{ie}{\hbar c} F_{\mu\nu}, \qquad (19.145)$$

where

$$F_{\mu\nu} = \frac{\partial A_\nu}{\partial x_\mu} - \frac{\partial A_\mu}{\partial x_\nu} \qquad (19.146)$$

is the electromagnetic tensor (remember that $F_{ij} = \varepsilon_{ijk} B_k$, $F_{4i} = iE_i$). Substituting and multiplying by $(-i\hbar)^2$, the preceding equation takes the form

$$(p_\mu - \frac{e}{c}A_\mu)^2 \Psi + m^2 c^2 \Psi - \frac{\hbar e}{2c}\sigma_{\mu\nu} F_{\mu\nu} \Psi = 0. \qquad (19.147)$$

19.5.1 Electric and magnetic moments of the electron

The preceding result shows that the Dirac equation for a particle subject to the action of an electromagnetic field is the relativistic generalization of the corresponding Pauli equation (see Section 14.2.1), which allows us to write

$$\frac{\hbar e}{4mc}\sigma_{\mu\nu}F_{\mu\nu} = -\frac{i\hbar e}{4mc}\gamma_\mu\gamma_\nu F_{\mu\nu} = -\frac{i\hbar e}{2mc}(\gamma_i\gamma_j F_{ij} + \gamma_i\gamma_4 F_{i4})$$
$$= \frac{\hbar e}{2mc}(\sigma_k B_k + i\gamma_i\gamma_4 i E_i), \tag{19.148}$$

or taking into account that $\gamma_i\gamma_4 = \gamma_i\rho_3 = \rho_2\sigma_i\rho_3 = \rho_2\rho_3\sigma_i = i\rho_1\sigma_i = i\alpha_i$,

$$\frac{\hbar e}{4mc}\sigma_{\mu\nu}F_{\mu\nu} = \frac{\hbar e}{2mc}(\boldsymbol{\sigma}\cdot\boldsymbol{B} - i\boldsymbol{\alpha}\cdot\boldsymbol{E}) = \boldsymbol{\mu}_{\text{mag}}\cdot\boldsymbol{B} + \boldsymbol{\mu}_{\text{elect}}\cdot\boldsymbol{E}, \tag{19.149}$$

where

$$\boldsymbol{\mu}_{\text{mag}} = -(\hbar e_0/2mc)\boldsymbol{\sigma} \tag{19.150}$$

and

$$\boldsymbol{\mu}_{\text{elect}} = i(\hbar e_0/2mc)\boldsymbol{\alpha} \tag{19.151}$$

are the electric and magnetic Dirac moments of the electron, respectively.

It is particularly interesting to check that Dirac's equation predicts the correct value of the intrinsic magnetic moment, including the factor $g = 2$ that we had to insert by hand in nonrelativistic theory. It also predicts an electric moment of the electron proportional to the velocity operator. Since the eigenvalues of σ are of the order of unity, while those of α are of the order of $|\dot{x}|/c$, in the nonrelativistic limit one has

$$|\langle\mu_{\text{mag}}\rangle| \sim \frac{\hbar e_0}{2mc} \gg \left|\frac{\langle\dot{x}\rangle}{c}\right|\frac{\hbar e_0}{2mc} \sim |\langle\mu_{\text{elect}}\rangle|, \tag{19.152}$$

and the interaction term reduces to the Pauli expression in the low-velocity limit, as expected.

Up to now, the properties of the Dirac electron have been obtained in a natural way, without the need to introduce ad hoc assumptions, so that Dirac's equation is considered to be an exact equation for the electron. However, in a second-quantization theory, the electron also couples to the fluctuations of the radiation field (which in fact always exist) and through them to itself, that is there is a self-interaction of the particle that contributes to modify the parameters that characterize it (mass, charge, etc.). One of the effects of self-interaction is the appearance of the anomalous magnetic moment, a contribution that is added to the proper magnetic moment predicted by Dirac's theory. As a result, the total magnetic moment of the electron is not that given by Dirac's theory for a particle, but carries a correction given by (see the explanation in Section 14.4)

$$\mu = \mu_0\left(1 + \frac{\alpha}{2\pi} + \ldots\right) \tag{19.153}$$

with $\alpha \approx 1/137$. This is an important result of QED, although because of the small size of the correction it is common to start from Dirac's equation to describe the electron and add the anomalous moment as a correction if necessary. In the case of nucleons, on the other hand, although the anomalous magnetic moment is of the order of the electronic one, the normal moment is much smaller (of the order of a thousandth of the electronic one, since $\mu \sim m^{-1}$), so in this case the anomalous magnetic moment cannot be introduced as a mere correction.

19.6 Solution of Dirac's Equation for the Hydrogen-Like Atom

In Section 19.2, the first-order Dirac equation for a four-component bi-spinor was transformed into a pair of first-order equations, each for a two-component spinor. A further separation of the angular and radial variables leads to a second-order equation corresponding to the Klein–Gordon equation plus additional terms representing corrections introduced by the coupling of the particle's spin to the external field. Finally, the angular and radial functions can be factored out.

In the stationary case this transformation leads to the radial equation for the hydrogen-like atom,

$$\left(\nabla_r^2 + \frac{\Lambda_D}{r^2} + \frac{2c_E}{r} - d_E\right) R = 0, \tag{19.154}$$

with

$$\nabla_r^2 = \frac{1}{r}\frac{\partial^2}{\partial r^2}r, \quad \Lambda_D = l(l+1) - \alpha^2 Z^2, \tag{19.155}$$

$$d_E = \frac{m^2 c^2}{\hbar^2}\left(1 - \mathcal{E}^2\right), \quad c_E = \frac{Zme^2}{\hbar^2}\mathcal{E}, \tag{19.156}$$

where

$$\mathcal{E} = \frac{mc^2 + E}{mc^2} \tag{19.157}$$

is the energy of the atom, including the rest energy, expressed in units of the rest energy, and $\alpha = e^2/\hbar c$.

19.6.1 Exact solution of the central-force problem

For the case of the Coulomb potential, Dirac's equation can be solved exactly. The result for the energy levels (including the proper energy), is given in Exercise 19.33,

$$E_{nj} = \frac{mc^2}{\sqrt{1 + \frac{\alpha^2 Z^2}{\left[n-j-\frac{1}{2}+\sqrt{k^2-\alpha^2 Z^2}\right]^2}}}. \tag{19.158}$$

where $n = k + l + 1$. It is interesting to compare this expression with the one obtained from the Klein–Gordon equation (19.29) with $k = k_{KG}$, in which j takes the place of k due to the absence of electron spin; with $k = k_D$ the correspondence is $k_{KG} + \frac{1}{2} \to k_D + \frac{1}{2} + (l - j)$). An expansion of the previous result in powers of $Z\alpha$ gives

$$E_{nj} = mc^2\left[1 - \frac{Z^2\alpha^2}{2n^2} - \frac{Z^4\alpha^4}{2n^4}\left(\frac{n}{j+\frac{1}{2}} - \frac{3}{4}\right) + \ldots\right]. \tag{19.159}$$

The first term after the proper mass represents the Schrödinger energy; the next represents the relativistic and spinorial correction to order α^4, which follows from Dirac's equation; this correction differs from (19.30) derived from the Klein–Gordon equation by substituting j for l. It is clear that this approximate expression is valid only when $k^2 > Z^2\alpha^2$, that is, for relatively light atoms, or highly excited states of heavy atoms. The solution (19.159) predicts the breaking of the degeneracy with respect to j, but not with respect to l.

Exercise E19.3. Solution of the radial Dirac equation for the H atom using the algebraic consistency method

Construct the exact energy eigenvalues for the Dirac H atom by means of the algebraic procedure developed in Section A.1.

Solution. The method of algebraic consistency conditions can be used to solve the H-atom eigenvalue problem without the need to know the explicit solution of the Dirac equation, as follows.

The consistency conditions hold whenever the operators S and A can be written in the form of Eqs. (A.94) and (A.95), where B_{n-} represents the Hamiltonian operator expressed in the appropriate dynamical variable.

In terms of $u = rR$, Eq. (19.154) acquires the same structure as the radial Schrödinger equation for the hydrogen-like atom, (A.99). They both correspond to the operator B_{1-}, although with a different set of parameters, so (19.154) is equivalent to

$$\left(-q\frac{d^2}{dq^2} + \frac{\Lambda_D}{q} + q\right)u = B_{1-}u = \frac{2c_E}{d_E^{1/2}}u, \qquad (19.160)$$

$$\Lambda_D = l(l+1) - \alpha^2, \qquad q = d_E^{1/2}r. \qquad (19.161)$$

Thus, the solution for the Schrödinger atom can be transferred to the Dirac atom,

$$b_k = b_0 + 2k, \qquad (19.162)$$

where b_0 is given by $b_0 = 1 + \sqrt{1 + 4\Lambda_D}$ according to Eq. (A.98). The minus sign is excluded, since $b_0 \geq 0$ for any l. From the preceding equations, after some algebra, we obtain

$$\mathcal{E}_k = \left[1 + \frac{4\alpha^2}{(b_0 + 2k)^2}\right]^{-1/2}, \qquad (19.163)$$

which is the exact result for the energy of the Dirac H atom in units of mc^2.

19.7 Approximate Versions of Dirac's Equation

We will now study the corrections to the Schrödinger (or rather to the Pauli) equation predicted by Dirac's equation. An expansion up to the second order in v/c allows us to derive the terms coming from both relativity and spin, which can then be treated as perturbations to the Pauli equation, as was done in Section 14.4.

Consider an electron in an external electromagnetic field; we write the wave function as $\Psi = e^{-iEt/\hbar}\Psi(r)$ and substitute into Eq. (19.135) to obtain the following system of equations for the spinors (see Eqs. (19.107) and (19.108) with $p_\mu \to p_\mu - eA_\mu/c$),

$$c\boldsymbol{\sigma}' \cdot (\boldsymbol{p} - \frac{e}{c}\boldsymbol{A})\Psi_B = (E - eA_0 - mc^2)\Psi_A, \quad (19.164)$$

$$c\boldsymbol{\sigma}' \cdot (\boldsymbol{p} - \frac{e}{c}\boldsymbol{A})\Psi_A = (E - eA_0 + mc^2)\Psi_B. \quad (19.165)$$

Solving for Ψ_B from the second equation and inserting the solution into the first gives a second-order equation of the van der Waerden type,

$$\boldsymbol{\sigma}' \cdot (\boldsymbol{p} - \frac{e}{c}\boldsymbol{A}) \frac{c^2}{E - eA_0 + mc^2} \boldsymbol{\sigma}' \cdot (\boldsymbol{p} - \frac{e}{c}\boldsymbol{A})\Psi_A = (E - eA_0 - mc^2)\Psi_A. \quad (19.166)$$

This is an exact Dirac's equation, and it is Lorentz invariant, although not explicitly covariant. In the nonrelativistic approximation we can consider $E \approx mc^2$ and expand up to the first order in $|v/c|$; with $E_0 = E - mc^2$ and $|E_0 - eA_0| \ll mc^2$, where E is the total energy, we get

$$\frac{c^2}{E - eA_0 + mc^2} = \frac{1}{2m} \frac{2mc^2}{2mc^2 + E_0 - eA_0} = \frac{1}{2m}\left(1 - \frac{E_0 - eA_0}{2mc^2} + \ldots\right). \quad (19.167)$$

Keeping for the moment only the first term $1/2m$, (19.166) reduces to

$$\frac{1}{2m}\left[\boldsymbol{\sigma}' \cdot (\boldsymbol{p} - \frac{e}{c}\boldsymbol{A})\right]^2 \Psi_A = (E_0 - eA_0)\Psi_A, \quad (19.168)$$

which is the Pauli equation (14.25) for the stationary case, with $V = eA_0$. The Pauli equation can then be seen as the lowest-order nonrelativistic approximation of the Dirac equation. If, in addition, the spin is suppressed, the previous equation reduces to that of Schrödinger. Note that this result endorses the substitution of $\boldsymbol{p}^2$ by $(\boldsymbol{\sigma} \cdot \boldsymbol{p})^2$ used in Section 14.2 as a method of introducing spin into Schrödinger's theory.

After some algebra carried out to second order, the Dirac Hamiltonian reduces to

$$H = \frac{\boldsymbol{p}^2}{2m} + V - \frac{\boldsymbol{p}^4}{8m^3c^2} - \frac{e\hbar^2}{8m^2c^2}(\nabla \cdot \boldsymbol{E}) - \frac{\hbar e}{4m^2c^2}\boldsymbol{\sigma} \cdot \boldsymbol{E} \times \boldsymbol{p}. \quad (19.169)$$

We see that to this order, Dirac's theory predicts three correction terms to Schrödinger's theory, namely,
(i) the relativistic correction to the kinetic energy, $\widehat{H}_1 = -\boldsymbol{p}^4/8m^3c^2$,
(ii) the Darwin term due to the *zitterbewegung* of the electron,

$$\widehat{H}_2 = -\frac{e\hbar^2}{8m^2c^2}(\nabla \cdot \boldsymbol{E}) = \frac{\hbar^2}{8m^2c^2}\nabla^2 V, \quad (19.170)$$

(iii) the term due to the coupling between orbital motion and spin,

$$\widehat{H}_3 = -\frac{\hbar e}{4m^2c^2}\boldsymbol{\sigma} \cdot \boldsymbol{E} \times \boldsymbol{p} = \frac{\hbar}{4m^2c^2}\boldsymbol{\sigma} \cdot (\nabla V) \times \boldsymbol{p}. \quad (19.171)$$

Note that $\widehat{H}_2$ has appeared here as a result of the coupling of the electron's electric moment to the electric field, generated by the orbital motion. This makes it possible to link the origin of the electron's electric moment to the *zitterbewegung*. Sometimes this correspondence is carried further by considering that even the magnetic moment, and hence the electron spin, is due to the *zitterbewegung*. This possibility has been seriously considered by several authors over the years, but no consensus has been reached.

Problems

P19.1 Solve the Klein–Gordon equation for an isotropic attractive potential of depth V_0 and radius a; find the continuity conditions at $r = a$. Obtain the minimum value of V_0 needed to bind a particle of mass m.

P19.2 Use the method suggested in Problem 19.7.1 and the approximate equivalence

$$\frac{2m}{\hbar^2}[E - V(r)] \to \frac{1}{\hbar^2 c^2}\left[(E - V(r))^2 - m^2 c^4\right], \tag{19.172}$$

to obtain the Klein–Gordon solution for the spherical uniform well from the solution of the corresponding Schrödinger problem.

P19.3 Show that the expectation values of E^2 and p^2 for a general packet solution of the Klein–Gordon equation, satisfy the equation $\langle E^2 \rangle = c^2 \langle p^2 \rangle + m^2 c^4$.

P19.4 Show explicitly that the matrices γ_μ and α_μ satisfy the appropriate anticommutation rules, Eqs. (19.36) and (19.52).

P19.5 In applications, the matrix γ_5, defined as the product of the four matrices γ_μ,

$$\gamma_5 = \gamma_1 \gamma_2 \gamma_3 \gamma_4 \tag{19.173}$$

is frequently used. Prove that $\hat{\gamma}_5$ is Hermitian, that it anticommutes with the matrices γ_μ,

$$\gamma_5 \gamma_\mu + \gamma_\mu \gamma_5 = 0 \quad (\mu = 1, 2, 3, 4), \tag{19.174}$$

and that

$$\gamma_5 = \begin{pmatrix} \mathbb{O} & -\mathbb{I} \\ -\mathbb{I} & \mathbb{O} \end{pmatrix}, \quad \gamma_5^2 = 1. \tag{19.175}$$

P19.6 Solve Problem 19.7.1, this time using Dirac's equation. Compare and discuss your results.

P19.7 Prove that $[\widehat{H}, \boldsymbol{a} \cdot \boldsymbol{\sigma}] = -2ic\boldsymbol{\alpha} \cdot (\boldsymbol{p} \times \boldsymbol{a})$ for $\boldsymbol{a}$ a fixed vector and $\widehat{H}$ the free-particle Dirac Hamiltonian.

P19.8 Consider the sign operator for the free-particle Dirac equation given by the equation

$$\hat{\Lambda} = \frac{\widehat{H}_D}{E_+} = \frac{c\boldsymbol{\alpha} \cdot \boldsymbol{p} + \beta mc^2}{E_+}, \quad E_+ = +c\sqrt{p^2 + m^2 c^2}. \tag{19.176}$$

Show that this operator is Hermitian and unitary, with eigenvalues $\lambda = \pm 1$, and that these eigenvalues are integrals of motion for the free particle. Construct with its help the projectors $\hat{\Pi}^\pm = \frac{1}{2}(1 \pm \hat{\Lambda})$ and show that these operators project the positive and negative energy states, respectively, for the free particle.

P19.9 Given the Dirac Hamiltonian with minimal coupling (in units $c = \hbar = 1$)

$$\widehat{H}_D = \boldsymbol{\alpha} \cdot \boldsymbol{\pi} + \beta m + e\phi, \tag{19.177}$$

where $\boldsymbol{\pi} = \boldsymbol{p} - \frac{e}{c}\boldsymbol{A}$, prove that

$$\dot{\boldsymbol{r}} = \boldsymbol{\alpha}, \quad \ddot{\boldsymbol{r}} = \dot{\boldsymbol{\alpha}} = 2i\left[\boldsymbol{\pi} - \boldsymbol{\alpha}(\widehat{H}_D - e\phi)\right], \tag{19.178}$$

$$\dot{\boldsymbol{\pi}} = e\boldsymbol{E} + e\boldsymbol{\alpha} \times \boldsymbol{B}. \tag{19.179}$$

Note that the Lorentz force is expressed in terms of $\boldsymbol{\alpha}$.

P19.10 Prove the equation

$$(\boldsymbol{\alpha} \cdot \boldsymbol{p})(\boldsymbol{\sigma} \cdot \boldsymbol{J}) = \hat{\rho}_1(\boldsymbol{\sigma} \cdot \boldsymbol{p})(\boldsymbol{\sigma} \cdot \boldsymbol{J}) = \rho_1 \boldsymbol{p} \cdot \boldsymbol{J} + i\boldsymbol{\alpha} \cdot (\boldsymbol{p} \times \boldsymbol{J}). \tag{19.180}$$

P19.11 The selection rules for dipole transitions that apply to Dirac's theory are $\Delta l = \pm 1$, $\Delta j = 0, \pm 1$. Determine the frequencies of the transitions allowed between the states with $n = 2$ and $n = 3$ for the H atom in Dirac's theory, and compare with the results predicted by Schrödinger's theory.

Bibliographical Notes

Dirac's classic book (2023), first published in 1930, is still a standard reference.

The idea that the electron spin is related to the *zitterbewegung* has been proposed independently by many authors, starting with Schrödinger (1935). This has been an ongoing topic of research over the years; see, for example, Dirac (2023), Huang (1952), Bhabha and Corben (1941), Barut and Zhangi (1984), Hestenes (1990), Rodrigues et al. (1993), G. Salesi (1993). The last word on this subject has not yet been said, however.

Appendix A: Mathematical Tools

A.1 Solving Second-Order Linear Homogeneous Differential Equations: Special Functions

The differential equations that generate the special functions of physics are of the general form

$$\frac{d^2y}{dx^2} + p(x)\frac{dy}{dx} + q(x)y = 0. \tag{A.1}$$

If y_1 and y_2 are two independent solutions of (A.1), then any other solution can be written in the form

$$y_3 = ay_1 + by_2, \tag{A.2}$$

where a, b are arbitrary constant coefficients, which are chosen so that y_3 satisfies the required boundary conditions; y_3 is called the general solution of Eq. (A.1) (it contains already two constants of integration).

Among the most important orthogonal polynomials for this course are the Hermite polynomials, the Legendre and associated Legendre polynomials, the Laguerre and associated Laguerre polynomials, and the cylindrical and spherical Bessel functions, which we will briefly review in what follows.

A.1.1 Hermite polynomials

The Hermite[1] equation is

$$H_n'' - 2xH_n' + 2nH_n = 0. \tag{A.3}$$

The normalization is chosen so that the coefficient of x^n in H_n is 2^n; it follows that

$$H_n(x) = (2x)^n - \frac{n(n-1)}{1!}(2x)^{n-2} + \frac{n(n-1)(n-2)(n-3)}{2!}(2x)^{n-4} + \ldots; \tag{A.4}$$

the last term is proportional to x or is a constant; H_n is even or odd if n is even or odd, respectively. Hermite polynomials have the following *generating function*,

$$F(x,t) = e^{2xt-t^2} = \sum_{n=0}^{\infty} \frac{H_n(x)}{n!} t^n. \tag{A.5}$$

---

[1] Charles Hermite (1822–1901) was a famous French mathematician; the term Hermitian applies to properties of several mathematical entities.

and the Rodrigues formula,

$$H_n(x) = (-1)^n e^{x^2} \frac{d^n e^{-x^2}}{dx^n}. \tag{A.6}$$

By deriving the generating function it is easy to show that

$$H'_n(x) = 2n H_{n-1}(x). \tag{A.7}$$

From this expression and the differential Hermite equation follows the recurrence relation

$$H_{n+1}(x) = 2x H_n(x) - 2n H_{n-1}(x). \tag{A.8}$$

Using the generating function, it is possible to check that the orthogonality relation is

$$\int_{-\infty}^{\infty} e^{-x^2} H_n(x) H_m(x) dx = 0 \quad (n \neq m), \tag{A.9}$$

which shows that $w = e^{-x^2}$ is the weight function; for this reason the Hermite functions $e^{-x^2/2} H_n(x)$ are sometimes introduced which are orthonormal with weight 1; the normalization factor is

$$\int_{-\infty}^{\infty} e^{-x^2} H_n^2(x) dx = 2^n n! \sqrt{\pi}. \tag{A.10}$$

The first Hermite polynomials are

$$\begin{array}{ll} H_0 = 1 & H_1 = 2x \\ H_2 = 4x^2 - 2 & H_3 = 8x^3 - 12x \\ H_4 = 16x^4 - 48x^2 + 12 & H_5 = 32x^5 - 160x^3 + 120x. \end{array} \tag{A.11}$$

A.1.2 Legendre polynomials and spherical harmonics

We turn to the solutions of Eq. (11.47), obtained for the angular part of the Schrödinger equation of a central problem,

$$\frac{1}{\sin\theta} \frac{d}{d\theta} \left(\sin\theta \frac{d\Theta}{d\theta} \right) + \left(\lambda - \frac{m^2}{\sin^2\theta} \right) \Theta = 0. \tag{A.12}$$

With the change of variable $x = \cos\theta$, this equation transforms into

$$(1-x^2)\frac{d^2\Theta}{dx^2} - 2x\frac{d\Theta}{dx} + \left(\lambda - \frac{m^2}{1-x^2} \right) \Theta = 0. \tag{A.13}$$

A particular case is the Legendre[2] equation, when $m = 0$,

$$(1-x^2)\frac{d^2\Theta}{dx^2} - 2x\frac{d\Theta}{dx} + \lambda\Theta = 0. \tag{A.14}$$

The regular solutions in the interval $(-1, +1)$ are obtained by a Taylor series expansion around $x = 0$, $\Theta = \sum_{k=0}^{\infty} a_k x^k$, which leads to the recurrence relation

$$a_{k+2} = \frac{k(k+1) - \lambda}{(k+1)(k+2)} a_k. \tag{A.15}$$

[2] Adrien-Marie Legendre (1752–1833) was an outstanding French mathematician.

For $a_1 = 0$, $a_0 \neq 0$, the solution is even and for $a_0 = 0$, $a_1 \neq 0$, it is odd; in both cases $(a_{k+2}/a_k) \to 1$ when $k \to \infty$, so that both diverge for $x = \pm 1$ ($\theta = 0$ or π). The only way to obtain regular solutions is by choosing λ such that one of the series becomes a polynomial. From the recurrence relation it follows that a regular solution exists if and only if

$$\lambda = l(l+1), \quad l = 0, 1, 2, \ldots. \tag{A.16}$$

The first Legendre polynomials, obtained directly from the preceding results, are the following,

$$\begin{aligned}
&P_0 = 1 & &P_1 = x \\
&P_2 = \tfrac{1}{2}(3x^2 - 1) & &P_3 = \tfrac{1}{2}(5x^3 - 3x) \\
&P_4 = \tfrac{1}{8}(35x^4 - 30x^2 + 3) & &P_5 = \tfrac{1}{8}(63x^5 - 70x^3 + 15x).
\end{aligned} \tag{A.17}$$

The normalization has been chosen such that

$$P_l(\pm 1) = (\pm 1)^l. \tag{A.18}$$

Using the Rodrigues formula

$$P_l(x) = \frac{1}{2^l l!} \frac{d^l}{dx^l}(x^2 - 1)^l, \tag{A.19}$$

we obtain after integrating by parts,

$$\int_{-1}^{1} P_l(x) P_{l'}(x) dx = \frac{2}{2l+1} \delta_{ll'}, \tag{A.20}$$

as well as the recurrence relation

$$(n+1)P_{n+1}(x) = (2n+1)x P_n(x) - n P_{n-1}(x). \tag{A.21}$$

The generating function leads to a formula of great practical value, namely

$$F(x, t) = \frac{1}{\sqrt{1 - 2xt + t^2}} = \sum_{n=0}^{\infty} P_n(x) t^n. \tag{A.22}$$

The regular solutions of Eq. (A.13) with $m \neq 0$ are the associated Legendre polynomials. Since the equation depends on m^2, changing the sign of m should not affect the solution; therefore if Θ_l^m is the solution for $m > 0$, the solution for $m < 0$ is taken as

$$\Theta_l^{-|m|}(x) = (-1)^m C_{lm} \Theta_l^{|m|}(x), \tag{A.23}$$

where C_{lm} is an arbitrarily chosen constant.

For $m > 0$ the associated Legendre polynomial $\Theta = P_l^m$ is given by

$$P_l^m(x) = (1 - x^2)^{m/2} \frac{d^m P_l}{dx^m}, \tag{A.24}$$

or alternatively,

$$\begin{aligned}
P_l^m(x) &= \frac{1}{2^l l!} (1 - x^2)^{m/2} \frac{d^{l+m}}{dx^{l+m}}(x^2 - 1)^l \\
&= \frac{(-1)^m (l+m)!}{2^l l! (l-m)!} (1 - x^2)^{-m/2} \frac{d^{l-m}}{dx^{l-m}}(x^2 - 1)^l.
\end{aligned} \tag{A.25}$$

It is clear that only for $m \leq l$ are the solutions regular and physically acceptable. The first associated Legendre polynomials are the following,

$$P_1^1(x) = \sqrt{1-x^2}, \quad P_2^1(x) = 3x\sqrt{1-x^2}, \quad P_2^2(x) = 3(1-x^2) \tag{A.26}$$

or in terms of the angle θ, with $x = \cos\theta$,

$$P_1^1 = \sin\theta, \quad P_2^1 = 3\sin\theta\cos\theta, \quad P_2^2 = 3\sin^2\theta. \tag{A.27}$$

Note that the polynomials are orthogonal with respect to l, but not to m,

$$\int_{-1}^{1} P_l^m P_{l'}^m(x) dx = \frac{2}{2l+1} \frac{(l+m)!}{(l-m)!} \delta_{ll'}. \tag{A.28}$$

Equation (11.45) defines the spherical harmonics as the functions normalized to unity obtained by multiplying an associated Legendre function and the function $e^{im\varphi}$. Taking into account the normalization of P_l^m, Eq. (A.28), we obtain the spherical harmonics as the following functions, for l a positive integer and m a positive or negative integer, $|m| \leq l$,[3]

$$Y_l^m(\theta, \varphi) = \sqrt{\frac{2l+1}{4\pi} \frac{(l-m)!}{(l+m)!}} (-1)^m P_l^m(\cos\theta) e^{im\varphi}. \tag{A.29}$$

The first spherical harmonics are

$$Y_0^0 = \frac{1}{\sqrt{4\pi}} \qquad Y_1^0 = \sqrt{\frac{3}{4\pi}} \cos\theta \qquad Y_1^{\pm 1} = \mp\sqrt{\frac{3}{8\pi}} \sin\theta e^{\pm i\varphi}$$

$$Y_2^0 = \sqrt{\frac{5}{16\pi}} (3\cos^2\theta - 1) \qquad Y_2^{\pm 1} = \mp\sqrt{\frac{15}{8\pi}} \sin\theta \cos\theta e^{\pm i\varphi} \tag{A.30}$$

$$Y_2^{\pm 2} = \sqrt{\frac{15}{32\pi}} \sin^2\theta e^{\pm 2i\varphi}.$$

A.1.3 Laguerre polynomials

The associated Laguerre[4] polynomials are the regular solutions of the differential equation

$$\rho u'' + [2(l+1) - \rho]u' + (\beta - l - 1)u = 0, \tag{A.31}$$

which is obtained for the radial part of the hydrogen-like wave function once the asymptotic solutions at the origin and infinity have been extracted. With the change of variable $\rho = 2x$ this equation transforms into

$$xu'' + 2(l+1-x)u' + 2(\beta - l - 1)u = 0. \tag{A.32}$$

According to the results given in the introduction, the point $x = 0$ is singular, so the general solution of (A.31) is singular at the origin, but there is a particular regular

---

[3] For $m < 0$ we use (A.23) with $C_{lm} = 1$.
[4] Edmond Nicolas Laguerre (1834–1886) was a noted French mathematician.

solution, which is constructed from a Taylor series of the form $u = \sum_{k=0}^{\infty} a_k x^k$. This gives the recurrence relation

$$a_{k+1} = 2 \frac{k+l+1-\beta}{(k+1)(k+2l+2)} a_k, \tag{A.33}$$

from which $a_{k+1}/a_k \to 2/k$ when $k \to \infty$. This is precisely the behavior of the coefficients of the power series of e^{2x}; therefore, the series diverges exponentially, unless it is cut off by reducing β to a positive integer,

$$\beta \equiv n = k + l + 1, \tag{A.34}$$

so that a_{n-l} and hence all a_k for $k > n - l$ are zero. The method gives us the associated Laguerre polynomials and the preceding relation is the most important result in connection with these polynomials.

In order to generalize a little, we write (A.31) in the form

$$xu'' + (p + 1 - x)u' + (q - p)u = 0, \tag{A.35}$$

which reduces to (A.31) with the identification $x = \rho$, $p = 2l + 1$, $q = n + l$, $\beta = n$. With $p = 0$ we get the Laguerre equation,

$$xu'' + (1 - x)u' + qu = 0. \tag{A.36}$$

By deriving it p times we obtain (A.35), therefore if $L_{q-p}^{p}(x)$ are the solutions of (A.35) and $L_q(x)$ those of (A.36), we have

$$L_{q-p}^{p}(x) = (-1)^p \frac{d^p}{dx^p} L_q(x); \tag{A.37}$$

in particular, $L_q^0 = L_q$ (the sign $(-1)^p$ is conventional and has been chosen so that the coefficient of x^n is $(-1)^n$). From (A.37) we get the Rodrigues formula

$$x^p L_n^p(x) = \frac{1}{n!} e^x \frac{d^n}{dx^n} \left(e^{-x} x^{n+p} \right), \tag{A.38}$$

where $L_n^p(x)$ is a polynomial only if $p = 0, 1, 2, \ldots$. Expanding in series we get

$$L_n^p(x) = \sum_{s=0}^{n} (-1)^s \binom{n+p}{n-s} \frac{x^s}{s!}, \tag{A.39}$$

where $\binom{n}{r} = \frac{n!}{r!(n-r)!}$ is a binomial coefficient. The Rodrigues formula leads also to the orthogonality relation,

$$\int_0^\infty e^{-x} x^p L_n^p(x) L_m^p(x) dx = \frac{\Gamma(n+p+1)}{n!} \delta_{nm} \tag{A.40}$$

(Re $p > 0$). The most common recurrence relations are

$$x L_n^{p+1} = (n + p + 1) L_n^p - (n + 1) L_{n+1}^p = (n + p) L_{n-1}^p - (n - x) L_n^p, \tag{A.41}$$

$$\frac{d}{dx} L_n^p(x) = -L_{n-1}^{p+1}(x). \tag{A.42}$$

The generating function is

$$e^{-\frac{tx}{1-t}} = (1-t)^{p+1} \sum_{n=0}^{\infty} L_n^p(x) t^n, \qquad (A.43)$$

which shows that the exponential on the left can be expanded in terms of associated Laguerre polynomials of arbitrary index p, including $p = 0$. The simplest Laguerre polynomials are

$$L_0^p = 1 \qquad L_1 = 1 - x \qquad L_2 = 1 - 2x + \tfrac{1}{2}x^2. \qquad (A.44)$$

A.1.4 Cylindrical Bessel functions

The (cylindrical) Bessel[5] functions are the solutions of the differential Bessel equation,

$$y'' + \frac{1}{x} y' + \left(1 + \frac{\nu^2}{x^2}\right) y = 0 \qquad (A.45)$$

where ν is an arbitrary (real or complex) parameter. One of the solutions of this equation is called a Bessel function of the first kind and is conventionally denoted as $J_\nu(x)$; if ν is not an integer, a second independent solution is $J_{-\nu}(x)$. In this case it is customary to introduce the Bessel function of the second kind or Neumann[6] function (also called Weber function), which is usually denoted in physics by $N_\nu(x)$ (and in mathematics by $Y_\nu(x)$) and defined as

$$N_\nu(x) = \frac{1}{\sin \pi \nu} \left(\cos \pi \nu J_\nu(x) - J_{-\nu}(x)\right), \quad |\arg x| < \pi. \qquad (A.46)$$

In terms of these functions, the Hankel functions or Bessel functions of the third kind are defined as

$$H_\nu^{(1,2)}(x) = J_\nu(x) \pm i N_\nu(x), \qquad (A.47)$$

where the index 1(2) corresponds to the sign $+(-)$. When $\nu = n$ is an integer, J_n and J_{-n} cease to be independent,

$$J_{-n}(x) = (-1)^n J_n(x), \quad n \text{ integer}. \qquad (A.48)$$

Applying the methods discussed in the introduction, it is possible to show that

$$J_\nu(x) = \left(\frac{x}{2}\right)^\nu \sum_{k=0}^{\infty} (-1)^k \frac{1}{k! \Gamma(\nu + k + 1)} \left(\frac{x}{2}\right)^{2k}. \qquad (A.49)$$

The generating function of $J_n(x)$ for integer n is

$$e^{\frac{x}{2}(t - \frac{1}{t})} = \sum_{n=-\infty}^{\infty} J_n(x) t^n, \qquad (A.50)$$

and the most common recurrence relations are ($Z_\nu \equiv Z_\nu(x)$ stands for any Bessel function)

$$x(Z_{\nu-1} + Z_{\nu+1}) = 2\nu Z_\nu, \qquad (A.51)$$

$$x Z_\nu' = x Z_{\nu-1} - \nu Z_\nu = \nu Z_\nu - x Z_{\nu+1}. \qquad (A.52)$$

The half-indexed Bessel functions reduce to a finite sum of circular functions,

---

[5] Friedrich Wilhelm Bessel (1784-1846) was German astronomer, mathematician, and physicist.
[6] Carl Neumann (1832-1925) was a German mathematician and theoretical physicist; he is one of the initiators of the theory of integral equations.

$$J_{n+1/2}(x) = -\sqrt{\frac{2x}{\pi}} x^n \left(\frac{d}{xdx}\right)^n \frac{\sin x}{x}, \qquad (A.53)$$

$$J_{-n-1/2}(x) = \sqrt{\frac{2x}{\pi}} x^n \left(\frac{d}{xdx}\right)^n \frac{\cos x}{x}. \qquad (A.54)$$

For $|x| \gg |\nu|$ the asymptotic behavior is given by

$$\begin{aligned}J_{\pm\nu}(x) = &\sqrt{\frac{2}{\pi x}} \cos\left(x \mp \frac{\pi}{2}\nu - \frac{\pi}{4}\right)\left[1 + \mathcal{O}\left(\frac{1}{x^2}\right)\right] \\ &- \sqrt{\frac{2}{\pi x}} \sin\left(x \mp \frac{\pi}{2}\nu - \frac{\pi}{4}\right)\left[\frac{\Gamma(\nu+3/2)}{\Gamma(\nu-1/2)} \cdot \frac{1}{2x} + \mathcal{O}\left(\frac{1}{x^3}\right)\right]. \end{aligned} \qquad (A.55)$$

When the variable x is imaginary, it is often convenient to introduce the Bessel functions with imaginary argument I_ν and K_ν, according to the formulas

$$I_\nu(x) = e^{-\frac{\pi}{2}\nu i} J_\nu(ix), \qquad (A.56)$$

$$K_\nu(x) = \tfrac{1}{2}\pi i e^{\frac{\pi}{2}\nu i} H_\nu^{(1)}(ix). \qquad (A.57)$$

For large values of x, I diverges exponentially, whereas K tends to zero,

$$I_\nu(x) \xrightarrow[x\to\infty]{} \frac{e^x}{\sqrt{2\pi x}}, \qquad (A.58)$$

$$K_\nu(x) \xrightarrow[x\to\infty]{} \sqrt{\frac{\pi}{2x}} e^{-x}. \qquad (A.59)$$

I_ν and K_ν are called *modified Bessel functions*.

A.1.5 Spherical Bessel functions

In Section 12.1 we obtained the following differential equation for the radial wave function of a free particle in a central potential,

$$y'' + \frac{2}{x} y' + \left(1 - \frac{l(l+1)}{x^2}\right) y = 0. \qquad (A.60)$$

By making the change of variable $y = Z/\sqrt{x}$, the resulting equation for the new function $Z(x)$ is

$$Z'' + \frac{1}{x} Z' + \left(1 - \frac{(l+\tfrac{1}{2})^2}{x^2}\right) Z = 0. \qquad (A.61)$$

Comparing with (A.45), we see that Z is a cylindrical Bessel function with half-integer index $\nu = l + 1/2$; for this reason, the solutions of (A.60), which can be written in the conventional form

$$j_l(x) = \sqrt{\frac{\pi}{2x}} J_{l+1/2}(x) \qquad (A.62)$$

(and analogous expressions for the other cylindrical Bessel functions), are called *spherical Bessel* functions.[7]

Spherical Bessel functions include those specifically called spherical Bessel functions of the first kind $j_l(x)$, of the second kind or Neumann functions $n_l(x)$, and of the third kind or Hankel functions $h_l^{(1)}(x)$, $h_l^{(2)}(x)$. They are related by

$$h_l^{(1,2)}(x) = j_l(x) \pm i n_l(x); \quad (A.63)$$

the pairs (j_n, n_n) o $(h_k^{(1)}, h_n^{(2)})$ are linearly independent solutions of (A.60) for any $l = n = 0, 1, 2, \ldots$. Being proportional to the half-indexed Bessel functions, the spherical Bessel functions can be expressed as the sum of a finite number of terms, each of which contains a circular function. The first spherical Bessel functions are

$$j_0 = \tfrac{1}{x} \sin x \qquad n_0 = -\tfrac{1}{x} \cos x$$

$$j_1 = \tfrac{1}{x^2} \sin x - \tfrac{1}{x} \cos x \qquad n_1 = -\tfrac{1}{x^2} \cos x - \tfrac{1}{x} \sin x \quad (A.64)$$

$$j_2 = \left(\tfrac{3}{x^3} - \tfrac{1}{x}\right) \sin x - \tfrac{3}{x^2} \cos x \quad n_2 = -\left(\tfrac{3}{x^3} - \tfrac{1}{x}\right) \cos x - \tfrac{3}{x^2} \sin x,$$

which shows that j_n are the regular solutions of (A.60), while n_n are singular at the origin. Their asymptotic behavior at the origin is

$$j_l(x) \simeq \frac{1}{(2l+1)!!} x^l, \quad n_l(x) \simeq -\frac{(2l-1)!!}{x^{l+1}} \quad (x \ll |l|), \quad (A.65)$$

and at $x \to \infty$,

$$j_l(x) \simeq \frac{1}{x} \cos\left[x - \frac{\pi}{2}(l+1)\right], \quad n_l(x) \simeq \frac{1}{x} \sin\left[x - \frac{\pi}{2}(l+1)\right] \quad (x \gg |l|). \quad (A.66)$$

In closed form, we have

$$j_l(x) = (-x)^l \left(\frac{1}{x}\frac{d}{dx}\right)^l \left(\frac{\sin x}{x}\right), \quad n_l(x) = -(-x)^l \left(\frac{1}{x}\frac{d}{dx}\right)^l \left(\frac{\cos x}{x}\right). \quad (A.67)$$

The generating function for the Bessel functions is

$$\frac{1}{x} \cos \sqrt{x^2 - 2xt} = \sum_{l=0}^{\infty} \frac{t^l}{l!} j_{l-1}(x), \quad (A.68)$$

and for the Neumann functions it is

$$\frac{1}{x} \sin \sqrt{x^2 + 2xt} = \sum_{l=0}^{\infty} \frac{(-t)^l}{l!} n_{l-1}(x). \quad (A.69)$$

The following recurrence relations hold for any of the functions $j_l(x)$, $n_l(x)$, $h_l^{(1,2)}(x)$,

$$(2l+1) f_l = x(f_{l-1} + f_{l+1}), \quad (A.70)$$
$$(2l+1) f_l' = l f_{l-1} - (l+1) f_{l+1}. \quad (A.71)$$

---

[7] The term "Bessel function" is used here to refer generically to any of the solutions of the corresponding differential equations.

A.2 Mathematical Identities

$$A \times (B \times C) = (A \cdot C) B - (A \cdot B) C$$
$$(A \times B) \times C = (A \cdot C) B - (B \cdot C) A$$
$$A \cdot (B \times C) = (A \times B) \cdot C = (C \times A) \cdot B$$
$$(A \times B) \cdot (C \times D) = (A \cdot C)(B \cdot D) - (A \cdot D)(B \cdot D)$$
$$(A \times B) \times (C \times D) = [A \cdot (B \times D)] C - [A \cdot (B \times C)] D$$
$$= [A \cdot (C \times D)] B - [B \cdot (C \times D)] A$$
$$\nabla \times (\nabla u) = 0, \quad \nabla \cdot (\nabla u) = \nabla^2 u$$
$$\nabla \cdot (\nabla \times A) = 0, \quad \nabla \times (\nabla \times A) = \nabla (\nabla \cdot A) - \nabla^2 A$$
$$\nabla (uv) = u \nabla v + v \nabla u$$
$$\nabla \cdot (uA) = (\nabla u) \cdot A + u \nabla \cdot A$$
$$\nabla \times (uA) = (\nabla u) \times A + u \nabla \times A$$
$$\nabla \cdot (A \times B) = B \cdot (\nabla \times A) - A \cdot (\nabla \times B)$$
$$\nabla (A \cdot B) = (A \cdot \nabla) B + (B \cdot \nabla) A + A \times (\nabla \times B) + B \times (\nabla \times A)$$
$$\nabla \times (A \times B) = (B \cdot \nabla) A - (A \cdot \nabla) B + A (\nabla \cdot B) - B (\nabla \cdot A)$$

$$F = \hat{n} (F \cdot \hat{n}) + \hat{n} \times (F \times \hat{n}) = F_\parallel + F_\perp$$
$$\int_V (\nabla \cdot F) \, dv = \oint_S F \cdot da \quad \text{(Gauss divergence theorem)}$$
$$\int_V (\nabla \times F) \, dv = -\oint_S F \times da \quad \text{(a variant of Gauss theorem)}$$
$$\int_V (\nabla u) \, dv = \oint_S u \, da \quad \text{(a variant of Gauss theorem)}$$
$$\int_S (\nabla \times F) \cdot da = \oint_C F \cdot dr \quad \text{(Stokes theorem)}$$
$$\int_S da \times \nabla \psi = \oint_C \psi \, dr \quad \text{(a variant of Stokes theorem)}$$
$$\int_V \left(\phi \nabla^2 \psi - \psi \nabla^2 \phi \right) dv = \oint_S (\phi \nabla \psi - \psi \nabla \phi) \cdot da \quad \text{(Green theorem)}$$
$$\int_V \left(\phi \nabla^2 \psi + \nabla \phi \cdot \nabla \psi \right) dv = \oint_S \phi \, da \cdot \nabla \psi \quad \text{(first Green identity)}$$

Noncommuting operators

Consider two noncommuting operators P, Q that belong to a Hilbert space of linear vector operators. To third order in their commutators, a Taylor expansion gives

$$\ln(e^P e^Q) = (P + Q) + \frac{1}{2}[P, Q] + \frac{1}{12}([P, [P, Q]] + [Q, [Q, P]]) - \frac{1}{24}[Q, [P, [P, Q]]]. \quad (A.72)$$

This is the (logarithm of the) *Baker–Campbell–Hausdorff formula*, including up to third-order commutators.

A.3 Curvilinear Coordinates

A.3.1 Spherical coordinates

$$x = r \sin \theta \cos \varphi, \quad y = r \sin \theta \sin \varphi, \quad z = r \cos \theta,$$
$$0 \leq r < \infty, \quad 0 \leq \theta \leq \pi, \quad 0 \leq \varphi \leq 2\pi.$$

$$\frac{\partial r}{\partial x_i} = \frac{x_i}{r}, \quad \frac{\partial x_i}{\partial r} = \frac{x_i}{r}.$$

$$\frac{\partial \theta}{\partial x} = \frac{1}{r}\cos\theta\cos\varphi, \quad \frac{\partial \theta}{\partial y} = \frac{1}{r}\cos\theta\sin\varphi, \quad \frac{\partial \theta}{\partial z} = -\frac{1}{r}\sin\theta.$$

$$\frac{\partial \varphi}{\partial x} = -\frac{1}{r\sin\theta}\sin\varphi, \quad \frac{\partial \varphi}{\partial y} = \frac{1}{r\sin\theta}\cos\varphi, \quad \frac{\partial \varphi}{\partial z} = 0.$$

$$\hat{\mathbf{a}}_r = \hat{\mathbf{i}}\sin\theta\cos\varphi + \hat{\mathbf{j}}\sin\theta\sin\varphi + \hat{\mathbf{k}}\cos\theta,$$

$$\hat{\mathbf{a}}_\theta = \hat{\mathbf{i}}\cos\theta\cos\varphi + \hat{\mathbf{j}}\cos\theta\sin\varphi - \hat{\mathbf{k}}\sin\theta,$$

$$\hat{\mathbf{a}}_\varphi = -\hat{\mathbf{i}}\sin\varphi + \hat{\mathbf{j}}\cos\varphi.$$

$$\hat{\mathbf{i}} = \hat{\mathbf{a}}_r\sin\theta\cos\varphi + \hat{\mathbf{a}}_\theta\cos\theta\cos\varphi - \hat{\mathbf{a}}_\varphi\sin\varphi,$$

$$\hat{\mathbf{j}} = \hat{\mathbf{a}}_r\sin\theta\sin\varphi + \hat{\mathbf{a}}_\theta\cos\theta\sin\varphi + \hat{\mathbf{a}}_\varphi\cos\varphi,$$

$$\hat{\mathbf{k}} = \hat{\mathbf{a}}_r\cos\theta - \hat{\mathbf{a}}_\theta\sin\theta.$$

$$\hat{\mathbf{a}}_r \times \hat{\mathbf{a}}_\theta = \hat{\mathbf{a}}_\varphi, \quad \hat{\mathbf{a}}_r \times \hat{\mathbf{a}}_\varphi = -\hat{\mathbf{a}}_\theta, \quad \hat{\mathbf{a}}_\theta \times \hat{\mathbf{a}}_\varphi = \hat{\mathbf{a}}_r,$$

$$\frac{\partial \hat{\mathbf{a}}_r}{\partial \theta} = \hat{\mathbf{a}}_\theta, \quad \frac{\partial \hat{\mathbf{a}}_\theta}{\partial \theta} = -\hat{\mathbf{a}}_r, \quad \frac{\partial \hat{\mathbf{a}}_\varphi}{\partial \theta} = 0,$$

$$\frac{\partial \hat{\mathbf{a}}_r}{\partial \varphi} = \hat{\mathbf{a}}_\varphi\sin\theta, \quad \frac{\partial \hat{\mathbf{a}}_\theta}{\partial \varphi} = \hat{\mathbf{a}}_\varphi\cos\theta, \quad \frac{\partial \hat{\mathbf{a}}_\varphi}{\partial \varphi} = -\hat{\mathbf{a}}_r\sin\theta - \hat{\mathbf{a}}_\theta\cos\theta$$

$$d\mathbf{r} = \hat{\mathbf{a}}_r\,dr + \hat{\mathbf{a}}_\theta r\,d\theta + \hat{\mathbf{a}}_\varphi r\sin\theta\,d\varphi$$

$$da = r^2\sin\theta\,d\theta\,d\varphi$$

$$dv = r^2\sin\theta\,dr\,d\theta\,d\varphi.$$

$$\mathrm{grad}\,u = \nabla u = \hat{\mathbf{a}}_r\frac{\partial u}{\partial r} + \hat{\mathbf{a}}_\theta\frac{1}{r}\frac{\partial u}{\partial \theta} + \hat{\mathbf{a}}_\varphi\frac{1}{r\sin\theta}\frac{\partial u}{\partial \varphi},$$

$$\mathrm{div}\,\mathbf{A} = \nabla\cdot\mathbf{A} = \frac{1}{r^2}\frac{\partial}{\partial r}r^2 A_r + \frac{1}{r\sin\theta}\frac{\partial}{\partial \theta}\sin\theta A_\theta + \frac{1}{r\sin\theta}\frac{\partial A_\varphi}{\partial \varphi},$$

$$\mathrm{rot}\,\mathbf{A} = \nabla\times\mathbf{A} = \hat{\mathbf{a}}_r\frac{1}{r\sin\theta}\left(\frac{\partial}{\partial \theta}\sin\theta A_\varphi - \frac{\partial A_\theta}{\partial \varphi}\right) +$$

$$+\frac{\hat{\mathbf{a}}_\theta}{r}\left(\frac{1}{\sin\theta}\frac{\partial A_r}{\partial \varphi} - \frac{\partial}{\partial r}r A_\varphi\right) + \frac{\hat{\mathbf{a}}_\varphi}{r}\left(\frac{\partial}{\partial r}r A_\theta - \frac{\partial A_r}{\partial \theta}\right),$$

$$\nabla^2 u = \frac{1}{r}\frac{\partial^2}{\partial r^2}ru + \frac{1}{r^2\sin\theta}\frac{\partial}{\partial \theta}\sin\theta\frac{\partial u}{\partial \theta} + \frac{1}{r^2\sin^2\theta}\frac{\partial^2 u}{\partial \varphi^2},$$

$$\nabla^2 \mathbf{A}\,|_r = \left(-\frac{2}{r^2} + \frac{2}{r}\frac{\partial}{\partial r} + \frac{\partial^2}{\partial r^2} + \frac{\cos\theta}{r^2\sin\theta}\frac{\partial}{\partial \theta} + \frac{1}{r^2}\frac{\partial^2}{\partial \theta^2} + \frac{1}{r^2\sin^2\theta}\frac{\partial^2}{\partial \varphi^2}\right)A_r$$

$$+\left(-\frac{2}{r^2}\frac{\partial}{\partial \theta} - \frac{2\cos\theta}{r^2\sin\theta}\right)A_\theta + \left(-\frac{2}{r^2\sin\theta}\frac{\partial}{\partial \varphi}\right)A_\varphi =$$

$$= \nabla^2 A_r - \frac{2}{r^2}A_r - \frac{2}{r^2}\frac{\partial A_\theta}{\partial \theta} - \frac{2\cos\theta}{r^2\sin\theta}A_\theta - \frac{2}{r^2\sin\theta}\frac{\partial A_\varphi}{\partial \varphi},$$

$$\nabla^2 \mathbf{A}\,|_\theta = \nabla^2 A_\theta - \frac{1}{r^2\sin^2\theta}A_\theta + \frac{2}{r^2}\frac{\partial A_r}{\partial \theta} - \frac{2\cos\theta}{r^2\sin^2\theta}\frac{\partial A_\varphi}{\partial \varphi},$$

$$\nabla^2 \mathbf{A}\,|_\varphi = \nabla^2 A_\varphi - \frac{1}{r^2\sin^2\theta}A_\varphi + \frac{2}{r^2\sin\theta}\frac{\partial A_r}{\partial \varphi} + \frac{2\cos\theta}{r^2\sin^2\theta}\frac{\partial A_\theta}{\partial \varphi}.$$

$$\nabla = \hat{\mathbf{a}}_r \frac{\partial}{\partial r} - \frac{i}{\hbar r^2} \mathbf{r} \times \hat{\mathbf{L}}$$

$$\nabla^2 = \frac{1}{r} \frac{\partial^2}{\partial r^2} r - \frac{\hat{\mathbf{L}}^2}{\hbar^2 r^2}.$$

A.3.2 Cylindrical coordinates

$$x = \rho \cos\varphi, \quad y = \rho \sin\varphi, \quad z = z,$$
$$0 \le \rho < \infty, \quad 0 \le \varphi \le 2\pi, \quad -\infty < z < \infty.$$
$$\hat{\mathbf{a}}_\rho = \hat{\mathbf{i}}\cos\varphi + \hat{\mathbf{j}}\sin\varphi, \quad \hat{\mathbf{a}}_\varphi = -\hat{\mathbf{i}}\sin\varphi + \hat{\mathbf{j}}\cos\varphi, \quad \hat{\mathbf{k}} = \hat{\mathbf{k}},$$
$$\hat{\mathbf{i}} = \hat{\mathbf{a}}_\rho \cos\varphi - \hat{\mathbf{a}}_\varphi \sin\varphi, \quad \hat{\mathbf{j}} = \hat{\mathbf{a}}_\rho \sin\varphi + \hat{\mathbf{a}}_\varphi \cos\varphi.$$
$$\hat{\mathbf{a}}_\rho \times \hat{\mathbf{a}}_\varphi = \hat{\mathbf{k}}, \quad \hat{\mathbf{a}}_\varphi \times \hat{\mathbf{k}} = \hat{\mathbf{a}}_\rho, \quad \hat{\mathbf{k}} \times \hat{\mathbf{a}}_\rho = \hat{\mathbf{a}}_\varphi,$$
$$\frac{\partial \hat{\mathbf{a}}_\rho}{\partial \varphi} = \hat{\mathbf{a}}_\varphi, \quad \frac{\partial \hat{\mathbf{a}}_\varphi}{\partial \varphi} = -\hat{\mathbf{a}}_\rho.$$
$$d\mathbf{r} = \hat{\mathbf{a}}_\rho \, d\rho + \hat{\mathbf{a}}_\varphi \, \rho d\varphi + \hat{\mathbf{k}} \, dz.$$
$$dv = \rho \, d\rho \, d\varphi \, dz.$$
$$\nabla u = \hat{\mathbf{a}}_\rho \frac{\partial u}{\partial \rho} + \hat{\mathbf{a}}_\varphi \frac{1}{\rho} \frac{\partial u}{\partial \varphi} + \hat{\mathbf{k}} \frac{\partial u}{\partial z},$$
$$\nabla \cdot \mathbf{F} = \frac{1}{\rho} \frac{\partial}{\partial \rho}(\rho F_\rho) + \frac{1}{\rho} \frac{\partial F_\varphi}{\partial \varphi} + \frac{\partial F_z}{\partial z},$$
$$\nabla \times \mathbf{F} = \frac{1}{\rho} \begin{vmatrix} \hat{\mathbf{a}}_\rho & \rho \hat{\mathbf{a}}_\varphi & \hat{\mathbf{k}} \\ \frac{\partial}{\partial \rho} & \frac{\partial}{\partial \varphi} & \frac{\partial}{\partial z} \\ F_\rho & \rho F_\varphi & F_z \end{vmatrix},$$
$$\nabla^2 u = \frac{1}{\rho} \frac{\partial}{\partial \rho}\left(\rho \frac{\partial u}{\partial \rho}\right) + \frac{1}{\rho^2} \frac{\partial^2 u}{\partial \varphi^2} + \frac{\partial^2 u}{\partial z^2},$$
$$\nabla^2 \mathbf{F}|_\rho = \nabla^2 F_\rho - \frac{1}{\rho^2} F_\rho - \frac{2}{\rho^2} \frac{\partial F_\varphi}{\partial \varphi},$$
$$\nabla^2 \mathbf{F}|_\varphi = \nabla^2 F_\varphi - \frac{1}{\rho^2} F_\varphi + \frac{2}{\rho^2} \frac{\partial F_\rho}{\partial \varphi},$$
$$\nabla^2 \mathbf{F}|_z = \nabla^2 F_z,$$
$$\nabla = \hat{\mathbf{a}}_\rho \frac{\partial}{\partial \rho} + \hat{\mathbf{a}}_\varphi \frac{1}{\rho} \frac{\partial}{\partial \varphi} + \hat{\mathbf{k}} \frac{\partial}{\partial z}.$$

A.3.3 Parabolic coordinates

$$x = \xi\eta\cos\varphi, \quad y = \xi\eta\sin\varphi, \quad z = \tfrac{1}{2}\left(\xi^2 - \eta^2\right),$$
$$0 \le \xi < \infty, \quad 0 \le \eta < \infty, \quad 0 \le \varphi \le 2\pi.$$
$$\frac{\partial \xi}{\partial x} = \frac{\eta \cos\varphi}{\xi^2 + \eta^2}, \quad \frac{\partial \xi}{\partial y} = \frac{\eta \sin\varphi}{\xi^2 + \eta^2}, \quad \frac{\partial \xi}{\partial z} = \frac{\xi}{\xi^2 + \eta^2}.$$
$$\hat{\mathbf{a}}_\xi = \hat{\mathbf{i}}\eta\cos\varphi + \hat{\mathbf{j}}\eta\sin\varphi + \hat{\mathbf{k}}\xi,$$

$$\hat{\mathbf{a}}_\eta = \widehat{\mathbf{i}}\xi \cos\varphi + \widehat{\mathbf{j}}\xi \sin\varphi - \hat{\mathbf{k}}\eta,$$
$$\hat{\mathbf{a}}_\varphi = -\widehat{\mathbf{i}}\xi\eta \operatorname{sen}\varphi + \widehat{\mathbf{j}}\xi\eta \cos\varphi.$$
$$dv = \xi\eta\left(\xi^2 + \eta^2\right) d\xi\, d\eta\, d\varphi.$$
$$\nabla^2 u = \frac{1}{\xi^2 + \eta^2}\left(\frac{1}{\xi}\frac{\partial}{\partial \xi}\xi\frac{\partial u}{\partial \xi} + \frac{1}{\eta}\frac{\partial}{\partial \eta}\eta\frac{\partial u}{\partial \eta}\right) + \frac{1}{\xi^2\eta^2}\frac{\partial^2 u}{\partial \varphi^2}.$$

A.4 Dirac's Delta

$$\delta(x) = 0,\ x \neq 0, \quad \int_{-\infty}^{\infty} f(x)\,\delta(x)\,dx = f(0),$$
$$\delta(ax - b) = \frac{1}{|a|}\delta\left(x - \frac{b}{a}\right),$$
$$f(x)\delta(x) = f(0)\delta(x), \quad x\delta(x) = 0,$$
$$f(x)\delta'(x) = f(0)\delta'(x) - f'(0)\delta(x),$$
$$x\delta^{(m)}(x) = -m\delta^{(m-1)}(x),$$
$$x^n \delta^{(m)}(x) = 0 \quad n \geq m+1,$$
$$\delta(x) = \frac{d}{dx}H(x), \quad H(x)\ \text{the Heaviside function}$$
$$\delta[f(x)] = \sum_n \frac{1}{|f'(x_n)|}\delta[x - x_n], \quad f(x_n) = 0,\ f'(x_n) \neq 0,$$
$$\delta\left(x^2 - a^2\right) = \frac{1}{2|a|}[\delta(x - a) + \delta(x + a)],$$
$$\delta((x - a)(x - b)) = \frac{1}{|a - b|}[\delta(x - a) + \delta(x - b)]$$
$$\lim_{\eta \to 0^+} \frac{1}{\eta}e^{-x^2/\eta^2} = \sqrt{\pi}\delta(x), \quad \lim_{\eta \to 0^+} \frac{\eta}{\eta^2 + x^2} = \pi\delta(x),$$
$$\delta(\omega) = \frac{1}{2\pi}\int_{-\infty}^{+\infty} e^{i\omega t}\,dt,$$
$$\delta_+(\omega) = \frac{1}{2\pi}\int_0^\infty e^{i\omega t}\,dt = \frac{i}{2\pi}\lim_{\varepsilon \to 0^+}\frac{1}{\omega + i\varepsilon} = \frac{i}{2\pi}\mathrm{P}\frac{1}{\omega} + \frac{1}{2}\delta(\omega),$$
$$\delta_-(\omega) = \frac{1}{2\pi}\int_0^\infty e^{-i\omega t}\,dt = -\frac{i}{2\pi}\lim_{\varepsilon \to 0^+}\frac{1}{\omega - i\varepsilon} = -\frac{i}{2\pi}\mathrm{P}\frac{1}{\omega} + \frac{1}{2}\delta(\omega),$$

where the operator P represents the principal value.

$$\nabla^2 |x| = 2\delta(x), \quad \nabla^2 \frac{1}{|\mathbf{r} - \mathbf{r}_0|} = -4\pi\delta(\mathbf{r} - \mathbf{r}_0).$$
$$\delta(\mathbf{r} - \mathbf{r}_0) = \frac{1}{r^2 \sin\theta}\delta(r - r_0)\,\delta(\theta - \theta_0)\,\delta(\varphi - \varphi_0), \quad x_0, y_0, z_0 \neq 0;$$
$$= \frac{1}{2\pi r^2 \sin\theta}\delta(r - r_0)\,\delta(\theta), \quad x_0, y_0 = 0,\ z_0 \neq 0;$$

$$= \frac{1}{4\pi r^2}\delta(r), \quad x_0, y_0, z_0 = 0,$$

$$\delta(\mathbf{r} - \mathbf{r}_0) = \frac{1}{|J(x_i, \xi_i)|}\delta\left(\xi_1 - \xi_1^0\right)\delta\left(\xi_2 - \xi_2^0\right)\delta\left(\xi_3 - \xi_3^0\right).$$

Transverse delta function:

$$\delta_{ij}^\perp(\mathbf{r} - \mathbf{r}_0) = \frac{1}{(2\pi)^3}\int d^3k \delta_{ij}^\perp e^{i\mathbf{k}\cdot(\mathbf{r}-\mathbf{r}_0)},$$

$$\delta_{ij}^\perp = \left(\delta_{ij} - \frac{k_i k_j}{k^2}\right).$$

A.5 The Gamma Function

$$\Gamma(z) \equiv \lim_{n\to\infty}\frac{1\cdot 2\cdot 3\cdots n}{z(z+1)(z+2)\cdots(z+n)}n^z, \quad z \neq 0, -1 -2, \ldots, \operatorname{Re} z > 0 \quad \text{(Euler)},$$

$$\Gamma(z) \equiv \int_0^\infty e^{-t}t^{z-1}dt, \quad \operatorname{Re} z > 0 \quad \text{(Euler)},$$

$$\frac{1}{\Gamma(z)} \equiv z e^{\gamma z}\prod_{n=1}^\infty \left(1 + \frac{z}{n}\right)e^{-z/n} \quad \text{(Weierstrass)};$$

γ is the Euler–Mascheroni constant $\gamma = 0.577216\ldots$
For $-(n+1) < \operatorname{Re}(z) \leq -n, z \neq 0, -1, -2, \ldots$, the Γ function is defined by the recurrence relation

$$\Gamma(z) = \frac{\Gamma(z+1)}{z} = \frac{\Gamma(z+n+1)}{z(z+1)\cdots(z+n)}.$$

The points $z = 0, -1, -2, \ldots$ are simple poles of $\Gamma(z)$.

$$\Gamma(z)\Gamma(1-z) = \frac{\pi}{\operatorname{sen}\pi z},$$

$$|\Gamma(1+i\beta)|^2 = \frac{\pi\beta}{\operatorname{senh}\pi\beta}, \quad \beta \text{ real}.$$

$$\Gamma(x+1) = x\Gamma(x).$$

$$\Gamma(n) = (n-1)!, \quad \Gamma\left(n+\tfrac{1}{2}\right) = \frac{(2n-1)!!}{2^n}\sqrt{\pi}, \quad \Gamma\left(\tfrac{1}{2}\right) = \sqrt{\pi},$$

where

$$(2n+1)!! = 1\cdot 3\cdot 5\cdots(2n+1) = \frac{(2n+1)!}{2^n n!},$$

$$(2n)!! = 2\cdot 4\cdot 6\cdots 2n = 2^n n!.$$

For $|z| \to \infty$,

$$\Gamma(z) \sim \sqrt{2\pi}z^{z-1/2}e^{-z}\left(1 + \frac{1}{12z} + \frac{1}{288z^2} - \cdots\right), \quad |\arg z| \leq \pi - \delta, \quad 0 < \delta \ll 1.$$

For x real, $x \to \infty$,

$$\Gamma(x) \sim \sqrt{2\pi}x^{x-1/2}e^{-x}.$$

A.6 Solving Eigenvalue Problems by Means of Algebraic Consistency Conditions

Throughout the book we have encountered situations where the conditions imposed on the function to be a physical solution of Schrödinger's equation lead to quantization. The conditions usually arise from the requirement that the wave function has the correct properties to correspond to a statistical amplitude of probability, and from the boundary conditions appropriate to the problem. In the case of bound systems, this procedure usually leads to quantization. Here we explore an alternative possibility that does not require knowing the solution of the Schrödinger equation for the problem, but instead requires considering certain algebraic relations between elements of two of the most important dynamical variables (operators) that should hold in any quantum mechanical system. More specifically, given the two reference variables, certain matrix elements of them are considered, which should be related by a set of specific (and simple) relations.

The operators are usually either bilinear forms of appropriate raising and lowering operators, or else auxiliary operators constructed by resorting to the factorization of the Hamiltonian. The quantization rules follow from the solution of the (generally elementary) condition obtained. A couple of examples will illustrate the clarity and simplicity of the procedure.

A.6.1 State-dependent and state-independent ladder operators

We consider the eigenvector basis $\{|k\rangle\}$ of a certain Hermitian operator P of interest with discrete spectrum and use this basis to construct a pair of operators $\eta^\dagger, \eta$ as follows,

$$\eta^\dagger = \sum_i C_i |i+1\rangle\langle i| = \sum_i C_i \eta_i^\dagger, \qquad (A.73)$$

$$\eta = \sum_i C_i^* |i\rangle\langle i+1| = \sum_i C_i^* \eta_i, \qquad (A.74)$$

where the constants C_i are to be determined, and

$$\eta_i^\dagger = |i+1\rangle\langle i|, \quad \eta_i = |i\rangle\langle i+1|. \qquad (A.75)$$

From this equation we note that

$$\eta_i^\dagger |n\rangle = |n+1\rangle \delta_{in}, \quad \eta_i |n\rangle = |n-1\rangle \delta_{i,n-1}. \qquad (A.76)$$

The raising and lowering operators $\eta_i^\dagger, \eta_i$ may depend explicitly or not on the state i on which they operate. We see that the basis $\{|k\rangle\}$ corresponds to the eigenvectors of the operators $\eta\eta^\dagger$ and $\eta^\dagger\eta$, and of the linear combinations of these operators,

$$S \equiv \eta\eta^\dagger + \eta^\dagger\eta = \{\eta, \eta^\dagger\}, \quad A \equiv \eta\eta^\dagger - \eta^\dagger\eta = [\eta, \eta^\dagger], \qquad (A.77)$$

which are respectively the anticommutator and the commutator of η and $\eta^\dagger$. Therefore, S and A satisfy the eigenvalue equations

$$S|k\rangle = s_k |k\rangle, \quad A|k\rangle = a_k |k\rangle, \tag{A.78}$$

whence

$$s_k + a_k = s_{k+1} - a_{k+1} \tag{A.79}$$

and

$$s_{k+1} - s_k = a_{k+1} + a_k. \tag{A.80}$$

These basic consistency conditions between the eigenvalues of S and A corresponding to adjacent states will play an important role in what follows. In particular, it is clear from Eqs. (A.78) that

$$s_k \geq |a_k|. \tag{A.81}$$

From Eqs. (A.73), (A.74), (A.77), and (A.78) we get

$$s_k = |C_k|^2 + |C_{k-1}|^2, \tag{A.82}$$

$$a_k = |C_k|^2 - |C_{k-1}|^2. \tag{A.83}$$

We will consider that the spectrum of interest is bounded from below. As is customary, we denote the ground state with $|k_{\min}\rangle = |0\rangle$ unless otherwise specified, so that $C_{-1} = 0$ and $s_0 = a_0$. Successive application of Eq. (A.80) leads then to the recurrence relation

$$s_k = a_k + 2\sum_{i=0}^{k-1} (-1)^{i-(k-1)} s_i, \tag{A.84}$$

establishing the relationship between the eigenvalues of A and those of S.

For some applications it is convenient to introduce a couple of auxiliary operators α, β, indirectly defined through the relations

$$\eta^\dagger = \frac{1}{\sqrt{2}}(\alpha - i\beta), \text{ and} \tag{A.85}$$

$$\eta = \frac{1}{\sqrt{2}}(\alpha + i\beta). \tag{A.86}$$

From Eqs. (A.77) it follows that

$$A = i[\beta, \alpha], \tag{A.87}$$

$$S = \alpha^2 + \beta^2. \tag{A.88}$$

In this form, when the ladder operators are known, the consistency conditions allow us to determine the spectrum of A in terms of the spectrum of S or vice versa. This is particularly useful when either one of these operators is identified with a dynamical variable of interest, as is the case in the following example.

A.6.1.1 The radial problem

There is a family of problems for which a factorization method allows one to construct appropriate state-independent operators A and S, which can be used to solve the eigenvalue problem with the help of the consistency relations. These are

N-dimensional radial problems, that is, central-force problems in N dimensions ($N = 1, 2, 3$) reduced to the radial variable only. The procedure is briefly presented for the 3D case ($N = 3$).

To study the spectrum associated with the radial factor, we introduce two operators that depend on a continuous dimensionless variable q as follows (where Λ is a real parameter)

$$B_{n\pm} = q^n \pm \left(\frac{-\Lambda}{q^n} + \frac{1}{q^{n-2}} \frac{d^2}{dq^2} \right), \tag{A.89}$$

with the properties

$$\left[q \frac{d}{dq}, B_{n\pm} \right] = n B_{n\mp}, \tag{A.90}$$

$$B_{n+} = 2q^n - B_{n-}. \tag{A.91}$$

This procedure is justified by the fact that for certain values of the exponent n, B_{n-} happens to take the form of an important quantum-mechanical radial Hamiltonian, as will be seen below. A careful examination of the above equations suggests the choice of

$$\alpha = B_{n+}, \tag{A.92}$$

$$i\beta = q \frac{d}{dq} - \left(q \frac{d}{dq} \right)^\dagger = 2q \frac{d}{dq} + (n-1)\mathbb{I}, \tag{A.93}$$

so that, after some simple transformations, one gets

$$S = \alpha^2 + \beta^2 = B_{n-}^2 + (n^2 - 1 - 4\Lambda)\mathbb{I}, \tag{A.94}$$

$$A = i[\beta, \alpha] = 2n B_{n-}. \tag{A.95}$$

These equations show that the eigenvalues of the operators A and S are related by the eigenvalues of the operator B_{n-}, which are the ones of interest. Denoting these eigenvalues by b_k, we obtain, from the consistency condition (A.80),

$$b_{k+1}^2 - b_k^2 = 2n(b_{k+1} + b_k). \tag{A.96}$$

Upon elimination of the common factor $b_{k+1} + b_k$ (assuming it is different from zero, which is true in the cases of interest), this equation simplifies into $b_{k+1} - b_k = 2n$, with solution

$$b_k = b_0 + 2nk. \tag{A.97}$$

For $n > 0$, which is the only case we shall study here, there is no upper bound. Equation (A.84) gives $2nb_0 = b_0^2 + n^2 - 1 - 4\Lambda$, whence

$$b_0 = n \pm \sqrt{1 + 4\Lambda} \geq 0. \tag{A.98}$$

We see that the proposed choice of the operators α, β has led to the general solution expressed in Eq. (A.97). The essence of the method was to obtain a nonlinear relationship between s_k and a_k, with $a_k \propto b_k$, and to use the consistency condition to determine b_k. In the following we apply the procedure to Schrödinger's hydrogen-like atom, and in exercise E19.3 to Dirac's hydrogen-like atom.

The hydrogen-like atom

The radial Schrödinger equation for the hydrogen-like atom is

$$\left[-\frac{\hbar^2}{2m}\frac{d^2}{dr^2} + \frac{\hbar^2 l(l+1)}{2mr^2} - \frac{Ze^2}{r}\right] u = Eu, \qquad (A.99)$$

where $u = rR$, $R(r)$ being the radial part of the wave function. Multiplying by r, in terms of the variable $q = \beta_E r$ with $\beta_E = \sqrt{2m|E|}/\hbar$, the radial Hamiltonian reduces, after some simple rearrangements, to the operator B_{n-} for $n = 1$,

$$\left(-q\frac{d^2}{dq^2} + \frac{l(l+1)}{q} + q\right) u = B_{1-}u = \frac{Ze^2}{\hbar}\sqrt{\frac{2m}{|E|}} u. \qquad (A.100)$$

To determine $b_0 = (Ze^2/\hbar)\sqrt{2m/|E_0|}$ we proceed, using (A.98) now with $n = 1$ and $\Lambda = l(l+1)$, thus obtaining $b_0 = 2(l+1)$ (the minus sign in Eq. (A.98) is ruled out since $b_0 \geq 0$ for any l). Using Eq. (A.97), we get $b_k = 2(l+k+1)$, which combined with $b_k = (Ze^2/\hbar)\sqrt{2m/|E_k|}$ gives, for the energy eigenvalues,[8]

$$E_{kl} = -\frac{Z^2 e^4 m}{2\hbar^2 (k+l+1)^2} = -\frac{Z^2 e^4 m}{2\hbar^2 \mathrm{n}^2}. \qquad (A.101)$$

Bibliographical Notes

Clear, elegant and comprehensive textbooks on mathematical methods in physics are Morse and Feshbach (1953), and Hassani (2013). Another recommended book is Margenau and Murphy (2009).

The handbooks of integrals and special functions by Abramowitz and Stegun (1972), and Gradshteyn and Ryzhik (2007) are a necessary and very useful general reference.

For early work on the algebraic consistency method, see de la Peña and Montemayor (1980). An elaboration of the method is given in Fernández and Castro (1984), and further applications are given in de la Peña et al. (2022).

---

[8] The principal quantum number $\mathrm{n} = k+l+\frac{N-1}{2}$, is an integer in the 1D and 3D cases, and a half-integer in the 2D case. Although central-force problems can be reduced to 1D (radial) problems regardless of their physical dimension N, the energy spectrum explicitly depends on the number of dimensions.

Physical Constants

Electron charge	e	-1.6021×10^{-19} C
Electron mass	m_e	$\begin{cases} 9.10938 \times 10^{-31} \text{ kg} \\ 0.510998 \text{ MeV}/c^2 \end{cases}$
Proton mass	m_p	$\begin{cases} 1836.152 \, m_e \\ 1.67262 \times 10^{-27} \text{ kg} \\ 938.272 \text{ MeV}/c^2 \\ 1.007276 \text{ u.a.m.} \end{cases}$
Velocity of light in vacuum	c	$\begin{cases} 2.99792458 \times 10^8 \text{ m/s} \\ 137.036 \text{ u.a.} \end{cases}$
Planck constant	$\begin{cases} \hbar \\ \\ \hbar c \\ \\ \hbar\omega_{\text{visible}} \end{cases}$	$\begin{cases} 1.0546 \times 10^{-34} \text{ J} \cdot \text{s} \\ 6.5821 \times 10^{-15} \text{ eV} \cdot \text{s} \\ 3.1613 \times 10^{-26} \text{ J} \cdot \text{m} \\ 1.9732 \times 10^{-11} \text{ MeV} \cdot \text{cm} \\ \sim 2 \text{ eV} \end{cases}$
Avogadro number	N_A	6.02214×10^{23} part/mol
Free space permeability	μ_0	$4\pi \times 10^{-7}$ H/m
Free space permittivity	$\varepsilon_0 = 1/\mu_0 c^2$	8.85418×10^{-12} F/m
Classical electron radius	$r_0 = e^2/4\pi\varepsilon_0 mc^2$	2.8179×10^{-15} m
Fine structure constant	$\alpha = e^2/\hbar c$	$1/137.036$
Compton electron wavelength	$\lambda_C = h/m_e c$	2.4263×10^{-12} m
Bohr radius	$a_0 = \lambda_C/\alpha$	0.5291 Å
Rydberg constant	R_H	109737.315 1/cm
Bohr magneton	$\mu_B = |e|\hbar/2m_0$	9.2740×10^{-24} J/tesla
Boltzmann constant	$\begin{cases} k_B \\ \\ k_B T_{\text{room temp}} \end{cases}$	$\begin{cases} 1.3806 \times 10^{-23} \text{ J/K} \\ 8.6173 \times 10^{-5} \text{ eV/K} \\ \sim 0.025 \text{ eV} \end{cases}$

References

Abramowitz, M., and Stegun, I. A., eds. (1972) Handbook of Mathematical Functions with Formulas, Graphs, and Mathematical Tables. Washington: National Bureau of Standards.

Afnan, I. R. (2011) Quantum Mechanics with Applications. Sharjah Airport International Free Trade Zone, UAE: Bentham Science Publishers.

Aharonov, Y., and Bohm, D. (1959) 'Significance of Electromagnetic Potentials in the Quantum Theory', Physical Review, 115(3), pp. 485–491. DOI: https://doi.org/10.1103/physrev.115.485.

Aitchison, I. J. R., MacManus, D. A., and Snyder, T. M. (2004) 'Understanding Heisenberg's "Magical" Paper of July 1925: A New Look at the Calculational Details', American Journal of Physics, 72, pp. 1370–1379. DOI: https://doi.org/10.1119/1.1775243.

Allahverdyan, A. E., Balian, R., and Nieuwenhuizen, T. M. (2013) 'Understanding Quantum Measurement from the Solution of Dynamical Models', Physics Reports, 525(1), pp. 1–166. DOI: https://doi.org/10.1016/j.physrep.2012.11.001.

Allen, J. F., and Misener, A. D. (1938) 'Flow of Liquid Helium II', Nature, 141(3558), p. 75. DOI: https://doi.org/10.1038/141075a0.

Anderson, C. D. (1932) 'The Apparent Existence of Easily Deflectable Positives', Science, 76(1967), pp. 238–239. DOI: https://doi.org/10.1126/science.76.1967.238.

Aspect, A., Dalibard, J., and Roger, G. (1982) 'Experimental Test of Bell's Inequalities Using Time-Varying Analyzers', Physical Review Letters, 49(25), pp. 1804–1807. DOI: https://doi.org/10.1103/physrevlett.49.1804.

Auletta, G., Fortunato, M., and Parisi, G. (2009) Quantum Mechanics. Cambridge, UK: Cambridge University Press.

Avis, D., Moriyama, S., and Owari, M. (2009) 'From Bell's Inequalities to Tsirelson's Theorem', IEICE Transactions on Fundamentals of Electronics, E92.A, pp. 1254–1627. DOI: https://doi.org/10.1587/transfun.E92.A.1254.

Bacciagaluppi, G., and Valentini, A. (2009) Quantum Theory at the Crossroads: Reconsidering the 1927 Solvay Conference. New York, NY: Cambridge University Press. DOI: https://doi.org/10.48550/arXiv.quant-ph/0609184.

Ballentine, L. E. (1970) 'The Statistical Interpretation of Quantum Mechanics', Reviews of Modern Physics, 42, pp. 358. DOI: https://doi.org/10.1103/RevModPhys.42.358.

Ballentine, L. E. (1972) 'Einstein's Interpretation of Quantum Mechanics', American Journal of Physics, 40(12), pp. 1763–1771. DOI: https://doi.org/10.1119/1.1987060.

Ballentine, L. E. (1998) Quantum Mechanics: A Modern Development. Singapore: World Scientific.

Balmer, J. J. (1885) 'Notiz über die Spektrallinien des Wasserstoffs', Verhandlungen der Naturforschenden Gesellschaft in Basel, 7(3), pp. 548–560.

Barnett, S. J. (1917) 'The Magnetization of Iron, Nickel, and Cobalt by Rotation and the Nature of the Magnetic Molecule', Physical Review, 10(1), pp. 7–21.

Barut, A. O., and Zanghi, N. (1984) 'Classical Model of the Dirac Electron', Physical Review Letters, 52(23), pp. 2009–2012. DOI: https://doi.org/10.1103/physrevlett.52.2009.

Basdevant, J.-L. (2023) Variational Principles in Physics. Cham: Springer.

Bell, J. S. (1964) 'On the Einstein Podolsky Rosen Paradox', Physics Physique Fizika, 1, pp. 195–200. DOI: https://doi.org/10.1103/PhysicsPhysiqueFizika.1.195.

Bertrand, J. (1873) 'Théorème relatif au mouvement d'un point attiré vers un centre fixe', Comptes Rendus de l'Académie des Sciences, 77, pp. 849–853.

Bethe, H. A., and Salpeter, E. E. (1957) Quantum Mechanics of One and Two Electron Atoms. Heidelberg: Springer Berlin.

Bhabha, H. J., and Corben, H. C. (1941) 'General Classical Theory of Spinning Particles in a Maxwell Field', Proceedings of the Royal Society A: Mathematical, Physical and Engineering Sciences, 178(974), pp. 273–314. DOI: https://doi.org/10.1098/rspa.1941.0056.

Bloch, F. (1946) 'Nuclear Induction', Physical Review, 70(7–8), pp. 460–474.

Bohm, D. (1989) Quantum Theory. Mineola, NY: Dover Publications.

Bohr, N. (1913) 'I. On the Constitution of Atoms and Molecules', Philosophical Magazine, 26(151), 1–25. DOI: https://doi.org/10.1080/14786441308634955.

Bohr, N. (1913) 'XXXVII. On the Constitution of Atoms and Molecules', Philosophical Magazine, 26(153), pp. 476–502.

Bohr, N. (1913) 'LXXIII. On the Constitution of Atoms and Molecules', Philosophical Magazine, 26(155), pp. 857–875. DOI: https://doi.org/10.1080/14786441308635031.

Born, M., and Einstein, A. (2005) Born–Einstein Letters 1916–1955, Friendship, Politics and Physics in Uncertain Times. New York, NY: Macmillan.

Born, M., and Oppenheimer, R. (1927) 'Zur Quantentheorie der Molekeln', Annalen der Physik, 389(20), pp. 457–484. DOI: https://doi.org/10.1002/andp.19273892002.

Bose, S. N. (1924) 'Plancks Gesetz und Lichtquantenhypothese', Zeitschrift für Physik, 26(1), pp. 178–181.

Bosov, N. G., and Prokhorov, A. M. (1954) 'Application of Molecular Beams to the Radio Spectroscopic Study of the Rotation Spectra of Molecules', JETP, 27, 4(10), pp. 431–438.

Boyer, T. H. (1969) 'Recalculations of Long-Range van der Waals Potentials', Physical Review, 180(1), pp. 19–24. DOI: https://doi.org/10.1103/physrev.180.19.

Boyer, T. H. (1980) 'A Brief Survey of Stochastic Electrodynamics', pp. 49–63 in Barut, A. O Foundations of Radiation Theory and Quantum Electrodynamics. New York, NY: Plenum.

Boyer, T. H. (2000) 'Does the Aharonov–Bohm Effect Exist?', Foundations of Physics, 30(6), pp. 893–905. DOI: https://doi.org/10.1023/a:1003602524894.

Boyer, T. H. (2023) 'The Classical Aharonov–Bohm Iinteraction as a Relativity Paradox', European Journal of Physics, 44(3), 035202. DOI: https://doi.org/10.1088/1361-6404/acc0e6.

Brillouin, L. (1926) 'La mécanique ondulatorie de Schrödinger: une méthode générale de resolution par approximations successives', Comptes Rendus de l' Académie des Sciences, Vol. 183, pp. 24–26, 1926.

Brink, D. M., and Satchler, G. R. (1968) Angular Momentum. Oxford: Clarendon Press.

Brody, T. A. (1993) The Philosophy Behind Physics. DOI: https://doi.org/10.1007/978-3-642-78978-6. Heidelberg: Springer.

Brown, R. (1928) 'A Brief Account of Microscopical Observations Made in the Months of June, July and August 1827, on the Particles Contained in the Pollen of Plants; and on the General Existence of Active Molecules in Organic and Inorganic Bodies', Philosophical Magazine, 4(21), pp. 161–173.

Bush, J. W. M. (2015) 'The New Wave of Pilot-Wave Theory', Physics Today, 68(8), pp. 47–53. DOI: https://doi.org/10.1063/pt.3.2882.

Bush, J. W. M., and Oza, A. U. (2020) 'Hydrodynamic Quantum Analogs', Reports on Progress in Physics, 84(1), 017001. DOI: https://doi.org/10.1088/1361-6633/abc22c.

Butkov, E. (1968) Mathematical Physics. Boston, MA: Addison-Wesley.

Casado, A., Guerra, S., and Plácido, J. (2019) 'From Stochastic Optics to the Wigner Formalism: The Role of the Vacuum Field in Optical Quantum Communication Experiments', Atoms, 7(3), 76. DOI: https://doi.org/10.3390/atoms7030076.

Casimir, H. B. G. (1948) 'On the Attraction between Two Perfectly Conducting Plates', Proceedings of the Royal Netherlands Academy of Arts and Sciences, 51, pp. 793–795.

Casimir, H. B. G., and Polder, D. (1948) 'The Influence of Retardation on the London-van der Waals Forces', Physical Review, 73(4), 360–372. DOI: https://doi.org/10.1103/physrev.73.360.

Cetto, A. M. (2022) 'Electron Spin Correlations: Probabilistic Description and Geometric Representation', Entropy, 24, 1439. DOI: https://doi.org/10.3390/e24101439.

Cetto, A. M., and de la Peña, L. (2012) 'Radiative Corrections for the Matter–Zeropoint Field System: Establishing Contact with Quantum Electrodynamics', Physica Scripta, T151, 014009. DOI: https://doi.org/10.1088/0031-8949/2012/t151/014009.

Cetto, A. M., de la Peña, L., and Valdés-Hernández, A. (2014) 'Emergence of Quantization: The Spin of the Electron', Journal of Physics: Conference Series, 504 (012007). DOI: https://doi.org/10.1088/1742-6596/504/1/012007.

Cetto, A. M., de la Peña, L., and Valdés, A. (2021) 'Relevance of Stochasticity for the Emergence of Quantization', European Physics Journal Special Topics, 230, pp. 923–929. DOI: https://doi.org/10.1140/epjs/s11734-021-00066-4.

CIAAW (2015) Periodic Table of the Isotopes. Available at https://ciaaw.org/periodic-table-isotopes.htm (Accessed: 2024).

Clauser, J. F., et al. (1969) 'Proposed Experiment to Test Local Hidden-Variable Theories', Physical Review Letters, 23(15), pp. 880–884. DOI: https://doi.org/10.1103/physrevlett.23.880.

Cohen, E., et al. (2019) 'Geometric Phase from Aharonov–Bohm to Pancharatnam–Berry and Beyond', Nature Reviews Physics. DOI: https://doi.org/10.1038/s42254-019-0071-1.

Cohen-Tannoudji, C., Diu, B., and Laloë, F. (2019) Quantum Mechanics, Volume 3: Fermions, Bosons, Photons, Correlations, and Entanglement. John Wiley & Sons.

Cohen-Tannoudji, C., Diu, B., and Laloë, F., (1977) Quantum Mechanics. Vol. I and II. Wiley-VCH.

Cohen-Tannoudji, C., Dupont-Roc, J., and Grynberg, G. (1997) Photons and Atoms. WILEY-VCH Verlag GmbH & Co. KGaA.

Cohen-Tannoudji, C., Dupont-Roc, J., and Grynberg, G. (1998) Atom-Photon Interactions: Basic Processes and Applications. John Wiley & Sons, Inc.

Compton, A. H. (1926) X-Rays and Electrons: An Outline of Recent X-Ray Theory. New York: D. Van Nostrand Company.

Condon, E. U., and Shortley, G. H. (1964) The Theory of Atomic Spectra. Cambridge: Cambridge University Press.

Corben, H. C., and Stehle, P. (1960) Classical Mechanics. New York: Wiley.

Davisson, C., and Germer, L. H. (1927) 'Diffraction of Electrons by a Crystal of Nickel', Physical Review, 30(6), 705–740. DOI: https://doi.org/10.1103/physrev.30.705.

Davydov, A. S. (2013) Quantum Mechanics. International Series in Natural Philosophy. Oxford: Pergamon Press.

de Broglie, L. (1927). 'La mécanique ondulatoire et la structure atomique de la matière et du rayonnement', Journal de Physique et le Radium, 8(5), pp. 225–241. DOI: https://doi.org/10.1051/jphysrad:0192700805022500.

de Broglie, L. (1925) 'Recherches sur la théorie des quanta', Annales de Physique, 3(10), pp. 22.

de la Peña, L. (1969) 'New Formulation of Stochastic Theory and Quantum Mechanics', Journal of Mathematical Physics, 10(9), pp. 1620–1630. DOI: https://doi.org/10.1063/1.1665009.

de la Peña, L. (2014) Introducción a la Mecánica Cuántica. Mexico: Fondo de Cultura Económica.

de la Peña, L. and Cetto, A. M. (1996) The Quantum Dice. An Introduction to Stochastic Electrodynamics. Dordrecht: Kluwer Acad. Publ.

de la Peña, L. and Cetto, A. M. (1997) 'Estimate of Planck's Constant from an Electromagnetic Mach Principle', Foundations of Physics Letters, 10(6), pp. 591–598. DOI: https://doi.org/10.1023/a:1022401403771.

de la Peña, L., and Cetto, A. M. (1975) 'Stochastic Theory for Classical and Quantum Mechanical Systems', Foundations of Physics, 5(2), pp. 355–370. DOI: https://doi.org/10.1007/bf00717450.

de la Peña, L., and Cetto, A. M. (1982) 'Does Quantum Mechanics Accept a Stochastic Support?', Foundations of Physics, 12(10), pp. 1017–1037. DOI: https://doi.org/10.1007/bf01889274.

de la Peña, L., and Cetto, A. M. (2024) 'Completing the Quantum Ontology with the Electromagnetic Zero-Point Field', in Castro, P., Bush, J.W.M., Croca, J. (eds.) Advances in Pilot Wave Theory. Boston Studies in the Philosophy and History of Science, 344. Springer, Cham. DOI: https://doi.org/10.1007/978-3-031-49861-9_10.

de la Peña, L., and Montemayor R. (1980) 'Raising and Lowering Operators and Spectral Structure: A Concise Algebraic Technique', American Journal of Physics, 48(10), pp. 855–860. DOI: https://doi.org/10.1119/1.12236.

de la Peña, L., Cetto, A. M., and Valdés, A. (2015) The Emerging Quantum. The Physics Behind Quantum Mechanics. Cham: Springer.

de la Peña, L., Cetto, A. M., and Valdés, A. (2022) 'Solution of Quantum Eigenvalue Problems by Means of Algebraic Consistency Conditions', European Journal of Physics, 43(015401). DOI: https://doi.org/10.1088/1361-6404/ac2ecd.

de la Peña, L., and Villavicencio, M. (2003) Problemas y ejercicios de mecánica cuántica. Mexico: Fondo de Cultura Económica.

Dewdney, C., and Hiley, B. J. (1982) 'A Quantum Potential Description of One-Dimensional Time-Dependent Scattering from Square Barriers and Square Wells', Foundations of Physics, 12(1), pp. 27–48. DOI: https://doi.org/10.1007/bf00726873.

Dirac, P. A. M. (1927) 'The Quantum Theory of the Emission and Absorption of Radiation', Proceedings of the Royal Society A: Mathematical, Physical and Engineering Sciences, 114(767), pp. 243–265. DOI: https://doi.org/10.1098/rspa.1927.0039.

Dirac, P. A. M. (2023) The Principles of Quantum Mechanics. Jenson Books Inc.

Donati, O., Missiroli, G. F., and Pozzi, G. (1973) 'An Experiment on Electron Interference', American Journal of Physics, 41, pp. 639–644. DOI: https://doi.org/10.1119/1.1987321.

Duane, W. (1923) 'The Transfer in Quanta of Radiation Momentum to Matter', Proceedings of the National Academy of Sciences, 9(5), pp. 158–164. DOI: 10.1073/pnas.9.5.158.

Ebert, M. R., and Reissig, M. (2018) Methods for Partial Differential Equations. Springer.

Edmonds, A. R. (1957) Angular Momentum in Quantum Mechanics. Princeton: Princeton University Press.

Ehrenberg, W., and Siday, R. E. (1949). The Refractive Index in Electron Optics and the Principles of Dynamics. Proceedings of the Physical Society. Section B, 62(1), 8–21. DOI: https://doi.org/10.1088/0370-1301/62/1/303.

Einstein, A. (1905) 'Über die von der molekularkinetischen Theorie der Wärme geforderte Bewegung von in ruhenden Flüssigkeiten suspendierten Teilchen', Ann. d. Physik, 322(8), pp. 549–560. DOI: https://doi.org/10.1002/andp.19053220806.

Einstein, A. (1905) 'Über einen die Erzeugung und Verwandlung des Lichtes betreffenden heuristischen Gesichtspunkt', Ann. d. Physik, 322(6), 132–148. DOI: https://doi.org/10.1002/andp.19053220607.

Einstein, A. (1907) 'Die Plancksche Theorie der Strahlung und die Theorie der spezifischen Wärme', Ann. d. Physik, 327(1), 180–190. DOI: https://doi.org/10.1002/andp.19063270110.

Einstein, A. (1916) 'Zur Quantentheorie der Strahlung', Mitteil. d. Phys. Ges. Zürich, 18, pp. 47–62.

Einstein, A. (1924) 'Quantentheorie des einatomigen idealen Gases', Akademie-Vorträge, pp. 237–244. DOI: https://doi.org/10.1002/3527608958.ch27.

Einstein, A. (1948) 'Quantum Mechanics and Reality', Dialectica, 2, pp. 320–324, (reproduced in the Born–Einstein correspondence). DOI: https://doi.org/10.1111/j.1746-8361.1948.tb00704.x.

Einstein, A. (1953) 'Elementary Considerations on the Interpretation of the Foundations of Quantum Mechanics', Scientific Papers Presented to Max Born on his retirement from the Tait Chair of Natural Philosophy in the University of Edinburgh, pp. 33–40.

Einstein, A., and de Haas, W. J. (1915) 'Experimenteller Nachweis der Ampéreschen Molekularströme', Deutsche Physikalische Gesellschaft, Verhandlungen.

Einstein, A., and Hopf, L. (1910) 'Über einen Satz der Wahrscheinlichkeitsrechnung und seine Anwendung in der Strahlungstheorie', Ann. d. Physik, 338(16), pp. 1096–1104. DOI: 10.1002/andp.19103381603.

Einstein, A., Podolsky, B., and Rosen, N. (1935) 'Can Quantum. Mechanical Description of Physical Reality Be Considered Complete?', Physical Review, 47(10), pp. 777–780. DOI: 10.1103/physrev.47.777.

Estermann, I., Frisch, O., and Stern, O. (1931) 'Beugung von Molekularstrahlen', Zeitschrift für Physik, 73, pp. 348–365.

Faget, J. (1961) 'Interferences d'ondes electroniques: application à une methode de microscopie electronique interferentielle', Revue Optical, 40, pp. 347–381.

Feenberg, E. (1932) 'The Scattering of Slow Electrons by Neutral Atoms', Physical Review, 40(1), 40–54. DOI: 10.1103/physrev.40.40.

Feinberg, G., and Sucher, J. (1968) 'Long-Range Forces from Neutrino-Pair Exchange', Physical Review, 166(5), pp. 1638–1644. DOI: 10.1103/physrev.166.1638.

Fermi, E. (1955) 'Lectures on Pions and Nucleons', reprinted in Rivista Nuovo Cimento, 31(1), pp. 1–73 (2008). DOI: 10.1393/ncr/i2008-10028-x.

Fermi, E., and Marshall, L. (1947) 'Interference Phenomena of Slow Neutrons', Physical Review Journals Archive, 71, pp. 666.

Fernández, F. (2020) Introduction to Perturbation Theory in Quantum Mechanics. CRC Press.

Feynman, R. P., Leighton, R. B., and Sands, M. (1971) Lectures on Physics. Vol III. Addison-Wesley.

Fock, V. (1930) 'Näherungsmethode zur Lösung des quantenmechanischen Mehrkörperproblems', Zeitschrift f. Physik, 61(1–2), pp. 126–148. DOI: https://doi.org/10.1007/bf01340294.

Fort, E., et al. (2010) 'Path-Memory Induced Quantization of Classical Orbits', PNAS, 107(41), pp. 17515–17520. DOI: https://doi.org/10.1073/pnas.1007386107.

Friedman, B. (2011) Principles and Techniques of Applied Mathematics. New York: Dover Publications, Inc.

Friedrich, H. (2013) Scattering Theory. Springer.

Gamow, G. (1928) 'Zur Quantentheorie des Atomkernes', Zeitschrift für Physik, 51, 204. DOI: https://doi.org/10.1007/BF01343196.

Gasiorowicz, S. (2003) Quantum Physics. John Wiley & Sons, Inc.

Gerry, Ch., and Knight, P. (2012) Introductory Quantum Optics. Cambridge: Cambridge University Press.

Ghirardi, G. C., Rimini, A., and Weber, T. (1986) 'Unified Dynamics for Microscopic and Macroscopic Systems', Physical Review D, 34(2), pp. 470–491. DOI: https://doi.org/10.1103/physrevd.34.470.

Goldberg, A., Schey, H. M., and Schwartz, J. L. (1967). 'Computer-Generated Motion Pictures of One-Dimensional Quantum-Mechanical Transmission and Reflection Phenomena', American Journal of Physics, 35, pp. 177–186. DOI: https://doi.org/10.1119/1.1973991.

Goldstein, H., Poole, Ch., and Safko, J. (2011) Classical Mechanics. 3rd ed. Uttar Pradesh: Pearson.

Göppert-Mayer, M. (1931) 'Über Elementarakte mit zwei Quantensprüngen', Annalen der Physik, 401(3), pp. 273–294. DOI: https://doi.org/10.1002/andp.19314010303.

Göppert-Mayer, M., and Jensen J. H. (1955) Elementary Theory of Nuclear Shell Structure. New York: John Wiley & Sons.

Gordon, J. P., Zeiger, H. J., and Townes, C. H. (1954) 'Molecular Microwave Oscillator and New Hyperfine Structure in the Microwave Spectrum of NH_3', Physical Review, 95(1), pp. 282–284.

Gradshteyn, I. S., and Ryzhik, I. M. (2007) Table of Integrals, Series, and Products. Amsterdam: Academic Press.

Grangier, P., Roger, G., and Aspect, A. (1986) 'Experimental Evidence for a Photon Anticorrelation Effect on a Beam Splitter: A New Light on Single-Photon Interferences', Europhysics Letters, 1, pp. 173–179. DOI: https://doi.org/10.1209/0295-5075/1/4/004.

Grundmann, M. (2021) 'The Physics of Semiconductors', Graduate Texts in Physics. DOI: https://doi.org/10.1007/978-3-030-51569-0.

Gurney, R. W., and Condon, E. U. (1928) 'Quantum Mechanics and Radioactive Disintegration', Nature, 122, 439–440. DOI: https://doi.org/10.1038/122439a0.

Hall, E. H. (1879) 'On a New Action of the Magnet on Electric Currents', American Journal of Mathematics, 2(3), pp. 287. DOI: https://doi.org/10.2307/2369245.

Harrison, W, A. (2000) Applied Quantum Mechanics. World Scientific. DOI: https://doi.org/10.1142/4485.

Hartree, D. R. (1928) 'The Wave Mechanics of an Atom with a Non-Coulomb Central Field. Part I. Theory and Methods', Mathematical Proceedings of the Cambridge Philosophical Society, 24(01), pp. 89. DOI: https://doi.org/10.1017/s0305004100011919.

Hassani, S. (2013) Mathematical Physics. Cham: Springer. DOI: https://doi.org/10.1007/978-3-319-01195-0.

Hassé, H. R. (1930) 'The Polarizability of the Helium Atom and the Lithium Ion', Math. Proc. Cambridge Philosophical Society, 26(04), 542. DOI: https://doi.org/10.1017/s0305004100016327.

Heisenberg, W. (1925). 'Über quantentheoretische Umdeutung kinematischer und mechanischer Beziehungen', Z Physik, 33(1), 879–893. DOI: https://doi.org/10.1007/bf01328377.

Heisenberg, W. (1927) 'Über den anschaulichen Inhalt der quantentheoretischen Kinematik und Mechanik', Zeitschrift für Physik, 43(3-4), pp. 172–198. DOI: https://doi.org/10.1007/bf01397280.

Heisenberg, W. (1951) '50 Jahre Quantentheorie', Die Naturwissenschaften, 38(3), 49–55. DOI: https://doi.org/10.1007/bf00589911.

Heisenberg, W. (1958) Physics & Philosophy. Great Britain: George Allen & Unwin LTD.

Heisenberg, W. (1958) The Physicist's Conception of Nature. London: Hutchinson Scientific and Technical.

Heitler, W., and London, F. (1927) 'Wechselwirkung neutraler Atome und homöopolare Bindung nach der Quantenmechanik', Zeitschrift f. Physik, 44(6–7), pp. 455–472. DOI: https://doi.org/10.1007/bf01397394.

Hermann, G. (1935) 'Die naturphilosophischen Grundlagen der Quantenmechanik', Die Naturwissenschaften, 23(42), pp. 718–721. DOI: 10.1007/bf01491142.

Hertz, H. (1887) 'Über sehr schnelle elektrische Schwingungen', Annalen der Physik und Chemie, 267(7), 421–448. DOI: https://doi.org/10.1002/andp.18872670707.

Hestenes, D. (1990) 'The zitterbewegung interpretation of quantum mechanics', Foundations of Physics, 20(10), pp. 1213–1232. DOI: https://doi.org/10.1007/bf01889466.

Hillery, M., et al. (1984) 'Distribution Functions in Physics: Fundamentals', Physics Reports, 106(3), pp. 121–167. DOI: https://doi.org/10.1016/0370-1573(84)90160-1.

Hitachi Brand Channel (2012) Observing Hidden Worlds - Part 1 - Hitachi [Video]. Available at: www.youtube.com/watch?v=mypzz99_MrM (Accessed: 2024).

Home, D. (1997) Conceptual Foundations of Quantum Physics. An Overview from Modern Perspectives. New York: Plenum Press. DOI: https://doi.org/10.1007/978-1-4757-9808-1.

Huang, K. (1952) 'On the Zitterbewegung of the Dirac Electron', American Journal of Physics, 20(8), pp. 479–484. DOI: https://doi.org/10.1119/1.1933296.

Hund, F. (1925) 'Zur Deutung verwickelter Spektren, insbesondere der Elemente Scandium bis Nickel', Zeitschrift für Physik, 33(1), pp. 345–371. DOI: https://doi.org/10.1007/bf01328319.

Israelachvili, J. N. (2011) Intermolecular and Surface Forces. Amsterdam: Elsevier.

IUPAC (2024) Periodic Table of Elements. Available at: https://iupac.org/what-%20we-do/periodic-table-of-elements/#a10 (Accessed: 2024).

James, R. W., Brindley, G. W., and Wood, R. G. (1929) 'A Quantitative Study of the Reflexion of X-Rays from Crystals of Aluminium', Proc. Roy. Soc. A: Mathematical, Physical and Engineering Sciences, 125(798), pp. 401–419. DOI: https://doi.org/10.1098/rspa.1929.0176.

Jaynes, E. T., and Cummings, F. W. (1963) 'Comparison of Quantum and Semiclassical Radiation Theories with Application to the Beam Maser', Proceedings of the IEEE, 51(1), pp. 89–109. DOI: https://doi.org/10.1109/proc.1963.1664.

Jönsson, C. (1961) 'Elektroneninterferenzen an mehreren künstlich hergestellten Feinspalten', Zeitschrift für Physik, 161, pp. 454–474. DOI: https://doi.org/10.1007/BF01342460.

José, J. V., and Saletan, E. J. (1998) Classical Dynamics. A Contemporary Approach. London: Cambridge University Press.

Kamerlingh O. H. (1911) 'Further Experiments with Liquid Helium. G. On the Electrical Resistance of Pure Metals, etc. VI. On the Sudden Change in the Rate at Which the Resistance of Mercury Disappears', Comm. Phys. Lab. Univ. Leiden, 124c.

Khersonskii, V. K., Moskalev, A. N., and Varshalovich, D. A. (1988) Quantum Theory of Angular Momentum. Singapore: World Scientific. DOI: https://doi.org/10.1142/0270.

Klitzing, K. v., Dorda, G., and Pepper, M. (1980) 'New Method for High-Accuracy Determination of the Fine-Structure Constant Based on Quantized Hall Resistance', Physical Review Letters, 45(6), pp. 494–497. DOI: https://doi.org/10.1103/physrevlett.45.494.

Kochen, S., and Specker, E. P. (1967) 'The Problem of Hidden Variables in Quantum Mechanics', Journal of Mathematics and Mechanics, 17(1), 59–87.

Kramers, H. A. (1926) 'Wellenmechanik und halbzahlige Quantisierung', Z. Physik, 39, pp. 828–840. DOI: https://doi.org/10.1007/BF01451751.

Kronig, L., and Penney, W. G (1931) 'Quantum Mechanics of Electrons in Crystal Lattices', Proceedings of the Royal Society of London. A, 130, pp. 499–513.

Lamb, W. E., and Scully, M. O. (1968) The photoelectric effect without photons. Center for Theoretical Studies.

Lamb, W. E., and Retherford, R. C. (1947) 'Fine Structure of the Hydrogen Atom by a Microwave Method', Physical Review, 72(3), pp. 241–243. DOI: https://doi.org/10.1103/physrev.72.241.

Lamoreaux, S. K. (2004) 'The Casimir Force: Background, Experiments, and Applications', Reports on Progress in Physics, 68(1), pp. 201–236. DOI: https://doi.org/10.1088/0034-4885/68/1/r04.

Landau, L.D. (1965) 'The Damping Problem Wave Mechanics', Collected Papers, pp. 8–18. DOI: https://doi.org/10.1016/b978-0-08-010586-4.50007-9.

Landau, L. D., and Lifshitz, E. M. (1976) Mechanics. New Delhi: Butterworth-Heinemann.

Landé, A. (1921) 'Über den anomalen Zeemaneffekt (Teil I)', Zeitschrift für Physik, 5(4), pp. 231–241. DOI: https://doi.org/10.1007/bf01335014.

Langer, R. E. (1937) 'On the Connection Formulas and the Solutions of the Wave Equation', Physical Review, 51, pp. 669–676. DOI: https://doi.org/10.1103/physrev.51.669.

Lewis, G. N. (1926) 'The Conservation of Photons', Nature, 118(2981), 874–875. DOI: https://doi.org/10.1038/118874a0.

London, F. (1926) 'Die Zahl der Dispersionselektronen in der Undulationsmechanik', Zeitschrift für Physik, 39(4), pp. 322–326. DOI: https://doi.org/10.1007/bf01322017.

London, F. (1938) 'On the Bose-Einstein Condensation', Physical Review, 54(11), pp. 947–954. DOI: https://doi.org/10.1103/physrev.54.947.

Longair, M. (2013) Quantum Concepts in Physics. New York: Cambridge University Press.

Lyman, T. (1914) 'The Spectrum of Hydrogen in the Region of Extremely Short Wave-Lengths', Memoirs of the American Academy of Arts and Sciences, New Series, 13(3), pp. 125–146.

Mandel, L., and Wolf, E. (1995) Optical coherence and quantum optics. Cambridge: Cambridge University Press

Mandel, L., Sudarshan, E. C. G., and Wolf, E. (1964) 'Theory of Photoelectric Detection of Light Fluctuations', Proceedings of the Physical Society, 84(3), pp. 435–444. DOI: https://doi.org/10.1088/0370-1328/84/3/313.

Margenau, H., and Murphy, G. M. (2009) The Mathematics of Physics and Chemistry. Read Books.

Marshall, T. W. (1963) 'Random Electrodynamics', Proceedings of the Royal Society A: Mathematical, Physical and Engineering Sciences, 276(1367), pp. 475–491. DOI: https://doi.org/10.1098/rspa.1963.0220.

Marshall, T. W. (1965) 'Statistical Electrodynamics', Mathematical Proceedings of the Cambridge Philosophical Society, 61(02), pp. 537. DOI: https://doi.org/10.1017/s0305004100004114.

Marshall, T., and Santos, E. (1988) 'Stochastic Optics: A Reaffirmation of the Wave Nature of Light', Foundations of Physics, 18(2), pp. 185–223. DOI: https://doi.org/10.1007/bf01882931.

McClendon, M., and Rabitz, H. (1988) 'Numerical Simulations in Stochastic Mechanics', Physical Review A, 37, pp. 3479.

Matteucci, G., and Pozzi, G. (1978) 'Two Further Experiments on Electron Interference', American Journal of Physics, 46, pp. 619–623. DOI: https://doi.org/10.1119/1.11099.

Meitner, L. (1922) 'Über die Entstehung der β-Strahl-Spektren radioaktiver Substanzen', Zeitschrift für Physik, 9(1), pp. 131–144. DOI: https://doi.org/10.1007/bf01326962.

Mendeleev, D. (1871) 'The Natural System of Elements and Its Application to the Indication of the Properties of Undiscovered Elements', Journal of the Russian Chemical Society, 3, pp. 25–56.

Merli, P. G., Missiroli, G. P., and Pozzi, G. (1976) 'On the Statistical Aspect of Electron Interference Phenomena', American Journal of Physics, 44, pp. 306–307. DOI: https://doi.org/10.1119/1.10184.

Mermin, D. N. (1989). 'What's Wrong with This Pillow?', Physics Today, 42(4), pp. 9–11.

Merzbacher, E. (1998) Quantum Mechanics. 3rd edn. New York: Wiley.

Millikan, R. A. (1914) 'A Direct Determination of "h"', Physical Review, 4(1), 73–75. DOI: 10.1103/physrev.4.73.2.

Millikan, R. A. (1916) 'A Direct Photoelectric Determination of Planck's "h"', Physical Review, 7(3), 355–388. DOI: https://doi.org/10.1103/physrev.7.355.

Milonni, P. W. (2013) The Quantum Vacuum: An Introduction to Quantum Electrodynamics. San Diego: Academic Press, Inc.

Mizobuchi, Y., and Ohtaké, Y. (1991) 'An Experiment to Throw More Light on Light', Physics Letters A, 153, pp. 403–406. DOI: https://doi.org/10.1016/0375-9601(91)90686-3.

Möllenstedt, G., and Bayh, W. (1962) 'Kontinuierliche Phasenschiebung von Elektronenwellen im kraftfeldfreien Raum durch das magnetische Vektorpotential eines Solenoids', Physik Journal, 18(7), 299–305. DOI: https://doi.org/10.1002/phbl.19620180702.

Morse, P. M. (1929) 'Diatomic Molecules According to the Wave Mechanics. II. Vibrational Levels', Physical Review, 34(1), pp. 57–64. DOI: https://doi.org/10.1103/physrev.34.57.

Morse, P. M., and Feshbach, H. (1953) Methods of Theoretical Physics. Vols. I and II. McGraw-Hill.

Moshinsky, M., and Smirnov, Y. F. (1996) The Harmonic Oscillator in Modern Physics. Harwood Academic Publishers.

Nelson, E. (1966) 'Derivation of the Schrödinger Equation from Newtonian Mechanics', Physical Review, 150(4), pp. 1079–1085. DOI: https://doi.org/10.1103/physrev.150.1079.

Nelson, E. (1967) Dynamical Theories of Brownian Motion. Princeton, NJ: Princeton University Press.

Nelson, E. (2012) 'Review of Stochastic Mechanics', Journal of Physics: Conferences Series, 361, 012011. DOI: https://doi.org/10.1088/1742-6596/361/1/012011.

Newton, R. G. (1982) Scattering Theory of Waves and Particles. New York, NY: Springer-Verlag.

Nicholson, J. W. (1912) 'The Spectrum of Nebulium', Monthly Notices of the Royal Astronomical Society, 72(1), pp. 49–64.

Nieto, M. M., and Simmons, L. M. (1979) 'Coherent States for General Potentials. I. Formalism', Physical Review D, 20(6), pp. 1321–1331. DOI: https://doi.org/10.1103/physrevd.20.1321.

Noether, E. (1915) 'Der Endlichkeitssatz der Invarianten endlicher Gruppen', Mathematische Annalen, 77(1), pp. 89–92. DOI: https://doi.org/10.1007/bf01456821.

Paschen, F. (1908) 'Zur Kenntnis ultraroter Linienspektra. I', Annalen der Physik, 332, pp. 537–570.

Pauli, W. (1925) 'Über den Zusammenhang des Abschlusses der Elektronengruppen im Atom mit der Komplexstruktur der Spektren', Zeitschrift Für Physik, 31(1), pp. 765–783. DOI: https://doi.org/10.1007/bf02980631.

Pauli, W. (1940) 'The Connection Between Spin and Statistics', Physical Review, 58(8), pp. 716–722. DOI: https://doi.org/10.1103/physrev.58.716.

Pauli, W. (1960) 'Wahrscheinlichkeit und Physik', Aufsätze und Vorträge über Physik und Erkenntnistheorie.

Pauling, L. (1960) The Nature of the Chemical Bond. New York, NY: Cornell University Press.

Peltier, J. C. A. (1834) Nouvelles Expériences sur la Caloricité des Courants Électriques. Paris.

Peshkin, M. and Tonomura, A. (1989) The Aharonov–Bohm Effect. New York, NY: Springer.

Philippidis, C., Dewdney, C., and Hiley, B. J. (1979) 'Quantum Interference and the Quantum Potential', Il Nuovo Cimento B Series 11, 52(1), pp. 15–28. DOI: https://doi.org/10.1007/bf02743566.

Pickering, E. (1896) 'The Spectrum of zeta Puppis', Astronomische Nachrichten.

Planck, M. (1899) 'Über irreversible Strahlungsvorgänge', Sitzungsberichte der Königlich Preußischen Akademie der Wissenschaften zu Berlin, 5, pp. 440–480.

Planck, M. (1901) 'Ueber das Gesetz der Energieverteilung im Normalspectrum', Annalen der Physik, 309(3), 553–563. DOI: https://doi.org/10.1002/andp.19013090310.

Planck, M. (1911) 'Eine neue Strahlungshypothese', Verhandlungen der Deutschen Physikalischen Gesellschaft, 13, pp. 138–148.

Planck, M. (1912) 'Über die Begründung des Gesetzes der schwarzen Strahlung', Annalen Der Physik, 342(4), pp. 642–656. DOI: https://doi.org/10.1002/andp.19123420403.

Planck, M. (1913) Vorlesungen über die Theorie der Wärmestrahlung. Leipzig: J. A. Barth.

Planck, M. (1958) Physikalische Abhandlungen und Vorträge. Vol. 2. Braunschweig: Vieweg & Sohn.

Power, E. A. (1965) Introductory Quantum Electrodynamics. New York: Elsevier.

Purcell, E. M., Torrey, H. C., and Pound, R. V. (1946) 'Resonance Absorption by Nuclear Magnetic Moments in a Solid', Phys. Rev., 69(1–2), pp. 37–38. DOI: https://doi.org/10.1103/physrev.69.37.

Rabi, I. I., and Cohen, V. W. (1933) 'The Nuclear Spin of Sodium', Physical Review, 43(7), pp. 582–583. DOI: https://doi.org/10.1103/physrev.43.582.

Ramsauer, C. (1921) 'Über den Wirkungsquerschnitt der Gasmoleküle gegenüber langsamen Elektronen', Annalen der Physik, 369(6), pp. 513–540.

Rayleigh, J. (1878) 'On the Propagation of Waves through Elastic Solids', Proceedings of the London Mathematical Society, 9, pp. 21–26.

Rayleigh, J. W. S. (1877) The Theory of Sound. London: Macmillan and Co.

Risken, H. (1996) The Fokker–Planck Equation: Methods of Solution and Applications. 2nd edn. Berlin: Springer.

Ritz, W. (1908) 'Recherches critiques sur l'Électrodynamique Générale', Annales de Chimie et de Physique, 13, 145–275, p. 172.

Ritz, W. (1909) 'Über eine neue Methode zur Lösung gewisser Variationsprobleme der mathematischen Physik', Journal für die reine und angewandte Mathematik (Crelle's Journal), 135, 1–61. DOI: https://doi.org/10.1515/crll.1909.135.1.

Rodrigues, W. A., et al. (1993) 'About zitterbewegung and electron structure', Physics Letters B, 318(4), pp. 623–628. DOI: https://doi.org/10.1016/0370-2693(93)90464-s.

Ross-Bonney, A. A. (1975) 'Does God Play Dice? A Discussion of Some Interpretations of Quantum Mechanics', Il Nouvo Cimento B, pp. 55–79. DOI: https://doi.org/10.1007/bf02721493.

Schrödinger, E. (1926) 'Quantisierung als Eigenwertproblem', Annalen der Physik, 384(4), 361–376. DOI: https://doi.org/10.1002/andp.19263840404.

Schrödinger, E. (1935) 'Die gegenwärtige Situation in der Quantenmechanik', Die Naturwissenschaften, 23(48), pp. 807–812. DOI: https://doi.org/10.1007/bf01491891.

Scully, M. O., and Zubairy, M. S. (1997) Quantum Optics. Cambridge: Cambridge University Press.

Seebeck, T. J. (1822) 'Über den Magnetismus der galvanischen Kette', Abhandlungen der Physikalischen Klasse der Königlich-Preußischen Akademie der Wissenschaftten aus den Jahren 1820–1821, pp. 289–346.

Selleri, F. (1983) Die Debatte um die Quantentheorie. Braunschweig: Vieweg. DOI: https://doi.org/10.1007/978-3-322-88796-2.

Slater, J. C. (1929) 'The Theory of Complex Spectra', Physical Review Journals Archive, 34, pp. 1293.

Sokolov, A. A., Loskutov, T. M., and Ternov, I. M. (1966) Quantum Mechanics. New York: Holt, Rinehart and Winston, Inc.

Sommerfeld, A. (1916) 'Zur Quantentheorie der Spektrallinien', Annalen der Physik, 365(17), pp. 1–94.

Sommerfeld, A. (1923) Atomic Structure and Spectral Lines. New York: Dutton and Company.

Srivastava, G. P. (2022) The Physics of Phonons. London: CRC Press.

Stark, J. (1914) 'Beobachtungen über den Effekt des elektrischen Feldes auf Spektrallinien. I. Quereffekt', Annalen der Physik, 348(7), 965–982. DOI: https://doi.org/10.1002/andp.19143480702.

Stern, O. (1921) 'Ein Weg zur experimentellen Prüfung der Richtungsquantelung im Magnetfeld', Zeitschrift für Physik, 7(1), pp. 249–253. DOI: https://doi.org/10.1007/bf01332793.

Stewart, J. Q. (1918) 'The Moment of Momentum Accompanying Magnetic Moment in Iron and Nickel', Physical Review, 11(2), pp. 100–120. DOI: https://doi.org/10.1103/physrev.11.100.

Stone, A. D. (2013) Einstein and the Quantum: The Quest of the Valiant Swabian. Oxfordshire: Princeton University Press.

Stueckelberg, E. C. G. (1934) 'Relativistisch invariante Störungstheorie des Diracschen Elektrons I. Teil: Streustrahlung und Bremsstrahlung', Annalen der Physik, 413(4), pp. 367–389. DOI: https://doi.org/10.1002/andp.19344130403.

Tamm, I. (1932) 'About a Possible State of the Electron Bound to a Crystal Surface', Physikalische Zeits. der Sowjetunion, 1, pp. 733.

Thomson, G. P. (1927) 'The Diffraction of Cathode Rays by Thin Films of Platinum', Nature, 120, pp. 802.

Thomson, J. J. (1904) 'XXIV. On the Structure of the Atom: An Investigation of the Stability and Periods of Oscillation of a Number of Corpuscles Arranged at Equal Intervals Around the Circumference of a Circle; with Application of the Results to the Theory of Atomic Structure', Philos. Mag., 7(39), pp. 237–265.

Tinkham, M. (1996) Introduction to Superconductivity. New York, NY: McGraw-Hill, Inc.

Tomonaga, S.-I. (1997) The Story of Spin. London: The University of Chicago Press.

Tong, D. Lectures on the Quantum Hall Effect. www.damtp.cam.ac.uk/user/tong/qhe.html (Accessed: 2025).

Tonomura, A., et al. (1989) 'Demonstration of Single-Electron Buildup of an Interference Pattern', American Journal of Physics, 57, pp. 117–120. DOI: https://doi.org/10.1119/1.16104.

Townsend, J. S. (1921) 'The Motion of Electrons in Gases', Philosophical Magazine, 42(250), pp. 873–891.

Uhlenbeck, G., and Goudsmit, S. (1925) 'Ersetzung der Hypothese vom unmechanischen Zwang durch eine Forderung bezüglich des inneren Verhaltens jedes einzelnen Elektrons', Naturwissenschaften, 13, pp. 953–954.

van der Waerden, B. L. (1968) Sources of Quantum Mechanics. New York, NY: Dover Publications, Inc.

van Kampen, N. G. (1981) Stochastic Processes in Physics and Chemistry. Amsterdam: North-Holland.

von Laue, M. (1952) 'Eine quantitative Prüfung der Theorie für die Interferenz-Erscheinungen bei Röntgenstrahlen', Naturwissenschaften 39, 368–372.

von Neumann, J. (1932) Mathematische Grundlagen der Quantenmechanik. Berlin: Springer.

von Neumann, J. (1927) 'Wahrscheinlichkeitstheoretischer Aufbau der Quantenmechanik', Göttinger Nachrichten, 1, pp. 245–272.

Wang, S. C. (1928) 'The Problem of the Normal Hydrogen Molecule in the New Quantum Mechanics', Physical Review, 31(4), pp. 579–586. DOI: https://doi.org/10.1103/physrev.31.579.

Webster, A. G. (2016) Partial Differential Equations of Mathematical Physics. 2nd edn. New York, NY: Dover Publications.

Wee, L. K. (2012) 2nd simulation of Brownian motion of a big particle. Available at https://weelookang.blogspot.com/2010/06/ejs-open-source-brownian-motion-gas.html (Accessed: 2024).

Weinberger, P. (2006) 'Revisiting Louis de Broglie's Famous 1924 Paper in the Philosophical Magazine', Phil. Mag. Lett., 86(7), pp. 405–410. DOI: https://doi.org/10.1080/09500830600876565.

Weisskopf, V. F. (1979) The formation of Cooper Pairs and the Nature of Superconducting Currents. Geneva: CERN.

Wick, D. (1995) The Infamous Boundary: Seven Decades of Heresy in Quantum Physics. New York, NY: Copernicus.

Wentzel, G. (1926) 'Eine Verallgemeinerung der Quantenbedingungen für die Zwecke der Wellenmechanik', Z. Physik, 38, pp. 518–529. DOI: https://doi.org/10.1007/BF01397171.

Wien, W. (1896) 'Über die Energieverteilung im Emissionsspektrum eines schwarzen Körpers"', Annalen der Physik, 58, 662–669.

Wigner, E. P. (1960) 'The Unreasonable Effectiveness of Mathematics in the Natural Sciences', Richard Courant Lecture in Mathematical Sciences. Communications on Pure and Applied Mathematics, 13(1), 1–14. DOI: https://doi.org/10.1002/cpa.3160130102.

Wikipedia (2012) Category: Animations of Quantum Tunneling. https://commons.wikimedia.org/wiki/Category:Animations_of_quantum_tunneling (Accessed: 2024)

Wikipedia (2023) Interpretations of Quantum Mechanics. https://en.wikipedia.org/w/index.php?title=Interpretations_of_quantum_mechanics&oldid=1188445223#Further_reading (Accessed: 2025)

Wilson, W. (1915) 'LXXXIII. The Quantum-Theory of Radiation and Line Spectra', Philos. Mag., 29(174), pp. 795–802.

Young, T. (1804) 'The Bakerian Lecture: Experiments and Calculations Relative to Physical Optics', Philosophical Transactions of the Royal Society of London, 94(0), pp. 1–16. DOI: https://doi.org/10.1098/rstl.1804.0001.

Zangwill, A. (2013) Modern Electrodynamics. Cambridge, UK: Cambridge University Press.

Zavoisky, Y. (1944) 'Spin-Magnetic Resonance in Paramagnetics', Fizicheskii Zhurnal, 9, pp. 211–245.

Zeeman, P. (1896) 'On the Influence of Magnetism on the Nature of the Light Emitted by a Substance', The London, Edinburgh and Dublin Philosophical Magazine and Journal of Science, 43(262), pp. 226–239.

Index

actinide, 445
Aharonov–Bohm effect, 326
alkali metal, 444, 462
alkaline earth metal, 444
Anderson, C. D., 567
angular momentum, 270
　addition, 291
　eigenfunctions, 274
　eigenvalues, 278
　intrinsic, 366
　matrix representation, 283
　notation, 278
　orbital, 270
　precession, 277, 286
　spin, 284, 289
anticommutator, 527
antiparticles, 564
atomic
　clock, 130
　orbital, 442
　shell, 440
　subshell, 440, 442
Auger, P., 447

band, 156
　conduction, 157
　valence, 157
BCS theory, 412
Bell, J. S., 11
　inequality, 386
　state, 295
　theorem, 11, 389
Bessel, F. W., 593
　cylindrical functions, 593
　equation, 593
　functions, 309
　spherical functions, 594
blackbody, 15
Blackett, P., 317
Bloch vector, 61
Blokhintsev, D. I., 10
Bogolyubov, N. N., 351
　canonical transformation, 361

Bohm, D., 75, 249
Bohr, N., 9, 440
　atom, 27, 30
　correspondence principle, 88
　magneton, 323, 364
Boltzmann, L., 15
　distribution, 17, 374
Born, M., 9, 44
　rule, 72
Born–Oppenheimer approximation, 465
Bose, N., 34, 176, 407, 413, 607
Bose–Einstein
　condensate, 34, 400, 413
　distribution, 406
　statistics, 34, 399, 528
boson, 398
bra, 95
Bragg, W. L., 36
　Bragg's Law, 36
Brillouin zone, 156
Brody, T. A., 1, 13
Brownian motion, 245

canonical, 46
Casimir, H. B. G., 253
　Casimir–Polder force, 476
　effect, 253
　force, 253, 476
causality, 11
center-of-mass frame, 300, 532
central potential, 302
centrifugal barrier, 281
characteristics, 186
chemical bond, 444, 454
　covalent, 444, 457
　homopolar, 454
　ionic, 444, 457
　valence, 444, 461
chemical forces, 463
chemical potential, 406, 430
CHSH inequality, 388
Clebsch–Gordan coefficients, 292, 293
coherence length, 412

coherent, 236
   state, 180, 236
   superposition, 179, 203
cold emission, 150
collective phenomena, 408
commutator, 47
   and Poisson bracket, 47
   canonical, 47, 266
   time-dependent, 238
completeness relation, 61, 82, 96, 167, 209
Compton, A. H., 24
   effect, 24
   frequency, 252, 518
   time, 251
   wavelength, 26, 576
Condon, E., 142
contact potential, 151
Cooper pair, 412
Copenhagen interpretation, 4, 9
correspondence rule
   Dirac, 47
   Weyl, 51
Coulomb gauge, 262, 321, 370, 493, 508
critical temperature, 411, 413
cross section
   differential, 506
   effective, 506
   total, 540
Curie–Langevin formula, 374
current density, 69, 181
cutoff frequency, *see also* Compton, A. H.

Davisson, C. J., 35
Davisson–Germer experiment, 35
de Broglie, L., 5, 33, 249
   wavelength, 33, 35, 65, 71, 163
Debye, P., 24
   temperature, 24, 471
decay
   alpha, 142
   radioactive, 144
   rate, 124, 125, 143
   spontaneous, 123
decoherence, 431, 433
degeneracy, 77, 102
   order of, 77
density
   current, 69
density matrix, 180, 415
   canonical ensemble, 430
   pure state, 425
   reduced, 432
   stationarity time, 424
density of modes, 17
detailed balance, 499, 515, 516
diamagnetism, 373
diatomic gas, 470
diatomic molecule, 465
   nuclear motion, 465

rotational spectrum, 308
selection rule, 465
vibrational spectrum, 466
dielectric constant, 348
diffusion coefficient, 246
dipole approximation, 16
Dirac equation, 566, 572
   adjoint, 572, 573
   algebraic consistency, 584
   anomalous magnetic moment, 582
   antimatter, 566
   approximations, 584
   bispinor, 568, 571
   Dirac sea, 566
   electric moment, 582
   electromagnetic field, 580
   electron spin, 574
   free particle, 568, 576
   gyromagnetic ratio, 582
   hydrogen-like atom, 583
   magnetic moment, 582
   oscillator, 568
   self-interaction corrections, 582
   spin components, 575
Dirac matrices, 568, 569
Dirac, P. A. M., 9, 506
   delta function, 167, 173
   notation, 95
dispersion, *see* scattering
dispersion relation, 410
Dulong–Petit law, *see also* specific heat

Ehrenfest, P., 18, 260
   theorem, 187, 193, 203
eigenstate, 55
   energy, 55
eigenvalue, 55, 76
   energy, 55
   equation, 77, 91, 101, 196
eigenvectors, 60
Einstein, A., 10, 13, 22–26, 34, 245, 498, 504
   A and B coefficients, 500
   Einstein, Podolsky and Rosen, 384
Einstein–de Haas experiment, 364
electron diffraction, 35, 67
electronic configuration, 442
empiricist interpretation, 9
energy, 139, 225
energy balance, 515
energy level
   splitting, 241
energy quantum, 19
energy splitting, 126
ensemble, 68
ensemble interpretation, *see* statistical interpretation
entangled state, 11, 297, 383, 384, 398, 413, 449
EPR, *see* magnetic resonance
ergodic principle, 18, 68, 89, 278

exchange
  degeneracy, 393
  energy, 403, 450
  operator, 403
  potential, 439
  splitting, 403
exclusion principle, 399
expansion theorem, 83, 98
expectation value, 90, 102

factorization method, 235
Faraday, M., 380
Fermi, E., 149
  energy, 149
  golden rule, 490
  level, 152
  surface, 149
Fermi–Dirac
  distribution, 148, 406
  statistics, *see* 158, 400, 528
fermion, 398
ferromagnetism, 374
Feynman, R. P., 171
  propagator, 171
Feynman–Hellman formula, 352
field, 12, 476
field commutators, 508
field quantization, 506, 519
Fifth Solvay Conference, 7
fine structure constant, 313
flow density, 322
fluctuation-dissipation relation, 261
fluctuations, 57, 68, 225
Fock, V. A., 437, 439
  space, 525
Fokker–Planck equation, 259
form factor, 545
Franck, J., 30
Franck–Condon principle, 469
free particle, 162
  Born normalization, 164
  densities, 175
  Dirac normalization, 166
  in 3D, 308
  propagator, 170
  wave packet, 164
Fresnel, A., 22

Gamow, G., 142
gauge transformation, 324
Gaussian distribution, 104, 189, 203, 223, 252
Gedankenexperiment, 67
Geiger, H., 27
Germer, L., 35
Ginzburg, V., 412
Glauber, R., 237
  representation, 415
Göppert Mayer, M., 477, 519
Goudsmit, S., 366

Green, G., 133
  function, 171, 185
Groenewold, H., 7
ground state
  atomic, 42
  energy, 39, 55
  harmonic oscillator, 54
ground-state energy, 522
group velocity, 410
guiding wave, 75
gyromagnetic ratio, 322, 363, 370
  anomalous, 364

Hall effect, 414
  fractional, 414
  quantum, 414
halogen, 444, 462
Hamilton, W. R., 134
Hamilton–Jacobi equation, 134, 249
Hamiltonian, 31
  density, 523
  dynamics, 47, 262
  operator, 53
  radial, 280
Hankel
  functions, 593, 595
harmonic oscillator, 54, 222
  canonical ensemble, 258
  coupled, 239, 343
  eigenfunctions, 227
  eigenvalues, 228
  in SED, 256
Hartree, D. R., 437
  equations, 438
Hartree–Fock
  equations, 439
  method, 438, 477
Hassé, H. R., 474
Heisenberg, W., 5, 44
  description, 191, 210
  equation of evolution, 53
  inequality, 40, 48, 56, 200, 201
  picture, 45, 58
Heitler–London
  method, 461
  wave function, 457
helium, 411
  atom, 445
  orthohelium, 450
  parahelium, 450, 453
  perturbative method, 448
  spectrum, 319, 446
  variational method, 449, 452
Hermite
  equation, 227
  polynomials, 227, 588
Hermitian, 45
  matrix, 45
  operator, 46, 91

Hertz, H., 23
hidden variable, 386
　deterministic, 388, 389
Hilbert, D., 7
　space, 60, 62, 83, 95
hole, *see* vacancy
Hund rule, 442, 450
hybridization, 464
hydrodynamic quantum analogs, 244, 267
hydrogen atom, 310
　Balmer series, 317
　Blackett series, 317
　Bohr energy, 313
　Bohr radius, 315
　eigenfunctions, 314
　energy eigenvalues, 313
　fine spectrum, 375
　hyperfine spectrum, 375
　ionization energy, 315
　lifetime, 317
　Lyman series, 316
　relativistic correction, 565
　Ritz–Paschen series, 317
　Schrödinger equation, 311
　selection rule, 315
hydrogen bond, 471

idealistic interpretation, 9
incoherent superposition, 180, 419
induced absorption, *see* stimulated absorption
induced emission, *see* stimulated emission
interaction description, 193, 211, 377, 425
intermolecular force, 471
　dipole-dipole, 471
　dispersion forces, 472
　long range, 471
　retarded dispersion force, 475
　van der Waals, 472
isotope effect, 412

Jacobi, C. G., 134
Jaynes–Cummings model, 509, 512
Jeans, J., 18
Jensen, H., 477
Jordan, P., 44
Josephson effect, 412

Kapitza, P., 412, 539
Kapitza–Dirac effect, 539
ket, 95
Kirchhoff, G., 16
　Kirchhoff's law, 16
Klein–Gordon equation, 561, 573
　hydrogen-like energy, 565
Kocher–Specker theorem, 197
Kronig–Penney model, 153

Lagrangian density, 523
Laguerre
　associated polynomials, 315, 593
　equation, 592
　polynomials, 313, 591
Lamb, W., 375
　shift, 375, 512, 517
Landé factor, 372
Landé, A., 363
Landau, L. D., 230, 412, 415
　level, 230, 268
Langevin equation, 256
Langevin formula, 374
lanthanide, 445
Larmor formula, 27
Larmor frequency, 230, 323, 372
Larmor radiation, 517
laser, 236, 502
lattice dynamics, 409
LCAO method, 455
LED, *see* light-emitting diode
Legendre, 275
　associated polynomials, 276, 590
　equation, 275, 589
　polynomials, 309, 590
lifetime, 124
light-emitting diode, 503
Liouville, J., 133
　equation, 423
London penetration depth, 412
long-wavelength approximation, 16, 495
Lord Rayleigh, 17
Lorentz distribution, 104, 556
Lorentz, H. A., 7

magnetic quantum number, 275, 322
magnetic resonance, 376
　electron, 376, 486
　nuclear, 376
Mandelstam, L. I., 204
Mandelstam–Tamm, 279
　inequality, 204, 274
many-body problem, 437
Markov approximation, 259
Markov process, 245
Marsden, E., 27
Marshall, T. W., 431
maser, 130
matrix
　mechanics, 45, 46
　transition elements, 265
maximal set of commuting operators, 198, 525
Maxwel–Boltzmann statistics, 400, 429
Maxwell, J. C., 14
mean life, 124, 144
mean lifetime, *see* mean life
mean-square dispersion, *see* variance
measurement theory, 431
Meissner effect, 411
Meitner, L., 447
　Meitner–Auger effect, 447
Mendeleev, D. I., 440

Mensing, L., 45
metal
 electron gas model, 148
metalloid, 442
Millikan, R., 23
minimal coupling, 262, 320, 493, 564, 580
mixture, 180
molecular clock, 130
molecular hydrogen, 456, 460
 ion, 454
molecular orbital, 456, 460
Morse, P., 466
 potential, 466
Moyal, J. E., 7
multipartite system, 392

Nelson, E., 245
Nernst, W., 24, 38
Neumann
 functions, 593, 595
NMR, see magnetic resonance
noble gas, 445, 462
noble metal, 444
Noether, E., 215
 theorem, 215
nonfactorizable, 398
nonlocality, 11, 12, 72, 110, 121, 142, 194, 249
nonmetal, 442
normal mode, 240
nuclear
 island of stability, 480
 magic numbers, 477
 shell model, 477
 spin–orbit interaction, 478
 structure, 477
 superheavy elements, 480

observable, 45
oganesson, 462
ontology, 6
operator, 46
 annihilation, 235, 507, 521, 528
 antilinear, 220
 antinormal order, 522
 antiunitary, 220
 chronological, 213
 creation, 235, 507, 521, 528
 displacement, 236
 evolution, 172
 helicity, 571, 579
 ladder, 601
 lowering, 57, 233, 285, 288
 normal order, 522
 number, 525
 orthogonal, 207
 parity, 219
 particle density, 525
 permutation, 393
 raising, 57, 233, 285, 288

 rotation, 217
 tensor, 371
 time inversion, 220
 unitary, 206
 vector, 218
Oppenheimer, R., 465
orthodox interpretation, see Copenhagen
  interpretation
orthogonal, 92
orthonormal, see orthogonal
oscillator, see harmonic oscillator

paramagnetism, 373
parity, 219
 intrinsic, 219
Parseval theorem, 83
Pauli, W., 7, 44, 365
 equation, 369, 585
 exclusion principle, 365, 399
 matrices, 61, 290
 Pauli–Schrödinger equation, 369
Pauling, L., 444
periodic table, 365, 442
Perrin, J., 35
perturbation theory, 331
 canonical transformation method, 351
 degenerate states, 339
 finite time, 487
 level repulsion, 336
 nondegenerate states, 332
 periodic, 491
 time-dependent, 483
 time-independent, 331
phase shift, 123, 145, 548
phase space, 266
phenomenological, 72, 244
phonon, 410
 acoustic, 410
 gas, 411
 optical, 410
photodetection, 522
photoelectric effect, 23, 504
photoemission, 522
photon, 22, 494
pilot wave, 249, 268
Planck, M., 12, 38
 distribution law, 19, 258, 406, 499
 Planck's constant, 20
 reduced Planck constant, 21
plane wave expansion, 309
Poisson bracket, 263
Poisson distribution, 47, 408
polar molecule, 471
 ammonia, 471
 water, 471
Popper, K., 10
potential
 short-range, 533
 well, 85

probability
  amplitude, *see* wave function, 73
  density, 74
Proca equation, 561
product, 81
  dot, 81
  inner, 81
  scalar, 81, 83
projection operator, 61, 100, 208
projector, *see* projection operator
propagator, 204, 524
  free particle, 171
  general case, 184
pure state, 180

QED, *see* quantum electrodynamics
quantization, 75
  conditions, 138, 139
  rule, 46
  Wilson–Sommerfeld rule, 34, 139
quantum
  electrodynamics, 6, 40, 171, 211, 507
  fluctuations, 5
  information, 383
  number, 198
  potential, 249
  quantum dot, 116, 160
  regime, 261
  statistics, 34
quantum dot, 116, 160, 384
quantum electrodynamics
  cavity, 519

Rabi, I., 376
  oscillations, 487
radiation
  absorption, 493
  emission, 493
  reaction, 16
radiative
  corrections, 506, 517
  terms, 260
Ramsauer–Townsend effect, 118, 558
randomness, *see* stochasticity
rare gas, *see* noble gas
Rarita–Schwinger equation, 561
Rayleigh–Jeans formula, 18, 499
Rayleigh–Ritz method, 353
Rayleigh–Schrödinger, *see* perturbation theory
realistic interpretation, 8
reduced mass, 319
reflection coefficient, 111, 141
relativistic electron, 576
  equations, 561
representation
  change of, 98
  Heisenberg, 45
  matrix, 46, 93, 101
  number, 525

p-representation, 50
Schrödinger, 71
theory, 95
x-representation, 49, 98, 102
resonance
  energy, 555
  width, 555
resonant
  response, 261, 265, 266
  transmission, 118, 126, 144
response coefficient, 265, 520
rigid rotor, 306
  time-dependent solution, 308
Ritz, W., 353
Rodrigues formula, 228, 276, 309
rotating-wave approximation, 377, 486, 512
rotation generator, 218
Rutherford, E., 20, 27, 547
Rydberg, J., 28
  atom, 452
  constant, 28, 29, 313, 319

Santos, E., 431
scattering, 531
  amplitude, 533, 534, 555
  Born approximation, 537, 551
  Breit–Wigner cross section, 556
  central potential, 542
  Coulomb, 543
  cross section, 533, 555
  delta potential, 541
  differential cross section, 536
  effective area, 536
  elastic, 531
  electric form factor, 546
  form factor, 544
  impenetrable sphere, 552
  lifetime, 557
  matrix, 119
  optical theorem, 550
  partial wave expansion, 547
  periodic potential, 539
  phase shift, 548
  resonance width, 556
  resonant, 117, 554
  Rutherford cross section, 542, 547
  S matrix, 550
  spherical barrier, 557
  weak potential, 539
Schrödinger, E., 5, 65, 575
  cat, 3, 384
  curvilinear coordinates, 282
  Darwin term correction, 585
  description, 192, 210
  minimal coupling, 321
  quasi-stationary states, 319
  radial, 281, 302
  relativistic correction, 585
  spin–orbit correction, 585

Schrödinger, E. (*cont.*)
  stationary equation, 71
  time-dependent equation, 177
second quantization
  bosons, 523
  fermions, 527
secular equation, 102, 341
SED, *see* stochastic electrodynamics
selection rule, 231, 266, 497
  angular momentum, 305, 465
  harmonic oscillator, 231
  hydrogen atom, 315
  rigid rotor, 307
self-ionization, 447
semiclassical approximation, 133
semiconductor, 153, 158, 442
  acceptor, 158
  diode, 159
  donor, 158
  impurities, 158
  n-type, 158
  p-n junction, 159
  p-type, 158
  photoconductor, 158
  transistor, 159
shielding effect, 448
singlet state, 294
Slater, J. C., 10
  determinant, 399
Sommerfeld, A., 30
space quantization, 275, 364
specific heat, 24
  diatomic gas, 470
  metal, 150
spectral decomposition, *see* completeness relation
spectral energy density, 16, 38, 177, 251
spectral line, 59
spectroscopic notation, 442
spectrum, 78
  continuous, 78
  discrete, 78
  vibrational-rotational, 468
spherical harmonics, *see* angular momentum, 302, 309
spin, 363, 366
  correlation, 388
  polarization, 427
spin–orbit interaction, 375
spin–spin interaction, 376
spin-statistics theorem, 398
spinor, 367, 369
spontaneous emission, 317, 498, 506, 514
  Einstein A coefficient, 498
  semiclassical approach, 500
squeezed state, 203
Stückelberg, E., 211
standard deviation, *see* variance

Standard Model, 531
Stark, J., 347
  effect, 347
state, 60
  metastable, 450
  mixture, 236, 415
  pure, 415
statistical interpretation, 13, 74
Stefan–Boltzmann law, 15
step potential, 109
Stern, O., 364
Stern–Gerlach experiment, 365
stimulated absorption
  Einstein B coefficient, 498
  probability, 496
stimulated emission
  Einstein B coefficient, 498
  probability, 496
stochastic electrodynamics, 40, 244, 251
stochastic optics, 431
stochastic quantum mechanics, 244, 245
stochasticity, 194
superconducting quantum interference device, 384, 412
superconductivity, 351, 411
superposition principle, 73
symmetry
  continuous, 216
  discrete, 219

Tamm, I. Y., 204
thermocouple, *see* contact potential
Thomas–Reiche–Kuhn sum rule, 59, 265
Thomson, J. J., 20, 27
  scattering, 26
't Hooft, G., 389
time delay, 145
time resolution, 250
transformation, 171
  canonical, 47, 50, 206
  continuous, 216
  continuous group, 207
  finite, 216
  generator, 217
  group, 216
  infinitesimal, 215
  inverse, 206
  orthogonal, 207
  unitary, 206
transition, 59
  dipole, 495, 497
  electric dipole, 231
  frequency, 59
  multipole, 497
  probability, 317
  radiative, 493, 513
  resonant, 493
transition metal, 444, 445, 462
transmission coefficient, 111, 141

triplet state, 294
tunneling, 120, 140, 142, 151
   microscope, 151

Uhlenbeck, G., 366
ultraviolet catastrophe, 18
uncertainty principle, *see* Heisenberg
   inequality

vacancy, 158
vacuum fluctuations, 253, 512, 517
vacuum state, 525
valence, 444, 454
   saturation, 461
van der Waals, J., *see* intermolecular forces
van der Waerden equation, 570
variance, 49, 197
variational method, 353, 473
   hydrogen atom, 355
   shielded Coulomb potential, 357
velocity
   diffusive, 90, 246
   flow, 69, 90, 181, 245
virial theorem, 195
Volta, E., 151
von Laue, M., 36
von Neumann, J., 10, 60, 386, 415
   equation, 423

wave function, 75, 156
   antisymmetric, 396
   collapse, 10, 12, 432
   symmetric, 396
wave packet, 163, 199, 223
   coherent, 224
   Gaussian, 203
   minimum dispersion, 202
Wien, W., 19
   displacement law, 19, 41, 499
Wigner, E., 7, 293
   distribution function, 430
   Wigner functions, 307
Wigner–Eckart theorem, 371
Wilson–Sommerfeld, 31
WKB method, *see* semiclassical approximation
work function, 23, 148, 153

Young, Th., 22
Yukawa, H.
   potential, 361, 543

Zeeman, P., 320
   anomalous effect, 320, 363, 370
   normal effect, 320
zero-point, 38
   energy, 139, 225
   field, 12, 476
zitterbewegung, 378, 572, 575

For EU product safety concerns, contact us at Calle de José Abascal, 56–1°,
28003 Madrid, Spain or eugpsr@cambridge.org.

www.ingramcontent.com/pod-product-compliance
Ingram Content Group UK Ltd.
Pitfield, Milton Keynes, MK11 3LW, UK
UKHW051857220126
467219UK00008B/133